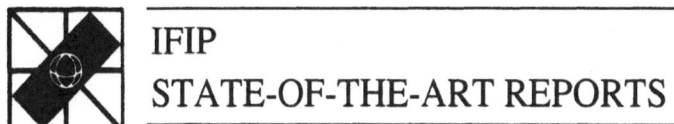

IFIP
STATE-OF-THE-ART REPORTS

Springer
Berlin
Heidelberg
New York
Barcelona
Hong Kong
London
Milan
Paris
Singapore
Tokyo

E. Astesiano H.-J. Kreowski
B. Krieg-Brückner (Eds.)

Algebraic Foundations
of Systems Specification

With 57 Figures

 Springer

Editors

Prof. Dr. Egidio Astesiano

Università di Genova
Dipartimento di Informatica e Scienze dell' Informazione (DISI)
I-16146 Genova, Italy
astes@disi.unige.it

Prof. Dr. Hans-Jörg Kreowski
Prof. Dr. Bernd Krieg-Brückner

Universität Bremen
Fachbereich Mathematik/Informatik
D-28334 Bremen, Germany
{kreo, bkb}@informatik.uni-bremen.de

ACM Computing Classification (1998): D.2.1, D.2.4, D.3, F.3.1-2

Library of Congress Cataloging-in-Publication Data applied for
Die Deutsche Bibliothek – CIP-Einheitsaufnahme
Algebraic foundations of systems specification/E. Astesiano ... (ed.).
Berlin; Heidelberg; New York; Barcelona; Hong Kong; London; Milan;
Paris; Singapore; Tokyo: Springer, 1999
(IFIP state-of-the-art reports)
ISBN-13: 978-3-642-64151-0   e-ISBN-13: 978-3-642-59851-7
DOI: 10.1007/978-3-642-59851-7

© 1999 IFIP International Federation for Information Processing,
Hofstrasse 3, A-2361 Laxenburg, Austria
Softcover reprint of the hardcover 1st edition 1999

Cover design: design & production GmbH, Heidelberg
Typesetting: Camera-ready by the editors
SPIN: 10519695   45/3142 – 5 4 3 2 1 0 – Printed on acid-free paper

Preface

The aim of software engineering is the provision and investigation of methods for the development of software systems of high quality with correctness as a key issue. A system is called correct if it does what one wants, if it meets the requirements. To achieve and to guarantee correct systems, the need of formal methods with rigorous semantics and the possibility of verification is widely accepted. Algebraic specification is a software engineering approach of this perspective.

When Liskov and Zilles, Guttag and the ADJ-group with Goguen, Thatcher, Wagner and Wright introduced the basic ideas of algebraic specification in the mid seventies in the U.S.A. and Canada, they initiated a very successful and still flourishing new area. In the late seventies, algebraic specification became a major research topic also in many European countries. Originally, the algebraic framework was intended for the mathematical foundation of abstract data types and the formal development of first-order applicative programs. Meanwhile, the range of applications has been extended to the precise specification of complete software systems, the uniform definition of syntax and semantics of programming languages, and to the stepwise development of correct systems from the requirement definitions to the running programs. The activities in the last 25 years have led to an abundance of concepts, methods, approaches, theories, languages and tools, which are mathematically founded in universal algebra, category theory and logic.

The development of the area of algebraic specification is continuously reflected in the proceedings of the international ADT-workshop series on *Algebraic Development Techniques* (formerly called *Specification of Abstract Data Types*) which started in 1982 and enjoyed 13 meetings so far. Algebraic specification concepts were employed in several national and international research projects in the last decade. In particular, they were the central subjects of interest in the ESPRIT Basic Research Working Group COMPASS (*Comprehensive Algebraic Approach to System Specification and Development*) running from 1989 to 1996. Many leading researchers in the field are members of the IFIP Working Group 1.3 (formerly 14.3) *Foundations of Systems Specification* which was established in 1992. The aims of the working group are "to support and promote the systematic development of the fundamental mathematical theory of systems specification" and "to investigate the theory of formal models for systems specification, development, transformation and verification".

This volume provides a state-of-the-art report of the area of algebraic specification. It consists of surveys of the major research topics in the area and is authored by leading experts in the field. The production of this book was an activity of the IFIP Working Group 1.3 in cooperation with the ESPRIT Basic Research Working Group COMPASS. In recent years, IFIP WG1.3 has been organizing and supporting the *Common Framework Initiative for Algebraic Specification and Development* (CoFI) that aims at the provision of a family of specification languages at different levels of software development with a widely accepted algebraic and logical basis. As a first step, the central language CASL has been designed for the specification of conventional software including requirement definition, design and architectural aspects. One may say that the presented state-of-the-art report provides the algebraic and logical foundations of the CoFI project.

In more detail, this volume is organized in the following way. Bernot and Gaudel motivate the use of formal methods and – in particular – of algebraic specification techniques from a software engineering point of view in Chapter 1. The basic notions and notations of algebraic specifications centered around many-sorted algebras and equational calculus are recalled by Sannella and Tarlecki in Chapter 2 while Cerioli, Mossakowski, and Reichel extend the foundational framework to a partial first-order logic in Chapter 3. The variety of algebraic approaches due to the abundance of possible logical systems and semantical choices is discussed in Chapters 4 and 5. Tarlecki introduces the concept of institutions as an abstract framework for formal specification, and Reichel surveys specification semantics from initial algebra semantics and loose semantics to the use of constraints and observability. The development of large software systems demands structuring principles that allow specifications and programs to be built in a stepwise and modular way. Chapters 6 and 7 are devoted to algebraic methods supporting structured and stepwise development. While Orejas surveys the possibilities of horizontal structuring by means of specification-building operations, parameterization and parameter passing as well as of modularization in Chapter 6, Ehrig and Kreowski discuss the spectrum of stepwise refinement and implementation concepts for vertical structuring in Chapter 7 including the issue of compatibility of horizontal and vertical structuring. Another necessary precondition for the practical use of formal methods is the availability of appropriate languages and tools. In Chapter 8, Sannella and Wirsing give an overview of the various proposed algebraic specification languages from the early candidates CLEAR and ACT ONE to the most recent attempt, the design of CASL as a standard specification language. Chapters 9 to 11 are concerned with term rewriting and proof theory as the foundations for evaluation and verification of algebraic specifications. In Chapter 9, Kirchner discusses various forms and aspects of term rewriting as a basis for interpreter semantics. As a major advantage of formal specification is the chance to guarantee correctness, theorem proving techniques and proof systems for algebraic specifications are extensively presented by Padawitz with respect to flat specifications in Chapter 10 and by Bidoit, Cengarle, and Hennicker with respect to structuring and refinement. The last three chapters deal with prominent applications domains. Ehrich relates algebraic specification with object orientation in Chapter 12. Astesiano, Broy, and Reggio introduce the reader to the algebraic specification of concurrent systems in Chapter 13. Finally, Basin and Krieg-Brückner provide a formalization of the development process in Chapter 14.

The main goal of this book is to disseminate fundamental knowledge about the area of algebraic specification and to expose the state of the art in a systematic way. The aim is to attract students, scientists, researchers and system developers interested in formal methods in general and algebraic specification in particular.

We would like to express our gratitude to all authors and reviewers for their patient cooperation, enthusiasm, and professional efforts. We are very grateful to Renate Klempien-Hinrichs who assisted us at various stages of the editing process in a marvelous way. We are also grateful to Frank Drewes and Gianna Reggio for their technical help with LaTeX matters. We would like to thank the publisher, in particular Hans Wössner and Ingeborg Mayer, Gabriele Fischer, Jacqueline Lenz and the copy editor Andrew Ross for their remarkable cooperation. We would also like to thank the IFIP authorities who supported the project, in particular Giorgio Ausiello, Wilfried Brauer, R.G. Johnson, and Plamen Nedkov.

Genova and Bremen,   *Egidio Astesiano*
May 1999   *Hans-Jörg Kreowski*
                                          Bernd Krieg-Brückner

Contents

1 The Role of Formal Specifications

Marie-Claude Gaudel[1] and Gilles Bernot[2]

[1] Université de Paris-Sud, LRI, CNRS UMR 8623, Bâtiment 490,
F-91405 Orsay Cedex, France
[2] Université d'Évry, LaMI, CNRS EP738, F-91025 Évry Cedex, France

Abstract. This introductory chapter aims at stating the context and the motivations of the rest of the book. The first section is a brief general reminder of the role of specifications in the software development process. Important concepts such as abstraction, refinement, validation, and verification are introduced informally. The second section gives a characterization of formal specifications, sketches a classification, and discusses the possibilities that they bring for software development. Section 3 is devoted to the use of formal specifications for requirement engineering and validation. Section 4 addresses the notions of refinement and verification. Section 5 discusses what kind of tools can be developed on the basis of formal specifications.

1.1 The role of specifications in software development

Software (and hardware) systems have to be developed following rigorous guidelines to have a chance of being dependable, finished on time, and easy to maintain. The aim of *software engineering* is to define criteria to evaluate software-intensive systems, as well as methods and techniques to develop them and satisfy these criteria. The search for these criteria, methods, and techniques requires a study of the characteristics of such systems and their development process.

The software development process can be seen as the construction of a sequence of more and more detailed descriptions of the software under development, leading to a final set of documents which contains an executable program and its documentation. This view is clearly oversimplified since it ignores the essential role of backtracking and iterations in such a process [Hum90, GMSB96], but it is sufficient for the purpose of this introductory chapter.

At every stage of the development process but the last one, the core of the description document is a *specification* of the future software, namely a definition of *what* it must do, without a complete description of *how* it will do it.

Because it plays a special role, it is interesting to distinguish the earliest specification document: we will call it the *global specification*, while the other ones are called intermediate or detailed specifications.

Clearly, the concept of specification is a cornerstone of software development: the global specification is the basis of the agreement between the developers and the users (some authors see it as a "contract"); global and intermediate specifications are essential for communication of precise information between the developers.

At each stage of the development, the current description of the future system must be checked against inconsistencies and omissions.

Moreover, it must be *validated* if it is the first one, or *verified* with respect to the previous one. Namely, it is necessary:

- to validate the global specification with respect to the requirements, the needs of the clients;
- to verify every intermediate specification with respect to the previous specification;
- to verify every piece of program with respect to its detailed specification;
- to verify the integrated final system with respect to its global specification.

Validation aims at establishing that the future system will fit for its operational mission. Verifications aim at establishing that it will satisfy the global specification. Validation, verification, consistency, and completeness checks are crucial activities since it has been recognized for a long time [Boe81] that the later a fault is discovered in the development process, the more expensive it is to correct.

The ability to use the specifications effectively for verification and validation is a major issue. For example, depending of the kind of developed system, a specification technique should provide some support for prototyping, correctness proofs, elaboration of test data, and failure detection when running these tests.

As said above, the role of the global specification is to establish as clearly as possible what has to be done. This must be understood and stated before deciding how to realize it. In practice, it is often difficult to write "unbiased" specifications at each level, i.e., specifications that stay within the boundaries of the *"what,"* without premature choices which mention unnecessary *"how"* aspects. However this effort of abstraction is profitable:

- *A priori*: before the software under consideration is finished, an abstract specification is a powerful reference document for discussions and elaboration of the design; besides, abstraction prevents local design choices which could result in a poor global design;
- *A posteriori*: a sequence of specifications, progressing from abstract to concrete, provides a trace of the design activity and makes easier the reuse, evolution, and maintenance of the software.

The transition from a specification to a more detailed one is often called a *refinement*. Most specification techniques provide some support for such refinements and the corresponding verifications. Verification may cause some iterations on the refinement until its result, namely the more detailed specification, is satisfactory with respect to the original specification.

Specifications are more abstract than the corresponding programs, but in many cases they are big enough to require some way for mastering size and complexity when writing them. As for programming languages, this is mainly provided by decomposition and modularity principles (see Chapters 6 and 8 of this book), or, more recently, by object-orientation (see Chapter 12). However, it is worth to note that the structure of a specification is mainly dictated by understanding the role of the future system. It is not the case of the structure of the implementation, which may be very different due to considerations such as efficiency, security, or reuse [FJ90, Gau92]. Thus, elaboration of the software architecture [SG96] may lead to a structure different from the organization of the specifications.

Like the choice of a programming language, the choice of a specification technique depends on the kind of system to be specified. For certain projects, it can even depend on the components to be specified: for example, certain components may require to specify real time aspects, while some other parts may require a careful specification of complex data structures without time constraints. A specification technique which would deal with all the possible aspects of all the possible systems would have a good chance to be too complex to be usable. This justifies the existence of a variety of languages and methods, each of them emphasizing particular aspects of some application domain and some class of target programs. As a consequence, in some cases it may be useful to write several specifications using different specification techniques for some component, in order to enlighten different aspects.

In the current state of practice, specifications are often written using diagrams or tables, enriched by comments in natural language [Dav93]. These diagrams and comments are the major media for dialog between the actors of the development. However, they often admit several interpretations. This plurality of interpretations results from *ambiguities* in the used notations. Such ambiguities of informal specifications do not necessarily prevent from producing software of good quality. However, they may raise some problems, especially when checking, validating, or verifying.

This book presents a class of *formal specification* techniques. Formal specification techniques make it possible to remove any ambiguities in the expression of the specifications of a program, in order to provide a sound basis for checking, validating, and verifying.

1.2   Formal specifications

A specification written in a formal specification language defines rigorously all the acceptable behaviors of the specified system. Besides, it is possible to perform some reasoning on the specification and its refinements.

A formal specification technique must provide at least three well defined facets: a *syntax*, some *semantics* and an *inference system*.

- The syntax exhaustively defines what a specifier is allowed to write to obtain a specification, as it is the case for programming languages. Thus the text of a formal specification has a structure which makes possible powerful computer aided treatments. In particular, a specification contains properties which are written as well formed formulas of some logic (these properties are often called axioms or sentences in this book).
- The semantics describes the models associated to a given specification. A model is a mathematical object which defines the behaviors of the acceptable realizations of the specification. "Acceptable" implies in particular that each model satisfies the properties of the specification. The role of the semantics is precisely to avoid ambiguities.
- The inference system serves to define the deductions that can be made from a formal specification. These deductions allow some formulas to be derived, which are consequences of the properties listed in the specification. Such derivations are mechanically checkable as developed in Section 4. So, the inference system can help to partly automate theorem proving, functional testing, prototyping, verification of refinements, etc.

This classical view of formal specifications leads to distinguish (at least) two classes of formal specification techniques. They are called respectively "model oriented" and "property oriented" (or "constructive" and "declarative").

In the *model oriented* approaches (such as VDM, Z, or B) the specifier builds a *unique* model, from a choice of built-in data structures and construction primitives that the specification language offers [Jon90], [Spi92], [Abr96]. Then, a program is *correct* with respect to the specification if the functions it provides have the *same behaviors* as in the specified model.

In the *property oriented* approaches, the specifier declares first a list of "function names" and by default there is an infinity of models that provide, in all the possible ways, a function for each name. Next, the specifier states several properties (i.e., formulas, which are often called "axioms" as they have not to be proved: they are required). Among all the previously mentioned models only a few of them satisfy the required properties; all the other models "do not satisfy the specification" and are discarded. Then, a program is *correct* with respect to a specification if it provides all the declared function names and defines a model that satisfies the specification. It is this class of specification techniques, also called *algebraic specifications*, which is studied in this book.

Among other classes of formal specification techniques, there are operational specifications. Starting from a set of elementary actions, these techniques describe the computations, i.e., the sequences of actions, that the system can perform. Some representatives of these techniques are Petri nets

[Rei85] and process algebras [Hen88]. Extensions of algebraic specifications on this direction are presented in the Chapter 13 of this book.

Formal specifications define unambiguously the correctness of a program with respect to its specification. They are indeed the only way to have a rigorous definition of correctness. Consequently, formal specifications must be used if correctness proofs are foreseen for some verifications. From a logical point of view, the notion of correctness of a program, without a formal specification, is a nonsense. Nevertheless, making use of formal specifications is a demanding process and should be suitably targeted. There are many cases where formal specifications would be a mere luxury.

In most projects where formal specification were used, informal specifications and formal ones were mixed. Some components and steps were formalized, some others not. The decision to use formal specifications mainly depends on the *criticality* of the component, in term of consequences of a fault (human lives, cost) and of the complexity of its requirements or of its development.

There are also some aspects of the development process which are beyond the scope of formal development methods and correctness proofs [Gau95]. The figure below shows the boundary of what can be formal in software development. The right hand side shows a purely formal process which goes from a high level abstract formal specification to the program. The left hand side of the figure mentions informal entities such as needs, opinions, or physical systems which belong to the real world. The formalization of such entities always induces some schematization [McD91], and is a delicate activity where some misunderstanding or error may occur.

The figure is slightly simplified since some validation against the actual needs may take place after each refinement, as developed in [Gau95], and some kinds of verification may skip some intermediate specifications.

Formal methods make it possible to perform a stepwise refinement process (from the global specification to the most concrete one) in a provably correct way. But the validation of the global specification with respect to the needs of the clients is a special case, because these needs are not formally specified (otherwise this specification itself would be the global specification and would have to be validated ...).

For the crucial parts of a software project, formal specifications oblige the specifier to treat a lot of particular cases that would have been forgotten or ambiguously specified otherwise. They also facilitate a "mutual validation" between two texts written according to formal syntactical rules (specification against program texts, or against lower level specifications).

In the remainder of this introductory chapter we develop some of the main contributions of formal specifications in general, and of algebraic specifications in particular, to sofware development activities. In Section 3, we discuss the new possibilities that formal specifications bring to requirement

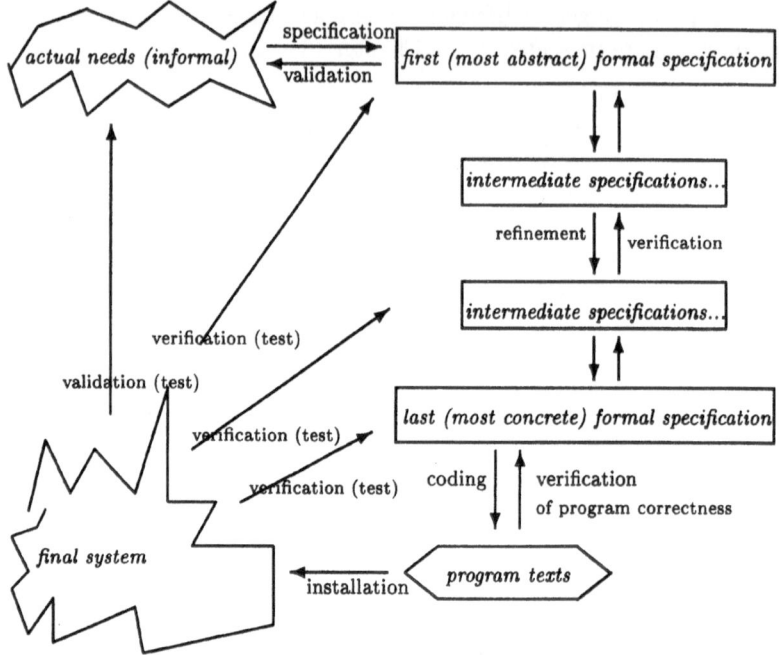

Fig. 1.1. Formal and informal aspects in software development

engineering and validation. In Section 4, we make more precise the notions of refinements and verifications. In Section 5, under the heading "mechanization of formal specifications", we briefly presents some of the development tools that formal specifications make possible. Finally, in Section 6, a quick tour of industrial use of formal specifications is presented.

1.3   Requirement engineering and validation

Even if it seems obvious, it is always useful to recall that a formal specification may be wrong. It may be wrong for two reasons: some misunderstanding of the users needs; some error in the expression of these needs. However, as said above, the notion of a formal correctness proof is meaningless in the case of a global specification since there is no formal correctness reference. Thus, the only possibility is to use "testing-like" methods. However, the great advantage of using formal specifications is that some tools exist to analyze and sometimes animate them. Such tools can be used to guide and support the validation activity.

Two classes of specification faults are of special interest in the case of formal specifications: *adequacy faults*, and *underspecifications* (for a more complete classification see for instance [Par89]). Adequacy faults arise when some of the properties expressed in the specification are in contradiction with the informal requirements. There is some underspecification when all the properties expressed in the specification are adequate, but some models of the specification correspond to unacceptable behaviors, i.e., the specification is not precise enough. A third class of specification faults are overspecifications. In these cases, the specification is too prescriptive, not abstract enough, and jumps to implementation choices which eliminate other acceptable, and maybe better, implementation possibilities.

In formal specifications, adequacy faults and underspecification may appear as inconsistencies or incompletenesses.

Inconsistency arises when contradictory properties are required. Thus there exists no implementation satisfying such a specification.

Incompleteness may (or may not) correspond to some underspecification: some properties, which are expressible in the specification language, cannot be stated as being either compatible or contradictory with the specification. Either some cases have been forgotten (underspecification) or it is just that these properties are irrelevant: it does not matter whether or not they are fulfilled by the future system.

Besides these fundamental faults, there are all the possible "typographic" faults: the cause of these faults is not a misunderstanding of the problem, but a mistake in writing the specification. It can come from a lack of attention (using one identifier in the place of another, etc.) or from difficulty in mastering the specification language. Clearly, as above, the resulting specification will be inadequate, inconsistent, or incomplete.

First, we discuss briefly how to prevent specification faults by applying some methodologic principles when developing a specification, then we give some hints on how to detect and remove such faults, making use of formal techniques.

Concerning the first kind of specification faults (those coming from a misunderstanding of the problem), an obvious methodological principle for avoiding them is to try to ensure simplicity and conciseness of the global specification. The main concept for achieving both simplicity and conciseness is abstraction [LZ74], [Neu89]. Abstraction is a key concept of algebraic specifications, the formal specification technique which is presented in this book. It is also present in most formal specification languages. However, this is not always sufficient to get abstract and concise specifications, just as modules in a programming language do not ensure that any programmer will write modular programs.

For expressing formally the required properties of a system, it is always necessary to formalize some aspects of the application domain or of the environment. This is one of the reasons why most case studies in formal specification result in apparently large specifications: it does not come from the formalization of the system itself but from the number of concepts which are necessary to the formal expression of the system functionalities. Here, as elsewhere in software engineering, structure and reuse improve the situation.

By reuse we mean reuse of theories: mathematicians do not reinvent Peano's axioms each time they use natural numbers. In computer science, a good example is the specification of compilers where accumulated knowledge is sufficient to guide and facilitate any new specification. Another less classical example of reusing pieces of formal specifications specific to an application domain is discussed in [GD90].

As indicated above, the only possible methods for validation are "testing-like" methods, in the sense that these methods can provide some evidence that the specification is wrong; they can only increase confidence in the claim that it is satisfactory. However, thanks to the deduction possibilities of formal techniques, these methods consist, most of the time, of proving or refuting consequences of the specification.

More precisely, given some conjectures (which play the role of "test data") that should be consequences of the formal specification, a theorem prover supporting the inference system of the specification technique is used to try to prove them. In [GG90] this technique is called "theory containment". Similarly, some properties which must not hold can be refuted. In [Rus93], the properties to be proved or refuted are called "challenges". This method is very powerful for detecting adequacy faults. The main difficulty is the invention of pertinent challenges which often requires a very good knowledge of the application domain.

For some formal specification languages, there is a possibility of detecting inconsistencies via some specific tool: for instance, such tools exist for algebraic specifications with equational or positive conditional axioms (see Chapter 10 of this book). It is also possible to give sufficient conditions for (some adapted notion of) completeness, which are mechanizable. *Sufficient completeness* [LG86] is such a criterion for algebraic specifications. As indicated above, the fact that a specification is consistent and complete does not mean that it is adequate. However, inconsistency reveals the presence of a fault, and incompleteness may correspond to some underspecification.

1.4   Refinements

Once a global formal specification has been stated, it has to be *refined* in order to incrementally obtain more and more concrete specifications and to make the implementation more and more precise. Each successive intermediate specification adds some implementation choices, until the last one is detailed enough to be easily translatable into a program.

In algebraic specifications, elementary refinement steps are called abstract implementations. A classical example is the representation of an abstract data type, such as set or collection, by another one, closer to a programming language, such as list or array. Another example is the choice of a precise recursive specification of a sorting algorithm, for instance a quicksort, to implement an abstract sorting operation, originally specified as returning any ordered permutation of a list. The specification of an abstract implementation consists of the definition of an abstraction function from the "implementing" data type into the "implemented" one, and the specification of a realization of each implemented operation, using the implementing data type and its operations.

The resulting (more) concrete specification must be correct with respect to the original specification, and this must be verified.

The definition of an adequate notion of *implementation correctness* is a crucial point in a formal specification technique. It should coincide with current practice and accept or reject the same realizations. But capturing a universal notion of implementation correctness turns out to be far from obvious. Different respectable definitions of implementation correctness may coexist for the same technique, depending on the application domain or on the operational context. A well-known example is the Lotos language [Bri89], and more generally process algebras.

A key point in such a definition is the fact that the implementation may be very different from the original specification, provided that these differences are not observable. Thus notions such as *observable equivalences* or *behavioral equivalences* play an essential role in such definitions. It is important to note that these notions are characterized by the kind of observations which are performable. They depend on the developed system and its environment. Thus the observability of the future system must be specified, or at least taken into account, in the verification activities. For instance, in the case of algebraic specifications, observations can be restricted to some observable data types, or to the results of some observable operations, or even to the fact that some observable formula holds or not.

To make a long story short, in almost all formal techniques the correctness criteria require that a refined specification "satisfies observationally" the properties stated in the original specification, the notion of observational satisfaction being the core of the correctness definition. These points are developed in Chapter 7.

The problem of verifying implementation correctness is made harder by the fact that in the development process, a specification evolves in two ways.

First, it is developed in a "horizontal" way, in the sense that the abstraction level is constant (and high). Most specification techniques provide structuring facilities. In this case the specification is developed piece by piece (or more likely as a collection of components which are written more or less concurrently). The result of this horizontal development is a structured specification, made of related components. *The aim of horizontal development is adequacy and completeness with respect to the user needs.*

The second evolution is the refinement process mentioned above, which is often called "vertical development" because the level of abstraction is decreasing. *The aim of vertical development is efficiency and correctness of the implementation.*

These two evolutions must be compatible, i.e., the composition of refined and correct pieces of specifications must be correct with respect to the composition of the original pieces.

Consequently, the definitions of "how to structure formal specifications" and "how to refine formal specifications" are often deeply interrelated according to the underlying specification theories.

1.5   Mechanization of formal specifications

Since there are precise rules to manipulate them, formal specifications constitute good raw material for mechanized treatments. Moreover, the soundness of these treatments can be justified with respect to their semantics. This makes it possible to support advanced aspects of computer aided software design, where each tool has a deep meaning which is sound with respect to some formal semantics.

There are four main kinds of activity where the mechanization of formal specifications can play a useful role, namely *theorem proving*, *prototyping*, *code generation*, and *testing*.

Theorem proving seems to be the first natural activity when formal methods are under consideration. Proving properties of a formal specification can be useful, as said in Section 3, for validation or, as developed in the previous section, to verify refinement steps. However, even if the underlying logic of a specification technique has a complete inference system, the truth of a formula with respect to a specification is not decidable in general as the proof tree can be infinite. So, *theorem provers* rely on *strategies*, and the various existing strategies make a significant difference between their possibilities.

The less powerful tools are *proof checkers*, which are already very useful to detect human errors in reasoning. Among more powerful tools, a distinction can be made between two classes of theorem provers: General theorem provers, such as HOL [GM93] or PVS [ORS92] among many others, can be adapted to specific formal specification techniques; Specialized ones, such as LP [GG90] for the LARCH language or MURAL [BFLM94] for VDM, are specific to a formal specification technique.

A general remark is that the more expressive is the underlying logic, the more difficult is the design of effective proof strategies. Thus interesting proofs can rarely be performed completely automatically. However, as claimed in [RvH93], the user of such a theorem prover has to struggle with its implacable logic, and this is very fruitful for debugging the specification.

Theorem proving for algebraic specifications is developed in Chapters 9, 10, and 11 of this book.

Prototyping is also an activity which is facilitated by formal specifications. Of course, ad hoc simulation tools can be developed. But techniques based on logic, such as rewriting techniques (see Chapter 9) or resolution algorithms, provide powerful bases for executing specifications written in some restricted logic.

Almost all specification languages have subparts which make it possible to write *executable specifications*. This is the case for algebraic specifications, when the axioms are restricted to equational clauses. Then, the prototyping activity can be seen as a sort of refinement process whose target is to write a specification within the boundaries of this sublanguage.

If the specification is executable, it can be tested and, since there is a formal syntax, some coverage criteria for the text of the specification, similar to those existing for programs, can be defined.

Code generation from formal specifications has been successfully studied for a long time, the pioneering project in this domain being the CIP project [BBB$^+$85]. Almost all formal specification environments have a code generator. Code generation considerably facilitates good programming disciplines and eliminates programming errors. In order to avoid problems of efficiency of the generated code, it is recommended to start from a rather detailed specification, obtained by successive refinements of the global specification, as discussed in the previous section.

However, the correctness of the translation may be difficult to state. Another approach to code generation, which avoid this problem, is to try to obtain a program directly from an abstract specification by synthesis. These methods are based on the expression of an induction principle attached to each data structure and they can be based on intuitionistic constructive logics. A classical approach consists in proving the realizability of a specification of the form: for each symbolic input I of a specified function f, there exists an acceptable output O for $f(I)$ (where "acceptable" means "that satisfies the specification"). Then, since the way to prove the property is constructive, it is possible to extract the computation of O from the proof. This computation is a correct realization of f and the program is correct by construction.

Such techniques are presented in Chapter 14.

Formal specifications can play a useful role for the testing activities as well [Bri89, BGM91, DF93]. Mainly two of the difficult aspects of testing can be aided: *test data selection* and *decision on success or failure* of a test experiment.

The test selection activity corresponds to functional testing since the choice of the test data is based on the specification, not on the code of the program under test. In the case of algebraic specifications [BGM91], test cases are extracted from each axiom of the specification and "interesting" subcases are determined by some case analysis based on an adapted resolution algorithm.

Moreover formal specifications can help the decision on success or failure of the selected test cases, since the correctness of the obtained results is defined by the specification. A formal specification provides formal *"oracle"* predicates characterizing this correctness. If the notions of *observability* in the specification and the program are sufficiently close, it becomes possible to automatically derive the oracle.

These possibilities bring interesting perspectives of intensive functional testing of critical software-based systems (see for instance [DGM93]).

Toward industrial use

The transfer of formal methods into industrial practice has been slow. There are several reasons for these difficulties. Some of them are: problems of training of the developers, lack of good tool support, lack of well-documented methodologies, and poor integration of these methods into the usual software development models.

However, formal methods are more and more used for critical software, clearly because the effort is justified and sometimes mandatory. Some certification agencies require them, and some standards recommend their use [BH94]. An interesting list of industrial projects where such methods were used can be found in [CGR93], with a discussion of the way these methods were used, and the positive and negative effects of their use. Other examples can be found in [GH90], [DGM93], [Siv96]. The proceedings of the Formal Methods Europe symposia also report practical experiences in the use of these methods (see [GW96], [FJL97], which are the last two volumes), and the numerous workshops on industrial applications of formal methods which are flourishing now.

2 Algebraic Preliminaries

Donald Sannella[1] and Andrzej Tarlecki[2]

[1] Laboratory for Foundations of Computer Science, University of Edinburgh, Edinburgh, Scotland
dts@dcs.ed.ac.uk http://www.dcs.ed.ac.uk/~dts/
[2] Institute of Informatics, Warsaw University and Institute of Computer Science, Polish Academy of Sciences, Warsaw, Poland
tarlecki@mimuw.edu.pl http://wwwat.mimuw.edu.pl/~tarlecki/

The purpose of this chapter is to present the basic definitions and results on which the following chapters rely. Most of this material is quite standard and for that reason the presentation will be concise. More detailed presentations with greater emphasis on motivation, exercises, and examples may be found in [EM85, Wir90, LEW96, ST].

The most basic assumption of work on algebraic specification is that a program is modeled as an *algebra*, that is, a set of data together with a number of functions over this set. The branch of mathematics which deals with algebras in a general sense (as opposed to the study of specific classes of algebras, such as groups and rings) is called *universal algebra* or sometimes *general algebra*. This chapter presents the basics of universal algebra, generalized to the *many-sorted* case as required to model programs which manipulate several kinds or *sorts* of data. Some extensions useful for modeling more complex programs are sketched at the end of the chapter.

2.1 Many-sorted sets

When using an algebra to model a program which manipulates several sorts of data, it is natural to partition the underlying set of values in the algebra so that there is one set of values for each sort of data. It is often convenient to manipulate such a family of sets as a unit in such a way that operations on this unit respect the "typing" of data values.

Let S be a set (of sorts). An *S-sorted set* is an S-indexed family of sets $X = \langle X_s \rangle_{s \in S}$, which is *empty* if X_s is empty for all $s \in S$. The empty S-sorted set is written $\emptyset$.

Let $X = \langle X_s \rangle_{s \in S}$ and $Y = \langle Y_s \rangle_{s \in S}$ be S-sorted sets. *Union, intersection, Cartesian product, disjoint union, inclusion (subset),* and *equality* of X and Y are defined as follows:

$$X \cup Y = \langle X_s \cup Y_s \rangle_{s \in S}$$
$$X \cap Y = \langle X_s \cap Y_s \rangle_{s \in S}$$
$$X \times Y = \langle X_s \times Y_s \rangle_{s \in S}$$

$X \uplus Y = \langle X_s \uplus Y_s \rangle_{s \in S}$ (where $X_s \uplus Y_s$ is the disjoint union of X_s and Y_s)
$X \subseteq Y$ iff $X_s \subseteq Y_s$ for all $s \in S$
$X = Y$ iff $X \subseteq Y$ and $Y \subseteq X$.

An *S-sorted function* $f : X \to Y$ is an S-indexed family of functions $f = \langle f_s : X_s \to Y_s \rangle_{s \in S}$; X is called the *domain* of f, and Y is called its *codomain*. An S-sorted function $f : X \to Y$ is an *identity* (an *inclusion, surjective, injective, bijective, ...*) if for every $s \in S$, the function $f_s : X_s \to Y_s$ is an identity (an inclusion, surjective, injective, bijective, ...). The identity S-sorted function on X will be written as $id_X : X \to X$.

If $f : X \to Y$ and $g : Y \to Z$ are S-sorted functions, then their *composition* $f;g : X \to Z$ is the S-sorted function defined by $(f;g)_s(x) = g_s(f_s(x))$ for $s \in S$ and $x \in X_s$.

Let $f : X \to Y$ be an S-sorted function and $X' \subseteq X$, $Y' \subseteq Y$ be S-sorted sets. The *image of X' under f* is the S-sorted set $f(X') = \langle \{f_s(x) \mid x \in X'_s\} \rangle_{s \in S} \subseteq Y$. The *coimage of Y' under f* is the S-sorted set $f^{-1}(Y') = \langle \{x \in X_s \mid f_s(x) \in Y'_s\} \rangle_{s \in S} \subseteq X$.

An *S-sorted binary relation on X*, written $R \subseteq X \times X$, is an S-indexed family of binary relations $R = \langle R_s \subseteq X_s \times X_s \rangle_{s \in S}$. For $s \in S$ and $x, y \in X_s$, $x R_s y$ (sometimes written $x R y$) means $\langle x, y \rangle \in R_s$. R is an *S-sorted equivalence (relation) on X* if it is reflexive ($x R_s x$), symmetric ($x R_s y$ implies $y R_s x$), and transitive ($x R_s y$ and $y R_s z$ implies $x R_s z$). The symbol $\equiv$ is often used for (S-sorted) equivalence relations.

Let $\equiv$ be an S-sorted equivalence on X. If $s \in S$ and $x \in X_s$, then the *equivalence class of x modulo* $\equiv$ is the set $[x]_{\equiv_s} = \{y \in X_s \mid x \equiv_s y\}$. The *quotient of X modulo* $\equiv$ is the S-sorted set $X/\equiv\ = \langle \{[x]_{\equiv_s} \mid x \in X_s\} \rangle_{s \in S}$.

Let $f : X \to Y$ be an S-sorted function. The *kernel of f* is the S-sorted equivalence relation $K(f) = \langle \{\langle x, y \rangle \in X_s \times X_s \mid f_s(x) = f_s(y)\} \rangle_{s \in S} \subseteq X \times X$.

Subscripts selecting components of S-sorted sets (functions, relations, ...) are often omitted where there is no danger of confusion.

2.2 Signatures and algebras

An algebra consists of named sets and named functions on these sets. The set of names associated with an algebra is called its signature. The signature of an algebra defines the *syntax* of the algebra; the algebra itself supplies the *semantics* by assigning interpretations to the names.

A *(many-sorted) signature* is a pair $\Sigma = \langle S, \Omega \rangle$, where S is a set (of sort names) and Ω is an $S^* \times S$-sorted set (of operation names). Here, S^* is the set of finite (including empty) sequences of elements of S. We will sometimes write $sorts(\Sigma)$ for S and $opns(\Sigma)$ for Ω. Σ is a *subsignature* of $\Sigma' = \langle S', \Omega' \rangle$ if $S \subseteq S'$ and $\Omega \subseteq \Omega'$.

Saying that $f\colon s_1 \times \cdots \times s_n \to s$ is in $\Sigma = \langle S, \Omega \rangle$ means that $s_1 \ldots s_n \in S^*$, $s \in S$, and $f \in \Omega_{s_1 \ldots s_n, s}$. Then f is said to have *arity* $s_1 \ldots s_n$ and *result sort* s. If $n = 0$, that is $f\colon \to s$, we use the abbreviation $f\colon s$.

Many-sorted signatures will be referred to as *algebraic* signatures when it is necessary to distinguish them from other kinds of signatures to be introduced later. The above definition of signature permits overloading, since it is possible to have different arities and result sorts for a single operation name.

In the rest of this section, let $\Sigma = \langle S, \Omega \rangle$ be a signature.

A *Σ-algebra* A consists of an S-sorted set $|A|$ of *carrier sets* (or *carriers*); and, for each $f\colon s_1 \times \cdots \times s_n \to s$ in Σ, a function (or *operation*) $(f\colon s_1 \times \cdots \times s_n \to s)_A \colon |A|_{s_1} \times \cdots \times |A|_{s_n} \to |A|_s$. The class of all Σ-algebras will be denoted by $Alg(\Sigma)$.

If $f\colon s_1 \times \cdots \times s_n \to s$ is in Σ for $n = 0$ (i.e., $f\colon s$), then $|A|_{s_1} \times \cdots \times |A|_{s_n}$ is a singleton set containing the empty tuple $\langle \rangle$, and so $(f\colon s)_A$ may be viewed as a constant denoting the value $(f\colon s)_A(\langle \rangle) \in |A|_s$. Notice that $(f\colon s_1 \times \cdots \times s_n \to s)_A$ is a *total* function; see Section 2.10 for several ways of extending the definitions to cope with partial functions. Note also that there is no restriction on the cardinality of $|A|_s$; in particular, $|A|_s$ may be empty (but not if, e.g., $\Omega_{\varepsilon, s} \neq \emptyset$).

We always write f_A in place of $(f\colon s_1 \times \cdots \times s_n \to s)_A$ where there is no possibility of confusion.

Example 2.1. Let $S1 = \{food, car\}$ and let $\Omega 1_{\varepsilon, food} = \{soup\}$, $\Omega 1_{\varepsilon, car} = \{vw\}$, $\Omega 1_{food, food} = \{boil\}$, $\Omega 1_{food\ car, car} = \{f\}$, and $\Omega 1_{w, s} = \emptyset$ for all other $w \in S1^*, s \in S1$. Then $\Sigma 1 = \langle S1, \Omega 1 \rangle$ is a signature which can be presented in tabular form as follows:

$\Sigma 1 =$ **sorts** *food*, *car*
    opns *soup*: *food*
       vw: *car*
       boil: *food* $\to$ *food*
       f: *food* $\times$ *car* $\to$ *car*

Let $|A1|_{food} = \{\oplus, \otimes\}$, $|A1|_{car} = \{a, b, c\}$, $soup_{A1} = \oplus \in |A1|_{food}$, $vw_{A1} = b \in |A1|_{car}$ and $boil_{A1}\colon |A1|_{food} \to |A1|_{food} = \{\oplus \mapsto \oplus, \otimes \mapsto \oplus\}$, and let $f_{A1}\colon |A1|_{food} \times |A1|_{car} \to |A1|_{car}$ be defined by the following table:

f_{A1}	a	b	c
$\oplus$	a	c	b
$\otimes$	b	c	c

This defines a $\Sigma 1$-algebra $A1$. Reference will be made to $\Sigma 1$ and $A1$ in examples throughout the rest of this chapter.   ■

Let A and B be Σ-algebras. B is a *subalgebra* of A if $|B| \subseteq |A|$, and if $f_B(b_1, \ldots, b_n) = f_A(b_1, \ldots, b_n)$ for any $f\colon s_1 \times \cdots \times s_n \to s$ in Σ and $b_1 \in |B|_{s_1}, \ldots, b_n \in |B|_{s_n}$. B is a *proper* subalgebra of A if it is a subalgebra of

A and $|B| \neq |A|$. A subalgebra of A is determined by an S-sorted subset $|B|$ of $|A|$ such that for each $f: s_1 \times \cdots \times s_n \to s$ in Σ and $b_1 \in |B|_{s_1}, \ldots, b_n \in |B|_{s_n}$, $f_A(b_1, \ldots, b_n) \in |B|_s$.

The intersection of any family of (carriers of) subalgebras of A is a (carrier of a) subalgebra of A. This implies that for any $X \subseteq |A|$, there is a least subalgebra of A that contains X, called the *subalgebra of A generated by X*. A is *reachable* if it has no proper subalgebra (equivalently, if A is generated by $\emptyset$). It follows that every algebra has a unique reachable subalgebra.

Example 2.2. Let $\Sigma 1 = \langle S1, \Omega 1 \rangle$ and $A1$ be as in Example 2.1. Define a $\Sigma 1$-algebra $B1$ by $|B1|_{food} = \{\oplus\}$, $|B1|_{car} = \{b, c\}$, $soup_{B1} = \oplus \in |B1|_{food}$, $vw_{B1} = b \in |B1|_{car}$, $boil_{B1}: |B1|_{food} \to |B1|_{food} = \{\oplus \mapsto \oplus\}$, and $f_{B1}: |B1|_{food} \times |B1|_{car} \to |B1|_{car} = \{\langle \oplus, b \rangle \mapsto c, \langle \oplus, c \rangle \mapsto b\}$. $B1$ is the subalgebra of $A1$ generated by $\emptyset$. That is, $B1$ is the reachable subalgebra of $A1$. ∎

2.3   Homomorphisms and congruences

A homomorphism between algebras is a function between the carrier sets which preserves the operations. Similarly, a congruence relation on an algebra is an equivalence which is preserved by the operations.

Throughout this section, let $\Sigma = \langle S, \Omega \rangle$ be a signature and let A and B be Σ-algebras.

A *Σ-homomorphism* $h: A \to B$ is an S-sorted function $h: |A| \to |B|$ such that for all $f: s_1 \times \cdots \times s_n \to s$ in Σ and $a_1 \in |A|_{s_1}, \ldots, a_n \in |A|_{s_n}$, $h_s(f_A(a_1, \ldots, a_n)) = f_B(h_{s_1}(a_1), \ldots, h_{s_n}(a_n))$.

Example 2.3. Let $\Sigma 1 = \langle S1, \Omega 1 \rangle$ and $A1$ be as in Example 2.1. Define a $\Sigma 1$-algebra $B1$ by $|B1|_{food} = |B1|_{car} = \{1, 2, 3\}$, $soup_{B1} = 1 \in |B1|_{food}$, $vw_{B1} = 2 \in |B1|_{car}$ and $boil_{B1}: |B1|_{food} \to |B1|_{food} = \{1 \mapsto 1, 2 \mapsto 3, 3 \mapsto 1\}$, where $f_{B1}: |B1|_{food} \times |B1|_{car} \to |B1|_{car}$ is defined by the following table:

f_{B1}	1	2	3
1	1	2	3
2	2	1	2
3	2	2	1

Let $h1: |A1| \to |B1|$ be the $S1$-sorted function such that $h1_{food} = \{\oplus \mapsto 1, \otimes \mapsto 3\}$ and $h1_{car} = \{a \mapsto 1, b \mapsto 2, c \mapsto 2\}$. It is easy to verify that $h1: A1 \to B1$ is a $\Sigma 1$-homomorphism by checking the following:

$$h1_{food}(soup_{A1}) = soup_{B1}$$
$$h1_{car}(vw_{A1}) = vw_{B1}$$
$$h1_{food}(boil_{A1}(\oplus)) = boil_{B1}(h1_{food}(\oplus))$$
$$h1_{food}(boil_{A1}(\otimes)) = boil_{B1}(h1_{food}(\otimes))$$

$$h1_{car}(f_{A1}(\oplus,a)) = f_{B1}(h1_{food}(\oplus), h1_{car}(a))$$
$$h1_{car}(f_{A1}(\oplus,b)) = f_{B1}(h1_{food}(\oplus), h1_{car}(b))$$
$$h1_{car}(f_{A1}(\oplus,c)) = f_{B1}(h1_{food}(\oplus), h1_{car}(c))$$
$$h1_{car}(f_{A1}(\otimes,a)) = f_{B1}(h1_{food}(\otimes), h1_{car}(a))$$
$$h1_{car}(f_{A1}(\otimes,b)) = f_{B1}(h1_{food}(\otimes), h1_{car}(b))$$
$$h1_{car}(f_{A1}(\otimes,c)) = f_{B1}(h1_{food}(\otimes), h1_{car}(c)).$$ ∎

The identity function on the carrier of a Σ-algebra is a Σ-homomorphism, and composition of Σ-homomorphisms yields another Σ-homomorphism.

Let $h\colon A \to B$ be a Σ-homomorphism, and let A' be a subalgebra of A. Let the *image of A' under h* be the Σ-subalgebra $h(A')$ of B such that $|h(A')| = h(|A'|)$, and $f_{h(A')}(h_{s_1}(a_1), \ldots, h_{s_n}(a_n)) = h_s(f_{A'}(a_1, \ldots, a_n))$ for each $f\colon s_1 \times \cdots \times s_n \to s$ in Σ and $a_1 \in |A'|_{s_1}, \ldots, a_n \in |A'|_{s_n}$. The *coimage* of a subalgebra B' of B under h is a subalgebra $h^{-1}(B')$ of A, defined analogously.

A Σ-homomorphism $h\colon A \to B$ is a *Σ-isomorphism* if it has an inverse, i.e., there is a Σ-homomorphism $h^{-1}\colon B \to A$ such that $h;h^{-1} = id_{|A|}$ and $h^{-1};h = id_{|B|}$. A homomorphism is an isomorphism iff it is bijective.

If there is an isomorphism from A to B, they are called *isomorphic* and we write $h\colon A \cong B$ or just $A \cong B$. Then $\cong$ (as a binary relation on Σ-algebras) is reflexive, symmetric, and transitive, and is therefore an equivalence relation.

Two isomorphic algebras are typically regarded as indistinguishable for all practical purposes: the only way in which they can differ is in the particular choice of data values in the carriers.

Example 2.4. Let $\Sigma1 = \langle S1, \Omega1 \rangle$ and $A1$ be as in Example 2.1. Define a $\Sigma1$-algebra $B1$ by $|B1|_{food} = \{\oplus, \otimes\}$, $|B1|_{car} = \{1, 2, 3\}$, $soup_{B1} = \otimes \in |B1|_{food}$, $vw_{B1} = 2 \in |B1|_{car}$, and $boil_{B1}\colon |B1|_{food} \to |B1|_{food} = \{\oplus \mapsto \otimes, \otimes \mapsto \otimes\}$, where $f_{B1}\colon |B1|_{food} \times |B1|_{car} \to |B1|_{car}$ is defined by the following table:

f_{B1}	1	2	3
$\oplus$	2	3	3
$\otimes$	1	3	2

Let $i1\colon |A1| \to |B1|$ be the $S1$-sorted function such that $i1_{food} = \{\oplus \mapsto \otimes, \otimes \mapsto \oplus\}$ and $i1_{car} = \{a \mapsto 1, b \mapsto 2, c \mapsto 3\}$. This defines a $\Sigma1$-homomorphism $i1\colon A1 \to B1$ which is a $\Sigma1$-isomorphism, so $A1 \cong B1$. ∎

A *Σ-congruence on A* is an (S-sorted) equivalence $\equiv$ on $|A|$ which respects the operations of Σ: for all operations $f\colon s_1 \times \cdots \times s_n \to s$ in Σ and values $a_1, a_1' \in |A|_{s_1}, \ldots, a_n, a_n' \in |A|_{s_n}$, if $a_1 \equiv_{s_1} a_1'$ and $\ldots$ and $a_n \equiv_{s_n} a_n'$ then $f_A(a_1, \ldots, a_n) \equiv_s f_A(a_1', \ldots, a_n')$.

The intersection of any family of Σ-congruences on A is a Σ-congruence on A, which implies that for any S-sorted binary relation R on $|A|$ there exists a least (with respect to $\subseteq$) Σ-congruence on A which includes R.

Let $\equiv$ be a Σ-congruence on A. The *quotient of A modulo* $\equiv$ is the Σ-algebra $A/\equiv$ such that $|A/\equiv| = |A|/\equiv$, and $f_{A/\equiv}([a_1]_{\equiv_{s_1}}, \ldots, [a_n]_{\equiv_{s_n}}) = [f_A(a_1, \ldots, a_n)]_{\equiv_s}$ for each $f \colon s_1 \times \cdots \times s_n \to s$ and $a_1 \in |A|_{s_1}, \ldots, a_n \in |A|_{s_n}$.

Example 2.5. Let $\Sigma1 = \langle S1, \Omega1 \rangle$ and $A1$ be as in Example 2.1, and let $\equiv = \langle \equiv_s \rangle_{s \in S1}$ be the $S1$-sorted congruence on $|A1|$ defined by $\equiv_{food} = \{\langle \oplus, \oplus \rangle, \langle \otimes, \otimes \rangle\}$ and $\equiv_{car} = \{\langle a, a \rangle, \langle b, b \rangle, \langle b, c \rangle, \langle c, b \rangle, \langle c, c \rangle\}$. $A1/\equiv$ is the $\Sigma1$-algebra defined by

$$|A1/\equiv|_{food} = \{\{\oplus\}, \{\otimes\}\}, \qquad |A1/\equiv|_{car} = \{\{a\}, \{b, c\}\},$$
$$soup_{A1/\equiv} = \{\oplus\} \in |A1/\equiv|_{food}, \qquad vw_{A1/\equiv} = \{b, c\} \in |A1/\equiv|_{car},$$
$$boil_{A1/\equiv} \colon |A1/\equiv|_{food} \to |A1/\equiv|_{food} = \{\{\oplus\} \mapsto \{\oplus\}, \{\otimes\} \mapsto \{\oplus\}\},$$

where $f_{A1/\equiv} \colon |A1/\equiv|_{food} \times |A1/\equiv|_{car} \to |A1/\equiv|_{car}$ is defined by the following table:

$f_{A1/\equiv}$	$\{a\}$	$\{b, c\}$
$\{\oplus\}$	$\{a\}$	$\{b, c\}$
$\{\otimes\}$	$\{b, c\}$	$\{b, c\}$

∎

The kernel of any Σ-homomorphism $h \colon A \to B$ is a Σ-congruence on A. Moreover, if $\equiv$ is a Σ-congruence on A, and $h_s(a) = [a]_{\equiv}$ for $s \in S$, $a \in |A|_s$, then $\langle h_s \colon |A|_s \to (|A|/\equiv)_s \rangle_{s \in S}$ is a Σ-homomorphism $h \colon A \to A/\equiv$. Therefore, a binary relation on $|A|$ is a Σ-congruence on A iff it is the kernel of a Σ-homomorphism from A.

Proposition 2.6. *Let $\equiv$ be a Σ-congruence on A. If $h \colon A \to B$ is a Σ-homomorphism such that $\equiv \subseteq K(h)$, then there is a unique Σ-homomorphism $g \colon A/\equiv \to B$ such that $h_s(a) = g_s([a]_{\equiv_s})$ for all $s \in S$ and $a \in |A|_s$.*

The above property characterizes quotient algebras up to isomorphism. It follows in particular that for any Σ-homomorphism $h \colon A \to B$, $A/K(h)$ is isomorphic to $h(A)$.

2.4   Term algebras

Throughout this section, let $\Sigma = \langle S, \Omega \rangle$ be a signature and let X be an S-sorted set (of variables), where $x \in X_s$ for $s \in S$ means that the variable x is of sort s (written $x \colon s$). Note that "overloading" of variable names is permitted here, since there is no requirement that X_s and $X_{s'}$ be disjoint for $s \neq s' \in S$.

The *Σ-algebra $T_\Sigma(X)$ of terms with variables X* is the Σ-algebra defined as follows:

- $|T_\Sigma(X)|$ is the least S-sorted set such that $x \in |T_\Sigma(X)|_s$ for all $s \in S$ and $x \in X_s$, and $f(t_1, \ldots, t_n) \in |T_\Sigma(X)|_s$ for all $f \colon s_1 \times \cdots \times s_n \to s$ in Σ and $t_1 \in |T_\Sigma(X)|_{s_1}, \ldots, t_n \in |T_\Sigma(X)|_{s_n}$,

- for all $f: s_1 \times \cdots \times s_n \to s$ in Σ and $t_1 \in |T_\Sigma(X)|_{s_1}, \ldots, t_n \in |T_\Sigma(X)|_{s_n}$,
  $f_{T_\Sigma(X)}(t_1, \ldots, t_n) = f(t_1, \ldots, t_n) \in |T_\Sigma(X)|_s$.

Note the distinction between syntactic term formation $f(t_1, \ldots, t_n)$ and the application of the operation named f. If $s \in S$ and $t \in |T_\Sigma(X)|_s$, then t is a Σ-*term of sort s with variables* X; the *free variables* of t is the set $FV(t) \subseteq X$ of variables that actually occur in t.

The Σ-*algebra of ground terms* is the Σ-algebra $T_\Sigma = T_\Sigma(\emptyset)$ of terms without variables. If $s \in S$ and $t \in |T_\Sigma|_s$, then t is a *ground Σ-term*.

Example 2.7. Let $\Sigma 1 = \langle S1, \Omega 1 \rangle$ be as in Example 2.1. Then $T_{\Sigma 1}$ is the $\Sigma 1$-algebra defined by

$$|T_{\Sigma 1}|_{food} = \{soup(), boil(soup()), boil(boil(soup())), \ldots\},$$

$$|T_{\Sigma 1}|_{car} = \{vw(), f(soup(), vw()), f(boil(soup()), vw()),$$
$$f(soup(), f(soup(), vw())), \ldots\}$$

where the operations of $T_{\Sigma 1}$ are the term-formation operations

$$soup_{T_{\Sigma 1}} = soup() \in |T_{\Sigma 1}|_{food}, \qquad vw_{T_{\Sigma 1}} = vw() \in |T_{\Sigma 1}|_{car},$$

$$boil_{T_{\Sigma 1}}: |T_{\Sigma 1}|_{food} \to |T_{\Sigma 1}|_{food} = \{soup() \mapsto boil(soup()),$$
$$boil(soup()) \mapsto boil(boil(soup())), \ldots\},$$

and similarly for $f: food \times car \to car$. ∎

It is implicitly assumed above that the result sort of each term is determined unambiguously. If the signature Σ and the set of variables X do not ensure this property, then appropriate sort decorations must be added to terms. We will henceforth assume that variables and constants (0-ary operations) of the same sort are distinct, which allows us to drop the parentheses "()" in terms like $boil(soup())$ above. So, in the example we would have:

$$|T_{\Sigma 1}|_{food} = \{soup, boil(soup), \ldots\},$$

$$|T_{\Sigma 1}|_{car} = \{vw, f(soup, vw), f(boil(soup), vw), \ldots\}$$

In examples we will also use infix notation for binary operations when convenient.

Let A be a Σ-algebra, $v: X \to |A|$ be an S-sorted function, $s \in S$, and $t \in |T_\Sigma(X)|_s$ be a Σ-term of sort s. The *value of t in A under the valuation v* is $v^\#(t) \in |A|_s$, defined as follows:

- $v^\#(x) = v(x)$ for all $s \in S$ and $x \in X_s$; and
- $v^\#(f(t_1, \ldots, t_n)) = f_A(v^\#(t_1), \ldots, v^\#(t_n))$ for all $f: s_1 \times \cdots \times s_n \to s$ in Σ and $t_1 \in |T_\Sigma(X)|_{s_1}, \ldots, t_n \in |T_\Sigma(X)|_{s_n}$.

Proposition 2.8. *For any Σ-algebra A and S-sorted function $v: X \to |A|$, $v^\#: T_\Sigma(X) \to A$ is the unique Σ-homomorphism that extends v, i.e., such that $v_s^\#(x) = v_s(x)$ for all $s \in S$, $x \in X_s$.*

20   Donald Sannella and Andrzej Tarlecki

It is easy to see that the value of a term $t \in |T_\Sigma(X)|$ depends only on the valuation of variables in $FV(t)$. In particular, the value of a ground term $t \in |T_\Sigma|$ does not depend on the valuation. Hence we write $t_A = \emptyset^{\#}(t)$, where $\emptyset: \emptyset \to |A|$ is the empty function, for the *value of t in A*.

The Σ-algebra A is *reachable* iff every element in $|A|$ is the value of a ground Σ-term, or equivalently, iff it is isomorphic to a quotient of T_Σ. It follows that there is a one-to-one correspondence between isomorphism classes of reachable Σ-algebras and congruences on T_Σ.

When the algebra A above is a term algebra $T_\Sigma(Y)$ for some S-sorted set Y, valuations are called *substitutions* (of terms in $T_\Sigma(Y)$ for variables), and the value of a term t under a substitution $\theta: X \to |T_\Sigma(Y)|$, written $t[\theta]$, is just the result of substituting $\theta(x)$ for all x in t in the usual sense. We write $t[x \mapsto u]$ for the result of replacing x in t by u, regarding $x \mapsto u$ as a shorthand for the obvious substitution which is the identity on all variables in X other than x (here, Y is $X \cup FV(u)$).

2.5 Signature morphisms

A signature morphism defines a mapping from the sort and operation names in one signature to those in another signature, in such a way that the arity and result sort of operations are respected.

More formally, let $\Sigma = \langle S, \Omega \rangle$ and $\Sigma' = \langle S', \Omega' \rangle$ be signatures. A *signature morphism* $\sigma: \Sigma \to \Sigma'$ is a pair $\sigma = \langle \sigma_{sorts}, \sigma_{opns} \rangle$ where $\sigma_{sorts}: S \to S'$ and $\sigma_{opns} = \langle \sigma_{w,s}: \Omega_{w,s} \to \Omega'_{\sigma^*_{sorts}(w), \sigma_{sorts}(s)} \rangle_{w \in S^*, s \in S}$ (where for $w = s_1 \ldots s_n \in S^*$, $\sigma^*_{sorts}(w) = \sigma_{sorts}(s_1) \ldots \sigma_{sorts}(s_n)$). Both σ_{sorts} and σ_{opns} (and its components $\sigma_{w,s}$ for all $w \in S^*, s \in S$) will be denoted by σ.

Signature morphisms as defined above will be referred to as *algebraic* signature morphisms when it is necessary to distinguish them from other kinds of signature morphisms to be introduced later.

Example 2.9. Let $\Sigma = \langle S, \Omega \rangle$ be the signature

sorts *warm, cold, vehicle*
opns *borscht: cold*
      heat: cold → warm
      heat: warm → warm
      h: warm × vehicle → vehicle

Let $\Sigma1 = \langle S1, \Omega1 \rangle$ be the signature in Example 2.1. Define $\sigma_{sorts}: S \to S1$ and $\sigma_{opns} = \langle \sigma_{w,s}: \Omega_{w,s} \to \Omega1_{\sigma^*_{sorts}(w), \sigma_{sorts}(s)} \rangle_{w \in S^*, s \in S}$ by $\sigma_{sorts} = \{ warm \mapsto food, cold \mapsto food, vehicle \mapsto car \}$, $\sigma_{\varepsilon, cold} = \{ borscht \mapsto soup \}$, $\sigma_{cold, warm} = \{ heat \mapsto boil \}$, $\sigma_{warm, warm} = \{ heat \mapsto boil \}$, $\sigma_{warm\, vehicle, vehicle} = \{ h \mapsto f \}$, and $\sigma_{w,s} = \emptyset$ for all other $w \in S^*, s \in S$. Then $\sigma: \Sigma \to \Sigma1$ is a signature morphism. ∎

In the rest of this section, let $\sigma\colon \Sigma \to \Sigma'$ be a signature morphism. This gives rise to a translation of Σ-terms to Σ'-terms, and of Σ'-algebras and homomorphisms to Σ-algebras and homomorphisms, as defined below. Note that the direction of translation of algebras and homomorphisms is "backwards" with respect to the direction of the signature morphism.

Let A' be a Σ'-algebra. The σ-*reduct of* A' is the Σ-algebra $A'|_\sigma$ such that $|A'|_\sigma|_s = |A'|_{\sigma(s)}$ for all $s \in S$, and $f_{A'|_\sigma} = \sigma(f)_{A'}$ for all $f\colon s_1 \times \cdots \times s_n \to s$ in Σ. Similarly, if $h'\colon A' \to B'$ is a Σ'-homomorphism, the σ-*reduct of* h' is the Σ-homomorphism $h'|_\sigma\colon |A'|_\sigma| \to |B'|_\sigma|$ such that $(h'|_\sigma)_s = h'_{\sigma(s)}$ for all $s \in S$.

If Σ is a subsignature of Σ', then we write $A'|_\Sigma$ for $A'|_\sigma$ where $\sigma\colon \Sigma \to \Sigma'$ is the obvious signature inclusion (and similarly for homomorphisms). Then $A'|_\Sigma$ is just A' with some carriers and/or operations removed.

Example 2.10. Let $\sigma\colon \Sigma \to \Sigma 1$ and $A1$ be as in Examples 2.9 and 2.1 respectively. Then $A1|_\sigma$ is the Σ-algebra such that $|A1|_\sigma|_{warm} = |A1|_\sigma|_{cold} = \{\oplus, \otimes\} = |A1|_{food}$, $|A1|_\sigma|_{vehicle} = \{a, b, c\} = |A1|_{car}$, $borscht_{A1|_\sigma} = \oplus = soup_{A1}$, $(heat\colon cold \to warm)_{A1|_\sigma} = \{\oplus \mapsto \oplus, \otimes \mapsto \oplus\} = boil_{A1}$, $(heat\colon warm \to warm)_{A1|_\sigma} = \{\oplus \mapsto \oplus, \otimes \mapsto \oplus\} = boil_{A1}$ and $h_{A1|_\sigma} = \{\langle\oplus, a\rangle \mapsto a, \langle\oplus, b\rangle \mapsto c, \dots\} = f_{A1}$.   ■

Let X be an S-sorted set of variables such that X_s and $X_{s'}$ are disjoint for $s \ne s' \in S$. Define $X' = \langle\bigcup_{\sigma(s)=s'} X_s\rangle_{s'\in S'}$. The *translation of a Σ-term* $t \in |T_\Sigma(X)|$ *by* σ is the Σ'-term $\sigma(t) \in |T_{\Sigma'}(X')|$ obtained by replacing each operation name f in t by $\sigma(f)$. (The disjointness assumption on X is for notational convenience only. It may be avoided by taking the disjoint union in the definition of X'.)

Example 2.11. Let $\sigma\colon \Sigma \to \Sigma 1$ be the signature morphism in Example 2.9, where $\Sigma = \langle S, \Omega\rangle$ and $\Sigma 1 = \langle S1, \Omega 1\rangle$. Let X be the S-sorted set of variables $x\colon cold, x'\colon warm, y\colon warm, z\colon vehicle$. The $S1$-sorted set of variables X' is then $x\colon food, x'\colon food, y\colon food, z\colon car$, and

$$\sigma(h(heat(x), h(x', z))) = f(boil(x), f(x', z)),$$

$$\sigma(h(x', h(heat(heat(borscht)), z))) = f(x', f(boil(boil(soup)), z)),$$

and so on.   ■

The following result states that the value of a term is invariant under change of signature.

Proposition 2.12. *Let X be an S-sorted set of variables such that X_s and $X_{s'}$ are disjoint for $s \ne s' \in S$, and $X' = \langle\bigcup_{\sigma(s)=s'} X_s\rangle_{s'\in S'}$. Let A' be a Σ'-algebra and $v'\colon X' \to |A'|$ be a valuation. Define $v\colon X \to |A'|_\sigma|$ by $v_s(x) = v'_{\sigma(s)}(x)$ for $s \in S$ and $x \in X_s$. Then for any Σ-term $t \in |T_\Sigma(X)|$, $v^\#(t) = (v')^\#(\sigma(t))$. In particular, if t is a ground term, then $t_{A'|_\sigma} = \sigma(t)_{A'}$.*

2.6 Equations

In the simple algebraic specifications considered in this chapter, equations are used as axioms to constrain the permitted behaviour of operations.

Throughout this section, let $\Sigma = \langle S, \Omega \rangle$ be a signature.

A Σ-equation $\forall X.\ t = t'$ consists of an S-sorted set X (of variables) such that X_s and $X_{s'}$ are disjoint for $s \neq s' \in S$, and two Σ-terms $t, t' \in |T_\Sigma(X)|_s$ for some sort $s \in S$. A Σ-equation $\forall \emptyset.\ t = t'$, sometimes abbreviated $t = t'$, is called a *ground (Σ-) equation*.

A Σ-algebra A *satisfies* (or, *is a model of*) a Σ-equation $\forall X.\ t = t'$, written $A \models_\Sigma \forall X.\ t = t'$, if for every ($S$-sorted) function $v \colon X \to |A|$, $v^\#(t) = v^\#(t')$.

A satisfies (or, is a model of) a set Φ of Σ-equations, written $A \models_\Sigma \Phi$, if $A \models_\Sigma \varphi$ for every equation $\varphi \in \Phi$. A class $\mathcal{A}$ of Σ-algebras satisfies a Σ-equation φ, written $\mathcal{A} \models_\Sigma \varphi$, if $A \models_\Sigma \varphi$ for every $A \in \mathcal{A}$. Finally, a class $\mathcal{A}$ of Σ-algebras satisfies a set Φ of Σ-equations, written $\mathcal{A} \models_\Sigma \Phi$, if $A \models_\Sigma \Phi$ for every $A \in \mathcal{A}$. We sometimes write $\models$ in place of $\models_\Sigma$ where Σ is obvious.

The explicit quantification over X in a Σ-equation $\forall X.\ t = t'$ is essential. For example, if $|A|_s = \emptyset$ but $X_s \neq \emptyset$ for some s in S, then A trivially satisfies any equation $\forall X.\ t = t'$. Thus variables in X may influence satisfaction even if they do not actually occur in t or t'.

Satisfaction of Σ-algebras is preserved under subalgebras and homomorphic images: if $A \models \varphi$ then φ is satisfied by any subalgebra of A and by any homomorphic image of A (and thus by any algebra isomorphic to A).

Let $\sigma \colon \Sigma \to \Sigma'$ be a signature morphism. The translation of Σ-terms to Σ'-terms defined above extends in the obvious way to a translation of Σ-equations to Σ'-equations. We will write $\sigma(\forall X.\ t = t')$ for $\forall X'.\ \sigma(t) = \sigma(t')$, where $X'_{s'} = \bigcup_{\sigma(s)=s'} X_s$ for each $s' \in S'$ as above.

An important result that brings together some of the main definitions above is as follows:

Lemma 2.13 (Satisfaction Lemma [BG80]). *If $\sigma \colon \Sigma \to \Sigma'$ is a signature morphism, φ is a Σ-equation, and A' is a Σ'-algebra, then $A' \models_{\Sigma'} \sigma(\varphi)$ iff $A'|_\sigma \models_\Sigma \varphi$.*

This states that the translations of syntax (terms, equations) and semantics (algebras) induced by signature morphisms are coherent with the definition of satisfaction. The proof follows from Proposition 2.12.

2.7 Presentations and theories

A signature, together with a set of equations over that signature, constitutes a simple form of specification. We refer to these as *flat* (meaning *unstructured*) specifications in order to distinguish them from the *structured* specifications to be introduced in later chapters.

Throughout this section, let Σ be a signature.

A *presentation* (also known as a *flat specification*) is a pair $\langle \Sigma, \Phi \rangle$ where Φ is a set of Σ-equations (called the *axioms* of $\langle \Sigma, \Phi \rangle$). A presentation $\langle \Sigma, \Phi \rangle$ is sometimes referred to as a Σ-*presentation*.

A *model* of a presentation $\langle \Sigma, \Phi \rangle$ is a Σ-algebra A such that $A \models_\Sigma \Phi$. $Mod_\Sigma(\Phi)$ is the class of all models of $\langle \Sigma, \Phi \rangle$. Taking $\langle \Sigma, \Phi \rangle$ to denote the semantic object $Mod_\Sigma(\Phi)$ is sometimes called taking its *loose semantics*.

Example 2.14. Let $Bool = \langle \Sigma Bool, \Phi Bool \rangle$ be the following presentation.

$Bool =$ **sorts** $\quad bool$
$\qquad\qquad$ **opns** $\quad true: bool$
$\qquad\qquad\qquad\quad false: bool$
$\qquad\qquad\qquad\quad \neg: bool \to bool$
$\qquad\qquad\qquad\quad \wedge: bool \times bool \to bool$
$\qquad\qquad$ **axioms** $\neg true = false$
$\qquad\qquad\qquad\quad \neg false = true$
$\qquad\qquad\qquad\quad \forall p{:}bool.\ p \wedge true = p$
$\qquad\qquad\qquad\quad \forall p{:}bool.\ p \wedge false = false$
$\qquad\qquad\qquad\quad \forall p{:}bool.\ p \wedge \neg p = false$

Define $\Sigma Bool$-algebras $A1$, $A2$, and $A3$ as follows:

$|A1|_{bool} = \{\star\}$
$true_{A1} = \star$
$false_{A1} = \star$
$\neg_{A1} = \{\star \mapsto \star\}$

$|A2|_{bool} = \{a, b, c\}$
$true_{A2} = a$
$false_{A2} = b$
$\neg_{A2} = \{a \mapsto b,$
$\qquad\quad b \mapsto a,$
$\qquad\quad c \mapsto c\}$

$|A3|_{bool} = \{1, 0\}$
$true_{A3} = 1$
$false_{A3} = 0$
$\neg_{A3} = \{1 \mapsto 0,$
$\qquad\quad 0 \mapsto 1\}$

$\wedge_{A1}$	$\star$
$\star$	$\star$

$\wedge_{A2}$	a	b	c
a	a	b	b
b	b	b	b
c	c	b	b

$\wedge_{A3}$	1	0
1	1	0
0	0	0

Each of these algebras is a model of *Bool*. (Reference will be made to *Bool* and to $A1$, $A2$, and $A3$ in later sections of this chapter.) ∎

For any class $\mathcal{A}$ of Σ-algebras, $Th_\Sigma(\mathcal{A})$ (the *theory* of $\mathcal{A}$) denotes the set of all Σ-equations satisfied by each Σ-algebra in $\mathcal{A}$:

$$Th_\Sigma(\mathcal{A}) = \{\varphi \mid \varphi \text{ is a } \Sigma\text{-equation and } \mathcal{A} \models_\Sigma \varphi\}.$$

The *closure* of a set Φ of Σ-equations is the set $Cl_\Sigma(\Phi) = Th_\Sigma(Mod_\Sigma(\Phi))$; Φ is *closed* if $\Phi = Cl_\Sigma(\Phi)$.

Proposition 2.15. *For any sets Φ and Ψ of Σ-equations and classes $\mathcal{A}, \mathcal{B}$ of Σ-algebras:*

1a. If $\Phi \subseteq \Psi$ then $Mod_\Sigma(\Psi) \subseteq Mod_\Sigma(\Phi)$.
1b. If $A \subseteq B$ then $Th_\Sigma(B) \subseteq Th_\Sigma(A)$.
2a. $\Phi \subseteq Th_\Sigma(Mod_\Sigma(\Phi))$.
2b. $A \subseteq Mod_\Sigma(Th_\Sigma(A))$.
3a. $Mod_\Sigma(\Phi) = Mod_\Sigma(Th_\Sigma(Mod_\Sigma(\Phi)))$.
3b. $Th_\Sigma(A) = Th_\Sigma(Mod_\Sigma(Th_\Sigma(A)))$.

A Σ-equation φ is a *semantic* (or *model-theoretic*) *consequence* of a set Φ of Σ-equations, written $\Phi \models_\Sigma \varphi$, if $\varphi \in Cl_\Sigma(\Phi)$ (equivalently, if $Mod_\Sigma(\Phi) \models_\Sigma \varphi$). We will write $\Phi \models \varphi$ instead of $\Phi \models_\Sigma \varphi$ where the signature Σ is obvious.

Proposition 2.16. *Semantic consequence is preserved by translation along signature morphisms: for any signature morphism $\sigma\colon \Sigma \to \Sigma'$, set Φ of Σ-equations, and Σ-equation φ,*

if $\Phi \models_\Sigma \varphi$ then $\sigma(\Phi) \models_{\Sigma'} \sigma(\varphi)$.

Proposition 2.17. *Let $\sigma\colon \Sigma \to \Sigma'$ be a signature morphism and let Φ' be a closed set of Σ'-equations. Then $\sigma^{-1}(\Phi')$ is a closed set of Σ-equations.*

A *theory* is a presentation $\langle \Sigma, \Phi \rangle$ such that Φ is closed. A presentation $\langle \Sigma, \Phi \rangle$ (where Φ need not be closed) *presents* the theory $\langle \Sigma, Cl_\Sigma(\Phi) \rangle$. A theory $\langle \Sigma, \Phi \rangle$ is sometimes referred to as a Σ-*theory*. For any theories $\langle \Sigma, \Phi \rangle$ and $\langle \Sigma', \Phi' \rangle$, a *theory morphism* $\sigma\colon \langle \Sigma, \Phi \rangle \to \langle \Sigma', \Phi' \rangle$ is a signature morphism $\sigma\colon \Sigma \to \Sigma'$ such that $\sigma(\varphi) \in \Phi'$ for every $\varphi \in \Phi$.

Example 2.18. Let Σ be the signature

$$\Sigma = \textbf{sorts } s, b$$
$$\textbf{opns } tt\colon b$$
$$ff\colon b$$
$$not\colon s \to b$$
$$and\colon s \times b \to b$$

and recall the presentation $Bool = \langle \Sigma Bool, \Phi Bool \rangle$ in Example 2.14. Define a signature morphism $\sigma\colon \Sigma \to \Sigma Bool$ by $\sigma_{sorts} = \{s \mapsto bool, b \mapsto bool\}$, $\sigma_{\varepsilon, b} = \{tt \mapsto true, ff \mapsto false\}$, $\sigma_{s, b} = \{not \mapsto \neg\}$, and $\sigma_{s\,b, b} = \{and \mapsto \wedge\}$. Let $\Phi = \{\forall x\colon s.\ and(x, and(x, not(x))) = ff, \forall x\colon s.\ and(x, ff) = ff\}$. Then $Cl_\Sigma(\Phi)$ includes Σ-equations that were not in Φ, such as the equation $\forall x, y\colon s.\ and(y, and(x, and(x, not(x)))) = ff$. The presentations $\langle \Sigma, Cl_\Sigma(\Phi) \rangle$ and $\langle \Sigma Bool, Cl_{\Sigma Bool}(\Phi Bool) \rangle$ are theories – the latter is the theory presented by $Bool$ – and $\sigma\colon \langle \Sigma, Cl_\Sigma(\Phi) \rangle \to \langle \Sigma Bool, Cl_{\Sigma Bool}(\Phi Bool) \rangle$ is a theory morphism. ∎

Proposition 2.19. *Let $\sigma\colon \Sigma \to \Sigma'$ be a signature morphism, Φ be a set of Σ-equations, and Φ' be a set of Σ'-equations. Then the following conditions are equivalent:*

1. σ is a theory morphism $\sigma\colon \langle \Sigma, Cl_\Sigma(\Phi) \rangle \to \langle \Sigma', Cl_{\Sigma'}(\Phi') \rangle$.
2. $\sigma(\Phi) \subseteq Cl_{\Sigma'}(\Phi')$.
3. For every $A' \in Mod_{\Sigma'}(\Phi')$, $A'|_\sigma \in Mod_\Sigma(\Phi)$.

2.8   Equational calculus

The set of consequences of a presentation $\langle \Sigma, \Phi \rangle$ has been defined in a model-theoretic way. In this section we present a calculus for deriving consequences of a set of equational axioms in a "syntactic" way. It turns out that these two notions of consequence coincide.

A Σ-equation φ is a *syntactic* (or *proof-theoretic*) *consequence* of Φ, written $\Phi \vdash_\Sigma \varphi$, if φ can be derived from Φ by application of the following inference rules:

Reflexivity:
$$\frac{}{\forall X.\, t = t} \quad t \in |T_\Sigma(X)|$$

Symmetry:
$$\frac{\forall X.\, t = t'}{\forall X.\, t' = t}$$

Transitivity:
$$\frac{\forall X.\, t = t' \qquad \forall X.\, t' = t''}{\forall X.\, t = t''}$$

Congruence:
$$\frac{\forall X.\, t_1 = t_1' \quad \cdots \quad \forall X.\, t_n = t_n'}{\forall X.\, f(t_1, \ldots, t_n) = f(t_1', \ldots, t_n')}$$

$$\text{for } f : s_1 \times \cdots \times s_n \to s \text{ and } t_i, t_i' \in |T_\Sigma(X)|_{s_i} \text{ for } i \leq n$$

Instantiation:
$$\frac{\forall X.\, t = t'}{\forall Y.\, t[\theta] = t'[\theta]} \quad \theta : X \to |T_\Sigma(Y)|$$

Example 2.20. Recall the presentation $Bool = \langle \Sigma Bool, \Phi Bool \rangle$ from Example 2.14. The following derivation proves $\Phi Bool \vdash_{\Sigma Bool} \forall p{:}bool.\ \neg(p \wedge \neg false) = \neg p$:

$$\frac{\dfrac{\dfrac{}{\forall p{:}bool.\ p = p} \qquad \dfrac{\dfrac{}{\neg false = true}}{\forall p{:}bool.\ \neg false = true}}{\dfrac{\forall p{:}bool.\ p \wedge \neg false = p \wedge true}{\forall p{:}bool.\ \neg(p \wedge \neg false) = \neg(p \wedge true)}} \qquad \dfrac{\forall p{:}bool.\ p \wedge true = p}{\forall p{:}bool.\ \neg(p \wedge true) = \neg p}}{\forall p{:}bool.\ \neg(p \wedge \neg false) = \neg p} \qquad \blacksquare$$

As mentioned above, $\vdash_\Sigma$ is both *sound* (only valid consequences may be derived) and *complete* (all valid consequences may be derived) for $\models_\Sigma$.

Theorem 2.21. *For any set Φ of Σ-equations and any Σ-equation φ, $\Phi \vdash_\Sigma \varphi$ if and only if $\Phi \models_\Sigma \varphi$.*

Simplifying the above calculus by omitting explicit quantifiers in equations yields an unsound system because algebras may have empty carrier sets. In particular, unused variables cannot always be removed from equations. The instantiation rule allows quantified variables to be eliminated when it is sound to do so [GM85].

2.9 Initial models

The class of algebras given by the loose semantics of a Σ-presentation always includes degenerate Σ-algebras with a single value of each sort in Σ, and usually includes unreachable Σ-algebras. Equational axioms are not sufficient to eliminate such obviously undesired models. One standard remedy is to take the so-called *initial semantics* of presentations.

Let A be a model of a presentation $\langle \Sigma, \Phi \rangle$. We say that A *contains junk* if it is not reachable, and that A *contains confusion* if it satisfies a ground Σ-equation that is not in $Cl_\Sigma(\Phi)$.

Example 2.22. Recall the presentation $Bool = \langle \Sigma Bool, \Phi Bool \rangle$ and its models $A1$, $A2$, and $A3$ given in Example 2.14. $A1$ contains confusion ($A1 \models_{\Sigma Bool}$ $true = false \notin Cl_{\Sigma Bool}(\Phi Bool)$) but not junk; $A2$ contains junk (there is no ground $\Sigma Bool$-term t such that $t_{A2} = c \in |A2|_{bool}$) but not confusion; $A3$ contains neither junk nor confusion. There are models of $Bool$ containing both junk and confusion.   ∎

A Σ-algebra $A \in Mod_\Sigma(\Phi)$ is an *initial model* of $\langle \Sigma, \Phi \rangle$ if for every $B \in Mod_\Sigma(\Phi)$ there is a unique Σ-homomorphism $h\colon A \to B$.

The initial models of an equational presentation are those that have no junk and no confusion. An initial model may be constructed as a quotient of the algebra T_Σ of ground Σ-terms by the least congruence generated by the axioms:

Theorem 2.23. $\langle \Sigma, \Phi \rangle$ *has an initial model.*

Proof sketch. An initial model of $\langle \Sigma, \Phi \rangle$ is the quotient $T_\Sigma/\equiv_\Phi$, where $\equiv_\Phi$ is the Σ-*congruence generated by* Φ: $t \equiv_\Phi t' \iff \Phi \models_\Sigma \forall \emptyset.\ t = t'$, for all $t, t' \in |T_\Sigma|$. The existence and uniqueness of a Σ-homomorphism from $T_\Sigma/\equiv_\Phi$ to any $B \in Mod_\Sigma(\Phi)$ follows from Proposition 2.6.   □

Example 2.24. The model $T_{\Sigma Bool}/\equiv_{\Phi Bool}$ of $Bool$ (see Example 2.14) is defined as follows:

$|T_{\Sigma Bool}/\equiv_{\Phi Bool}|_{bool} = \{[true]_{\equiv_{\Phi Bool}}, [false]_{\equiv_{\Phi Bool}}\}$

$true_{T_{\Sigma Bool}/\equiv_{\Phi Bool}} = [true]_{\equiv_{\Phi Bool}}$

$false_{T_{\Sigma Bool}/\equiv_{\Phi Bool}} = [false]_{\equiv_{\Phi Bool}}$

$\neg_{T_{\Sigma Bool}/\equiv_{\Phi Bool}} = \{[true]_{\equiv_{\Phi Bool}} \mapsto [false]_{\equiv_{\Phi Bool}}, [false]_{\equiv_{\Phi Bool}} \mapsto [true]_{\equiv_{\Phi Bool}}\}$

$\wedge_{T_{\Sigma Bool}/\equiv_{\Phi Bool}}$	$[true]_{\equiv_{\Phi Bool}}$	$[false]_{\equiv_{\Phi Bool}}$
$[true]_{\equiv_{\Phi Bool}}$	$[true]_{\equiv_{\Phi Bool}}$	$[false]_{\equiv_{\Phi Bool}}$
$[false]_{\equiv_{\Phi Bool}}$	$[false]_{\equiv_{\Phi Bool}}$	$[false]_{\equiv_{\Phi Bool}}$

where

$[true]_{\equiv_{\Phi Bool}} = \{true, \neg false, \neg(false \wedge true), \neg(false \wedge \neg false), \dots\},$

$[false]_{\equiv_{\Phi Bool}} = \{false, \neg true, \neg(true \wedge true), \neg(true \wedge \neg false), \dots\}.$

This is an initial model of *Bool* by the proof sketched for Theorem 2.23. $\Sigma Bool$-homomorphisms from $T_{\Sigma Bool}/\equiv_{\Phi Bool}$ to $A1$, $A2$, and $A3$ are as follows:

$h1\colon T_{\Sigma Bool}/\equiv_{\Phi Bool} \to A1$
$h1_{bool} = \{[true]_{\equiv_{\Phi Bool}} \mapsto \star, [false]_{\equiv_{\Phi Bool}} \mapsto \star\}$

$h2\colon T_{\Sigma Bool}/\equiv_{\Phi Bool} \to A2$
$h2_{bool} = \{[true]_{\equiv_{\Phi Bool}} \mapsto a, [false]_{\equiv_{\Phi Bool}} \mapsto b\}$

$h3\colon T_{\Sigma Bool}/\equiv_{\Phi Bool} \to A3$
$h3_{bool} = \{[true]_{\equiv_{\Phi Bool}} \mapsto 1, [false]_{\equiv_{\Phi Bool}} \mapsto 0\}$ ∎

Taking a presentation to denote the (non-empty) class of its initial models is called taking its *initial semantics*. The initiality property identifies a model of $\langle \Sigma, \Phi \rangle$ up to isomorphism: any two initial models are isomorphic, and any model isomorphic to an initial model is itself initial. We therefore refer to *the* initial model of a presentation.

Example 2.25. $A3$ is an initial model of *Bool* (see Example 2.14) since it is isomorphic to $T_{\Sigma Bool}/\equiv_{\Phi Bool}$. On the other hand, $A1$ and $A2$ are not isomorphic to $A3$ and hence are not initial models. This can be checked directly as well: for example, $\nexists h\colon A1 \to A2$ and $\nexists h\colon A1 \to A3$. ∎

2.10 Variations on a theme

The simple specification framework presented above is the classical one in the field of algebraic specifications. A wide variety of modifications have been made to increase its expressive power and to take account of the various features of software systems which it does not handle adequately. This section is devoted to a sketch of some of these modifications; details may be found in the cited references.

2.10.1 Conditional equations

Equational axioms can be generalized to (positive) conditional equational axioms of the form $\forall X.\ t_1 = t_1' \wedge \ldots \wedge t_n = t_n' \Rightarrow t_0 = t_0'$. A Σ-algebra A *satisfies* such an axiom if for every (S-sorted) function $v\colon X \to |A|$, if $v^{\#}(t_1) = v^{\#}(t_1')$ and $\ldots$ and $v^{\#}(t_n) = v^{\#}(t_n')$, then $v^{\#}(t_0) = v^{\#}(t_0')$. With these changes, most results still apply with appropriate minor modifications. For example, any presentation $\langle \Sigma, \Phi \rangle$, where Φ is a set of conditional Σ-equations, has an initial model which can be constructed in a similar way as in the proof of Theorem 2.23 (see, e.g., [MT92a]). There is also a sound and complete proof system for conditional equational consequence [Sel72].

2.10.2   Partial algebras

An obvious way to generalize the standard definition of an algebra is to allow partial functions as interpretations of operation names. Homomorphisms between such *partial algebras* are required to preserve definedness of operations, and (as usual) their results when these are defined. Term evaluation is defined as in ordinary algebras, except that terms need not have defined values. An equation $\forall X.\ t = t'$ is satisfied in a partial algebra A when for all valuations $v\colon X \to |A|$, the values of t and t' under v either are defined and equal, or are both undefined. Additional axioms are required to assert definedness: $\forall X.\ D(t)$ holds in A when the value of t is defined under all valuations of X in A. Every presentation $\langle \Sigma, \Phi \rangle$, where Φ is a set of Σ-equations and definedness formulas, has an initial model which contains no junk, is *minimally defined* (i.e., the value of a ground term t is defined only if $\Phi \models_\Sigma \forall \emptyset.\ D(t)$), and contains no confusion, i.e., the values of two ground terms t, t' are defined and equal only if $\Phi \models_\Sigma \forall \emptyset.\ t = t'$.

This is one possible approach to the specification of partial algebras, following [BW82b]. There are various other choices for the basic definitions [Rei87, Bur86].

2.10.3   Error algebras

To model operations that may produce erroneous or exceptional results, we can partition each of the carrier sets of an algebra into an *error* part and an *OK* part. Operations in signatures are classed as *safe* or *unsafe*, where the former are required to yield OK values when applied to OK arguments. Homomorphisms are required to preserve OK-ness. Like operations, variables in equations are classed as safe or unsafe; the former range over OK values only, while the latter range over all values. Again, all presentations have initial models, in which operations propagate errors unless otherwise specified.

The details of this approach may be found in [GDLE84]. Again, there are many other approaches, see for instance [Gog78] or [BBC86a].

2.10.4   Order-sorted algebras

In order to model sort inclusion and coercions, signatures may be enriched with an order relation on the set of sorts. An *order-sorted Σ-algebra* A is required to respect the sort ordering in the *order-sorted signature* Σ: if $s \leq s'$ in Σ then we require that $|A|_s \subseteq |A|_{s'}$. Overloading is forced by requiring operations to be applicable to values from subsorts of their argument sorts and to yield results in supersorts of their result sorts. Under certain conditions terms are guaranteed to have least sorts and unambiguous values. Then once more, all presentations have initial models and there is a version of the equational calculus that is sound and complete for order-sorted satisfaction.

For details see [GM85]. Alternative approaches are [Gog84, Poi84, Smo86]; see also [Mos93, GD94a, CHKM97].

2.10.5 First-order predicate logic

Signatures may be modified to enable them to include (typed) *predicate names* in addition to operation names, e.g., $\leq: nat \times nat$. Atomic formulas are then formed by applying predicates to terms; in *first-order predicate logic with equality*, the predicate $=: s \times s$ is implicitly available for any sort s. Formulas are built from atomic formulas using the usual logical connectives and quantifiers. Algebras are modified to include relations on their carriers to interpret predicate names (giving what are sometimes called *relational structures*). Homomorphisms are required to preserve predicates as well as operations. The satisfaction of a *sentence* (a formula without free variables) by an algebra is as usual in first-order logic. Presentations involving predicates and first-order axioms do not always have initial models or even reachable models. Details of first-order predicate logic for use in algebraic specifications may be found in, e.g., [GB92].

2.10.6 Higher-order functions

Higher-order functions (which take functions as parameters and/or return functions as results) can be accommodated by interpreting certain sort names as (subsets of) function spaces. Given a set S of (base) sorts, let $S^{\rightarrow}$ be the closure of S under formation of function types: $S^{\rightarrow}$ is the smallest set such that $S \subseteq S^{\rightarrow}$ and for all $s_1, \ldots, s_n, s \in S^{\rightarrow}$, $s_1 \times \cdots \times s_n \rightarrow s \in S^{\rightarrow}$. Then a *higher-order signature* Σ is a pair $\langle S, \Omega \rangle$ where Ω is an $S^{\rightarrow}$-indexed set of operation names. This determines an ordinary signature $\Sigma^{\rightarrow}$ comprised of the sort names $S^{\rightarrow}$ and the operation names in Ω (as constants of sorts in $S^{\rightarrow}$) together with operation names $apply: (s_1 \times \cdots \times s_n \rightarrow s) \times s_1 \times \cdots \times s_n \rightarrow s$ for every $s_1, \ldots, s_n, s \in S^{\rightarrow}$. A *higher-order Σ-algebra* is just an ordinary (total) $\Sigma^{\rightarrow}$-algebra, and analogously for the definitions of higher-order Σ-homomorphism, higher-order Σ-equation, higher-order presentation, etc. A higher-order Σ-algebra A is *extensional* if for all sorts $s_1 \times \cdots \times s_n \rightarrow s \in S^{\rightarrow}$ and values $f, g \in |A|_{s_1 \times \cdots \times s_n \rightarrow s}$, $f = g$ whenever $apply_A(f, a_1, \ldots, a_n) = apply_A(g, a_1, \ldots, a_n)$ for all $a_1 \in |A|_{s_1}, \ldots, a_n \in |A|_{s_n}$. In an extensional algebra A, every carrier $|A|_{s_1 \times \cdots \times s_n \rightarrow s}$ is isomorphic to a subset of the function space $|A|_{s_1} \times \cdots \times |A|_{s_n} \rightarrow |A|_s$. Higher-order equational presentations always have initial extensional reachable models. See [MTW88b] for details, and for alternative approaches see, e.g., [Poi86, Mei92].

2.10.7 Polymorphic types

Programming languages such as Standard ML [Pau96] can be used to define *polymorphic types* such as $\alpha\ list$ and *polymorphic values* such as the function $head: \forall \alpha.\ \alpha\ list \rightarrow \alpha$. To specify such types and functions, signatures are modified to contain *type constructors* in place of sort names. Terms built using these type constructors and *type variables* (such as α above) are the *polymorphic types* of the signature. The set Ω of operation names is then indexed by non-empty sequences of polymorphic types, where $f \in \Omega_{t_1...t_n,t}$ means $f \colon \forall FV(t_1) \cup ... \cup FV(t_n) \cup FV(t). \ t_1 \times \cdots \times t_n \to t$. There are various choices for algebras over such signatures. The most straightforward is to require each algebra A to incorporate a (single-sorted) *algebra of carriers*, $Carr(A)$, having sets which interpret types as values and an operation to interpret each type constructor. Then, for each operation $f \in \Omega_{t_1...t_n,t}$ and for each instantiation of type variables $i \colon V \to |Carr(A)|$, A has to provide a function $f_{A,i} \colon i^{\#}(t_1) \times \cdots \times i^{\#}(t_n) \to i^{\#}(t)$. Various conditions may be imposed to ensure that the interpretation of polymorphic operations is *parametric*, by requiring $f_{A,i}$ and $f_{A,i'}$ to be appropriately related for different type variable instantiations i, i'. Axioms contain (universal) quantifiers for type variables in addition to quantifiers for ordinary variables, as in System F [GLT89]; alternatively, type-variable quantification may be left implicit, as in Extended ML [KST97].

2.10.8   Non-deterministic functions

Non-deterministic functions may be handled by interpreting operation names in algebras as relations or, equivalently, as set-valued functions. Homomorphisms are required to preserve possible values of functions: for any homomorphism $h \colon A \to B$ and operation $f \colon s_1 \times \cdots \times s_n \to s$, if a is a possible value of $f_A(a_1, \ldots, a_n)$ then $h_s(a)$ is a possible value of $f_B(h_{s_1}(a_1), \ldots, h_{s_n}(a_n))$. Universally quantified inclusions between sets of possible values may be used as axioms: $t \subseteq t'$ means that every possible value of t is a possible value of t'. See [Nip86, Huß89, BS93, BK98] for details.

2.10.9   Continuous algebras

Following [Sco76], partial functions may be specified as least solutions of recursive equations. To accommodate this, we can use *continuous algebras*, i.e., ordinary (total) Σ-algebras with carriers that are complete partially ordered sets (so-called *cpos*) and operation names interpreted as *continuous functions* on these sets. The "bottom" element $\perp$ of the carrier for a sort, if it exists, represents the completely undefined value of that sort. The order on carriers induces an order on (continuous) functions in the usual fashion. A homomorphism between continuous algebras is required to be continuous as a function between cpos. For details see, e.g., [GTWW77]. It is possible to define a language of axioms that allows direct reference to least upper bounds of chains and/or to the order relation itself (see, e.g., [TW86]).

3 From Total Equational to Partial First-Order Logic

Maura Cerioli[1], Till Mossakowski[2], and Horst Reichel[3]

[1] DISI – Dipartimento di Informatica e Scienze dell'Informazione
Università di Genova, Via Dodecaneso 35, Genova 16146, Italy
cerioli@disi.unige.it   http://www.disi.unige.it/person/CerioliM/
[2] Universität Bremen, Fachbereich Mathematik und Informatik
Postfach 330440, D-28334 Bremen, Germany
till@informatik.uni-bremen.de
http://www.informatik.uni-bremen.de/~till/
[3] TU Dresden, Fakultät Informatik, Institut für Theoretische Informatik
D-01062 Dresden, Germany
reichel@tcs.inf.tu-dresden.de
http://wwwtcs.inf.tu-dresden.de/~reichel/english-index.html

Abstract. The focus of this chapter is the incremental presentation of partial first-order logic, seen as a powerful framework where the specification of most data types can be directly represented in the most natural way. Both model theory and logical deduction are fully described.

Alternatives to partiality, like (variants of) error algebras and order-sortedness, are also discussed, emphasizing their uses and limitations.

Moreover, both the total and the partial (positive) conditional fragments are investigated in detail, and in particular the existence of initial (free) models for such restricted logical paradigms is proved.

Finally some more powerful algebraic frameworks are sketched.

Equational specifications, introduced in last chapter, are a powerful tool to represent the most common data types used in programming languages and their semantics. Indeed, Bergstra and Tucker have shown in a series of papers (see [BT87] for a complete exposition of results) that a data type is semicomputable if and only if it is (isomorphic to) the initial model of a finite set of equations over a finite set of symbols. However this result has two main limitations.

The first point is that the initial approach is appropriate only if the specifying process of the data type has been completed, because it defines one particular realization (up to isomorphism), instead of a class of possible models, still to be refined. In particular, if the data type has partial functions, the treatment of the "erroneous" elements must already be fixed in all detail.

The second, more problematic, point is that, since the expressive power of the logic used to axiomatize the data types is so poor, quite often it is not possible to define the intended data type through its abstract properties, but it is necessary to describe one of its possible implementations. Technically speaking, in order to define a data type, auxiliary types and operators can be needed, drastically decreasing the level of abstraction of the specification and reducing its readability and naturalness. Consider the following example, showing an artificial but simple data type, that cannot be specified by a finite set of equations. Other, far more interesting, data types, like the algebra of regular sets, cannot be expressed in this way either, but the proof that more powerful logics are needed is made too complicated by their richer structure.

Example 3.1. We want to specify a data type having sorts for the natural numbers and for their quotient identifying all odd numbers, with the usual constructors for the natural numbers and an operation associating each number with its equivalence class. Thus, the signature of the type should be the following.

> **sig** $\Sigma_{\text{Nat}} =$
> > **sorts** $\text{nat}, \text{nat}/\equiv$
> > **opns** $\text{zero}: \;\rightarrow \text{nat}$
> > > $\text{succ}: \text{nat} \rightarrow \text{nat}$
> > > $nat_\equiv: \text{nat} \rightarrow \text{nat}/\equiv$

Let us see whether we can give a finite set E of equations on this signature in a way that the initial model of such a specification is (isomorphic to) our intended data type.

First of all note that our set E cannot contain any nontrivial equation of sort **nat**. Indeed, using $f^k(\ldots)$ to denote the iterative application of any function f a number k times, an equation of sort **nat** can have (up to symmetry) four forms.

$\text{succ}^k(\text{zero}) = \text{succ}^n(\text{zero})$ – either trivially valid or not valid in our intended model, as the term $\text{succ}^k(\text{zero})$ is interpreted by the number k.

$\text{succ}^k(\text{zero}) = \text{succ}^n(x)$ – cannot be satisfied by a nontrivial model, because the left-hand side is interpreted as a constant and the right-hand side changes its value depending on the interpretation of x; in particular in our intended model it does not hold if $k+1$ is substituted for x.

$\text{succ}^k(x) = \text{succ}^n(x)$ – either trivial or does not hold in our intended model if 0 is substituted for x.

$\text{succ}^k(x) = \text{succ}^n(y)$ with distinct variables x and y – does not hold if 0 is substituted for x and $k+1$ for y.

Let us analyze, analogously, the nontrivial equalities of sort **nat**/$\equiv$ to see whether they can belong to E. Since terms of sort **nat**/$\equiv$ are given by the application of $nat_\equiv$ to terms of sort **nat**, we again have four cases.

$nat_\equiv(\text{succ}^k(\text{zero})) = nat_\equiv(\text{succ}^n(\text{zero}))$ – not valid in our intended model if either k or n is even (unless $k = n$, in which case it is trivial), while it is valid whenever both k and n are odd.

$nat_{\equiv}(\text{succ}^k(\text{zero})) = nat_{\equiv}(\text{succ}^n(x))$ – not valid in our intended model, because $nat_{\equiv}(\text{succ}^k(\text{zero}))$ is interpreted as a constant (either k, if k is even, or the class of all odd numbers), while the value of $nat_{\equiv}(\text{succ}^n(x))$ changes, and in particular $nat_{\equiv}(\text{succ}^n(\text{zero}))$ and $nat_{\equiv}(\text{succ}^{n+1}(\text{zero}))$ have different interpretations, the classes of an even and an odd number respectively, so $nat_{\equiv}(\text{succ}^k(\text{zero})) = nat_{\equiv}(\text{succ}^n(x))$ cannot be true if both 0 and 1 can be substituted for x.

$nat_{\equiv}(\text{succ}^k(x)) = nat_{\equiv}(\text{succ}^n(x))$ – either trivial or does not hold in our intended model, because if either k or n is even it is not satisfied after substituting 0 for x, otherwise it is not satisfied substituting 1 for x.

$nat_{\equiv}(\text{succ}^k(x)) = nat_{\equiv}(\text{succ}^n(y))$ with distinct variables x and y, – does not hold if k is substituted for x and $n + 2k + 2$ for y.

Therefore, all nontrivial equations in E must have the form

$$nat_{\equiv}(\text{succ}^{2k+1}(\text{zero})) = nat_{\equiv}(\text{succ}^{2n+1}(\text{zero})).$$

Any such equation can only identify the result of the interpretation of $nat_{\equiv}$ on the two odd numbers $2k + 1$ and $2n + 1$. Thus, if E is finite, only a finite number of identities between terms of the form $nat_{\equiv}(\text{succ}^{2k+1}(\text{zero}))$ can be inferred.

Therefore, there is not a finite equational specification of the required data type using the signature Σ_{Nat}. However, if we enrich the signature, we can define the data type, using the extra symbols. Let us consider the following specification.

> **spec** Odd = **enrich** Σ_{Nat} **by**
> **sorts**   bool
> **opns**   true,false: $\rightarrow$ bool
>   odd: nat $\rightarrow$ bool
>   cond: bool $\times$ nat $\times$ nat $\rightarrow$ nat
> **vars** x, y : nat
> **axioms** cond(true, x, y) = x
>   cond(false, x, y) = y
>   odd(zero) = false
>   odd(succ(zero)) = true
>   odd(succ(succ(x))) = odd(x)
> $nat_{\equiv}(x)$ = cond(odd(x), $nat_{\equiv}$(succ(zero)), $nat_{\equiv}(x)$)   ∎

Roughly speaking, equational specifications are sufficiently expressive to define initially any semicomputable (total) data type, because recursive functions can be described using recursion and conditional choice. However recursion is implicitly embedded in the equational framework, as recursive definitions *are* given by equalities, and Booleans and conditional choice can be implemented equationally, as in the previous example. Thus, the intuition here is that if the logic used in specifications is poor, for instance, equational, complex data types can still be expressed by implementing *inside* the data type a "Boolean" sort with operations to represent logical connectives, and by translating any assertion ϕ at the metalevel into an equation between the Boolean term corresponding to ϕ and the constant value true.

Since all logical connectives can be described equationally, all theories of predicate calculus (without quantifiers) can be translated into an equational specification with *hidden sorts and operations*, that is, adding auxiliary symbols to the data type, that should be not exported to the users of the specification. However, the resulting specification lacks abstraction, because that which is logically a statement on the data type has been implemented by an equation between elements of the data type itself. In other words, the equational specification is actually an *implementation* of the natural axiomatic description of the data type. In particular this implies that we have to fix the data type of Booleans to have just two elements. We cannot do this within positive conditional logics (logics which have a nearly-executable proof theory).

This chapter will be devoted to introducing an algebraic framework sufficiently expressive in order to represent the most common data types *directly*. As we have seen, the first obstacle to overcome is the limitation of the formulas that can be used to specify the data types. Thus, we need a richer logic, but not too rich. Indeed, we want to keep the logical language easy to read and implement, in order to have tools for rapid prototyping of the data types. Moreover, we do not want to lose the initial semantic approach. Thus, our formulas should be able to describe only classes of algebras having an initial object.

A far more challenging problem is the specification of partial functions. Many data types in practice have partial operations whose result on some input is "erroneous". Sometimes such errors can be avoided simply by using a better typing, as it is the case, in a programming language with declarations

```
type   my_array=array[1..k] of T
var    i:integer;
       A: my_array;
```

for expressions of the form $A[i]$ if i assumes values outside the array range, but that could be forbidden by declaring i of type $1..k$.

Even if a better typing is not possible (or not convenient), most errors can be detected statically and hence axioms to identify them with some "error element", representing an error message, can be given.

However, when a partial recursive function that has no recursive domain has to be specified, it is obviously not possible to detect the errors introduced by its application. Hence there does not exist a (total) specification of the function identifying all its erroneous applications with some "error". Note that partial recursive functions without recursive domains are needed, for instance, whenever describing the semantics of (universal) programming languages. Hence any algebraic approach has to deal with them in some way – otherwise, it can be used only to describe the data types of a program but not to verify properties of the programs using them.

As usual, the easier the theory, the harder its use. Thus, in the total equational framework, having nice, intuitive semantics and efficient rewriting techniques, specifications of complex data types are often hard to find (if at all). On the other hand, making the framework more powerful can make its theory too complex and hence hinder its understanding by users. Here we will incrementally introduce a very expressive partial framework, showing how and when its features are needed or are simply convenient, so that users can restrict themselves to one of its subtheories, if dealing with sufficiently easy problems.

3.1   Conditional axioms

Following the guideline of Example 3.1, we need a way to impose equations on only those values that satisfy a condition. This was implemented in that example by introducing the operation cond, corresponding to a conditional choice, and then imposing the axiom

$$nat_{\equiv}(x) = \mathrm{cond}(\mathrm{odd}(x), nat_{\equiv}(\mathrm{succ}(\mathrm{zero})), nat_{\equiv}(x))$$

which corresponds, logically, to requiring

$$nat_{\equiv}(x) = nat_{\equiv}(\mathrm{succ}(\mathrm{zero})) \text{ if } \mathrm{odd}(x).$$

Thus, we move from *equational* to *equational conditional*, or simply conditional, specifications. Therefore, in the following the axioms will have the form

$$t_1 = t_1' \wedge \ldots \wedge t_n = t_n' \Rightarrow t = t'$$

which is *satisfied* by a valuation if the consequence $t = t'$ is satisfied whenever all the premises $t_i = t_i'$ are satisfied and *holds* in a model iff it is satisfied by all valuations in that model (see Definition 2.7.1 for the formal details).

Although, as shown in [BT87], conditional specifications, as well as equational ones, need hiding sorts and operations to define all semicomputable total data types, they are strictly more expressive than equational axioms, because the data type introduced in Example 3.1, which cannot be axiomatized by a finite set of equations on its signature, can be easily defined by the following conditional axiom

$$\phi_{\mathrm{odd}} \quad nat_{\equiv}(x) = nat_{\equiv}(\mathrm{succ}(\mathrm{zero})) \Rightarrow nat_{\equiv}(\mathrm{succ}(\mathrm{succ}(x))) = nat_{\equiv}(x)$$

Note that the above specification works, because all even numbers are distinct from odd numbers. Indeed, let us consider the same problem, but with $nat/\equiv$, the quotient identifying all odd numbers *and* 0. Then the axiom ϕ_{odd} is incorrect, because if the classes of 0 and 1 coincide when instantiating x on 0, we identify all integers.

The point is that the informal specification of $nat_\equiv$ is based on the distinction between odd and even numbers, but our signature does not have the syntactical means to express this concept. Thus, although by using conditional axioms we actually enrich the expressive power of our logic, the logic we obtain is still too poor, because the atoms we can use to build axioms are only equations, while we would need symbols to state whether a number is even or odd. Of course we can always use the same trick, implementing a Boolean sort with a odd Boolean function, but it is much clearer to add a facility to our specification framework, providing symbols for *predicates*.

Definition 3.2. A *first-order signature* Σ is a triple (S, Ω, Π) where

- (S, Ω) is a many-sorted signature;
- Π is an S^*-sorted family (of *predicate symbols*).

Given a first-order signature $\Sigma = (S, \Omega, \Pi)$, the Σ-terms on an S-sorted family of variables X, denoted by $T_\Sigma(X)$, are the many-sorted term algebra $T_{(S,\Omega)}(X)$ on the many-sorted signature underlying Σ. ◇

Example 3.3. A reasonable signature for Example 3.1 then is the following.

    sig $\Sigma_{\text{Nat}\,P}$ = **enrich** Σ_{Nat} **by**
      preds is_odd: nat

The signature $\Sigma_{\text{Nat}\,P}$ completely captures our intuition that we want to enrich the natural numbers by the new sort of their quotient and that to describe the equivalence relation we discriminate odd from even numbers and hence we need a symbol stating whether a number is odd. ∎

In the rest of this section, let $\Sigma = (S, \Omega, \Pi)$ be a first-order signature. In each Σ-structure predicate symbols are interpreted by their truth set.

Definition 3.4. A Σ-structure consists of

- an (S, Ω) many-sorted algebra A, called the *underlying many-sorted algebra*;
- for each $p: s_1 \times \cdots \times s_n \in \Pi$ a subset p_A of $|A|_{s_1} \times \cdots \times |A|_{s_n}$, representing the *extent* of p in A, that is, the tuples of elements on which the predicate is true.

Given two Σ-structures A and B, a *homomorphism of Σ-structures* from A into B is a truth-preserving homomorphism of many-sorted algebras between the underlying many-sorted algebras, that is, a homomorphism $h: A \to B$ such that, if $(a_1, \ldots, a_n) \in p_A$, then $(h_{s_1}(a_1), \ldots, h_{s_n}(a_n)) \in p_B$ for all $p: s_1 \times \cdots \times s_n \in \Pi$ and all $a_i \in |A|_{s_i}$ for $i = 1, \ldots, n$.

Let C be a class of Σ-structures. A Σ-structure I is *initial* in C iff $I \in C$ and for each $A \in C$ a unique homomorphism of Σ-structures $!_A: I \to A$ exists.

Given a Σ-structure A and an S-sorted family of variables X, *variable valuations* and *term evaluations* for $T_\Sigma(X)$ in A are, respectively, variable valuations and term evaluations for $T_\Sigma(X)$ in the many-sorted algebra underlying A.   ◇

Notice the difference between enriching a signature by a Boolean sort and some operations, as in the equational presentation of Example 3.1, and by predicates. In the former case the models are *larger* than the models in which we are interested, in the sense that sets and functions have been added to their structure. In the latter the models are *richer*, because though they are the same algebras, as collections of sets and functions, the language which we use to handle them is more expressive (and consequently we now need to know how to interpret some other conditions in them). Thus, for instance, we have the same number of elements, but we know each element more clearly and hence are able to state the (un)truth of some property about them.

For instance, consider once again Example 3.1. Then a model of the specification Odd is the algebra

algebra $N_{eq} =$
   Carriers
      $|N_{eq}|_{nat} = N$
      $|N_{eq}|_{nat/\equiv} = 2N \cup \{\bar{1}\}$
      $|N_{eq}|_{bool} = \{T, F\}$
   Functions
      $zero_{N_{eq}} = 0$
      $true_{N_{eq}} = T$
      $false_{N_{eq}} = F$
      $succ_{N_{eq}}(n) = n+1$
      $plus_{N_{eq}}(n, m) = n + m$
      $nat{\equiv}_{N_{eq}}(n) = \begin{cases} n, & \text{if there is } k \text{ s.t. } n = 2k \\ \bar{1}, & \text{otherwise} \end{cases}$
      $cond_{N_{eq}}(b, n, m) = \begin{cases} n, & \text{if } b = T \\ m, & \text{otherwise} \end{cases}$
      $odd_{N_{eq}}(n) = \begin{cases} F, & \text{if there is } k \text{ s.t. } n = 2k \\ T, & \text{otherwise} \end{cases}$

and N_{eq} consists of the algebra we require, that is, Σ_{Nat}-reduct plus a set and a group of functions. Using predicates instead, we get the first-order structure

algebra $N_p =$
   Carriers
      $|N_p|_{nat} = N$
      $|N_p|_{nat/\equiv} = 2N \cup \{\bar{1}\}$
   Functions
      $zero_{N_p} = 0$
      $succ_{N_p}(n) = n+1$
      $plus_{N_p}(n, m) = n + m$

$$nat\equiv_{N_p}(n) = \begin{cases} n, & \text{if there is } k \text{ such that } n = 2k \\ \bar{1}, & \text{otherwise} \end{cases}$$

Predicates

$$\texttt{is\_odd}_{N_p} = \{n \mid \exists k \in N \text{ such that } n = 2k + 1\}$$

which is exactly as we wanted, enriched by the information as to which elements of its carrier are odd.

Notice that with positive conditional axioms and predicates, we cannot talk about the falsehood of predicates, while the approach of enriching a signature by a Boolean sort and treating predicates as operations onto this sort does not have such a restriction. Thus, apparently the latter approach is richer. But the capability to use negative information has the drawback that, if the Boolean sort has to contain only the interpretations of **true** and **false**, the untruth, as well as the truth, of each relation has to be stated. Thus, in particular, semicomputable relations cannot be conditionally axiomatized, because their falseness cannot be recursively axiomatized.

Of course, it is possible to ensure that the Boolean sort has just two elements, using a more complex (first-order) axiom, without specifying the actual result of the application of the functions representing relations. However, in this case the simpler proof theory of positive conditional axioms cannot be used and the existence of free extensions is not guaranteed, that is, we may lose the capability of extending the models of some data type in a uniform way, because the new operations yield Boolean terms whose interpretation could be true as well as false.

Another merit of the approach using predicates is the easy specification of inductively defined relations using initial or free semantics (or initiality or freeness constraints). This is possible only with the predicate approach, which combines the property of the existence of initial models (and free extensions) with the flexibility of homomorphisms (which have to preserve truth, but not falsehood). With this, it is possible to specify the minimal relation satisfying some set of axioms. Using free extensions, transitive closure, for example, can be specified by just stating that the transitive closure contains the original relation and it is transitive.[1] Note that initiality constraints are a second-order principle, so proof theory here becomes more complex, since an induction principle is needed.

Let us now formally define conditional axioms and their validity.

Definition 3.5. Let Σ be the first-order signature (S, Ω, Π) and X be an S-sorted family of variables. The set of Σ-*atoms* on X is

$$At(\Sigma, X) = \{t = t' \mid t, t' \in |T_\Sigma(X)|_s\} \cup \{p(t_1, \dots, t_n) \mid$$
$$p \colon s_1 \times \cdots \times s_n \in \Pi \text{ and } t_i \in |T_\Sigma(X)|_{s_i}, i = 1, \dots, n\}$$

---

[1] When trying to specify transitive closure in a purely functional style, we either get additional truth values in the free extension, or equate **true** and **false**. Adding a first-order axiom stating that there are exactly two truth values destroys the existence of free extensions. These problems arise because homomorphisms are not so flexible: they have to preserve both truth and falsehood.

The set of Σ-*conditional axioms* on X is

$$Cond(\Sigma, X) = \{\forall X.\epsilon_1 \wedge \cdots \wedge \epsilon_n \Rightarrow \epsilon_{n+1} \mid \epsilon_i \in At(\Sigma, X), i = 1, \dots, n+1\}$$

$\diamond$

In other words, conditional axioms are positive Horn-Clauses, built using the predicates in Π and the equality symbol. As for many-sorted algebras, quantification is explicit to avoid inconsistent deductions in the case of possibly empty carriers.

Definition 3.6. Given a Σ-structure A, we say that A *satisfies* a conditional axiom $\forall X.\varphi \in Cond(\Sigma, X)$ (denoted by $A \models_\Sigma \forall X.\varphi$) if all valuations v for X in A *satisfy* φ (denoted by $v \Vdash \varphi$), where satisfaction of a conditional axiom by a valuation is defined by the following rules:

- $v \Vdash t = t'$ iff $v^\#(t) = v^\#(t')$
- $v \Vdash p(t_1, \dots, t_n)$ iff $(v^\#(t_1), \dots, v^\#(t_n)) \in p_A$
- $v \Vdash \epsilon_1 \wedge \cdots \wedge \epsilon_n \Rightarrow \epsilon_{n+1}$ iff $v \Vdash \epsilon_{n+1}$ or there is an ϵ_i such that $v \not\Vdash \epsilon_i$

A *presentation* consists of a first-order signature Σ and a set AX of Σ-conditional axioms. The class of models of a presentation $Sp = \langle \Sigma, AX \rangle$, denoted by $Mod(Sp)$, consists of all those Σ-structures satisfying the axioms in AX. $\diamond$

Exercise 3.7. Generalize the notion of signature morphism, reduct, and sentences translation, and prove the satisfaction lemma for first-order structures with conditional axioms.

Most of the theory of many-sorted algebras carries over to Σ-structures with no difficulty, though the behavior of the model theory of many-sorted first-order structures resembles more that of the model theory for *partial* algebras/first-order structures (which is discussed in the next section) than that for total algebras.

The model categories of total algebras are of a more "algebraic" flavor, while model categories of first-order structures and partial algebras have a more "topologically algebraic" flavor. In [AHS90], a precise mathematical background is given for this. The main point is that both for first-order structures and partial algebras, there are bijective homomorphisms that are not isomorphisms, while for total algebras, bijective homomorphisms are always isomorphisms. Thus the notions of full and closed homomorphisms, and their interconnection with relative and closed substructures (Proposition 3.38), already make sense for first-order structures. But since first-order structures are a special case of partial first-order structures, we refer to Section 3.3.1 for details.

However, we present a sound and complete calculus for conditional axioms that we will "borrow" for the partial case as well, and show how the calculus itself defines the initial (free) model for a presentation. Indeed, conditional axioms, as in the case without predicates, define *quasivarieties* and hence their model classes always admit initial models, which may be "computed" by a proof calculus.

Definition 3.8. Let $\Sigma = (S, \Omega, \Pi)$ be a first-order signature. The $\vdash$ inference system consists of the following axioms and inference rules, where we assume that, as usual, ϵ and η, possibly decorated, are atoms over Σ, φ is a conditional axiom over Σ, Φ is a countable set of conditional axioms over Σ, X and Y are S-sorted family of variables, and t, t', t'', t_i, t'_i are Σ-terms.

Congruence Axioms

$$\Phi \vdash \forall X.t = t$$
$$\Phi \vdash \forall X.t = t' \Rightarrow t' = t$$
$$\Phi \vdash \forall X.t = t' \wedge t' = t'' \Rightarrow t = t''$$
$$\Phi \vdash \forall X.t_1 = t'_1 \wedge \ldots \wedge t_n = t'_n \Rightarrow f(t_1, \ldots, t_n) = f(t'_1, \ldots, t'_n)$$
$$\Phi \vdash \forall X.t_1 = t'_1 \wedge \ldots \wedge t_n = t'_n \wedge p(t_1, \ldots, t_n) \Rightarrow p(t'_1, \ldots, t'_n)$$

Proper Axioms

$$\Phi \vdash \forall X.\varphi \qquad \text{for } \forall X.\varphi \in \Phi$$

Substitution

$$\frac{\Phi \vdash \forall X.\varphi}{\Phi \vdash \forall Y.\varphi[\theta]} \quad \text{for } \theta : X \longrightarrow |T_\Sigma(Y)|$$

Here, $\varphi[\theta]$ is the formula φ with each term in it translated by $\theta^{\#}$.

Cut Rule

$$\frac{\Phi \vdash \forall X.\epsilon_1 \wedge \cdots \wedge \epsilon_n \Rightarrow \eta_i \qquad \Phi \vdash \forall Y.\eta_1 \wedge \cdots \wedge \eta_k \Rightarrow \epsilon}{\Phi \vdash \forall X \cup Y.\eta_1 \wedge \cdots \wedge \eta_{i-1} \wedge \epsilon_1 \wedge \cdots \wedge \epsilon_n \wedge \eta_{i+1} \wedge \ldots \wedge \eta_k \Rightarrow \epsilon}$$

$\diamond$

Proposition 3.9. *The calculus introduced in Definition 3.8 is sound, that is, if $\Phi \vdash \forall X.\varphi$ and $M \models \Phi$, then $M \models \forall X.\varphi$.*

Exercise 3.10. Using the congruence axioms and the cut rule, show that $\Phi \vdash \forall X.\epsilon \Rightarrow \epsilon$ for all atoms ϵ on X.

As for the equational case, the proposed calculus here is also complete with respect to equations without variables and the proof is by building a reachable model satisfying exactly the deduced equations. Therefore, since the calculus is sound too, that model is initial, satisfying the *no-junk & no-confusion* conditions.

Definition 3.11. Let $Sp = \langle \Sigma, AX \rangle$ be a presentation and $\Theta \subseteq At(\Sigma, X)$ be a finite set of atoms. Then the structure $F_{Sp}(X.\Theta)$ has as underlying algebra $T_\Sigma(X)/{\equiv_\Theta}$, where $t \equiv_\Theta t'$ if and only if $AX \vdash \forall X. \bigwedge \Theta' \Rightarrow t = t'$ for some $\Theta' \subseteq \Theta$, and a predicate $p\colon s_1 \times \cdots \times s_n \in \Pi$ is interpreted as

$$\{\,([t_1]_{\equiv_\Theta}, \ldots, [t_n]_{\equiv_\Theta}) \mid AX \vdash \forall X. \bigwedge \Theta' \Rightarrow p(t_1, \ldots, t_n) \text{ for some } \Theta' \subseteq \Theta\,\}$$

where we denote by $[t]_{\equiv_\Theta}$ the equivalence class with respect to $\equiv_\Theta$ of any term t.   ⋄

It can be immediately verified that $\equiv_\Theta$ is a many-sorted congruence, because of the first four axioms and the Cut Rule. Moreover, because of the fifth axiom and the Cut Rule, the interpretation of p in $F_{Sp}(X.\Theta)$ is well defined.

Lemma 3.12. *Each valuation $v\colon Y \longrightarrow F_{Sp}(X.\Theta)$ can be factorized (in general not uniquely) through $[\_]_{\equiv_\Theta}\colon T_\Sigma(X) \longrightarrow F_{Sp}(X.\Theta)$ as follows*

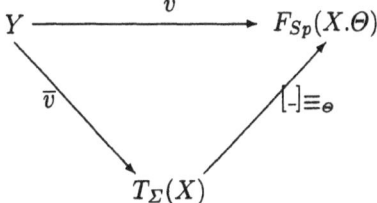

Moreover, for any $\bar{v}$ s.t. $[\_]_{\equiv_\Theta} \circ \bar{v} = v$ and any atom $\epsilon \in At(\Sigma, Y)$, we have $v \Vdash \epsilon$ if and only if $AX \vdash \forall X. \bigwedge \Theta' \Rightarrow \bar{v}^\#(\epsilon)$ for some $\Theta' \subseteq \Theta$.

Theorem 3.13. *Using the notation of Definition 3.11,*

1. *$F_{Sp}(X.\Theta)$ is a Sp-algebra.*
2. *The valuation $\iota\colon X \longrightarrow F_{Sp}(X.\Theta)$ given by $\iota(x) = [x]_{\equiv_\Theta}$ satisfies Θ and is universal with respect to this property, i.e., for any valuation $v\colon X \longrightarrow A$ into a Sp-algebra A satisfying Θ, there exists exactly one homomorphism $\tilde{v}\colon F_{Sp}(X.\Theta) \to A$ with $\tilde{v} \circ \iota = v$*

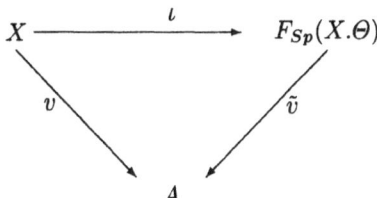

3. *If $\Theta = \emptyset$ then $F_{Sp}(X.\Theta)$ and ι are the free objects in the model class of AX.*

The calculus proposed can deduce all conditional formulas valid in all models in a stronger form, that is, with possibly fewer premises. We call this property *practical completeness*. Obviously any practically complete system can be made complete in the usual sense by adding a weakening rule of the form

$$\frac{\Phi \vdash \forall X. \Theta \Rightarrow \epsilon}{\Phi \vdash \forall X. \Theta' \Rightarrow \epsilon} \quad \text{for } \Theta \subseteq \Theta'$$

which we do not include, as the resulting system would be less efficient.

Theorem 3.14. *The calculus is practically complete, that is, if*

$$\Phi \models \forall X. \bigwedge \Theta \Rightarrow \epsilon,$$

then

$$\Theta' \subseteq \Theta \text{ exists such that } \Phi \vdash \forall X. \bigwedge \Theta' \Rightarrow \epsilon.$$

If $\Theta = \emptyset$, $F_{Sp}(X.\Theta)$ is called the *free Sp-structure over* X, written $F_{Sp}(X)$. Moreover, if $X = \emptyset$, $F_{Sp}(X)$ is called the *initial Sp*-structure, written I_{Sp}.

Let us look at an incremental use of conditional specifications to describe a data type as the initial model of a presentation.

Example 3.15. Let us consider the problem of the definition of a very primitive dynamic data type that is a subset of the CCS language. The *Calculus of Communicating Systems* has been introduced to study in isolation the problems due to concurrency and to describe reactive systems. It provides primitives to express the interactions between the components of complex systems. The intuition is that there is a set of *agents* that can perform *actions* either individually or in cooperation with each other. The description of the activity of an agent at a fixed instant cannot be given by a function because the agents are able to perform nondeterministic choices. Therefore, the activity is defined as a *transition* predicate stating how an agent, performing an action, evolves in another.

The starting point is the specification of the possible actions which we assume is given by the specification Action[2]. Although such a specification can be as complex as needed by the concrete problem of concurrency we want to describe, at this level the only interesting feature is that each action determines a *complementary* action, representing the subject/object viewpoint of two interacting agents. For instance, the complementary action of *sending* a message is *receiving* it (and vice versa). Moreover, there is an *internal* action τ given by the composition of an action with its complement. This represents the abstraction of a system composed of two agents interacting between them, that performs a change of its internal state without effects on the external world. Here and in the following we use the notation _ to denote the place of operands in an infix notation.

```
spec Action_ =
    sorts   Act
    opns    τ: → Act
            bar: Act → Act
    vars    a : Act
    axioms bar(τ) = τ
            bar(bar(a)) = a
```

---

[2] In this case, as in most complex examples, the data type in which we are interested is defined using simpler data types, that, of course, should be not modified. The elementary data preservation is guaranteed, in this and in the following examples, by two stronger properties: no operations with an elementary type as result are added and all the equations (and predicate assertions) in the consequences of the axioms are of nonelementary sort (involve nonelementary predicates).

As examples of actions, we can think of instructions like **send** or **receive**.

Now we add the sort **Agents** with the idle agent, which cannot perform any action, operations for prefixing an action, parallel composition, and nondeterministic choice. The dynamic aspects of the data type are captured by a *transition* predicate.

spec CCS_ = **enrich** Action **by**
    sorts Agents
    opns λ: $\to$ Agents
            _._: Act $\times$ Agents $\to$ Agents
            _||_, _+_: Agents $\times$ Agents $\to$ Agents
    preds _ $\overset{\_}{\Rightarrow}$ _: Agents $\times$ Act $\times$ Agents
    vars a : Act; p, q, p', q' : Agents
    axioms $p + \lambda = p$
              $p||\lambda = p$
              $p + q = q + p$
              $p||q = q||p$
              $a.p \overset{a}{\Rightarrow} p$
              $p \overset{a}{\Rightarrow} p' \Rightarrow p + q \overset{a}{\Rightarrow} p'$
              $p \overset{a}{\Rightarrow} p' \Rightarrow p||q \overset{a}{\Rightarrow} p'||q$
              $p \overset{a}{\Rightarrow} p' \wedge q \overset{\text{bar}(a)}{\Rightarrow} q' \Rightarrow p||q \overset{\tau}{\Rightarrow} p'||q'$

The axioms stating equalities between agents capture the properties of the operations between agents, but note that there are agents with the same transition capability that are not identified, for instance, $(a.\lambda + b.\lambda)||c.\lambda$ and $(a.\lambda||c.\lambda) + (b.\lambda||c.\lambda)$, both of which can perform either a or b but then are in a situation where c is the only move available.

Thus, the given axioms leave open many different semantics defined at a metalevel in terms of the action capabilities of agents. But more restrictive axioms could be imposed as well to describe more tightly the operations on agents, for example, a distributive law $(p + p')||q = (p||q) + (p'||q)$ would impose the equivalence of terms disregarding the level where the nondeterministic choice has taken place.

In our specification there is no way to force an action to be performed instead of another in some context, if both choices are available for that component. This capability is usually achieved by *hiding* some action in an agent, so that it cannot be individually performed but is activated only by a parallel interaction with an agent capable of its complementary action. To axiomatize this construct we must be able to say whether two actions are equal or not, in order to allow all actions but the restricted one. Note that the use of equality is not sufficient, because we want to express properties with inequalities in premises such as $a \neq b \wedge p \overset{a}{\Rightarrow} p' \Rightarrow p_{|b} \overset{a}{\Rightarrow} p'_{|b}$, saying that if p has the capability of making an action a and a is not the action b we want to hide, then the restriction of p can perform a as well. This is a limitation of conditional axioms: whenever the negation of a property is needed in the premises of an axiom, it must be introduced as a new symbol and axiomatized. Equivalently, the property (in this case the equality) must be expressed not as a predicate, but as a Boolean function, introducing Boolean sort and operations as well.

Thus, let us assume that the specification of actions is actually richer than first proposed and includes a predicate $\mathtt{different} : \mathtt{Act} \times \mathtt{Act}$. Then we can enrich the agent specification

> **spec** CCS = **enrich** CCS_ **by**
>   **opns**   _|_: **Agents** × **Act** → **Agents**
>   **vars** $a, b : \mathtt{Act}; p, p' : \mathtt{Agents}$
>   **axioms** $p \overset{a}{\Rightarrow} p' \wedge \mathtt{different}(a, b) \Rightarrow p_{|b} \overset{a}{\Rightarrow} p'_{|b}$

Negation of equality (and more generally of atomic sentences) is not the only kind of logical expression that is not allowed by the conditional framework.

For instance, let us suppose that we want to define a predicate specifying that an agent is allowed to perform at most one action. Then basically we would like to give the following specification.

> **spec** $\mathrm{CCS_1}$ = **enrich** CCS **by**
>   **preds**   _must do_: **Agents** × **Act**
>   **vars** $a, b : \mathtt{Act}; p, p' : \mathtt{Agents}$
>   **axioms** p must do $a \Leftrightarrow (\forall p' : \mathtt{Agents}.\forall b : \mathtt{Act}.p \overset{b}{\Rightarrow} p' \Rightarrow a = b)$

But this is not, nor can be reduced to, a conditional specification. Indeed, it has no initial model, because the minimality of the transition predicate conflicts with the minimality of the predicate must do.

The lack of an initial model, as well as the need for a more powerful logic, is quite common whenever functions and predicates are described through their properties rather than by an algorithm computing them. This situation is unavoidable in the phase of the *requirement* specification of a data type, when the implementation details are still left underspecified. (The fixing of details is usually done in the *design* phase.) Indeed, the given description of the predicate must do, for instance, does not depend on the structure of the agents, which could still be changed leaving the specification unaffected, nor does it suggest a way to compute/verify whether it holds on a given agent and action. However, exploiting the information we have on the definition of the transition predicate, we can also specify the same predicate using conditional axioms:

> **spec** $\mathrm{CCS_2}$ = **enrich** CCS **by**
>   **preds**   _must do_: **Agents** × **Act**
>   **vars** $a : \mathtt{Act}; p, q : \mathtt{Agents}$
>   **axioms** $a.p$ must do a
> p must do $a \wedge q$ must do $a \Rightarrow p\|q$ must do a
> p must do $a \wedge q$ must do $a \Rightarrow p + q$ must do a

The latter specification has the expected initial model. However, it also has models that do not satisfy the specification $\mathrm{CCS_1}$ (as some agent in them has more transitions than those strictly required by the specification CCS). Moreover, the definition of the predicate must do is only correct for this particular description of the agents, and should be updated if, for instance, a new combinator is added for agents. Therefore, CCS_1 is much more flexible and can be used during the requirement phase, while CCS_2 can be adopted as a solution only for the design phase. ■

Let us look at one more example of an (initial) specification of data types: the specification of finite maps. Since finite maps are the basis for the abstract description of stores and memories, this data type is obviously crucial for the description of each imperative data type.

Example 3.16. Let us assume that we have specifications of locations (Sp_L, with main sort loc) and values (Sp_V, with main sort value) for a given type of our imperative language. We want to define the specification of the *store* data type. Since we want to update a store introducing a new value at a given location, we need the capability of looking up whether two locations are equal or not, as in Example 3.15. Therefore, we assume that a Boolean function representing equality is implemented in our specification of locations.

> **spec** $Sp_L =$
> > **sorts**   loc, bool...
> > **opns** $T, F \colon \ \rightarrow$ bool
> > $eq \colon$ loc $\times$ loc $\rightarrow$ bool ...

Notice that the axioms describing *eq* cannot be given without further assumption on the form of locations. Indeed, it is quite easy to guarantee that *eq* yields T over equal elements by requiring the axioms $x = y \Rightarrow eq(x, y) = T$ and $eq(x, y) = T \Rightarrow x = y$. However, in order to obtain that *eq* yields F over distinct elements, we need to add an axiom of the form $(\neg x = y) \Rightarrow eq(x, y) = F$, which is not positive conditional. Another possibility is to add the axiom[3] $\forall b \colon$ bool$.b = T \lor b = F$, that, in connection with the sentences axiomatizing the truth of *eq*, suffices to identify to F the result of *eq* over distinct elements, but this axiom is not conditional either. Moreover, in both cases we need to state that T and F represent distinct elements of sort bool, that is, we need an axiom of the form $\neg T = F$, which is not conditional. There is no general way of fully axiomatizing equality using conditional axioms, though it is often possible to do it in particular cases taking advantage of the structure of the terms of the argument sort.

Using *eq*, we can impose the extensional equality on stores requiring that the order of updates of different locations is immaterial and that only the last update for each variable is recorded.

> **spec** $Stores_1 = $ **enrich** Sp_L, Sp_V **by**
> > **sorts**   store
> > **opns**   empty$\colon \ \rightarrow$ store
> >   update$\colon$ store $\times$ loc $\times$ value $\rightarrow$ store
> > **vars** $x, y \colon$ loc; $v_1, v_2 \colon$ value; $s \colon$ store

---

[3] Adding the axiom is equivalent to restricting the model class to those models having (at most) a two-valued boolean carrier set and developing the theory of conditional specifications for those model classes.

$$\text{axioms } \mathtt{update}(\mathtt{update}(s, x, v_1), x, v_2) = \mathtt{update}(s, x, v_2)$$
$$eq(x, y) = F \Rightarrow \mathtt{update}(\mathtt{update}(s, x, v_1), y, v_2) =$$
$$\mathtt{update}(\mathtt{update}(s, y, v_2), x, v_1)$$

The initial model of $\mathtt{Stores}_1$ has the intended stores as elements of sort $\mathtt{store}$, but no tools to retrieve the stored values. In order to introduce an operation $\mathtt{retrieve\colon store \times loc \longrightarrow value}$, we should first define the result of retrieving a value from a location that has not been initialized in that store. If we simply give the specification

> **spec** $\mathtt{Stores}_2$ = **enrich** $\mathtt{Stores}_1$ **by**
> **opns** $\mathtt{retrieve\colon store \times loc \to value}$
> **vars** $x : \mathtt{loc}; v : \mathtt{value}; s : \mathtt{store}$
> **axioms** $\mathtt{retrieve}(\mathtt{update}(s, x, v), x) = v$

then in its initial model the application $\mathtt{retrieve}(s, x)$ to $s = \mathtt{empty}$, cannot be reduced to a primitive value, but is a new element of sort $\mathtt{value}$. This is a patent violation of any elementary principle of modularity and unfortunately does not depend on the initial approach or the particular specification.

Indeed, in all (total) models of $\mathtt{Stores}_2$, a value for (the interpretation of) $\mathtt{retrieve}(\mathtt{empty}, x)$ must be supplied that logically should represent an error. Hence if Sp_V defines only "correct" values, either a new "error" value is introduced, violating the modularity principle, or an arbitrary correct value is given as a result of $\mathtt{retrieve}(\mathtt{empty}, x)$, against the logic of the problem. A (quite unsatisfactory) solution is to require that all sorts of all specifications provide an "error" so that when modularly defining a function on data types already specified, if that function is incorrect on some input, the result can be assigned to the "error" element. This solution has two main limitations. Specifications which are possibly simple and perfectly correct have to be made far more complicated. This is caused by the introduction of error elements that can appear as arguments of the specification operators, requiring axioms for error propagation and disrupting the axioms for "correct" values. Secondly, if the errors are provided by the basic specifications, then they are classified depending on the needs of the original specifications; hence they do not convey any distinction between different errors due to the newly introduced operators. Therefore, the different origins of errors get confused and it is a complex task to define a sensible system of "error messages" for the user.

The point is that stores are inherently *partial* functions and hence the specification of their application should be partial as well. In the following sections we will see how, relaxing the definition of Σ-structure by allowing the interpretation of some function symbols to be partial, the specification of most partial data types is simplified.   ∎

Bibliographical notes. After proposing the equational specification of abstract data types [GTW78] (see also [EM85]), the well-known ADJ group soon recognized the need for conditional axioms [TWW81]. Conditional axioms with predicates (but without full equality) are also used in logic programming and Prolog [Llo87]. A combination of both points of view, that is, conditional axioms with equations *and* predicates, appears in the Eqlog language [GM86b]. The proof theory of this is studied in [Pad88a].

3.2   Partial data types

The need for a systematic treatment of partial operations is clear from practice. One must be able to handle *errors* and *exceptions*, and account for *non-terminating operations*. There are several approaches to deal with these in literature, none of which appears to be fully satisfactory. *S. Feferman* [Fef92]

3.2.1   Different motivations for partiality

Partial operations, besides being a useful tool to represent functions not yet completely specified during the design refinement process, are needed to represent partial recursive functions. In practice, partiality arises from situations that can be roughly partitioned into the following categories:

- a normal, total abstract data type, like the positive natural numbers, is enriched by a partial function, like subtraction. A canonical case is the axiomatic introduction of the inverse of some constructor. Most of the examples take place in this category, such as the well-known case of *stacks*, where stacks are built by the total functions *empty* and *push*. Then, *pop* and *top* are defined on them (i.e., the result of the application of these operations is either an "error" or a term built from the primitive operations);
- the partial functions that have to be specified are the "constructors" of their image set. For example, consider the definition of *lists* without repetitions of elements, or of *search trees*; in both cases the new data type is built by partial inserting operations. This is not uncommon especially for hardware design or at a late stage of projects, when *limited* or *bounded* data types have to be defined, for instance, *integer* subranges (where *successor* and *predecessor* operations are the constructors – they result in an error when applied to the bounds of the subrange), or bounded *stacks* (where *push*ing an element on a full *stack* yields an error).
- a partial recursive function with nonrecursive domain, for example, an interpreter of a programming language, has to be specified;
- a semidecidable predicate p has to be specified, such as in concurrency theory the *transition* relation on processes, or the *typing* relation for higher-order languages. Thus, representing p as a Boolean function f_P, it is possible to axiomatize the truth (recursively), but not whether f_P yields *false* on some inputs and hence f_P is partial (or its image is larger than the usual Boolean values set);

The last case has already been solved by explicitly adding predicates to our signature, as in the last section. Thus, let us focus on the others.

Programming on data types. Consider the following situation: we have
defined a data type by a minimal set of (total) functions providing a method
of construction for each element of our data type that are hence called *constructors*. Then we want to enrich it by some (possibly partial) functions
that are *programmed* in terms of the constructors, in the sense that their
application to primitive elements either also reduces to a term built by the
constructors, or is an error.

A particular case is the specification of the constructor inverses, from now
on called *selectors*. Consider, for instance, as running example, the (Peano-
style) specification of the natural numbers, by *zero* and *successor*.

> **spec Nat =**
> **sorts**   nat
> **opns**   zero: $\rightarrow$ nat
>   succ: nat $\rightarrow$ nat

Suppose now that we want to define the inverse of succ, that is the *predecessor* prec. Then the unique problem comes from the application of prec
to elements that are not in the succ image, that is, to zero.

Within a total approach, we have the following possibilities to cope with
this. Either we introduce a new sort, representing the domain of prec, that
does not contain zero, so that prec(zero) is no longer a well-formed term.
Or we let prec(zero) denote an *erroneous* element err, and hence an error-
management mechanism has to be provided as well, for instance, identifying
all the applications of the operations of the data type to the error(s) with
err.

Partial constructors. A quite different problem from that introduced in the
last subsection is the definition of data types whose constructors themselves
are partial functions.

A paramount example of this case can be found in the formal languages
field. Each production of a context-free grammar, with the form

$$s ::= w_1 s_1 \ldots s_n w_{n+1},$$

where the possibly decorated s's are nonterminal symbols, and each w_i is a
string of terminal symbols, corresponds to a total function from $s_1 \times \ldots \times s_n$
into s. Thus, context-free grammars can be represented by total signatures,
but attributed grammars cannot, because the applicability of production
rules, i.e., of constructors, may be partial. The same also applies to grammars
for languages whose operators are assigned a priority. Indeed, in the case of
plus and times on integers, for instance, so that a string $x + y \times z$ unambiguously represents $x + (y \times z)$, the rule for times cannot apply to terms
having some plus in the outermost position. Therefore, the interpretation on
the times operator must be partial.

Other very relevant examples may be found, for instance, during the implementation phase, where, due to the machine limits, data types are limited.

Indeed, in these cases, the constructors of infinite data type sets, like *successor* and *predecessor* for integers, or *push* for stacks, are not defined on the extreme values and hence become partial.

Partial recursive functions with nonrecursive domains. Examples are typically found in the specification of system programs. When specifying an interpreter of a programming language that is expressive enough to describe all Turing computable functions, we get a partial recursive function. Indeed, the interpreter is recursive, but it loops for ever on some inputs. Since the Halting Problem is undecidable, it is undecidable whether the interpreter eventually yields an output or not. Thus the interpreter is a partial recursive function with nonrecursive domain.

Another example is an automatic theorem proving system that, given a proof goal, searches for a proof tree proving that goal. Now for a Turing-complete logic, such as first-order logic, provability is undecidable in general. So our automatic theorem-proving system will be a partial recursive function with a nonrecursive domain.

3.2.2 Capturing partiality within total frameworks

We now discuss how good the occurrences of partiality listed above can be treated within a total framework.

Static elimination of errors with order-sorted algebra. Let us consider again the example of natural numbers. Essentially we want to enrich Nat with a new sort **pos** (representing the domain of **prec**) and the function **prec** itself. But we also have the intuition that the domain of **prec** is a subset of **nat**. In particular, we would like to be able to apply **prec** to all strictly positive elements of sort **nat**. Thus, in a pure many-sorted style we should also add the embedding of pos into Nat, producing the following specification.

$$\textbf{spec } \textbf{Nat}^{\textbf{pos}} =$$
$$\quad \textbf{sorts} \quad \text{nat}, \text{pos}$$
$$\quad \textbf{opns} \quad \text{zero}: \to \text{nat}$$
$$\quad\quad\quad \text{succ}: \text{nat} \to \text{pos}$$
$$\quad\quad\quad \text{prec}: \text{pos} \to \text{nat}$$
$$\quad\quad\quad e: \text{pos} \to \text{nat}$$
$$\quad \textbf{axioms } \forall x: \text{nat}.\text{prec}(\text{succ}(x)) = x$$
$$\quad\quad\quad \forall x, y: \text{pos}.e(x) = e(y) \Rightarrow x = y$$

Thus, for instance, the expression $0 + 1 + 1 - 1$ is represented by the term $\text{prec}(\text{succ}(e(\text{succ}(\text{zero}))))$, while we would expect $\text{prec}(\text{succ}(\text{succ}(\text{zero})))$. Therefore, it is much more preferable to enrich the theory by explicitly allowing the *subsorts*, that is, having sorts that must be interpreted in each model as subsets of the interpretation of other sorts. Thus, for instance, the size of each model is not unduly enlarged by a subsort declaration. This can be relevant in the implementation phase. The rules for term formation are accordingly relaxed, so that any function requiring an argument of the supersort can also accept an argument of the subsort.

The resulting theory of *order-sorted* algebras, which will be more extensively used in the next subsection, is much deeper and more complicated. We refer to [GM87a, GM92, MG93, Yan93, Mos93, Poi90] for a formal presentation of the order-sorted approach. Here, we informally use the following specification

> **spec** Nat^{osa} =
>> **sorts**   pos $\leq$ nat
>> **opns**   zero: $\to$ nat
>>   succ: nat $\to$ pos
>>   prec: pos $\to$ nat
>> **axioms** $\forall x :$ nat.prec(succ(x)) = x

with the convention that if t is a term of sort pos then it is a term of sort nat as well. In other words the above specification is a more convenient presentation of Nat^{pos} but it has an equivalent semantics.

Then two main problems can be seen. Firstly, as the domain of prec is *statically* described by means of the signature, it cannot capture our intuitive identifications of terms as different representations of the same value, that depend on the deductive mechanism of the specification and hence are, so to speak, *dynamic*. Thus, the term prec(prec(succ(succ(zero)))) is incorrect because the result type of prec is nat, while prec expects an argument of sort pos, even if prec(succ(succ(zero))) can be deduced to be equal to succ(zero). Hence, intuitively it should have sort pos. This problem cannot be avoided by any approach based on a static elimination of error elements, that is, on a refinement of the typing of functions at the signature definition level.

Consider now the problem of specifying a partial constructor, say, the push operation for bounded stacks. This specification is, of course, parametric on the definition of the element data type which we assume is given by the specification Elem, with a sort elem, and is based also on a specification of natural numbers with an order relation $\leq$ on them, in order to define the depth of a stack.

> **spec** Nat$_\leq$ =
>> **sorts**   nat
>> **opns**   zero: $\to$ nat
>>   succ: nat $\to$ nat
>> **preds**   _ $\leq$ _: nat $\times$ nat ...
>> **vars** $x, y :$ nat
>> **axioms** zero $\leq x$
>> $x \leq y \Rightarrow$ succ(x) $\leq$ succ(y)

The problem is that only a fixed number of push of iterations should be allowed. Of course it is always possible, though awkward, to introduce ad hoc constructors, for instance, $n - 1$ constructors, each one representing the creation of a stack with exactly i elements, for $i = 1, \ldots, n - 1$, where n is the maximal allowed depth of a stack. This approach would be not only unnatural, but also nonparametric with respect to the maximum n; indeed if the value n were to be changed at a further stage of design, functions would need to be added to the specification as well as axioms.

A more promising approach is to specify the domain of the partial constructor push:

> **spec BoundedStacks = enrich Nat$_\leq$, Elem by**
> **sorts**   nonfullbstack $\leq$ bstack
> **opns**   max: $\to$ nat
> **opns**   empty: $\to$ bstack
>   push: elem $\times$ nonfullbstack $\to$ bstack
>   depth: bstack $\to$ nat
> **axioms max = ...**
>   depth(empty) = zero
> $\forall x$: elem, s : nonfullbstack.depth(push(x, s)) = succ(depth(s))

But now we have the problem similar to that with nested applications of succ and prec. A term such as

> push(x, push(y, empty))

is not wellformed because push expects a nonfullbstack, but only delivers a bstack.

The situation becomes even worse when considering partial recursive functions with nonrecursive domains: the specification of a nonrecursive subsort seems to be difficult, if not impossible, in the order-sorted approach.

Dynamic treatment of errors: retracts. In OBJ3 [GMW+93], the problem of ill-formedness of prec(prec(succ(succ(zero)))) is solved by automatically adding retracts r : nat > pos which can be removed using retract equations

> $\forall s$:pos.r : nat > pos(s) = s

and which are irreducible in the case of ill-typed terms. Then in a term like prec(prec(succ(succ(zero)))), a retract is added:

> prec(r : nat > pos(prec(succ(succ(zero)))))

which, by the retract equation, has the intended semantics. Now terms such as prec(zero) are parsed as an irreducible term:

> prec(r : nat > pos(zero))

which can be seen as an error message. In a similar vein, by adding a conditional retract equation

> $\forall s$: bstack.succ(depth(s)) $\leq$ max $\Rightarrow$ r : bstack > nonfullbstack(s) = s

multiple pushes on a bounded stack are allowed, provided that they do not exceed the bound; otherwise, we get an irreducible term serving as an error message.

Discussion. The main problem with the retract approach is the following: terms involving irreducible retracts introduce useful new error elements, but thereby change the semantics of the specification. In [MG93] it is shown that specifications with retracts have an initial semantics given by an injective homomorphism from the initial algebra of the specification without retracts to the initial algebra of the specification with retracts[4]. This hiatus between the signature of the models, and in particular that of the initial one, which has no retracts in it, and the signature of the terms used in the language can be eliminated if retracts are allowed to be truly *partial* functions. This solution needs a framework where order-sorted algebra is combined with partiality, as, for instance, in the language CASL in the course of definition within the CoFI initiative (see Section 3.3.5).

Dynamic treatment of errors: sort constraints. An alternative approach uses (conditional) sort constraints [MG93, Yan93]. A sort constraint expresses that some term, which syntactically belongs to some sort s, is always interpreted in such a way that it already belongs to a sort $s' \leq s$[5].

Now let us add a sort constraint[6]

$\forall n{:}\mathtt{nat}.\mathtt{prec}(\mathtt{succ}(\mathtt{succ}(n))) : \mathtt{pos}$

to the specification.

If well-typedness of terms is now defined by taking sort constraints into account as well [GMJ85], the term

$\mathtt{prec}(\mathtt{prec}(\mathtt{succ}(\mathtt{succ}(\mathtt{zero}))))$

is welltyped because $\mathtt{prec}(\mathtt{succ}(\mathtt{succ}(n)))$ is of sort **pos** by the sort constraint.

Similarly, by adding a conditional sort constraint

$\forall s : \mathtt{bstack}.\mathtt{succ}(\mathtt{depth}(s)) \leq \mathtt{max} \Leftrightarrow s : \mathtt{nonfullbstack}$

we also have the possibility of multiple pushes on a bounded stack (provided that they do not exceed the bound).

If the partial function to be introduced is not the inverse of one of the constructors, then the domain of the partial function has to be introduced and axiomatized explicitly. This can be done with sort constraints, which allow us to specify subsorts to consist of all the values that satisfy a given predicate. As an artificial, but simple, example let us consider the division by 2, which is only well defined for even numbers, corresponding to the following specification in an order-sorted simplified notation.

---

[4] Likewise, there is a free construction embedding any algebra of a specification without retracts into an algebra of the same specification with retracts.

[5] Actually, s and s' only need to belong to the same connected component.

[6] In CASL there are *membership* axioms that can be used to express sort constraints, but do not influence parsing.

spec Nat^2 =
    sorts   even $\leq$ nat
    opns   _ mod 2 : nat $\rightarrow$ nat
           _ div 2 : even $\rightarrow$ nat
    axioms zero mod 2 = zero
            succ(zero) mod 2 = succ(zero)
            $\forall x$: nat.succ(succ(x)) mod 2 = x mod 2
            $\forall x$: nat.x : even $\Leftrightarrow$ x mod 2 = zero
            zero div 2 = zero
            $\forall x$: even.succ(succ(x)) div 2 = succ(x div 2)

Note that the equivalence defining **even** can be expressed with a conditional sort constraint

$$\forall x : \text{nat}.x \text{ mod } 2 = \text{zero} \Rightarrow x : \text{even}$$

together with an ordinary axiom

$$\forall x : \text{even}.x \text{ mod } 2 = \text{zero}$$

With this method we can even specify nonrecursive domains of partial recursive functions.

If the partial function to be introduced is not unary, the order-sorted approach does not immediately apply, because the domain should be a subsort of a *product* sort and usually the sort set does not include products. One then has to define products explicitly by a tupling operation together with projections, and specify the domain of the operation to be a subsort of the product using sort constraints.

Discussion. Allowing sort constraints to have influence on wellformedness, as proposed in [GMJ85], means that type checking can generate proof obligations, which in general can be resolved only dynamically by doing some theorem proving.[7] So parsing of terms becomes quite complex. Some work in this direction has been done in [Yan93], but there is not yet a fully worked out theory.

The contribution of order-sorted algebra with sort constraints is to allow a separation of those parts of type-checking which can be done statically from those which can only be done dynamically. This distinction is lost in the approach of partial algebras introduced below. Therefore, it is worthwhile to combine the order-sorted and the partial approach, as is done in the CASL language (see Section 3.3.5).

---

[7] But we cannot expect all definedness conditions to be resolved statically because definedness is undecidable in general. And indeed, within the usual approaches to partial algebras, terms may not denote and definedness can only be checked dynamically with theorem proving.

Underspecification. Another possible way to deal with prec(zero) is to use this term as a denotation for error. Underspecification means that the value of prec(zero) is just left unspecified - it can be any natural number [GS95b, MS97]. This approach seems to work quite well in some cases: one cannot prove any interesting properties about prec(zero), so one cannot really use prec(zero) as a normal natural number. However, there is much danger of expressing more properties than wanted – it is even possible to get inconsistent specifications [LW98]. Moreover, this method cannot be combined with initial semantics: with initial semantics, prec(zero) would generate a new data value.

Error elements. While underspecification leaves the value of prec(zero) unspecified, the error element approach introduces a new error value as the denotation for prec(zero). This is clearly inconsistent with a modular approach, as one or more new element(s) interpreting the error(s) have to be added to the models of the original specification (see [Poi87] for an argumentation against the phenomenon that error elements in basic types are introduced by hierarchically building more complex specifications). Moreover, having added (at least) one new element, the application of the data type functions to it has to be specified as well. In the pioneering *Error algebras*, introduced by the ADJ group in [GTW78], in order to achieve a reasonable uniformity, one constant symbol for each sort is added and all the errors have to reduce to it by introduction and propagation axioms. Of course the naive application of error propagation can cause problems. Indeed, consider, for instance, the definition of natural numbers with product. Then, by instantiating the error propagation axiom for the product, $err * x = err$, with zero and the standard basis of the inductive product definition, $x * zero = zero$, with err, we deduce zero = err.

Thus, in [GTW78] a uniform technique to avoid these inconsistencies is introduced, basically consisting of a distinction of axioms into *axioms for correct elements* and *error propagation axioms*. Since error algebras are described as equational specifications of many-sorted algebras, the resulting specifications are quite heavy. However, by using predicates and conditional formulas, their usage is improved, because the implementation of the Boolean type, with its connectives, is no longer needed. Still, for each function of n arguments, n error propagation axioms have to be stated, each constructor requires a correctness propagation axiom and each error introduction must be detected by an appropriate axiom. Thus, specifications in this style cannot be concise. Moreover, each axiom stating properties on the correct elements, that is, the proper axioms of the data type, must be *guarded* by the predicates stating the correctness of their input in the premises.

For instance, one of the more classical examples, that is, the specification of natural numbers with sum and product, using a predicate to state that a term is correct, would be as follows.

spec Nat* =
    sorts   nat
    opns   zero, err : $\rightarrow$ nat
              succ : nat $\rightarrow$ nat
              plus, minus, times : nat $\times$ nat $\rightarrow$ nat
    preds OK, IsErr : nat
    vars x, y : nat
    axioms OK(zero)
             $OK(x) \Rightarrow OK(succ(x))$
             $IsErr(x) \Rightarrow IsErr(succ(x))$
             $IsErr(err)$
             $OK(x) \Rightarrow plus(x, zero) = x$
             $OK(x) \wedge OK(y) \Rightarrow plus(x, succ(y)) = succ(plus(x, y))$
             $IsErr(x) \Rightarrow IsErr(plus(x, y))$
             $IsErr(x) \Rightarrow IsErr(plus(y, x))$
             $OK(x) \Rightarrow times(x, zero) = zero$
             $OK(x) \wedge OK(y) \Rightarrow times(x, succ(y)) = plus(x, times(x, y))$
             $IsErr(x) \Rightarrow IsErr(times(x, y))$
             $IsErr(x) \Rightarrow IsErr(times(y, x))$
             $OK(x) \Rightarrow minus(x, zero) = x$
             $OK(x) \wedge OK(y) \Rightarrow minus(succ(x), succ(y)) = minus(x, y)$
             $IsErr(minus(zero, succ(x)))$
             $IsErr(x) \Rightarrow IsErr(minus(x, y))$
             $IsErr(x) \Rightarrow IsErr(minus(y, x))$
             $IsErr(x) \Rightarrow x = err$

Note that removing the last axiom, the incorrect terms are all distinct and thus can serve as very informative error messages.

The error algebra approach can also be used to specify partial constructors. In the bounded stack example, pushing too many elements on the same stack results in one *error* element:

spec BoundedStacks = **enrich** Nat$_\leq$, Elem **by**
    sorts   bstack
    opns   max : $\rightarrow$ nat
           empty, err : $\rightarrow$ bstack
           Bpush : elem $\times$ bstack $\rightarrow$ bstack
           depth : bstack $\rightarrow$ nat
    preds OK, Is_Err : bstack
    vars x : elem; s : bstack
    axioms max = ...
           $OK(empty)$
           $depth(empty) = zero$
           $depth(err) = succ(max)$
           $succ(depth(s)) \leq max \Rightarrow OK(Bpush(x, s))$
           $OK(Bpush(x, s)) \Rightarrow depth(Bpush(x, s)) = succ(depth(s))$
           $max \leq depth(s) \Rightarrow Is\_Err(Bpush(x, s))$
           $Is\_Err(s) \Rightarrow s = err$

The first axiom fixes the actual maximal size of the bounded stacks, while the last one identify all errors.

The specification of partial recursive functions with nonrecursive domain is possible within the error algebra approach, but in semicomputable models it leads inevitably to infinitely many error values.[8]

Discussion. Many other approaches flourished from the original error algebras, refining the basic idea of cataloging the elements of the data type but using a more powerful algebraic framework to express the specifications. See, e.g., the *exception algebras* in [BBC86a], where both the elements of the algebras and the terms are labeled to capture the difference between errors and exceptions, or the *clean algebras* in [Gog87], where an order-sorted approach is adopted to catalog the elements of algebras. Among the approaches that use error elements in some way to model partiality within a total framework, there are Equational Type Logic [MSS90, MSS92], unified algebras [Mos89a], and based algebras [Kre87, KM95]. In spite of their greater power and embellishments, these approaches share the original difficulties of interaction with the modular definition of data types. Indeed the *non-ok* elements of basic types have to be designed *a priori* to support error messages, or exceptions caused by other modules that use the basic ones.

Thus, error algebras (and variations on the theme) are more suitable for specifying a completely defined system than for refining a project or represent (parts of) a library of specifications *on the shelf*. Quite often in the last stages of the refinement process even functions that are partial from a philosophical point of view (for instance, the constructors of bounded stacks, bounded integer, search trees, ordered lists and other bounded resources or finite domains) are implemented as total functions, identifying incorrect applications with error messages, that are special algebra elements, or using more sophisticated approaches, where there is a notion of state and error handling means that a special state is entered (see Chapter 14). However, we cannot delay the semantics of the data until every detail has been decided, because of methodological reasons: we want to separate *requirement specifications* that are close to the informal understanding of human beings from *design specifications* that are close to implementations.

Therefore, we have to find a way of dealing with the requirement specification of partial functions and in particular of partial constructors. In the above example we saw the *design* specification of bounded stacks, which cannot be refined further. For instance, all errors have been identified and can no longer be distinguished in order to obtain a more informative error message system. In a total approach it is impossible (or, rather, unnatural and inconvenient) to give the *requirement* specification of bounded stacks. Therefore,

---

[8] If there were only finitely many errors, one could use them to recursively enumerate the complement of the domain. The domain itself can be recursively enumerated as well. Both recursive enumerations lead to a decision procedure for the domain, which contradicts the assumption that the domain is nonrecursive.

in the next section we will introduce a more powerful framework, based on the (possibly) *partial* interpretation of function symbols and see how it can be used to describe easily this and other specifications.

3.3 Partial first-order logic

When developing a model theory for partial first-order structures, it is not obviously clear which way to proceed as there are possibilities for different choices at various points. Sometimes different choices have severe technical implications, sometimes they are more or less only a matter of taste. See [CJ91, Far91] for an overview of different approaches.

The introduction of symbols denoting partial functions has the effect that not all terms can be interpreted in each model (unless wellformedness of terms is made dependent on the model where they have to be interpreted, which does not seem very useful). Thus, the valuations of terms is inherently partial, but the question reamins as to whether such partiality should propagate to formulas built over terms.

We follow here the two-valued approach developed and motivated by Burmeister [Bur82] and others [Bee85, Far91, Far95, Fef95, Par95]. In a two-valued logic of partial functions, formulas which contain some nondenoting term are interpreted as either true or false. The main motivation for this principle is that for *specification* of software systems, we need definiteness, i.e., we want to know whether some property does or does not hold, and do not want to have a "may-be". Moreover, the logic which we will develop is a so-called *negative logic* [GL97], that is, atomic formulas containing some nondenoting term are interpreted as false.[9] Later on, we briefly sketch a three-valued logic [CJ91, Jon90] where the valuations for formulas may be partial as well, that is, a formula containing some nondenoting term is interpreted as neither true nor false, but has some third truthvalue assigned to it.

3.3.1 Model theory

This section is devoted to the study of categories of partial first-order structures. Since many definitions and results can be stated in a uniform language, using category theory, and hold for many algebraic frameworks, we will try to clarify the basic nature of these categories and discuss the existence of very simple constructions, to allow an experienced reader to apply the available theories to our framework. However, as the proposed constructions have a quite natural and intuitive counterpart, essentially generalizing analogous constructions in (indexed) set theory, those who are not familiar with categories and their applications can find our theory useful, by simply ignoring the categorical terminology.

---

[9] Strictly speaking, this holds only if one counts strong equations as nonatomic formulas. But strong equations are a derived notion in our formalism.

Partial first-order structures differ from (total) structures in the interpretation of function symbols being possibly *partial* functions, i.e., they are not required to yield a result on each possible input. Thus, total functions are a particular case of partial functions that happen to be defined on all the elements of their source. It is convenient anyway to discriminate as soon as possible between total and partial functions of a data type, because knowing a function (symbol) to be (interpreted as) total simplifies its treatment not only from an intellectual point of view of designing the data type, but also, for example, when applying rewriting techniques or proof deductions. Therefore, we already distinguish at the signature level between function symbols that must be interpreted as total and function symbols that are allowed to denote partial operations (but can, obviously, be total as well in some model).

Definition 3.17. A *partial signature* $\Sigma = \langle S, \Omega, P\Omega, \Pi \rangle$ consists of a set S, denoted by *sorts*(Σ), of *sorts*, two componentwise disjoint $S^* \times S$-sorted families Ω and $P\Omega$, denoted respectively by *opns*(Σ) and *popns*(Σ), of *total* and *partial function symbols* and an S^*-sorted family Π, denoted by *preds*(Σ), of *predicate symbols*.

Given a partial signature $\Sigma = \langle S, \Omega, P\Omega, \Pi \rangle$ and an S-sorted family X of variables, the Σ-*terms* are the terms on the first-order signature $(S, \Omega \cup P\Omega, \Pi)$, as introduced in Definition 3.2.   ◇

Thus, terms on a partial signature are defined as usual, disregarding the distinction between total and partial functions. Therefore, the same symbol cannot be used for a total and a partial function with the same arity, as such an overloading would introduce a semantic ambiguity.

Example 3.18. Let us look at a signature for non-negative integers, with partial operations, like predecessor, subtraction and division, and a predicate stating whether a number is a multiple of another.

> **sig** $\Sigma_{\mathtt{Nat}} =$
> **sorts**   nat
> **opns**   zero: $\to$ nat
>   succ: nat $\to$ nat
>   plus, times: nat $\times$ nat $\to$ nat
> **popns** prec: nat $\to$ nat
>   minus, div, mod: nat $\times$ nat $\to$ nat
> **preds** multiple: nat $\times$ nat

Analogously, a signature for stacks with possibly partial interpretation of top and pop on an empty stack, based on a signature $\Sigma_{\mathtt{Elem}}$, describing the type elem of the elements for the stack, is the following.

> **sig** $\Sigma_{\mathtt{Stack}} =$ **enrich** $\Sigma_{\mathtt{Elem}}$ **by**
> **sorts**   stack
> **opns**   empty: $\to$ stack
>   push: elem $\times$ stack $\to$ stack
> **popns** pop: stack $\to$ stack

```
           top: stack ─◦→ elem
preds   is_in: elem × stack
           is_empty: stack                                              ∎
```

Since function symbols are partitioned into total and partial and both families are classified depending on their input/output types, the same symbol can appear many times in the same signature, possibly making the term construction ambiguous. From now on, we assume that terms are not ambiguous, i.e., the overloading of function symbols is not introducing problems (or, if the overloading *is* problematic, a different, unambiguous notation for terms has been adopted, for instance, substituting for each function symbol a pair consisting of its name and its type).

Notation 3.19. For each partial function $pf \colon X \longrightarrow\!\!\!\!\!\circ\ Y$, we will denote its *domain* by dom pf, that is, the subset of X defined by

$$\text{dom } pf = \{x \mid x \in X \text{ and } pf(x) \in Y\}.$$

In the total case, signature morphisms map sorts to sorts, operation symbols to operation symbols, and predicate symbols to predicate symbols. In the partial case, having partitioned the operation symbols into total and possibly partial ones, total operation symbols must be mapped to total operation symbols, in order to be able to define reduct functors on models. For partial operation symbols, we have two possibilities: they can be required to be mapped either to partial operation symbols or to the union of total and partial operation symbols. The latter choice is more appropriate for using signature morphisms to represent refinement and therefore has been adopted in CASL.

Exercise 3.20. Generalize the notion of signature morphism for partial signatures, both for the case that partial operation symbols are required to be mapped to partial operation symbols and for the case that partial operation symbols are allowed to be mapped to either partial or total operation symbols.

As the interpretations of function symbols in $P\Omega$ can be undefined on some input, not all (meta)expressions denote values in the carriers of a partial first-order structure. Thus, the meaning of an equality between expressions that can be nondenoting becomes ambiguous; indeed it is arbitrary to decide whether an equality implicitly states the existence of the denoted element, or holds also if both sides are undefined (assuming the viewpoint that (nonexisting) undefined elements are indistinguishable), or is satisfied whenever the two sides do not denote different elements. Therefore, we will use different equality symbols for the different concepts and in particular we will use $e \overset{\mathrm{E}}{=} e'$ (*existential equality*) to state that both sides denote the same value, $e \overset{\mathrm{W}}{=} e'$ (*weak equality*) to state that if both sides denote a value, then the two values coincide, and $e \overset{\mathrm{S}}{=} e'$ (*strong equality*) to state that either both sides denote the same value or both do not denote any value. Thus, in particular, $e\overset{E}{=}e$ is equivalent to e denoting a value, and hence is usually represented by $e\downarrow$ (and its negation becomes $e\uparrow$).

It is interesting to note that assuming either existential or strong equality as primitive, the other notions can be derived; indeed $e\downarrow$ for an expression e of sort s can be expressed simply as $e\overset{E}{=}e$ or as "there exists some x with $e\overset{S}{=}x$" with x of sort s nonfree in e. Then, using $e\downarrow$ as syntactic sugar for the formulas of our metalanguage mentioned above, we have the following table.

using:	$\overset{E}{=}$ becomes	$\overset{W}{=}$ becomes	$\overset{S}{=}$ becomes
$e\overset{E}{=}e'$	$e\overset{E}{=}e'$	$(e\downarrow\wedge e'\downarrow)\supset e\overset{E}{=}e'$	$(e\downarrow\vee e'\downarrow)\supset e\overset{E}{=}e'$
$e\overset{S}{=}e'$	$e\downarrow\wedge e\overset{S}{=}e'$	$(e\downarrow\wedge e'\downarrow)\supset e\overset{E}{=}e'$	$e\overset{S}{=}e'$

On the other hand, weak equality is too weak to describe the other kinds of equality because, in particular, it is not possible to state the definedness of an expression using only weak equalities.[10] But using weak equality and definedness assertions, it is possible to represent both existential and strong equalities, as follows.

$\overset{E}{=}$ becomes	$\overset{W}{=}$ becomes	$\overset{S}{=}$ becomes
$e\downarrow\wedge e'\downarrow\wedge e\overset{W}{=}e'$	$e\overset{W}{=}e'$	$(e\downarrow\vee e'\downarrow)\supset(e\downarrow\wedge e'\downarrow\wedge e\overset{W}{=}e')$

Definition 3.21. Given a partial signature $\Sigma = \langle S, \Omega, P\Omega, \Pi\rangle$, a *partial Σ-structure* A is a triple consisting of

- an S-sorted family $|A|$ of *carriers*;
- a family $\{\mathcal{I}_A^{w,s}: \Omega_{w,s}\cup P\Omega_{w,s}\longrightarrow\mathrm{PFun}_{w,s}\}_{(w,s)\in S^*\times S}$ of *function interpretations*, where $\mathrm{PFun}_{s_1,\ldots,s_n,s}$ is the set of all partial functions from $|A|_{s_1}\times\cdots\times|A|_{s_n}$ into $|A|_s$, such that $\mathcal{I}_A^{w,s}(f)$ is total for each $f\colon s_1\times\cdots\times s_n\longrightarrow s\in\Omega$. In the following, if no ambiguity arises, $\mathcal{I}_A^{w,s}(f)$ is denoted by f_A for each $f\in\Omega_{w,s}\cup P\Omega_{w,s}$.
- a family $\{\mathcal{J}_A^{s_1,\ldots,s_n}: \Pi_{s_1,\ldots,s_n}\longrightarrow\wp(|A|_{s_1}\times\cdots\times|A|_{s_n})\}_{s_1,\ldots,s_n\in S^*}$ of *predicate interpretations*. In the following, if no ambiguity arises, $\mathcal{J}_A^w(p)$ is denoted by p_A for each $p\colon s_1\times\cdots\times s_n\in\Pi$.

Moreover, given partial Σ-structures A and B, a *homomorphism* of partial Σ-structures from A into B is an S-sorted family h of (total) functions $h_s\colon|A|_s\longrightarrow|B|_s$ such that

- $h_s(f_A(a_1,\ldots,a_n))\overset{E}{=}f_B(h_{s_1}(a_1),\ldots,h_{s_n}(a_n))$ for all $f\colon s_1\times\cdots\times s_n\longrightarrow s\in\Omega$ and all $a_i\in|A|_{s_i}$ for $i=1,\ldots,n$;

---

[10] For the object level (we introduce $\overset{W}{=}$ for weak equality in our formal logical language later), this can be seen as follows. In any trivial first-order structure with singleton carriers, all weak equalities are true, disregarding the definedness of the involved expressions.

- if $pf_A(a_1, \ldots, a_n){\downarrow}$ then $h_s(pf_A(a_1, \ldots, a_n)) \overset{E}{=} pf_B(h_{s_1}(a_1), \ldots, h_{s_n}(a_n))$
  for all $pf\colon s_1 \times \cdots \times s_n \relbar\joinrel\mapsto s \in P\Omega$ and all $a_i \in |A|_{s_i}$ for $i = 1, \ldots, n$;
- if $(a_1, \ldots, a_n) \in p_A$, then $(h_{s_1}(a_1), \ldots, h_{s_n}(a_n)) \in p_B$ for all $p\colon s_1 \times \cdots \times s_n \in \Pi$ and all $a_i \in |A|_{s_i}$ for $i = 1, \ldots, n$.

The category $Mod(\Sigma)$ has partial Σ-structures as objects and homomorphisms of partial Σ-structures as arrows, with the obvious composition and identities.   ◇

Therefore, in order to define a partial Σ-structure, we must provide:

- the S-sorted family $|A|$ of carriers;
- for each $f\colon s_1 \times \cdots \times s_n \longrightarrow s \in \Omega$, the *interpretation* of f in A, that is, a total function $f_A\colon |A|_{s_1} \times \cdots \times |A|_{s_n} \longrightarrow |A|_s$;
- for each $pf\colon s_1 \times \cdots \times s_n \relbar\joinrel\mapsto s \in P\Omega$, the *interpretation* of pf in A, that is, a partial function $pf_A\colon |A|_{s_1} \times \cdots \times |A|_{s_n} \relbar\joinrel\mapsto |A|_s$;
- for each $p\colon s_1 \times \cdots \times s_n \in \Pi$, a subset p_A of $|A|_{s_1} \times \cdots \times |A|_{s_n}$, representing the *truth values* of p in A.

Notice that, although the interpretation function for partial and total function symbols is the same, it is often convenient to distinguish between partial and total function symbols, as in the definition of homomorphism above, where the definedness condition can be dropped for the total symbols.

Let us look at two examples of partial Σ-structures on the signatures introduced by the Example 3.18.

Example 3.22. The natural numbers with the "obvious" interpretation of function symbols is a partial $\Sigma_{\mathbf{Nat}}$-structure.

algebra N =
    Carriers
        $|N|_{\mathbf{nat}} = N$
    Functions
        $\mathbf{zero_N} = 0$
        $\mathbf{succ_N}(n) = n + 1$
        $\mathbf{plus_N}(n, m) = n + m$
        $\mathbf{times_N}(n, m) = n * m$
$$\mathbf{prec_N}(n) = \begin{cases} n - 1, & \text{if } n > 0 \\ \text{undefined,} & \text{otherwise} \end{cases}$$
$$\mathbf{minus_N}(n, m) = \begin{cases} n - m, & \text{if } n \geq m \\ \text{undefined,} & \text{otherwise} \end{cases}$$
$$\mathbf{div_N}(n, m) = \begin{cases} n \text{ div } m, & \text{if } m \neq 0 \\ \text{undefined,} & \text{otherwise} \end{cases}$$
$$\mathbf{mod_N}(n, m) = \begin{cases} n \text{ mod } m, & \text{if } m \neq 0 \\ \text{undefined,} & \text{otherwise} \end{cases}$$
    Predicates
        $\mathbf{multiple_N} = \{(n, m) \mid n \text{ mod } m \overset{E}{=} 0\} \cup \{(0, 0)\}$

Another first-order structure on the same signature is, for instance, the following, where "error messages" have been added to the carrier.

algebra $N_E =$

Carriers

$|N_E|_{\text{nat}} = N \cup E$ for $E = \{\texttt{underflow}, \texttt{err\_minus}, \texttt{division\_by\_0}\}$

Functions

$\texttt{zero}_{N_E} = 0$

$\texttt{succ}_{N_E}(n) = \begin{cases} n+1, & \text{if } n \in N \\ n, & \text{otherwise} \end{cases}$

$\texttt{plus}_{N_E}(n, m) = \begin{cases} n+m, & \text{if } n, m \in N \\ n, & \text{if } n \in E \\ m, & \text{otherwise} \end{cases}$

$\texttt{times}_{N_E}(n, m) = \begin{cases} n*m, & \text{if } n, m \in N \\ n, & \text{if } n \in E \\ m, & \text{otherwise} \end{cases}$

$\texttt{prec}_{N_E}(n) = \begin{cases} n-1, & \text{if } n \in N, n > 0 \\ \texttt{underflow}, & \text{if } n = 0 \\ n, & \text{otherwise} \end{cases}$

$\texttt{minus}_{N_E}(n, m) = \begin{cases} n-m, & \text{if } n, m \in N, n \geq m \\ n, & \text{if } n \in E \\ m, & \text{if } n \in N, m \in E \\ \texttt{err\_minus}, & \text{otherwise} \end{cases}$

$\texttt{div}_{N_E}(n, m) = \begin{cases} n \text{ div } m, & \text{if } n, m \in N, m \neq 0 \\ n, & \text{if } n \in E \\ m, & \text{if } n \in N, m \in E \\ \texttt{division\_by\_0}, & \text{otherwise} \end{cases}$

$\texttt{mod}_{N_E}(n, m) = \begin{cases} n \text{ mod } m, & \text{if } n, m \in N, m \neq 0 \\ n, & \text{if } n \in E \\ m, & \text{if } n \in N, m \in E \\ \texttt{division\_by\_0}, & \text{otherwise} \end{cases}$

Predicates

$\texttt{multiple}_{N_E} = \{(n, m) \mid n, m \in N \text{ and } n \text{ mod } m \overset{E}{=} 0\} \cup \{(0, 0)\}$

It is clear that the embedding of the Σ_{Nat}-structure N into N_E is a homomorphism. ∎

Exercise 3.23. Following the guideline of the previous example, define several Σ_{Stack}-algebras and relate them by homomorphisms, when possible.

Homomorphisms of partial first-order structures are truth-preserving *weak* homomorphisms using the notation of Definition 2.7.28. There are several other possible notions of homomorphism, basically due to the combinations of choices for the treatment of predicates (truth-preserving, truth-reflecting, or both) and partial functions (the condition $pf_A(a_1, \ldots, a_n) \downarrow$ can be dropped, or substituted by $pf_B(h_{s_1}(a_1), \ldots, h_{s_n}(a_n)) \downarrow$), that are used in the literature (see, e.g., [Bur82, Rei87]). The definition adopted here guarantees that initial (free) models (if any) in a class are *minimal*, following the *no-junk & no-confusion* principle from [MG85].

Exercise 3.24. Generalize the notion of reduct for partial first-order structures for both kinds of signature morphisms as defined in Exercise 3.20.

In each algebraic formalism, like group or ring theories or total algebras, there is a notion of subobject that is (up to isomorphism) a subset of the set underlying the algebraic structure, inheriting the interpretation of the operations from its superobject, that is, the interpretation of each operation in the subobject is a function of the required class (e.g., the binary operation of a group is associative) and its graph is a subset of the graph of the interpretation of the same operation in the superobject. In the case of partial first-order structures, the same intuition applies; but since some operation symbols are interpreted as (possibly) partial functions, there are different possible generalizations. Let us, indeed, compare the interpretation pf_B of a partial symbol pf in a subobject B of a partial first-order structure A with the interpretation pf_A of the same partial symbol in A. The weakest possible requirement is that pf_B is a partial function and $graph(pf_B) \subseteq graph(pf_A)$. Thus the application of pf_B to an appropriate tuple of elements from the carrier sets of B may be undefined, though the application of pf_A to the same tuple results in a value and this, in turn, may cause some element, which in A is the interpretation of a term, to become "junk".

Several possible definitions of subobject have been explored and used in the literature. For model theory the most interesting are the *weak* and *closed* substructures that we present and investigate here. A few more notions are presented shortly in Section 3.3.1 and their applications are sketched there and in the following.

Definition 3.25. Let A be a partial Σ-structure; then a partial Σ-structure A_0 is a *weak substructure* (a *subobject*) of A iff

- $|A_0| \subseteq |A|$;
- $f_A(a_1, \ldots, a_n) \overset{\mathrm{E}}{=} f_{A_0}(a_1, \ldots, a_n)$ for all $f: s_1 \times \cdots \times s_n \longrightarrow s \in \Omega$ and all $a_i \in |A_0|_{s_i}$ for $i = 1, \ldots, n$;
- if $pf_{A_0}(a_1, \ldots, a_n) \downarrow$, then $pf_{A_0}(a_1, \ldots, a_n) \overset{\mathrm{E}}{=} pf_A(a_1, \ldots, a_n)$ for all $pf: s_1 \times \cdots \times s_n \rightarrowtail s \in P\Omega$ and all $a_i \in |A_0|_{s_i}$ for $i = 1, \ldots, n$;
- $p_{A_0} \subseteq p_A$ for all $p: s_1 \times \cdots \times s_n \in \Pi$.

A weak substructure Σ-structure A_0 of A is a *closed substructure* (see [Bur82]) of A iff

- $pf_{A_0}(a_1, \ldots, a_n) \overset{\mathrm{S}}{=} pf_A(a_1, \ldots, a_n)$ for all $pf: s_1 \times \cdots \times s_n \rightarrowtail s \in P\Omega$ and all $a_i \in |A_0|_{s_i}$ for $i = 1, \ldots, n$;
- $(a_1, \ldots, a_n) \in p_{A_0}$ iff $(a_1, \ldots, a_n) \in p_A$ for all $p: s_1 \times \cdots \times s_n \in \Pi$ and all $a_i \in |A_0|_{s_i}$ for $i = 1, \ldots, n$.

The *embedding* $e: A_0 \hookrightarrow A$ of a weak substructure A_0 into A is the homomorphism of partial Σ-structures whose components are set embeddings.

A Σ-structure without proper closed substructures is said to be *reachable*.

◇

Thus, many different weak substructures of one Σ-structure sharing the same carriers exist, because partial function and predicate interpretations can be weakened. On the other hand, each subset of the carriers closed under functional application defines a closed substructure.

Example 3.26. Let us consider the following homogeneous signature

> **sig** $\Sigma =$
>> **sorts**　s
>> **opns**　$c: \rightarrow s$
>>> $f: s \rightarrow s$
>> **popns**　$pc: \negmedspace\rightarrow s$
>>> $pf: s \times s \negmedspace\rightarrow s$
>> **preds**　$p: s \times s \times s$

and the following Σ-structure A

> **algebra** $A =$
>> **Carriers**
>>> $|A|_s = \{0, \dots, \mathbf{max}\}$
>> **Functions**
>>> $c_A = 0$
>>> $f_A(x) = x$
>>> $pc_A = \mathbf{max}$
>>> $pf_A(x,y) = \begin{cases} x, & \text{if } x = y \\ \text{undefined}, & \text{otherwise} \end{cases}$
>> **Predicates**
>>> $p_A = \{(x,y,z) \mid x = y \text{ or } x = z \text{ or } y = z\}$

Then any subset of the range $\{0, \dots, \mathbf{max}\}$ including 0 can be the carrier of several weak substructures. For instance, let us consider the singleton $\{0\}$, then the following Σ-structures A_1 and A_2 are both weak substructures of A:

> **algebra** $A_1 =$
>> **Carriers**
>>> $|A_1|_s = \{0\}$
>> **Functions**
>>> $c_{A_1} = 0$
>>> $f_{A_1}(x) = x$
>>> $pc_{A_1} = \text{undefined}$
>>> $pf_{A_1}(x,y) = \begin{cases} x, & \text{if } x = y \\ \text{undefined}, & \text{otherwise} \end{cases}$
>> **Predicates**
>>> $p_{A_1} = \emptyset$

> **algebra** $A_2 =$
>> **Carriers**
>>> $|A_2|_s = \{0\}$
>> **Functions**
>>> $c_{A_2} = 0$

$$f_{A_2}(x) = x$$
$$pc_{A_2} = \text{undefined}$$
$$pf_{A_2}(x, y) = \text{undefined}$$
Predicates
$$p_{A_2} = \{(0, 0, 0)\}$$

and moreover the two weak substructures are not related by homomorphisms in either way. There does not exist a closed substructure of A with $\{0\}$ as carrier, because the interpretation of pc in A is defined but its value does not belong to $\{0\}$. But each subset X of $\{0, \dots, \text{max}\}$ including 0 and max defines a *unique* closed substructure A_X of A, consisting of:

algebra $A_X =$
    Carriers
$$|A_X|_s = X$$
    Functions
$$c_{A_X} = 0$$
$$f_{A_X}(x) = x$$
$$pc_{A_X} = \text{max}$$
$$pf_{A_X}(x, y) = \begin{cases} x, & \text{if } x = y \\ \text{undefined}, & \text{otherwise} \end{cases}$$
    Predicates
$$p_{A_X} = \{(x, y, z) \mid x = y \text{ or } x = z \text{ or } y = z, x, y, z \in X\}$$ ∎

As expected in a categorical setting, there is an obvious relation between the notion(s) of subobject and the domain of special kinds of homomorphisms. Indeed, weak substructures are (standard) subobjects, that is, domains of monomorphisms.

Proposition 3.27. *Let A and B be partial Σ-structures. A homomorphism of partial Σ-structures $h\colon A \to B$ is a monomorphism, that is, $h_1; h = h_2; h$ implies $h_1 = h_2$ for all $h_1, h_2\colon A_0 \to A$, if and only if h_s is injective for all $s \in S$.*

Analogously, closed substructures are regular subobjects, that is, domains of equalizers. Notice that the image of a homomorphism is not, in general, a closed substructure, because the interpretation of a partial function in the target can be more defined than in the source.

Exercise 3.28. Prove that the image of a homomorphism is a weak substructure of the homomorphism target and show an example of a homomorphism whose image is not a closed substructure of the target. Moreover, show that the image of a closed homomorphism is a closed substructure of the homomorphism target.

If a family of functions satisfies the homomorphism conditions for a Σ-structure, then it is a homomorphism into any closed substructure including its image.

66   Maura Cerioli, Till Mossakowski, and Horst Reichel

Lemma 3.29. *Let A_0 be a closed substructure of a partial Σ-structure A and $h\colon B \to A$ be a homomorphism such that the image of B is an S-indexed subset of $|A_0|$. Then $h\colon B \to A_0$ is a homomorphism.*

The category of partial Σ-structures has equalizers. That is, given two parallel homomorphisms $h, h'\colon A \to B$, there exists a homomorphism $e\colon E \to A$ such that e *equalizes* h and h', i.e., $e; h = e; h'$, and moreover e is *universal*, that is, every other homomorphism equalizing h and h' factorizes in a unique way through e, i.e., $k; h = k; h'$ for some $k\colon K \to A$ implies that there exists a unique $k'\colon K \to E$ such that $k = k'; e$. Indeed, it is possible to "restrict" the domain of such h and h' to the elements on which they yield the same result.

Proposition 3.30. *Let $h, h'\colon A \to B$ be parallel homomorphisms of partial Σ-structures; then the equalizer of h and h' is the (embedding of the) closed substructure E of A whose carriers are defined by*

$$|E|_s = \{a \mid a \in |A|_s \text{ and } h_s(a) \overset{\mathrm{E}}{=} h'_s(a)\}$$

for all $s \in S$ (into A).

Thus, the domains of equalizers are closed substructures[11], and hence closed substructures are "regular subobjects".

While monomorphisms are all injective functions, as in set theory, epimorphisms are not required to be surjective.

Definition 3.31. A homomorphism $h\colon A \to B$ is called *generating* (or *dense*), if the smallest closed substructure of B containing the image of h is B itself.

◇

Proposition 3.32. *Let $h\colon A \to B$ be a homomorphism of partial Σ-structures. B is generated by h iff h is an epimorphism, that is, $h; k = h; k'$ implies $k = k'$ for all $k, k'\colon B \to C$.*

It is worth noting that bijective homomorphisms are not required to be isomorphisms, that is, their inverse may not exist. Consider the signature Σ with one sort, no total function, one partial constant, and no predicates at all. Then let us call A the Σ-structure on such signature with a singleton carrier and the interpretation of the constant defined, and let us call B its weak substructure, with the same carrier, but with the interpretation of the constant undefined. Then the embedding of B into A is a bijective homomorphism, but there does not exist any homomorphism from A into B, as the

---

[11] The converse is also true, that is, for any given substructure E of a Σ-structure A there is a pair of homomorphisms whose equalizer is E itself. However, the result is unnecessary for the model theory we wish to present here and the construction of such homomorphisms is not elementary. Indeed, they are basically the embedding of A into a Σ-structure B, built by duplicating the elements of A, and then identifying the elements of E, and adding new values to denote the results of total functions on mixed inputs from both copies of A.

constant is defined in A but not in B. Therefore, in the category of partial Σ-structures, bimorphisms (i.e., monic epimorphisms) are not required to be isomorphisms.

As usual in most algebraic approaches, a notion of *congruence* is introduced to represent the kernels of homomorphisms.

Definition 3.33. Let A be a partial Σ-structure; a *congruence* on A is an S-sorted family $\equiv$ of subsets $\equiv_s$ of $|A|_s \times |A|_s$ satisfying the following conditions:

- if $a \equiv_s a'$, then $a' \equiv_s a$ (*symmetry*);
- if $a \equiv_s a'$ and $a' \equiv_s a''$, then $a \equiv_s a''$ (*transitivity*);
- if $a_i \equiv_{s_i} a'_i$ for $i = 1, \dots, n$, then $f_A(a_1, \dots, a_n) \equiv_s f_A(a'_1, \dots, a'_n)$ for all $f: s_1 \times \cdots \times s_n \longrightarrow s \in \Omega$ (*total function closure*);
- if $a_i \equiv_{s_i} a'_i$ for $i = 1, \dots, n$ and both $pf_A(a_1, \dots, a_n) \equiv_s pf_A(a_1, \dots, a_n)$[12] and $pf_A(a'_1, \dots, a'_n) \equiv_s pf_A(a'_1, \dots, a'_n)$, then $pf_A(a_1, \dots, a_n) \equiv_s pf_A(a'_1, \dots, a'_n)$ for all $pf: s_1 \times \cdots \times s_n \relbar\joinrel\rightharpoonup s \in P\Omega$ (*partial function weak closure*).

Given a congruence $\equiv$ on a partial Σ-structure A the *domain* of $\equiv$ is the S-family $\{a \mid a \in |A|_s$ and $a \equiv_s a\}$; if the domain of $\equiv$ coincides with the whole carrier, then $\equiv$ is called *total*.

Given a congruence $\equiv$ on a partial Σ-structure A, the *quotient* of A by $\equiv$ is the partial Σ-structure $A/\equiv$ defined by:

- $|A/\equiv|_s = \{[a]_\equiv \mid a \equiv_s a\}$ for all $s \in S$;
- $f_{A/\equiv}([a_1]_\equiv, \dots, [a_n]_\equiv) = [f_A(a_1, \dots, a_n)]_\equiv$, for all $f: s_1 \times \cdots \times s_n \longrightarrow s \in \Omega$ and all $[a_i]_\equiv \in |A/\equiv|_{s_i}$ for $i = 1, \dots, n$;
- $pf_{A/\equiv}(x_1, \dots, x_n) \overset{\text{E}}{=} [pf_A(a_1, \dots, a_n)]_\equiv$ if $a_i \in x_i$ exist for $i = 1, \dots, n$ such that $pf_A(a_1, \dots, a_n) \equiv_s pf_A(a_1, \dots, a_n)$; otherwise it is undefined, for all $pf: s_1 \times \cdots \times s_n \relbar\joinrel\rightharpoonup s \in P\Omega$ and all $x_i \in |A/\equiv|_{s_i}$ for $i = 1, \dots, n$;
- $(x_1, \dots, x_n) \in p_{A/\equiv}$ iff $(a_1, \dots, a_n) \in p_A$ for some $a_i \in x_i$ for $i = 1, \dots, n$, for all $p: s_1 \times \cdots \times s_n \in \Pi$ and all $x_i \in |A/\equiv|_{s_i}$ for $i = 1, \dots, n$.

Given a total congruence $\equiv$ on a partial Σ-structure A, we will denote by $nat_\equiv$ the homomorphism from A into $A/\equiv$ associating each element with its equivalence class in $\equiv$.

Let h be a homomorphism of partial Σ-structures from A into B. Then the *kernel* $K(h)$ of h is the total congruence on A defined by $a\, K(h)\, a'$ iff $h(a) = h(a')$ for all $a, a' \in |A|_s$ and all $s \in S$.   ◇

It can be immediately verified that $A/\equiv$ is actually a partial Σ-structure for any given congruence $\equiv$ on a partial Σ-structure A. Indeed the conditions on weak partial and total functional closure ensure the unambiguous definition of function interpretation.

---

[12] Note that this implicitly implies $pf_A(a_1, \dots, a_n)\downarrow$.

Moreover, $K(h)$ is symmetric, reflexive, and transitive, by definition, and the conditions of functional closure are guaranteed by the definition of homomorphism. Therefore, $K(h)$ is a total congruence. The first homomorphism theorem holds for our definition of congruence.

Proposition 3.34. *Given a homomorphism of partial Σ-structures h from A into B, there exists a unique $h_K(h) \colon A/K(h) \to B$ with $h = nat_{K(h)}; h_K(h)$.*

Different notions of subobjects. Having introduced two definitions of substructure, it can be supposed that there are others as well. And indeed between weak and closed subobjects there are various reasonable possibilities that have been proposed for algebras in the literature.

The first requirement that we can impose on a weak substructure B of a partial first-order structure A is that the application of pf_B to an appropriate tuple of elements from the carrier sets of B is defined whenever this is possible, that is, if the application of pf_A to the same tuple results in a value within the carrier set of B. This observation leads to the notion of *relative* substructure, where pf_B (regarded as a graph) is the largest subset of pf_A included in the (appropriate cartesian product of the) carrier sets of B. This concept of substructure is particularly relevant because it allows us to regard a bounded implementation of a data type as a substructure of its unbounded abstraction (see, e.g., [Grä79]). The important property here is that each subset of the carriers induces a unique relative substructure.

While closed substructures satisfy some upward closure principle with respect to partial function application, *normal* substructures satisfy some *downward* closure principle. Normal substructures are important for term evaluation (the downward closure here means that all subterms of an interpretable term are interpretable).

To help the experienced reader to gain an overview of how we will generalize the different notions of subalgebra to the case of first-order structures, the key idea is that predicates are thought of as partial functions in a singleton set, identifying definedness of such functions to predicate truth.

Definition 3.35. A weak substructure Σ-structure A_0 of A is a *relative substructure* (see [Bur82]) of A iff

- for all $pf \colon s_1 \times \cdots \times s_n \xrightarrow{\;\;} s \in P\Omega$ and all $a_i \in |A_0|_{s_i}$, for $i = 1, \ldots, n$, if $pf_A(a_1, \ldots, a_n) \in |A_0|_s$, then $pf_{A_0}(a_1, \ldots, a_n) \stackrel{\mathrm{E}}{=} pf_A(a_1, \ldots, a_n)$, and
- $(a_1, \ldots, a_n) \in p_{A_0}$ iff $(a_1, \ldots, a_n) \in p_A$ for all $p \colon s_1 \times \cdots \times s_n \in \Pi$ and all $a_i \in |A_0|_{s_i}$, for $i = 1, \ldots, n$.

A relative substructure Σ-structure A_0 of A is a *normal substructure* (see [Sch70]) of A iff for all $pf \colon s_1 \times \cdots \times s_n \xrightarrow{\;\;} s \in P\Omega$ and all $a_i \in |A|_{s_i}$, for $i = 1, \ldots, n$, if $pf_A(a_1, \ldots, a_n) \in |A_0|_s$ then $pf_{A_0}(a_1, \ldots, a_n) \stackrel{\mathrm{E}}{=} pf_A(a_1, \ldots, a_n)$[13]. ◇

---

[13] Note that this also means that $a_i \in |A_0|_{s_i}$.

As in the case of closed substructure, a sorted set completely determines the relative (or normal) substructure having the set as carrier. Moreover, each subset of the carriers determines a unique relative substructure.

Instead of classifying the different notions of subobject through more or less complex categorical constructions[14], as we did for closed substructures that are the domain of equalizers, we can also see them as standard subobjects in a subcategory, where homomorphisms are restricted to those satisfying some extra requirement. We can also characterize the different kinds of substructures using different specializations of homomorphisms.

Definition 3.36. Let $\Sigma = \langle S, \Omega, P\Omega, \Pi \rangle$ be a partial signature, A and B be partial Σ-structures.

A homomorphism h from A into B is called *full* [Bur82] iff the following conditions hold:

- if $pf_B(h_{s_1}(a_1), \ldots, h_{s_n}(a_n)) = b \in |B|_s$ and there exists $a \in |A|_s$ such that $h_s(a) = b$ then there exist $a'_i \in |A|_{s_i}$, for $i = 1, \ldots, n$, such that $h_{s_i}(a_i) = h_{s_i}(a'_i)$ and $pf_A(a'_1, \ldots, a'_n) \in |A|_s$ for all $pf \in P\Omega_{w,s}$, where $w = s_1 \times \cdots \times s_n$, and all $a_i \in |A|_{s_i}$ for $i = 1, \ldots, n$;
- if $(h_{s_1}(a_1), \ldots, h_{s_n}(a_n)) \in p_B$ then there exist $a'_i \in |A|_{s_i}$, for $i = 1, \ldots, n$, such that $h_{s_i}(a_i) = h_{s_i}(a'_i)$ and $(a'_1, \ldots, a'_n) \in p_A$.

A full homomorphism h from A into B is called *normal* iff the following condition holds:

- if there exists $a \in |A|_s$ such that $h_s(a) \stackrel{\mathrm{E}}{=} f_B(b_1, \ldots, b_n)$ then there exist $a_i \in |A|_{s_i}$, for $i = 1, \ldots, n$, such that $h_{s_i}(a_i) = b_i$ and $f_A(a_1, \ldots, a_n) \downarrow$ for all $f \in \Omega_{w,s} \cup P\Omega_{w,s}$, where $w = s_1 \times \cdots \times s_n$, and all $b_i \in |B|_{s_i}$, for $i = 1, \ldots, n$;

A homomorphism h from A into B is *closed* [Bur82] iff the following conditions hold:

- if $pf_B(h_{s_1}(a_1), \ldots, h_{s_n}(a_n)) \downarrow$ then $pf_A(a_1, \ldots, a_n) \downarrow$ for all $pf \in P\Omega_{w,s}$, where $w = s_1 \times \cdots \times s_n$, and all $a_i \in |A|_{s_i}$, for $i = 1, \ldots, n$;
- if $(h_{s_1}(a_1), \ldots, h_{s_n}(a_n)) \in p_B$, then $(a_1, \ldots, a_n) \in p_A$ for all $p \colon s_1 \times \cdots \times s_n \in \Pi$ and all $a_i \in |A|_{s_i}$, for $i = 1, \ldots, n$.

See Table 3.1.   ◇

It is worth noting that for any congruence $\equiv$ on a partial first-order structure A, the homomorphism $nat_\equiv$ is a full surjection. Moreover, in case that the congruence is not total, $nat_\equiv$ can be defined to start from the relative subalgebra induced by $(\{a | a \equiv_s a\})_{s \in S}$.

---

[14] For instance, relative substructures are initial subobjects, that is, domains of initial monomorphisms [AHS90] (in a different terminology, they would be called cartesian monomorphisms [Bor94]) with respect to the obvious forgetful functor.

Kind	partial operations	predicates
hom.	$h(graph\ pf_A) \subseteq graph\ pf_B$	$h(p_A) \subseteq p_B$
full	$h(graph\ pf_A) = graph\ pf_B \cap h(A)$	$h(p_A) = p_B \cap h(A)$
normal	$h(graph\ f_A) = graph\ f_B \cap |B|_{s_1,\dots,s_n} \times h_s(|A|_s)$	$h(p_A) = p_B \cap h(A)$
closed	$dom\ pf_A = h^{-1}(\ dom\ pf_B)$	$p_A = h^{-1}(p_B)$

Table 3.1.

Proposition 3.37. *Any closed homomorphism is full.*

Proposition 3.38. *A weak substructure is a relative (normal, closed) sub-structure iff the embedding is a full (normal, closed) homomorphism.*

Corollary 3.39. *Any closed substructure is relative.*

Term evaluation and initiality. Standard term algebras (as defined in the total case) can be endowed with (possibly infinite) choices of predicate interpretations in order to get partial first-order structures (with predicates). But, the usual inductive definition of term evaluation is not a homomorphism, disregarding the interpretation of predicates, because generally it is a *partial* function. However, the domain of term evaluation is, by definition, a normal substructure of the term algebra and term evaluation is the unique closed homomorphism extending variable evaluation with a normal substructure as domain (with respect to the signature where predicates that do not play any role in term evaluation are dropped[15]). For a similar presentation, factorizing term evaluation in two steps (the definition of the domain as a substructure satisfying appropriate conditions and the definition of term evaluation as a closed homomorphism), see [Wol90].

A *term Σ-structure* over X consists of the (total) term algebra $T_\Sigma(X)$ and an interpretation $p_{T_\Sigma(X)} \subseteq |T_\Sigma(X)|_{s_1} \times \cdots \times |T_\Sigma(X)|_{s_n}$ of each $p: s_1 \times \cdots \times s_n \in \Pi$.

In particular we will denote the term Σ-structure with $p_{T_\Sigma(X)} = \emptyset$ for all $p: s_1 \times \cdots \times s_n \in \Pi$ by $T_\Sigma(X)$, and if X is the empty family of variables, $T_\Sigma(X)$ will be simply denoted by T_Σ.

A *variable valuation* v for X in A is any S-sorted family of partial functions $v_s: X_s \longrightarrow |A|_s$; given a variable valuation v for X in A, the *term evaluation* $v^\#: T_\Sigma(X) \longrightarrow A$ is inductively defined by:

- $v_s^\#(x) \overset{S}{=} v(x)$ for all $x \in X_s$ and all $s \in S$;
- $v_s^\#(f(t_1, \dots, t_n)) \overset{S}{=} f_A(v_{s_1}^\#(t_1), \dots, v_{s_n}^\#(t_n))$ for all $f \in \Omega_{w,s} \cup P\Omega_{w,s}$, where $w = s_1 \times \cdots \times s_n$, and all $t_i \in |T_\Sigma(X)|_{s_i}$, for $i = 1, \dots, n$;

---

[15] If we allow predicates to persist in the signature, then, since closed homomorphisms, as all homomorphisms, both preserve and reflect truth, we would need an *ad hoc* definition of predicate interpretation in the term structure for each algebra where we wisht to evaluate terms.

Given a term t, we say that t is *v-interpretable*, if $t \in$ dom $v^{\#}$ and in this case we call $v^{\#}(t) \in |A|_s$ the *value of t in A under the valuation v*.

In particular if X is the empty family of variables, then v is the empty map for each partial Σ-structure A and we will denote $v_s^{\#}$ by $eval_A$ and its application to a term t by t_A.

Whenever $v^{\#}$ is surjective, we say that the Σ-structure A is *generated by v*, and if X is the empty set (and hence v is the empty map), A is simply said to be *term-generated*.

Proposition 3.40. *A homomorphism h from A into B is generating iff B is generated by h (regarded as a valuation of the family $|A|$ of variables).*

It is straightforward to verify, by induction on the definition of $v^{\#}$, the following technical lemma, whose proof is left as an exercise to the reader.

Lemma 3.41. *Let $\Sigma = \langle S, \Omega, P\Omega, \Pi \rangle$ be a partial signature, A be a partial Σ-structure, X be an S-sorted family of variables, and v be a valuation for X in A.*

- *For each term t, if $v^{\#}(t) \downarrow$, then $h(v^{\#}(t)) \overset{\mathrm{E}}{=} (v; h)^{\#}(t)$ for all homomorphisms $h: A \to B$.*
- *For each valuation v' for X in A, if $v^{\#}(x) \overset{\mathrm{W}}{=} v'^{\#}(x)$ for all variables $x \in X$, then $v^{\#}(t) \overset{\mathrm{W}}{=} v'^{\#}(t)$ for all terms t.*

Notation 3.42. In the sequel let us fix a partial signature $\Sigma = \langle S, \Omega, P\Omega, \Pi \rangle$.

Closed substructures are closed with respect to term evaluation, while weak substructures are not.

Exercise 3.43. Prove that if the valuation of a family of variables is contained in the carriers of a closed substructure, the evaluation of any term with respect to such a valuation in the Σ-structure and in the closed substructure are strongly equal (while for weak substructures, they are only weakly equal).

Corollary 3.44. *A Σ-structure is term-generated if and only if it is reachable (i.e., it does not have proper closed substructures).*

Definition 3.45. Let $\mathcal{A}$ be a class of partial first-order structures over a partial first-order signature $\Sigma = \langle S, \Omega, P\Omega, \Pi \rangle$ and X be an S-sorted family of variables.

Then a partial first-order structure $F \in \mathcal{A}$ is *free* for X in $\mathcal{A}$ iff there exists a total valuation e for X in F such that for all $A \in \mathcal{A}$ and all total valuations v for X in A a unique homomorphism $h_v : F \to A$ exists such that $v = e; h_v$.

A free partial first-order structure $F \in \mathcal{A}$ for the empty family of variables in $\mathcal{A}$ is called *initial* in $\mathcal{A}$.   ◇

Lemma 3.46. *Let $\mathcal{A}$ be a class of partial first-order structures over a partial first-order signature $\Sigma = \langle S, \Omega, P\Omega, \Pi \rangle$ closed under closed substructures. Then a free first-order structure for an S-sorted family X of variables in $\mathcal{A}$, if any, is generated by X.*

3.3.2 Partial logic

We now want to generalize the concept of describing classes of algebras by axioms introduced in Chapter 2 and extended at the beginning of this chapter, by allowing conditional axioms to be built starting not only from equalities, but also from predicate symbols applied to tuples of terms.

The presence of partial functions introduces the possibility of terms which do not denote in all structures. This phenomenon causes different possible generalizations of the concept of equation to the partial case. Thus, the proliferation of equality symbols that we introduced at the metalevel also reflects on formulas, having three different kinds of atomic formulas representing existential, weak, and strong equalities between terms, besides atomic predicate formulas.

Moreover, in Section 3.1, we only allowed universally quantified conditional axioms, since they have the nice properties that initial models and relatively fast theorem provers exist. Here, we move to full first-order logic. This increased expressiveness of axioms allows us to write requirement specifications (to be interpreted loosely) which are more succinct and more related to informal requirements than specifications with conditional equations can be, as illustrated at the end of Example 3.15. Moreover, there are interesting data types that cannot be directly expressed within the conditional fragment; see, for instance, Section 3.4 below. Since the existence of initial models is needed sometimes (e.g., for initial constraints or for design specifications), we later identify those fragments of first-order logic that still have initial models.

We now want to use terms for building formulas, which will eventually serve as axioms in specifications.

Definition 3.47 (First-order formula). Let $\Sigma = \langle S, \Omega, P\Omega, \Pi \rangle$ be a signature. We inductively define for all S-sorted sets X in parallel the set $Form(\Sigma, X)$ of Σ-formulas in variables X. $Form(\Sigma, X)$ is the least S-sorted set containing

- $t_1 \overset{e}{=} t_2$ for $t_1, t_2 \in |T_\Sigma(X)|_s$
- $p(t_1, \ldots, t_n)$ for $p \colon s_1 \times \cdots \times s_n \in \Pi$ and $t_i \in |T_\Sigma(X)|_{s_i}$, $i = 1, \ldots, n$
- F (read: false)
- $(\varphi \wedge \psi)$ and $(\varphi \Rightarrow \psi)$ for $\varphi, \psi \in Form(\Sigma, X)$
- $(\forall Y.\varphi)$ for $\varphi \in Form(\Sigma, X \cup Y)$, Y an S-sorted set

If there is no ambiguity, the brackets around $(\varphi \wedge \psi)$, etc., can be omitted.

We define the following abbreviations:

$(\neg \varphi)$ stands for $(\varphi \Rightarrow F)$

$(\varphi \vee \psi)$ stands for $\neg(\neg\varphi \wedge \neg\psi)$

$(\varphi \Leftrightarrow \psi)$ stands for $(\varphi \Rightarrow \psi) \wedge (\psi \Rightarrow \varphi)$

$D(t)$ stands for $t \overset{e}{=} t$

$t_1 \overset{s}{=} t_2$ stands for $(D(t_1) \vee D(t_2)) \Rightarrow t_1 \overset{e}{=} t_2$

$t_1 \overset{w}{=} t_2$ stands for $(D(t_1) \wedge D(t_2)) \Rightarrow t_1 \overset{e}{=} t_2$

$(\exists Y.\varphi)$ stands for $\neg(\forall Y.\neg\varphi)$ ◇

Definition 3.48 (First-order axiom). A *first-order axiom* over a signature Σ is a pair (X, φ), written $X.\varphi$, where $\varphi \in Form(\Sigma, X)$.
A first-order axiom $X.\varphi$ is called *closed*, if $X = \emptyset$. ◇

The usual definition of the free variables of a term or a formula now becomes easy.

Definition 3.49. Given a term $t \in T_\Sigma(X)$, the set $FV(t)$ of *free variables of t* is the least set $X' \subseteq X$ such that $t \in T_\Sigma(X')$.
Likewise, given a formula $\varphi \in Form(\Sigma, X)$, the set $FV(\varphi)$ of *free variables of φ* is the least set $X' \subseteq X$ such that $\varphi \in Form(\Sigma, X')$. ◇

It is common practice to omit from the definition of axioms, the family X of variables over which the formula is defined and just write φ, where X is recovered as $FV(\varphi)$. However, at least for a semantics based on total valuations, it is essential to also allow axioms $X.\varphi$ where $FV(\varphi)$ is a proper subset of X, which may behave differently from $FV(\varphi).\varphi$, see Section 3.3.3.
To be able to understand formally a model of a specification, we now have to define satisfaction of first-order axioms by first-order structures. According to Feferman [Fef95], there are basically two different ways to define satisfaction: in the *logics of existence*, satisfaction is defined using *partial* variable valuations, i.e., variables may be undefined. On the other hand, in the *logics of definedness*, satisfaction is defined using *total* variable valuations, i.e., variables are always defined. Note that in both cases, we quantify over the defined only – bound variables are always defined. Therefore, quantification is treated by extension of the valuations which have to be defined for the quantified variables.

Definition 3.50 (Satisfaction). A partial Σ-structure A *satisfies* a first-order axiom $X.\varphi$ *with respect to total valuations* (written $A \models_\Sigma^t X.\varphi$), if all total valuations $v \colon X \longrightarrow |A|$ satisfy φ.
A partial Σ-structure A *satisfies* a first-order axiom $X.\varphi$ *with respect to partial valuations* (written $A \models_\Sigma^p X.\varphi$), if all (partial or total) valuations $v \colon X \longrightarrow |A|$ satisfy φ.
Satisfaction of a formula $\varphi \in Form(\Sigma, X)$ by a (possibly partial) valuation $v \colon X \longrightarrow |A|$ is defined inductively over the structure of φ:

- $v \Vdash_\Sigma t_1 \overset{e}{=} t_2$ iff $v^\#(t_1) \overset{E}{=} v^\#(t_2)$
- $v \Vdash_\Sigma p(t_1, \ldots, t_n)$ iff $v^\#(t_1) \downarrow$ and $\cdots$ and $v^\#(t_n) \downarrow$ and $(v^\#(t_1), \ldots, v^\#(t_n)) \in p_A$
- not $v \Vdash_\Sigma F$
- $v \Vdash_\Sigma (\varphi \wedge \psi)$ iff $v \Vdash_\Sigma \varphi$ and $v \Vdash_\Sigma \psi$
- $v \Vdash_\Sigma (\varphi \Rightarrow \psi)$ iff $v \Vdash_\Sigma \varphi$ implies $v \Vdash_\Sigma \psi$
- $v \Vdash_\Sigma (\forall Y.\varphi)$ iff for all valuations $\xi \colon X \cup Y \longrightarrow |A|$ which
  - extend v on $X \setminus Y$ (i.e., $\xi(x) \overset{S}{=} v(x)$ for all $x \in X_s \setminus Y_s, s \in S$), and
  - are defined on Y (i.e., $\xi(y) \downarrow$ for $y \in Y_s, s \in S$)
  we have $\xi \Vdash_\Sigma \varphi$. ◇

Thus, we treat quantification by extensions of valuations to the quantified variables. By requiring ξ to be an extension of v on $X \setminus Y$ only, variables in Y are treated as fresh variables: their value under v is disregarded within φ.

Satisfaction of arbitrary first-order axioms with respect to total valuations can be reduced to that of closed axioms, because the satisfaction of a quantified formula is equivalent to that of its universal closure.

Exercise 3.51. Prove that if a formula φ is satisfied by some valuation v then it is satisfied by all valuations coinciding with v on the free variables of φ.

Exercise 3.52. Prove that for a partial Σ-structure A and $\varphi \in Form(\Sigma, X)$

$$A \models_\Sigma^t X.\varphi \text{ if and only if } A \models_\Sigma^t \emptyset.\forall X.\varphi \text{ if and only if } A \models_\Sigma^p \emptyset.\forall X.\varphi$$

The counterexample $A \models_\Sigma^p \emptyset.\forall x : s.x \overset{e}{=} x$ but $A \not\models_\Sigma^p \{x : s\}.x \overset{e}{=} x$, as not $v \Vdash_\Sigma x \overset{e}{=} x$ for v the totally undefined partial valuation, shows that open formulas are interpreted by $\models^p$ quite differently.

Proposition 3.53. *Let X and Y be two S-sorted variable systems and $\varphi \in Form(\Sigma, X \cap Y)$ be a first-order formula. Then for any partial Σ-structure A,*

$$A \models_\Sigma^p Y.\varphi \text{ if and only if } A \models_\Sigma^p X.\varphi$$

while the corresponding property for $\models^t$ does not hold, unless all carrier sets of A are nonempty.

Definition 3.54 (Semantical consequence). A first-order axiom $X.\varphi$ is said to *follow semantically with respect to total valuations* (with respect to partial valuations) from a set of first-order axioms M, written $M \models_\Sigma^t \varphi$ ($M \models_\Sigma^p \varphi$), if for all total (total or partial) valuations $v \colon X \cup \bigcup_{Y.\psi \in M} Y \longrightarrow |A|$ into partial Σ-structures A we have:

if $v|_Y \Vdash_\Sigma \psi$ for all $Y.\psi \in M$, then $v|_X \Vdash_\Sigma \varphi$ ◇

Proposition 3.53 can be easily extended to semantical consequence.

Proposition 3.55. *Let X and Y be two S-sorted variable systems and $M \cup \{\varphi\} \subseteq Form(\Sigma, X \cap Y)$ be first-order formulas. Then*

$$M \models_\Sigma^p Y.\varphi \text{ if and only if } M \models_\Sigma^p X.\varphi$$

while the corresponding property for $\models^t$ does not hold.

The peculiarity of $\models^t$ shown in Propositions 3.53 and 3.55 is not introduced by the extension of logical power, because it is already present in the total many-sorted equational fragment, where it leads to inconsistent calculi unless quantification is very carefully treated. For this reason, in most of the literature[16] on total algebras empty carrier sets are not allowed or they

---

[16] Indeed, the only references developing full many-sorted first-order logic with possibly empty carriers we found are [AR94, Hai53, KGS88, KN94].

are required not to be connected to the nonempty carriers by any function symbol. In [HO80] syntactical conditions on signatures are given, guaranteeing that the empty carriers cannot introduce problems. Not only are such conditions no longer significant for the partial case, but in the context of specification, there may very well be the situation where some data set is empty, for instance, during the requirement phase, before decisions on some kinds of elements have been completed. Thus, we do not require that the models of a specification have nonempty carriers.

Both Exercise 3.52 and Proposition 3.53 (the latter together with its companion 3.55) do hold for one-sorted, total first-order logic. When generalizing to the partial many-sorted case, we cannot keep both true. So we have to choose between the equivalence of formulas to their universal closure (which holds for $\models^t$) and invariance under changes of the variable system (which holds for $\models^p$).[17] While many treatments of partial logics [Bur82, Bee85, Rei87] are guided by the former, we prefer the latter, partly because of the easier Substitution Lemma for $\models^p$ (see Lemmas 3.62 and 3.63 below). The price for this preference is a slightly more complex manipulation of quantification. But the invariance under changes of the variable system of $\models^p$ allows us now to drop the variable system.

Notation 3.56. Referring to $\models^p_\Sigma$, we may drop the variable system from formulas and treat φ as an abbreviation of $FV(\varphi).\varphi$.

As in Section 2.2, we define a presentation to be a pair $\langle \Sigma, AX \rangle$ where AX is a set of Σ-first-order axioms. A *model* of a presentation $\langle \Sigma, AX \rangle$ is a partial Σ-structure A such that $A \models^p_\Sigma AX$. $Mod_\Sigma(AX)$ is the class of all models of $\langle \Sigma, AX \rangle$.

Definition 3.57. A presentation $\langle \Sigma, AX \rangle$ is called *semantically inconsistent*, if $Mod_\Sigma(AX)$ is empty. Otherwise, it is called *semantically consistent*.

$\diamond$

Proposition 3.58. *For Σ-first-order formulas $\varphi_1, \ldots, \varphi_n, \psi$, the following are equivalent:*

1. $M \cup \{ \varphi_1, \ldots, \varphi_n \} \models^p_\Sigma \psi$;
2. $M \models^p_\Sigma \varphi_1 \wedge \cdots \wedge \varphi_n \Rightarrow \psi$;
3. $\langle \Sigma, M \cup \{ \varphi_1, \ldots, \varphi_n, \neg\psi \} \rangle$ *is inconsistent.*

Example 3.59. An easy example of an inconsistent presentation is given requiring $A \wedge \neg A$ for some formula A without free variables.

---

[17] Of course, we can keep both true if we define the semantics of quantification over partial extensions of valuations, so that quantified variables, as free variables, need not denote a value. But then, in practice, we have to add many definedness conditions to quantified axioms.

Another possibility to keep both conditions true, for total satisfaction, is to restrict the models to those with nonempty carriers, but this is too restrictive from a methodological point of view during the initial phases, as already mentioned.

    spec *INCONSISTENT =*
       sorts   *s*
       preds *p: s*
       axioms $(\forall x : s.p(x)) \wedge (\neg \forall x : s.p(x))$ ∎

It is interesting to note that the following, quite similar, specification is *not* inconsistent.

    spec *PECULIAR =*
       sorts   *s*
       preds *p: s*
       axioms $\forall x : s.(p(x) \wedge \neg p(x))$

Indeed it has the empty structure as a model, because if the carrier of sort s is empty, a total valuation for $\{x\}$ does not exist in it and hence $\forall x : s.A$ is satisfied disregarding the formula A.

3.3.3 Proof theory

Whereas model theory introduced in the previous section lays the foundation for specification of data types (understood as partial algebras), proof theory is essential for deriving in a syntactical, computable way the semantical consequences of a specification. The consequences may not only reveal wanted or unwanted behavior of the specified system, but possibly also the inconsistency of the specification.

    We present here two natural deduction-style proof calculi for partial first-order logic. The first one was developed by Burmeister [Bur82] to capture $\models^t$. Burmeister's calculus covers only the one-sorted case. Here, in accordance with previous sections, we generalize it to the many-sorted case (again allowing carriers to be empty), which forces us to carefully keep track of variables [GM85, GM86a]. The second calculus captures $\models^p$ and follows the ideas of Scott [Sco79]. In this calculus, because of Proposition 3.55, we can omit the variable system.

    Both calculi are based on a notion of substitution. The usual notion (see Exercise 1.4.9) can be easily generalized to the partial case.

Definition 3.60. A function $\theta : X \longrightarrow |T_\Sigma(Y)|$ is called a *substitution*. Given a substitution $\theta : X \longrightarrow |T_\Sigma(Y)|$ and an S-sorted variable system Z, we denote by $\theta \backslash Z : X \cup Z \longrightarrow T_\Sigma(Y \cup Z)$ the substitution being the identity on Z and being θ on $X \backslash Z$.   ◇

Substitutions can be applied to terms as well as to formulas, where in the case of formulas, the application is not defined in all cases because of possible naming conflicts between substituted and quantified variables.

Definition 3.61. The term $t[\theta] \in T_\Sigma(Y)$ resulting from applying the substitution θ to a term $t \in T_\Sigma(X)$ is defined by

$$t[\theta] = \theta^{\#}(t)$$

The formula $\varphi[\theta] \in Form(\Sigma, Y)$, which, if defined, results from applying the substitution θ to a formula $\varphi \in Form(\Sigma, X)$ is defined inductively over φ:

- $(t_1 \overset{e}{=} t_2)[\theta] \overset{E}{=} t_1[\theta] \overset{e}{=} t_2[\theta]$
- $p(t_1, \ldots, t_n)[\theta] \overset{E}{=} p(t_1[\theta], \ldots, t_n[\theta])$
- $F[\theta] \overset{E}{=} F$
- $(\varphi \wedge \psi)[\theta] \overset{S}{=} (\varphi[\theta]) \wedge (\psi[\theta])$
- $(\varphi \Rightarrow \psi)[\theta] \overset{S}{=} (\varphi[\theta]) \Rightarrow (\psi[\theta])$
- $(\forall Z.\varphi)[\theta] \overset{S}{=} \begin{cases} \forall Z.(\varphi[\theta \setminus Z]), & \text{if } \forall x \in X_s, s \in S. \\ & (x[\theta] \neq x \text{ and } x \in FV(\forall Z.\varphi) \\ & \text{implies } Z \cap FV(x[\theta]) = \emptyset) \\ \text{undefined}, & \text{otherwise} \end{cases}$ ◇

The last case, causing $(\forall Z.\varphi)[\theta]$ to be undefined in the case of name conflicts, prevents a free variable in $x[\theta]$ from being bound by the quantification over Z. This restriction is important to keep the intended semantics of substitutions. This semantics is reflected by the following Lemma from [Bur82].

Lemma 3.62 (Substitution Lemma for $\models^t$). *Let A be a partial Σ-structure, $v: Y \longrightarrow A$ be a total valuation, $\theta: X \longrightarrow |T_\Sigma(Y)|$ be a substitution, and $\varphi \in Form(\Sigma, X)$ a formula. Under the conditions that*

- *$v^\# \circ \theta: X \longrightarrow A$ is also a total valuation (i.e., for all $x \in X_s, s \in S$ we have $v \Vdash_\Sigma D(\theta(x))$), and*
- *$\varphi[\theta]$ is defined,*

we have

$$v^\# \circ \theta \Vdash_\Sigma \varphi \text{ if and only if } v \Vdash_\Sigma \varphi[\theta]$$

Compared with the usual Substitution Lemma for total logics, we have to make the additional assumption here that the terms being substituted are defined. On the other hand, the Substitution Lemma for $\models^p$ keeps the simplicity of substitution in the total case.

Lemma 3.63 (Substitution Lemma for $\models^p$). *Let A be a partial Σ-structure, $v: Y \longrightarrow A$ be a (total or partial) valuation, $\theta: X \longrightarrow |T_\Sigma(Y)|$ be a substitution, and $\varphi \in Form(\Sigma, X)$ a formula. Under the condition that $\varphi[\theta]$ is defined, we have*

$$v^\# \circ \theta \Vdash_\Sigma \varphi \text{ if and only if } v \Vdash_\Sigma \varphi[\theta]$$

The more complicated Substitution Lemma for $\models^t$ also complicates the rules of the calculus dealing with substitution, while the other rules can be taken directly from total first-order logic.

A calculus for $\models^t$. Let Φ, Φ_1, Φ_2, and Φ_3 be finite sets of formulas in $Form(\Sigma, X)$. The application of substitution to such sets is understood elementwise. We introduce the following rules of derivation:

Assumption

$$\frac{}{\Phi \vdash^t_{\Sigma,X} \varphi} \quad \varphi \in \Phi$$

∧-introduction

$$\frac{\Phi_1 \qquad \vdash^t_{\Sigma,X} \varphi}{\Phi_2 \qquad \vdash^t_{\Sigma,X} \psi}{\Phi_1 \cup \Phi_2 \vdash^t_{\Sigma,X} (\varphi \wedge \psi)}$$

∧-left elimination

$$\frac{\Phi \vdash^t_{\Sigma,X} (\varphi \wedge \psi)}{\Phi \vdash^t_{\Sigma,X} \psi}$$

∧-right elimination

$$\frac{\Phi \vdash^t_{\Sigma,X} (\varphi \wedge \psi)}{\Phi \vdash^t_{\Sigma,X} \varphi}$$

Tertium non datur

$$\frac{\Phi_1 \cup \{\varphi\} \qquad\qquad \vdash^t_{\Sigma,X} \psi}{\Phi_2 \cup \{(\varphi \Rightarrow F)\} \vdash^t_{\Sigma,X} \psi}{\Phi_1 \cup \Phi_2 \qquad\qquad \vdash^t_{\Sigma,X} \psi}$$

Absurdity

$$\frac{\Phi \vdash^t_{\Sigma,X} F}{\Phi \vdash^t_{\Sigma,X} \psi}$$

Cut

$$\frac{\Phi_1 \qquad \vdash^t_{\Sigma,X} \quad \varphi_1 \wedge \cdots \wedge \varphi_n \Rightarrow \psi_i}{\Phi_2 \qquad \vdash^t_{\Sigma,Y} \quad \psi_1 \wedge \cdots \wedge \psi_k \Rightarrow \epsilon}{\Phi_1 \cup \Phi_2 \vdash^t_{\Sigma,X \cup Y} \psi_1 \wedge \cdots \wedge \psi_{i-1} \wedge \varphi_1 \wedge \cdots \wedge \varphi_n \wedge \psi_{i+1} \wedge \ldots \wedge \psi_k \Rightarrow \epsilon}$$

⇒-introduction

$$\frac{\Phi \cup \{\varphi_1, \ldots, \varphi_n\} \vdash^t_{\Sigma,X} \psi}{\Phi \qquad\qquad\qquad \vdash^t_{\Sigma,X} \varphi_1 \wedge \cdots \wedge \varphi_n \Rightarrow \psi}$$

∀-elimination

$$\frac{\Phi \vdash^t_{\Sigma,X} \quad (\forall Y.\varphi)}{\Phi \vdash^t_{\Sigma,X \cup Y} \varphi}$$

∀-introduction

$$\frac{\Phi \vdash^t_{\Sigma,X \cup Y} \varphi}{\Phi \vdash^t_{\Sigma,X} \quad (\forall Y.\varphi)} \quad \text{if } Y \cap FV(\Phi) = \emptyset$$

Reflexivity

$$\frac{}{\Phi \vdash^t_{\Sigma,X} x \stackrel{e}{=} x} \quad \text{for } x \in X_s$$

Congruence

$$\frac{\Phi \vdash^t_{\Sigma,X} \varphi}{\Phi \vdash^t_{\Sigma,X \cup Y} (\bigwedge_{x \in X_s} x \stackrel{e}{=} \theta(x)) \Rightarrow \varphi[\theta]} \quad \text{for } \theta : X \longrightarrow |T_\Sigma(Y)| \text{ with } \varphi[\theta] \downarrow$$

Substitution

$$\frac{\Phi \quad \vdash^t_{\Sigma,X} \varphi}{\Phi[\theta] \vdash^t_{\Sigma,Y} (\bigwedge_{x \in X_s} D(\theta(x))) \Rightarrow \varphi[\theta]}$$
$$\text{for } \theta : X \longrightarrow |T_\Sigma(Y)| \text{ with } \varphi[\theta] \downarrow \text{ and } \Phi[\theta] \downarrow$$

Function Strictness

$$\frac{\Phi \vdash^t_{\Sigma,X} t_1 \stackrel{e}{=} t_2}{\Phi \vdash^t_{\Sigma,X} D(t)} \quad t \text{ some subterm of } t_1 \text{ or } t_2$$

Predicate Strictness

$$\frac{\Phi \vdash^t_{\Sigma,X} p(t_1, \dots, t_n)}{\Phi \vdash^t_{\Sigma,X} D(t_i)} \quad \text{for } p : s_1 \times \cdots \times s_n \in \Pi$$

Totality

$$\frac{\Phi \vdash^t_{\Sigma,X} \bigwedge_{i=1,\dots,n} D(t_i)}{\Phi \vdash^t_{\Sigma,X} D(f(t_1, \dots, t_n))} \quad \text{for } f : s_1 \times \cdots \times s_n \longrightarrow s \in \Omega$$

Here $D(t)$ is syntactical sugar for $t \stackrel{e}{=} t$.

A *derivation* is a finite sequence of judgments of form $\Phi \vdash^t_{\Sigma,X} \varphi$ such that each member of the sequence is either an axiom or obtained from previous members of the sequence by application of a rule. A *derivation of a formula* $X.\varphi$ is a derivation whose last member is the judgment $\emptyset \vdash^t_{\Sigma,X} \varphi$.

Burmeister states the following theorem in [Bur82].

Theorem 3.64. *The calculus is sound and complete, i.e.,*

$$\Phi \models^t_\Sigma X.\varphi \text{ if and only if } \Phi \vdash^t_{\Sigma,X \cup \bigcup_{Y.\psi \in \Phi} Y} \varphi$$

Derived rules for the defined connectives and quantifiers can be found in [Her73]. If we want a calculus with definedness and strong equality as basic notions, we have to add the following derived rules:

Reflexivity

$$\frac{}{\Phi \vdash^t_{\Sigma,X} t \stackrel{s}{=} t} \quad \text{for } t \in |T_\Sigma(X)|_s$$

Equality1

$$\frac{\Phi \vdash^t_{\Sigma,X} t_1 \overset{e}{=} t_2}{\Phi \vdash^t_{\Sigma,X} t_1 \overset{s}{=} t_2}$$

Equality2

$$\frac{\Phi \vdash^t_{\Sigma,X} D(t_1)}{\Phi \vdash^t_{\Sigma,X} t_1 \overset{s}{=} t_2}$$
$$\frac{}{\Phi \vdash^t_{\Sigma,X} t_1 \overset{e}{=} t_2}$$

Equality3

$$\frac{\Phi \vdash^t_{\Sigma,X} D(t_1) \Rightarrow F}{\Phi \vdash^t_{\Sigma,X} D(t_2) \Rightarrow F}$$
$$\frac{}{\Phi \vdash^t_{\Sigma,X} t_1 \overset{s}{=} t_2}$$

and replace $\overset{e}{=}$ by $\overset{s}{=}$ in **Congruence**. Moreover, the old **Reflexivity** can be dropped.

All rules of the calculus up to **Congruence** are taken from a calculus for total first-order logic [Her73] (except that we index judgments with signatures and variables here, which is necessary to cover the case of empty carriers). On the other hand, the last four rules are entirely new. **Function Strictness** and **Predicate Strictness** state that atomic formulas are interpreted as "existentially strict", that is, their truth entails definedness of all terms occurring in them. **Totality** states that the application of a total function to defined terms is defined.

Finally, the **Substitution** Rule also occurs in the calculus for total first-order logic, but has to be modified for the partial case: it contains additional assumptions

$$\{ D(\theta(x)) \mid x \in X_s \}$$

in the derived judgment, which state that the terms to be substituted are defined. This is a syntactical version of the definedness condition in the Substitution Lemma for $\models^t$.

The calculus (especially when extended with suitable derived rules[18]) is useful for proofs whose structure follows the reasoning of a mathematician. But for automated theorem proving, more efficient proof calculi are used, like analytic tableaux, resolution, and the connection structure method [G+93]. One crucial source of efficiency is the use of unification (i.e., finding a substitution under which two given formulas become equal). However in the above calculus, the rule of substitution is restricted to the case where the items

---

[18] Note that the derived rule for the introduction of existential quantifiers has to leave the variable system intact. In particular, from $\vdash_{\Sigma,\{x:s\}} x \overset{e}{=} x$ we can only derive $\vdash_{\Sigma,\{x:s\}} \exists x : s.x \overset{e}{=} x$, but not $\vdash_{\Sigma,\emptyset} \exists x : s.x \overset{e}{=} x$.

being substituted are defined. This causes difficulties, at least, when using the well-known techniques and results based on unification.

A calculus for $\models^p$. This is a further strong argument in favor of $\models^p$, which can also be captured by a proof calculus. This calculus consists of the rules **Assumption, ∧-introduction, ∧-left elimination, ∧-right elimination, Tertium non datur, Absurdity, Predicate Strictness, Function Strictness** and **Totality** which are obtained from the corresponding rules in the above calculus by dropping the variable system as an index for $\vdash$, and the following additional rules:

∀-elimination

$$\frac{\varPhi \vdash^p_\Sigma (\forall Y.\varphi)}{\varPhi \vdash^p_\Sigma (\bigwedge_{y \in Y_s, s \in S} D(y)) \Rightarrow \varphi}$$

∀-introduction

$$\frac{\varPhi \vdash^p_\Sigma (\bigwedge_{y \in Y_s, s \in S} D(y)) \Rightarrow \varphi}{\varPhi \vdash^p_\Sigma (\forall Y.\varphi)} \quad \text{if } Y \cap FV(\varPhi) = \emptyset$$

Symmetry

$$\frac{}{\varPhi \vdash^p_\Sigma x \overset{e}{=} y \Rightarrow y \overset{e}{=} x}$$

Substitution

$$\frac{\varPhi \quad \vdash^p_\Sigma \varphi}{\varPhi[\theta] \vdash^p_\Sigma \varphi[\theta]}$$

$$\text{for } \theta : X \longrightarrow |T_\Sigma(Y)| \text{ with } \varphi[\theta] \downarrow \text{ and } \varPhi[\theta] \downarrow$$

Again, $D(t)$ is syntactical sugar for $t \overset{e}{=} t$.

Theorem 3.65. *The calculus is sound and complete, i.e.,*

$$\varPhi \models^p_\Sigma \varphi \text{ if and only if } \varPhi \vdash^p_\Sigma \varphi$$

This calculus has a simple substitution rule **Substitution**, while the quantifier rules **∀-elimination** and **∀-introduction** now have to take account of definedness. Reflexivity of $\overset{e}{=}$ no longer holds and has to be replaced by symmetry, which is strictly weaker because it follows from reflexivity together with **Congruence**. (Transitivity follows by **Congruence** in either case.)

Translating partial to total first-order logic. The success of total first-order logic is based on the fact that it is expressive enough to be called a *universal logic* in [MT93] but not too expressive, so there is a sound and complete calculus. First-order logic being universal means that there are translations from many-sorted, higher-order, dynamic, modal, etc., logics to first-order

logic. We will now describe a translation from partial first-order logic (denoted by PFOL) to total first-order logic, from now on denoted by FOL. This
translation allows us to take any deductive system for total first-order logic
and re-use it for PFOL (with either$\models^t$ or $\models^p$). This is a particular case of
the *borrowing* technique proposed in [CM97].

Of course, since PFOL is a superset of FOL, it shares the universal character of FOL. On the other hand, it becomes clear that no essential expressive
power is added by the passage from FOL to PFOL, but, as we shall see, we
gain much notational convenience.

We use here the standard FOL introduced in many textbooks. That is,
a FOL-signature consists of a PFOL-signature with exactly one sort symbol
and no partial operation symbols. Σ-structures have a nonempty carrier, in
order to obtain a simple calculus. Such a calculus for FOL consists of the rules
**Assumption, ∧-introduction, ∧-left elimination, ∧-right elimination,
Tertium non datur, Absurdity, ∀-elimination, ∀-introduction, Reflexivity, Substitution,** and **Congruence.** For the rules **∀-elimination,
∀-introduction, Reflexivity,** and **Congruence,** the variable system has
to be dropped and $\stackrel{e}{=}$ has to be replaced by $=$. This gives us an entailment
relation $\vdash_{\Sigma}^{FOL}$.

The idea is now to translate PFOL to FOL and then re-use the easier
calculus for FOL via this translation. In particular, the world of automated
theorem provers for FOL can thus be adapted for PFOL.

The only translation from PFOL to FOL fully representing partial structures and homomorphisms is the representation of partial operations by their
graph relations, sketched by Burmeister in [Bur82]. Substitution of terms containing partial operation symbols is avoided, but instead applications of partial operation symbols have to be expanded into long, existentially quantified
conjunctions: a term $pf(t_1, \ldots, t_n)$ is translated to the formula

$$\alpha(pf(t_1, \ldots, t_n)) =$$
$$\exists x_1 : s_1, \ldots, x_n : s_n, x : s.$$
$$(R^{pf}(x_1, \ldots, x_n, x) \wedge \alpha(x_1 \stackrel{e}{=} t_1) \wedge \cdots \wedge \alpha(x_n \stackrel{e}{=} t_n) \wedge \alpha(x \stackrel{e}{=} t))$$

This makes the treatment of partial operations even more cumbersome than
in Burmeister's calculus.

But there is another translation from PFOL to FOL along the lines of
Scott's ideas [Sco79]. Although model categories are not represented faith-
fully, proof theory is, so it fits our purposes here.

A PFOL-signature $\Sigma = \langle S, \Omega, P\Omega, \Pi \rangle$ is translated to a FOL-signature
$\Phi(\Sigma) = (\{ * \}, \Omega' \uplus P\Omega', \Pi')$, where Ω' and $P\Omega'$ result from Ω and $P\Omega$ by
replacing all sorts by $*$, and $\Pi' = \Pi \cup \{ \equiv_s : s \times s \mid s \in S \}$. To the translated
signature, there has to be added the set of axioms $C(\Sigma)$ consisting of

- $x \equiv_s y \Rightarrow y \equiv_s x$ for $s \in S$
- $x \equiv_s y \wedge y \equiv_s z \Rightarrow x \equiv_s z$ for $s \in S$

- $x_1\equiv_{s_1} y_1 \wedge \cdots \wedge x_n\equiv_{s_n} y_n \Rightarrow f(x_1, \ldots, x_n)\equiv_s f(y_1, \ldots, y_n)$ for $f\colon s_1 \times \cdots \times s_n \longrightarrow s \in \Omega$
- $x_1\equiv_{s_1} y_1 \wedge \cdots \wedge x_n\equiv_{s_n} y_n \wedge pf(x_1, \ldots, x_n)\equiv_s pf(x_1, \ldots, x_n)$
  $\Rightarrow pf(x_1, \ldots, x_n)\equiv_s pf(y_1, \ldots, y_n)$ for $pf\colon s_1 \times \cdots \times s_n \overset{}{\longrightarrow} s \in P\Omega$
- $x_1\equiv_{s_1} y_1 \wedge \cdots \wedge x_n\equiv_{s_n} y_n \wedge p(x_1, \ldots, x_n) \Rightarrow p(y_1, \ldots, y_n)$ for $p\colon s_1 \times \cdots \times s_n \in \Pi$
- $f(x_1, \ldots, x_n)\equiv_s f(x_1, \ldots, x_n) \Rightarrow x_i\equiv_{s_i} x_i$ for $f\colon s_1 \times \cdots \times s_n \longrightarrow s \in \Omega$
- $pf(x_1, \ldots, x_n)\equiv_s pf(x_1, \ldots, x_n) \Rightarrow x_i\equiv_{s_i} x_i$ for $pf\colon s_1 \times \cdots \times s_n \overset{}{\longrightarrow} s \in P\Omega$
- $p(x_1, \ldots, x_n) \Rightarrow x_i\equiv_{s_i} x_i$ for $p\colon s_1 \times \cdots \times s_n \in \Pi$

stating that $\equiv$ is a strict partial congruence, and partial operations and predicates are strict.

A Σ-axiom φ (in PFOL) is translated to the $\Phi(\Sigma)$-axiom $\alpha_\Sigma(\varphi)$:

- $\alpha_\Sigma(t_1\overset{e}{=}t_2) = t_1\equiv_s t_2$ for $t_1, t_2 \in T_\Sigma(X)_s$
- $\alpha_\Sigma(p(t_1, \ldots, t_n)) = p(t_1, \ldots, t_n)$
- $\alpha_\Sigma(F) = F$
- $\alpha_\Sigma(\varphi \wedge \psi) = \alpha_\Sigma(\varphi) \wedge \alpha_\Sigma(\psi)$
- $\alpha_\Sigma(\varphi \Rightarrow \psi) = \alpha_\Sigma(\varphi) \Rightarrow \alpha_\Sigma(\psi)$
- $\alpha_\Sigma(\forall Y.\varphi) = (\forall Y. \bigwedge_{\{y\in Y_s, s\in S\}} y\equiv_s y) \Rightarrow \alpha_\Sigma(\varphi)$

A $\langle \Phi(\Sigma), C(\Sigma)\rangle$-structure B (in FOL) is translated to the partial Σ-structure $\beta_\Sigma(B)$ (in PFOL) with

$$\beta_\Sigma(B) = (B|_{\Sigma \longrightarrow \Phi(\Sigma)})/\equiv_B$$

where $B|_{\Sigma \longrightarrow \Phi(\Sigma)}$ is the reduct of B along the obvious map $\Sigma \longrightarrow \Phi(\Sigma)$, i.e. B interpreted as a partial Σ-structure, and the quotient is taken as in Definition 3.33.

This translation now has the following crucial properties.

Proposition 3.66. *For each Σ-axiom φ and each total valuation $v\colon X \longrightarrow B$ into a $\langle \Phi(\Sigma), C(\Sigma)\rangle$-structure B, we have*

$$nat \equiv_B \circ v \Vdash_\Sigma^{PFOL} \varphi \text{ if and only if } v \Vdash_{\Phi(\Sigma)}^{FOL} \alpha_\Sigma(\varphi)$$

Proposition 3.67. *For each partial Σ-structure A, there is a $\langle \Phi(\Sigma), C(\Sigma)\rangle$-structure B with $\beta_\Sigma(B) = A$, such that for each valuation $\rho\colon X \longrightarrow A$ there exists a total valuation $v\colon X \longrightarrow B$ with $\rho = nat \equiv_B \circ v$. Moreover if ρ is total as well, then $v(x)\equiv_{s,B} v(x)$ for all $x \in X_s, s \in S$.*

Theorem 3.68 (Borrowing proof calculus for PFOL from FOL).

$$\Phi \models_\Sigma^t X.\varphi \text{ iff } C(\Sigma) \cup \alpha_\Sigma(\Phi) \cup \{x\equiv_s x \mid x\colon s \in X \cup \bigcup_{Y.\psi\in\Phi} Y\} \vdash_{\Phi(\Sigma)}^{FOL} \alpha_\Sigma(\varphi)$$

$$\Phi \models_\Sigma^p \varphi \text{ iff } C(\Sigma) \cup \alpha_\Sigma(\Phi) \vdash_{\Phi(\Sigma)}^{FOL} \alpha_\Sigma(\varphi)$$

The translation of PFOL to FOL can also take advantage of a special theorem prover for FOL which copes very well with partial congruences. The partial congruences generated by the translation can be treated in a similar way to equality, with the results of Bachmair and Ganzinger [BG94b, CGW95].

3.3.4 Conditional logic with existential premises

Although having full first-order logic at hand to describe a specification allows us in many cases to give a concise and intuitive axiomatization, there are several data types that are quite easily and naturally described within a far smaller fragment of PFOL, consisting only of the conditional axioms.

There are essentially two advantages of using such restricted language: on one side, if the form of the axioms used in the deduction is restricted, better theorem provers are available, taking advantage, for instance, of paramodulation [Pad88a] and conditional term-rewriting techniques (see Chapter 1 and [DJ90, Klo92]).

On the semantic side, the existence of an initial model for such classes of specifications is guaranteed. Moreover, the initial model is characterized as the *minimal* first-order structure satisfying the axioms of the specifications; employing it yields an economy of thought.

Let us first look at an example of partial conditional specifications, before their formal definition and the proof of existence of initial (free) models for such a class of specifications.

Example 3.69. Using the partial framework, the specification of *stacks* with their constructors and selectors becomes easy and elegant.

> **sig** Σ_{Stack} = enrich Σ_{Elem} **by**
> **sorts**   stack
> **opns**   empty: $\to$ stack
>   push: elem $\times$ stack $\to$ stack
> **popns** pop: stack $\to$ stack
>   top: stack $\to$ elem

The minimal specification of stacks on this signature is given by the axioms identifying pop and top as (partial) inverses of the constructor push

> **spec Stack** = enrich Σ_{Stack} **by**
> **vars** e : elem; s stack
> **axioms** $\text{pop}(\text{push}(e, s)) \overset{s}{=} s$
> $\text{top}(\text{push}(e, s)) \overset{s}{=} e$

As we will see, all *positive conditional* specifications, i.e., specifications with axioms that are implications whose premises are (first-order equivalent to) a set of existential equalities and predicate applications, have an initial model characterized by the *no-junk & no-confusion* properties. Therefore, in particular, the above specification of stacks has the following initial model I.

> **algebra** $I =$
> **Carriers**
> $|I|_{\text{elem}} = X$
> $|I|_{\text{stack}} = X^*$
> **Functions**
> $\text{empty}_I = \lambda$
> $\text{push}_I(x, s) = x \cdot s$

$$\text{pop}_I(s) = \begin{cases} s', & \text{if } s = x \cdot s' \\ \text{undefined}, & \text{otherwise} \end{cases}$$

$$\text{top}_I(s) = \begin{cases} x, & \text{if } s = x \cdot s' \\ \text{undefined}, & \text{otherwise} \end{cases}$$

It is interesting to note that the specification Stack is the most abstract interpretation of stacks and can be further specialized to obtain an *implementation* where more details have been fixed. For instance, the operation pop on the empty stack could be recovered on the empty stack, as in many standard total approaches, by enriching Stack with the following axiom

$$\text{pop}(\text{empty}) \overset{s}{=} \text{empty}$$

Since this axiom also is positive conditional, the enriched specification has an initial model too, that is, the Σ_{Stack}-structure I with the interpretation of function pop modified into

$$\text{pop}_I = \begin{cases} s', & \text{if } s = x \cdot s' \\ \lambda, & \text{otherwise} \end{cases}$$

More refined error recovery (or detection) techniques can be implemented as well, by enriching Stack in different ways.

In any case the (initial) models of the enrichments are already models of the specification Stack and hence any property proved for Stack holds for them too, allowing incremental tests.   ∎

Definition 3.70. A *positive conditional* formula is a wellformed first-order formula $\varphi \in Form(\Sigma, X)$ of the form

$$\varphi = \forall X. \epsilon_1 \wedge \cdots \wedge \epsilon_n \Rightarrow \epsilon$$

where ϵ is any atom and each ϵ_i is either a predicate application or an existential equality.

A specification Sp is called *positive conditional* if it has the same model class as a specification $Sp' = \langle \Sigma, AX \rangle$ and each $\varphi \in AX$ is a positive conditional formula.   ◇

A particular case of partial positive conditional specifications is total conditional specifications. Indeed, a total first-order signature is a partial first-order signature with the empty family of partial function symbols $popns(\Sigma) = \emptyset$ and, moreover, each partial first-order structure is a total first-order structure too. Thus, the distinction between different kinds of equalities is immaterial and hence the model class of a total conditional specification is the same as the model class of the partial positive conditional specification having the "same" axioms, where each $=$ symbol has been replaced by $\overset{e}{=}$.

Another important class of positive conditional specifications that can be easily recognized is that whose axioms are conditional and each strong equality in the premises is guarded by a definedness assertion on either side of the equality and, analogously, each weak equality is guarded on both sides of the equality, because such a formula has the same models as the given conditional formula where all equalities in the premises have been substituted by existential equalities.

Exercise 3.71. Show that a specification $Sp = \langle \Sigma, AX \rangle$ is positive conditional if each $\varphi \in AX$ has the form

$$\varphi = \forall X. \epsilon_1 \wedge \cdots \wedge \epsilon_n \Rightarrow \epsilon$$

where ϵ and all ϵ_i are atoms and the following two conditions are satisfied:

- if ϵ_i is the strong equality $t \overset{s}{=} t'$, then there exists $j \in 1, \ldots, n$ such that ϵ_j is $D(t)$ or $D(t')$;
- if ϵ_i is the weak equality $t \overset{w}{=} t'$, then there exist $j, k \in 1, \ldots, n$ such that ϵ_j is $D(t)$ and ϵ_k is $D(t')$.

Using initial semantics of specifications with positive conditional formulas to describe a data type intuitively corresponds to using inductive definition. A particular, but very common, case is the axiomatization of a data type where the carriers are built by some total functions, called *constructors*, and then other, possibly partial, operations are defined on such elements by simply imposing the equality of their applications to terms built by the constructors.

An instance of this methodology of data definition is the previous example of the stacks that are built by pushing elements on the empty stack, where the evaluation of a pop (top) reduces by the axioms to the evaluation of simpler terms without pop (top). Let us look at another example, that is, the specification of the minus between non-negative integers.

Example 3.72. The basic specification of non-negative integers is the usual (total) one, given by the absolutely free constructors "zero" and "successor".

sig $Sp_{\mathbf{Nat}}{}^T =$
    sorts   nat
    opns   zero: $\rightarrow$ nat
            succ: nat $\rightarrow$ nat

On this signature we want to define, for example, the predecessor and minus operations.

spec $Sp_{\mathbf{Nat}} = $ **enrich** $Sp_{\mathbf{Nat}}{}^T$ **by**
    popns   prec: nat $\multimap$ nat
            minus: nat $\times$ nat $\multimap$ nat
    vars $x, y :$ nat
    axioms   prec(succ(x))$\overset{s}{=}x$
             minus(x, zero)$\overset{s}{=}x$
             minus(succ(x), succ(y))$\overset{s}{=}$minus(x, y)

The above specification follows the intuition that the new operations are *programs* on a data type built by zero and successor, inductively defined by means of the constructors. Indeed, a term starting with a **prec** or a

minus symbol can be deduced defined iff it reduces to a term of the form $\text{succ}^k(\text{zero})$ because the axioms are strong equalities. Thus, for instance, $\text{prec}(\text{zero})$ cannot be deduced to be defined (and indeed in the initial model, it is undefined) and represents an erroneous *call* of prec.

It is also worth noting that the given axioms allow the intuitively intended identifications for minimal models, which are those partial first-order structures where each element of the carrier is denoted by a term of the form $\text{succ}^k(\text{zero})$. ∎

The existence of initial (free) models for positive conditional axioms is due to the particular structure of the model class, as in the total (both equational and conditional) case.

Briefly the first step to prove the existence of an initial (free) model I is to show that if it exists, then it is termgenerated. This property is due to the fact that the model class of a positive conditional specification is closed under closed substructure, so that the term-generated part of I is a model too and hence, as the initial (free) model is in a sense the *smallest* (with respect to the partial order induced by homomorphism existence), I and its term-generated part must coincide.

Thus, I is (isomorphic to) a term-algebra quotient. Moreover, since homomorphisms preserve existential equalities and predicate assertions, the congruence defining I must be *minimal*, i.e., it must be the intersection of all the kernels of term-evaluation in a model of the specification.

The last step is to prove that the quotient of the term algebra with respect to the intersection of all the kernels of term-evaluation in a model of the specification, is actually a model too, i.e., it satisfies the axioms. This point too relies on the form of the axioms. Indeed, if the premises of an axiom hold in such a quotient, then they must hold in each model. Hence the consequence also holds in all models, so that it holds in the quotient, which is, therefore, a model.

Theorem 3.73. *Let Sp be a positive conditional specification over a partial first-order signature* $\Sigma = \langle S, \Omega, P\Omega, \Pi \rangle$ *and* X *be an S-sorted family of variables. Then there is a free Sp-model for* X *that is (isomorphic to) the quotient* F *of the term algebra* $T_\Sigma(X)$ *with the following interpretation of predicate symbols:*

$$(t_1, \ldots, t_n) \in p_{T_\Sigma(X)} \text{ iff } A \models \forall X.p(t_1, \ldots, t_n) \text{ for all models } A \text{ of } Sp.$$

by the following congruence $\equiv$*:*

$$t \equiv t' \text{ iff } A \models \forall X.t \overset{e}{=} t' \text{ for all models } A \text{ of } Sp.$$

We also can use the translation from PFOL to FOL to obtain initial models in the partial conditional case.

Proposition 3.74. *Let* $\langle \Sigma, AX \rangle$ *be a presentation and* I *an initial model in* $Mod(\langle \Phi(\Sigma), C(\Sigma) \cup \alpha(AX) \rangle)$ *Then* $\beta(I)$ *is an initial model in* $Mod(\langle \Sigma, AX \rangle)$.

A sound and practically complete calculus for the conditional fragment of partial first-order logic for $\models^t$ consists of the rules **Assumption, Cut, $\forall$-elimination, Reflexivity, Congruence, Substitution, Function Strictness, Predicate Strictness** and **Totality**. For $\models^p$, it consists of the same rules modified for $\models^p$, except that **Reflexivity** is replaced by **Symmetry**.

But like full partial first-order logic, positive conditional specifications are also reducible, from a deductive point of view, to the usual total first-order specifications which again turn out to be conditional, so that automatic tools and techniques developed for the conditional total case can be *borrowed* for the partial.

The key point of such a reduction technique is the translation of a positive conditional specification into a corresponding total conditional specification, whose models satisfy the same atomic formulas (up to translation). This is a particular case of the *borrowing* technique proposed in [CM97] and a sugared version of the borrowing for PFOL, where definedness predicates are also allowed.

Definition 3.75. Let $\Sigma = \langle S, \Omega, P\Omega, \Pi \rangle$ be a partial signature.

- Let Σ^T denote the total first-order signature

$$\langle S, \Omega \cup P\Omega, \Pi \cup \{D_s : s, \overset{e}{=}_s : s \times s\}_{s \in S} \rangle$$

and AX^T denote the following set of total conditional formulas on Σ^T:

$$D_{s_1}(x_1) \wedge \cdots \wedge D_{s_n}(x_n) \Rightarrow D_s(f(x_1, \ldots, x_n))$$
$$\text{for all } f : s_1 \times \cdots \times s_n \longrightarrow s \in \Omega$$
$$D_s(f(x_1, \ldots, x_n)) \Rightarrow D_{s_i}(x_i)$$
$$\text{for all } f : s_1 \times \cdots \times s_n \longrightarrow s \in \Omega \cup P\Omega$$
$$x \overset{e}{=}_s y \Rightarrow D_s(x)$$
$$D_s(x) \Rightarrow x \overset{e}{=}_s x$$
$$D_s(x) \wedge x = y \Rightarrow x \overset{e}{=}_s y$$
$$x \overset{e}{=}_s y \Rightarrow x = y$$
$$x \overset{e}{=}_s y \Rightarrow y \overset{e}{=}_s x$$
$$x \overset{e}{=}_s y \wedge y \overset{e}{=}_s z \Rightarrow x \overset{e}{=}_s z$$

- Let us call *strictly positive conditional* a formula over a finite set of variables X having the form $\forall X. \epsilon_1 \wedge \cdots \wedge \epsilon_n \Rightarrow \epsilon_{n+1}$ with each ϵ_i a predicate application, an existential equality, or a definedness assertion. For each strictly positive conditional formula $\varphi = \forall X. \epsilon_1 \wedge \cdots \wedge \epsilon_n \Rightarrow \epsilon_{n+1}$, let $\alpha(\varphi)$ denote the following total conditional axiom:

$$\wedge_{x \in X_s} D_s(x) \wedge \epsilon_1 \wedge \cdots \wedge \epsilon_n \Rightarrow \epsilon_{n+1}$$

- For each total first-order structure A modeling $\langle \Sigma^T, AX^T \rangle$, let $\beta(A)$ denote the following partial first-order structure B:
  - $|B|_s = D_{sA}$ for all $s \in S$.

- f_B is the restriction of f_A to the carriers of B for all $f : s_1 \times \cdots \times s_n \longrightarrow s \in \Omega$.
- pf_B is the restriction of pf_A to the carriers of B for all $pf : s_1 \times \cdots \times s_n \rightarrow s \in P\Omega$; in particular if $(a_1, \ldots, a_n) \in D_{s_1 A} \times \ldots \times D_{s_n A}$ but $pf_A(a_1, \ldots, a_n) \notin D_{sA}$, then $pf_B(a_1, \ldots, a_n)$ is undefined.
- p_B is the restriction of p_A to the carriers of B for all $p : s_1 \times \cdots \times s_n \in \Pi$. $\diamond$

Thus, each total first-order structure satisfying AX^T corresponds to the partial first-order structure where the "undefined" elements have been dropped. This correspondence reflects on the logic too, in the sense that the reduction of a total to a partial first-order structure satisfies the same strictly positive conditional formulas (up to the α translation).

Lemma 3.76. *Using the notation of Definition 3.75, the following* satisfaction condition *holds for all total first-order structures A and all strictly positive conditional formulas φ:*

$$\beta(A) \models^t \varphi \iff A \models \alpha(\varphi)$$

Exercise 3.77. Show that each positive conditional specification has the same model class as a specification whose axioms are all strictly positive conditional formulas.

Therefore, for each partial positive conditional specification $Sp = \langle \Sigma, AX \rangle$ the model class of Sp satisfies a strictly positive conditional formula iff the model class of the total conditional specification $Sp^T = (\Sigma^T, \alpha(AX) \cup AX^T)$ satisfies its translation along α. Hence each deductive system (theorem prover) for total conditional specification can be used to verify the validity of strictly positive conditional formulas in the model classes of partial positive conditional specification. Moreover, this result can be extended to any class of formulas that can be effectively translated into strictly positive conditional form without affecting their validity. Indeed, in this case the validity verification splits into

- a preliminary coding of the formula into strictly positive conditional,
- a translation of this form into a total conditional formula via α,
- an application of any (conditionally complete) deduction system for Sp^T.

Theorem 3.78. *Let $Sp = \langle \Sigma, AX \rangle$ be a partial positive conditional specification (with axioms all in strictly positive conditional form) and φ be a strictly positive conditional formula.*
  Using the notation of Definition 3.75, $Mod_\Sigma(AX) \models^t \varphi$ iff $\alpha(AX) \cup AX^T \vdash \alpha(\varphi)$, where $\vdash$ is given in Definition 3.8.

Example 3.79. Let us consider again the problem of store specification presented in Example 3.16.

Here, as normal in a context with predicates, we assume that the specification of locations also defines some predicates, instead of their implementations as Boolean functions as in Example 3.16. Using the predicates `AreEqual` and `AreDifferent` to check the equality between locations rather than any of the predefined equalities of our logic, allows greater freedom. Indeed, the user can axiomatize those predicates in such a way that their interpretation in some model is not the identity.

Notice that, since the (assertion of the) negation of the equality between locations is needed in the premises of some axioms, its negation also has to be axiomatized as a (different) predicate, because we want to get a *positive conditional* specification. Otherwise, using a more expressive fragment of partial first-order logic, we could just have the predicate `AreEqual`.

Therefore, let us assume that the specification Sp_L of locations includes the following:

> **spec** Sp_L =
>> **sorts**   loc...
>> **preds**   AreEqual, AreDifferent : loc × loc ...

Then we can enrich Sp_L and the specification Sp_V of values, with the main sort **value** to get the store specification.

> **spec** Stores = **enrich** Sp_L, Sp_V **by**
>> **sorts**   store
>> **opns**   empty : $\rightarrow$ store
>>   update : store × loc × value $\rightarrow$ store
>> **popns**   retrieve : store × loc $-\!\!\!\rightarrow$ value
>> **vars** x, y : loc; v, v_1, v_2 : value; s : store
>> **axioms** AreEqual(x, y) $\Rightarrow$ update$(s, x, v)\overset{s}{=}$update(s, y, v)
>>   AreEqual(x, y) $\Rightarrow$
>>   update$($update$(s, x, v_1), y, v_2)\overset{s}{=}$update$(s, x, v_2)$
>>   AreDifferent(x, y) $\Rightarrow$ update$($update$(s, x, v_1), y, v_2)\overset{s}{=}$
>>   update$($update$(s, y, v_2), x, v_1)$
>>   AreEqual(x, y) $\Rightarrow$ retrieve$($update$(s, x, v), y)\overset{s}{=}v$

Notice that, owing to the partial setting, in the `Stores` initial model the application `retrieve`(s, x) to stores s where x has never been updated, for instance, if s is `empty`, does not yield any value, because it cannot be deduced defined. Hence the problem of hierarchical consistency seen in the total case does not apply here.

In [Rei87], a much more sophisticated specification for stores, where for instance it is possible to remove an association from a store, is given in a different setting. The reader is encouraged to rephrase it using positive conditional data types.

The inference system for `Stores` is total conditional using the specification

> **spec** Stores^T = **enrich** Sp_L^T, Sp_V^T **by**
>> **sorts**   store

opns empty: $\to$ store
update: store $\times$ loc $\times$ value $\to$ store
retrieve: store $\times$ loc $\to$ value
vars x, y : loc; v, v_1, v_2 : value; s : store
axioms $D(\text{empty})$
$D(s) \land D(x) \land D(v) \Rightarrow D(\text{update}(s, x, v))$
$D(\text{update}(s, x, v)) \Rightarrow D(s)$
$D(\text{update}(s, x, v)) \Rightarrow D(x)$
$D(\text{update}(s, x, v)) \Rightarrow D(v)$
$D(\text{retrieve}(s, x)) \Rightarrow D(s)$
$D(\text{retrieve}(s, x)) \Rightarrow D(x)$
$D(x) \land D(y) \land D(s) \land D(v) \land \text{AreEqual}(x, y) \Rightarrow$
  $\text{update}(s, x, v) = \text{update}(s, y, v)$
$D(x) \land D(y) \land D(s) \land D(v_1) \land D(v_2) \land \text{AreEqual}(x, y) \Rightarrow$
  $\text{update}(\text{update}(s, x, v_1), y, v_2) = \text{update}(s, x, v_2)$
$D(x) \land D(y) \land D(s) \land D(v_1) \land D(v_2) \land \text{AreDifferent}(x, y) \Rightarrow$
  $\text{update}(\text{update}(s, x, v_1), y, v_2) =$
  $\text{update}(\text{update}(s, y, v_2), x, v_1)$
$D(x) \land D(s) \land D(v) \land \text{AreEqual}(x, y) \Rightarrow$
  $\text{retrieve}(\text{update}(s, x, v), y) = v$

where Sp_L^T and Sp_V^T are the corresponding translations where for example axioms like $\text{AreDifferent}(x, y) \Rightarrow D(x)$ have been added. ∎

Another case where partial specifications are useful is the specification of *bounded* data types, where the *constructors* themselves are partial functions. For instance, let us consider the specification of bounded stacks, parametric on a positive constant *max* representing the maximum number of elements that can be stacked.

Example 3.80.

spec BoundedStacks = **enrich** Nat$_\leq$, Σ_{Elem} **by**
  sorts bstack
  opns empty: $\to$ bstack
    max: $\to$ nat
  popns Bpush: elem $\times$ bstack $\twoheadrightarrow$ bstack
    pop: bstack $\twoheadrightarrow$ bstack
    top: bstack $\twoheadrightarrow$ elem
  vars x : elem; s : bstack
  axioms depth(empty) = zero
    $\text{succ}(\text{depth}(s)) \leq \text{max} \Rightarrow \text{depth}(\text{Bpush}(x, s)) = \text{succ}(\text{depth}(s))$
    $D(\text{Bpush}(x, s)) \Leftrightarrow \text{succ}(\text{depth}(s)) \leq \text{max}$
    $D(\text{Bpush}(x, s)) \Rightarrow \text{pop}(\text{Bpush}(x, s)) \stackrel{e}{=} s$
    $D(\text{Bpush}(x, s)) \Rightarrow \text{top}(\text{Bpush}(x, s)) \stackrel{e}{=} x$ ∎

Let us finally look at a motivating example of a partial recursive function with nonrecursive domain (that hence cannot be described by a total specification identifying all "erroneous applications"). It is (a fragment of) the definition of the semantics for an imperative language based on environments and states. Here we try to capture the keypoints of this complex example, leaving the details to the interested reader. In particular we are not considering the declarative part of the language, nor the environment aspects. Thus, commands can be simply represented as functions from states to states (and expressions as functions from states to values).

Example 3.81. Let us assume the specification of the states is given; then commands are *partial* functions among them. It is worth noting that we are not interested in deducing the extensional equality between commands, because not only is it unnecessary in order to describe the language semantics, but it is also too restrictive to capture, for instance, complexity criteria that are interesting for imperative language semantics. Therefore, we do not need (a representation of) partial higher-order and hence can restrict ourselves to positive conditional types, while extensionality implicitly requires a more powerful and technically more complex fragment of the logic (see, e.g., [AC92, AC95] for an extended treatment or Section 3.4 of this chapter for an introductory discussion of the subject).

Here we are more interested in *command constructs* than in basic commands like skip, read, and write, that are more relevant to the specification of states and the language data types than to proper commands. In particular we are interested in the while_do command, as a source of possible nontermination and hence as a paramount example of a partial recursive function with nonrecursive domain.

Therefore, we also assume the specification of a BExp data type is given, where the Boolean expressions of our imperative language should be interpreted, with the obvious constants, axiomatized by the following specification. The actual constructs for the Boolean expressions are not important and have been omitted. The relevant point of Boolean expressions for the command specification is that any such expression can be evaluated in a state and produce a truth value, if the evaluation terminates.

```
spec BExp = enrich States by
    sorts   bool, BoolExps
    opns    true, false: → bool ...
    popns   BEval: BoolExps × state -⊖→ bool ...
```

Let us finally look at the specification for the kernel of the language semantics concerning the command constructs.

```
spec ImpLang = enrich States, BExp by
    sorts   commands
    opns    skip: → commands ...
            if_then_else: BoolExps × commands × commands → commands
            while_do: BoolExps × commands → commands
            conc: commands × commands → commands ...
    popns   CEval: commands × state -⊖→ state ...
    vars    c, c′ : commands; s : state; b : BoolExps
```

axioms $\text{CEval}(c, s) \overset{e}{=} s' \Rightarrow \text{CEval}(\text{conc}(c, c'), s) \overset{s}{=} \text{CEval}(c', s')$

$\quad\quad \text{BEval}(b, s) \overset{e}{=} \text{true} \Rightarrow \text{CEval}(\text{if\_then\_else}(b, c, c'), s) \overset{s}{=} \text{CEval}(c, s)$

$\quad\quad \text{BEval}(b, s) \overset{e}{=} \text{false} \Rightarrow$

$\quad\quad\quad \text{CEval}(\text{if\_then\_else}(b, c, c'), s) \overset{s}{=} \text{CEval}(c', s)$

$\quad\quad \text{BEval}(b, s) \overset{e}{=} \text{false} \Rightarrow \text{CEval}(\text{while\_do}(b, c), s) \overset{e}{=} s$

$\quad\quad \text{BEval}(b, s) \overset{e}{=} \text{true} \wedge \text{CEval}(c, s) \overset{e}{=} s' \Rightarrow$

$\quad\quad\quad \text{CEval}(\text{while\_do}(b, c), s) \overset{s}{=} \text{CEval}(\text{while\_do}(b, c), s')$

Here, as in programming languages, the conditional choice if_then_else is *nonstrict*, in the sense that $\text{CEval}(\text{if\_then\_else}(b, c, c'), s)$ can result in a value even if the evaluation of some subterm in the same state does not. For instance, if the Boolean expression yields true, then the evaluation of the second branch can be undefined. However the interpretation of all functions of the signature are *strict*. Indeed, the nonstrictness has been achieved by using *functions* rather than *ground elements* as interpretation for commands and expressions.

If *real* nonstrictness is needed, as is sometimes the case, for instance, in the design phase, then partial logic is inadequate and more powerful frameworks should be applied (see, e.g., Section 3.4 for a discussion of the problem and references).   ∎

3.3.5   Subsorting in partial first-order logic

Although some applications of subsorting can be represented better by using partiality and predicates, there are still important cases where data types are naturally supersets or subsets of other data types. For this reason and to support the many users of order-sorted specification languages, it may be convenient to enrich the language used to describe specifications by subsort declaration. Indeed, this viewpoint has been adopted by the language CASL[19] in the course of definition under the CoFI initiative (see, e.g., [Mos97]) and the semantics of the subsorting constructs can be easily described using the theory presented so far.

Indeed, the idea is to encode an order-sorted signature into a partial, using explicit *embeddings* to represent the subsorting relationship. Then, formulas and models of partial logic are used for the subsorted framework as well. However, in order-sorted formalisms, overloading of function (predicate) symbols usually carries a semantics, in the sense that using the same name for functions on sorts connected by the subsort relation requires the corresponding interpretations of the function to be *compatible*. Hence we have to restrict the models to those satisfying such a property by means of a set of formulas. Therefore, we represent an order-sorted signature by a partial positive conditional presentation, from which we borrow formulas and models.

---

[19] See   http://www.brics.dk/Projects/CoFI/Documents/CASL/Summary/index.html for a summary of the CASL language.

The notion of subsorted signatures proposed here extends the notion of order-sorted signatures as given by Goguen and Meseguer [GM92], by allowing not only total function symbols, but also partial function symbols and predicate symbols. The resulting logic is called *subsorted partial first-order logic (SubPFOL)*.

Definition 3.82. A *subsorted signature* $\Sigma = (S, \Omega, P\Omega, \Pi, \leq_S)$ consists of a partial first-sorted signature $(S, \Omega, P\Omega, \Pi)$ together with a reflexive transitive *subsort relation* $\leq_S$ on the set S of sorts. The relation $\leq_S$ extends pointwise to sequences of sorts.

For a subsorted signature, $\Sigma = (S, \Omega, P\Omega, \Pi, \leq_S)$, the *overloading relations* $\sim_\Omega$ and $\sim_P$, for function and predicate symbols, are defined by:

- $f : w_1 \rightarrow_1 s_1 \sim_\Omega f : w_2 \rightarrow_2 s_2$ iff there exist $w \in S^*$ such that $w \leq w_1, w_2$ and a common supersort of s_1 and s_2; for each $f : w_1 \rightarrow_1 s_1, f : w_2 \rightarrow_2 s_2 \in \Omega \cup P\Omega$ (where $\rightarrow_1, \rightarrow_2 \in \{ \rightarrow, \rightarrowtail \}$);
- $p : w_1 \sim_P p : w_2$ iff there exists $w \in S^*$ such that $w \leq w_1, w_2$; for each $p : w_1, p : w_2 \in \Pi$.

A *signature morphism* $\sigma : \Sigma \rightarrow \Sigma'$ is a many-sorted signature morphism that preserves the relations $\leq_S$, $\sim_\Omega$ and $\sim_P$.   ◇

Note that two profiles of an overloaded constant declared with two different sorts are in the overloading relation iff the two sorts have a common supersort.

In the following, due to the possibility of name conflicts for functions and predicates, we will use $f_{\langle w,s \rangle}$ instead of f for a (total or partial) function symbol in term formation, in order to make terms unambiguous.

Definition 3.83. With each subsorted signature $\Sigma = (S, \Omega, P\Omega, \Pi, \leq_S)$ we associate a many-sorted signature $\Sigma^\#$, which is the extension of the underlying many-sorted signature $(S, \Omega, P\Omega, \Pi)$ with

1. a total *injection* function symbol $\mathrm{inj} \colon s \longrightarrow s'$, for each $s \leq_S s'$,
2. a partial *projection* function symbol $\mathrm{pr} \colon s' \rightarrowtail s$, for each $s \leq_S s'$, and
3. a unary *membership* predicate symbol $\in^s \colon s'$, for each $s \leq_S s'$.

We assume that the symbols used for injections, projections, and membership are not used elsewhere in Σ.

Subsorted Σ-sentences are ordinary many-sorted $\Sigma^\#$-sentences.

Subsorted Σ-models are ordinary many-sorted $\Sigma^\#$-models satisfying the following set of axioms $J(\Sigma)$:

identity $\forall x : s.\mathrm{inj}_{\langle s,s \rangle}(x) \stackrel{\mathrm{e}}{=} x$;

embedding-injectivity
$\forall x, y : s.\mathrm{inj}_{\langle s,s' \rangle}(x) \stackrel{\mathrm{e}}{=} \mathrm{inj}_{\langle s,s' \rangle}(y) \Rightarrow x \stackrel{\mathrm{e}}{=} y$ for $s \leq_S s'$;

transitivity $\forall x : s.\mathrm{inj}_{\langle s',s'' \rangle}(\mathrm{inj}_{\langle s,s' \rangle}(x)) \stackrel{\mathrm{e}}{=} \mathrm{inj}_{\langle s,s'' \rangle}(x)$ for $s \leq_S s' \leq_S s''$;

projection $\forall x : s.\mathrm{pr}_{\langle s',s \rangle}(\mathrm{inj}_{\langle s,s' \rangle}(x)) \overset{\mathrm{s}}{=} x$ for $s \leq_S s'$;

projection-injectivity $\forall x, y : s.\mathrm{pr}_{\langle s',s \rangle}(x) \overset{\mathrm{e}}{=} \mathrm{pr}_{\langle s',s \rangle}(y) \Rightarrow x \overset{\mathrm{e}}{=} y$ for $s \leq_S s'$;

membership $\forall x : s'.(\in^s_{s'} x \Leftrightarrow D_s(\mathrm{pr}_{\langle s',s \rangle}(x)))$ for $s \leq_S s'$;

function monotonicity

$$\forall x_1 : s_1 \ldots x_n : s_n.$$
$$\mathrm{inj}_{\langle s^1,s \rangle}(f_{\langle w_1,s^1 \rangle}(\mathrm{inj}_{\langle s_1,s_1^1 \rangle}(x_1),\ldots,\mathrm{inj}_{\langle s_n,s_n^1 \rangle}(x_n))) \overset{\mathrm{s}}{=}$$
$$\mathrm{inj}_{\langle s^2,s \rangle}(f_{\langle w_2,s^2 \rangle}(\mathrm{inj}_{\langle s_1,s_1^2 \rangle}(x_1),\ldots,\mathrm{inj}_{\langle s_n,s_n^2 \rangle}(x_n)))$$

for $f : w_1 \to_1 s^1 \sim_\Omega f : w_2 \to_2 s^1$, where $w_1 = s_1^1 \times \cdots \times s_n^1$, $w_2 = s_1^2 \times \cdots \times s_n^2$, $w = s_1 \times \cdots \times s_n$ with $w \leq w_1, w_2$ and $s^1, s^2 \leq s$;

predicate monotonicity

$$\forall x_1 : s_1 \ldots x_n : s_n. \, p_{w_1}(\mathrm{inj}_{\langle s_1,s_1^1 \rangle}(x_1),\ldots,\mathrm{inj}_{\langle s_n,s_n^1 \rangle}(x_n)) \Leftrightarrow$$
$$p_{w_2}(\mathrm{inj}_{\langle s_1,s_1^2 \rangle}(x_1),\ldots,\mathrm{inj}_{\langle s_n,s_n^2 \rangle}(x_n))$$

for $p : w_1 \sim_P p : w_2$, where $w_1 = s_1^1 \times \cdots \times s_n^1$, $w_2 = s_1^2 \times \cdots \times s_n^2$ and $w = s_1 \times \cdots \times s_n$ with $w \leq w_1, w_2$;

Σ-homomorphisms are $\Sigma^{\#}$-homomorphisms.   ◇

Let us look at a simple application of the subsorting formalism proposed here.

Example 3.84. Let us extend the natural number specification to the (positive) rational numbers.

spec $Sp_{\mathrm{rat}} =$ **enrich** Σ_{Nat} **by**
    sorts   nat $\leq$ rat
    opns   plus, times $\ldots$: rat $\times$ rat $\to$ rat
    popns $\_/\_$: nat $\times$ nat $\overset{\mathrm{e}}{\to}$ rat
    axioms $\forall x : \mathrm{nat}.\mathrm{inj}_{\langle nat,rat \rangle}(x) \overset{\mathrm{e}}{=} x/\mathrm{succ}(\mathrm{zero})$
            $\forall x, y : \mathrm{nat}.D(x/y) \Rightarrow D(x/\mathrm{succ}(y))$
            $\forall x_1, y_1, x_2, y_2 : \mathrm{nat}.D(x_1/y_1) \wedge D(x_2/y_2) \wedge$
               $\mathrm{times}(x_1, y_2) \overset{\mathrm{e}}{=} \mathrm{times}(x_2, y_1) \Rightarrow x_1/y_1 \overset{\mathrm{e}}{=} x_2/y_2$
            $\forall x_1, y_1, x_2, y_2 : \mathrm{nat}.\mathrm{plus}(x_1/y_1, x_2/y_2) \overset{\mathrm{s}}{=}$
               $\mathrm{plus}(\mathrm{times}(x_1, y_2), \mathrm{times}(x_2, y_1))/\mathrm{times}(y_1, y_2) \ldots$

The models of Sp_{rat} are as we would expect – the carrier of sort nat is isomorphic to a subset of rat, allowing implementations where natural numbers are represented differently with respect to rational numbers for efficiency reasons, and the first and third axioms identify multiple representations of the same number.   ∎

In standard order-sorted approaches, signatures are required to satisfy conditions in order to ensure that the usual functional notation for terms is semantically unambiguous, without requiring the clarification of multiply defined functions and explicit injections as here. Indeed, note that in the first axiom of the example above, an explicit injection has been introduced to produce a wellformed equation. Thus, in our approach we gain simplicity for the theory, because we do not have to impose regularity, monotonicity, and other similar conditions on signatures that hinder modular constructions of complex data types. On the other hand we pay for this because the language of sentences is much less user-friendly than in standard order-sorted approaches. However, it is possible to eat the cake and have it too, by introducing a more liberal language in order to describe the axioms and then parse its expressions, yielding unambiguous formulas in the standard algebraic style.

This two-level approach has been adopted in CASL, the CoFI language, and is based on the idea that terms and axioms at the user level are liberally built, as in standard order-sorted approaches. This is done by regarding terms of the subsort as terms of the supersort as well (while projections/retracts have to be explicitly stated) and using function symbols with or without their qualification. Then, such terms are expanded in all possible ways to terms in a standard, fully qualified, functional style. Finally, a term is successfully parsed if all its expansions are deducible as equivalent from the axioms in Definition 3.83, otherwise it is rejected and the user is asked to disambiguate it.

The details may be found in [KKM98, Mos98a, CHKM97].

A calculus for $SubPFOL$ can be obtained from the calculus for $\models^t$ by adding the axioms $J(\Sigma)$ as axioms to the calculus. If the model theory is restricted to models with nonempty carriers (as it is done in CASL), we also have to drop the variable subscripts from the derivability relation $\vdash$.

A different version of subsorted partial logic, which has a more restrictive treatment of subsorting, but which can also deal with higher-order functions, is developed in [Far93].

Bibliographical notes. Full many-sorted and order-sorted first order logic was introduced by Oberschelp [Obe62]. A survey over the conditional fragment of order-sorted logic can be found in [GM92]. Many-sorted logic is studied extensively in [MT93]. The empty carrier problem, which is ignored by most authors, is treated by Hailperin [Hai53] for the first-order case (the work on so-called free logics is also relevant [Lam91, KGS88], Chapter 9.2) and by Goguen and Meseguer [GM85] for the many-sorted case.

The two-valued partial first-order logic *of definedness* (using $\models^t$) was introduced by Burmeister [Bur82, Bur86] and Beeson [Bee85] and generalized to categorical logic by Knijnenburg and Nordemann [KN94]. The partial first-order logic of *existence* (using $\models^p$) follows ideas of Scott [Sco79], see also [Mog88]. $\models^t$ and $\models^p$ are compared by Beeson in [Bee86], see also [Fef95].

There is a calculus and a Henkin-style completeness theorem for partial higher-order logic in [Far91], while Burmeister has a calculus for one-sorted partial logic [Bur82]. The translation from partial to total first-order logic is described by Scott [Sco79]. This translation generates partial congruence relations, which can be treated in a similar way to equality with the results of Bachmair and Ganzinger [BG94b]. There have been two workshops on theorem proving with partial functions.[20] The restriction to the positive conditional case is studied by Reichel and others in [Rei87, BR83, AC95, BW82b, Cer93, Mos96a, Tar85].

3.4 More advanced problems

Although the positive conditional fragment of partial first-order logic is powerful enough for most data type specifications, there are a few cases where it is insufficient to represent the intended data type directly, or where even full partial first-order logic is too poor. In the following paragraphs we will examine some of the most relevant and common problems.

Partial higher-order specifications. As we have noted before, higher-order partial data-types are quite common in programming languages, for instance, to represent environment and stores in the imperative paradigm, or to describe functional (or procedural) parameters.

Since their use is reasonably restricted, it is not necessary to have *real* higher-order logic, but it is sufficient to consider a particular case of first-order specifications. The main intuition is that the set of sort is not unstructured, but is constructed by a subset B of *basic* sorts and a (polymorphic) operation building the functional type. That is, the set of sort is a subset of the set $S^{\rightarrow}$ inductively defined by the following rules:

$$B \subseteq S^{\rightarrow} \qquad s_1, \dots, s_n, s \in S^{\rightarrow} \Rightarrow (s_1 \times \cdots \times s_n \rightarrow s) \in S^{\rightarrow}$$

Of course the sort $(s_1 \times \cdots \times s_n \rightarrow s)$ represents the type of functions with arguments in the cartesian product $s_1 \times \cdots \times s_n$ and the result of sort s. Accordingly, in the signature, an explicit *apply* operation is provided, taking as input an element of a functional sort and the arguments for it, and yielding the result of the application of the function to its input.

Therefore, the set S of sorts must be downwardclosed, that is, if $(s_1 \times \cdots \times s_n \rightarrow s_{n+1}) \in S$, then all $s_i \in S$.

Since we are interested in the specification of partial functions, the application must be a partial function[21], even if the other operations of the signature are total (consider, for example, a total function delivering partial functions as result). For instance, in the previous example of the specification of a kernel of a programming language semantics, the only partial operation was the application of commands and Boolean expressions to their input,

---

[20] See http://www.cs.bham.ac.uk/~mmk/partiality/.

[21] Only if we introduce two different function space constructors, one for total and one for partial functions, can we choose a total application for total function spaces.

the current state, denoted by CEval (for commands) and BEval (for Boolean expressions), because all the command constructors were total functions.

The intuition that the elements of a functional sort $(s_1 \times \cdots \times s_n \to s_{n+1}) \in S$ are actually (isomorphic to) functions, is expressed not only by the application function, but also by the *extensionality principle*, requiring that two functions yielding the same result on each possible input must be the same.

Although the specification of higher-order types for programming languages can usually be described within the positive conditional fragment, the axiomatization of the extensionality property requires a more powerful logic. Indeed, the natural form of the extensionality axiom (for simplicity in the case of unary functions) is

$$\star \quad \forall f, f' : (s \to s).(\forall x : s \ apply(f, x) \overset{s}{=} apply(f', x)) \ \Rightarrow \ f \overset{e}{=} f'$$

Note that the equality in the premises is strong, capturing the idea that f and f' should have the same definition domain and yield the same result on applications within their domain.

A particularly interesting case is that of term-generated models. In that case, the premise of the extensionality axiom is equivalent to the infinitary conjunction of all its possible instantiations on (defined) terms. Therefore, for term-generated models, extensionality can be reduced to an *infinitary* conditional axiom

$$\star^t \quad \wedge_{t \in T_{\Sigma_\bullet}} apply(f, t) \overset{s}{=} apply(f', t) \ \Rightarrow \ f \overset{e}{=} f'$$

But notice that the equalities in the premises are not existential, nor can they be substituted by existential equalities. Thus, even in this simplified case, partial higher-order specifications do not reduce to positive conditional specifications. Indeed, in the general case, even total equational specifications of partial higher-order algebras do not have an initial model in the class of all extensional models, that is the models satisfying the axiom $\star$, nor in the class of all term-extensional models, that is the models satisfying the axiom $\star^t$ (see e.g. [AC92]).

It is worth noting that moving from a finitary to an infinitary logic, although obviously affecting computational issues for the involved deductive systems, does not change the nature of the problem. Indeed, the same results as for total conditional types are achieved in [Mei92, Möl87, MTW88a] where the same problem is tackled for total types. Moreover, in a partial context, the same problems existing for term-extensional higher-order algebras, are already present for *strongly conditional* partial specifications, that are conditional specifications whose axioms admit (unguarded) strong equalities in their premises (see, e.g., [AC95]).

Using the extensionality requirement as a unique nonpositive conditional axiom of a specification, it is possible to describe awkward data types, for instance, with all nontrivial models having finite carriers of bounded cardinality. We refer to [AC92] for an analysis and exposition of the problem of higher-order partial types.

The theory sketched so far can be extended in several ways.

A first possible extension is to add λ-abstraction. Given a term t, the term

$$\lambda x : s.t$$

denotes an anonymous function f defined by

$$\forall x : s.f(x) \overset{s}{=} t$$

Thus, typed λ-abstraction works well when combined with partiality [Far91]. However note that to be able to interpret λ-abstraction consistently, we require that all λ-definable functions exist in our models. Such models are called *generalized Henkin models* [Hen50]. This requirement can be achieved by adding a complete set of combinators (of combinatory logic) as operations to the signature and the corresponding axioms in order to define their semantics as default for any specification. Another possibility is to require that functional sorts are always interpreted with the full[22] function space. This requirement goes beyond the power of first-order logic. Moreover, due to Gödel's famous Incompleteness Theorem [Göd31], such a logic has no complete, finitely axiomatized proof system.

A second, orthogonal issue is the addition of predicate types. This is useful, for example, for modeling predicate transformers. To achieve this, we have to generalize the sort structur, by introducing a predicate-type constructor. Moreover, as we added explicit application functions for the functional sorts, we have to add explicit application predicates for the predicate types.

In this context, since the application of any predicate to some undefined argument yields false, the application of a term of a predicate type to (appropriately typed) terms yields false if the evaluation of any argument term and that of the predicate term is undefined. This approach is proposed for higher-order CASL [HKM98].

While, for the argument terms, it is natural to choose the same interpretation as in the first-order case (requiring any undefined argument to lead to false), it may seem strange that undefined predicate terms (which may now be applications of a possibly partial function) yield false as well. In order to avoid possible conflicts, it is possible to avoid undefined predicate terms by dividing the set of types into two kinds: kind $*$ containing the types of higher-order functions that may eventually (when applied to enough arguments) deliver a truth value, and kind ι containing all the other types. Now all functions of a $*$-type are required to be total. This is achieved by introducing, for each type of kind $*$, a canonical value corresponding to "everywhere

---

[22] Whatever this "full" means; in general, this depends on the meta set theory.

false". This canonical value is returned when a function of a ∗-type is applied to an undefined argument. This approach is used in LUTINS [Far91, Far93].

Another quite natural generalization of the first-order reduction of functional specifications for predicates is to introduce a special sort of Boolean values, and regard the predicates as functions onto that sort. This is a more powerful approach, as it is now easy to describe logical connectives and use formulas in term formation. To our knowledge, such a logic has not been investigated in the literature. Note that, in this case, a *nonstrict* framework as presented in the next section is needed to perform a reduction to first-order logic, because the application of Boolean-valued functions is not strict: it yields false for undefined arguments.

For the combination of partiality with λ-calculus, combinatory logic, and polymorphism, see [Mog88, Fef95].

Nonstrictness. A completely orthogonal problem is that of nonstrictness, because it concerns not the logic used to specify data types, but the semantic side, that is, the class of acceptable models. Indeed, in partial logic the interpretation of both total and partial operation symbols in the models are *strict*, that is they can only produce a result if all their inputs are provided correctly.

This is not the case, for instance, for the *conditional choice*, like the if_then_else in many programming languages, which can result in a correct value even if one of the branches does not, because only one branch is actually evaluated.

The classical trick for representing such functions is lifting their domain to function spaces.

Consider, for example, the case of if true then c_1 else c_2, where c_1 and c_2 are commands, then its evaluation on a state s is the evaluation of c_1 on s disregarding the value, if any, of the evaluation of c_2. But, even if the evaluation of c_2 on s is incorrect, the interpretation of c_2 is still a well-defined element of a functional sort and hence, technically, the interpretation of if_then_else is strict. The same technique, although less obviously, can be applied, for instance, to Boolean and and or with lazy valuation, as follows.

> **spec Bool** =
> **sorts** bool, BoolExps, dummy
> **opns** · : → dummy
> T, F: → bool
> true, false: → BoolExps
> BEval: BoolExps × dummy → bool
> and, or: BoolExps × BoolExps → BoolExps ...
> **vars** x, y : BoolExps
> **axioms** BEval(true, ·) $\stackrel{e}{=}$ T
> BEval(false, ·) $\stackrel{e}{=}$ F
> BEval(x, ·) $\stackrel{e}{=}$ F $\Rightarrow$ and(x, y) $\stackrel{e}{=}$ false
> BEval(x, ·) $\stackrel{e}{=}$ T $\Rightarrow$ and(x, y) $\stackrel{e}{=}$ y ...

$$\mathtt{BEval}(x, \cdot) \overset{e}{=} \mathtt{T} \Rightarrow \mathtt{or}(x, y) \overset{e}{=} \mathtt{true}$$
$$\mathtt{BEval}(x, \cdot) \overset{e}{=} \mathtt{F} \Rightarrow \mathtt{or}(x, y) \overset{e}{=} y \ \ldots$$

where, of course, the specification becomes interesting only if some *partial* constructors for Boolean expressions are provided.

Instead of trying to implement nonstrictness inside a strict framework, it is also possible to weaken the requirements on the semantic models to obtain a richer class providing *true* nonstrictness, so to speak.

A first possibility is to consider the case of *monotonic* nonstrictness, that is, requiring that if a function can produce a result with some of its arguments undefined, then, whatever is substituted for them, the result should be the same. This approach deals perfectly well with the so-called *don't care* parameters, such as the uninteresting branch in conditional choices or superfluous data for suspended valuations. However it cannot be used for error recovery, because in that case different "undefined" cases should result in different correctly recovered data.

In [AC96], an algebraic paradigm for the nonstrict don't care case is presented, which is based on the idea of *partial product*. The intuition is that, while in the standard strict case the arguments of an n-ary function are n-tuples that are functions from the range $[1 \ldots n]$ into the carriers, here *partial* n-tuples that are partial functions from the range $[1 \ldots n]$ into the carriers, are allowed as well.

Starting from this new point of view, the standard algebraic theory is developed. But it is important to note that the monotonicity requirement implicitly introduces disjunctive axioms. Indeed if we know that $pf(a)$ is defined, for some constant a and unary function pf, then we have that a is defined or $pf(x)$ is defined for any value of x (including the undefined). Thus, $D(pf(a))$ is equivalent to $D(a) \vee D(pf(x))$.

Therefore, the theory of equational nonstrict data types is more or less equivalent to the theory of disjunctive nonstrict data types which are studied in [AC96], where necessary and sufficient conditions for the existence of initial models are given.

When moving from algebras to first-order structures, nonstrictness extends to predicates in the following way: a predicate is strict if it yields false whenever some of its arguments is undefined, while nonstrict predicates can be true even for undefined arguments. A nonstrict partial first-order logic along these lines is developed in [GL97]. It uses total variable valuations (corresponding to our $\models^t$), and has a special constant denoting undefined. A monotonicity requirement is not imposed.

A completely different point of view on nonstrictness is presented in [Cer95], where the intuition is that nonstrictness comes from evaluational issues and is not inherent to the underlying datatype. Thus, the idea is to keep standard partial algebras as models, but each one is equipped with a total congruence on terms, representing the simplifications that are performed on terms before the actual evaluation.

In this approach error recovery follows the intuition that errors never occur, because terms can be simplified before their evaluation. Thus, a term can be simplified into a perfectly correct term, even if it contains some subterm that is incorrect. For instance, a term denoting an integer value, of the form zero $* t$ can be simplified to zero, and then evaluated to the 0 value even if t is incorrect, allowing in this way a strategy for error recovery.

The form of the axioms allow the definition of subtle error-recovery strategy, or lazy/suspended evaluation.

The description operator. A definite description operator allows a term to be constructed from a formula describing the properties of a unique value.

The presence of partial functions makes it straightforward to introduce a definite description operator [LvF67, Far91]. Assume that we have an arbitrary first-order formula φ; then $\imath x : s.\varphi$ (read "the x for which φ") is a new term of sort s. φ should be the implicit description of some value, such that φ is true if and only if this value is substituted for x. The intended meaning is that $\imath x : s.\varphi$ denotes this unique value, if it exists, while it is undefined, if no such value exists, or if there is more than one. Thus, we have to add the following semantical rule:[23]

$$v^{\#}(\imath x : s.\varphi) = \begin{cases} \xi(x), & \text{if there is a unique } \xi : X \cup \{x : s\} \longrightarrow A \\ & \text{extending } v \text{ on } X \backslash \{x : s\} \\ & \text{for which } \xi \Vdash \varphi \\ \text{undefined,} & \text{otherwise} \end{cases}$$

The description operator is characterized by the axiom

$$\forall y : s.(y \overset{e}{=} \imath x : s.\varphi \Leftrightarrow \forall x : s.(\varphi \Leftrightarrow x \overset{e}{=} y))$$

which has to be added to the calculus.

For example, the division function can now be easily defined in terms of multiplication:

$$\forall x, y : real.(x/y \overset{s}{=} \imath z : real.x \overset{e}{=} y * z)$$

where $x/0$ is undefined, as expected.

In the absence of a division operation, the term $\imath z : real.x \overset{e}{=} y * z$ is not equivalent to a term without $\imath$. Therefore, Corollary 3.44 no longer holds: there may be term-generated structures which are not reachable.

Three-valued logic. When reasoning about programs, we cannot avoid the use of some three-valued logic. For example, the condition in a while-loop is of the three-valued Boolean type, because it may either be true, false, or undefined due to some infinite computation or exception.

---

[23] Syntax of terms and formulas as well as their semantics (i.e., $v^{\#}$ and $v \Vdash \_$) have to be defined in parallel now.

On the other hand, when writing specifications, we want to specify *properties* of data types and programs, which may hold or not hold, but without a third possibility. For example, a definedness predicate delivering undefined when the argument is undefined does not make sense. The outer level of specifications is therefore two-valued, but somewhere we have to have the possibility to express properties of three-valued programs.

Basically, there are two points where three-valuedness can be introduced.

1. Incorporate three-valuedness into the logic itself. Thus, there is a distinction between the false and the undefined. The fact that some term is nondenoting is propagated to the formulas containing the term. Thus, valuations of terms *and* formulas have to be partial. The latter causes the need for a third truthvalue, say $\bot$, which is assigned to formulas containing a nondenoting term. Then, nonstrict predicates may also yield $\bot$ when applied to denoting terms. The connectives can be extended to deal with $\bot$ in several ways. A natural choice is Kleene's three-valued logic [Kle52], which is guided by a strictness (or continuity with respect to a natural associated topology) principle [Cle91], in which:

$\wedge$	t	$\bot$	f
t	t	$\bot$	f
$\bot$	$\bot$	$\bot$	f
f	f	f	f

$\vee$	t	$\bot$	f
t	t	t	t
$\bot$	t	$\bot$	$\bot$
f	t	$\bot$	f

$\neg$	
t	f
$\bot$	$\bot$
f	t

$\Rightarrow$	t	$\bot$	f
t	t	$\bot$	f
$\bot$	t	$\bot$	$\bot$
f	t	t	t

It is used, e.g., in the specification languages SPECTRUM [BFG$^+$93] and VDM [Jon90]. The latter reference also describes an easy calculus for the Kleene logic.

But, as mentioned above, at the level of specifications there is the need for other two-valued connectives which are nonstrict (or noncontinuous) when incorporated in the three-valued world. For example, the definedness predicate

D	
t	t
$\bot$	f
f	t

is nonstrict but is nevertheless needed for specifications. Another example is an implication such as

$$\forall x, y, z : nat.\, x = z/y \;\Rightarrow\; x * y = z$$

which is valid without a restriction on y in two-valued logic, but not in three-valued logic. To make it valid in three-valued logic, we have to use a nonstrict implication which identifies false and $\bot$.

But nonstrict connectives complicate the calculus by introducing the need for a more complex case analysis.

See [Cle91] for an overview of multi-valued logics.

2. Use a two-valued logic as introduced above and shift the issue of three-valuedness to the object level. This can be done with a specification like

> **spec** *ThreeValued* =
>> **sorts** Bool3
>> **preds** $\overset{3}{=}$: Bool3 × Bool3
>> **opns** t, f, ⊥: → Bool3
>>> $\neg^3$: Bool3 → Bool3
>>> $\wedge^3, logor^3, \Rightarrow^3$: Bool3 × Bool3 → Bool3
>> **vars** x, y : Bool3
>> **axioms** $\forall x$: Bool3.$x \overset{e}{=} t \vee x \overset{e}{=} f \vee x \overset{e}{=} \bot$
>>> $t \vee^3 \bot \overset{e}{=} t$
>>> . . .
>>>
>>> $(x \overset{3}{=} y) \overset{e}{=} t \Leftrightarrow x \overset{e}{=} y$
>>> $(x \overset{3}{=} y) \overset{e}{=} f \Leftrightarrow (D(x) \wedge D(y) \wedge \neg x \overset{e}{=} y)$
>>> $(x \overset{3}{=} y) \overset{e}{=} \bot \Leftrightarrow (\neg D(x) \vee \neg D(y))$
>>> . . .

A quantified formula $\forall X.\varphi$ is expressed by first translating φ inductively to a term $\overset{3}{\varphi}$: Bool3 and then taking $\forall \overset{3}{X}.\varphi$ to be

$$\imath b : \text{Bool3.} \, ((\forall X. \overset{3}{\varphi} \overset{e}{=} t) \Rightarrow b \overset{e}{=} t)$$
$$\wedge ((\forall X. \neg \overset{3}{\varphi} \overset{e}{=} \bot \wedge \exists X. \overset{3}{\varphi} \overset{e}{=} f) \Rightarrow b \overset{e}{=} f)$$
$$\wedge ((\exists X. \overset{3}{\varphi} \overset{e}{=} \bot) \Rightarrow b \overset{e}{=} \bot)$$

The need for a case analysis when performing theorem proving is made explicit in the definitions here.

Note. A preliminary and somehow extended version of this chapter, including proofs, is available as technical report DISI-TR-96-25 of DISI - University of Genova (Italy) at `ftp://ftp.disi.unige.it/person/CerioliM/IFIP96.ps.gz`.

4 Institutions: An Abstract Framework for Formal Specifications

Andrzej Tarlecki

Institute of Informatics, Warsaw University and Institute of Computer Science, Polish Academy of Sciences, Warsaw, Poland
tarlecki@mimuw.edu.pl   http://wwwat.mimuw.edu.pl/~tarlecki/

In this chapter we present some basic concepts and results of the theory of institutions, introduced by Goguen and Burstall to formally capture the informal notion of a logical system viewed from a model-theoretic perspective. We also sketch some possibilities of linking this to more proof-oriented concepts. We argue that the theory of institutions provides an appropriate framework for much of the work on formal software specification and development, as presented in this volume. Many standard logical systems used in particular versions of the algebraic specification paradigm may be viewed as institutions; some examples are given explicitly here, some others are hinted at. Developing (as much as possible) the ideas common to different versions of the algebraic specification paradigm in the framework of an arbitrary institution, and in particular providing a theory of formal specification and software development parameterized by an arbitrary institution rather than having a particular logical system built in, should be beneficial both by helping to avoid repetitive work and by bringing the concepts and results to an appropriate level of mathematical and practical abstraction.

4.1 Working with an arbitrary logical system

As already argued in Chapter 1, a "universal specification formalism" is a myth: there exist many specification formalisms and this diversity is practically necessary to adequately capture various needs of specific application areas, systems and programming paradigms, and even the taste and customs of particular users. To a novice, the wealth of possibilities to choose from when attempting a formal specification and development task is as much a problem as an advantage, since a large investment in effort is required to master each specific formalism. Fortunately, a closer look at the range of possibilities shows that within this apparent confusion, numerous specification formalisms share common ideas, concepts, theoretical results and practical tools. The choice then becomes easier, and somewhat less crucial; the effort required to learn a new formalism does not disappear, but at least many of the basics can be learnt once and for all. It is a non-trivial task, however, to make this uniformity apparent and mathematically precise. Work in this direction is worthwhile for at least two reasons. On one hand, a good presentation of common features of various specification formalisms should potentially enable them to be transferred to other, perhaps newly developed formalisms "for free". This is a very practical concern: specification formalisms are numerous, new ones are being created all the time, and we should as much as possible avoid tedious, repetitive work if we can just borrow its expected results from a related formalism. On the other hand, a precise account of some common features of a number of formalisms typically requires the features themselves to be presented in a more abstract way, resulting in more insight and a better understanding of the concepts involved by unwinding them from the often messy intricacies of particular formalisms. The details are thrown away, and the essence is exposed!

At the heart of any algebraic (and not only algebraic) specification formalism there is a *logical system* allowing the user to write *axioms* describing the properties of the software system to be developed. The software itself is formally represented (via the semantics of the programming language in use) as a *model* of the system, and a notion of *satisfaction* determines whether an axiom *holds* in a model, and hence whether it holds for the system the model represents, or not. In the most traditional approach to algebraic specifications, models are just standard many-sorted algebras, axioms are equations, and their satisfaction in an algebra is defined in the usual way, see Chapter 2. We have already sketched, in Section 2.10, a number of other useful logical systems, each motivated by the need to conveniently capture some common system feature. Specification formalisms based on most (if not all) of them, as well as on many others, may be found in the literature. It is often the case that the most essential differences between specification formalisms lie in the underlying logical system they use.

It therefore seems natural and most effective to attempt development or at least presentation of a specification formalism in a way which would be as much as possible independent from the actual choice of the underlying logical system. This should bring extra flexibility, separation of orthogonal issues (on one hand the choice of features independent from the underlying logical system, and on the other hand the logical system itself) and in a sense modularization of the work that has to be done. If we are successful with such an approach, the outcome would be a specification formalism parameterized by an arbitrary logical system. Then, to obtain a specification formalism well suited for a given task, we would just have to instantiate this to a specific logical system chosen for the task.

These remarks apply not just to specification formalisms as such, but to many other areas of research in computer science: we should strive to work as long and as much as possible independently from the underlying logical system we would like to eventually adopt. The benefits are obvious: reusability, abstraction, deeper understanding; however, so are the dangers and shortcomings. A considerable part of the necessary work will not be possible at an overly high level of abstraction. To give an almost trivial example: when thinking about an arbitrary logical system we probably want to abstract away from the syntactic details of the axioms – and these details are crucial for the developments of some tools, like type checkers, theorem provers, etc., which have to deal with the syntax. But it would be premature to say that each such tool has to be developed separately for each different logical system. They typically would contain many common parts, which, together with the underlying concepts, results and strategies, can be developed once and then reused. Quite similarly, some results or even features we would like to incorporate (say, in a specification formalism or a mathematical theory under development) often depend on a crucial property of the underlying logical system that only some systems enjoy (like the existence of initial models of any theory, which equational logic ensures but first-order logic does not). Again, if a property like this is necessary, then we should identify exactly what is needed and formulate this as a requirement on permitted logical systems, rather than switching immediately to a particular one.

Finally, a word of warning: once the abstract development is finished, to make it applicable in a specific context we will still need to provide a logical system which is appropriate for the task at hand. This need not be easy; in fact, for many programming paradigms and application areas (like, in our view, object-oriented programming and distributed programming, in spite of much important work and numerous proposals) the development of the appropriate logical system is the most basic research problem yet to be solved!

4.2   Institutions

A preliminary step in the direction of developing specification formalisms independent from their underlying logical systems, necessary if the specification formalisms are indeed to be formal, is an abstract mathematical definition of what a logical system is. Many approaches are possible here, and many views have been taken. It seems, however, that the closest to the spirit and technical needs of the specification formalisms within the algebraic specification paradigm is the proposal due to Goguen and Burstall to formalize the concept of a logical system as an institution [GB92].

As presented in Chapter 2 and hinted at in the previous section, the basic components of a logical system are models (algebras) and sentences (axioms), linked by a satisfaction relation – these are therefore also the basic components of the notion of institution, as formally defined below. In addition, the presentation of these concepts in Chapter 2 has been parameterized by an arbitrary algebraic signature. In classical logic such a parameterization would normally be quite sufficient. However, for our purposes, where the change of signature occurs so frequently in the process of software specification and development, this parameterization by signatures must be incorporated into the very concept of a logical system, and an appropriate account of the notion of an algebraic signature morphism (see Section 2.5 as a tool for moving from one signature to another must be given. In a sense, explicit parameterization by signatures and an appropriate basis for moving from one signature to another within a single logical system is one of the main contributions of the theory of institutions.

Definition 4.1 (Institution [GB92]).
An *institution* **INS** consists of:

- a category $\mathbf{Sign_{INS}}$ of *signatures*;
- a functor $\mathbf{Sen_{INS}} \colon \mathbf{Sign_{INS}} \to \mathbf{Set}$, giving a set $\mathbf{Sen}(\Sigma)$ of Σ-*sentences* for each signature $\Sigma \in |\mathbf{Sign_{INS}}|$;
- a functor $\mathbf{Mod_{INS}} \colon \mathbf{Sign_{INS}^{op}} \to \mathbf{Cat}$, giving a category $\mathbf{Mod}(\Sigma)$ of Σ-*models* for each signature $\Sigma \in |\mathbf{Sign_{INS}}|$; and
- for each signature $\Sigma \in |\mathbf{Sign_{INS}}|$, a *satisfaction relation* $\models_{\mathrm{INS},\Sigma} \subseteq |\mathbf{Mod_{INS}}(\Sigma)| \times \mathbf{Sen_{INS}}(\Sigma)$

such that for any signature morphism $\sigma \colon \Sigma \to \Sigma'$, Σ-sentence $\varphi \in \mathbf{Sen_{INS}}(\Sigma)$ and Σ'-model $M' \in |\mathbf{Mod_{INS}}(\Sigma')|$:

$$M' \models_{\mathrm{INS},\Sigma'} \mathbf{Sen_{INS}}(\sigma)(\varphi) \iff \mathbf{Mod_{INS}}(\sigma)(M') \models_{\mathrm{INS},\Sigma} \varphi$$
$$[\textit{Satisfaction condition}] \qquad \diamond$$

When working within an institution, we will freely use standard logical terminology, and for example say that a Σ-model $M \in |\mathbf{Mod}(\Sigma)|$ *satisfies* a Σ-sentence $\varphi \in \mathbf{Sen}(\Sigma)$, where $\Sigma \in |\mathbf{Sign}|$ is a signature, or that φ *holds* in M, whenever $M \models_{\mathrm{INS},\Sigma} \varphi$. When there is no danger of confusion, we will omit the subscript **INS** when referring to the components of an institution **INS**. Similarly, the subscript Σ on the satisfaction relations will often be omitted. For any signature morphism $\sigma \colon \Sigma \to \Sigma'$, the function $\mathbf{Sen}(\sigma) \colon \mathbf{Sen}(\Sigma) \to \mathbf{Sen}(\Sigma')$ will be denoted simply by σ and referred to as the σ-*translation* of Σ-sentences to Σ'-sentences. Then the functor $\mathbf{Mod}(\sigma) \colon \mathbf{Mod}(\Sigma') \to \mathbf{Mod}(\Sigma)$ will be denoted by $\_|_\sigma$ and referred to as the σ-*reduct* of Σ'-models and morphisms to Σ-models and morphisms, respectively. For any signature Σ, the satisfaction relation extends naturally to sets of Σ-sentences and classes of Σ-models. Namely, for any set $\Phi \subseteq \mathbf{Sen}(\Sigma)$ of Σ-sentences, Σ-sentence $\varphi \in \mathbf{Sen}(\Sigma)$, Σ-model $M \in |\mathbf{Mod}(\Sigma)|$, and class $\mathcal{M} \subseteq |\mathbf{Mod}(\Sigma)|$ of Σ-models, we will write $M \models \Phi$, $\mathcal{M} \models \varphi$ and $\mathcal{M} \models \Phi$ with the obvious meanings (for instance, $M \models \Phi$ means $M \models \psi$ for all $\psi \in \Phi$).

The concept of an institution as given above is very general, and especially in the light of the examples below, one might wonder if any logical systems do *not* form an institution. However, some of the restrictions imposed by the above definition are worth discussing.

Most obviously, there is the explicit requirement that the satisfaction condition holds. Very informally, relying on the intuitions induced by the examples below and the presentation in Chapter 2, this constrains the situation where a "smaller" signature Σ is embedded into a "larger" signature

Σ' via a signature morphism $\sigma \colon \Sigma \to \Sigma'$. Then to check whether a sentence $\varphi \in \mathbf{Sen}(\Sigma)$ over the smaller signature Σ holds in a model $M \in |\mathbf{Mod}(\Sigma')|$ over the larger signature Σ', we can either translate the sentence φ to the larger signature, and check whether it holds in the model M there, or we can reduce the model M to the smaller signature Σ and check whether it satisfies the sentence φ there. The satisfaction condition states that both ways always yield the same result, thus capturing the requirement that the satisfaction of a sentence does not depend on the interpretation of the symbols in the signature that do not explicitly occur in the sentence. When we think of σ as an arbitrary renaming rather than an embedding, this also captures the requirement that the satisfaction of a sentence does not depend on the actual names of the model components used in the signature.

Other requirements are hidden in the structure of the notions involved: **Mod** and **Sen** are required to be functorial. That is, translations of signatures and reducts of models under a signature morphism must be independent from the possible decompositions of the morphism considered. This may be thought rather restrictive, especially when we realize that for sentences this means that given two consecutive signature morphisms $\sigma \colon \Sigma \to \Sigma'$ and $\sigma' \colon \Sigma' \to \Sigma''$ and a Σ-sentence $\varphi \in \mathbf{Sen}(\Sigma)$, the two translations $(\sigma; \sigma')(\varphi) \in \mathbf{Sen}(\Sigma'')$ and $\sigma'(\sigma(\varphi)) \in \mathbf{Sen}(\Sigma'')$ must be identical, while what we really typically care about is that they would be equivalent in some suitable sense (which would have to imply that they hold in exactly the same Σ''-models).

Nevertheless, there is a huge variety of logical systems that satisfy these mild restrictions and can easily be put into the mold of an institution. Here are some typical examples.

*Example 4.2 (Equational logic **EQ**).*
The institution **EQ** of equational logic is defined as follows:

- The category $\mathbf{Sign_{EQ}}$ is the category **AlgSig** of algebraic signatures and their morphisms, see Section 2.5.
- The functor $\mathbf{Sen_{EQ}} \colon \mathbf{AlgSig} \to \mathbf{Set}$ gives (see Section 2.6):
  - the set of Σ-equations for each $\Sigma \in |\mathbf{AlgSig}|$; and
  - the σ-translation function taking Σ-equations to Σ'-equations for each signature morphism $\sigma \colon \Sigma \to \Sigma'$.
- The functor $\mathbf{Mod_{EQ}} = \mathbf{Alg} \colon \mathbf{AlgSig}^{op} \to \mathbf{Cat}$ gives:
  - the category $\mathbf{Alg}(\Sigma)$ of Σ-algebras and Σ-homomorphisms for each $\Sigma \in |\mathbf{AlgSig}|$ (see Section 2.3); and
  - the reduct functor $\_\!|_\sigma \colon \mathbf{Alg}(\Sigma') \to \mathbf{Alg}(\Sigma)$ mapping Σ'-algebras and Σ'-homomorphisms to Σ-algebras and Σ-homomorphisms for each signature morphism $\sigma \colon \Sigma \to \Sigma'$ (see Section 2.5).
- For each $\Sigma \in |\mathbf{AlgSig}|$, the satisfaction relation $\models_{\mathbf{EQ}, \Sigma} \subseteq |\mathbf{Alg}(\Sigma)| \times \mathbf{Sen_{EQ}}(\Sigma)$ is the usual relation of satisfaction of a Σ-equation by a Σ-algebra, as defined in Section 2.6.

The Satisfaction Lemma (Lemma 2.13) ensures that the required satisfaction condition holds and so that the above definition indeed yields an institution.

∎

Example 4.3 (Partial equational logic **PEQ***).*
The institution **PEQ** of partial equational logic is defined as follows (see Section 2.10.2):

- **Sign**$_{\mathbf{PEQ}}$ is **AlgSig** again.
- **Sen**$_{\mathbf{PEQ}}$: **AlgSig** → **Set** gives:
  - the set of Σ-equations and Σ-definedness formulas for each $\Sigma \in$ |**AlgSig**|; and
  - the σ-translation function taking Σ-equations and Σ-definedness formulas to Σ'-equations and Σ'-definedness formulas for each signature morphism $\sigma\colon \Sigma \to \Sigma'$, defined using the σ-translation of terms, as expected.
- **Mod**$_{\mathbf{PEQ}}$: **AlgSig**^{op} → **Cat** gives:
  - the category **PAlg**(Σ) of partial Σ-algebras and weak Σ-homomorphisms for each $\Sigma \in$ |**AlgSig**|; and
  - the reduct functor $\_|_{\sigma}\colon$ **PAlg**$(\Sigma') \to$ **PAlg**(Σ) defined similarly as in the total case for each signature morphism $\sigma\colon \Sigma \to \Sigma'$.
- For each $\Sigma \in$ |**AlgSig**|, the satisfaction relation $\models_{\mathbf{PEQ},\Sigma}\ \subseteq$ |**PAlg**(Σ)|× **Sen**$_{\mathbf{PEQ}}(\Sigma)$ is the satisfaction of Σ-equations and Σ-definedness formulas by partial Σ-algebras.

Checking the satisfaction condition for this institution requires a proof analogous to that of the standard Satisfaction Lemma (Lemma 2.13).   ∎

Example 4.4 (First-order logic with equality **FOEQ***).*
The institution **FOEQ** of first-order logic with equality is defined as follows (see Section 2.10.5):

- **Sign**$_{\mathbf{FOEQ}}$, from now on denoted by **FOSig**, is the category of *first-order signatures*:
  - A *first-order signature* Θ is a triple $\langle S, \Omega, \Pi \rangle$, where S is a set (of *sort names*), $\Omega = \langle \Omega_{w,s} \rangle_{w \in S^*, s \in S}$ is a family of sets (of *operation names*, indexed by arities and result sorts), and $\Pi = \langle \Pi_w \rangle_{w \in S^*}$ is a family of sets (of *predicate* or *relation names*, indexed by arities).
  - A *first-order signature morphism* $\theta\colon \langle S, \Omega, \Pi \rangle \to \langle S', \Omega', \Pi' \rangle$ consists of a function $\theta_{sorts}\colon S \to S'$, an $S^* \times S$-indexed family of functions $\theta_{opns} = \langle (\theta_{opns})_{w,s}\colon \Omega_{w,s} \to \Omega'_{\theta^*_{sorts}(w), \theta_{sorts}(s)} \rangle_{w \in S^*, s \in S}$, and an S^*-indexed family of functions $\theta_{preds} = \langle (\theta_{preds})_w\colon \Pi_w \to \Pi'_{\theta^*_{sorts}(w)} \rangle_{w \in S^*}$.

- **Sen$_{FOEQ}$: FOSig → Set** gives:
  - For each first-order signature $\Theta = \langle S, \Omega, \Pi \rangle$, **Sen$_{FOEQ}$**$(\Theta)$ is the set of all closed (i.e., without unbound occurrences of variables) *first-order formulas* built out of atomic formulas using the standard propositional connectives ($\vee, \wedge, \Rightarrow, \neg$) and quantifiers ($\forall, \exists$).[1] The *atomic formulas* are equalities of the form $t = t'$, where t and t' are $\langle S, \Omega \rangle$-terms of the same sort, and atomic predicate formulas of the form $p(t_1, \dots, t_n)$, where $p \in \Pi_{s_1 \dots s_n}$ and $t_1, \dots, t_n$ are terms of sorts s_1, $\dots$, s_n, respectively.
  - For each first-order signature morphism $\theta \colon \Theta \to \Theta'$, **Sen$_{FOEQ}$**$(\Theta)$ is the obvious translation of first-order Θ-sentences to first-order Θ'-sentences determined by the renaming θ of sort, operation and predicate names in Θ to the corresponding names in Θ'.
- **Mod$_{FOEQ}$: FOSig^{op} → Cat**, from now on denoted by **FOStr**, gives:
  - For each first-order signature $\Theta = \langle S, \Omega, \Pi \rangle$, the category **FOStr**$(\Theta)$ of *first-order Θ-structures*:
    * A *first-order Θ-structure* $A \in |\textbf{FOStr}(\Theta)|$ consists of a carrier set $|A|_s$ for each sort name $s \in S$, a function $f_A \colon |A|_{s_1} \times \dots \times |A|_{s_n} \to |A|_s$ for each operation name $f \in \Omega_{s_1 \dots s_n, s}$, and a relation $p_A \subseteq |A|_{s_1} \times \dots \times |A|_{s_n}$ for each predicate name $p \in \Pi_{s_1 \dots s_n}$. In the following we write $p_A(a_1, \dots, a_n)$ for $\langle a_1, \dots, a_n \rangle \in p_A$.
    * For any first-order Θ-structures A and B, a *first-order Θ-morphism* between them, $h \colon A \to B$, is a family of functions $h = \langle h_s \colon |A|_s \to |B|_s \rangle_{s \in S}$ which preserves the operations and predicates.
  - For each first-order signature morphism $\theta \colon \Theta \to \Theta'$, we have the *$\theta$-reduct functor* **FOStr**$(\theta) \colon \textbf{FOStr}(\Theta') \to \textbf{FOStr}(\Theta)$ defined similarly as reduct functors corresponding to algebraic signature morphisms.
- For each first-order signature $\Theta \in |\textbf{FOSig}|$, the satisfaction relation $\models_{FOEQ, \Theta} \subseteq |\textbf{FOStr}(\Theta)| \times \textbf{Sen}_{FOEQ}(\Theta)$ is the usual relation of satisfaction of first-order sentences in first-order structures, determined by the usual interpretation of $\vee, \wedge, \Rightarrow$, and $\neg$ as disjunction, conjunction, implication, and negation, respectively, of $\forall$ and $\exists$ as universal and existential quantifiers, respectively, of equalities $t = t'$ as identity of the values of t and t', and of atomic predicate formulas $p(t_1, \dots, t_n)$ as membership of the tuple of the values of $t_1, \dots, t_n$ in the relation named p in the structure.

The satisfaction condition for this institution may be proved as an appropriate generalization of the Satisfaction Lemma (Lemma 2.13). ∎

It is not much more difficult to define, for example, the institution of partial first-order equational logic, or any other institution formalizing one

---

[1] Similarly as for equalities – cf. Section 2.6 – we will assume that in each first-order formula variables of different sorts are distinct.

of the many standard variants of the classical notions. Institutions for the logical systems hinted at in Section 2.10 may be defined in a similar manner.

The notion of an institution is general enough, however, to admit many non-standard and perhaps somewhat unexpected examples. We refer to [ST] for some examples of institutions which depart considerably from standard systems based on usual logics and some of their variants as those presented above.

To hint at the wealth of possibilities the notion of an institutions leaves open, let us mention some trivial examples:

- There is an empty institution based on the empty category of signatures: no signatures, hence no sentences and no models.
- Given any category **Sign** and functor **Mod**: $\mathbf{Sign}^{op} \to \mathbf{Cat}$, a trivial institution with signatures **Sign**, with models given by **Mod**, and with no sentences may be constructed. Similarly, for any functor **Sen**: $\mathbf{Sign} \to \mathbf{Set}$, a trivial institution with signatures **Sign**, with sentences given by **Sen**, and with no models may be constructed.
- Given any category **Sign** and functors **Sen**: $\mathbf{Sign} \to \mathbf{Set}$ and **Mod**: $\mathbf{Sign}^{op} \to \mathbf{Cat}$, two trivial institutions may be constructed: one is obtained by making all sentences hold in any model, and the other by making none hold.

It should also be realized that there is nothing in the formal definition of an institution to support the usual informal intuition that in some sense sentences are syntactic objects and models are more on the semantic side of the world. For example, for any category of signatures **Sign** and model functor **Mod**: $\mathbf{Sign} \to \mathbf{Cat}$,[2] we can define an institution where for each signature $\Sigma \in |\mathbf{Sign}|$, Σ-"sentences" are just sets of Σ-models, their translations are obtained as the coimage with respect to the reduct functors, and their satisfaction is simply the membership relation (it is easy to check that the satisfaction condition holds under these definitions). A construction of models as sets of sentences is also possible – we will return to this in Section 4.5.

Given some institutions, there are many standard ways in which we can modify them, thus building new institutions out of the ones already defined. For instance, given an institution **INS** we can form its "closure under conjunction" by considering the institution $\mathbf{INS}^{\wedge}$ with the same category of signatures and the same model functor, but with sentences which are sets (or arbitrary conjunctions) of **INS**-sentences and with satisfaction defined using **INS**-satisfaction in "conjunctive" way. Similarly, given an institution **INS** we can form an institution in which signatures are **INS**-signatures additionally equipped with sets of **INS**-sentences (and signature morphisms are required to preserve these sets), sentences are the same as in **INS**, models are those

---

[2] We additionally assume here that for each $\Sigma \in |\mathbf{Sign}|$, the category $\mathbf{Mod}(\Sigma)$ is small.

INS-models that satisfy the sentences added to the signature, and the satisfaction relation is again inherited from **INS**. Such formal modifications may be viewed as operations on the class of institutions and studied separately; examples are numerous, see [ST].

4.3 Flat specifications in an arbitrary institution

In this section we will attempt to illustrate how some standard concepts and results may be developed independently from the details of the underlying logical system formalized as an institution. What we present here is rather simple: we just re-develop the basic concepts related to presentations, as formulated for equational logic in Section 2.7. We repeat the developments presented there without further motivation and examples. What is instructive here is that this is indeed possible for an arbitrary institution, without any knowledge whatsoever about the actual details of signature components, the structure of models and sentences, and the exact definition of satisfaction. It turns out that the satisfaction condition is all we need to know to obtain the desired results.

Let then $\mathbf{INS} = \langle \mathbf{Sign}, \mathbf{Sen}, \mathbf{Mod}, \langle \models_\Sigma \rangle_{\Sigma \in |\mathbf{Sign}|} \rangle$ be an arbitrary institution, fixed throughout this section.

Consider a signature $\Sigma \in |\mathbf{Sign}|$.

A *presentation* (also known as a *flat specification*) is a pair $\langle \Sigma, \Phi \rangle$ where $\Phi \subseteq \mathbf{Sen}(\Sigma)$ is a set of Σ-sentences (called the *axioms* of $\langle \Sigma, \Phi \rangle$). A presentation $\langle \Sigma, \Phi \rangle$ is sometimes referred to as a Σ-*presentation*.

A *model* of a presentation $\langle \Sigma, \Phi \rangle$ is a Σ-model $M \in |\mathbf{Mod}(\Sigma)|$ such that $M \models_\Sigma \Phi$.[3] $Mod_\Sigma(\Phi)$, written also as $Mod(\langle \Sigma, \Phi \rangle)$, is the class of all models of $\langle \Sigma, \Phi \rangle$. Taking $\langle \Sigma, \Phi \rangle$ to denote the semantic object $Mod_\Sigma(\Phi)$ is sometimes called taking its *loose semantics*.

For any class $\mathcal{M} \subseteq |\mathbf{Mod}(\Sigma)|$ of Σ-models, the *theory* of $\mathcal{M}$, written as $Th_\Sigma(\mathcal{M})$, is the set of all Σ-sentences satisfied by each Σ-model in $\mathcal{M}$:

$$Th_\Sigma(\mathcal{M}) = \{\varphi \in \mathbf{Sen}(\Sigma) \mid \mathcal{M} \models_\Sigma \varphi\}.$$

The *closure* of a set Φ of Σ-sentences is the set $Cl_\Sigma(\Phi) = Th_\Sigma(Mod_\Sigma(\Phi))$; Φ is *closed* if $\Phi = Cl_\Sigma(\Phi)$.

Proposition 4.5. *For any sets* $\Phi, \Psi \subseteq \mathbf{Sen}(\Sigma)$ *of* Σ-*sentences and classes* $\mathcal{M}, \mathcal{N} \subseteq |\mathbf{Mod}(\Sigma)|$ *of* Σ-*models:*

1. If $\Phi \subseteq \Psi$ *then* $Mod_\Sigma(\Psi) \subseteq Mod_\Sigma(\Phi)$.

---

[3] Note the overloading of the term "model": it is used to refer both to objects in the category $\mathbf{Mod}(\Sigma)$ and to *models* of a specification. This is not entirely incompatible, as the objects of $\mathbf{Mod}(\Sigma)$ are *models* of the specification $\langle \Sigma, \emptyset \rangle$, but may require some caution. We follow the terminology of [GB84, GB92] and hope that this will not cause confusion.

2. *If $M \subseteq N$ then $Th_\Sigma(N) \subseteq Th_\Sigma(M)$.*
3. *$\Phi \subseteq Th_\Sigma(Mod_\Sigma(\Phi))$ and $M \subseteq Mod_\Sigma(Th_\Sigma(M))$.*
4. *$Mod_\Sigma(\Phi) = Mod_\Sigma(Th_\Sigma(Mod_\Sigma(\Phi)))$ and*
   $Th_\Sigma(M) = Th_\Sigma(Mod_\Sigma(Th_\Sigma(M)))$.

A Σ-sentence $\varphi \in \mathbf{Sen}(\Sigma)$ is a *semantic* (or *model-theoretic*) *consequence* of a set $\Phi \subseteq \mathbf{Sen}(\Sigma)$ of Σ-sentences, written $\Phi \models_\Sigma \varphi$, if $\varphi \in Cl_\Sigma(\Phi)$ (equivalently, if $Mod_\Sigma(\Phi) \models_\Sigma \varphi$). As before, the subscript Σ will be omitted when obvious.

Proposition 4.6. *Semantic consequence is preserved by translation along signature morphisms: for any signature morphism $\sigma: \Sigma \to \Sigma'$, set $\Phi \subseteq \mathbf{Sen}(\Sigma)$ of Σ-sentences, and Σ-sentence $\varphi \in \mathbf{Sen}(\Sigma)$,*

if $\Phi \models_\Sigma \varphi$ then $\sigma(\Phi) \models_{\Sigma'} \sigma(\varphi)$.

Proposition 4.7. *Let $\sigma: \Sigma \to \Sigma'$ be a signature morphism and let Φ' be a closed set of Σ'-sentences. Then $\sigma^{-1}(\Phi')$ is a closed set of Σ-sentences.*

A *theory* is a presentation $\langle \Sigma, \Phi \rangle$ such that Φ is closed. A presentation $\langle \Sigma, \Phi \rangle$ (where Φ need not be closed) *presents* the theory $\langle \Sigma, Cl_\Sigma(\Phi) \rangle$. A theory $\langle \Sigma, \Phi \rangle$ is sometimes referred to as a Σ-*theory*. For any theories $\langle \Sigma, \Phi \rangle$ and $\langle \Sigma', \Phi' \rangle$, a *theory morphism* $\sigma: \langle \Sigma, \Phi \rangle \to \langle \Sigma', \Phi' \rangle$ is a signature morphism $\sigma: \Sigma \to \Sigma'$ such that $\sigma(\varphi) \in \Phi'$ for every $\varphi \in \Phi$. Theories and theory morphisms with identities and composition inherited from the category **Sign** of signatures form a category, which we will denote by $\mathbf{Th_{INS}}$ (omitting the index **INS** when no confusion arises).

Proposition 4.8. *Let $\sigma: \Sigma \to \Sigma'$ be a signature morphism, $\Phi \subseteq \mathbf{Sen}(\Sigma)$ a set of Σ-sentences, and $\Phi' \subseteq \mathbf{Sen}(\Sigma')$ a set of Σ'-sentences. Then the following conditions are equivalent:*

1. *σ is a theory morphism $\sigma: \langle \Sigma, Cl_\Sigma(\Phi) \rangle \to \langle \Sigma', Cl_{\Sigma'}(\Phi') \rangle$.*
2. *$\sigma(\Phi) \subseteq Cl_{\Sigma'}(\Phi')$.*
3. *For every $M' \in Mod_{\Sigma'}(\Phi')$, $M'|_\sigma \in Mod_\Sigma(\Phi)$.*

The above proposition allows us to define a functor $\mathbf{Mod_{Th}}: \mathbf{Th}_{\mathbf{INS}}^{op} \to \mathbf{Cat}$, which maps any theory $T = \langle \Sigma, \Phi \rangle$ to the category $\mathbf{Mod}(T)$ defined as the full subcategory of $\mathbf{Mod}(\Sigma)$ with the class of objects $|\mathbf{Mod}(T)| = Mod_\Sigma(\Phi)$, and any theory morphism $\sigma: T \to T'$ to (the appropriate restriction of) the σ-reduct functor $\_|_\sigma: \mathbf{Mod}(T') \to \mathbf{Mod}(T)$.

Many further standard properties of theories investigated in the realm of classical model theory may be formulated for an arbitrary institution as well. For instance, a presentation $\langle \Sigma, \Phi \rangle$ is *consistent* if it has a model (that is, $Mod_\Sigma(\Phi) \neq \emptyset$) and it is *complete* if it is a maximal consistent presentation (that is, if it is consistent and every presentation $\langle \Sigma, \Phi' \rangle$ where Φ' properly contains Φ is inconsistent).

Proposition 4.9. *Any complete presentation is a (consistent) theory.*

We have avoided giving proofs in the above presentation, but the reader is encouraged to check that they are just the same – and just as simple – as for equational or first-order logic presentations.

4.4 Institutions with composable signatures

One of the main topics of study in the area of formal software specification and development is modularity, as applied to both specifications and the software systems they specify. In both cases, one of the major practical issues is that of the size and complexity of objects one deals with. The only way to cope with these is by building objects (specifications and systems) in some systematic, structured way, out of simpler and hence easier-to-master components.

In the theory of specifications this was first identified in the seminal paper [BG77] on "putting theories together to make specifications". It was proposed there, and assumed as a standard in much subsequent work, see for instance Chapter 6, that the categorical concept of colimit provides an appropriate abstract basis for combining theories, and hence specifications, in a systematic way. However, if this is to work, the category of theories in the underlying institution has to be (finitely) cocomplete – which is not always the case. In such situations, when some special property of the underlying institution is needed, we suggested that an appropriate requirement should be identified and used in further work. In this case, the required property follows from a condition which is much easier to check.

Let $\mathbf{INS} = \langle \mathbf{Sign}, \mathbf{Sen}, \mathbf{Mod}, \langle \models_\Sigma \rangle_{\Sigma \in |\mathbf{Sign}|} \rangle$ be an arbitrary institution.

Theorem 4.10 (Burstall and Goguen [BG80]).
If the category **Sign** *of signatures in* **INS** *is cocomplete, then so is the category* **Th** *of theories in* **INS**.

Proof sketch. Let D be a diagram in **Th** with objects $D_n = \langle \Sigma_n, \Phi_n \rangle$ for $n \in N$. Let D' be the corresponding diagram in **Sign**, with objects $D'_n = \Sigma_n$ for $n \in N$. By the assumption, D' has a colimit, say $\langle \alpha_n \colon \Sigma_n \to \Sigma \rangle_{n \in N}$. Let $\Phi = Cl_\Sigma(\bigcup_{n \in N} \alpha_n(\Phi_n))$. Then $\langle \alpha_n \colon \langle \Sigma_n, \Phi_n \rangle \to \langle \Sigma, \Phi \rangle \rangle_{n \in N}$ is a colimit of D in **Th**. $\qquad\square$

The proof sketched above shows that in fact a stronger property holds: the category of theories has all the colimits that the category of signatures has, and moreover, the obvious functor mapping theories to their underlying signatures *reflects colimits*, cf. [GB92].

Corollary 4.11. *If the category* **Sign** *of signatures in* **INS** *is finitely cocomplete, then so is the category* **Th** *of theories in* **INS**.

Notice that the above theorem applies to *any* institution, regardless of the exact nature of its signatures, sentences, models, etc., and of the means used to construct it. It therefore provides a convenient and very general way of checking whether a logical system presented as an institution has a cocomplete category of theories. For instance, knowing that the category **AlgSig** of algebraic signatures is cocomplete, we immediately conclude that the categories **Th$_{EQ}$** and **Th$_{PEQ}$** of theories of equational and partial equational logic, respectively, are cocomplete. Similarly, to prove that the category **Th$_{FOEQ}$** of theories of first-order logic with equality is cocomplete, it is enough to check that the category **FOSig** of first-order signatures is cocomplete, which indeed is the case.

The importance of colimits, and in particular pushouts, for putting equational theories together is strongly supported by the fact that colimits of theories always have a counterpart construction at the level of models. For pushouts, this is presented in the form of the so-called Amalgamation Lemma (see, e.g., [EM85] and the general formulation below). Very roughly, this states that individual models of the component theories can be put together to form a model of the combined theory.

Example 4.12. Let Σ be an algebraic signature with one sort s, a constant $a: s$, and a unary operation $f: s \to s$. Let Σ_1 extend the signature Σ by a new constant $b: s$, let Σ_2 extend Σ by a new constant $c: s$, and let $\iota_1: \Sigma \to \Sigma_1$ and $\iota_2: \Sigma \to \Sigma_2$ be the algebraic signature inclusions. Then the pushout of ι_1 and ι_2 in the category **AlgSig** of algebraic signatures consists of the algebraic signature Σ' extending Σ by both $b: s$ and $c: s$ together with signature inclusions $\iota_1': \Sigma_1 \to \Sigma'$ and $\iota_2': \Sigma_2 \to \Sigma'$; informally, we think of Σ' as the result of putting together the two extensions Σ_1 and Σ_2 of their common part Σ.

Let then T be a Σ-theory in the institution **EQ** of equational logic, presented by a single equation $a = f(f(f(a)))$, T_1 be a Σ_1-theory presented by $\{f(a) = b, f(f(b)) = a\}$, and T_2 be a Σ_2-theory presented by $\{f(c) = a, f(f(a)) = c\}$. It is easy to check that $\iota_1: T \to T_1$ and $\iota_2: T \to T_2$ are theory morphisms in the category **Th$_{EQ}$** of theories in **EQ**. By the construction in the proof of Theorem 4.10 (of course, this can also be checked directly), their pushout is given as the Σ'-theory T' presented by $\{f(a) = b, f(f(b)) = a, f(c) = a, f(f(a)) = c\}$ (which is the same as presented by $\{f(a) = b, f(b) = c, f(c) = a\}$) together with theory morphisms $\iota_1': T_1 \to T'$ and $\iota_2': T_2 \to T'$.

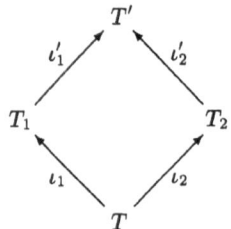

Consider a Σ_1-algebra $A_1 \in \mathbf{Alg}(\Sigma_1)$ and a Σ_2-algebra $A_2 \in \mathbf{Alg}(\Sigma_1)$, where $|A_1|_s = |A_2|_s = \{0,1,2\}$, $f_{A_1}(i) = f_{A_2}(i) = (i+1) \bmod 3$ for $i \in \{0,1,2\}$, $a_{A_1} = a_{A_2} = 1$, $b_{A_1} = 2$, and $c_{A_2} = 0$. Then A_1 is a model of T_1, A_2 is a model of T_2, and moreover, they share a common part $A_1|_{\iota_1} = A_2|_{\iota_2}$ which is a model of T. There exists then a unique Σ'-algebra $A' \in \mathbf{Alg}(\Sigma')$ such that $A'|_{\iota'_1} = A_1$ and $A'|_{\iota'_2} = A_2$, and moreover A' is a model of T'. The Amalgamation Lemma for the institution $\mathbf{EQ}$ of equational logic ensures that there is a bijective correspondence between models of T' and pairs of models of T_1 and T_2, respectively, that share a model of T as a common part. ∎

It is hardly surprising that the Amalgamation Lemma is a property which holds for some, but certainly not for all logical systems. Again, the "game" is to try to identify an elegant condition which we can impose on the institutions we want to consider and which would ensure that the Amalgamation Lemma holds.

An institution $\mathbf{INS}$ has *(finitely) composable signatures* if its category of signatures $\mathbf{Sign}$ is (finitely) cocomplete and its model functor $\mathbf{Mod}: \mathbf{Sign}^{op} \to \mathbf{Cat}$ is (finitely) continuous, mapping (finite) colimits in $\mathbf{Sign}$ to limits in $\mathbf{Cat}$.

Lemma 4.13 (Amalgamation Lemma).
Assume that an institution $\mathbf{INS}$ has finitely composable signatures. Consider a pushout in the category $\mathbf{Sign}$ of signatures:

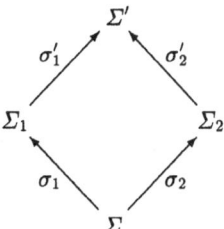

Then, for any two models $M_1 \in |\mathbf{Mod}(\Sigma_1)|$ and $M_2 \in |\mathbf{Mod}(\Sigma_2)|$ such that $M_1|_{\sigma_1} = M_2|_{\sigma_2}$, there exists a unique model $M' \in |\mathbf{Mod}(\Sigma')|$ such that $M'|_{\sigma'_1} = M_1$ and $M'|_{\sigma'_2} = M_2$.

Similarly, for any two model morphisms $f_1: M_{11} \to M_{12}$ in $\mathbf{Mod}(\Sigma_1)$ and $f_2: M_{21} \to M_{22}$ in $\mathbf{Mod}(\Sigma_2)$ such that $f_1|_{\sigma_1} = f_2|_{\sigma_2}$, there exists a unique model morphism $f': M'_1 \to M'_2$ in $\mathbf{Mod}(\Sigma')$ such that $f'|_{\sigma'_1} = f_1$ and $f'|_{\sigma'_2} = f_2$.

Proof sketch. First, notice that since the model functor $\mathbf{Mod}: \mathbf{Sign}^{op} \to \mathbf{Cat}$ is finitely continuous, it maps the above pushout of signatures to a pullback in the category $\mathbf{Cat}$ of the corresponding categories of models. Then the result follows from the construction of pullbacks in $\mathbf{Cat}$. □

The Amalgamation Lemma is equivalent to the requirement that the model functor **Mod** of **INS** maps pushouts in the category **Sign** of signatures to pullbacks in the category **Cat** of all categories. We will refer to institutions which satisfy this property as *institutions with semi-composable signatures*. This is weaker than the condition that the institution has finitely composable signatures, and in fact there are institutions where the model functor preserves pushouts, but does not preserve arbitrary (finite) colimits of signatures. Typical examples of institutions with semi-composable but not finitely composable signatures may be obtained by constructing single-sorted versions of institutions listed in Section 4.2.

Corollary 4.14 (Amalgamation Lemma for theories).
Let **INS** *be an institution with semi-composable signatures. Consider a pushout in the category* **Th** *of theories:*

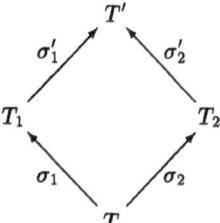

Then, for any two models $M_1 \in Mod(T_1)$ *and* $M_2 \in Mod(T_2)$ *such that* $M_1|_{\sigma_1} = M_2|_{\sigma_2}$, *there exists a unique model* $M' \in Mod(T')$ *such that* $M'|_{\sigma_1'} = M_1$ *and* $M'|_{\sigma_2'} = M_2$.

Proof sketch. The result follows from the Amalgamation Lemma for signatures (Lemma 4.13) by the construction of pushouts in **Th**, as sketched in the proof of Theorem 4.10, and then by the satisfaction condition for the underlying institution **INS**.   □

A similar argument leads to the following result:

Corollary 4.15. *Let* **INS** *be an institution with (finitely) composable signatures. Then the functor* $Mod_{Th}: Th^{op} \to Cat$, *mapping theories to their model categories and theory morphisms to the corresponding reduct functors, is (finitely) continuous.*

The above two corollaries give a formal basis for the methodology of software development which links modularization of software systems with the way the theories that specify them are (de)composed. Namely, in order to realize a specification given as the theory T' which consists essentially of two parts T_1 and T_2, sharing a common part T, it is enough to produce realizations M_1 of T_1 and M_2 of T_2, sharing a common part which realizes T, and then Corollary 4.14 ensures that M_1 and M_2 can be unambiguously put together to yield a realization M' of T'. We should stress again that this important property was derived here from a much simpler property for signatures in a way which does not depend on the details of the underlying institution at all.

4.5 Entailment relations

The notion of an institution as introduced in Definition 4.1 adopts a model-theoretic view of what logical systems are: the central notion is that of the satisfaction of a sentence by a model. Although we believe that this is indeed the most appropriate as a fundamental concept, in practical applications it is often most crucial to be able to prove properties of systems, either directly from the system definition or from its specification. Therefore, a more *proof-theoretic* view of logic is important as well. In a sense, this can be derived from the institutional perspective, and in fact much of the basic material presented in Section 4.3 where we introduced the notions of presentation, theory, semantic consequence, etc., leads in this direction. In this section we will start by taking the concept of *logical entailment* (or *consequence relation*, see [Avr92] for a survey) as the main topic of study.

An *entailment relation on a set S of sentences* is a relation $\vdash \subseteq \wp(S) \times S$ such that:

(Reflexivity) $\{\varphi\} \vdash \varphi$;
(Transitivity) If $\Phi_i \vdash \varphi_i$ for $i \in \mathcal{I}$ and $\Phi \cup \{\varphi_i\}_{i \in \mathcal{I}} \vdash \psi$, then $\Phi \cup \bigcup_{i \in \mathcal{I}} \Phi_i \vdash \psi$;
(Weakening) If $\Phi \vdash \psi$, then $\Phi \cup \Phi' \vdash \psi$,

where $\wp(S)$ is the power set of S, $\mathcal{I}$ is a set of indices, $\varphi, \psi, \varphi_i \in S$ are sentences, and $\Phi, \Phi', \Phi_i \subseteq S$ are sets of sentences, for $i \in \mathcal{I}$.

Semantic consequence relations, as defined for an arbitrary institution in Section 4.3 (just before Prop. 4.6), provide numerous important examples:

Proposition 4.16. *Let* **INS** $= \langle$**Sign, Sen, Mod,** $\langle \models_\Sigma \rangle_{\Sigma \in |\mathbf{Sign}|} \rangle$ *be an institution. For any signature* $\Sigma \in |\mathbf{Sign}|$, *the semantic consequence relation* $\models_\Sigma \subseteq \wp(\mathbf{Sen}(\Sigma)) \times \mathbf{Sen}(\Sigma)$ *satisfies the reflexivity, transitivity and weakening conditions, that is,* $\models_\Sigma$ *is an entailment relation on* $\mathbf{Sen}(\Sigma)$.

Another, perhaps more typical and expected source of examples is that of the usual natural deduction proof systems. Most roughly, given a set of sentences S, a natural deduction proof system P consists of a number of *proof rules* which schematically define elementary entailments. We say then that a sentence $\varphi \in S$ is derivable from a set $\Phi \subseteq S$ in a proof system P if φ can be obtained from Φ by application of (instances of) the proof rules of P, which we then write as $\Phi \vdash_P \varphi$ (as usual omitting the index P when no confusion can arise). In other words, $\vdash_P \subseteq \wp(S) \times S$ is the least entailment relation which contains all the elementary entailments determined by the rules in P. A simple example is the equational calculus as presented in Section 2.8.

Proposition 4.17. *For any algebraic signature $\Sigma \in |\mathbf{AlgSig}|$, the proof-the-oretic consequence relation $\vdash_\Sigma$ between sets of Σ-equations and Σ-equations is reflexive, transitive, and closed under weakening, that is, $\vdash_\Sigma$ is an entail-ment relation on $\mathbf{Sen_{EQ}}(\Sigma)$.*

One important topic of study is the mutual relationship between semantic and proof-theoretic consequence relations.

Let $\vdash \subseteq \wp(\mathcal{S}) \times \mathcal{S}$ and $\vdash' \subseteq \wp(\mathcal{S}) \times \mathcal{S}$ be two entailment relations on a common set $\mathcal{S}$ of sentences. We say that $\vdash$ is *sound* for $\vdash'$ if $\vdash \subseteq \vdash'$ and that $\vdash$ is *complete* for $\vdash'$ if $\vdash' \subseteq \vdash$. Thus, Theorem 2.21 states that in the equational logic, for any algebraic signature $\Sigma \in |\mathbf{AlgSig}|$, the proof-theoretic consequence relation $\vdash_\Sigma$ is sound and complete for the semantic consequence relation $\models_\Sigma$. This is in fact a typical example of the use of the concepts of soundness and completeness as introduced above: the semantic consequence relations of an institution are viewed as target entailment relations which we want to simulate using some proof-theoretic means. The soundness of proof-theoretic consequence for semantic consequence (if $\Phi \vdash_\Sigma \varphi$ then $\Phi \models_\Sigma \varphi$) states that whatever can be proved is indeed true; the completeness of proof-theoretic consequence for semantic consequence (if $\Phi \models_\Sigma \varphi$ then $\Phi \vdash_\Sigma \varphi$) states that whatever is true, can be proved.

Just as when introducing institutions (see Section 4.2) we argued that it is important to model the concept of a logical system incorporating a collection of signatures and taking proper account of their change, similarly here, we study not so much individual entailment relations, but rather their families indexed by signatures, again taking care of the possibility of moving from one signature to another.

Definition 4.18 (Entailment system [Mes89, HST94]).
An *entailment systems* **ES** consists of:

- a category $\mathbf{Sign_{ES}}$ of *signatures*;
- a functor $\mathbf{Sen_{ES}} : \mathbf{Sign_{ES}} \to \mathbf{Set}$, giving a set $\mathbf{Sen}(\Sigma)$ of Σ-*sentences* for each signature $\Sigma \in |\mathbf{Sign_{ES}}|$;
- an *entailment relation* $\vdash_{\mathbf{ES},\Sigma} \subseteq \wp(\mathbf{Sen_{ES}}(\Sigma)) \times \mathbf{Sen_{ES}}(\Sigma)$ for each sig-nature $\Sigma \in |\mathbf{Sign_{ES}}|$;

such that the entailment relations are preserved under translation of sen-tences with respect to signature morphisms, that is, for any signature mor-phism $\sigma \colon \Sigma \to \Sigma'$, Σ-sentence $\varphi \in \mathbf{Sen_{ES}}(\Sigma)$ and set $\Phi \subseteq \mathbf{Sen_{ES}}(\Sigma)$ of Σ-sentences,

    if $\Phi \vdash_\Sigma \varphi$ then $\mathbf{Sen_{ES}}(\sigma)(\Phi) \vdash_{\Sigma'} \mathbf{Sen_{ES}}(\sigma)(\varphi)$.   ◇

Here, as usual, $\mathbf{Sen_{ES}}(\sigma)(\Phi)$ is the image of Φ with respect to the function $\mathbf{Sen_{ES}}(\sigma)$. In the following, similarly as for institutions, the index **ES** will be omitted when no confusion can arise, and for any signature morphism $\sigma \colon \Sigma \to \Sigma'$, the function $\mathbf{Sen}(\sigma) \colon \mathbf{Sen}(\Sigma) \to \mathbf{Sen}(\Sigma')$ will be denoted simply by σ and referred to as the σ-*translation* of Σ-sentences to Σ'-sentences.

Very informally, the requirement that entailment relations in an entailment system are preserved under translation along signature morphisms states that the entailment relations in the system have been defined in some way uniformly with respect to the symbols in signatures and their renamings as incorporated in the notion of signature morphisms. For entailment systems defined proof-theoretically, this typically requires that the proof rules provided over particular signatures are closed under translation along signature morphisms. For entailment systems defined via model-theoretic satisfaction, this is given directly by the institutional satisfaction condition:

Corollary 4.19. *Let* $\mathbf{INS} = \langle \mathbf{Sign}, \mathbf{Sen}, \mathbf{Mod}, \langle \models_\Sigma \rangle_{\Sigma \in |\mathbf{Sign}|} \rangle$ *be an institution. Then*

$$\mathbf{ES_{INS}} = \langle \mathbf{Sign}, \mathbf{Sen}, \langle \models_\Sigma \rangle_{\Sigma \in |\mathbf{Sign}|} \rangle$$

is an entailment system, where for $\Sigma \in |\mathbf{Sign}|$, $\models_\Sigma \subseteq \wp(\mathbf{Sen}(\Sigma)) \times \mathbf{Sen}(\Sigma)$ *is the semantic consequence relation in* $\mathbf{INS}$.

Proof sketch. For $\Sigma \in |\mathbf{Sign}|$, $\models_\Sigma \subseteq \wp(\mathbf{Sen}(\Sigma)) \times \mathbf{Sen}(\Sigma)$ is indeed an entailment relation by Proposition 4.16. It is preserved under translation along signature morphisms by Proposition 4.6 (and hence essentially by the satisfaction condition for $\mathbf{INS}$). □

Given an entailment system $\mathbf{ES} = \langle \mathbf{Sign}, \mathbf{Sen}, \langle \vdash_\Sigma \rangle_{\Sigma \in |\mathbf{Sign}|} \rangle$, a $(\Sigma\text{-})pres$-*entation* in $\mathbf{ES}$ is a pair $\langle \Sigma, \Phi \rangle$ where $\Sigma \in |\mathbf{Sign}|$ is a signature and $\Phi \subseteq \mathbf{Sen}(\Sigma)$ is a set of Σ-sentences. The *closure* of a set Φ of Σ-sentences in $\mathbf{ES}$ is the set $Cl_\Sigma(\Phi) = \{\varphi \in \mathbf{Sen}(\Sigma) \mid \Phi \vdash_\Sigma \varphi\}$; Φ is *closed* in $\mathbf{ES}$ if $\Phi = Cl_\Sigma(\Phi)$. A $(\Sigma\text{-})theory$ in $\mathbf{ES}$ is a presentation $\langle \Sigma, \Phi \rangle$ such that Φ is closed. For any theories $\langle \Sigma, \Phi \rangle$ and $\langle \Sigma', \Phi' \rangle$, a *theory morphism* $\sigma \colon \langle \Sigma, \Phi \rangle \to \langle \Sigma', \Phi' \rangle$ is a signature morphism $\sigma \colon \Sigma \to \Sigma'$ such that $\sigma(\varphi) \in \Phi'$ for every $\varphi \in \Phi$. Theories and theory morphisms with identities and composition inherited from the category $\mathbf{Sign}$ of signatures form a category, which we will denote by $\mathbf{Th_{ES}}$ (omitting the index $\mathbf{ES}$ when no confusion arises).

The above terminology for entailment systems deliberately coincides with the terminology introduced in Section 4.3 for institutions: the institutional concepts of presentation, closure, theory, and theory morphism for an institution $\mathbf{INS}$ coincide with those introduced above for the semantic entailment system $\mathbf{ES_{INS}}$ determined by $\mathbf{INS}$ as in Corollary 4.19.

Some simple facts about institutional theories and their morphisms, which therefore hold for semantic entailment systems determined by institutions, generalize to arbitrary entailment systems. For example:

Proposition 4.20. *Consider an arbitrary entailment system* $\mathbf{ES} = \langle \mathbf{Sign}, \mathbf{Sen}, \langle \vdash_\Sigma \rangle_{\Sigma \in |\mathbf{Sign}|} \rangle$. *Let* $\sigma \colon \Sigma \to \Sigma'$ *be a signature morphism*, $\Phi \subseteq \mathbf{Sen}(\Sigma)$ *a set of* Σ-sentences, *and* $\Phi' \subseteq \mathbf{Sen}(\Sigma')$ *a set of* Σ'-sentences.

- *If* Φ' *is a closed set of* Σ'-sentences, *then* $\sigma^{-1}(\Phi')$ *is a closed set of* Σ-sentences.

- *The following conditions are equivalent:*
    1. σ *is a theory morphism* $\sigma \colon \langle \Sigma, Cl_{\Sigma}(\Phi) \rangle \to \langle \Sigma', Cl_{\Sigma'}(\Phi') \rangle$.
    2. $\sigma(\Phi) \subseteq Cl_{\Sigma'}(\Phi')$.

The most typical situation in the practice of formal software specification and development is that we have a logical system which includes both model theory, that is, a notion of model for each signature and a satisfaction relation between models and sentences – an institution – and proof theory, presenting an entailment relation on sentences for each signature – an entailment system. The model theory is then used as a basis for semantic definitions of the specification formalism, together with the related methodological concepts of refinement and implementation, thus providing an ultimate reference for the correctness of methods and tools provided for the formalism. The proof system and any extensions that might be required are then used in the development process to prove the verification conditions inherent in the semantic definitions. Ideally, we would like to achieve the situation where the proof system would be capable of deriving only true facts and all true facts; that is, it should be sound and complete for the semantic consequence relation determined by the institution. Unfortunately, in many practically important cases this is not achievable. We cannot give up safety, though, so the proof system is always required to be sound. This leads to the final definition of this section, putting together the concepts of institution and of entailment system.

Definition 4.21 (General logic [Mes89]).

A *general logic* **GL** is a system $\langle \mathbf{Sign}, \mathbf{Sen}, \mathbf{Mod}, \langle \models_{\Sigma} \rangle_{\Sigma \in |\mathbf{Sign}|}, \langle \vdash_{\Sigma} \rangle_{\Sigma \in |\mathbf{Sign}|} \rangle$ where

- $\mathbf{INS_{GL}} = \langle \mathbf{Sign}, \mathbf{Sen}, \mathbf{Mod}, \langle \models_{\Sigma} \rangle_{\Sigma \in |\mathbf{Sign}|} \rangle$ is an institution; and
- $\mathbf{ES_{GL}} = \langle \mathbf{Sign}, \mathbf{Sen}, \langle \vdash_{\Sigma} \rangle_{\Sigma \in |\mathbf{Sign}|} \rangle$ is an entailment system

such that for each signature $\Sigma \in |\mathbf{Sign}|$, the entailment relation $\vdash_{\Sigma}$ is sound for the semantic consequence relation $\models_{\Sigma}$ of the institution $\mathbf{INS_{GL}}$.   ◇

Somewhat marginally for the main topic of this volume, we would like to end this section with some remarks which shed an interesting light on the relationship between views of logic based on satisfaction and on entailment, respectively. Recall that at the end of Section 4.2 we mentioned a construction of an institution out of a model functor only, where sentences are given simply as sets of models. A dual construction of models as sets of sentences is also possible. More interestingly, however, there is a natural construction of an institution out of an entailment system (see [Mes89]):

Let $\mathbf{ES} = \langle \mathbf{Sign}, \mathbf{Sen}, \langle \vdash_{\Sigma} \rangle_{\Sigma \in |\mathbf{Sign}|} \rangle$ be an entailment system. For any signature $\Sigma \in |\mathbf{Sign}|$, define a category $\mathbf{Mod_{ES}}(\Sigma)$ of Σ-"models" so that[4]

---

[4] This is in fact a very simple case of the *comma category* construction, see, e.g., [Mac71].

- the objects of $\mathbf{Mod_{ES}}(\Sigma)$ are pairs $\langle\langle \Sigma', \Phi'\rangle, \iota\rangle$, where $\langle \Sigma', \Phi'\rangle$ is a theory (in **ES**) and $\iota\colon \Sigma \to \Sigma'$ is a signature morphism;
- a morphism from $\langle\langle \Sigma_1, \Phi_1\rangle, \iota_1\colon \Sigma \to \Sigma_1\rangle$ to $\langle\langle \Sigma_2, \Phi_2\rangle, \iota_2\colon \Sigma \to \Sigma_2\rangle$ in $\mathbf{Mod_{ES}}(\Sigma)$ is a theory morphism $\sigma\colon \langle \Sigma_1, \Phi_1\rangle \to \langle \Sigma_2, \Phi_2\rangle$ such that $\iota_1; \sigma = \iota_2$.

Then, for any signature morphism $\sigma\colon \Sigma_0 \to \Sigma$, let $\mathbf{Mod_{ES}}(\sigma)\colon \mathbf{Mod_{ES}}(\Sigma) \to \mathbf{Mod_{ES}}(\Sigma_0)$ be the obvious functor given by precomposition with σ. This yields a functor $\mathbf{Mod_{ES}}\colon \mathbf{Sign}^{op} \to \mathbf{Cat}$. For any signature $\Sigma \in \mathbf{Sign}$, Σ-sentence $\varphi \in \mathbf{Sen}(\Sigma)$ and Σ-"model" $\langle\langle \Sigma', \Phi'\rangle, \iota\colon \Sigma \to \Sigma'\rangle \in |\mathbf{Mod_{ES}}(\Sigma)|$, define $\langle\langle \Sigma', \Phi'\rangle, \iota\colon \Sigma \to \Sigma'\rangle \models_{\mathbf{ES},\Sigma} \varphi$ to stand for $\iota(\varphi) \in \Phi'$. Then

$$\mathbf{INS_{ES}} = \langle \mathbf{Sign}, \mathbf{Sen}, \mathbf{Mod_{ES}}, \langle\models_{\mathbf{ES},\Sigma}\rangle_{\Sigma\in|\mathbf{Sign}|}\rangle$$

is an institution. Moreover, the semantic entailment system $\mathbf{ES_{INS_{ES}}}$ determined by this institution is just **ES**.

The construction sketched above gives in some sense an equivalence between entailment systems and a large class of institutions, which shows that at this very abstract level the views of logic based on satisfaction and model theory on one hand and on entailment and proof theory on the other hand are closer to each other than one usually tends to think.

4.6 Specifications in an arbitrary institution

A logical system underlying a specification framework must incorporate the following components, as indeed included in any institution:

- signatures: informally, each signature provides a vocabulary of a software system, the set of names for data types, operations, procedures, etc., the system offers for the user;
- models: informally, the meaning of each software system, via a formal semantics of the underlying programming language, forms a particular model of the logic;
- sentences with the notion of their satisfaction in models: informally, each sentence describes a property of models, and hence of software systems they represent.

Thus, an arbitrary institution

$$\mathbf{INS} = \langle \mathbf{Sign}, \mathbf{Sen}, \mathbf{Mod}, \langle\models_\Sigma\rangle_{\Sigma\in|\mathbf{Sign}|}\rangle$$

– let's consider it fixed throughout this section – provides a basic framework in which one can build specifications of software systems semantically represented by models of **INS**. Of course, the basic building blocks of specifications in any specification formalism are presentations (cf. Section 4.3), where a system's signature and a set of sentences to be satisfied by the system are given. Typically, however, a specification formalism will offer much more: tools to build complex specifications in a modular way, to specify parameterized systems, to enhance the specification power of the underlying logic exploiting some particular details of a class of institutions, etc. The most trivial examples are given below; see, for instance, Chapters 5, 6, and 8 for more elaborate examples and discussion, and [ST] for a more complete presentation of the "institutional" view of software specification we briefly advocate here.

Whatever exactly the specifications are – in any specification formalism they will be syntactic objects of some kind – their ultimate role is to describe a class of software systems which are permissible realizations of the specification. If we agree that software systems are adequately represented as models of the institution, which is the basic assumption for the developments presented here, then this means that ultimately a specification must describe a signature and a class of models over this signature, called the *models of the specification* (see [ST] for further, more detailed argument in this direction). Consequently, any specification formalism over **INS** will determine a class of specifications *Spec*, and then, for any specification $SP \in Spec$, its signature $Sig[SP] \in |\mathbf{Sign}|$ and the collection of its models $Mod[SP] \subseteq |\mathbf{Mod}(Sig[SP])|$. If $Sig[SP] = \Sigma$, we will sometimes call SP a Σ-specification and denote the collection of all Σ-specifications by $Spec_\Sigma$.

Even at this level of abstraction, working within an arbitrary institution, it is possible to provide a few rudimentary ways of building specifications. For example:

- Any presentation $\langle \Sigma, \Phi \rangle$ is a specification with the following semantics:
  $Sig[\langle \Sigma, \Phi \rangle] = \Sigma$
  $Mod[\langle \Sigma, \Phi \rangle] = Mod_\Sigma(\Phi) = \{M \in |\mathbf{Mod}(\Sigma)| \mid M \models \Phi\}$
  This directly imposes some requirements on the models, as captured by the axioms of the presentation.
- For any signature $\Sigma \in |\mathbf{Sign}|$, given any Σ-specifications SP_1 and SP_2, their union $SP_1 \cup SP_2$ is a specification with the following semantics:
  $Sig[SP_1 \cup SP_2] = \Sigma$
  $Mod[SP_1 \cup SP_2] = Mod[SP_1] \cap Mod[SP_2]$
  This combines the requirements imposed on models by the specifications SP_1 and SP_2.
- For any signature morphism $\sigma: \Sigma \to \Sigma'$ and Σ-specification SP, **translate** SP **by** σ is a specification with the following semantics:
  $Sig[\textbf{translate } SP \textbf{ by } \sigma] = \Sigma'$
  $Mod[\textbf{translate } SP \textbf{ by } \sigma] = \{M' \in |\mathbf{Mod}(\Sigma')| \mid M'|_\sigma \in Mod[SP]\}$
  Typically, this renames symbols in the old signature of the specification without changing the way in which their interpretation in models is constrained, and adds new symbols to the signature without constraining in any way their interpretation in the models.
- For any specification morphism $\sigma: \Sigma \to \Sigma'$ and Σ'-specification SP', **derive from** SP' **by** σ is a specification with the following semantics:
  $Sig[\textbf{derive from } SP' \textbf{ by } \sigma] = \Sigma$
  $Mod[\textbf{derive from } SP' \textbf{ by } \sigma] = \{M'|_\sigma \mid M' \in Mod[SP']\}$

This hides some symbols of the signature, and hence removes their interpretation from the models, and renames the remaining symbols without changing the way in which their interpretation in models is constrained.

We have in fact defined above a number of operations on specifications (union, translate, derive) – we will refer to such operations as *specification-building operations* – which semantically amount to certain functions on classes of models. The above operations, although extremely simple, already provide flexible mechanisms for expressing basic ways of putting specifications together and so for building specifications in a structured manner. Further, perhaps rather more convenient-to-use operations may be built on this basis. See for instance [ST88a] for examples, definitions of other similarly abstract operations, and much further discussion.

The semantics of specifications provides a natural way to equip the class of specifications with the structure of a category. The key idea is to notice that one of the conditions given in Proposition 4.8 can easily be expressed here.

Given two specifications $SP, SP' \in Spec$, by a specification morphism $\sigma \colon SP \to SP'$ we mean a signature morphism $\sigma \colon Sig[SP] \to Sig[SP']$ in **Sign** such that for all models $M' \in Mod[SP']$, we have $M'|_\sigma \in Mod[SP]$. This yields a category **Spec** of specifications and their morphisms, with composition and identities inherited from **Sign**.

Proposition 4.22. *If the category* **Sign** *of signatures is finitely cocomplete and the class of specifications is closed under the translate and union operations, then the category* **Spec** *is finitely cocomplete as well.*

Proof sketch. The proof of Theorem 4.10 may easily be adapted: just use the translate operation to translate component specifications to the combined signature, and then the union operation to combine the translated specifications there.   □

In fact, if the class of specifications is closed under the translate and union operations then, just as before, the category **Spec** has all the finite colimits that the category of signatures has, and the obvious functor mapping specifications to their signatures reflects colimits. To generalize the above result to arbitrary (infinite) colimits, a more general form of the union operation is needed, to allow one to form the union of an arbitrary set of specifications over the same signature.

There is also an obvious functor $\mathbf{Mod_{Spec}} \colon \mathbf{Spec}^{op} \to \mathbf{Cat}$, which for any specification $SP \in |\mathbf{Spec}|$ yields the full subcategory $\mathbf{Mod_{Spec}}(SP)$ of $\mathbf{Mod}(Sig[SP])$ with the class of objects $Mod[SP]$ and maps any specification morphism to the corresponding reduct functor. Similarly as for theories, we can prove here the Amalgamation Lemma for specifications and a more general continuity result for the $\mathbf{Mod_{Spec}}$ functor.

Corollary 4.23 (Amalgamation Lemma for specifications).
*Let **INS** be an institution with semi-composable signatures, and let the class of specifications be closed under the union and translate operations. Consider a pushout in the category **Spec** of specifications:*

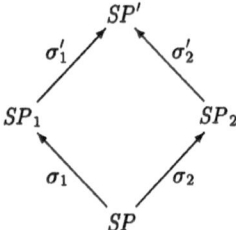

Then, for any two models $M_1 \in Mod[SP_1]$ and $M_2 \in Mod[SP_2]$ such that $M_1|_{\sigma_1} = M_2|_{\sigma_2}$, there exists a unique model $M' \in Mod[SP']$ such that $M'|_{\sigma_1'} = M_1$ and $M'|_{\sigma_2'} = M_2$.

Corollary 4.24. *If the institution **INS** has finitely composable signatures and the class of specifications is closed under the union and translate operations, then the functor $\mathbf{Mod_{Spec}} \colon \mathbf{Spec}^{op} \to \mathbf{Cat}$ is finitely continuous.*

In fact, the functor $\mathbf{Mod_{Spec}} \colon \mathbf{Spec}^{op} \to \mathbf{Cat}$ adequately describes many important features of the specification formalism. An indexed category, which such a functor in fact forms, cf. [TBG91], may be taken as a basic notion to underly the study and development of some of the theory of specifications, even more general and less detailed than an institution – see the concept of *specification frame*, or *specification logic* in [EBC91].

The semantics of specifications also enables us to introduce the concepts of semantic consequence and of theory for specifications.

Given a specification $SP \in Spec$ with $Sig[SP] = \Sigma$ and a Σ-sentence $\varphi \in \mathbf{Sen}(\Sigma)$, we say that φ is a *semantic consequence* of SP, written $SP \models_\Sigma \varphi$, if $Mod[SP] \models \varphi$ (that is, if every model of SP satisfies φ). Then, by the *theory* of SP we mean $Th[SP] = Th_\Sigma(Mod[SP])$. Notice that these notions are proper generalizations of semantic consequence and of the theory of a presentation in the underlying institution, as defined in Section 4.3.

We can also generalize the developments of Section 4.5 and study entailment between specifications and sentences. The most important task here is to develop proof-theoretic entailment relations $\vdash_\Sigma \subseteq Spec_\Sigma \times \mathbf{Sen}(\Sigma)$ that would soundly approximate semantic consequences $\models_\Sigma \subseteq Spec_\Sigma \times \mathbf{Sen}(\Sigma)$ for $\Sigma \in |\mathbf{Sign}|$. Of course, any such proof system would have to somehow incorporate a proof system for the underlying institution. This is captured by the following rule (we omit the index Σ on entailment relations, for brevity):

$$\frac{\text{for } i \in \mathcal{I}, SP \vdash \varphi_i \qquad \{\varphi_i\}_{i \in \mathcal{I}} \vdash \varphi}{SP \vdash \varphi}$$

Then, there would be a number of rules for the specification-building operations introduced in the formalism; for instance:

$$\frac{SP_1 \vdash \varphi}{SP_1 \cup SP_2 \vdash \varphi} \qquad \frac{SP_2 \vdash \varphi}{SP_1 \cup SP_2 \vdash \varphi}$$

$$\frac{SP \vdash \varphi}{\textbf{translate } SP \textbf{ by } \sigma \vdash \sigma(\varphi)}$$

$$\frac{SP \vdash \sigma(\varphi)}{\textbf{derive from } SP \textbf{ by } \sigma \vdash \varphi}$$

Soundness and *completeness* of such systems can be studied, under the usual definition of these concepts. For example, when added to a sound proof system for the underlying institution, the above proof rules yield a sound proof system for consequences of specifications. Unfortunately, completeness does not carry over (it does for institutions with *interpolation property* [Cen94, Bor98], and for arbitrary institutions for the very limited language of specifications built from presentations using the union and translate operations only). In fact, it can be shown that there is no similarly general, compositional proof system for specification formalisms which include all three operations presented above. The problem is that the derive operation may yield a specification denoting a class of models which is not definable by a theory of the underlying logic.

Example 4.25. [ST97].
Consider the institution **EQ** of equational logic. Let Σ_0 be an algebraic signature with two sorts s, s' and three constants $a: s$ and $b, c: s'$, and let $SP_0 = \langle \Sigma_0, \emptyset \rangle$. Let then Σ_1 be the algebraic signature resulting from Σ_0 by removing the constant a, $\iota: \Sigma_1 \to \Sigma_0$ be the algebraic signature inclusion, and let $SP_1 = \textbf{derive from } SP_0 \textbf{ by } \iota$. Finally, let $SP = SP_1 \cup \langle \Sigma_1, \{\forall x{:}s.\, b = c\} \rangle$. In a somewhat more intuitive notation, this may be presented as follows:

$$SP \left\{ \begin{array}{l} \textbf{enrich} \\ SP_1 \left\{ \begin{array}{l} \textbf{hide opns } a \textbf{ in} \\ SP_0 \left\{ \begin{array}{l} \textbf{sorts } s, s' \\ \textbf{opns } a : s \\ \qquad\quad b, c : s' \end{array} \right. \end{array} \right. \\ \textbf{by axioms } \forall x{:}s.\, b = c \end{array} \right.$$

(**hide** is a special case of the derive operation and a general form of **enrich** may be expressed using the translate – not needed here – and union operations.)

The example relies on the following well-known fact [GM85]: $\forall x{:}s.\, b = c$ does not imply $b = c$, although it implies $b = c$ for Σ_1-algebras with non-empty carrier of sort s, see Section 2.8.

Now, $Mod[SP_0]$ is the class of all Σ_0-algebras and $Mod[SP_1]$ consists of all Σ_1-algebras that are reducts of Σ_0-algebras, obtained by removing the constant a. Consequently, $Mod[SP_1]$ contains all and only those Σ_1-algebras that have a non-empty carrier of sort s. Then, selecting from $Mod[SP_1]$ the algebras that satisfy $\forall x{:}s.\, b = c$ yields the class $Mod[SP]$ – and all these algebras satisfy $b = c$ (since for the algebras in $Mod[SP_1]$, $b = c$ follows from $\forall x{:}s.\, b = c$). Thus, $SP \models b = c$.

On the other hand, the theory of SP_0 is trivial, and so is the theory of SP_1 (there are no equations capable of expressing the fact that a carrier is non-empty). Thus, the additional axiom $\forall x{:}s.\, b = c$ in the context of the theory of SP_1 does not entail the equation $b = c$.   ∎

The above example shows that it is indeed impossible to give a complete proof system where the properties of a specification would be derived directly from the properties of its immediate components only. A non-compositional proof system for the specification formalism involving presentations and the union, translate and derive operations, based on the idea that before deriving consequences of a specification one can "massage" it into some normal form, can be found in [Wir93, Far92]. The transformation of specifications required in such systems is typically justified by simple algebraic laws, which can be derived directly from the semantics of the specification-building operations, for instance:

$$Mod[\textbf{translate } (SP_1 \cup SP_2) \textbf{ by } \sigma] =$$
$$Mod[(\textbf{translate } SP_1 \textbf{ by } \sigma) \cup (\textbf{translate } SP_2 \textbf{ by } \sigma)].$$

Identities like the one above, or inclusions like:

$$Mod[\textbf{derive from } (SP_1 \cup SP_2) \textbf{ by } \sigma] \subseteq$$
$$Mod[(\textbf{derive from } SP_1 \textbf{ by } \sigma) \cup (\textbf{derive from } SP_2 \textbf{ by } \sigma)],$$

would also be used in a proof calculus for yet another generalized form of entailment, needed to show that constraints imposed on models by one specification are stricter than those imposed by another, which is typically an essential part of proving correctness of program development steps. For two Σ-specifications SP and SP' this can be written as $SP \models_\Sigma SP'$, which is just another way, perhaps more appealing in this context, of writing $Mod[SP] \subseteq Mod[SP']$. Developing a corresponding proof calculus is a topic of ongoing research, with [Wir93] and [Far92] providing important contributions. For further discussion of various aspects of proof systems for (structured and unstructured) specifications, see Chapters 10 and 11.

4.7   Further remarks

The basic idea behind the notion of institution was first used by Burstall and Goguen in the semantics of the CLEAR specification language [BG80], see the concept of a "language" there. In the form presented here it was formulated in [GB84], cf. also [GB92]. Various minor modifications of the notion of institution have been used by the same authors, for example considering sets rather than categories of models and categories rather than sets of sentences (with sentence morphisms modeling some form of entailment in the latter case). Mayoh [May85] proposed a generalization of the concept to a situation where a category of logical values, rather than just the two element set of boolean values, provides meanings for satisfaction of sentences in models; this was later reformulated in [GB86, ST]. A careful analysis of the satisfaction condition in the definition of institution, resulting in a number of somewhat weaker notions, was given in [SS93].

Similar ideas about an abstract view of logics may be found in work on so-called abstract model theory [Bar74, BF85], but the goals and methods used there are quite different. It is possible to develop some "very abstract" model theory in the framework of an arbitrary institution, see for instance [Tar86a, ST] (and some definitions towards the end of Section 4.3) for hints in this direction.

Most of the simple material on presentations, theories and semantic consequence in Section 4.3 comes from the original paper on institutions [GB84, GB92], which also contained the most standard examples of institutions; see [ST] for further details and examples, including some of a rather non-standard character. The role of the Amalgamation Lemma has been emphasized in the field of algebraic specifications in connection with so-called parameterized specifications, see [EM85] for a monographic presentation. It was formulated in the framework of an arbitrary institution and linked to the continuity of the model functor in [ST88a] and explored in [Tar85, Tar86b, Tar86a], see also the property of "exactness" in [DGS93].

The concept of an entailment relation is of course rather standard in work on proof theory, but in this context it was perhaps first introduced by Meseguer [Mes89] (cf. also [HST94]) where the notions of entailments sytem and of general logic were proposed as well. Fiadeiro and Sernadas [FS88] introduced the concept of π-institution as a formal counterpart of the notion of a logical system based on a proof-theoretic view, with the closure operator as the central notion. Of course, a proof-theoretic view of logical systems is inherent in much work on general logical frameworks based on various type theories such as Edinburgh LF [HHP93].

The model-theoretic semantics for specifications in the style given here was first explicitly advocated in work on the ASL specification language, where also operations like those defined in Section 4.6 were introduced in the context of a specific logical formalism [SW83, Wir86]. This was then lifted to an arbitrary institution in [ST88a]. The proof rules for the specification-building operations were given in [ST88a] as well, based on earlier work in [SB83]; see [HST94] for further analysis of proof search in this context, and [Wir93, Far92, Cen94, Bor98] for work on proofs for structured specifications. These topics are presented in more detail in Chapter 11.

One important idea we have not touched on here at all is that once the general notion of logical system is formalized as an institution, we can further equip it with a notion of *institution morphism* [GB92] and study the resulting category of institutions. This category was proved to be complete in [Tar86a], cf. also [TBG91], thus potentially providing a framework for building logical systems in a structured manner. For this, however, it seems more appropriate to work with some presentations of institutions, as given in [GB86] in the form of *parchments*, introduce parchment morphisms and use the existing limits in the resulting category [ST, Mos96b, MPT97]. Another intriguing possibility is to employ institution morphisms (or *semi-morphisms*) to build specifications which span a number of institutions, to move specifications from one institution to another, to implement specifications from one institution by specifications from another, etc., see [ST88b, Tar96, ST]. It is also possible to transfer proof systems between institutions linked by institution morphisms, see [GB92, ST], although perhaps for this purpose other concepts such as institution maps [Mes89, CM97], simulations [AC93] or representations [Tar87] are more appropriate. An overview of the related problems and results is given in [Tar96].

5 Specification Semantics

Horst Reichel

TU Dresden, Fakultät Informatik, Institut für Theoretische Informatik
D-01062 Dresden, Germany
reichel@tcs.inf.tu-dresden.de
http://wwwtcs.inf.tu-dresden.de/ALG/english-index.html

The most basic assumption of work on algebraic specification is that software systems are modeled as many-sorted algebras, abstracting from the concrete details of algorithms and code, and focusing on their functional behavior.

This assumption raises the question of whether a software system should be specified by a single *prototype algebra* or by a class of *algebras with equal external behavior*. What external behavior means precisely has to be defined by the semantics of the specification language. In both cases the similarity type of the algebras, called a signature, gives some very abstract information: the names of the data types and the names, the arities, and result sorts of the operations defined by the system. A single prototype algebra has the advantage that it supplies a constructive understanding of the behavior of the specified software system. For the description of the prototype algebra we can use very abstract mathematical terms to obtain both a precise and abstract description.

However, every single model has the serious disadvantage that only some properties of its external behavior are clearly visible. Some other interesting properties may be very hard to detect in the given model but much easier in another one with the same external behavior. This remains true if one considers isomorphic algebras as algebras with equal external behavior. Isomorphism is the narrowest interpretation of equal external behavior which will be discussed in the following in more detail.

We compare the quality of two different prototype algebras, considering the example of term algebras introduced in Section 2.4. The carrier $|T_\Sigma(X)|$ could be defined as the least S-sorted set of words such that $x \in |T_\Sigma(X)|_s$ for all $s \in S$ and $x \in X_s$, and $t_1 \ldots t_n f \in |T_\Sigma(X)|_s$ for all $f : s_1 \times \ldots \times s_n \to s$ in Σ and $t_1 \in |T_\Sigma(X)|_{s_1}, \ldots, t_n \in |T_\Sigma(X)|_{s_n}$. The interpretation of an operation name now becomes the appending of the operation name to the end of the argument sequence. This representation of terms is sometimes called postfix polish notation. It is a very compact representation of terms, without any technical symbols, and is very useful for evaluating terms by means of a stack machine. But for a human reader this prototype algebra of terms is not very convenient. Human readers prefer the representation with technical symbols, as in Section 2.4, which represents more clearly the tree-like structure of terms.

Using a class of algebras opens up the possibility of defining the class axiomatically, where the axioms express the characteristic properties that ensure the same external behavior of all algebras in the class.

In the following we will investigate the precise kinds of algebra classes and axioms which may appear in algebraic specifications, and the formal meaning of algebras displaying the same external behavior.

If we focus on the axiomatic specification only, then this does not imply that specifications based on a single prototype algebra are not worth considering.

So far we have argued that the semantics of a specification is in any case a class of algebras that may consist of a single algebra. But from a more general point of view there are also other possibilities of defining the meaning of specifications.

As suggested in [ST97] one may attempt to provide a semantics of specifications on (at least) three different levels:

- *Presentation semantics*: a specification denotes a signature and a finite, or at least recursive or recursively enumerable, set of axioms over this signature. In this case semantics is defined by purely syntactic means.
- *Theory semantics*: a specification denotes a signature and a set of axioms over this signature which is closed under logical consequences. In this case the meaning of a specification is no longer strictly syntactic. In addition to the presentation semantics, theory semantics performs the closure under logical consequence.
- *Model-class semantics*: a specification denotes a signature and a class of algebras over this signature.

Since in any logical system with a well-defined model theory there is a natural mapping from representations to theories and from theories to model classes, ultimately specifications determine classes of algebras.

For any signature Σ, there is a canonic translation from classes of Σ-algebras to sets of Σ-axioms and vice versa, which associates with each set of axioms the class of algebras that satisfy them, and in the other direction each class of algebras with the set of axioms that hold in them. There is a one-to-one correspondence between the class of sets of axioms closed under logical consequence and the class of suitably closed classes of algebras (i.e., definable by sets of axioms). This implies that the theory semantics as a semantic domain for specifications is less expressive than either the presentation semantics or the model-class semantics, whereas presentation semantics and model-class semantics are incomparable.

We agree with the intention of [ST97], considering the model-class semantics as the most convenient formal semantics of specifications, and will only be considering with different kinds of model-class semantics in this chapter.

In the first section we discuss the formalization of algebras with equal external behaviors by isomorphic algebras. Central concepts in this section are initiality and finality. Both concepts are higher-order concepts that allow

definitions of isomorphism classes. By means of examples we show that initial
algebras can be used to describe generation schemata of infinite sets whereas
final algebras represent schemata of observations and the behavior of systems.

Since isomorphism classes are not convenient in all cases, especially in the
early stages of software design, we discuss in the second section different kinds
of loose semantics, i.e., classes of algebras closed under isomorphism, but any
two algebras of a specified class do not have to be isomorphic. Basic concepts
include loose semantics, where a specified class consists of all algebras satisfying the set of axioms used, reachable loose semantics, where algebras with
junk elements, i.e., elements which are not representable as values of terms
over the signature used, are excluded from the specified classes, and finally
constructor-based loose semantics, where the representability of no-junk elements with respect to a distinguished subset of so-called constructors of the
set of operations is required.

The third section deals with the specification of parameterized objects.
This generalizes the semantics of monomorphic data types and systems specifications to parameterized polymorphic specifications. This step is essential
for breaking down large specifications in manageable pieces. The combination
of loose, and initially or finally constraint parts in specifications improves the
expressibility considerably.

The fourth section investigates in detail observable equivalence of algebras
on the one hand and observable equivalence of internal states of systems on
the other. These concepts are very useful for specifications of systems with
internal states and state transitions, in contrast to specifications of functions
and types of complicated structured data values.

The final section sketches a different formal approach to data and systems
specifications which is based on recursive type equations. It turns out that
least solutions are related to initial algebras and greatest solutions to final
coalgebras. Coalgebras are suitable mathematical formalizations of reactive
systems and their behavior.

5.1   Isomorphism class semantics

A first attempt to formalize the concept of equal functional behavior of algebras is the notion of *isomorphic algebras*, see Chapter 2. Intuitively two
isomorphic Σ-algebras differ only in the denotation of their elements. If A and
B are isomorphic then there is a bijective encoding $h : |A| \to |B|$ compatible
with the operations of A and B, i.e., for all $f : s_1 \times \ldots \times s_n \to s$ in Σ and
$a_1 \in |A|_{s_1}, \ldots, a_n \in |A|_{s_n}, h_s(f_A(a_1, \ldots, a_n)) = f_B(h_{s_1}(a_1), \ldots, h_{s_n}(a_n))$.
This implies that $h : A \to B$ is a Σ-homomorphism that has an inverse
representing the decoding.

Well-known examples of isomorphic algebras result from the different representations of natural numbers, taking, for instance, the decimals on the one
hand and the binary representation on the other. Decimals are well suited for

human beings whereas duals are well suited for digital computers. Isomorphic representations of term algebras have been discussed above.

An *isomorphism class* of algebras is a class where any two algebras of the class are isomorphic and any algebra isomorphic to an algebra of the class is also a member of the class.

It is well known from mathematical logic that there is no finite axiomatization of the isomorphism class of natural numbers within first-order predicate logic. The induction axiom of the Peano axiomatization of natural numbers is not expressible in first-order predicate logic. This raises the question of how to describe isomorphism classes in a finite manner. One way could be a finite description of one of the models within the intended isomorphism class. This leads to specifications based on a single prototype algebra, which do not represent explicitly the intended, or required, properties of correct implementations represented by the algebras of the isomorphism class.

Another way would be to use higher-order logics. In the following we refrain from using a full higher-order logic. Instead just two specific higher-order concepts, well known in the field of Universal Algebra, will allow us to characterize algebras unique up to isomorphism. These are the notions of *initial algebras* and *final algebras* within a class $\mathcal{A}$ of Σ-algebras, see Chapter 1. Since the existence of initial and final algebras is not guaranteed in arbitrary classes of models, in the following definition we will use model classes $Mod_\Sigma(AX)$ of conditional Σ-equations AX in which the existence of initial algebras is guaranteed.

Definition 5.1. A Σ-algebra T is called *initial* in a class $\mathcal{A}$ of Σ-algebras if for every A in $\mathcal{A}$ there is exactly one homomorphism $h : T \to A$, and, dually, it is called *final* (or *terminal*) if for every A in $\mathcal{A}$ there is exactly one homomorphism $h : A \to T$.   ◇

Evidently, if in a class $\mathcal{A}$ of Σ-algebras there is an initial algebra, or respectively a final algebra, then it is unique up to isomorphism.

Definition 5.2. For each set of conditional Σ-equations AX

$$IMod_\Sigma(AX) = \{A \mid A \text{ initial in } Mod_\Sigma(AX)\}$$

denotes the class of all initial Σ-algebras $A \in Mod_\Sigma(AX)$.   ◇

Since any two algebras in $IMod_\Sigma(AX)$ are isomorphic and each isomorphic algebra is initial as well $IMod_\Sigma(AX)$ is an isomorphism class.

Let us consider the following presentation:

$Nat =$
    sorts nat
    opns $0 \colon nat$
             $suc \colon nat \to nat$
             $+ \colon nat, nat \to nat$
    axioms $\forall n \colon nat.\, 0 + n = n$
                $\forall m, n \colon nat.\, suc(m) + n = suc(m + n)$

The natural numbers with zero as a distinguished constant, with the successor function, and with addition as a binary operation, form a model in $IMod(Nat)$, whereas N_k, the natural numbers modulo k, do not form a model in $IMod(Nat)$. N_k is only a model in $Mod(Nat)$. The class $Mod(Nat)$ contains both the natural numbers and N_k, and also a final algebra which is a degenerated algebra whose carrier set consists of a one-element set.

Let us now consider the following presentation:

$Aut(In, Out) =$
    sorts *in, out, states*
    opns *c*: *in* (for each $c \in In$)
           k: *out* (for each $k \in Out$)
           next: *in, states* → *states*
           print: *states* → *out*

where In, Out are two arbitrary, but fixed, sets serving respectively as input and output alphabets.

$Mod(Aut(In, Out))$ is the class of all abstract automata whose input and output alphabet equal respectively In and Out.

In this case there is an initial algebra I in $Mod(Aut(In, Out))$, but it is degenerated, because of $|I|_{states} = \emptyset$. Now there is an interesting final algebra T in $Mod(Aut(In, Out))$ given by

$$|T|_{states} = Out^{In^*},$$

i.e., the final model is given by the set of all mappings $\varphi : In^* \to Out$. For an arbitrary model $A \in Model(Aut(In, Out))$, the unique homomorphism $h^A : A \to T$ is given by $[h^A(s)](w) = print_A(next_A^w(s))$ where $next_A^w : |A|_{states} \to |A|_{out}$ is inductively defined by $next_A^\varepsilon(s) = s$ and $next_A^{wx}(s) = next_A(x, next_A^w(s))$ and where ε denotes the empty string of input symbols.

This demonstrates that there are both model classes with degenerated final models and interesting initial models, and model classes with degenerated initial models and interesting final ones.

In the following we will argue that specifications based on initial algebras are very convenient if the generation principle of values of the type of interest is of primary concern. Specifications based on final semantics are better suited to cases where the external behavior is the point.

We will discuss the relations between initial and final algebras using the example of lists:

$Lists =$
    sorts *elem, lists*
    opns *nil*: → *lists*
           cons: *elem, lists* → *lists*

If we consider the class $Mod(Lists)$ of all *Lists*-algebras, then both the initial and the final *Lists*-algebras exist and both are degenerated. The initial *Lists*-algebra I is given by $|I|_{elem} = \emptyset$ and $|I|_{lists} = \{nil\}$, and the final

LIST-algebra T is given by $|T|_{elem} = |T|_{lists} = \{*\}$. Why are both algebras degenerated?

The initial *Lists*-algebra I degenerates because the given set of operations is not strong enough to generate the intended values. It lacks ground terms of sort *elem* and, therefore, the carrier $|I|_{elem}$ becomes empty. Consequently the initial *Lists*-algebra I represents the set of lists for an empty set of list elements, which consists of the empty list *nil* only. If we extend *Lists* in the following way

> *Lists1* = *Lists* **plus**
>   **opns** $a, b, c, d: \rightarrow elem$

then the initial *Lists1*-algebra $I1$ would be given by $|I1|_{elem} = \{a, b, c, d\}$ and $|I1|_{lists}$ now becomes the intended set of all finite lists (or sequences) of elements in $|I1|_{elem}$.

The final *Lists1*-algebra coincides with the final *Lists*-algebra, i.e., it is still degenerated. The final *Lists*-algebra and *Lists1*-algebra both degenerate since there are no *output operations* to make observations. This implies that any two lists, considered as internal states of an automaton, are indistinguishable and therefore identified within the final algebras. To overcome the degeneration in this case we extend the presentation *Lists1* as follows

> *Lists2* = *Lists1* **plus**
>   **sorts**   *bool*
>   **opns** $true, false: \rightarrow bool$
> $empty: lists \rightarrow bool$
>   **axioms** $\forall \emptyset: empty(nil) = true$
> $\forall x \in lists. \forall l \in lists. empty(cons(x, l)) = false$

Let us look first at the initial *Lists2*-algebra $I2$, which can also be described as an extension of the initial *Lists1*-algebra. $I2$ is given by $|I2|_{bool} = \{true, false\}$ and $empty_{I2}(l) = true$ iff l is the empty list. What about the final *Lists2*-algebra? It is still degenerated! Why does the output operation not prevent the degeneration of the final algebra? The degeneration is caused by the fact that in the final algebra $T2$, the constants in the carrier set $|T2|_{bool}$, coincide and so the output operation becomes ineffective. If we consider the final algebra $T2_{\neq}$ in

$$Mod(Lists2, true \neq false) = \{A \in Mod(LIST2) \mid true_A \neq false_A\}$$

then $|T2_{\neq}|_{elem}$ is again a singleton set, but the other two carrier sets

$$|T2_{\neq}|_{bool} = \{true, false\} \quad \text{and} \quad |T2_{\neq}|_{lists} = \{nil, notnil\}$$

are not. The interpretation of the output operation $empty_{T2_{\neq}}$ is given by $empty_{T2_{\neq}}(nil) = true$ and $empty_{T2_{\neq}}(notnil) = false$. Although the final algebra $T2_{\neq}$ is not degenerated, it is not of interest.

To obtain a more interesting example of a final algebra we will use the following presentation of an equational theory. To avoid problems with the partiality of some operations we will use order-sorted algebras, see Section 2.10.4.

$Lists^* =$
    sorts $elem, lists, nelists$
    subsorts $nelists < lists$
    opns $nil\colon\ \to lists$
               $cons\colon elem, lists \to nelists$
               $a, b, c, d\colon\ \to elem$
               $hd\colon nelists \to elem$
               $tl\colon nelists \to lists$
    axioms $\forall x \in elem. \forall l \in nelists. hd(cons(x, l)) = x$
               $\forall x \in elem. \forall l \in nelists. tl(cons(x, l)) = l$

The declaration

    subsorts $nelists < lists$

implies $|A|_{nelists} \subset |A|_{lists}$ for each model A of $Lists^*$. If we had used the sort $lists$ only, the operations hd, tl would have been partial, since they are not defined in a natural way on the empty list. Therefore we have introduced the subsort $nelists$ to distinguish the subset of nonempty lists in each model.

The initial algebra I in $Mod(Lists^*)$ is given by the finite lists with elements in $\{a, b, c, d\}$ where hd_I selects for nonempty list the head element. and tl_I produces the list reduced by the head element. $IMod(Lists^*)$ represents the abstract data type of finite lists over the basic set $\{a, b, c, d\}$.

Let T denote the final algebra in $Mod(Lists^*)$. Then $|T|_{elem}$ is again a singleton set, but $|T|_{lists}$ is isomorphic to the set of natural numbers adjoint with a maximal element $\mathbb{N}^\infty = \mathbb{N} \cup \{\infty\}$ and $|T|_{nelists} = |T|_{lists} \setminus \{0\}$. $cons_T$ represents the successor function with $cons_T(\infty) = \infty$, hd_T is the unique constant function to the singleton set and tl_T represents the predecessor function with $hd_T(\infty) = \infty$.

The unique homomorphism $h : I \to T$ associates with each finite list its length, because in the final model the output function hd cannot distinguish different elements of a list, it can only count the length of a list. The elements of the final model may also be seen as finite and infinite lists with elements in a one-element set.

If we consider the following class of $Lists^*$-algebras

$$Mod(Lists^*; \{a, b, c, d\}) = \{A \in Mod(Lists^*) \mid |A|_{elem} = \{a, b, c, d\}\}$$

then I is also initial in $Mod(Lists^*; \{a, b, c, d\})$ but the algebra T is no longer a model in $Mod(Lists^*; \{a, b, c, d\})$. If in $Mod(Lists^*; \{a, b, c, d\})$ there is a final algebra $T1$, then the output function hd_{T1} will be able to distinguish different elements because $|T1|_{elem} = \{a, b, c, d\}$.

An arbitrary algebra A in $Mod(Lists^*; \{a, b, c, d\})$ may be seen as an automaton where the set $|A|_{lists}$ $(\supset |A|_{nelists})$ operates as a set of states, and the operation $hd_A : |A|_{nelists} \to |A|_{elem}$ acts as a output function, and $tl_A : |A|_{nelists} \to |A|_{lists}$ as a state transition function. The operation $cons_A : |A|_{elem} \times |A|_{lists} \to |A|_{nelists}$ produces a state $cons_A(x, a)$ such that

$x = hd_A(cons_A(x, a))$ and $a = tl_A(cons_A(x, a))$, i.e., some kind of predecessor state is built.

Having this interpretation in mind, the description of the final algebra in $Mod(Aut(In, Out))$ above is very helpful. Since we now have both terminating and nonterminating automata in $Mod(Lists^*; \{a, b, c, d\})$, the set of behaviors realized by an automaton A in $Mod(Lists^*; \{a, b, c, d\})$ is given by

$$|T1|_{lists} = \{\alpha : D \to \{a, b, c, d\} \mid D = \mathbb{N} \text{ or } D = \{0, \ldots, n-1\}, n \geq 0\}$$

and $|T1|_{nelists}$ is the subset of mappings with a nonempty domain. The unique homomorphism $h : I \to T1$ associates with a list $l = [x_1, \ldots, x_n]$ the mapping $\alpha_l : \{0, \ldots, n-1\} \to \{a, b, c, d\}$ with $\alpha_l(i) = x_{i+1}$.

Finally we constrain $Mod(Lists^*; \{a, b, c, d\})$ to its reachable algebras:

$$Mod(Lists^*; \{a, b, c, d\}; tg) = \{\ A \in Mod(Lists^*; \{a, b, c, d\}) \mid$$
$$A \text{ is reachable }\}$$

The algebra I is initial in the class $Mod(Lists^*; \{a, b, c, d\}; tg)$ and surprisingly also final in the constraint model class $Mod(Lists^*; \{a, b, c, d\}; tg)$. The unique homomorphism $f : I \to A$ for any algebra $A \in Mod(Lists^*; \{a, b, c, d\}; tg)$ can easily be described, since I is isomorphic to the term algebra of ground terms and therefore the homomorphism f associates with each term t its value t_A in A.

Recall that an algebra A in $Mod(Lists^*; \{a, b, c, d\})$ is called reachable if the unique homomorphism $f : I \to A$ is surjective. This implies that for each $a \in |A|_{lists}$, there is a natural number n such that $tl_A^n(a) = nil_A$, i.e., the n-fold iteration of the state transition function leads to the distinguished state nil_A. The unique homomorphism $h : A \to I$ associates with each $a \in |A|_{nelists}$, the list of results of the run in A, interpreted as an automaton, starting in a, i.e.,

$$h_{nelists}(a) = [a_0, \ldots, a_n]$$

with $a_0 = hd_A(a), a_{i+1} = hd_A(tl_A^i(a))$, for $0 \leq i \leq n-1$, and n equals the length of computation starting in a. Finally, since A is assumed to be reachable, the only element nil_A in $|A|_{lists} \setminus |A|_{nelists}$ is mapped by h_{lists} to the empty list.

If one looks more thoroughly at the conditions used, it is not too surprising that the initial and final algebras coincide since the two constraining conditions imply that $Mod(Lists^*; \{a, b, c, d\}; tg)$ is an isomorphism class. The first condition on the model class $Mod(Lists^*)$ resulting in $Mod(Lists^*; \{a, b, c, d\})$ implies that each homomorphism $f : I \to A$ with $A \in Mod(Lists^*; \{a, b, c, d\})$ is injective. The next condition on $Mod(Lists^*; \{a, b, c, d\})$, which defines $Mod(Lists^*; \{a, b, c, d\}; tg)$, implies that the unique homomorphism $f : I \to A$ is surjective, for each $A \in Mod(Lists^*; \{a, b, c, d\}; tg)$. These conditions algebra together make the unique homomorphism from the initial algebra I to the final algebra T an isomorphism.

The considerations above were aimed at a better understanding of the expressivity of initial and final algebras. In short : *initial algebras describe generation patterns whereas final algebras describe observation patterns*. In an initial algebra as many elements as possible are distinguished whereas in a final algebra all elements are identified if there is no distinguishing observation.

In the following we will demonstrate that isomorphism classes do not suit all application domains. We will illustrate this by means of the well-known concept of *stacks*. Sorts and operations for a specification of stacks can be given by the following presentation:

Stacks =
    sorts   *elem, states, nestates*
    subsorts *nestates < states*
    opns   *empty*: → *states*
               push: *elem, states* → *nestates*
               pop: *nestates* → *states*
               top: *nestates* → *elem*
    axioms $\forall x \in elem. \forall s \in states. top(push(x, s)) = x$
               $\forall x \in elem. \forall s \in states. pop(push(x, s)) = s$

Notice that the presentation *Stacks* is just a renaming of the presentation *Lists* with the exception of the constants of sort *elem*, so that there is no real need to introduce *Stacks*. However, the notations in *Stacks* are more intuitive for the following discussion.

It is well known that stacks can be implemented by pairs consisting of a pointer and *elem*-array. Let S be an algebra with $|S|_{elem}$ an arbitrary set of elements,

$$|S|_{states} = \{\langle n, a\rangle \mid n \in \mathbb{N}, a : \mathbb{N} \setminus \{0\} \to |S|_{elem}\},$$

$|S|_{nestates}$ is the subset of pairs where the first element, i.e., the pointer, is a natural number different from 0. The constant nil_S can be implemented by an arbitrary pair with the pointer value 0. The operations are implemented such that

$push_S(x, \langle n, a\rangle) = \langle n + 1, a[n + 1] := x\rangle$
$pop_S(\langle n + 1, a\rangle) = \langle n, a\rangle$
$top_S(\langle n + 1, a\rangle) = a(n + 1)$

The algebra S seems to be a correct stack implementation. However, the required equation $pop(push(x, s)) = s$ is not valid in S. The right-hand side results for $s = \langle n, a\rangle$ in $\langle n, a[n + 1] := x\rangle$, so that the array components of the two stack states, $pop_S(push_S(x, \langle n, a\rangle))$ and $\langle n, a\rangle$, may be different.

This example demonstrates that not in every situation does one want to describe either a generation pattern, by an initial algebra, or an observation pattern, by a final algebra. In the case of stacks one wants to describe on the one hand that the operations *nil* and *push* generate all (reachable) states

of a stack, and on the other that the states of a stack are considered to
be different if they can be distinguished by means of combinations of the
operations *nil, push, pop, top*.

There are other points that make isomorphism-class semantics insufficient.
If one specifies a system or a program in a top-down manner, one would
like to start with a very rough specification that will be constrained step
by step in subsequent design stages. Such a specification process will only
result in an isomorphism class in the final step, if at all. The semantics of
the intermediate stages will be represented by classes of algebras that are not
necessarily isomorphic.

5.2   Loose semantics

In zhe early stages of specifications, the behavior of a system and its mathematical structures are not completely known. In this situation algebras representing the system are considered to be equivalent if they have characteristic
properties in common.

An axiomatic specification is given by a signature Σ and a set AX of
axioms of a specification logic. In general many different specification logics are in use. The most popular ones are many-sorted equational logic (see
Sections 2.7 and 2.8)and first-order predicate logic with or without equality,
but there are also order-sorted equational logic, higher-order logics, logics for
partial algebras and also logics for higher-order partial algebras, etc. This has
led to the notion of institution [GB92] which extracts common concepts and
properties from the specific logics in use. The use of institutions in specification languages is described in Chapter 4. We will use many-sorted equational
logic as a representative of a specification logic in the following. However, the
definitions will not depend on the specification logic chosen.

Definition 5.3. The *loose semantics* of a specification $\langle \Sigma, AX \rangle$ is given by
the following class of algebras

$$[\![\Sigma, AX, loose]\!] = \{A \mid A \in Alg(\Sigma),\ A \models_\Sigma AX\}. \qquad \diamond$$

This terminology is not generally accepted. In [Wir90] and many other papers, an abstract data type is always an isomorphism class and a specification
with loose semantics describes a class of abstract data types.

Specifications languages with loose semantics include for instance LARCH
[GHW85], PLUSS [Gau86], and SPECTRUM [BFG+93]. An algebra in the
model class of a loose specification is just a possible implementation of the
sorts and operation names in Σ satisfying the required properties described
by AX.

As an example we sketch a loose specification of finite and cofinite (being
the complement of a finite subset) sets of natural numbers. The key word
loose indicates the loose semantics of the specification.

$NatSet$ = **loose**
>**sorts** $nat, bool, sets$
>**opns** $true, false: \rightarrow bool$
>$zero: \rightarrow nat$
>$suc: nat \rightarrow nat$
>$add: nat, nat \rightarrow nat$
>$empty, all: \rightarrow sets$
>$s\_add, s\_rem: nat, sets \rightarrow sets$
>$join, meet: sets, sets \rightarrow sets$
>$member: nat, sets \rightarrow bool$
>**axioms** $\forall s \in sets. join(empty, s) = s$
>$\forall s \in sets. meet(all, s) = s$
>$\forall s_1, s_2 \in sets. join(s_1, s_2) = join(s_2, s_1)$
>$\forall s_1, s_2 \in sets. meet(s_1, s_2) = meet(s_2, s_1)$
>$\forall x \in nat. \forall s \in sets. s\_rem(x, s\_add(x, s)) = s$
>$\forall x \in nat. \forall s \in sets. member(x, s) = true \Rightarrow$
>$\qquad s\_add(x, s\_rem(x, s)) = s$
>$\forall x \in nat. s\_add(x, all) = all$
>$\forall x \in nat. s\_rem(x, empty) = empty$

This loose specification is not very realistic. We have only written down some sample properties. There are, for instance, no requirements concerning the operations on nat and $bool$, and the requirements on operations on $sets$ are not complete in any sense. The loose specification $NatSet$ is sound in the sense that its denotation contains the intended algebra. However, it also contains algebras that have nothing to do with the intended meaning and would usually not be accepted as correct implementations of the desired data type.

In many situations it would be more reasonable to define the semantics of a loose specfication $SPEC$, not by the class $[\![SPEC, loose]\!]$ of algebras, but by the full subcategory of the category of all Σ-algebras, whose set of objects is $[\![SPEC, loose]\!]$. In contrast to the case of isomorphism-class semantics, the addition of the homomorphisms between the models of the required properties makes sense. The category of models carries much more information than the class of models themselves. One can ask for initial or final models, representing distinct implementations, or one can ask for interesting closure properties, like closure under finite colimits.

If $SPEC = \langle \Sigma, AX \rangle$ is a loose specification, $\mathbb{C}(SPEC)$ will denote the full subcategory of the category of Σ-algebras defined by the models in $[\![SPEC, loose]\!]$ and will be called the $SPEC$-category.

As mentioned for the loose specification $NatSet$, the specified class of algebras may contain incorrect realizations of the intended system or data type. A typical situation is the existence of *junk elements* that cannot be represented as values of terms. This is the reason why some authors, for instance, [Wir90], define the semantics of a loose specification as follows.

Definition 5.4. The *reachable loose semantics* of a specification $\langle \Sigma, AX \rangle$ is the following class of Σ-algebras

$$[\![\, \Sigma, AX, gen \,]\!] = \{ A \mid A \text{ is a reachable } \Sigma\text{-algebra with } A \models_\Sigma AX \}. \qquad \diamond$$

Recall that a Σ-algebra A is called *reachable* (*term-generated, minimal*), if there are no proper subalgebras of A. This implies that the unique homomorphism from the Σ-algebra of ground terms to A is surjective. Reachable Σ-algebras are called Σ-*computation structures* (or Σ-*data types*) in [Wir90].

The use of reachable loose specifications makes inductive reasoning available. If A is any Σ-algebra in the reachable loose specification, then for every element $a \in |A|$, there is at least one ground term $t \in T_\Sigma$, whose value in A equals a, i.e., $t_A = a$. As described in Chapter 2, each Σ-data type A in a reachable loose specification is isomorphic to a suitable quotient algebra $T_\Sigma / \equiv_A$ of the Σ-algebra of ground terms T_Σ, where $\equiv_A$ is the kernel of the unique homomorphism from T_Σ to A. If the congruence relation $\equiv_A$ is decidable, every computation in the Σ-algebra A can be equivalently simulated purely syntactically in the Σ-algebra of ground terms. This topic is discussed in the Chapter 9.

In the example *NatSet* it becomes obvious that not all operations are necessary to represent arbitrary values by means of ground terms. For elements in $|A|_{nat}$, one would only need ground terms built up with the operations *zero* and *suc*, and for elements in $|A|_{sets}$, the use of the operation *meet* in ground terms is not necessary. This observation is typical for many specifications and leads to specifications with a distinguished subsignature of so-called *constructors*. The reachability constraint in that case requires that any element has to be representable as the value of a ground term built up from constructors only.

Let Σ_0 be a subsignature of the signature Σ and $\iota : \Sigma_0 \to \Sigma$ the embedding signature morphism (see Section 2.5), then we denote the ι-reduct $A|_\iota$ of a Σ-algebra A by A_{Σ_0}. A Σ-algebra A is called Σ_0-*reachable*, if there is no proper Σ-subalgebra A' of A with $A'|_{\Sigma_0} = A|_{\Sigma_0}$.

Definition 5.5. The *constructor-based loose semantics* (or Σ_0-*based loose semantics*) of a specification $\langle \Sigma, AX \rangle$ with respect to a given subsignature Σ_0 of Σ is the following class of Σ-algebras

$$[\![\, \Sigma, \Sigma_0, AX \,]\!] = \{ A \mid A \in Alg(\Sigma) \text{ is } \Sigma_0\text{-reachable} \}. \qquad \diamond$$

In a constructor-based loose specification, the declaration part of operations is split into a declaration part of constructors and a declaration part of additional operations.

Let us return to the example of stacks and give a constructor-based loose specification which refines the storable elements to natural numbers. In this specification we introduce, besides the data type *nat* of natural numbers, a data type $nat^\perp$ of natural numbers with an additional default value $\perp$, in order to make the topoperation into a total operation. The operation

$c : nat \rightarrow nat^{\perp}$ embeds the natural numbers (without default value) into the natural numbers with default value.

$NatStack^{\perp}$	**loose**
sorts	$nat, nat^{\perp}, states$
constructors	$zero: \;\; \rightarrow nat$
	$suc: nat \rightarrow nat$
	$\perp: \;\; \rightarrow nat^{\perp}$
	$c: nat \rightarrow nat^{\perp}$
	$empty: \;\; \rightarrow states$
	$push: nat, states \rightarrow states$
opns	$top: states \rightarrow nat^{\perp}$
	$pop: states \rightarrow states$
axioms	$\forall n \in nat. \forall s \in states. top(push(n, s)) = c(n)$
	$\forall n \in nat. \forall s \in states. pop(push(n, s)) = s$
	$top(empty) = \perp$
	$pop(empty) = empty$

An algebra A with $|A|_{nat} = \mathbb{N} \cup \{\infty\}$ would not satisfy the reachability condition, since ∞ cannot be reached by starting with $zero$ and applying the successor operation finitely often. If stacks are implemented by lists in an algebra L, then $|L|_{states}$ must not contain an infinite list. Again in this case the reachability condition would be violated.

The implementation S of stacks by pairs of pointers and $nat^{\perp}$-arrays, as described above, is not an algebra of the constructor-based loose specification $NatStack$. The axiom

$$\forall n \in nat. \forall s \in states. \; pop(push(n, s)) = s$$

is still not valid. It would be satisfied if the interpretations of $empty$ and pop are modified as follows:

$$emptys = \langle 0, a_{\perp} : \mathbb{N} \rightarrow \mathbb{N} \cup \{\perp\} \rangle$$
$$pop_S(\langle n + 1, a \rangle) = \langle n, a[n + 1] := \perp \rangle$$

and if

$$|S|_{states} = \{\langle n, a \rangle \mid a : \mathbb{N} \rightarrow \mathbb{N} \cup \{\perp\} \text{ and } a(m) = \perp \text{ for all } m > n\}$$

The modified interpretation of the sort $states$ is necessary to satisfy the reachability condition. Setting the value in $a[n+1]$ to the default value is only necessary to satisfy the axiom in a very strong sense, which is not justified in practice.

To overcome this defect of loose semantics, Sannella and Tarlecki [ST88b] introduced the concept of *abstractor specifications*. The notion of *abstractor* allows us to abstract from a class of models to an enclosing class of algebras by means of a given equivalence relation $\equiv$ on the class of all Σ-algebras. *Observational equivalence relations* between algebras are important examples of such equivalence relations. The basic idea behind such relations is that two algebras are considered to be observationally equivalent if they cannot be distinguished by a predefined set of observations. In the example of stacks, the set of observations is given by terms of the form

$$top(s),\ top(pop(s)), \ldots , top(pop(\ldots (pop(s)) \ldots)) \ldots .$$

Definition 5.6. An *abstractor specification* $(\Sigma, \Sigma_0, AX, \equiv)$ is given by a specification (Σ, Σ_0, AX) of a constructor-based loose specification together with an equivalence relation $\equiv$ on the class of all Σ-algebras. Its semantics is given by the following class of Σ-algebras :

$$[\![\, \Sigma, \Sigma_0, AX, \equiv \,]\!] \ = \ \{A \in Alg(\Sigma) \mid \text{there exists}$$
$$B \in [\![\, \Sigma, \Sigma_0, AX \,]\!] \text{ with } A \equiv B \ \} \qquad \qquad \diamond$$

In Section 5.4 we will discuss the topic of *behavioral and abstractor specifications* in more detail.

There is another shortcoming of constructor-based loose specifications and also abstractor specifications. In both cases implementations have to be algebras of the same signature. It is sometimes convenient to cmbine algebras of different signatures into the model class of a loose data type. To overcome this restriction in specifications of abstract data types, several authors have suggested *ultra loose* semantics of abstract data types [AR93a, Pep92, CR94].

In the case of ultra loose semantics, also called *very abstract specifications*, an implementation of a constructor-based ultra loose specification (Σ, Σ_0, AX) (or an abstractor specification $(\Sigma, \Sigma_0, AX, \equiv)$) consists of a pair $\langle A, \Sigma' \rangle$, where $A \in Alg(\Sigma')$ and Σ' is a signature extension of Σ, such that the Σ-reduct $A \mid_{\Sigma}$ is in $[\![\, \Sigma, \Sigma_0, AX \,]\!]$ (or $[\![\, \Sigma, \Sigma_0, AX, \equiv \,]\!]$). Thus, the signature in a specification of an ultra loose abstract data type describes only the minimal sets of types and operations available for the specified data type.

5.3   Constraints

The semantic concepts introduced in the previous section are inconvenient for specifications of *parameterized objects*. The specification *NatStack* above defines only one instance of the general concept of a stack where the elements that may be stored (by means of *push*), released (by means of *pop*), and read (by means of *top*) are natural numbers. However, the general concept of a stack is independent of the type of elements. A stack may be seen as a parameterized object that associates with an arbitrary set of elements the corresponding stack algebra.

Due to Sannella and Tarlecki [ST92], it is important to distinguish between *parameterized specifications* and specifications of parameterized objects. Parameterized objects may also be understood as *type constructors*. Parameterized specifications will be thoroughly considered in the Chapters 6 and 7. We will concentrate here on the semantics of specifications of parameterized objects.

A stack as a parameterized object may be represented by a function

$$stack : Set \rightarrow Mod(Stacks)$$

where *Set* denotes the class of sets and *Mod(Stacks)* denotes the class of *Stack*-algebras as described above, such that for each set M the *Stack*-algebra $stack(M)$ is given by

$$|stack(M)|_{elem} = M,$$
$$|stack(M)|_{states} = \text{set of stack states over } M,$$
$$|stack(M)|_{nestates} = \text{set of nonempty stack states over } M.$$

Different implementations of the parameterized concept of stack would yield different functions $\varphi : Set \rightarrow Mod(Stacks)$. Thus, a specification of parameterized stacks defines a set of functions mapping each *Triv*-model M to a *Stacks*-model that expands M, where

$$Triv = \textbf{sorts } elem.$$

In general, a specification SP of parameterized objects is denoted by a set of functions mapping models of a parameter specification SP_{par} to models of the result specification SP_{res}. See [ST92] for more details.

In the example of stacks, *Triv* is the specification of parameter objects, *Stack* is the specification of result objects and, for instance, the denotation $[\![SP^{ini}]\!]$ of an *initial specification of parameterized stacks* would be given by

$$[\![SP^{ini}]\!] = \{\varphi : Set \rightarrow Mod(Stacks) \mid \varphi(M) \text{ is a free extension of } M\}.$$

An algebra $A \in Mod(Stacks)$ is called a free extension of M if $A|_{Triv} = M$ and, for every $B \in Mod(Stacks)$ and every $TRIV$-homomorphism $h : M \rightarrow B|_{Triv}$, there is exactly one homomorphism $h^* : A \rightarrow B$ with $h^*|_{Triv} = h$.

Not only can the initial semantics of specifications of nonparametric objects be extended to specifications of parameterized objects, but so can the final semantics and the different kinds of loose semantics.

For example, the denotation $[\![SP^{gen}]\!]$ of a *reachable specification of parameterized stacks* would be given by

$$[\![SP^{gen}]\!] = \{\varphi : Set \rightarrow Mod(Stacks) \mid \varphi(M) \text{ is a quotient of a}$$
$$Stacks\text{-algebra freely generated by } M \text{ and } \varphi(M)|_{Triv} = M\}$$

Alternatively, the specification of parameterized objects can be denoted by a class of algebras. This approach uses as the denotation for a specification SP, the set of all images of functions in $[\![SP]\!]$

$$\{\varphi(A) \in Mod(SP_{res}) \mid A \in Mod(SP_{par}), \varphi \in [\![SP]\!]\}.$$

For finite descriptions the resulting classes of algebras require additional higher order tools, usually called *constraints*.

We follow the presentation of constraints in Chapter 7 of [EM90].

For a given specification $SPEC$ (without constraints), as presented above, we introduce different sets $Constr_\#(SPEC)$ of constraints, where, $\#$ from $\{INIT, TGEN, GEN, FGEN, FIN\}$, indicates the *kind of constraints*. For any kind of constraints over a specification $SPEC$ and any $SPEC$-algebra A we define a satisfaction relation

$$\models \; \subseteq \; Mod(SPEC) \times Constr_\#(SPEC)$$

If $SPEC$ is a specification and C a set of constraints over $SPEC$, then the pair $(SPEC, C)$ is called a *specification with constraints* and the corresponding class of models is defined by

$$Mod(SPEC, C) = \{A \in Mod(SPEC) \mid A \models c \text{ for each } c \in C\}$$

If in $(SPEC, C)$ the specification $SPEC = (SPEC^*, C^*)$ is already constrained, we assume $(SPEC, C) = (SPEC^*, C \cup C^*)$.

In the following *subspecifications* of specifications will be used to represent constraints syntactically. For a specification $SPEC = \langle \Sigma, AX \rangle$, a second specification $SPEC_0 = \langle \Sigma_0, AX_0 \rangle$ is called a subspecification of $SPEC$, written $SPEC_0 \sqsubseteq SPEC$, if Σ_0 is a subsignature of Σ and the set of axioms AX_0 is a subset of AX.

Definition 5.7. For a given specification $SPEC$ the set of *initiality constraints* is defined by

$$Constr_{INIT}(SPEC) = \{init(SPEC_0) \mid SPEC_0 \sqsubseteq SPEC\}.$$

An algebra $A \in Mod(SPEC)$ satisfies an initiality constraint $init(SPEC_0)$ written

$$A \models init(SPEC_0)$$

if the $SPEC_0$-reduct $A|_{SPEC_0}$ of A is an initial $SPEC_0$-algebra.   ◇

If $SPEC = (\Sigma, AX)$ and AX is a set of conditional equations, then $IMod(AX)$ as defined above coincides with the class of all $SPEC$-algebras satisfying the initiality constraint $init(SPEC)$.

Constraining model classes by means of initiality constraints extends the initial semantics of Definition 5.1. In contrast to Definition 5.1, initiality constraints allow us to restrict the interpretation of parts of a loose specification to the initial case. For instance, in the specifications

$Bool =$
    sorts *bool*
    opns *true, false:* $\rightarrow bool$

$Nat =$
    sorts *nat*
    opns *zero:* $\rightarrow nat$
            suc: nat $\rightarrow nat$

and

$Equiv = Bool$ **plus** Nat **plus**
    opns $eq: nat, nat \rightarrow bool$
    axioms $\forall x \in nat. \, eq(x, x) = true;$
              $\forall x, y \in nat. \, eq(x, y) = eq(y, x);$
              $\forall x, y, z \in nat. \, eq(x, y) = true \wedge eq(y, z) = true \Rightarrow$
                $eq(x, z) = true;$

by the constraint specification

$$Mod(Equiv, \{init(Bool), init(Nat)\})$$

only the interpretation of the operation $eq : nat, nat \rightarrow bool$ is not affected by the two initiality constraints. It specifies the class of equivalence relations on natural numbers, more precisely the class of characteristic functions of equivalence relations on natural numbers. Without the initiality constraints the carrier set $|A|_{bool}$ was allowed to collapse to any nonempty set, for instance, a one-element set. Therefore the characteristic function eq would become trivial, contradicting our intuition. Similarly, the carrier set $|A|_{nat}$ could be an arbitrary nonempty set if the initiality constraint $init(Nat)$ would not have been stated.

The next class of constraints, called *generating constraints*, generalizes reachable loose specifications.

Definition 5.8. For a given specification $SPEC$ the set of *generating constraints* is defined by

$$Constr_{GEN}(SPEC) = \{gen(SPEC_0, SPEC_1) \mid$$
$$SPEC_0 \sqsubseteq SPEC_1 \sqsubseteq SPEC\}.$$

An algebra $A \in Mod(SPEC)$ satisfies a generating constraint, written

$$A \models gen(SPEC_0, SPEC_1)$$

if the Σ_1-reduct $A|_{\Sigma_1}$ of A is Σ_0-reachable, where $\Sigma_i, i = 0, 1$, are the signatures of the specifications $SPEC_i, i = 0, 1$. $\diamond$

The special case of a reachable loose specification is given by the generating constraint $gen(\langle \Sigma_\emptyset, \emptyset \rangle, SPEC)$, which constrains $Mod(SPEC)$ to those $SPEC$-algebras A generated by the only $\Sigma_\emptyset$-algebra of the empty signature $\Sigma_\emptyset$ consisting of the empty set of sort names and the empty set of operation names. This implies that A is reachable.

The semantics of *term-generating constraints* of a specification $SPEC$

$$Constr_{TGEN}(SPEC) = \{tgen(SPEC_0) \mid SPEC_0 \sqsubseteq SPEC\}$$

can be reduced to generating constraints by interpreting each constraint $tgen(SPEC_0)$ as an abbreviation of $gen(\langle \Sigma_\emptyset, \emptyset \rangle, SPEC_0)$.

Definition 5.9. For a given specification $SPEC$ the set of *free generating constraints* is defined by

$$Constr_{FGEN}(SPEC) = \{fgen(SPEC_0, SPEC_1) \mid \\ SPEC_0 \sqsubseteq SPEC_1 \sqsubseteq SPEC\}.$$

An algebra $A \in Mod(SPEC)$ satisfies a free generating constraint, written

$$A \models fgen(SPEC_0, SPEC_1),$$

if for any $B \in Mod(SPEC_1)$ and any homomorphism $h : A|_{\Sigma_0} \to B|_{\Sigma_0}$, there exists exactly one homomorphism $h^* : A|_{\Sigma_1} \to B$ with $h = h^*|_{\Sigma_0}$. ⋄

Clearly, initiality constraints can be expressed equivalently by free generating constraints. For any $SPEC$-algebra A we have:

$$A \models init(SPEC_0) \text{ if and only if } A \models fgen(\langle \Sigma_\emptyset, \emptyset \rangle, SPEC_0).$$

Definition 5.10. For a specification $SPEC$ the set of *finality constraints* is defined by

$$Constr_{FIN}(SPEC) = \{fin(SPEC_0, SPEC_1) \mid \\ SPEC_0 \sqsubseteq SPEC_1 \sqsubseteq SPEC\}.$$

A $SPEC$-algebra A satsfies a finality constraint, written

$$A \models fin(SPEC_0, SPEC_1),$$

if for every $SPEC_1$-algebra B and any homomorphism $h : B|_{SPEC_0} \to A|_{\Sigma_0}$, there is exactly one homomorphism $h^* : B \to A|_{\Sigma_1}$ with $h = h^*|_{SPEC_0}$, where Σ_i denotes the signature of $Spec_i$ for $i = 0, 1$. ⋄

Constraints were first considered by Reichel [Rei80]. His notion of *canon* corresponds to a finite set of free generating constraints. Similar constraints were called *data constraints* in the semantics definition of CLEAR [BG80] and simply *constraints* in LOOK [ETLZ82]. The term generating constraints first appeared in [SW82] and [WPP+83] as *hierarchy constraints*. Generating constraints have been studied in [EWT83] and [WE87].

Finality constraints have yet to be used in specification languages. Analogous to free generating constraints, finality constraints extend final semantics of data types to the case of parameterized objects.

We will illustrate the different kinds of constraints using the parameterized data type of infinite lists, usually called *streams*. Consider:

$FairStreams = Nat$ **plus**
　　sorts　　$items, streams$
　　opns　　$encode: nat \to items$
　　　　　　$cons: items, streams \to streams$
　　　　　　$hd: streams \to items$
　　　　　　$tl: streams \to streams$
　　　　　　$iter: nat, streams \to streams$

$$f: streams \rightarrow nat$$

axioms $\forall x \in items. \forall l \in streams. hd(cons(x,s)) = x;$
$\qquad \forall x \in items. \forall l \in streams. tl(cons(x,s)) = s;$
$\qquad \forall s \in streams. iter(zero,s) = s;$
$\qquad \forall x \in nat. \forall s \in streams. iter(suc(x),s) = tl(iter(x,s))$
$\qquad \forall s \in streams. hd(iter(f(s),s)) = encode(zero)$

with the following subspecifications

$Spec_0 = Nat$ **plus**
$\qquad$ **sorts** $\quad items$
$\qquad$ **opns** $\quad encode: nat \rightarrow items$

$Spec_1 = Nat$ **plus**
$\qquad$ **opns** $\quad iter: nat, streams \rightarrow streams$
$\qquad$ **axioms** $\forall s \in streams. iter(zero,s) = s;$
$\qquad\qquad\qquad \forall x \in nat. \forall s \in streams. iter(suc(x),s) = tl(iter(x,s))$

$Spec_2 = Nat$ **plus**
$\qquad$ **sorts** $\quad streams$
$\qquad$ **opns** $\quad cons: items, streams \rightarrow streams$

$Spec_3 = Nat$ **plus**
$\qquad$ **sorts** $\quad streams$
$\qquad$ **opns** $\quad hd: streams \rightarrow items$
$\qquad\qquad\qquad tl: streams \rightarrow streams$

The generating constraint

$$gen(Nat, Spec_0)$$

constrains the model class of *FairStreams* to those algebras for which the operation $encode_A : |A|_{nat} \rightarrow |A|_{items}$ is surjective. All other sorts and operations are not affected.

The free generating constraint

$$fgen(Nat, Spec_0)$$

turns the mapping $encode_A : |A|_{nat} \rightarrow |A|_{items}$ into a bijection.

The finality constraint

$$fin(Nat, Spec_0)$$

restricts the model class in such a way that $|A|_{items}$ is a set with one element.

The initiality constraint

$$init(Spec_1)$$

implies that for each algebra A of the resulting class we have:

- $|A|_{nat}$ is the set of natural numbers (up to isomorphism);

- $encode_A : |A|_{nat} \rightarrow |A|_{items}$ is a bijection;
- $iter_A(n, s) = \underbrace{tl(\ldots tl(s) \ldots)}_{n \ times}$

The same class of algebras is generated by the following two constraints

$$\{init(Nat), fgen(Nat, Spec_1)\}$$

The free generating constraint

$$fgen(Spec_0, Spec_2)$$

implies that $|A|_{streams} = \emptyset$ for each algebra A in the resulting class, since there is no constant of sort *streams*, which is necessary for a reasonable inductive definition.

Finally we discuss the consequences of the following two constraints:

$$\{init(Spec_1), fin(Spec_0, Spec_2)\}$$

The effect of the initiality constraint is described above. In addition, the finality constraint implies that, for each algebra A of the resulting class, the carrier $|A|_{streams}$ is (up to isomorphism) the set of streams with elements of $|A|_{items}$, i.e., mappings $s : \mathbb{N} \rightarrow |A|_{items}$, such that the element $encode_A(zero_A) \in |A|_{items}$ appears infinitely often in each stream of $|A|_{streams}$. In the terminology of temporal logic one would say that $encode_A(zero_A)$ appears *always sometimes* in each stream $s \in |A|_{streams}$.

If *Streams* is the specification resulting from *FairStreams* by removing the operation name $f : streams \rightarrow nat$ and also the axiom
$\forall s \in streams.hd(iter(f(s), s)) = encode(zero)$,
then the set of constraints

$$\{init(Spec_1), fin(Spec_0, Spec_2)\}$$

has weaker consequences. Now the finality constraint implies that the carrier set $|A|_{streams}$ becomes the set of all $|A|_{items}$-streams, i.e.,

$$|A|_{streams} = \{s : \mathbb{N} \rightarrow |A|_{items}\}.$$

These examples demonstrate that initiality and finality represent semantically different concepts. By means of initiality constraints one can describe generation schemata whereas finality constraints can be used to describe the observable effects of state transitions. Each generation schema degenerates if there are no starting points, given by constant function symbols, see $fgen(Spec_0, Spec_2)$, and each observation schema degenerates if there are no distinguishing results of observations, see $fin(Nat, Spec_1)$.

The kind of constraints considered so far are well suited for top-down development of specifications. We will not discuss tools for structuring specifications and specification-building operations here, since these are topics of Chapter 6.

Developing specifications in a bottom-up manner and re-using specification modules have influenced Bidoit [Bid88] to develop another kind of semantics, called *stratified loose semantics*. Here a specification module is an enrichment $(\Sigma, \Omega, AX$ use $Spec)$ of a given specification $Spec$, where Σ is a signature comprising the signature Σ_{Spec}, Ω is a distinguished set of constructors, and AX is a set of axioms over Σ. The semantics $[\![\, \Sigma, \Omega, AX$ use $Spec\,]\!]$ of a specification module [1] can be based on the class of parameterized objects [ST92], mapping $Spec$-models to those Σ-algebras which are finitely generated with respect to Ω, satisfy all axioms of AX, and whose Σ_{Spec}-reduct is a $Spec$-model.

A parameterized object of $[\![\, \Sigma, \Omega, AX$ use $Spec\,]\!]$ reflects the implementation choices intrinsic to the specification module $(\Sigma, \Omega, AX$ use $Spec)$ and also the implementation choices that are related to the whole specification and parameterized by the implementation choices of the subspecification $Spec$.

5.4   Observability

In the Section 5.2 an observational equivalence relation between algebras was used to define abstractor specifications. In this section we will discuss in more detail relations that may be used as observational equivalence relations, and look for other ways to describe the model classes $[\![\, \Sigma, \Sigma_0, AX, \equiv\,]\!]$ of abstractor specifications.

The notion of observational equivalence of algebras stems from automata theory and process theory where the related concepts are *bisimulation* and *testing equivalences* [NH84, Mil89]. Within algebraic specifications observability has its roots in final semantics. It was first introduced in [GGM76] and later used by Reichel [Rei81] and Goguen and Meseguer [GM82]. Observability was mainly related to the formalization of implementations of abstract data types. For a survey on that topic see [ONS93] and for more details consult Section 7.2 of this book.

The concept of institution, see Chapter 4, stimulates a redefinition of the satisfaction relation for a logic with equality in such a way that the model class $[\![\, \Sigma, \Sigma_0, AX, \equiv\,]\!]$ is exactly the class of those models that satisfy a given set of axioms behaviorally. The approach in [Rei81] was based on this idea. An equation is behaviorally satisfied by an algebra A if, for each substitution, the resulting left- and right-hand sides of an equation evaluate to states that are behaviorally equivalent in the sense of abstract automata, a concept which will be developed in the following.

Recently a interesting paper [BHW95] has thoroughly investigated and compared both approaches, calling them *behavioral* and *abstractor specifications*. In the following we present notions and results from [BHW95], extending the results of different authors.

---

[1] For a precise semantics definition, see [Bid88].

The remainder of this section uses a first-order language with infinitary formulas over a signature Σ, rather than an arbitrary institution. We exclude algebras with empty carriers.

Definition 5.11. Let Σ be a signature and $X = (X_s)_{s \in S}$ an arbitrary, but fixed, family of countably infinite sets of variables. The set of well-formed Σ-*formulae* is inductively defined by

1. If t, r are terms of sort s, then $t = r$ is a Σ-formula called equation.
2. If Φ, Ψ are Σ-formulas, then $\neg \Phi$ and $\Phi \wedge \Psi$ are Σ-formulas.
3. If Φ is a Σ-formula, then $\forall x : s.\Phi$ is a Σ-formula.
4. If $\{\Phi_i | i \in I\}$ is a countable family of Σ-formulas, then $\bigwedge_{i \in I} \Phi_i$ is a Σ-formula.

A Σ-sentence is a Σ-formula which contains no free variables. All other logical connectives are derived as usual.   ◇

The concept of partial Σ-congruences for general Σ-algebras (as defined below) captures the behavioral equivalence of internal states more generally than in automata theory.

Definition 5.12. A *partial* Σ-*congruence* on a Σ-algebra A is a family $R = (R_s)_{s \in S}$ of nonempty binary relations $R_s \subseteq |A|_s \times |A|_s$ such that each R_s is symmetric and transitive, and such that the family R is compatible with the signature Σ, i.e., for all Σ-operations f with arity $s_1 \times \ldots \times s_n \to s$ and for all $a_1, b_1 \in |A|_{s_1}, \ldots, a_n, b_n \in |A|_{s_n}$, the following holds: If $a_1 R_{s_1} b_1, \ldots, a_n R_{s_n} b_n$, then $f_A(a_1, \ldots, a_n) R_{s_n} f_A(b_1, \ldots, b_n)$.   ◇

One should have in mind that a family R of binary relations $R_s \subseteq |A|_s \times |A|_s$ is a partial Σ-congruence of a Σ-algebra A if and only if there is a subalgebra A_0 on A such that $R_s \subseteq |A_0|_s \times |A_0|_s$, for all sorts $s \in S$, and R is an ordinary Σ-congruence on A_0. Therefore any partial Σ-congruence $\approx_A$ on a Σ-algebra A will be considered as a pair $\approx_A = \langle A_0, \sim_{A_0} \rangle$ where A_0 is a subalgebra of A and $\sim_{A_0}$ is a Σ-congruence on A_0. The quotient algebra $A_0 / \sim_{A_0}$ will be denoted in the following by $A/ \approx_A$.

We now define the notion of satisfaction with respect to a given partial Σ-congruence.

Definition 5.13. Let A be a Σ-algebra and $\approx_A = \langle A_0, \sim_{A_0} \rangle$ be a partial Σ-congruence on A. Moreover, let $t, r \in |T_\Sigma(X)|_s$ be two terms of sort s, Φ, Ψ two Σ-formulas, $\{\Phi_i | i \in I\}$ a countable family of Σ-formulas, and $\alpha : X \to |A_0|$ an assignment. The *satisfaction relation w.r.t.* $\approx_A$, denoted by $\models_{\approx_A}$, is defined as follows:

1. $A, \alpha \models_{\approx_A} t = r$ holds if $\alpha^\#(t) \sim_{A_0} \alpha^\#(r)$, i.e. if the values of t and r in A under the assignment $\alpha : X \to |A_0|$ are equivalent w.r.t. $\sim_{A_0}$.
2. $A, \alpha \models_{\approx_A} \neg \Phi$ holds if $A, \alpha \models_{\approx_A} \Phi$ does not hold and $A, \alpha \models_{\approx_A} \Phi \wedge \Psi$ holds if both $A, \alpha \models_{\approx_A} \Phi$ and $A, \alpha \models_{\approx_A} \Psi$ hold.

3. $A, \alpha \models_{\approx_A} \forall x : s.\Phi$ holds if for all assignments $\beta : X \to |A_0|$ with $\beta(y) = \alpha(y)$ for all $y \neq x$, $A, \beta \models_{\approx_A} \Phi$ holds.
4. $A, \alpha \models_{\approx_A} \bigwedge_{i \in I} \Phi_i$ holds if for all $i \in I$ $A, \alpha \models_{\approx_A} \Phi_i$ holds.
5. $A \models_{\approx_A} \Phi$ holds if $A, \alpha \models_{\approx_A} \Phi$ holds for all assignments $\alpha : X \to |A_0|$.

◇

Recall that for the approach of behavioral specifications it is assumed that algebras do not have empty carrier sets. If this assumption were dropped, any Σ-algebra with an empty carrier set would satisfy (behaviorally or not) any Σ-formula.

Like the satisfaction relation, the generation principle of algebras changes with respect to a partial congruence.

Definition 5.14. Let Σ be a signature, Σ_0 a subsignature of Σ, A a Σ-algebra, and $\approx_A = (A_0, \sim_{A_0})$ a partial Σ-congruence on A. A Σ-algebra A satisfies a reachability (or generating) constraint $gen(\Sigma_0, \Sigma)$, written

$$A \models_{\approx_A} gen(\Sigma_0, \Sigma),$$

if, for any Σ_0-sort s and $a \in |A_0|_s$, there exists a constructor term $t \in T_{\Sigma_0}(X)$ and an assignment $\alpha : X \to |A_0|$ such that $\alpha^{\#}(t) \sim_{A_0} a$.

◇

For the generalized satisfaction relation and generating constraints, the following relation to the standard interpretation is proved in [BHW95].

Theorem 5.15. Let Σ be a signature, $(\approx_A)_{A \in Alg(\Sigma)}$ be a family of partial Σ-congruences, and $gen(\Sigma_0, \Sigma)$ a generating constraint over Σ. Then for all Σ-algebras A, the following holds:

1. $A \models_{\approx_A} \Phi$ if and only if $A/\approx_A \models \Phi$,
2. $A \models_{\approx_A} gen(\Sigma_0, \Sigma)$ if and only if $A/\approx_A \models gen(\Sigma_0, \Sigma)$.

Observational equalities between elements of an algebra are important examples of partial congruences. Let $\Sigma = (S, \Omega)$ be a signature, $Obs \subseteq S$ a distinguished set of *observable sorts*, and $In \subseteq S$ a second distinguished subset of so-called *input sorts* such that Σ is sensible w.r.t. In, i.e., for each $s \in S \setminus In$, there exists a term of sort s which is built from function symbols of Σ and variables of the nonempty sets X_s with $s \in In$. Two objects of a Σ-algebra are considered to be observationally equal, if they cannot be distinguished by experiments with observable results, where values of input sort can be used as inputs. This is formalized by the notion of *observable context*, which is any term $c \in T_{\Sigma}(X_{In} \cup Z)_s$, with $s \in Obs$, that contains (besides input variables) exactly one variable $z_s \in Z$. Thereby, the S-sorted family X_{In} of input variables is defined by $(X_{In})_s =_{def} \emptyset$ if $s \notin In$, $(X_{In})_s =_{def} X_s$ if $s \in In$ (where $X = (X_s)_{s \in S}$ is, as generally assumed, the family of countably infinite sets X_s of variables of sort s) and $Z = (\{z_s\})_{s \in S}$ is an S-sorted family of singleton sets $\{z_s\}$ where z_s is a variable of sort s not occurring in $(X_{In})_s$ for all $s \in S$.

Now, for any Σ-algebra A, the partial Σ-congruence

$$\approx_{Obs,In,A} = \langle A_0, \sim_{A_0} \rangle$$

is defined as follows.

- $|A_0|_s = \{a \in |A|_s \mid$ there exists a term $t \in T_\Sigma(X_{In})_s$ and an assignment $\alpha : X_{In} \to |A|$ such that $\alpha^\#(t) = a\}$ and, for any $f \in \Omega$, the operation f_{A_0}
  is defined by the restriction of f_A to A_0.
- $a \sim_{A_0} b$ holds for any $s \in S$ and $a, b \in |A_0|_s$ if, for all observable Σ-contexts $c \in T_\Sigma(X_{In} \cup \{z_s\})$ and for all assignments $\alpha : X_{In} \to |A|$, we have $\alpha_a^\#(c) = \alpha_b^\#(c)$, where $\alpha_a(z_s) = a, \alpha_b(z_s) = b$, and $\alpha_a(x) = \alpha_b(x) = \alpha(x)$ for all $x \in X_{In}$.

In this definition A_0 is the smallest subalgebra of A (with nonempty carrier sets) which is generated by the operations over the values of input sorts. In particular, if $In = \emptyset$ then A_0 is the finitely generated, smallest subalgebra of A. If $t = r$ is an equation, then it follows that

$$A \models_{\approx_{Obs,In,A}} t = r \text{ if and only if } A \models c[\sigma(t)] = c[\sigma(r)]$$

for all observable contexts c and for all substitutions σ which replace the variables occurring in t and r by arbitrary terms in $T_\Sigma(X_{In})$. This means that the equation $t = r$ is behaviorally satisfied if all its observable instances are satisfied in the standard sense.

If one chooses $In = S$, then $\approx_{Obs,S,A} = \langle A_0, \sim_{A_0} \rangle$ is a total Σ-congruence, i.e., $A = A_0$, and the relation $\sim_{A_0}$ becomes the behavioral equivalence relation used in [Rei87] and in [HWB97], and the satisfaction relation w.r.t. $\approx_{Obs,S,A}$ coincides with the behavioral satisfaction relations used in both these papers.

If $In = Obs$, then $\approx_{Obs,Obs,A} = \langle A_0, \sim_{A_0} \rangle$ is a partial Σ-congruence such that A_0 consists of all values that can be generated over observable elements of A by operations of Ω. The corresponding satisfaction relation w.r.t. $\approx_{Obs,Obs,A}$ is that which is used in [ONS93] for the behavioral satisfaction of equations. The advantage in this case is that nonobservable junk will not be considered for the satisfaction of formulas, and hence cannot cause problems, for instance, with respect to the correctness of implementations.

Definition 5.16. A *behavioral specification*

$$SPEC = (\Sigma, \Sigma_0, AX, \approx)$$

is given by a signature Σ, a subsignature Σ_0 describing the constructors, a set AX of axioms, and a family $\approx$ of partial Σ-congruences. A behavioral specification defines the following class of Σ-algebras:

$$[\![\Sigma, \Sigma_0, AX, \approx]\!] =$$
$$\{A \in Alg(\Sigma) \mid A \models_{\approx_A} \Phi \text{ for all } \Phi \in AX, A \models_{\approx_A} gen(\Sigma_0, \Sigma)\} \qquad \diamond$$

Now the question arises whether abstractor specifications and behavioral specifications may be compared or whether they define incomparable classes of Σ-algebras.

Important examples of abstractor specifications arise if one considers two algebras to be observationally equivalent if they cannot be distinguished by a predefined set of observations given by an appropriate set of terms. The construction of the set of terms can again be parameterized by two distinguished sets $Obj \subseteq S$ and $In \subseteq S$ of sorts for a given signature $\Sigma = (S, \Omega)$ such that Σ is sensible w.r.t. In. We consider the set of terms of observable sorts which may contain variables of input sorts. Two Σ-algebras A and B are called *observationally equivalent w.r.t. Obs and In*, denoted by

$$A \equiv_{Obs,In} B,$$

if there exists an S-sorted family Y_{In} of sets $(Y_{In})_{s \in S}$ of variables with $(Y_{In})_s = \emptyset$ for all $s \notin In$, $(Y_{In})_s \neq \emptyset$ for all $s \in In$ and if there are two assignments $\alpha : Y_{In} \to |A|$ and $\beta : Y_{In} \to |B|$ with surjective mappings $\alpha_s : (Y_{In})_s \to |A|_s$ and $\beta_s : (Y_{In})_s \to |B|_s$ for all $s \in In$, such that for all terms $t, r \in T_\Sigma(Y_{In})_s$ of observable sort $s \in Obs$ the following holds:

$$\alpha^\#(t) = \alpha^\#(r) \text{ if and only if } \beta^\#(t) = \beta^\#(r)$$

If one chooses $In = \emptyset$, two algebras are equivalent w.r.t. $\equiv_{Obs,\emptyset}$ if they satisfy the same equations between observable ground terms. Hence, in this case the equivalence relation $\equiv_{Obs,\emptyset}$ determines a behavioral abstraction in the sense of [ST88b]. The choice $In = Obs$ instantiates the relation $\equiv_{Obs,Obs}$ to the behavioral equivalence of algebras in the sense of [NO88, Sch87b], and finally one finds the behavioral equivalence of [Rei87] if one takes $In = S$.

There is a straightforward way to construct an equivalence $\equiv_\approx$ of algebras from a given family $\approx = (\approx_A)_{A \in Alg(\Sigma)}$ by setting for any $A, B \in Alg(\Sigma)$:

$$A \equiv_\approx B \text{ if } A/\approx_A \text{ and } B/\approx_B \text{ are isomorphic.}$$

An equivalence relation $\equiv \subseteq Alg(\Sigma) \times Alg(\Sigma)$ is called *factorizable* if there exists a family $\approx = (\approx_A)_{A \in Alg(\Sigma)}$ of partial Σ-congruences with $\equiv = \equiv_\approx$.

Corollary 5.17. *For any set Obs of observable sorts and any set In of input sorts, the equivalence $\equiv_{Obs,In}$ is factorizable by the family $\approx_{Obs,In}$ of partial Σ-congruences.*

A family $\approx = (\approx_A)_{A \in Alg(\Sigma)}$ is called *regular* if for any $A \in Alg(\Sigma)$ the quotient algebra $A/\approx_A$ is *fully abstract* w.r.t. $\approx$. A Σ-algebra B is called fully abstract w.r.t. $\approx$ if for $\approx_B = \langle B_0, \sim_{B_0} \rangle$ we have that $B_0 = B$ and $\sim_{B_0}$ is the identity relation in the carrier sets of B. The partial Σ-congruences $\approx_{Obs,In}$ as describes above are regular.

The following result relates behavioral specifications to abstractor specifications for regular families of partial congruences.

Theorem 5.18. *For any regular family $\approx$ of partial Σ-congruences, the class of models $[\![\,\Sigma, \Sigma_0, AX, \approx\,]\!]$ coincides with the closure of the class of fully abstract models out of $[\![\,\Sigma, \Sigma_0, AX\,]\!]$, under the equivalence relation $\equiv$ (which is factorizable by $\approx$).*

For a further characterization of the equivalence of behavioral and abstractor specifications, we need the following notation. For any class $C \subseteq Alg(\Sigma)$ of Σ-algebras

$$C/\approx = \{A/\approx_A \mid A \in C\}$$

denotes the class of all observable quotient algebras of C.

Theorem 5.19. *Let $\approx$ be a family of partial Σ-congruences such that for any two Σ-algebras A and B their quotient algebras $A/\approx_A$ and $B/\approx_B$ are also isomorphic. Then for any specifications (Σ, Σ_0, AX) the following conditions are equivalent:*

1. $[\![\,\Sigma, \Sigma_0, AX, \approx\,]\!] = [\![\,\Sigma, \Sigma_0, AX, \equiv_\approx\,]\!]$
2. $[\![\,\Sigma, \Sigma_0, AX\,]\!] \subseteq [\![\,\Sigma, \Sigma_0, AX, \approx\,]\!]$
3. $[\![\,\Sigma, \Sigma_0, AX\,]\!]/\approx \subseteq [\![\,\Sigma, \Sigma_0, AX\,]\!]$

This theorem gives for the general case a necessary and sufficient condition under which behavioral and abstractor specifications are semantically equivalent, and generalizes a corresponding result of [NO88]. In the special case where axioms are built up from conditional equations only condition (3) holds because the standard model class is closed under observable quotient construction. This case of the previous theorem has been proved in [Rei87].

5.5   Data types and systems as fixed points

In this last section we sketch a different approach to data type and systems specifications: *recursive type equations.*

As a motivating example, we consider the set A^* of finite lists of elements in A. One can easily see that the mapping

$$\varphi : A^* \to A \times A^* \cup \{*\}$$

with $\varphi(\varepsilon) = *$ and $\varphi(cons(x,l)) = \langle x, l \rangle$ is a bijection. The inverse mapping is given by $\varphi^{-1}(*) = \varepsilon$ and $\varphi^{-1}(\langle x, l\rangle) = cons(x,l)$. Thus, the set A^* is a fixed point of the recursive set equation $M \simeq A \times M \cup \{*\}$.

In general, we call a set X a fixed point of an equation

$$X \simeq A \times X \cup \{*\}$$

if it solves (up to bijection) the recursive set equation.

Is A^* (up to bijection) the only fixed point? Certainly not. The set

$$A^\infty = \{l : \mathbb{N}^+ \to A\} \cup \{l : \{1, \dots, n\} \to A \mid n \geq 0\}$$

is another fixed point, since the mapping

$$\psi : A^\infty \to A \times A^\infty \cup \{*\}$$

defined by $\psi(l : \emptyset \to A) = *$ and $\psi(l : dom(l) \to A) = \langle head(l), tail(l) \rangle$ is a bijection, where $head(l) = l(1)$ and $tail(l) = l'$ with $l'(n) = l(n+1)$, and $dom(l') = \{1, \dots, n-1\}$ if $dom(l) = \{1, \dots, n\}, n \geq 0$.

In general a recursive type equation

$$M \simeq \mathcal{R}(M)$$

is given by a set construction $\mathcal{R}$ yielding the right-hand side, which may be formally represented by a functor $\mathcal{R} : Set \to Set$ from the category of sets to itself, see [BW95].

In this categorical setting, the notion of an algebra gives rise to the notion of an $\mathcal{R}$-algebra, and to a categorically dual notion of an $\mathcal{R}$-coalgebra.

Definition 5.20. 1. For any category $\mathbb{C}$ an endofunctor $\mathcal{R} : \mathbb{C} \to \mathbb{C}$ is called a *signature functor* over $\mathbb{C}$. An $\mathcal{R}$-*algebra* is a pair $\langle A, a \rangle$, where A is an object of $\mathbb{C}$ and $a : \mathcal{R}(A) \to A$ a morphism of $\mathbb{C}$. An $\mathcal{R}$-*homomorphism* $h : \langle A, a \rangle \to \langle B, b \rangle$ is given by a morphism $h : A \to B$ of $\mathbb{C}$ such that $a; h = \mathcal{R}(h); b$. The category of all $\mathcal{R}$-algebras and $\mathcal{R}$-homomorphisms is denoted by $Alg(\mathcal{R})$.

2. An $\mathcal{R}$-*coalgebra* is a pair $\langle A, a \rangle$, where A is an object of $\mathbb{C}$ and $a : A \to \mathcal{R}(A)$ is a morphism of $\mathbb{C}$. An $\mathcal{R}$-*cohomomorphism* $h : \langle A, a \rangle \to \langle B, b \rangle$ is given by a morphism $h : A \to B$ of $\mathbb{C}$ such that $a; \mathcal{R}(h) = h; b$. The category of all $\mathcal{R}$-coalgebras and $\mathcal{R}$-cohomomorphisms is denoted by $CoAlg(\mathcal{R})$.   ◇

For more details on $\mathcal{R}$-algebras see [AT89, BW95], and on $\mathcal{R}$-coalgebras and their use for systems specification see [Rut96, JR97].

The right-hand side of the recursive set equation above is obtained from the functor $\mathcal{R}_A : Set \to Set$ with

$$\mathcal{R}_A(M) = A \times M \cup \{*\}.$$

Any fixed point of $M \simeq \mathcal{R}_A(M)$ is both an $\mathcal{R}_A$-algebra and an $\mathcal{R}_A$-coalgebra. The $\mathcal{R}_A$-coalgebra structure of the finite lists is given by the mapping $\varphi : A^* \to A \times A^* \cup \{*\} = \mathcal{R}_A(A^*)$, and the $\mathcal{R}_A$-algebra structure is given by the inverse mapping $\varphi^{-1} : \mathcal{R}_A(A^*) \to A^*$.

In the same way the set A^∞ defines an $\mathcal{R}_A$-algebra and an $\mathcal{R}_A$-coalgebra.

The relations to the initiality and finality constraints introduced above are given by the following [JR97].

Corollary 5.21. *Let $\mathcal{R} : Set \to Set$ be a functor. An initial $\mathcal{R}$-algebra, if it exists, is a least fixed point of $M \simeq \mathcal{R}(M)$, and each final $\mathcal{R}$-coalgebra, if it exists, is a greatest fixed point of $M \simeq \mathcal{R}(M)$.*

A second very useful reference to initial $\mathcal{R}$-algebras and final $\mathcal{R}$-coalgebras is [CS92, CS95]. These papers extend the approach of T. Hagino [Hag87]. R. Cockett and T. Fukushima [CF92], have developed an experimental programming language, called CHARITY, which implements both initial algebras and final coalgebras, and supports their combined application.

6 Structuring and Modularity

Fernando Orejas

Dept. L. S. I., Univ. Polit. Catalunya
Barcelona, Spain
orejas@lsi.upc.es

6.1 Introduction

Algebraic specification was originally introduced as a method to formally describe abstract data types. This means that the concern for structuring and modularity was present from the very beginning in this approach. In this sense, it is no surprise that the first algebraic specification languages including the (now considered) most basic structuring constructs were developed so early [BG77, GT77].

The algebraic approach originally concentrated on the use of equational logic as the basic formalism for specification. This use was successful in many senses. In particular, the good behavior of equational logic with respect to a number of semantic constructs has helped to facilitate the study of new specification concepts and constructs. At the same time, the categorical techniques used to obtain these results have allowed the generic study of specification constructs with independence of any specific formalism. In this sense, research in the foundations of algebraic specification methods departed from equational logic a number of years ago and has concentrated, to a large extent, on the generic study of specification constructs that can be used in connection with any "reasonable" specification approach.

In this chapter, we describe the main techniques for the semantic definition of some of the most used structuring and modular constructs. Our main aim will be to study the generic, "institution-independent", version of each construct. However, in order to provide intuition, in most cases, we will first study these constructions in connection with equational logic.

The rest of the chapter is structured as follows: in Section 2 we present some basic material on institutions, in order to fix some notation and terminology. Section 3 introduces the most basic structuring operations. In Section 4, we present the most used approaches to parameterization and parameter passing. Finally, in Section 5, modules and modular systems are studied.

6.2 Preliminaries

In this section we briefly present the abstract framework that we will use throughout the chapter to present all constructions and results.

Our specifications will be built over an arbitrary *institution* $\mathcal{L} = (Sig, Sen,$ $Mod, \models)$ [GB84, GB92], unless explicitly stated. Moreover, we assume that $\mathcal{L}$ is equipped with some notion of signature inclusion (see, e.g., [DGS91]). As usual, Sig denotes the category of signatures of $\mathcal{L}$; $Sen\colon Sig \to Set$ denotes the functor that maps every signature Σ into the set of all Σ-sentences and every signature morphism into the corresponding mapping translating sentences from one signature into sentences of the other; $Mod\colon Sig \to Cat^{op}$ denotes the functor mapping every signature Σ into the category of all Σ-structures and every signature morphism into its associated forgetful functor; finally, $\models$ denotes th! e satisfaction relation of the given institution. $\mathcal{L}$ is an *exact (semiexact)* institution iff Sig has finite colimits (pushouts) and, in addition, Mod transforms finite colimits (pushouts) in Sig into limits (pullbacks) in Cat.

Given an institution $\mathcal{L}$, one can define specifications over $\mathcal{L}$ in two different ways. On the one hand, specifications can be defined *syntactically* as pairs consisting of a signature and a set of axioms. On the other hand, a more *semantic* approach consists in defining specifications as pairs formed by a signature and a set of models over this signature [ST88a]. These classes of specifications can be made into categories by defining appropriate notions of specification morphisms as follows:

1. $SyntCatSpec_{\mathcal{L}}$ (or just $SyntCatSpec$ if $\mathcal{L}$ is clear from the context) is the syntactic category of specifications whose objects are all pairs (Σ, E), where Σ is a signature in Sig and E is a set of Σ-sentences (called the *axioms* of the specification), i.e. $E \subseteq Sen(\Sigma)$, and whose morphisms $h\colon (\Sigma, E) \to (\Sigma', E')$ are signature morphisms $h\colon \Sigma \to \Sigma'$ satisfying $Sen(h)(E) \subseteq E'$.

2. $SemCatSpec_{\mathcal{L}}$ (or just $SemCatSpec$ if $\mathcal{L}$ is clear from the context) is the semantic category of specifications whose objects are all pairs (Σ, M), where M is a set of Σ-models, i.e. $M \subseteq Mod(\Sigma)$, and whose morphisms $h\colon (\Sigma, M) \to (\Sigma', M')$, are signature morphisms $h\colon \Sigma \to \Sigma'$ satisfying $U_h(M') \subseteq M$, where U_h, usually called the *forgetful functor* associated with h, is a more standard notation for $Mod(h)$.

In either case, given a specification SP, we will assume that $Signature(SP)$ denotes the signature of the specification SP: if $SP = (\Sigma, E) \in SyntCatSpec$ (resp. $SP = (\Sigma, M) \in SemCatSpec$) then $Signature(SP) = \Sigma$. Similarly, we will assume that if $SP = (\Sigma, E) \in SyntCatSpec$ (resp. $SP = (\Sigma, M) \in SemCatSpec$) then $Axioms(SP) = E$ (resp. $Models(SP) = M$).

Remark 6.1.

1. The syntactic category of specifications can also be defined in terms of a slightly more general notion of morphism, in particular, by defining specification morphisms as signature morphisms satisfying $E' \models Sen(h)(E)$.

2. Inclusions of signatures can be extended to define inclusions of specifica-
tions in an obvious way: $(\Sigma, E) \hookrightarrow (\Sigma', E')$ (resp. $(\Sigma, M) \hookrightarrow (\Sigma', M')$)
iff $\Sigma \hookrightarrow \Sigma'$ and $E \subseteq E'$ (resp. $U_h(M') \subseteq M$).
3. It may be noted that if the given category of signatures, $\underline{Sig}$, has pushouts
then $\underline{SyntCatSpec}$ and $\underline{SemCatSpec}$ also have pushouts. In particular,
given specifications $SP0 = (\Sigma 0, E0)$, $SP1 = (\Sigma 1, E1)$, $SP2 = (\Sigma 2, E2)$
(resp. $SP0 = (\Sigma 0, M0)$, $SP1 = (\Sigma 1, M1)$, $SP2 = (\Sigma 2, M2)$) and morphisms $f1 \colon SP0 \to SP1$ and $f2 \colon SP0 \to SP2$ in $\underline{SyntCatSpec}$ (resp.
$\underline{SemCatSpec}$) one can build a pushout as in Figure 6.1 below, where the
pushout object $SP3 = (\Sigma 3, Sen(g1)(E1) \cup Sen(g2)(E2))$ (resp. $SP3 = (\Sigma 3, M3)$, with $M3 = \{A3 \in Mod(\Sigma 3)/U_{g1}(A3) \in M1 \text{ and } U_{g2}(A3) \in M2\}$), and $\Sigma 3$, $g1$ and $g2$ are defined in terms of the pushout of signatures
shown in Figure 6.2.

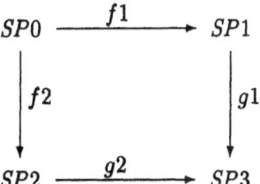

Fig. 6.1. Pushout of specifications

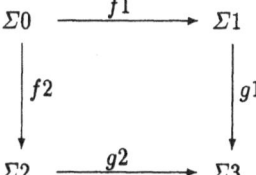

Fig. 6.2. Pushout of signatures

4. The (syntactic) category of specifications over $\mathcal{L}$, together with the *model
functor* $Mod \colon \underline{SyntCatSpec}_{\mathcal{L}} \to \underline{Cat}^{op}$ defined by
   - for every specification $SP = (\Sigma, E)$, $Mod(SP)$ is the full subcategory
   of $Mod(\Sigma)$ including all objects that satisfy all axioms in E, i.e.
   $A \in Mod(SP)$ iff $A \in Mod(\Sigma)$ and $\forall \alpha \in E$ $A \models_\Sigma \alpha$;
   - for every $h \colon (\Sigma, E) \to (\Sigma', E')$, $Mod(h) \colon Mod(\Sigma', E') \to Mod(\Sigma, E)$
   is the functor that maps every A in $Mod(\Sigma', E')$ (and, therefore, in
   $Mod(\Sigma')$) into $U_h(A)$ and every $f \colon A1 \to A2$ in $Mod(\Sigma', E')$ into
   $U_h(f)$. In both cases, the satisfaction condition ensures that $U_h(A)$
   and $U_h(f)$ are in $Mod(\Sigma, E)$;

   form an indexed category (see, e.g., [BGT91]), called a *specification frame*
   in [CEOB, JOE95]. Moreover, if $\mathcal{L}$ is semiexact then the associated specification frame has *pushouts* and *amalgamations* in the sense of [EPO89].

When working at the syntactic level, i.e. within $\underline{SyntCatSpec}$, one way
of assigning meaning to specifications is by defining a mapping Sem from
$\underline{SyntCatSpec}$ into $\underline{SemCatSpec}$, since this allow us to consider that the
class of models of any specification SP is $Models(Sem(SP))$. One possible
(obvious) definition of such a mapping would be

$$Sem(SP) = (Signature(SP), Mod(SP))$$

for every $SP = (\Sigma, E)$ in $\underline{SyntCatSpec}$. However, often this is not considered fully satisfactory. In particular, in a number of approaches the semantics of SP is defined in terms of a specific subclass of the class of all Σ-models satisfying E. For example, a standard approach in the framework of equational logic consists in defining the semantics of SP in terms of the class of initial models of SP. As a consequence, we will assume that the underlying institution is equipped with a mapping $Sem_{\mathcal{L}} \colon \underline{SyntCatSpec}_{\mathcal{L}} \rightarrow \underline{SemCatSpec}_{\mathcal{L}}$ (or just Sem if $\mathcal{L}$ is implicit from the context) such that for every $\overline{SP} = (\Sigma, E)$ in $\underline{SyntCatSpec}_{\mathcal{L}}$ $Signature(Sem(SP)) = \Sigma$ and $Models(Sem(SP)) \subseteq Mod(\overline{SP})$. Typically, if for every SP in $\underline{SyntCatSpec}_{\mathcal{L}}$ $Models(Sem(SP))$ consists of, at most, one model up to isomorphism, then Sem is said to be a *monomorphic semantics*, otherwise Sem will be called *polymorphic* or *loose*.

Remark 6.2.

1. It can be noted that $\underline{SemCatSpec}$ can be considered more expressive than $\underline{SyntCatSpec}$, since in general there may exist classes of Σ-models M such that there does not exist any set of Σ-axioms E with $Sem(\Sigma, E) = (\Sigma, M)$.
2. Some typical examples of semantic mappings for equational logic are: initial algebra semantics (where $Models(Sem(SP))$ is the class of all initial SP-algebras), loose semantics (where $Models(Sem(SP)) = Mod(SP)$ or, in some approaches, the class of all finitely generated SP-models), final algebra semantics (where $Models(Sem(SP))$ is the class of all algebras which are final in the category of non-trivial SP-models), etc.
3. It must be noted that, in general, Sem cannot be defined in an obvious way as a (contravariant) functor. For instance, this happens in the case of initial algebra semantics since, given $h \colon (\Sigma, E) \rightarrow (\Sigma', E')$, $U_h(T_{(\Sigma', E')})$ may differ from $T_{(\Sigma, E)}$.

6.3   Basic operations for building specifications

In this section we introduce the most basic operations for building specifications in a structured manner. In particular we introduce four operations that, possibly with some variations, can be found in almost every (algebraic) specification language.

When defining the semantics of a specification-building operation, two approaches may be considered, depending on the given framework. In particular, in some languages, the kind of operations provided are not considered to add any expressive power to the underlying specification formalism. This means that the given operations are used for allowing the stepwise construction of specifications and to help in understanding large specifications by imposing some structure on them. However, in this case, every structured specification built over the given operations would be equivalent to a (large) unstructured specification. This approach is the basis of the specification language ACT ONE [EM85] and, to some extent, of Clear and OBJ [BG77, GW88]. In this context, a quite natural semantics for a specification building operation would be defined at the specification level (i.e. within *SyntCatSpec*). This means that an operatio! n would be seen as a function mapping basic specifications (and perhaps other "things", such as morphisms) into basic specifications (i.e. signatures and axioms).

In some other frameworks, specification-building operations are seen as constructions that add expressive power to the underlying specification formalism. This means that a specification built using such operations may be not equivalent to any basic specification. This approach is the basis of the specification language ASL [SW83, Wir86]. In this context, the semantics for a specification building operation should be defined at the model level (i.e. within *SemCatSpec*). This means that an operation would be seen as a function mapping semantic specifications (in the sense of signatures and model classes) into semantic specifications.

Nevertheless, it must be noted that in the former case, when operations do not add expressive power, nothing prevents from defining their semantics at the model level. In particular, in some cases like in ACT ONE [EM85] both semantic definitions are provided, since they are considered complementary, and proved compatible.

As said in the introduction, each operation will be studied, first, considering equational logic as the underlying specification formalism and, then, considering its general institution-independent version.

6.3.1   Enrichments

This is one of the two most basic operation for building a specification incrementally. The idea is that we add more "elements" (e.g. sorts, operations) to a given specification. For instance, given the following boolean specification:

$$BOOL = \textbf{sorts } bool$$
$$\textbf{opns } true: \ \rightarrow bool$$
$$false: \ \rightarrow bool$$

one may define the natural numbers over the booleans as follows:

$$NAT = \textbf{enrich } BOOL \textbf{ by}$$
$$\textbf{sorts } nat$$
$$\textbf{opns } 0: \ \rightarrow nat$$
$$suc: nat \rightarrow nat$$
$$\leq: nat \ nat \rightarrow bool$$
$$\textbf{vars } X, Y: nat$$
$$\textbf{eqns } 0 \leq X = true$$
$$X \leq Y = suc(X) \leq suc(Y)$$
$$suc(X) \leq Y = false$$

In particular, when dealing with equational logic, standard syntax for enrichments is the following one:

enrich SP **by** $(\Sigma 1, E1)$

where SP is a specification and $Signature(SP) \cup \Sigma 1$ is a signature and $E1$ is a set of equations over $Signature(SP) \cup \Sigma 1$.

At the specification level, the meaning of an enrichment is the specification (Σ, E), where $\Sigma = Signature(SP) \cup \Sigma 1$ and $E = Axioms(SP) \cup E1$. At the model level, enrichments are given meaning by, first, associating a given semantic construction with inclusions of specifications and, then, defining the class of models of the enrichment $SP1 = $ **enrich** SP **by** $(\Sigma 1, E1)$ as the result of applying to the models of SP the construction associated with the inclusion $h: SP \hookrightarrow SP1$. For example, some standard choices in this sense would be:

1. The free functor associated with the given inclusion of specifications. In that case:

$$Models(\textbf{enrich } SP \textbf{ by } (\Sigma 1, E1)) = \{F_h(A)/A \in Models(SP)\}$$

2. A mapping associating, to every A in $Models(SP)$, the class of all its "proper extensions". In that case, $Models(\textbf{enrich } SP \textbf{ by } (\Sigma 1, E1))$ is defined as:

$$\{A \in Mod(\Sigma \cup \Sigma 1)/A|_\Sigma \in Models(SP) \text{ and } A \models E1\}$$

where $\Sigma = Signature(SP)$, $A|_\Sigma$ is a shorthand for $U_h(A)$ and h is the signature inclusion $h: \Sigma \hookrightarrow \Sigma \cup \Sigma 1$

Remark 6.3.

1. Some authors distinguish between *enrichments* and *extensions*, where the former are supposed not to add any new sort to the enriched specification.
2. Often enrichments are required to be a *conservative extension* of the enriched part, in the sense that:

$$Models(\textbf{enrich } SP \textbf{ by } (\Sigma 1, E1))|_{SP} = Models(SP)$$

where $\Sigma = Signature(SP)$. It may be noted that none of the above definitions guarantee this, unless some additional conditions are satisfied such as hierarchy consistency and sufficient completeness (e.g. see [EM85]), used in connection with initial algebra semantics. For instance, for the last definition we have that:

$$\{A \in Mod(\Sigma \cup \Sigma 1)/A|_\Sigma \in Models(SP) \text{ and } A \models E1\} \subseteq Models(SP)$$

but the converse is not true, in general.

3. In the framework of initial algebra semantics, free functor and specification level semantics for enrichments are compatible, in the sense that given the enrichment:

   enrich SP **by** $(\Sigma 1, E1)$

   with $SP = (\Sigma, E)$, we have that:

   $F_h(T_{SP}) = T_{SP1}$

   where $SP1 = (\Sigma \cup \Sigma 1, E \cup E1)$.

The standard (institution-independent) generalization of the notion of enrichment consists in considering that an enrichment is defined in terms of an arbitrary signature morphism (or just an inclusion) together with a set of axioms, i.e. the syntax for this operation may be:

enrich SP **by** (f, E)

where f is a morphism from $Signature(SP)$ to a given $\Sigma 1$ and E is a set of $\Sigma 1$-axioms. Then, at the specification level, the result of that enrichment would obviously be the specification having $\Sigma 1$ as signature and having as axioms the union of $E1$ and the translation, via f, of the axioms of SP. That is the semantics of **enrich** SP **by** (f, E) is $(\Sigma 1, Sen(f)(Axioms(SP)) \cup E1)$. On the other hand, in order to define the meaning of enrichments at the model level, we may note that one can associate the following class of models to the given enrichment:

$\{A \in Mod(\Sigma 1)/U_f(A) \in Models(SP) \text{ and } A \models E\}$

Let $SP1$ be the specification consisting of the signature $\Sigma 1$ and that class of models. Obviously, we could define $SP1$ as the meaning of the enrichment. However, a number of semantic constructions that are often used to give meaning to enrichments would not fit into this definition (e.g. free functor semantics). In order to provide a truly general definition, we must note, first, that the signature morphism f can also be considered as a specification morphism from SP into $SP1$ and, then, we must assume that the underlying institution is equipped with a mapping (used to give meaning to enrichments) that associates a "construction" $\kappa_h \colon Models(SP) \to 2^{Models(SP')}$ (or $\kappa_h \colon Models(SP) \to Models(SP')$) to every specification morphism $h \colon SP \to SP'$. In that case, the semantics of **enrich** SP **by** (f, E) would be defined:

$Models(\textbf{enrich } SP \textbf{ by } (f, E)) = \{A'/\exists A \in Models(SP) \text{ such that } A' \in \kappa_f(A)\}$

Remark 6.4.

1. As mentioned for the equational case, the two most typical constructions $\kappa 1_h$ and $\kappa 2_h$, associated with a morphism $h \colon SP \to SP'$, used to define

the meaning of enrichments are the free construction associated with h, i.e. $\kappa 1_h = F_h$, and the loose extension defined:

$$\kappa 2_h(A) = \{A' \in Mod(SP')/U_h(A') = A\}$$

for every $A \in Mod(SP)$. The former is typically used in connection with the initial approach while the latter is used in connection with the loose approach.

2. Obviously, specification and model-level semantics are compatible if and only if:

$$Models(\Sigma 1, Axioms(SP) \cup E1) = \{A'/\exists A \in Models(SP) \text{ with } A' \in \kappa_f(A)\}$$

This happens, for instance, in the initial and in the "pure" loose approach where the constructions used are $\kappa 1_h$ and $\kappa 2_h$ defined above.

3. It may be noticed that, for the same loose construction κ_h as above, the two semantics may fail to be compatible if the semantics of specifications is not the "pure" loose one, i.e. if $Models(SP) = Mod(SP)$ does not hold in general. For instance, this happens when considering the class of all finitely generated SP-models as the meaning of SP. In that case, we may consider that the enrichment operation adds expressive power to our specification language (see [BBTW81]).

4. Other constructors that have been used in the literature to define the semantics of enrichments are associated with other kinds of (monomorphic) semantics, such as final semantics (see e.g. [Gan83, Wan79]) or behavioral semantics (see e.g. [Rei81, NO88])

6.3.2   Union

This is the other most basic operation for building specifications incrementally. The idea is that we build a new specification by putting together two smaller ones, which may share some subspecification. For instance, the above specification of the natural numbers could have been built by first specifying the booleans as above, then defining the natural numbers with just the most basic operations, i.e.:

$$NAT0 = \textbf{sorts } nat$$
$$\textbf{opns } 0: \; \rightarrow nat$$
$$suc: nat \rightarrow nat$$

and, finally, enriching the *union* of the two specifications with an ordering relation:

$$NAT = \textbf{enrich } NAT0 + BOOL \textbf{ by}$$
$$\textbf{opns } \leq: nat \; nat \rightarrow bool$$
$$\textbf{vars } X, Y: nat$$
$$\textbf{eqns } 0 \leq X = true$$
$$X \leq Y = suc(X) \leq suc(Y)$$
$$suc(X) \leq Y = false$$

A standard syntax for the operation of union may be the following one:

$$SP1 +_{SP0} SP2$$

where $SP0$ is a subspecification of $SP1$ and $SP2$ denoting the common parts of $SP1$ and $SP2$. In the example above $SP0$ is the empty specification.

When dealing with equational logic, if $SPi = (\Sigma i, Ei)$, for $0 \le i \le 2$, at the specification level, the meaning of $SP1 +_{SP0} SP2$ may be simply defined as the specification $(\Sigma 1 \cup \Sigma 2, E1 \cup E2)$, provided that $\Sigma 1 \cap \Sigma 2 = \Sigma 0$. If we allow $\Sigma 1 \cap \Sigma 2$ to be different than $\Sigma 0$, then a set-theoretic definition of the union is slightly more complex and implies some kind of renaming of the elements (sorts or operations symbols) which are in $\Sigma 1 \cap \Sigma 2$ but not in $\Sigma 0$. However, in this general case, the categorical definition of the union of specifications is quite simple. In particular, the result of $SP1 +_{SP0} SP2$ can be defined as the pushout of the diagram in Figure 6.3, where $f1$ and $f2$ denote the inclusions $SP0 \hookrightarrow SP1$ and $SP0 \hookrightarrow SP2$, respectively.

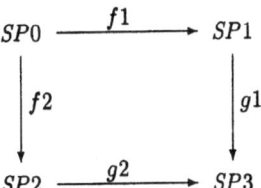

Fig. 6.3. Pushout of specifications

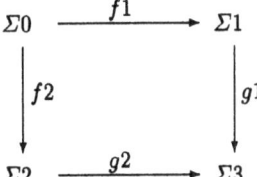

Fig. 6.4. Pushout of signatures

At the model level, the semantics of $SP1 +_{SP0} SP2$ is a pair $(\Sigma 3, M3)$, where $\Sigma 3$ can be defined, as above, in terms of a pushout (on the category of signatures) as in Figure 6.4.

The most obvious way of defining $M3$ is by considering all $\Sigma 3$ models such that their "reduct" via $g1$ and $g2$ is a model of $SP1$ and $SP2$, respectively, i.e.:

$$\{A3 \in Mod(\Sigma 3)/U_{g1}(A3) \in Models(SP1) \text{ and } U_{g2}(A3) \in Models(SP2)\}$$

This is equivalent to considering that the resulting models in $M3$ are built by *gluing* or *amalgamating* all the models of $SP1$ and $SP2$. That is, we may define the amalgamation of every pair of models $A1 \in Models(SP1)$ and $A2 \in Models(SP2)$ such that $A1|_{\Sigma 1} = A2|_{\Sigma 2} \in Models(SP0)$ as the algebra $A3$, denoted $A1 +_{A0} A2$ (where $A0 = A1|_{\Sigma 1} = A2|_{\Sigma 2}$) as:

For every $s1 \in S1$ $A3_{g1(s1)} = A1_{s1}$ and for every $\sigma 1: s1...sn \to sn \in \Sigma 1$ $g1(\sigma 1)_{A3} = \sigma 1_{A1}$ and, similarly, for every $s2 \in S2$ $A3_{g2(s2)} = A2_{s2}$ and for every $\sigma 2: s1...sn \to sn \in \Sigma 2$ $g2(\sigma 2)_{A3} = \sigma 2_{A2}$, $M3$ can equivalently be defined as

$$\{A1 +_{A0} A2/A1 \in Models(SP1), A2 \in Models(SP2) \text{ and } A0 \in Models(SP0)\}$$

In this case we say that $M3$ is the amalgamation of $Models(SP1)$ and $Models(SP2)$ with respect to $Models(SP0)$, denoted

$$M3 = Models(SP1) +_{Models(SP0)} Models(SP2)$$

Remark 6.5.

1. In the case of equational logic, the specification-level semantics and the model-level semantics are compatible (see the amalgamation lemma in [EM85]). This means that if $SP3$ is the specification-level semantics of $SP1 +_{SP0} SP2$ then $Models(Sem(SP3))$ is equal to

$$Models(Sem(SP1)) +_{Models(Sem(SP0))} Models(Sem(SP2))$$

2. The amalgamation operation defined over classes of models coincides with the result of the pullback diagram in Figure 6.5.

Mod(SP0)
U_f1
Mod(SP1)
U_f2
U_g1
Mod(SP2)
U_g2
Mod(SP3)

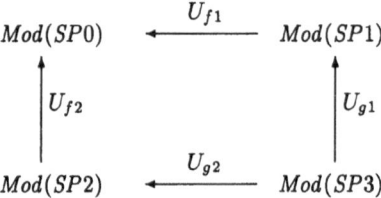

Fig. 6.5. Pullback of model classes

It must be noted that, as a consequence, the model-level semantics of the union operation may also be defined in terms of a pushout since, given a signature pushout, as in Figure 6.4, the diagram in Figure 6.6 is also a pushout.

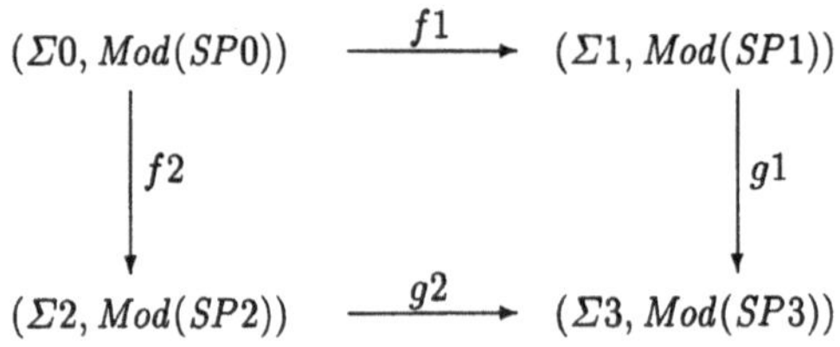

(Σ0, Mod(SP0))
f1
(Σ1, Mod(SP1))
f2
g1
(Σ2, Mod(SP2))
g2
(Σ3, Mod(SP3))

Fig. 6.6. Pushout in <u>*SemCatSpec*</u>

The previous constructions can be easily generalized to any arbitrary institution $\mathcal{L}$, provided that the corresponding category of signatures has pushouts (see Remark 6.1.3). In particular, the institution-independent version of the union operation is defined in terms of two signature morphisms (or just inclusions), i.e. the syntax for this operation is:

$$SP1 +_{(SP0,f1,f2)} SP2$$

where $fi: SP0 \to SPi$ (with $i = 1, 2$) are specification morphisms. Then, at the specification level, the semantics is again defined by the corresponding pushout construction, as in Figure 6.3.

At the model level, we may define the semantics of the union operation as a direct generalization of the equational case, i.e. $Models(SP1 +_{(SP0,f1,f2)} SP2)$ is equal to:

$$\{A3 \in Mod(\Sigma 3)/U_{gi}(A3) \in Models(SPi)(i = 1, 2)\}$$

or equivalently it can be defined as a pushout as in Figure 6.6.

Moreover, generalizing the notation used in the equational case, given models $A0 \in Models(SP0)$, $A1 \in Models(SP1)$ and $A2 \in Models(SP2)$ such that $U_{f1}(A1) = U_{f2}(A2) = A0$, the model $A3 \in Models(SP3)$, satisfying that $U_{g1}(A3) = A1$ and $U_{g2}(A3) = A2$ will be denoted $A3 = A1 +_{A0} A2$ and called the amalgamation of $A1$ and $A2$ with respect to $A0$. Similar notation and terminology will be used for model classes and morphisms.

In many cases, the specification-level semantics and the model-level semantics of the union operation are compatible if the underlying institution is semiexact. In particular, this obviously happens in the purely loose case. It also happens, for instance when dealing with initial algebra semantics in the following sense.

Proposition 6.6. *Given a pushout diagram as in Figure 6.3 over a semiexact institution $\mathcal{L}$, if $U_{f1}(T_{SP1}) = U_{f2}(T_{SP2}) = T_{SP0}$ then $T_{SP3} = T_{SP1} +_{T_{SP0}} T_{SP2}$.*

Proof. This proposition is a consequence of the (so-called) extension lemma (e.g., see [EM85]). In particular, we have to prove that the model $A3 \in Mod(SP3)$ satisfying that $U_{g1}(A3) = T_{SP1}$ and $U_{g2}(A3) = T_{SP2}$ is initial in $Mod(SP3)$. Let $A3' \in Mod(SP3)$, by initiality of T_{SP0}, T_{SP1} and T_{SP2}, we know that there are unique morphisms $h1: T_{SP1} \to U_{g1}(A3')$ in $Mod(SP1)$ and $h2: T_{SP2} \to U_{g2}(A3')$ in $Mod(SP2)$ such that $U_{f1}(h1) = U_{f2}(h2): T_{SP0} \to U_{f2}(U_{g2}(A3'))$. But this means that there is a (unique) morphism $h3: A3 \to A3'$ in $Mod(SP3)$ satisfying that $U_{g1}(h3) = h1$ and $U_{g2}(h3) = h2$. This implies that $h3$ is the unique morphism from $A3$ to $A3'$ since, if $h3': A3 \to A3'$ then $U_{g1}(h3')$ must be equal to $h1$ by initiality of T_{SP1} and $U_{g2}(h3')$ must be equal to $h2$ by initiality of T_{SP2}. □

6.3.3 Derive

Often, when building a specification, we may have to define some *auxiliary* sorts and operations that are needed in order to adequately describe the intended behavior of a given system. This may be caused by a lack of expressivity of the specification formalism or just for readability reasons. For instance, in [TWW82] it is proved that auxiliary operations may be needed for specifying some computable data types when using equational logic with initial semantics. However, in order to avoid an overpopulation of sorts and operations we may want, at some point, to *hide* these auxiliary elements. This is achieved by means of the operation *derive*. For instance, suppose that we have defined the following specification *NAT*1:

$$NAT1 = \textbf{enrich } NAT \textbf{ by}$$
$$\textbf{opns } 1: \ \rightarrow nat$$
$$+: nat \ nat \rightarrow nat$$
$$*: nat \ nat \rightarrow nat$$
$$\textbf{vars } X, Y: nat$$
$$\textbf{eqns } 1 = suc(0)$$
$$0 + X = X$$
$$X + suc(Y) = suc(X + Y)$$
$$0 * X = 0$$
$$X * suc(Y) = X + (X * Y)$$

Now, we may consider that *suc* is not useful anymore, since $suc(n)$ can be substituted by $n + 1$, if needed. In addition, in order to deal with too-large specifications, we may want to "forget" about *suc*. We may do that by stating:

$$NAT2 = \textbf{derive } \Sigma NAT2 \textbf{ from } NAT1$$

where $\Sigma NAT2$ would be a signature including all the operations in *NAT*1 but *suc*.

It must be noted that the kind of "hiding" provided by the *derive* operation is not quite equivalent to the one provided by module constructions of some programming languages. In particular, when some operation is considered hidden or "private" to a module in a given programming language (e.g. Ada) this means that the "user" of the module should not know about the existence of such operation and this also means that the (formal or informal) specification of the module should provide an adequate description of the contents of the module without "mentioning" the private operation. However, the "user" of a specification may need to know, at some point, the specification of the operations that are "hidden" by a *derive* operation in order to understand the whole specification, since there may be not a more abstract description of a given specification unit that allows one to understand the elements of that unit without "knowing about" the auxiliary operations. For instance, in the above example, one would be unable to understand how the plus operation is defined without knowing about the existence of *suc*.

When dealing with equational logic, a standard syntax for the derive operation may be the following one:

$$\textbf{derive } \Sigma 1 \textbf{ from } SP$$

where $\Sigma 1$ is a signature included in $Signature(SP)$. In particular, the sorts and operations not in $\Sigma 1$ are considered to be auxiliary.

At the specification level, given the specification $SP = (\Sigma, E)$ the meaning of **derive** $\Sigma 1$ **from** SP is the specification $(\Sigma 1, E1)$, where $E1$ is the set of all $\Sigma 1$-equations which are a logical consequence of SP. At the model level, the most obvious definition is the specification $(\Sigma 1, Models(SP)|_{\Sigma 1})$.

Remark 6.7.

1. Obviously, the specification and the model-level semantics for the derive operation are compatible.
2. It must be noted that, in general, the specification-level semantics of the operation **derive** $\Sigma 1$ **from** SP may be not an "acceptable" specification if we assume that specifications must be finite.

The standard (institution-independent) generalization of the derive operation consists in considering that derivations are defined in terms of an arbitrary signature morphism (or just an inclusion)

derive f **from** SP

where f is a morphism from a given $\Sigma 1$ to $Signature(SP)$. The semantics of this operation is just the straightforward generalization of the semantics given for the equational case. This means that, at the specification level, the meaning of **derive** f **from** SP is the specification $(\Sigma 1, E1)$ where

$$E1 = \{\alpha/SP \models Sen(f)(\alpha)\}$$

and, at the model level, its meaning is the specification $(\Sigma 1, U_f(Models(SP)))$.

6.3.4   Rename

The last operation that we consider is the rename operation used to change the names of some element (sorts and operations) of a given specification. This operation is needed, for instance, to avoid confusion when putting together two specifications, developed independently one from the other, sharing some names for different elements of the specification. A standard syntax of this operation is:

rename SP **by** f

where f is an isomorphism from $Signature(SP)$ into another signature $\Sigma 1$. The semantics of this operation can be explained easily in terms of the semantics of a derive operation, since we may consider that:

rename SP **by** f = **derive** f^{-1} **from** SP

where f^{-1} is the inverse of f.

6.4   Parameterizations and parameter passing

A key concept for the structuring and reusability of specifications (and programs) is the notion of parameterization. In the algebraic specification literature several different approaches for the study of parameterized specifications have been considered. In this paper we will present the two most well-known ones. The first notion is a generalization of the notion of enrichment. In particular, it is based in considering that a parameterization is just an enrichment where the enriched part is not fixed [BG77, TWW82, Ehr82, EKT+84, GM82]. In this case, instantiation of a parameterization consists in "fixing" the enriched part. The second approach is based on considering that a parameterized specification is any (generic) specification expression built over some specification variables [SW83, Wir86, SST92, Cen94]. In this case, instantiating a parameterization consists in substituting these variables by any (adequate) specification. Another existing approach [ETLZ82, CO88, OSC89, Mos89b], which we will not study in this paper, is based on considering that any (adequately structured) specification can be considered parameterized, without explicitly defining the parameters. In this case, instantiating a specification consists in substituting "part" of the structured specification by another specification.

6.4.1   Parameterizations as generic enrichments

The most classical approach to parameterizations [TWW82] was heavily influenced by the intuition underlying the notion of a parameterized data type, i.e. a data type built over some generic or varying "part". According to this intuition, one may consider that a parameterized specification is some kind of "generic enrichment" where the enriched (generic) part describes the admissible arguments that can be used to build specific instances of the parameterization. A standard example using equational logic of a parameterized specification would be the following one describing a parameterized specification of Sets of Values:

VAL = **enrich** $BOOL$ **by**
$\quad$ **sorts** val
$\quad$ **opns** $eq \colon val\ val \to bool$
$\quad$ **vars** $X, Y \colon val$
$\quad$ **axms** $eq(X, X) = true$
$\qquad\qquad X \neq Y \Rightarrow eq(X, Y) = false$

SET = **enrich** VAL **by**
$\quad$ **sorts** set
$\quad$ **opns** $\emptyset \colon\ \to set$
$\qquad\qquad add \colon set\ val \to set$
$\qquad\qquad \in \colon set\ val \to bool$
$\quad$ **vars** $S \colon set;\ X, Y \colon val$

axms $add(add(S, X), Y) = add(add(S, Y), X)$
$add(add(S, X), X) = add(S, X)$
$X \in \emptyset = false$
$X \in add(S, Y) = X \in Y \vee eq(X, Y)$

where the specification *VAL* is called the *formal parameter* specification and the specification *SET* is called the *body* of the parameterization. As said above, *VAL* describes the admissible arguments over which we may define specific instances of the data type of sets.

This means that the syntax of a parameterized specification can be seen as an inclusion of specifications, $PAR \subseteq BODY$, or more generally, as a pair of specifications $(PAR, BODY)$ where $BODY$ is an enrichment over PAR, i.e. $BODY =$ **enrich** PAR **by** $(\Sigma 1, E1)$. To define the semantics of a parameterization of this kind the most obvious approach consists in generalizing the constructions used for defining enrichments. In particular, the two possible choices that were considered for enrichments generalize in this case to defining the meaning of the parameterized specification $(PAR, BODY)$ as:

1. The free functor associated with the given inclusion of specifications,
2. The mapping associating each A in *Models*(SP) with the class of all its proper extensions.

The institution-independent generalization of these parameterizations is again obvious. Syntactically, a parameterized specification of a parameterization would just be a specification morphism (or just an inclusion) $h: PAR \to BODY$. Semantically, a parameterization would just denote a construction $\kappa_h: Models(PAR) \to 2^{Models(BODY)}$ (or $\kappa_h: Models(SP) \to Models(SP')$) that we would consider provided by the underlying institution.

Remark 6.8.

1. In most cases, one would expect constructions associated with parameterizations to be not just arbitrary mappings but functors. In particular, this is the case for the free functor approach to parameterizations [TWW82].
2. One can easily generalize the above approach to parameterizations with several parameters. In particular, in the case of exact institutions, parameterizations with n parameters can be considered equivalent to parameterizations with a single parameter, obtained as a colimit of the n parameters.

6.4.2 Parameter passing for generic enrichments

The operation of instantiating a parameterized specification to build a new specification is called parameter passing. In particular, to instantiate a parameterized specification we need an actual parameter specification and a specification morphism (fitting morphism) between the formal and the actual parameter specifications. This morphism should bind the sorts and operations from the formal and actual parameter specifications that we want to identify in the instantiation. It must be noted that requiring the fitting morphism to be a specification morphism (and not just a signature morphism) guarantees that the actual parameter specification satisfies the "requirements" imposed in the formal parameter specification.

For example, the set specification could be instantiated to create a specification of (finite) sets of natural numbers as follows: first we must have a specification of natural numbers satisfying the requirements specified by the formal parameters. In this case, we can do so by defining a specification *NATEQ* including an equality operation:

$NATEQ = $ **enrich** NAT **by**
   opns $\doteq: nat\ nat \to bool$
   vars $X, Y: nat$
   axms $(X \doteq Y) = (X \leq Y) \wedge (X \geq Y)$
         $(X \doteq X) = true$
         $X \neq Y \Rightarrow (X \doteq Y) = false$

And, second, we must define a specification morphism binding the sorts and operations of the specification *VAL* to the corresponding sorts and operations of *NATEQ*:

$h: VAL \to NATEQ$

$h(val) = nat$

$h(eq) = (\doteq)$

otherwise h is the identity

It is trivial to see that h is indeed a specification morphism. A simple syntax for the operation of parameter passing may be

$PSP[ACT]_f$

where *PSP* is the given parameterized specification, $PSP = (PAR, BODY)$, *ACT* is the actual parameter specification and $f: PAR \to ACT$ is the corresponding fitting morphism.

To define the semantics of parameter passing, at the specification level, one may observe that the result that one may expect after performing the operation $PSP[ACT]_f$ could be seen as a *combination* of the specifications *ACT* and *BODY* after identifying the corresponding sorts and operations from the formal and actual parameter specifications which are bound by f. For instance, in the above example, the result would be the specification

enrich $NATEQ$ **by**
sorts *set*
opns $\emptyset: \to set$
      $add: set\ nat \to set$

```
        ∈: set nat → bool
vars  S: set;  X, Y: nat
axms  add(add(S, X), Y) = add(add(S, Y), X)
      add(add(S, X), X) = add(S, X)
      X ∈ ∅ = false
      X ∈ add(S, Y) = X ∈ Y ∨ eq(X, Y)
```

This means that the result at the specification level of $PSP[ACT]_f$ can be defined as the specification RES in terms of the pushout diagram in Figure 6.7, where g denotes the inclusion $PAR \hookrightarrow BODY$.

Fig. 6.7. Parameter passing   **Fig. 6.8.** Parameter passing in <u>Sig</u>

Similarly, at the model level, the semantics of $PSP[ACT]_f$ can be defined as a specification $(\Sigma RES, MRES)$, where ΣRES can be defined, as above, in terms of a pushout (on the category of signatures), as in Figure 6.8, where ΣPAR, $\Sigma BODY$ and ΣACT denote, respectively, the signatures of the specifications PAR, $BODY$ and ACT.

On the other hand, reasonably, $MRES$ should be defined as the result of "applying" the mapping (or construction) denoted (at the model level) by the parameterization to the models of the actual parameter specification. However, this is not directly possible since this mapping is only defined over $Models(PAR)$. However, we may consider that each A in $Models(ACT)$ consist of two "parts", one of which is in $Models(PAR)$ (the PAR reduct of A), and the "rest" of the model. One may then consider that the meaning of applying the parameterization over A consists in applying its associated construction over the PAR reduct of A leaving the rest untouched. More precisely, if the inclusion $PAR \hookrightarrow BODY$ is denoted by h and the construction $\kappa_h: Models(PAR) \to 2^{Models(BODY)}$ is considered to be the meaning of the given parameterization, then this is equivalent to defining $MRES$ as the amalgamation of $Models!(ACT)$ and

$$MPSP = \{A \in \kappa_h(A')/A' \in Models(PAR)\}$$

with respect to $Models(PAR)$ and the corresponding morphisms or, as noted above (see Remark 6.5.2), as the result of a pullback at the model level.

Remark 6.9.

1. Parameterizations are often asked to be (strongly) persistent. This means that the construction denoted by the parameterization "preserves" the parameters, i.e. $U_g(\kappa_g(A)) = A$, for every A in $Mod(PAR)$).
2. Parameter passing for parameterizations with several parameters can be handled without too many problems. Essentially, the key idea is that one would need an arbitrary (finite) colimit for defining parameter passing at the specification level and a limit for the model level.
3. Parameter passing can be easily generalized to *parameterized* parameter passing, i.e. when the actual parameter is a parameterized specification. For instance, given the parameterization $LIST(X)$, parameterized parameter passing would allow us to define the parameterization $LIST(LIST(X))$, by instantiating $LIST(X)$ over itself. In particular, parameterized parameter passing can be simply defined at the specification level by means of the parameter passing diagram in Figure 6.9, i.e. given the parameterization $g\colon PAR \hookrightarrow BODY$, the parameterized actual parameter $g0\colon PAR' \hookrightarrow BODY'$ and the fitting morphism $f\colon PAR \to BODY'$, the result of parameter passing in this case would be the parameterized specification $g' \circ g0\colon PAR' \to RES$. Parameterized parameter passing at the model level would be a similar generalization of standard parameter passing.

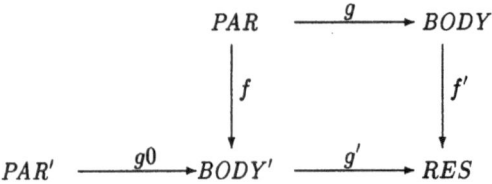

Fig. 6.9. Parameterized parameter passing

One may ask about the conditions ensuring the compatibility of the semantics of parameter passing at the model and specification levels. However this depends not only on the properties of the underlying institution but also on the kind of constructor used to give meaning to parameterizations. The following two theorems describe two situations where we have compatibility results. The first one states the compatibility of parameter passing for (pure) loose semantics when working in a semiexact institution $\mathcal{L}$ and is a direct consequence of the semiexactness of $\mathcal{L}$.

Theorem 6.10. *Let $PSP = PAR \hookrightarrow BODY$ be a parameterized specification over a semiexact institution $\mathcal{L}$, such that for every specification SP*

in $SyntCatSpec_{\mathcal{L}}$ *we have that* $Models(Sem_{\mathcal{L}}(SP)) = Mod(SP)$, *and let* ACT *be an actual parameter specification with respect to the fitting morphism* $f\colon PAR \to ACT$. *Then*

$$Models(RES) = MRES$$

where RES *and* $MRES$ *are obtained in the corresponding pushout and pull-back diagrams as above.*

The second theorem states the compatibility of parameter passing for (strongly persistent) parameterized specifications with initial semantics, also when working in a semiexact institution and is a generalization of Remark 6.3.3 and a consequence of the extension lemma (see [EM85]).

Theorem 6.11. *Let* $PSP = h\colon PAR \hookrightarrow BODY$ *be a parameterized specification over a semiexact institution* $\mathcal{L}$, *such that* F_h *is strongly persistent (i.e.* $U_h(F_h(A)) = A$, *for every* A *in* $Mod(PAR)$), *and let* ACT *be an actual parameter specification with respect to the fitting morphism* $f\colon PAR \to ACT$. *Then*

$$T_{RES} = F_h(T_{ACT})$$

where RES *is defined as above, with* κ_h *being the free functor* F_h.

6.4.3   Parameterizations as arbitrary generic expressions

The second approach to parameterization is quite more general. It is based on considering that any expression defined over a (specification) variable can be seen as a parameterization. This approach is due to Sannella and Wirsing and was first defined in the context of the ASL specification language [SW83, Wir86]. The specific syntax chosen to denote this kind of parameterization is a (typed) λ-notation, where the type of the specification variable would correspond to what we called the formal parameter specification in the previous approach. The idea is that the formal parameter specification can be seen as a type pattern that actual parameters must follow. For instance, the (parameterized) set specification seen above would now be written:

$$
\begin{aligned}
PSET = \lambda X\colon\ &VAL.\ \textbf{enrich } X \textbf{ by}\\
&\textbf{sorts}\quad set\\
&\textbf{opns}\quad \emptyset\colon\ \to set\\
&\qquad\quad add\colon set\ val \to set\\
&\qquad\quad \in\colon set\ val \to bool\\
&\textbf{vars}\quad S\colon set;\ V1, V2\colon val\\
&\textbf{axms}\quad add(add(S, V1), V2) = add(add(S, V2), V1)\\
&\qquad\quad add(add(S, V1), V1) = add(S, V1)\\
&\qquad\quad V1 \in \emptyset = false\\
&\qquad\quad V1 \in add(S, V2) = V1 \in S \vee eq(V1, V2)
\end{aligned}
$$

However, according to this approach, one may define parameterizations based on more complex specification expressions. In particular, parameterizations are assumed to have the form:

$$\lambda X : SP.\epsilon(X)$$

where $\epsilon(X)$ denotes any arbitrary expression, built over the specification variable X, including as operations not only enrichments, but also combinations, derivations, and renamings.

In order to define the syntax of this kind of parameterization in a precise manner, one may consider (following the Module Algebra approach [BHK90]) the specification-building operations as operators of an abstract signature $\Sigma SPEC$ over two sorts (specifications and morphisms). In this sense, specification expressions are just terms in $T_{\Sigma SPEC}$ and the two kinds of semantics defined for these operations in terms of the syntactic and semantic category of specifications can be seen just as the definition of two $\Sigma SPEC$-algebras. Then one can also define specification expressions over variables as terms in $T_{\Sigma SPEC}(X)$, and variable substitution can be defined in the usual way.

Remark 6.12. One of the advantages of this approach, besides its generality, is the simplicity for defining higher-order parameterizations. In particular, if in the definition of a parameterization:

$$\lambda X : SP.\epsilon(X)$$

we allow SP or $\epsilon(X)$ to be arbitrary λ-expressions then PSP would be a higher-order parameterized specification. For instance, given the specifications:

$ELEM = $ **sorts** $elem$

$PSEQ = \lambda X : ELEM.$ **enrich** X **by**
　　　　　sorts seq
　　　　　opns $emptyseq: \rightarrow seq$
　　　　　$app: seq\ elem \rightarrow seq$

defining a parameterized specification of sequences of a given type of elements, the following specifications:

$ELEM1 = $ **enrich** $BOOL$ **by**
　　　　sorts $elem$
　　　　opns $p: elem \rightarrow bool$

$PFILTER = \lambda X : ELEM1.$
　　　　　$\lambda Y : PSEQ.$**enrich** $Y[X]_{id}$ **by**
　　　　　　opns $filt: seq \rightarrow seq$
　　　　　　vars $S: seq;\ V: val$
　　　　　　axms
　　　　　　$filt(empty_seq) = empty_seq$
　　　　　　$p(V) = true \Rightarrow filt(app(S,V)) = app(filt(S),V)$
　　　　　　$p(V) = false \Rightarrow filt(app(S,V)) = S$

could define a specification describing how to "filter" the elements of a sequence satisfying a given property p.

However, in the following we will restrict our presentation to first-order parameterizations. The reason is that, in our opinion, there are certain aspects, namely parameter passing or the possible dependencies between formal parameters, that should be clarified.

There are several possible approaches for defining the semantics of this kind of parameterization. Firstly, we may consider the definition of a specification level semantics. This was not done in the previous case because parameterizations were already defined in terms of the syntactic category of specifications (i.e. as morphisms in that category). Moreover, in this case, this kind of semantics is especially needed for defining parameter passing at the specification level, as we will see below.

As in the case of generic enrichments, at the specification level, a λ-parameterization should denote the relation between the formal parameter specification and the specification denoted by the body. In the simple case of generic enrichments, this is just a morphism. In this case this is slightly more complex. For instance, given the λ-expression:

$$\lambda X \colon SP.X +_\emptyset X$$

we may consider that the meaning of this parameterization is some kind of mapping that, given any arbitrary argument $SP0$, yields as a result a specification consisting of two copies of $SP0$. In this case, the relation between the formal parameter and the body specification can be denoted by two morphisms binding the formal parameter SP with the two copies included in $SP +_\emptyset SP$, as in Figure 6.10.

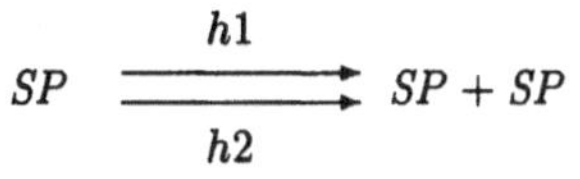

$$SP \underset{h2}{\overset{h1}{\Longrightarrow}} SP + SP$$

Fig. 6.10. "Double" morphism

On the other hand, consider the λ-expression:

$$PSP = \lambda X \colon SP.\textbf{derive (enrich } X \textbf{ by } (f1, E1)) \textbf{ from } g$$

In this case, the relations between the various specifications involved can be seen in the diagram in Figure 6.11, where $SP1$ is the specification denoted by **enrich** SP **by** $f1$, SP' is $SP1 \cup E$, i is the inclusion from $SP1$ into SP', and $SP2$ is the specification denoted by **derive(enrich** SP **by** $(f1, E1))$ **from** g. In particular, in this case, we may think that the meaning at the specification level of PSP can be explained in terms of the specification $SP1$ that represents the specification "built" in the parameterization and the two

$$SP \xrightarrow{\quad f1 \quad} SP1 \xrightarrow{\quad i \quad} SP' \xleftarrow{\quad g \quad} SP2$$

Fig. 6.11. Enrich-Derive specification

morphisms $i \circ f$ and g representing, respectively, the bindings between the formal parameter specification and the result of PSP; i.e. the semantics in this case can be denoted by the diagram in Figure 6.12.

$$SP \xrightarrow{\quad i \circ f1 \quad} SP' \xleftarrow{\quad g \quad} SP2$$

Fig. 6.12. Enrich-Derive specification

In general, we may consider that, for any arbitrary parameterized specification, $\lambda X : SP.\epsilon(X)$, the result is "obtained" as some derivation out of a "large" specification that includes any number of times the formal parameter specification. This means that we may consider that the specification-level meaning of a parameterized specification consists of a family of morphisms

$$< f_i : SP \to SP'/1 \le i \le n >$$

(with $n \ge 0$), from the formal parameter specification into certain specification SP', and a morphism g from the specification denoted by the result specification into SP'. However, if **derive** is not considered as an operation for defining λ-expressions, then the specification-level semantics of λ-parameterizations can be simplified. In particular, in this case, the semantics can be seen just as the family $< f_i : SP \to SP'/1 \le i \le n >$. On the other hand, in our opinion this kind of operation should only be considered in a second layer of a specification language. Being specific, hiding parts of a specification should only be possible through the use of a notion of module with well-defined import and export interfaces describing the "visible" parts of a specification, as we will see in the following section. For these reasons, we will only consider specification expressions not including the **derive** operati! on. Being specific, the specification-level semantics of a parameterized specification $[\![\lambda X : SP.\epsilon[X]]\!]_{Spec}$ can be defined as a triple $(SP, SP', \mathcal{F})$, where SP and SP' are the *formal parameter* and the *body* specifications, respectively, and $\mathcal{F}$ is a family of n morphisms, with $n \ge 0$, from SP to SP', i.e. by the diagram in Figure 6.13.

In particular, depending on the form of ϵ, $[\![\lambda X : SP.\epsilon[X]]\!]_{Spec}$ is defined as follows:

1. If $\epsilon[X] = SP0$, where $SP0$ is a "constant" expression, then

$$[\![\lambda X : SP.\epsilon[X]]\!]_{Spec} = (SP, SP0, \emptyset)$$

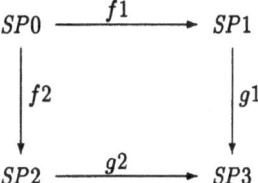

$$SP \quad \xrightarrow[\quad fn \quad]{\overset{f1}{\underset{\vdots}{\longrightarrow}}} \quad SP'$$

Fig. 6.13. Specification-level semantics of a parameterized specification

Fig. 6.14. Pushout of specifications

2. If $\epsilon[X] = X$ then $[\![\lambda X \colon SP.\epsilon[X]]\!]_{Spec} = (SP, SP, \{id \colon SP \to SP\})$.
3. If $\epsilon[X] = $ **enrich** $\epsilon'[X]$ **by** (f, E), $[\![\lambda X \colon SP.\epsilon'[X]]\!]_{Spec} = (SP, SP', \mathcal{F}')$
   and $f \colon SP' \to SP''$ then $[\![\lambda X \colon SP.\epsilon[X]]\!]_{Spec} = (SP, SP'' \cup E, \{i \circ f \circ$
   $g \colon SP \to (SP'' \cup E)/g \in \mathcal{F}'\})$, where i denotes the inclusion $SP'' \to$
   $SP'' \cup E$.
4. If $\epsilon[X] = \epsilon1[X] +_{(\epsilon0[X], f1, f2)} \epsilon2[X]$, $[\![\lambda X \colon SP.\epsilon i[X]]\!]_{Spec} = (SP, SPi, \mathcal{F}i)$
   $(0 \le i \le 2)$ and $f1$ and $f2$ are morphisms from $SP0$ to $SP1$ and $SP2$,
   resp., then $[\![\lambda X \colon SP.\epsilon[X]]\!]_{Spec} = (SP, SP3, \mathcal{F}3)$, where $\mathcal{F}3 = (\{g1 \circ$
   $h1 \colon SP \to SP3/h1 \in \mathcal{F}1\} \cup \{g2 \circ h2 \colon SP \to SP3/h2 \in \mathcal{F}2\})$ and $SP3$,
   $g1$, and $g2$ are defined by the pushout diagram in Figure 6.14.
5. If $\epsilon[X] = $ **rename** $\epsilon'[X]$ **by** f, $[\![\lambda X \colon SP.\epsilon'[X]]\!]_{Spec} = (SP, SP', \mathcal{F}')$ and
   $f \colon SP' \to SP''$ then

$$[\![\lambda X \colon SP.\epsilon[X]]\!]_{Spec} = (SP, SP'', \{f \circ g \colon SP \to SP''\}/g \in \mathcal{F}'\})$$

Remark 6.13.

1. The explicit handling of the **derive** operation in the context of λ-param-
   eterizations was dealt in [Bau92].
2. Note that the body of a parameterization can be independent of the
   formal parameter (e.g. as in1). In this case the corresponding family of
   morphisms would be empty.

In order to define the model-level semantics of λ-parameterizations we
have several possibilities according to different intuitions. In particular, if one
considers that a parameterized specification is a specification of a parameterized data type, then one may consider that the semantics of a parameterized
specification is a functor that builds, for any possible argument, the corresponding instance of the given parameterized data type. This is fine if the
underlying framework is monomorphic, i.e. if every specification describes a
unique model up to isomorphism. However, if we are working on a loose framework, we have several choices for defining the semantics of parameterizations.
First, we may consider that a loose parameterized specification describes not
a single functor (up to natural isomorphism) but a class of functors, in just
the same way that a loose (non-parameterized) specification describes not a

single model but a class of models. The second possibility! consists in considering that a parameterized specification describes a construction that maps every possible argument model into a class of models (i.e. the result is not just a single instance). Finally, we may consider that a parameterized specification is a construction that maps specifications (i.e. classes of models) into specifications (classes of models). A discussion of these different intuitions (in particular between the first and the last ones) and their different rôles in the software specification and design process can be found in [SST92], where these considerations were first aired. In [JOE95] a further study can be found. In this chapter we will just study the last approach. The reader is addressed to [SST92, JOE95] for a detailed study of the first approach to the semantics of parameterizations. Similar techniques can be used to study the second possible approach, which was studied in [Bid88, Bid89].

Therefore we consider that the second-level semantics of parameterized specifications is a function mapping model classes into model classes. In particular, given a parameterized specification $\lambda X : SP.\epsilon[X]$, this function can be defined, for every class of SP-models $\mathcal{M}$, as the evaluation at the model level of the expression ϵ over $\mathcal{M}$:

Definition 6.14. Given a parameterized specification $\lambda X : SP.\epsilon[X]$, its semantics at the model level, $[\![\lambda X : SP.\epsilon[X]]\!]_{Mod}$, is a mapping from $2^{Models(SP)}$ to $2^{Models(\epsilon[X/SP])}$ defined for each $\mathcal{M} \subseteq Models(SP)$ as

$$[\![\lambda X : SP.\epsilon[X]]\!]_{Mod}(\mathcal{M}) = \epsilon[X/\mathcal{M}]$$

where $\epsilon[X/\mathcal{M}]$ denotes the *evaluation* of ϵ over $\mathcal{M}$ (by abuse of notation) defined:

1. If $\epsilon[X] = SP0$ then $\epsilon[X/\mathcal{M}] = Models(SP0)$
2. If $\epsilon[X] = X$ then $\epsilon[X/\mathcal{M}] = \mathcal{M}$
3. If $\epsilon[X] =$ **enrich** $\epsilon'[X]$ **by** (f, E), and f is a morphism from $\epsilon'[X/SP]$ to SP', then $\epsilon[X/\mathcal{M}] = \{A \in Models(SP')/A \models E \quad and \quad U_f(A) \in \epsilon'[X/\mathcal{M}]\}$.
4. If $\epsilon[X] = \epsilon1[X] +_{(\epsilon0[X], f1, f2)} \epsilon2[X]$, then $\epsilon[X/\mathcal{M}] = \epsilon1[X/\mathcal{M}] +_{\epsilon0[X/\mathcal{M}]} \epsilon2[X/\mathcal{M}]$
5. If $\epsilon[X] =$ **rename** $\epsilon'[X]$ **by** f and $f : \epsilon'[X/SP] \to SP'$ then

$$\epsilon[X/\mathcal{M}] = \{A \in Models(SP')/U_f(A) \in \epsilon'[X/\mathcal{M}]\} \qquad \diamond$$

It may be noted that the semantics of the enrichment operation, for simplicity, is defined in terms of loose extensions (see Remark 6.4.1 above). In this sense, one could generalize the above definition by giving semantics to enrichments in terms of arbitrary constructors.

We can see that, for adequate institutions, both semantics are compatible:

Proposition 6.15. Let $\lambda X : SP.\epsilon[X]$ be a parameterized specification over a semiexact institution $\mathcal{L}$, with $[\![\lambda X : SP.\epsilon[X]]\!]_{Spec} = (SP, SP', \mathcal{F})$, and $\mathcal{M} \subseteq Models(SP)$ we have:

$$[\![\lambda X : SP.\epsilon[X]]\!]_{Mod}(\mathcal{M}) = \{A \in Models(SP')/\forall f \in \mathcal{F} \; U_f(A) \in \mathcal{M}\}$$

Proof. If ϵ is defined by cases 1, 2, 3 and 5 the proposition is a direct consequence of the definition of the specification and the model-level semantics. Case 4 is a consequence of the semiexactness of $\mathcal{L}$. $\qquad \square$

Proposition 6.16. *For every parameterized specification* $\lambda X : SP.\epsilon[X]$ *we have that:*

1. $[\![\lambda X : SP.\epsilon[X]]\!]_{Mod}$ *is monotonic, in the sense that for every* $\mathcal{M}1, \mathcal{M}2 \subseteq Models(SP)$, $\mathcal{M}1 \subseteq \mathcal{M}2$ *implies:*

$$[\![\lambda X : SP.\epsilon[X]]\!]_{Mod}(\mathcal{M}1) \subseteq [\![\lambda X : SP.\epsilon[X]]\!]_{Mod}(\mathcal{M}2)$$

2. *In general* $[\![\lambda X : SP.\epsilon[X]]\!]_{Mod}$ *is not additive, in the sense that for every* $\mathcal{M} \subseteq Models(SP)$

$$[\![\lambda X : SP.\epsilon[X]]\!]_{Mod}(\mathcal{M}) \neq \cup_{A \in \mathcal{M}}[\![\lambda X : SP.\epsilon[X]]\!]_{Mod}(\{A\})$$

3. *If* $[\![\lambda X : SP.\epsilon[X]]\!]_{Spec} = (SP, SP', \mathcal{F})$ *and* $Card(\mathcal{F}) \leq 1$ *then we have that* $[\![\lambda X : SP.\epsilon[X]]\!]_{Mod}$ *is additive.*

Proof. The first property is a direct consequence of the definition of the model-level semantics and of the fact that amalgamation is monotonic and the third one is a consequence of the fact that for standard pushouts amalgamation and multiple amalgamation coincide. For the second property it is enough to consider the following counterexample. Let SP and SP' be the following specifications:

$$SP = \textbf{sorts } s \qquad\qquad SP' = \textbf{sorts } s'$$
$$\textbf{opns } a, b: \ \rightarrow s \qquad\qquad \textbf{opns } a', b': \ \rightarrow s'$$

let A and B be two SP-algebras defined: $A_s = \{0, 1\}$, $a_A = 0$, $b_A = 1$ and $B_s = \{0\}$, $a_B = 0 = b_B$ and consider the parameterized specification $\lambda X : SP.X + (\textbf{enrich } X \textbf{ by } f)$, where f is the morphism from SP to SP' associating s', a', and b' with s, a, and b, respectively.

Now, the algebra C defined $C_s = \{0, 1\}$, $C_{s'} = 0$, $a_C = 0$, $b_C = 1$, $a'_C = 0 = b'_C$ is in $F(A, B)$ but not in $F(A) \cup F(B)$, where $F = [\![\lambda X : SP.X + (\textbf{enrich } X \textbf{ by } f)]\!]_{Mod}$. $\qquad \square$

Remark 6.17. The fact that the semantic definition above is not additive implies that, in general, this semantics and the other possible model-theoretic semantics (see the above discussion) are not equivalent. However this equivalence holds when considering only generic enrichments, as a consequence of Proposition 6.16.3.

6.4.4 Parameter passing for λ-parameterizations

As for the case of generic enrichments, we may consider that to instantiate a parameterized specification we need an actual parameter specification and a fitting morphism between the formal and the actual parameter specifications. As a consequence, parameter passing for λ-parameterizations may be denoted by:

$$(\lambda X : SPC.\epsilon[X])(SPC_{act})_h$$

where $\lambda X : SPC.\epsilon[X]$ is the given parameterized specification, SPC_{act} is the actual parameter, and h is the parameter passing morphism $h : SPC \rightarrow SPC_{act}$.

Now, due to the "multiple occurrence" of the formal parameter specification in the body specification, modeled by the family of morphisms between these specifications, one cannot use pushouts directly to define parameter passing at the specification level. Instead one has to use a generalization of this construction that we have called "multiple pushouts" [JOE95].

Definition 6.18. Given morphisms $f1, ..., fn : SP0 \rightarrow SP1$ $(n \geq 0)$ and given $f : SP0 \rightarrow SP2$ in *SyntCatSpec*, the diagram in Figure 6.15 is called a multiple pushout of $(f1, ..., fn)$ and f if we have

1. (Graded Commutativity): $g \circ fi = gi \circ f$ $(i = 1, ..., n)$
2. (Universal Property): For each object $SP3'$ and morphisms $g', g1, ..., gn$ with $g' \circ fi = gi' \circ f$, $(i = 1, ..., n)$ there is a unique morphism h such that $h \circ g = g'$ and $h \circ gi = gi'$ $(i = 1, ..., n)$.   ◇

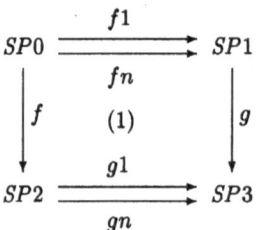

Fig. 6.15. Multiple pushout

Remark 6.19.

1. As in the case of pushouts (see Remark 6.1.3), to have multiple pushouts in *SyntCatSpec* (and also in *SemCatSpec*) it is enough to have multiple pushouts in *Sig*.
2. Note that $SP3$ is not the colimit object in diagram (1) since commutativity of diagram (1) would also mean $g \circ fi = g \circ fj$ and $gi \circ f = gj \circ f$ for all $i, j = 1, ..., n$ which is not required for graded commutativity in the case $n > 1$.

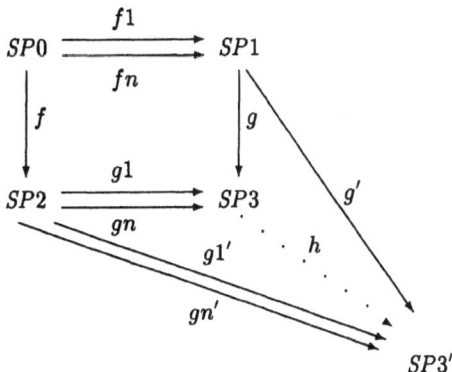

Fig. 6.16. Universal property of multiple pushouts

3. In the case $n = 1$ a multiple pushout is a pushout.
4. In the case $n = 0$ the multiple pushout object $SP3$ is equal to $SP1$ with $g = 1_{SP1}$.

Example 6.20. Consider the following specifications and morphisms:

$SP0 = $ **sorts** $s0$

$SP1 = $ **sorts** $s1, s2$
      opns $f1: s1 \to s1$
            $f2: s2 \to s2$

$h1: SP0 \to SP1$
with $h1(s0) = s1$

$h2: SP0 \to SP1$
with $h2(s0) = s2$

Now, if we define:

$SP2 = $ **sorts** $s0$
     opns $a: \ \to s0$
           $c: s0 \to s0$

together with the inclusion $i0: SP0 \to SP2$, we can "compute" the multiple pushout in Figure 6.17, where $SP3$ would be (up to isomorphism) the specification:

$SP3 = $ **sorts** $s1, s2$
      opns $f1: s1 \to s1$
             $f2: s2 \to s2$
             $a1: \ \to s1$
             $c1: s1 \to s1$
             $a2: \ \to s2$
             $c2: s2 \to s2$

and the two morphisms, $h1'$ and $h2'$, would map s, a, c to $s1, a1, c1$ and to $s2, a2, c2$, respectively. ∎

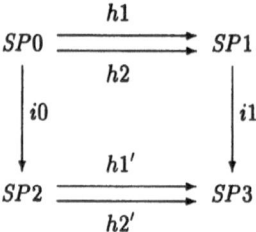

Fig. 6.17. Multiple pushout

As the following fact shows, in order to be able to define parameter passing for this kind of parameterization, the category of signatures (and, consequently, the corresponding syntactic category of specifications) of the underlying institution must have pushouts and (finite) coproducts or, more generally, finite colimits.

Fact 6.21. *If SyntCatSpec has finite coproducts and pushouts then it also has multiple pushouts.*

Proof. Consider the diagram in Figure 6.18 and let $\sqcup_n SP0$ be the coproduct of n copies of $SP0$ with injections $i1, ..., in$ and similarly $\sqcup_n SP2$ with injections $j1, ..., jn$. Then there is a unique $f0: \sqcup_n SP0 \to SP1$ such that we have graded commutativity of diagram (1) below and similar in diagram (2) with $\sqcup_n f$. Now $SP3$ can be constructed as the pushout in (3).

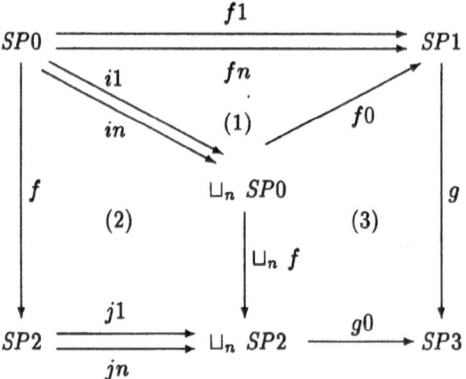

Fig. 6.18. Multiple pushout in terms of coproducts and pushouts

With $gi = g0 \circ ji$ for $i = 1, ..., n$ and g as above, $SP3$ becomes a multiple pushout object of $f1, ..., fn$ and fi. Graded commutativity of the combined diagram follows from graded commutativity of (1) and (2) and commutativity of (3). The universal property of multiple pushouts follows from the universal properties of pushouts and coproducts. In the case $n = 0$, the objects $\sqcup_n SP0$ and $\sqcup_n SP2$ are both initial, which implies $g = 1_{SP1}$. □

Remark 6.22. Note that the multiple pushout $SP3$ of $f1, ..., fn$ and f for $n \geq 1$ can also be constructed in terms of the colimit diagram of Figure 6.19.

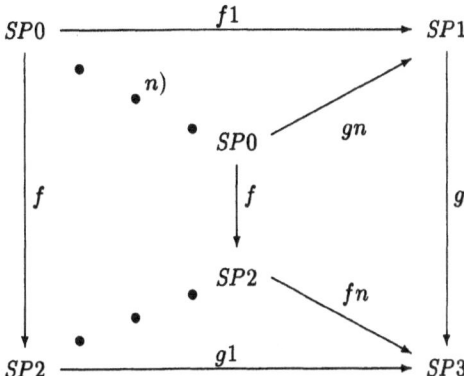

Fig. 6.19. Multiple pushout as a colimit

Therefore, assuming that the underlying institution has multiple pushouts, one may define parameter passing for λ-parameterizations, at the specification level, as follows:

Definition 6.23. The result at the specification level of the parameter passing expression $(\lambda X : SP.\epsilon[X])(ACT)_h$ is the specification RES, denoted

$$[\![(\lambda X : SP.\epsilon[X])(ACT)_h]\!]_{Spec}$$

obtained as the result of the multiple pushout diagram in Figure 6.20.   ◇

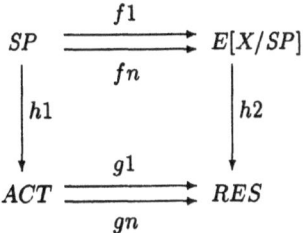

Fig. 6.20. Parameter passing

At the model level, similarly to the simpler case of generic enrichments, the semantics of $PSP[ACT]_f$ is defined in terms of the corresponding multiple pushout in $\underline{SemCatSpec}$. In particular, in this case, the resulting specification RES is the pair $(\Sigma RES, MRES)$, where ΣRES is the result of the corresponding pushout of signatures and $MRES$ consists of all models $A3$ in $Mod(\Sigma RES)$ such that there is an $A1$ in $E[X/Models(SP)]$ and a family of models $< A2_1, ..., A2_n >$ in $Models(ACT)$ such that $U_{h2}(A3) = A1$ and $U_{gi}(A3) = A2_i$ for $i = 1, ..., n$ (i.e. the *multiple pullback* of the associated model classes, denoted $E[X/Models(SP)) \oplus_{Models(SP)} Models(ACT))$.

Definition 6.24. The result at the model level of the parameter passing expression $(\lambda X : SP.\epsilon[X])(ACT)_h$, denoted $[\![(\lambda X : SP.\epsilon[X])(ACT)_h]\!]_{Mod}$, is defined as the *multiple amalgamation* with respect to the above multiple pushout diagram of the following model classes:

$$[\![(\lambda X : SP.\epsilon[X])]\!]_{Mod}(\mathcal{M}) \oplus_{\mathcal{M}} Models(ACT)$$

where $\mathcal{M} = U_h(Models(ACT))$.   ◇

One can generalize the compatibility result obtained for generic enrichments to the case of λ-parameterizations in the pure loose case.

Theorem 6.25. *Let* $\lambda X : SP.\epsilon[X]$ *be a parameterized specification over an exact institution* $\mathcal{L}$, *such that for every specification SP in* $\underline{SyntCatSpec}_{\mathcal{L}}$ *we have that* $Models(Sem_{\mathcal{L}}(SP)) = Mod(SP)$, *and let* $\overline{ACT}$ *be an actual parameter specification with respect to the fitting morphism* $h: SP \to ACT$. *Then*

$$Models(Sem([\![(\lambda X : SP.\epsilon[X])(ACT)_h]\!]_{Spec})) = [\![(\lambda X : SP.\epsilon[X])(ACT)_h]\!]_{Mod}$$

Proof. The theorem is a direct consequence, on one hand, of the exactness of $\mathcal{L}$ and, on the other hand, of the fact that the construction of the model class, $MRES$, of

$$[\![(\lambda X : SP.\epsilon[X])(ACT)_h]\!]_{Mod}$$

which can be shown to be equivalent to the limit of the diagram in Figure 6.21.   □

Remark 6.26. It must be noted that, in this context, it makes no sense to study compatibility for a free functor semantics to parameterization. Such kinds of results make sense only if the intuitions underlying the model level semantics for parameterizations would be different (see the discussion above, before Definition 6.14). In this sense, in [JOE95] compatibility results are presented for the (loose) functorial approach to parameterizations.

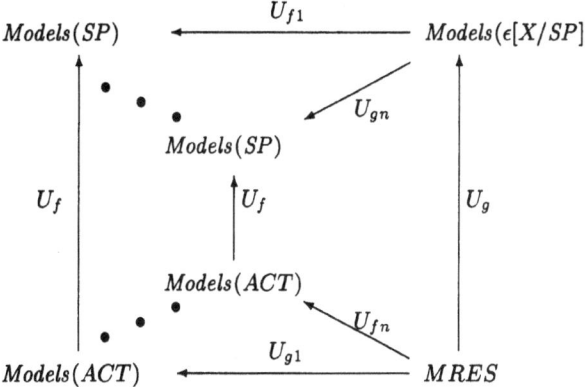

Fig. 6.21. Parameter passing at the model level

6.5 Modules and modular systems

The structuring operations introduced in Section 3 (and other similar ones) can be of great help for the systematic construction of (medium size) specifications. However, when building large specification, or even for understanding not-so-large specifications, the use of this kind of operation can be of little help. The main problem is that the specification fragments that one builds do not benefit from the principle of "separation of concerns". For instance, when defining (or trying to understand) an enrichment one must know what is the enriched specification. To avoid this kind of problems a number of notions of specification modules have been proposed in the literature, i.e. specification units satisfying the principle of separation of concerns, which means that can be understood by themselves. In this section, we will first study the syntax and semantics of a specific notion [EM90] and then we will see how these modules can be combined to define a large ! (modular) specification.

6.5.1 Modules: syntax and semantics

Essentially, a module is some kind of specification unit that not only consists of a specification of the "objects" specified in the unit, but also includes a specification of what is needed "to know" about the rest of the system. As we will see below, this second specification plays a similar role to formal parameter specifications in parameterized specifications. Actually, parameterizations as studied in Section 4 can be seen as specific kinds of modules. In this sense, the module notion below can be seen as a parameterized specification where some parts of it may remain hidden.

Definition 6.27. A module specification over an institution $\mathcal{L}$ consists of four specifications and four specification morphisms, $MOD = (PAR, EXP,$

IMP,BOD, e, s, i, v), that can be represented by the commuting diagram in
Figure 6.22, where *PAR* is called the parameter specification, *EXP* is the
export interface specification, *IMP* is the import interface specification, *BOD*
is the body specification, and the morphisms e, s, i, v represent the relations
among these specifications.   ◇

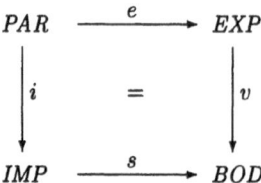

Fig. 6.22. Module

Remark 6.28.

1. This notion of module is the basis of the ACT TWO specification language and was inspired by Ada packages. In particular, the parameter
   part would correspond to the formal parameter specification of generic
   packages, the *EXP* specification would correspond to the *package specification*, the *BOD* specification would correspond to the *package body* and
   the *IMP* specification would, in some sense, correspond to the *with* declaration, where it is stated which "elements" (e.g. sorts and operations)
   defined externally are used inside the module. However, in Ada packages,
   the *with* clause declares the names of the modules where these external elements are defined, whereas *IMP* can be seen as a specification of
   some requirements that these external elements must satisfy. As a consequence, the binding relating the sorts and operations in *IMP* with actual
   sorts and operations defined in other modules should be done through
   an explicit fittin! g morphism (see Definition 6.33 of *module composition*
   below). This makes this module concept slightly more flexible than Ada
   packages. However, as a consequence, *IMP* becomes also a kind of formal
   parameter specification (which can be instantiated through module composition), making it difficult to establish a conceptual difference between
   PAR and *IMP*.
2. Another notion of module quite close to this one, although the semantics
   and the way of combining them are quite different, is the one used in EML
   [ST89]. The main difference would be the lack of the *PAR* component.
3. A second intuition on this module notion, which we like better, consists in
   seeing a module as a general form of a parameterized specification (generic
   enrichment), where *IMP* is then the formal parameter specification, *BOD*
   is the body of the parameterization, *EXP* defines the external interface
   of the parameterization (i.e. some elements in *IMP* or in *BOD* may be

considered hidden or invisible outside the module), and *PAR* just states the common parts of *IMP* and *EXP*.

Therefore, in order to define the semantics of a specification module, one must consider a module as a parameterized specification in which one may want to keep part of the body hidden. In this sense, one can define the meaning of a module as a generalization of the semantics given to generic enrichments studied in Section 4.1. However, here we will restrict our consideration to just one possible definition in terms of free functors (which would be assumed to exist in the underlying institution). This is actually the semantics given in the series of papers studying this notion [BEP87, EM90, EP91]. Nevertheless, it would be straightforward to study other possible definitions.

Definition 6.29. The (functorial) semantics of a module *MOD*, M_{MOD}, is the functor

$$M_{MOD} = U_v \circ F_s \colon Models(IMP) \to Models(EXP)$$

where $F_s \colon Models(IMP) \to Models(BOD)$, $U_v \colon Models(BOD) \to Models(EXP)$ are the free and forgetful functors corresponding to s and v respectively. The module specification *MOD* is called (internally) correct if the free functor F_s is strongly persistent, i.e. $U_s \circ F_s = ID_{Models(IMP)}$. ◇

The semantics of a module MOD, M_{MOD}, can be seen as a transformation from import data types in *Models(IMP)* to export data types in *Models(EXP)*. Internal correctness of *MOD* means that each imported data type I is protected within the free construction F_s, i.e. $U_s \circ F_s(I) = I$.

Remark 6.30.

1. In [EM90] another semantics was defined which was actually considered more adequate. In particular, according to that semantics the meaning of a module would be:

$$U_v \circ Reach_v \circ F_s \colon Models(IMP) \to Models(EXP)$$

where $Reach_v$ is the reachability functor associated with the morphism v, i.e. $Reach_v(A)$ is the subalgebra of A obtained after eliminating from A all values which are not generated from the values in $v(SEXP)$ and the operations in *BOD*, where *SEXP* is the set of sorts of *EXP*.
2. In [EBCO91] the semantics of modules was defined in terms of behavioral semantics. In general, as said above, one could associate with modules the appropriate generalization of any reasonable semantics that one could associate with parameterized specifications.

Example 6.31. The example used in [EM90] to illustrate this module notion is the specification of an airport scheduling system:

$APS\text{-}PAR$ = **enrich** $BOOL$ **by**

    sorts *flight-num*   /* flight number */
               dest   /* destination */
               dep   /* departure */
               plane-num   /* plane number */
               type
               seats

    opns *F-eq*: *flight-num flight-num* → *bool*
               P-eq: *plane-num plane-num* → *bool*
               no-depart: → *dep*

    eqns ...

$APS\text{-}EXP$ = **enrich** $APS\text{-}PAR$ **by**

    sorts *aps*   /* airport schedule */
    opns *create*: → *aps*
               sched: *flight-num dest dep plane type seats aps* → *aps*
               search :: *flight-num aps* → *bool*
               return :: *flight-num aps* → *dep*
               change :: *flight-num dest aps* → *aps*

    eqns ...

$APS\text{-}IMP$ = **enrich** $APS\text{-}PAR$ **by**

    sorts *fs*   /* flight schedule */
               ps   /* plane schedule */
    opns *create-fs*: → *fs*
               create-ps: → *ps*
               add-fs: *flight-num dest dep fs* → *fs*
               reserve-ps: *plane-num type seats ps* → *ps*
               search-fs :: *flight-num fs* → *bool*
               search-ps :: *plane-num ps* → *bool*
               return-fs :: *flight-num fs* → *dep*
               change-fs :: *flight-num dep fs* → *fs*

    eqns ...

$APS\text{-}BOD$ = **enrich** $(APS\text{-}EXP +_{APS\text{-}PAR} APS\text{-}IMP)$ **by**

    opns *tupling*: *fs ps* → *aps*
    eqns ...

In this example, airport schedules are constructed in the body of the module by means of a hidden function TUP (it is not in EXP), joining (imported) flight and plane schedules, respectively. ∎

6.5.2   Modular systems: explicit interconnection of modules

Once we have defined the syntax and semantics of our module notion, we have to show how can these units be used in defining (large) systems. For this we have two different approaches. The first one consists in having a number of operations for explicitly interconnecting the given modules. Then, a system composed by a number of modules may be seen (syntactically) just as an expression defined over the given interconnection operations. This approach is the one considered in ACT TWO. The second approach is based on considering that a modular system is just a collection of modules that are implicitly interconnected by references, for instance, made "inside" the modules. This approach is followed in most programming languages including a module notion. At the specification level, this approach is followed in EML [ST89] and is also the basis of [NOS95]. Let us now study both approaches in some detail.

As we did for the case of basic specifications, in order to define a given operation for putting together modular specifications, we may establish its meaning just at the specification level (by defining the result of the operation as another module), just at the model level (by defining the meaning of the operation in terms of the meaning of both modules), or at both levels showing the compatibility of the two levels. To do so, we will follow the same approach as above, namely defining syntactic and semantic categories of modular specifications, and then defining the meaning of a given operation in terms of both categories.

Definition 6.32. Given an institution $\mathcal{L}$, $\underline{SyntCatMSpec}_{\mathcal{L}}$ (or, if $\mathcal{L}$ is clear from the context, just $\underline{SyntCatMSpec}$) is the syntactic category of modular specifications in $\mathcal{L}$,whose objects are modules $MOD = (PAR, EXP, IMP, BOD, e, s, i, v)$ and whose morphisms $h: MOD0 \to MOD1$ are 4-tuples of specification morphisms $h = (h_P: PAR0 \to PAR1, h_E: EXP0 \to EXP1, h_I: IMP0 \to IMP1, h_B: BOD0 \to BOD1)$ such that all diagrams in Figure 6.23 commute and such that the following condition, stating the compatibility of the semantics of both modules, holds:

$$U_{h_B} \circ F_{s0} = F_{s1} \circ U_{h_I}$$

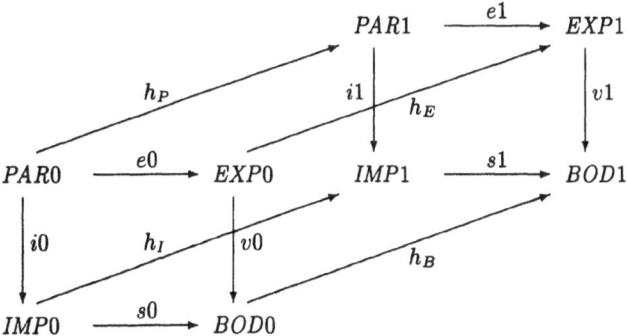

Fig. 6.23. Module morphism

$SemCatMSpec_{\mathcal{L}}$ (or just $SemCatMSpec$ if $\mathcal{L}$ is clear from the context) is the semantic category of modular specifications whose objects are all functors $H: Models(IMP) \rightarrow Models(EXP)$, for any IMP and EXP in $SemCatSpec$, and whose morphisms are natural transformations.   ⋄

Several operations have been defined to explicitly interconnect specification modules (see e.g. [EM90]). Here, we will just consider the two most basic ones: *composition* and *union*. The composition operation is related to parameter passing for parameterized specifications. The idea is to match the import and export interfaces of two modules, so that one module can import what the other exports:

Definition 6.33. Given module specifications

$$MODj = (PARj, EXPj, IMPj, BODj, ej, ij, sj, vj)$$

for $j = 1, 2$ and a pair $h = (h1, h2)$ of specification morphisms satisfying $e2 \circ h2 = h1 \circ i1$, the composition of $MOD1$ and $MOD2$ via h can be defined, at the specification level, as the module specification:

$$MOD3 = MOD1 \circ_h MOD2$$

defined by the outer square in the diagram in Figure 6.24, with $EXP3 = EXP1$, $IMP3 = IMP2$, $PAR3 = PAR1$, and $BOD3$ constructed as a pushout object in subdiagram (4).

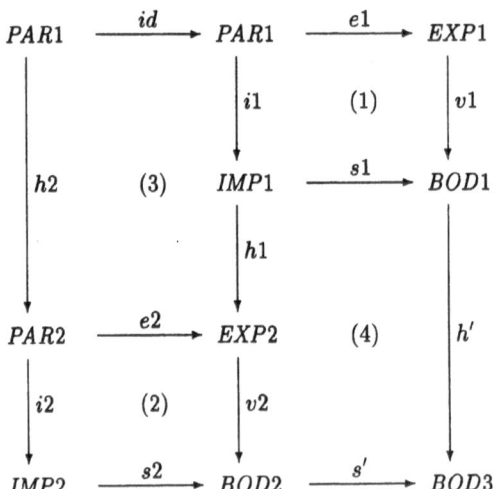

Fig. 6.24. Module Composition

At the model level, the semantics of the composition operation can be defined as the functor $M_{MOD1} \otimes_h M_{MOD2} \colon Mod(IMP2) \to Mod(EXP1)$ defined:

$$M_{MOD1} \otimes_h M_{MOD2} = M_{MOD1} \circ U_{h1} \circ M_{MOD2} \qquad \diamond$$

Remark 6.34.

1. The specification morphisms $h1 \colon IMP2 \to EXP2$ and $h2 \colon PAR1 \to PAR2$ match the import interface $IMP1$ of $MOD1$ with the export interface $EXP2$ of $MOD2$, and $PAR1$ with $PAR2$. The remaining import interface $IMP2$ becomes the import of $MOD3$ and the remaining export interface $EXP1$ becomes the export of $MOD3$. The new body $BOD3$ can be considered as the union of $BOD1$ and $BOD2$ with shared subpart $IMP1$. The specification morphisms of $MOD3$ are explicitly given by: $e3 = e1$, $i3 = i2 \circ h2$, $s3 = s' \circ s2$, and $v3 = h' \circ v1$, where s' and h' are defined by the pushout (4).

2. In the above definition we have considered that the parameter part of the two modules are explicitly bound through the morphism $h2$, i.e. $h2 \colon PAR1 \to PAR2$. We may relax this restriction if the underlying category of specifications has pullbacks. In this case, composition would be defined just by giving the fitting morphism $h1$, and square (3) in the above composition diagram would be a pullback.

Theorem 6.35. *Given correct module specifications MOD1 and MOD2 and h as above, defined over a semiexact institution $\mathcal{L}$, such that the composition $MOD3 = MOD1 \circ_h MOD2$ is defined then we have*

1. *MOD3 is correct*
2. *$M_{MOD3} = M_{MOD1} \otimes_h M_{MOD2}$*

Proof. Since $MOD1$ and $MOD2$ are correct we have strongly persistent free functors F_{s1} and F_{s2}. The semiexactness of $\mathcal{L}$ implies:

3. strong persistency of $F_{s'}$, and
4. $F_{s1} \circ U_{v2 \circ h1} = U_{h'} \circ F_{s'}$.

From (3) we conclude strong persistency of $F_{s3} = F_{s'} \circ F_{s2}$. F_{s3} is a free construction with respect to U_{s3} because free constructions are closed under composition. Hence we have correctness of $MOD3$ (1) and (4) implies (2):

$$M_{MOD3} = U_{v3} \circ F_{s3} =$$

(by def of $h3$ and $s3$)

$$= U_{v1} \circ U_{h'} \circ F_{s'} \circ F_{s2} =$$

(by condition (3) above)

$$= U_{v1} \circ F_{s1} \circ U_{h1} \circ U_{v2} \circ F_{s} =$$

(by def of M_{MOD1} and M_{MOD2})

$$= M_{MOD2} \circ U_{h1} \circ M_{MOD1} \qquad \square$$

The operation of *union* of modules is the generalization of the *combine* operation defined for "basic" specifications. The aim of the operation is to allow one to "put together" two modules that may share a common part, which is also expressed as a module.

Definition 6.36. Given module specifications

$$MODj = (PARj, EXPj, IMPj, BODj, ej, ij, sj, vj)$$

for $j = 0, 2$ and module morphisms $h1\colon MOD0 \to MOD1$ and $h2\colon MOD0 \to MOD2$, the union of $MOD1$ and $MOD2$ sharing $MOD0$ via $f1$ and $f2$, $MOD1 \cup_{(MOD0, f1, f2)} MOD2$, can be defined, at the specification level, as the pushout of the module diagram in Figure 6.25.

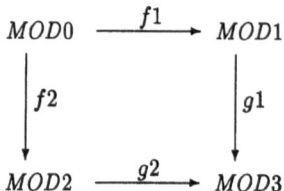

Fig. 6.25. Pushout of modules

At the model level, the semantics of the union operation can be defined as the corresponding pullback.   ◇

Remark 6.37. As shown in the definition, at the specification level, the semantics of the union of modules can only be defined if we have pushouts in $SyntCatMSpec$, which is a consequence of having pushouts in $SyntCatSpec$. In particular, the pushout in Figure 6.25 can be defined as the result of combining, respectively, the PAR, IMP, EXP, and $BODY$ specifications, and the connecting morphisms, of the modules $MOD1$ and $MOD2$, with respect to the corresponding specifications and morphisms of $MOD0$.

If the underlying institution is semiexact then it is easy to prove the following theorem:

Theorem 6.38. *Given correct module specifications $MOD0$, $MOD1$, and $MOD2$, and $h1$ and $h2$ as above, defined over a semiexact institution $\mathcal{L}$, we have*

1. *$MOD3$ is correct*
2. *$M_{MOD3} = M_{MOD1} +_{(M_{MOD0}, f1, f2)} M_{MOD2}$*

6.5.3   Modular systems: implicit interconnection of modules

As said above, the second approach is based on considering that a modular system is just a collection of modules that are, in some way, implicitly interconnected. Here, we will introduce this approach by presenting a slight variation of the formulation studied in [NOS95]. In particular, a modular system is seen as a collection of modules together with some *global description* of the system, e.g. the facilities or operations offered by the system. In our case we regard this description just as a signature. The modules are bound to this global signature by means of fitting (signature) morphisms, matching the services or operations imported and exported by a module with the global operations offered by the system. In this context, these signature morphisms play the role of (implicitly) interconnecting the modules of a given system.

Definition 6.39. A modular software system S is a pair, $(\Sigma, MSPEC)$, where $MSPEC$ is a (finite) set of modules on Σ, i.e. a set of triples consisting of a module and two signature monomorphisms $(MOD, h_I, h_E$, with $MOD = (PAR, EXP, IMP, BOD, e, i, s, v)$, $h_I \colon Signature(IMP) \to \Sigma$ and $h_E \colon Signature(EXP) \to \Sigma)$ such that the diagram in Figure 6.26 commutes.

◇

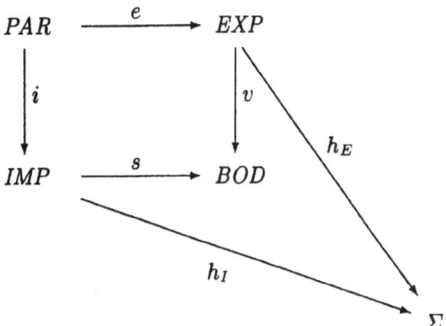

Fig. 6.26. Module component

This diagram states that, if PAR includes the common elements of IMP and EXP, then these elements must be bound to the same elements of the global signature through h_I and h_E. On the other hand, h_I and h_E should be monomorphisms, since it makes no sense that, for instance, a given operation offered by a system is internally defined in terms of two different operations within the same module.

Note that a modular system of this kind may be "incomplete" in the sense that there may be "elements" in the global signature which are not "defined" in (exported by) any module in the system. As a consequence, the meaning of a system of this kind may be seen as the class of all models (over the global signature) such that the different "parts" of the system corresponding to the given modules have the form specified by these modules. This is equivalent to regarding the modules as *constraints* over the system. That is, we say that the models of a system must *satisfy* the modules of the system. This idea is a generalization of the one used in [ETLZ82, CO88, OSC89, Bau91] where generating constraints (or data constraints [Rei80, BG80]) were used to deal with "incomplete" specifications.

Definition 6.40. Given a signature Σ, a module $MOD = (PAR, EXP, IMP, BOD, e, i, s, v)$ and two signature morphisms $h_I\colon Signature(IMP) \to \Sigma$, $h_E\colon Signature(EXP) \to \Sigma$, a model $A \in Mod(\Sigma)$ *satisfies MOD with respect to* $< h_I, h_E >$, denoted $A \models_{<h_I,h_E>} MOD$ iff $U_{h_I}(A) \in Mod(IMP)$ and, in addition, $U_{h_E}(A) = M_{MOD}(U_{h_I}(A))$. Given a modular system $S = (\Sigma, MSPEC)$, the semantics of S, denoted $[\![S]\!]$, is defined:

$$[\![S]\!] = \{A \in Mod(\Sigma)/\forall(MOD, h_I, h_E) \in MSPEC\, A \models_{<h_I,h_E>} MOD\} \quad \diamond$$

In [NOS95] a number of operations over a similar class of modular systems are studied, paying special attention to operations of refinement and implementation. Here, we will just provide some comparison between the two approaches studied above, since implementation notions are studied in a different chapter of this book. In particular, we will see how the effects of the operations of union and composition can be, in some sense, achieved by "putting together" the modules involved.

Theorem 6.41. *Given module specifications over a semiexact institution*

$$MODj = (PARj, EXPj, IMPj, BODj, ej, ij, sj, vj)$$

for $j = 0, 2$, module morphisms $g1\colon MOD0 \to MOD1$ and $g2\colon MOD0 \to MOD2$, and given a signature Σ and morphisms $h_{Ij}\colon Signature(IMPj) \to \Sigma$, $h_{Ej}\colon Signature(EXPj) \to \Sigma$, for $j = 0, 2$, such that the diagrams in Figures 6.27 and 6.28 commute, we have that for every $A \in Mod(\Sigma)$,

$$A \models_{<h_{I3},h_{E3}>} MOD1 \cup_{(MOD0,f1,f2)} MOD2 \;\; iff \;\; A \models_{<h_{Ij},h_{Ej}>} MODj$$

for every $j = 0, 2$, where h_{I3} and h_{E3} are the unique (universal) morphisms going from the pushout objects IMP3 and EXP3 into Σ, respectively.

Proof. Let $MOD3 = MOD1 \cup_{(MOD0,f1,f2)} MOD2$, $A \models_{<h_{Ij},h_{Ej}>} MODj$, for every $j = 0, 2$, if and only if $U_{h_{I3}} \in Mod(IMP3)$ since, for every $j = 0, 2$, $U_{h_{Ij}} \in Mod(IMPj)$ and $U_{h_{I3}}(A) = U_{h_{I1}}(A) +_{U_{h_{I0}(A)}} U_{h_{I2}}(A)$. In addition, for every $j = 0, 2$, $U_{h_{Ej}}(A) = M_{MODi}(U_{h_{Ij}}(A))$ if and only if

$$U_{h_{E3}}(A) = U_{h_{E1}}(A) +_{U_{h_{E0}(A)}} U_{h_{E2}}(A) =$$

$$M_{MOD1}(U_{h_{I1}}(A)) +_{M_{MOD0}(U_{h_{I0}}(A)} M_{MOD2}(U_{h_{I2}}(A))$$

but, by Theorem 6.38, this is equal to $M_{MOD3}(U_{h_{I3}}(A))$. $\qquad\square$

Fig. 6.27. Conditions for implicit module union (1)

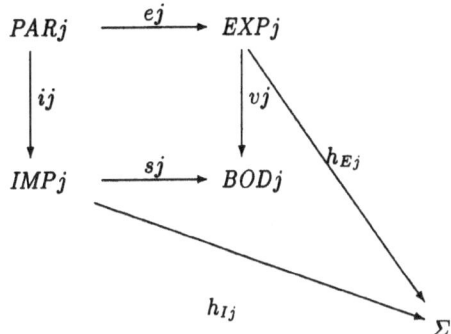

Fig. 6.28. Conditions for implicit module union (2)

Theorem 6.42. *Given module specifications over a semiexact institution*

$$MODj = (PARj, EXPj, IMPj, BODj, ej, ij, sj, vj)$$

$j = 1, 2$, $g = (g1, g2)$ *such that the composition* $MOD3 = MOD1 \circ_g MOD2$ *is defined, and given a signature* Σ *and morphisms* $h_{Ij} \colon Signature(IMPj) \to \Sigma$, $h_{Ej} \colon Signature(EXPj) \to \Sigma$, *for* $j = 1, 2$, *such that the diagrams in Figures 6.29 and 6.30 commute, we have that for every* $A \in Mod(\Sigma)$,

$$A \models_{<h_{I3}, h_{E3}>} MOD3 \text{ if for every } j = 1, 2 \ A \models_{<h_{Ij}, h_{Ej}>} MODj$$

where $h_{I3} = h_{I2}$ *and* $h_{E3} = h_{E1}$.

Proof. Let $A \in Mod(\Sigma)$ then $U_{h_{I3}}(A) \in Mod(IMP3)$ since, by definition, $IMP3 = IMP2$ and $h_{I3} = h_{I2}$. Additionally,

$$U_{h_{E3}}(A) = U_{h_{E1}}(A) = M_{MOD1}(U_{h_{I1}}(A)) = M_{MOD1}(U_{g1}(U_{h_{E2}}(A))) =$$
$$M_{MOD1}(U_{g1}(M_{MOD2}(U_{h_{I2}}(A)))) = M_{MOD3}(U_{h_{I3}}(A)) \qquad \square$$

Remark 6.43. Note that the converse is not true in general. For instance, consider two modules where $EXP1$ and $IMP2$ are the empty specifications. Then any Σ-model would trivially satisfy the resulting module $MOD1 \circ_h MOD2$, but depending on how $EXP2$ is defined, not every Σ-model would satisfy $MOD2$. If composition were defined in a slightly different way, by defining $EXP3$ as the union (appropriate pushout) of $EXP1$ and $EXP2$, then the converse of the theorem would hold.

200   Fernando Orejas

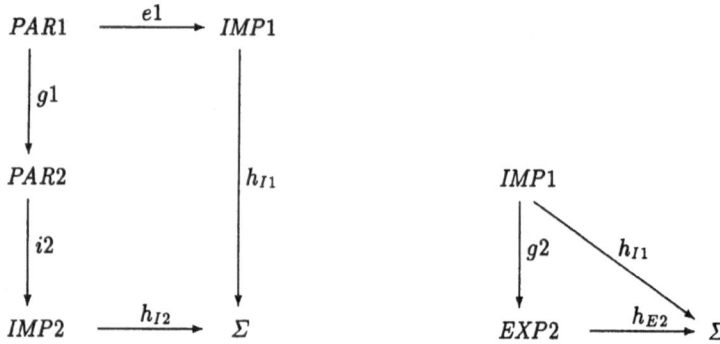

Fig. 6.29. Conditions for implicit module composition (1)

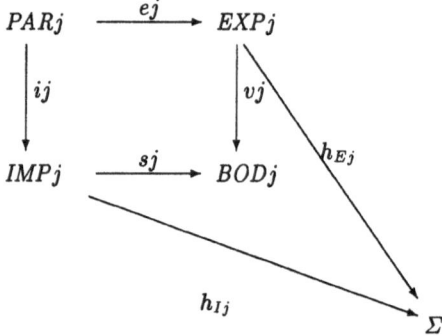

Fig. 6.30. Conditions for implicit module composition (2)

Acknowledgements. This paper is the result of the work of many people (who are mentioned in the paper and can be found in the references) and of discussions the author had in the last fifteen (?) years especially with Hartmut Ehrig, Michel Bidoit, Joseph Goguen, José Meseguer, Don Sannella, Andrzej Tarlecki, and Martin Wirsing. This work has been partially supported by the Esprit Basic Working Group COMPASS II (ref. 6112) and by the Spanish CICYT project COSMOS (ref. TIC95-1016-C02-01).

7 Refinement and Implementation

Hartmut Ehrig and Hans-Jörg Kreowski

[1] Technische Universität Berlin, Fachbereich Informatik
   Franklinstraße 28/29, D-10587 Berlin
   ehrig@cs.tu-berlin.de
[2] Universität Bremen, Fachbereich Mathematik/Informatik
   Postfach 330440, D-28334 Bremen
   kreo@informatik.uni-bremen.de

7.1 Introduction

Successful developments of data processing systems eventually result in systems that are programmed in existing programming languages, run on existing computers and serve certain purposes. The intended tasks and purposes of a data processing system may be put together in the requirements definition. Then the final program implements the intended system and is said to be correct if it meets the requirements.

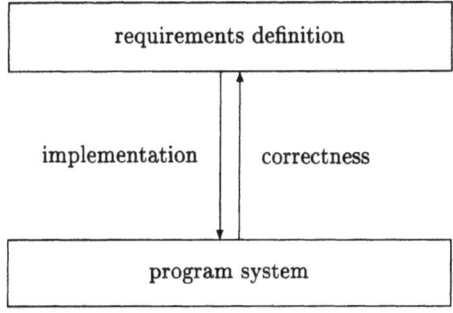

Correct programs are the goal of systems development, but this is hard to achieve and to guarantee in general, because requirements and programs are usually written in quite different languages which are difficult to compare on the semantic level. Moreover, systems development often involves many people and takes considerable time so that it is unreasonable to start the correctness proofs at the very end of the development. To overcome this difficult situation and to bridge the wide gap between requirements and programs, the idea of stepwise refinement may help. As illustrated in Fig. 7.1, it means putting in various additional layers of specifications between the requirements definition and the final program system where each pair of successive layers reflects certain design decisions, but are still close together so that correctness proofs become easier. We call a refinement step between algebraic specifications an implementation as often done in the literature.

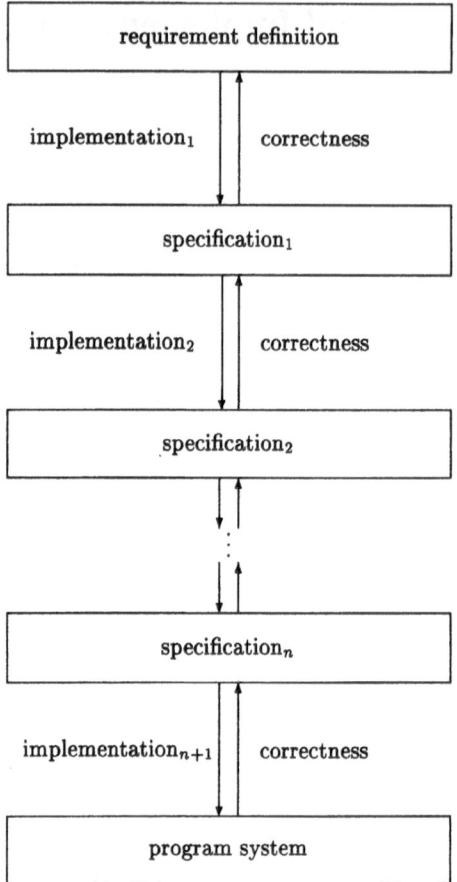

Fig. 7.1. Stepwise implementation

The questions are: How can two specifications be chosen, related, or constructed to form a refinement step? What does it mean semantically? And, in particular, which conditions must be met for a refinement step to be called correct?

In the literature, one encounters an amazing number of answers to these questions (e.g., [GTW78, Ehr78, EKP78, GHM78, Ore81, GM82, SW82, EK83, Gan83, KA84, Poi84, Ore85, BBC86b, BMPW86, ST88b, Ber89, Hen89], which is far from being a complete list). There are good reasons why so many approaches have been introduced. Firstly, different types of semantics for algebraic specifications (like loose, behavioural, and initial) must be taken into account and need different handling. Secondly, there are quite different views on how much two specifications which form a refinement step may be allowed to differ from each other. In particular, purely semantic approaches are opposed to others that involve syntactic constructions to relate the specifications of a refinement. Thirdly, the notion of correctness has various facets which can be emphasized in various ways.

Nevertheless, a few major phenomena can be identified that are addressed in the literature.

First of all, one has to observe that requirements definitions usually admit various solutions (if they are consistent at all) while the final program is a particular solution. For example, one may require some data container which allows one to store data, remove it or check its presence. Then clearly there are many solutions including sets, multisets, strings, trees, arrays, etc. The choice between the possible solutions is an important task of refinement and many design decisions narrow down the class of admitted models until eventually only one model is left. For example, one may decide that the order in which data are put into the container is irrelevant. Then strings and trees are no longer proper solutions. Furthermore, if multiple occurrences of data are forbidden, multisets are also ruled out.

However even if one already has a single data type solving a given problem, it may be necessary, or reasonable, to proceed in the implementation process because the present solution may be too inefficient or the concepts involved may not be available in the programming language used. Consider, for example, the sorting of strings specified by insertion sorting, that is, by placing the elements of the given string one after the other in its proper slot. In the worst case, one has to perform $\frac{n^2-n}{2}$ comparisons if n is the length of the given string. But there are much better solutions like merge sort where the number of comparisons is bounded by $n \cdot \log n$. Or consider stacks and queues as intended data types. If the programming language offers only integers, products and arrays, one has to represent stacks and queues in terms of the concepts available (e.g. stacks as pairs of arrays and pointers, being non-negative integers, and queues as triples of integers, arrays, and integers where the second integer is never smaller than the first). In such situations, one obtains a so-called abstraction function that maps representing data to represented data. This function may map many-to-one, meaning that an abstract data element is multiply represented. For example, a pair (A, i) of an array A and a non-negative integer i represents the stack $A[0] \cdots A[i-1]$ (with $A[i-1]$ being the top element), such that two such pairs (A, i) and (A', i) represent the same stack if A and A' differ only in entries with indices beyond $0, \ldots, i-1$. Moreover, the abstraction function may be partial, meaning that certain data is present as "junk", representing nothing. For example, a pair of an array and a negative number does not represent any stack.

This survey is organized in the following way. Section 7.2 provides a detailed discussion of some typical examples of implementation. In Section 7.3, a general notion of implementation based on the semantic concepts of constructors and abstractors is introduced, and several approaches in the literature are related to this framework, while syntactic aspects are considered in Section 7.4. In particular, constructors and abstractors are described by syntactic means yielding a variety of notions of implementations with a full syntactic description. Section 7.5 is concerned with the composition of implementations where different possibilities are studied. In Sections 7.6 and 7.7, the compatibility of implementation concepts with horizontal structuring of specifications, including parameterized specifications and modules, is considered. Moreover, the reader is refered to Chapter 11 for a discussion of proof systems for refinements and to Chapter 14 for an alternative framework for formalizing the system development process.

7.2 Examples

In this section, some typical examples of implementation of algebraic specifications by algebraic specifications are discussed. They were mentioned informally in the introduction but more detail is given here to illustrate the formal notions of implementations in the following sections.

7.2.1 Containers, strings, and sets

Consider the specification of a data container that is initially empty, allows the insertion and removal of data, and provides emptiness and membership tests.

$Container = Data$
sorts *container*
opns   *empty*: *container*
           $put - in -: data \times container \to container$
           $get - outof -: data \times container \to container$
           $- isempty: container \to bool$
           $- isin -: data \times container \to bool$
vars $x, y \in data$
           $c \in container$
axioms *get x outof empty = empty*
           get x outof (put x in c) = get x outof c
           get x outof (put y in c) = put y in (get x outof c) **if** $x \neq y$
           empty isempty
           $\neg((put\ x\ in\ c)\ isempty)$
           $\neg(x\ isin\ empty)$
           $x\ isin\ (put\ y\ in\ c) = (x \equiv y) \vee (x\ isin\ c)$

$Data = Bool$
sorts *data*
opns $\equiv, \neq: data \times data \to bool$
vars $x, y, z \in data$

axioms $x \equiv x$
$\qquad x \equiv y$ **if** $y \equiv x$
$\qquad x \equiv z$ **if** $(x \equiv y) \wedge (y \equiv z)$
$\qquad x \neq y = \neg(x \equiv y)$

Intuitively, it is clear that many data types, like strings, multisets, and sets, can serve as data containers and can be seen as implementations of data containers. For example, adding the constraints that *Container*-algebras are free constructions over their *Data*-reduct semantically yields lists of data, and adding further proper axioms as well yields sets of data.

$List = Container$
constrs free over *Data*

$Set = Container$
axioms *put* x *in* (*put* x *in* c) = *put* x *in* c
$\qquad$ *put* x *in* (*put* y *in* c) = *put* y *in* (*put* x *in* c)
constrs free over *Data*

In this way, *Set*-algebras and *List*-algebras are *Container*-algebras. If one considers *List* and *Set* as implementations of *Container*, it means semantically to choose particular models. If one adds the two axioms and the constraint of *Set* one after the other to *Container*, the semantic effect is to reduce the class of admitted models, reflecting certain design decisions.

However this view of implementation is not realistic if one thinks of strings and sets as standard data types that are predefined and re-usable. In such a case, one cannot assume that the specification of lists, for instance, uses the same names as the specification of containers. To overcome this discrepancy between implemented and implementing specifications, one may regard renamings as implementations.

$List_1$ **implements** *Container* **by**
sortsdefs *container* **is** *list*
opnsdefs *empty* **is** *empty*
$\qquad\qquad$ *put* $-$ *in* $-$ **is** *ladd*
$\qquad\qquad$ *get* $-$ *outof* $-$ **is** *rem*
$\qquad\qquad - $ *isempty* **is** $-$ *isempty*
$\qquad\qquad - $ *isin* $-$ **is** $-$ *isin* $-$

where the following specification of strings is used.

$List_1 = List_0$
opns $\quad ladd: data \times list \to list$
$\qquad\quad - isempty: list \to bool$
$\qquad\quad - isin -: data \times list \to bool$
axioms $ladd(x,l) = cat(in(x),l)$
$\qquad\quad empty\ isempty$
$\qquad\quad \neg(ladd(x,l)\ isempty)$
$\qquad\quad \neg(x\ isin\ empty)$
$\qquad\quad x\ isin\ ladd(y,l) = (x \equiv y) \vee (x\ isin\ l)$

$List_0 = Data$
sorts $list$
opns $empty\colon list$
             $in\colon data \to list$
             $cat\colon list \times list \to list$
             $rem\colon data \times list \to list$
vars $x, y \in data$
             $l, l', l'' \in list$
axioms $cat(empty, l) = cat(l, empty) = l$
             $cat(l, cat(l', l'')) = cat(cat(l, l'), l'')$
             $rem(x, empty) = empty$
             $rem(x, in(y)) = in(y)$ **if** $x \neq y$
             $rem(x, in(x)) = empty$
             $rem(x, cat(l, l')) = cat(rem(x, l), rem(x, l'))$

The renaming defines a signature morphism $h\colon \Sigma(Container) \to \Sigma(List_1)$ with the identity on the $Data$-part. Then the h-reduct $U_h(A)$ of any $List_1$-algebra A can be shown to be a $Container$-algebra. Hence looking at $List_1$ as an implementation of $Container$ through h means semantically to reduce the class of models using the forgetful functor given by h.

The same effect can be obtained by specifying the signature morphism as an enrichment of $List_1$ and $\Sigma(Container)$.

$List_1$ **implements** $Container = List_1 + \Sigma(Container)$
opns $copy\colon list \to container$
axioms $empty = copy(empty)$
             $put\ x\ in\ copy(l) = copy(ladd(x, l))$
             $get\ x\ outof\ copy(l) = copy(rem(x, l))$
             $copy(l)\ isempty = l\ isempty$
             $x\ isin\ copy(l) = x\ isin\ l$

Given a $List_1$ **implements** $Container$-algebra by choosing $A_{container}$ as A_{list} and the $Container$-operations as $empty_A$, $radd_A$, rem_A, $- isempty_A$, and $- isin -_A$ in this order, the reduct of this algebra to the $\Sigma(Container)$-part yields $U_h(A)$ as above.

The enrichment in $List_1$ **implements** $Container$ is a pure copying of sorts and operations. One may go a step further and allow an arbitrary enrichment as part of an implementation. For example, the enrichment of $List_0$ yielding $List_1$ may be done implicitly in the implementation (and not explicitly as before).

$List_0$ **implements** $Container = List_0 + \Sigma(Container)$
opns $copy\colon list \to container$
axioms $empty = copy(empty)$
             $put\ x\ in\ copy(l) = copy(cat(in(x), l))$
             $get\ x\ outof\ copy(l) = copy(rem(x, l))$
             $copy(l)\ isempty = l\ isempty$
             $x\ isin\ copy(l) = x\ isin\ l$

There is a small but important difference. In the former case, the left addition implementing the $put - in$-operation had to be specified explicitly. In the latter case, the left addition is kept implicit.

The idea behind this kind of implementation is that the specifications available do not provide all the concepts needed, but enough to specify them.

7.2.2   Sorting

The intended properties of sorting on strings are easily formulated if a sortedness test is available in *List*:

> $Sort = List$
> **opns** $sort: list \rightarrow list$
> **vars** $x \in data, u \in list$
> **axioms** $sort(u)$ *issorted*
> $\qquad\qquad x\ isin\ sort(u) = x\ isin\ u$

Note that this specification of sorting is loose even if one keeps the *List*-part persistent because the second axiom requires which elements must be present in the sorted list and which not, but nothing is required concerning the multiplicity of elements. Obviously this specification is not executable because a term $sort(l)$ for some string l cannot be transformed by the axioms. However many famous implementations of sorting exist. One is insertion sorting:

> $InsertionSort = List$
> **opns** $sort: list \rightarrow list$
> $\qquad\qquad insert: data \times list \rightarrow list$
> **vars** $x, y \in data, v \in list$
> **axioms** $insert(x, empty) = x$
> $\qquad\qquad insert(x, yv) = xyv$ **if** $x \le y$
> $\qquad\qquad insert(x, yv) = y\ insert(x, v)$ **if** $\neg(x \le y)$
> $\qquad\qquad sort(empty) = empty$
> $\qquad\qquad sort(xu) = insert(x, sort(u))$

Insertion sorting shows that there is an equational realization of the sorting problem. It is an executable specification because each term $sort(t)$, for a *List*-term t without variables, can be transformed into a *List*-term t' by applying a sequence of equations applications such that the equations t' *issorted* and $x\ isin\ t' = x\ isin\ t$ hold, showing that t' is a sorting of t. In the worst case, one must apply a number of equations that is quadratic in terms of the length of t. This is unacceptably inefficient. In this respect merge sort is a much better implementation.

> $MergeSort = List$
> **opns** $sort, left, right: list \rightarrow list$
> $\qquad\qquad merge: list \times list \rightarrow list$
> **vars** $x, y \in data, u, v \in list$

axioms $left(\lambda) = \lambda, \quad right(\lambda) = \lambda$
$\qquad\qquad left(x) = x, \quad right(x) = \lambda$
$\qquad\qquad left(xuy) = x\ left(u), \quad right(xuy) = right(u)y$
$\qquad\qquad merge(\lambda, v) = merge(v, \lambda) = v$
$\qquad\qquad merge(xu, yv) = x\ merge(u, yv)$ **if** $x \leq y$
$\qquad\qquad merge(xu, yv) = y\ merge(xu, v)$ **if** $\neg(x \leq y)$
$\qquad\qquad sort(\lambda) = \lambda$
$\qquad\qquad sort(x) = x$
$\qquad\qquad sort(xuy) = merge(sort(left(xuy)), sort(right(xuy)))$

The relationship of *MergeSort* to the requirements in *Sort* is exactly the same as that described above for *InsertionSort*, but the number of equations one must apply is now of the order $n \cdot \log n$ if n is the length of the input list. The reason is that the operations *left* and *right* decompose a string w in such a way that

$$w = left(w)right(w) \text{ and } |left(w)| \geq |right(w)| \geq |left(w)| - 1.$$

Hence each application of the last equation induces two further sortings of lists half as long as the original list. This stops after $\log n$ steps, and the merges required in each step do not need the application of more than n equations.

7.2.3   Queues

An essential feature of implementation is the notion of an abstraction function which may be illustrated by the implementation of queues by arrays with two pointers (mimicking the classical implementation of stacks by arrays with pointers). To make things easier, the data type *Queue* has a constant initial queue, and enqueue and dequeue operations, but no reading is specified. This allows one to avoid exception handling.

$\quad Queue = Data$
sorts$\quad queue$
opns$\quad init: queue$
$\qquad\quad enq: data \times queue \rightarrow queue$
$\qquad\quad deq: queue \rightarrow queue$
vars$\quad x, y \in data$
$\qquad\quad q \in queue$
axioms$\ deq(init) = init$
$\qquad\quad deq(enq(x, init)) = init$
$\qquad\quad deq(enq(x, enq(y, q))) = enq(x, deq(enq(y, q)))$

The specification of *arrays* uses integers as indices and the specification *Data* for entries. It provides an initial *array* and an assignment of data to indices as an operation, but no access to the entries because this is not needed in the implementation.

$Array = Int + Data$
sorts   *array*
opns $init: array$
      $-(-):= -: array \times int \times data \rightarrow array$
vars $A \in array$
      $i, j \in int$
      $x, y \in data$
axioms $(A(i):= x)(j):= y = $ *if* $i \equiv j$ *then* $A(i):= y$ *else* $(A(j):= y)(i):= x$

Now *Queue* can be implemented by arrays with two pointers (given by integers) as an enrichment of *Array* and the signature of *Queue*. The initial queue is represented by the initial array with both pointers set to 0. The enqueuing is provided by an assignment using the second pointer as an index and by incrementing the second pointer. The dequeuing leads to an increment of the first pointer if it is smaller than the second, but has no effect otherwise.

Array with 2 Pointers **implements** $Queue = Array + \Sigma(Queue)$
opns $(-, -, -): int \times array \times int \rightarrow queue$
vars $A \in array$
      $i, j \in int$
      $x \in data$
axioms $init = (0, init, 0)$
      $enq(x, (i, A, j)) = (i, A(j):= x, succ(j))$
      $deq((i, A, j)) = (succ(i), A, j)$ **if** $i < j$
      $deq((i, A, i)) = (i, A, i)$

To explain the semantic relation between *Queue* and *Array with 2 Pointers* as established by this implementation, we choose particular standard models. Let the finite set $X = \{x_1, \dots, x_k\}$ be the domain of type *data* (but any other carrier will do if it is not empty). Then consider all strings over X, i.e., the set X^*, as the domain of type *queue*, the empty string λ as *init*, the right addition of an element to a string as *enq*, and the removal of the leftmost element as *deq*, if there is an element, while *deq* of λ yields λ. Note that this is an explicit version of the free *Queue*-algebra over X. For *int* we take the usual set $\mathbb{Z}$ of integers. The domain of type *array* can then be defined by all partial mappings from $\mathbb{Z}$ to X with finite support, i.e.,

$$T_{array} = \{arr: \mathbb{Z} \rightarrow X \mid sup(arr) = \{i \in \mathbb{Z} \mid arr(i) \text{ defined}\} \text{ is finite}\}.$$

The initial array is the totally undefined mapping $\emptyset$. The assignment $arr(i) := x$ of $x \in X$ to array *arr* at index i yields a new array defined by $sup(arr(i):= x) = sup(arr) \cup \{i\}$ and

$$(arr(i):= x)(j) = \begin{cases} x & \text{for } i = j, \\ arr(j) & \text{otherwise.} \end{cases}$$

This provides an explicit version of the free *Array*-algebra over X and $\mathbb{Z}$.

The idea of the implementation *Array with 2 Pointers* **implements** *Queue* is to represent queues by triples consisting of an array and two pointers each, and to realize the queue operations as operations on these triples. The only implementing operation $(-, -, -)$ accordingly specifies the domain queue as the product domain $\mathbb{Z} \times T_{array} \times \mathbb{Z}$, and the implementing axioms state the following:

(1) the inital queue is represented by the triple $(0, \emptyset, 0)$,
(2) *enq* assigns the new value to the location determined by the second pointer and increases the second pointer by 1, and
(3) *deq* increases the first pointer if it is smaller than the second and leaves everything unchanged if both pointers are equal (whereas nothing is required if the first pointer is larger than the second).

If one now considers those triples that can be obtained by compositions of queue operations, one gets the significant representations

$$REPR = \{(i, arr, j) \mid 0 \le i \le j, \ arr = \emptyset \text{ if } j = 0, \text{ and}$$
$$sup(arr) = \{0, \dots, j-1\} \text{ otherwise}\}.$$

Hence, there are many triples that are never used if one operates on queues with queue operations only. These triples are junk. Moreover, each queue has many representations – actually an infinite number. For example, the queue of length 1 with entry x can be obtained by $n+1$ enqueue steps for some $n \ge 0$, followed by n dequeue steps, only if the value of the last enqueue step is x, because the queue axioms identify all those terms. However, on the implementation level, they are pairwise different because their first indices reflect the number of dequeue steps.

abstract queues:
$$deq(enq(x, enq(y, init))) = enq(x, deq(enq(y, init)))$$
$$= enq(x, init)$$

implemented queues:
$$\begin{aligned} deq(enq(x, enq(y, init))) &= deq(enq(x, enq(y, (0, \emptyset, 0)))) \\ &= deq(enq(x, (0, \emptyset(0) := y, 1))) \\ &= deq(0, (0, (\emptyset(0) := y)(1) := x, 2)) \\ &= (1, (\emptyset(0) := y)(1) := x, 2) \\ &\neq (0, \emptyset(0) := x, 1) \\ &= enq(x, (0, \emptyset, 0)) \\ &= enq(x, init) \end{aligned}$$

If one associates each significant triple with the abstract queue it represents, one gets the abstraction function of the implementation above, *abstr* : $REPR \to X^*$. It is defined as follows:

(i) $abstr(i, arr, i) = \lambda$ \quad for $i \ge 0$
(ii) $abstr(i, arr, j) = arr(i) \cdots arr(j-1)$ \quad for $0 \le i \le j$

The abstraction function may be considered as a partial mapping from $\mathbb{Z} \times T_{array} \times \mathbb{Z}$ to X^* with $REPR$ as the domain of definition.

7.3   Implementation concepts

As motivated in the introduction and in Section 7.2, a general notion of an implementation of a specification SP_1 by a specification SP_2 may consist of a constructive part and an abstraction part (see Section 7.3.5). The constructive part enriches SP_2 to SP such that one finds the data sorts and operations of SP_1 (i.e. its signature) in SP immediately or after some renaming. Moreover, a constructor (Section 7.3.1) from SP_2 to SP (together with the forgetful functor corresponding to the renaming) states which $\Sigma(SP_1)$-models are obtained from the SP_2-models. The constructor may associate a class of SP-models with each SP_2-model such that the implementation may have a loose semantics. Functional constructors, which are used in most approaches, are special cases. Moreover, the constructive part of an implementation may be trivial, meaning that the enrichment as well as the constructor are identities. The abstraction part is given by an abstractor (Section 7.3.3) stating which $\Sigma(SP_1)$-models are acceptable as realizations of any SP_1-model. Such an implementation can be considered as correct if all constructed models are acceptable and each SP_1-model has an acceptable constructed model. The implementation concept was introduced by Sannella and Tarlecki [ST88b] and – and in a slightly more general variant – by Orejas, Navarro, and Sanchez [ONS93, ONS96] to cover most of the notions one encounters in the literature.

7.3.1   Constructor

Let SP_0 and SP be specifications. A *constructor* from SP_0 to SP is a mapping κ from SP_0-models to SP-model classes, denoted by $\kappa: Mod(SP_0) \to Mod(SP)$.

A constructor κ is called functional if the model class $\kappa(A)$ for each $A \in Mod(SP_0)$ is a singleton set (up to isomorphism).

Constructors are obviously closed under sequential composition. Let κ be a constructor from SP_0 to SP and κ' a constructor from SP to SP_1. Then $\kappa' \circ \kappa$ given by

$$\kappa' \circ \kappa(A) = \bigcup_{A' \in \kappa(A)} \kappa'(A')$$

for all $A \in Mod(SP_0)$, is a constructor from SP_0 to SP_1.

7.3.2   Examples of constructors

1. Any mapping from SP_0-models to SP-models may be seen as a functional constructor. To meet the formal definition, one must assume that an SP-model A represents the singleton SP-model class $\{A\}$. In particular, any functor $F: Mod(SP_0) \to Mod(SP)$ is a constructor in this way.

2. Consider a specification morphism $i: SP_0 \rightarrow SP$ with the forgetful functor $U_i: Mod(SP) \rightarrow Mod(SP_0)$ and let $A_0 \in Mod(SP_0)$. Then the class of all SP-models A with $U_i(A) = A_0$, denoted by $translate_i(A_0)$, defines a constructor $translate_i: Mod(SP_0) \rightarrow Mod(SP)$.

3. If U_i has a left adjoint functor $T_i: Mod(SP_0) \rightarrow Mod(SP)$, this defines a functional constructor.

4. Consider equational specifications in particular. Then the class of all quotients of A_0, denoted by $quotient(A)$, defines a constructor $quotient: Mod(SP_0) \rightarrow Mod(SP_0)$.

5. Let E be a set of equations. Then the quotient $A_0/\equiv_E$ of A_0 through the congruence $\equiv_E$ induced by E yields a functional constructor $ident_E: Mod(SP_0) \rightarrow Mod(SP_0 + E)$ where $SP_0 + E$ is obtained from SP_0 by adding E to the set of equations.

   Note that the natural homomorphism $nat: A_0 \longrightarrow A_0/\equiv_E$ provides a surjective abstraction function for each $A_0 \in Mod(SP_0)$. One may say that A_0 represents the quotient algebra up to multiple data representations which are identified by the constructor.

6. Let $j: SP \rightarrow SP_0$ be a specification morphism, $U_j: Mod(SP_0) \rightarrow Mod(SP)$ the forgetful functor, T_{SP} the initial model, and $init_{A_0}: T_{SP} \rightarrow U_j(A_0)$ for $A_0 \in Mod(SP_0)$ the initial SP-morphism. Then the image of T_{SP} in $U_j(A_0)$ under $init_{A_0}$, denoted by $init_{A_0}(T_{SP})$, defines a functional constructor $restrict_j: Mod(SP_0) \rightarrow Mod(SP)$.

   The restriction of A_0 with respect to j yields the smallest SP-algebra of the j-reduct of A_0, i.e., the subalgebra that is generated by its operations. Intuitively, the construction removes all "junk" from A_0 if the SP-operations are the interesting and accessible part of the specification.

7. Consider specifications SP' and SP'' as well as a specification morphism $j: \Sigma(SP') \longrightarrow SP''$. Let E' be the equations of SP'. Then the composition $ident_{E'} \circ restrict_j: Mod(SP'') \longrightarrow Mod(SP')$ is a functional constructor which takes an SP''-algebra A'', restricts its j-reduct to the subalgebra $\overline{A}$ generated by the SP'-operations, and factorizes it according to the SP'-equations. By construction, one gets a partial abstraction function from A'' (or $U_j(A'')$) to $\overline{A}/\equiv_{E'}$ with $\overline{A}$ as the domain of definition.

   The composition of restriction and identification may be performed the other way round yielding $restrict_{j'} \circ ident_{E''}: Mod(SP'') \longrightarrow Mod(SP')$ where the set of equations E'' contains the translated equations with respect to j and $j': SP' \longrightarrow SP'' + E''$ extends j including the equations E'.

7.3.3 Abstractor

Let SP_0 be a specification. An *abstractor* on SP_0 is a mapping α of SP_0-models to $\Sigma(SP_0)$-model classes, denoted by $\alpha: Mod(SP_0) \rightarrow Mod(\Sigma(SP_0))$, subject to the *reflexivity* and *transitivity* conditions:

- $A_0 \in \alpha(A_0)$ for all $A_0 \in Mod(SP_0)$,
- $A_1 \in \alpha(A_2)$ and $A_2 \in \alpha(A_3)$ implies $A_1 \in \alpha(A_3)$,
  for all $A_1 \in Mod(\Sigma(SP_0))$ and $A_2, A_3 \in Mod(SP_0)$.

7.3.4   Examples of abstractors

1. Let $A_0 \in Mod(SP_0)$. Then the class of all $\Sigma(SP_0)$-models A with $f: A \to A_0$ for some $\Sigma(SP_0)$-morphism f, denoted by $abstr(A_0)$, defines an abstractor $abstr: Mod(SP_0) \to Mod(\Sigma(SP_0))$. The identity on A_0 and the composition $f_2 \circ f_1: A_1 \to A_3$ of $\Sigma(SP_0)$-morphisms $f_1: A_1 \to A_2$ and $f_2: A_2 \to A_3$ are $\Sigma(SP_0)$-morphisms. Hence $abstr$ satisfies the reflexivity and transitivity conditions.

2. If the morphism f above is required to satisfy some condition C in addition, one gets an abstractor $abstr_C: Mod(SP_0) \to Mod(\Sigma(SP_0))$ provided that identities satisfy C and compositions satisfy C if the components do. A typical property of this kind is surjectivity in cases where morphisms are based on functions. Another important case is where the abstractor mappings are required to be isomorphisms such that $abstr_{iso}(A_0)$ consists of all models isomorphic to $A_0 \in Mod(SP_0)$.

3. The constructor *quotient* defined in Section 7.3.2.4 is an abstractor, because each equational model can be seen as a quotient of itself (through the identity congruence) and a quotient of a quotient of some model is again a quotient of this model.

4. Let SP_0 again be an equational specification and E a subset of its set of equations. Let $A_0 \in Mod(SP_0)$. Then the class $abstr_E(A_0)$ of all $\Sigma(SP_0)$-models A with $A/\equiv_E \cong A_0$, where $\equiv_E$ is the congruence on A induced by E, defines an abstractor $abstr_E: Mod(SP_0) \to Mod(\Sigma(SP_0))$. Reflexivity and transitivity are obvious because $A_0/\equiv_E \cong A_0$ and $A_1/\equiv_E \cong A_2$ and $A_2/\equiv_E \cong A_3$ imply $A_1/\equiv_E \cong (A_1/\equiv_E)/\equiv_E \cong A_2/\equiv_E \cong A_3$. Note that this abstractor corresponds to the constructor $ident_E$ given in Section 7.3.2.5.

5. There is also an abstractor corresponding to $restrict_j$ in Section 7.3.2.6. Let SP_0 be a specification such that an initial $\Sigma(SP_0)$-algebra T, i.e., a term algebra in many cases, exists. Let $init_A: T \to A$ for any $A \in Mod(\Sigma(SP_0))$ be the corresponding initial morphism and $init_A(T)$ its image. Then the class $abstr_R(A_0)$ of all A, such that $init_A(T)$ is isomorphic to $A_0 \in Mod(SP_0)$, defines an abstractor $abstr_R: Mod(SP_0) \to Mod(\Sigma(SP_0))$.

6. Both the latter abstractors can be combined either way, leading to the abstractors RI_E and $I_E R$ (or RI and IR if E is clear from the context). $RI_E(A_0)$ consists of all $\Sigma(SP_0)$-algebras A such that $init_A(T)/\equiv_E$ is isomorphic to A_0. Similarly, $I_E R(A_0)$ consists of all $\Sigma(SP_0)$-algebras A' such that $init_{A''}(T)$ is isomorphic to A_0 with $A'' = A'/\equiv_E$. In the former case, the representing algebra A is first restricted to all data accessible by terms, i.e., by operations of the specification. This removes the "junk".

Multiple representations are then identified. The latter case eliminates multiple representations first and removes the "junk" afterwards.

7.3.5   Implementation

Let SP_1 and SP_2 be specifications. An *implementation* of SP_1 by SP_2 is a system $Impl = (SP, i, h, \kappa, \alpha)$ where SP is a specification, called the *intermediate specification*, $i\colon SP_2 \to SP$ an enrichment, $h\colon \Sigma(SP_1) \to \Sigma(SP)$ is a signature morphism, $\kappa\colon Mod(SP_2) \to Mod(SP)$ a constructor, and $\alpha\colon Mod(SP_1) \to Mod(\Sigma(SP_1))$ an abstractor.

Impl is said to be *correct* if it is *consistent* and *complete*, i.e.,

(consistent) $U_h(\kappa(Mod(SP_2))) \subseteq \alpha(Mod(SP_1))$ and

(complete) $U_h(\kappa(Mod(SP_2))) \cap \alpha(A) \neq \emptyset$ for every $A \in Mod(SP_1)$.

The enrichment provides the implementing specification SP_2 with all syntactic features that are additionally needed to cover the signature of the implemented specification SP_1. If one assumes that all necessary sorts and operations are already available in SP_2, one can choose SP_2 itself as the intermediate specification (with the identity as a trivial enrichment). Constructor and abstractor relate the models of the specifications involved to each other. The constructor transforms every algebra of the implementing type into algebras of the intermediate specification. Their h-reducts may be seen as potential representations of algebras of the implemented type while the abstractor describes the acceptable representations.

The notion of correctness relates potential and acceptable representations in a suitable way. Consistency in this context means that every potential representation is acceptable, and completeness requires that every algebra of the implemented specification is realized by an implementing algebra, up to abstraction.

If one deals with loose specifications, an implementation may be a refinement step reflecting a design decision. In this case, certain solutions of the given data processing problem are favored while others are ruled out. On the semantic level, this means that the model class of the implementing specification should be smaller than that of the implemented specification. In other words, such implementations should be considered to be correct if they are consistent. To require completeness in this case would be counterproductive.

If one deals with tight specifications, meaning that a specification SP_0 has a single model A_{SP_0} (which is unique up to isomorphism), the notion of correctness becomes much easier. An implementation $Impl(SP, i, h, \kappa, \alpha)$ of SP_1 by SP_2 is correct, if κ maps A_{SP_2} to A_{SP} and the h-reduct $U_h(A_{SP})$ is an acceptable representation of A_{SP_1}, i.e., $U_h(A_{SP}) \in \alpha(A_{SP_1})$.

Several authors (e.g., Ehrig, Kreowski, and Padawitz [EKP78], Lipeck [Lip83], Sannella and Tarlecki [ST88b]) have discussed implementation concepts that merely relate the model classes of two specifications (or two signatures) by mappings or functors from the implementing to the implemented algebras. Let $F: Mod(SP_2) \longrightarrow Mod(SP_1)$ be such a mapping. This can then be described as an implementation in the sense above. One may take the disjoint union $SP_1 + SP_2$ as the intermediate specification and the two inclusions $in_i: SP_i \longrightarrow SP_1 + SP_2$ for $i = 1, 2$, as enrichment and renaming respectively (where in_1 is considered a signature morphism). Now the pair $(F(A), A)$ for every $A \in Mod(SP_2)$ defines an $SP_1 + SP_2$-algebra meaning that F induces a functional constructor $\kappa(F): Mod(SP_2) \longrightarrow Mod(SP_1 + SP_2)$. Note that the in_1-reduct of $(F(A), A))$ is $F(A)$. Using the trivial abstractor, one obtains the *functorial implementation*, $Impl(F) = (SP_1 + SP_2, in_1, in_2, \kappa(F), abstr_{iso})$. By definition, this is always consistent. It is also complete if and only if F has an image in every isomorphism class of $Mod(SP_1)$.

7.3.6   Particular Implementation Concepts

In the following we will discuss some of the implementation approaches in the literature, using the framework introduced above. They use the constructors and abstractors discussed above in one way or other.

Hoare 1972

In his pioneering paper, Hoare [Hoa72] presents some fundamental ideas about implementations of abstract data types. He defined implementations by means of a partial abstraction functions that map representing data of the implementing concrete type to the represented data of the implemented abstract type. While a concrete value never represents several abstract values, the mapping is not assumed to be one-to-one because an abstract value may be multiply represented. The mapping may be partial because the implemented type is allowed to have junk data not needed for representation. The abstraction function is required to be homomorphic preserving the operations of the implemented type. But the implementing type is not forced to satisfy the axioms of the abstract type. The abstraction function describes how the concrete type simulates the abstract type. Hoare called the domain of definition of the abstraction function *representation domain* and the equivalence relation induced by the abstraction function *representation equivalence*.

The implementation of queues by arrays with 2 pointers in Section 7.2.3 reflects the spirit of Hoare's implementation concepts (who used the quite similar implementation of stacks by pairs of arrays and pointers as illustration).

Hoare's approach can be described in the given framework. He assumes that the implementing type is already enriched by the abstract operations such that the enrichment is the identity and the signature morphism from the signature of the abstract type SP_1 to the signature of the concrete type SP_2 is just the inclusion. Respectively, the constructor can be chosen as the identity on the SP_2-models. Only the abstractor is not trivial. An SP_2-model A_2 (or its SP_1-reduct) is related to an SP_1-model if and only if there is a partial abstraction function $f\colon A_2 \to A_1$ with $f(A_2) = A_1$. This is denoted by $A_2 \in abstr_{HOARE}(A_1)$. An implementation in Hoare's sense is correct if the abstractor is not empty for any SP_1-model, i.e. if there exists always an abstraction function.

Goguen, Thatcher, and Wagner 1976

Goguen, Thatcher and Wagner did not only initiate the area of algebraic specification with their seminal paper [GTW76, GTW78], but introduced also the first notion of implementation within this framework. To implement a specification SP_1 by a specification SP_2, the latter must be enriched by the abstract operations (after some possible renaming of abstract sorts by concrete ones) using so-called *derivors*. Together with every SP_1-operation $op\colon s_1 \ldots s_n \to s$, one adds an equation of the form $op(x_1, \ldots, x_n) = t$ where, for $i = 1, \ldots, n$, x_i is a variable of sort s_i and t is an s-sorted $\Sigma(SP_1)$-term with variables $x_1, \ldots, x_n$. In other words, each abstract operation is defined as a derived, composite operation of the implementing specification. This is reflected by the constructor associated to this enrichment. It is chosen as the free construction, which keeps every SP_2-algebra invariant adding just the derived operations. The abstractor makes use of congruence relations corresponding to representation equivalences: $A \in abstr_{GTW}(A_1)$ if and only if there is a congruence $\equiv$ on A_2 such that $A_1 \subseteq A/\equiv$. Note that the GTW-abstractor identifies multiple representation before it separates significant data from junk whereas the HOARE-abstractor does it the other way round.

An implementation in the sense of Goguen, Thatcher and Wagner is correct if, for every SP_1-algebra, there is an SP_2-algebra A such that – after adding the derived operations – a quotient of A contains A_1.

If one considers the operation $copy\colon list \to container$ as a renaming, the implementations of the data type *Container* in Section 7.2.1 are examples of derivor implementations. The derivor of the *Container*-operation $put - in -$ as a composition of the $List_0$-operations cat and in is typical for this restricted type of implementation while all other operations are derived by even simpler renamings. The other examples in Section 7.2 are not derivor implementations because they contain operations that are implemented by means of recursion.

Orejas [Ore81, Ore85] generalized the notion of derivor implementation by allowing recursive and infinite derivors meaning that a deriving term t may contain abstract operations or may be infinite. In this case, the constructor is given by means of least fixed points. Moreover, *partial representations* are discussed in [Ore81] to deal with bounded implementations like the implementation of stacks by bounded arrays such that overflow may happen.

Ehrich 1982

Another important generalization of the GTW-approach was introduced by Ehrich [Ehr82] who dropped the restrictions on the enrichment. An implementation of SP_1 by SP_2 in his sense consists of an intermediate specification

SP and two specification morphisms $h\colon SP_2 \to SP$ and $i\colon SP_1 \to SP$. The free construction associated with h is the constructor, and the abstractor is essentially chosen as $abstr_{GTW}$ defined above. That i is required to be a specification morphism rather than a signature morphism between the underlying signatures is not essential if one deals with equational specifications (as Ehrich did). In this case, an signature morphism $i\colon \Sigma(SP_1) \to \Sigma(SP)$ allows to translate the SP_1-equations and to add them to SP instead of assuming that SP does contain them already. And the factorization effects of the translated equations can be integrated into the congruences of the abstractor likewise.

Ehrich's approach can be seen as formalization of the rather informal approach introduced by Guttag, Horowitz and Musser [GHM78]. All the examples in Section 7.2 can be seen as implementations in Ehrich's sense because they are given by enrichments including renamings as a special case (see Sections 7.4.2 and 7.4.3 for more details).

Ehrig, Kreowski, Mahr, and Padawitz 1982

Compared with Ehrich's approach, Ehrig, Kreowski, Mahr, and Padawitz [EKMP82] proposed a slightly modified notion of implementation. The enrichment of the implementing specification SP_2 yielding the intermediate specification SP is done in two steps. First, the sorts of SP_1 are added together with so-called sorts-implementing operations (serving as constructors of the abstract sorts). Second, the operations of SP_1 are added together with operations-implementing equations (specifying the abstract operations in terms of the implementing ones). Consequently, the inclusion of $\Sigma(SP_1)$ into $\Sigma(SP_2)$ can be chosen as the signature morphism of the implementation. But an arbitrary signature morphism, i.e., a renaming of sorts and operations would also do. All examples in Section 7.2 can be seen as EKMP-implementations because they are given by renamings and enrichments that can be separated into sorts-implementing and operations-implementing parts (see Section 7.4.3 for more details).

Semantically, the enrichment is again reflected by the free construction as constructor. It is called SYNTHESIS in [EKMP82] because it covers the constructive part of an implementation. To enrich SP_2 in two steps, allows to formulate and study extra correctness conditions. An example is the OP-completeness, requiring that the free construction of the second step does not extend the data domains. This means that – at the intermediate level – the abstract operations are fully defined on data representations provided by the first enrichment step and do not generate new data. It should be noted that the 2-step enrichment could be integrated into Ehrich's approach without any trouble.

The significant difference between Ehrich's approach and the EKMP-implementation is the choice of abstractors. In [EKMP82], the semantics of an implementation is composed of SYNTHESIS (as introduced above) followed by RESTRICTION and IDENTIFICATION. While the latter factorizes a

$\Sigma(SP_1)$-algebra through the congruence induced by the SP_1-equations, RE-STRICTION maps an SP-algebra to the reachable $\Sigma(SP_1)$-subalgebra of its $\Sigma(SP_1)$-reduct, i.e. the subalgebra generated by the abstract operations. In this way, meaningful data representations are separated from junk before multiple representations are identified. The composition of RESTRICTION and IDENTIFICATION can be interpreted as an abstractor as explained in Section 7.3.4.

If one uses IDENTIFICATION instead of an arbitrary representation equivalence in the definition of an implementation, the semantics is uniquely determined by the syntactic component such that an implementation of SP_1 by SP_2 can be seen as a partial program with the implementing specification SP_2 as parameter. Syntactic aspects are further discussed in Section 7.4.

An implementation is correct in this framework if, for every SP_1-algebra A_1, there is an SP_2-algebra A_2 such that

$$\text{IDENTIFICATION} \circ \text{RESTRICTION} \circ \text{SYNTHESIS}(A_2) = A_1.$$

In the case of correctness, the semantic construction induces a partial representation function from SYNTHESIS(A_2) to A_1. Therefore, the EKMP-approach is closely related to Hoare's approach. In both cases, the restriction to significant data and the removal of junk precedes the identification of multiple representations.

In contrast to that, the abstractor in Ehrich's approach as well as in the GTW-approach identifies multiple representations before it restricts to the meaningful ones. Besides the different assumptions concerning the enrichment part of an implementation, the order of restriction and identification establishes the major difference between the approaches discussed so far. If one considers equational specifications with initial algebra semantics, one can show that the representation equivalence of a correct implementation in the sense of Hoare, GTW and Ehrich coincides always with the congruence induced by the abstract equations and corresponding to IDENTIFICATION in the EKMP-approach. Hence there is no choice in this respect. Moreover, let an implementation in the GTW-approach or in Ehrich's approach be correct meaning that IDENTIFICATION followed by RESTRICTION yield the initial SP_1-algebra T_{SP_1} if applied to the initial SP-algebra T_{SP}:

$$\text{RESTRICTION} \circ \text{IDENTIFICATION}(T_{SP}) = T_{SP_1}.$$

Then the implementation is also correct in the EKMP-approach, i.e.

$$\text{IDENTIFICATION} \circ \text{RESTRICTION}(T_{SP}) = T_{SP_1}.$$

The converse is not always true as one can see from the example discussed in Section 7.5.5. In this respect, the use of the RI-abstractor is more general than the IR-abstractor. On the other hand, correctness based on the IR-abstractor is easier to prove because it can be characterized by a proof-theoretic condition. More details concerning these both abstractors and the associated notions of correctness can be found in Ehrich and Lipeck [EL80], Ehrig, Kreowski, Mahr, and Padawitz [EKMP82], and Orejas, Navorro, and Sánchez [ONS93, ONS96].

The troublemaker with respect to correctness proofs and lack of compositionality (cf. Section 7.5.5) is the possible generation of junk by the sorts-implementing operations. Poigné [Poi84], Bernot, Bidoit, and Choppy [BBC86b, Ber89], and, more recently, Sánchez [San92] have tried to avoid or circumvent the problem by employing order sorted algebras and partial algebras respectively.

Sannella and Tarlecki [SW82] and Broy, Möller, Pepper, and Wirsing [BMPW86] advocate implementation concepts similar to the EKMP-approach, but explicitly for the specification framework with loose semantics.

An interesting and quite powerful alternative to representation equivalences that identify multiple representations (like the congruence described by IDENTIFICATION) is an abstractor based on observational or behavioral equivalence between implementing algebras and implemented ones (cf. Chapter 5 for the notion of behavioral equivalence). In this case, even many abstract values may be represented by a single concrete value provided that the abstract values are observationally equivalent. Many researchers have proposed implementation approaches using behavioral equivalence instead of identification (see, e.g., Goguen and Meseguer [GM82], Hennicker [Hen89, Hen91b], Nivela [Niv87], Sánchez [San92], and Sannella and Tarlecki [ST87, ST88b]).

Sannella and Tarlecki 1988

In another crucial paper [ST88b] (mentioned already at the end of Section 7.3.5 and in the last paragraph), Sannella and Tarlecki investigated implementation concepts in the framework of algebraic specification from a general point of view including, in particular, the refinement of requirement specifications with loose semantics. As illustrated by the examples in Section 7.2.1, the major aim of implementation on this semantical level is to cut down the class of accepted models until one reaches a single solution eventually. To achieve this, all kinds of constructors are welcome including – among others – SYNTHESIS, RESTRICTION and IDENTIFICATION as used above. But there is no longer any particular order of application required.

The implementation concept of this survey stems from Orejas, Navarro and Sánchez [ONS93, ONS96], but is much inspired by Sannella and Tarlecki who introduced the notions of constructor and abstractor implementations. The constructors are assumed to be functional meaning that they map every algebra to an algebra rather than a class of algebras. The considered abstractors are also slightly more restrictive in that they are required to relate algebras in a symmetric way like the abstractors based on behavioral equivalence. The RI- and IR-abstractors fail to satisfy the symmetry condition in general. Therefore, RESTRICTION and IDENTIFICATION are handled as constructors. Ignoring these technical differences, a constructor implementation in the sense of Sannella and Tarlecki is an implementation with the trivial abstractor $abstr_{iso}$ while abstractor implementations coincide more or less with the general notion of implementation.

7.4   Syntactic aspects

Although the enrichment and the signature morphism are syntactic features, the notion of implementation as introduced above is rather semantic because constructor and abstractor are described as relations on models, not by syntactic means. In this section, we discuss various suggestions in the literature as to how a notion of implementation with a complete syntactic description may look. In most cases, constructor and abstractor are defined in some unique way using the syntactic part of an implementation. However many variants exist.

7.4.1   Renaming

From the point of view of implementation, the easiest approach is to assume that all constructive work is done at the layer of the implementing specification. This means that the signature of the implementation specification is already available as part of the signature of the implementing specification, i.e., $\Sigma(SP_1) \subseteq \Sigma(SP_2)$, provided that the institution considered provides a notion of subsignatures. In particular, one may require $\Sigma(SP_1) = \Sigma(SP_2)$, which can be done in any institution.

A slightly more general view is to assume that everything needed is specified in the implementing specification, but may have other names. This reasoning leads to the notion of a *renaming* being a signature morphism $h\colon \Sigma(SP_1) \to \Sigma(SP_2)$.

A renaming $h\colon \Sigma(SP_1) \to \Sigma(SP_2)$ induces an implementation $Impl(h) = (SP_2, id, h, id, abstr)$ where the implementing specification provides the intermediate specification, the enrichment and the constructor are the respective identities, and $abstr$ is the standard abstractor for SP_1 given in Section 7.3.4.1. By definition, $Impl(h)$ is correct if, for every $A_1 \in Mod(SP_1)$, there are $A_2 \in Mod(SP_2)$ and a $\Sigma(SP_1)$-morphism $f\colon U_h(A_2) \to A_1$, and, for every $A_2 \in Mod(SP_2)$, there are $A_1 \in Mod(SP_1)$ and a $\Sigma(SP_1)$-*morphism* $f\colon U_h(A_2) \to A_1$.

Clearly, the abstractor considered can be replaced by any other abstractor that can be associated with specifications in some standard way. This results in modified notions of correctness. An obvious candidate is the abstractor $abstr_{surjective}$ that requires surjectivity of the $\Sigma(SP_1)$-*morphisms* involved in the notion of correctness (provided that morphisms in the institution considered are based on mappings). Correctness in this case intuitively means that any SP_1-model is fully simulated by a corresponding SP_2-model including all data.

Another quite restrictive candidate is the abstractor $abstr_{iso}$ that requires isomorphisms in the notion of correctness. In this case, $Mod(SP_1)$ and $U_h(Mod(SP_2))$ are equivalent if $Impl(h)$ is correct.

A typical example of the renaming kind of implementation is the first implementation of *Container* by *List*$_1$ in Section 7.2.1, which defines a signature morphism h from $\Sigma(Container)$ to $\Sigma(List_1)$. The h-reduct $U_h(A)$ of a *List*$_1$-algebra yields a *Container*-algebra by forgetting the data inclusion and concatenation of A. This means that the implementation is consistent. Conversely, let T_h be the free functor corresponding to h and C be a *Container*-algebra. Then $T_h(C)$ is a $\Sigma(List_1)$-algebra, and there is a surjective homomorphism from the h-reduct $U_hT_h(C)$ to C such that the implementation is also complete.

Moreover, the enrichments *List* and *Set* of *Container* may be seen as examples of this kind, where even the names are kept. *List*-algebras and *Set*-algebras are *Container*-algebras such that both examples are consistent. However not all *Container*-algebras are *List*- or *Set*-algebras such that the implementations are not complete with respect to the isomorphism abstractor.

7.4.2   Enrichment and renaming

To assume a signature morphism $h\colon \Sigma(SP_1) \to \Sigma(SP_2)$ behind an implementation intuitively means that the declarations of the implemented specification are declarations of the implementing specification up to renaming. To consider an intermediate specification SP, an enrichment $i\colon SP_2 \to SP$, and a signature morphism $h\colon \Sigma(SP_1) \to \Sigma(SP)$, as in the general notion of implementation is looking at the situation from a different angle. The assumption in this case is that the implementing specification chosen may not provide all the necessary concepts, and some additional specification is needed before the declarations can be related. Clearly, this view contains that in the previous subsection as a special case, since the identity on SP_2 is an enrichment – the most trivial one.

If one combines the syntactic features SP, i, and h with the constructor *translate*$_i$ and the standard abstractor *abstr*, one gets an induced implementation

$$Impl(SP, i, h) = (SP, i, h, translate_i, abstr).$$

By definition, such an implementation is correct if, for every $A_1 \in Mod(SP_1)$, there are $A \in Mod(SP)$ and a $\Sigma(SP_1)$-morphism $f\colon U_h(A) \to A_1$ and, for every $A \in Mod(SP)$, there are $A_1 \in Mod(SP_1)$ and a $\Sigma(SP_1)$-morphism $f\colon U_h(A) \to A_1$, and U_i is surjective. The last condition makes sure that all SP_2-models have got corresponding SP-models. The other two conditions say that SP_1-models are simulated by corresponding SP-models.

As in Section 7.4.1, the abstractor in $Impl(SP, i, h)$ may be replaced by any other that can be associated with a specification. Analogously, the constructor may be replaced by any other that can be associated with an enrichment in a standard way.

Examples of the enrichment and renaming kind of implementations are $List_1$ **implements** *Container* and $List_0$ **implements** *Container* in Section 7.2.1. They are discussed in more detail in the next subsection.

The implementation given by an enrichment and a renaming is essentially the notion introduced by Ehrich [Ehr82] (see Section 7.3.6).

7.4.3   Implementation of sorts and operations in two steps

The idea of enrichment in the previous subsection is to provide those components on the implementing level that are needed to implement a given specification, but are not yet available in theimplementing specification chosen. In the extreme case, one may extend the implementing specification by the whole specification to be implemented. The implementing and the implemented specifications then have nothing to do with each other, and the implementing specification is effectively not used.

If one wants to avoid this somewhat counterintuitive situation, one may split the enrichment into two steps. The first step extends the implementing specification by extra sorts and sorts-implementing operations; the second step adds operations and operations-implementing axioms. These notions allow the formulation of further requirements on the semantic relation between the implementing and the implemented specifications.

More formally, we may assume that the given specifications SP_1 and SP_2 share a common subspecification SP_0 and that they are combinations of SP_0 with further sorts, operations, and axioms:

$$SP_i = SP_0 + (S_i, OP_i, E_i) \quad \text{for } i = 1, 2.$$

Hence, the specification SP_2 must be extended by sorts and operations which correspond to S_1 and OP_1. The easiest way to achieve this is to add S_1 and OP_1 themselves. S_1 may be implemented (using SP_2) by *sorts-implementing operations* $OP(S_1)$ which yields the *sort implementation*

$$SORTIMPL = SP_2 + (S_1, OP(S_1)).$$

On top of $SORTIMPL$, OP_1 can be implemented by *operations-implementing axioms* $E(OP_1)$ which yields the *operation implementation*

$$OPIMPL = SORTIMPL + (OP_1, E(OP_1)).$$

Intuitively, this means constructing the sorts of SP_1 from sorts of SP_2 by sort constructors and to specify the operations of SP_1 in terms of operations of SP_2 and sorts-implementing operations.

The composition of sort implementation and operation implementation leads to an enrichment given by the inclusion in_2: $SP_2 \to OPIMPL$. Moreover, the inclusion in_1 from $\Sigma(SP_1)$ to $\Sigma(OPIMPL)$ is a signature morphism such that $IMPL(OPIMPL, in_2, in_1)$ is a syntactic description of an implementation in the sense of the previous subsection. In particular, the meaning of correctness is fixed in this way.

However one can say more. For this purpose, we consider the framework of equational specifications. Moreover, let T_{in_3} be the free construction given by the inclusion in_3: $SP_2 \to SORTIMPL$, and T_{in_4} the free construction given by the inclusion in_4: $SORTIMPL \to OPIMPL$. If all sorts-implementing operations have sorts of SP_1 as ranges (meaning that they can be considered as constructors), then the domain of type $s_1 \in S_1$ of the free model $T_{in_3}(A)$ for $A \in Mod(SP_2)$ consists of constants and trees, the nodes of which are labeled by tuples of data from the domains of A. In other words, the intermediate specification provides explicit data representations, which are called SP_2-colored trees. Furthermore, one can now require, as an additonal correctness condition, that the operation implementation defines the abstract operations in terms of the data-representation level given by $SORTIMPL$, i.e., each $\Sigma(SP_1)$-term is $OPIMPL$-equivalent to a $\Sigma(SORTIMPL)$-term. This condition is called OP-completeness.

The example $List_1$ **implements** $Container$ in Section 7.2.1 is of the type of implementations considered here. The enriching operation $copy$ is a sort-implementing operation which is interpreted as a mapping from A_{list} to $A_{container}$ in an arbitrary algebra A of the sort-implementing type. This mapping is a bijection if A is the free algebra of a $List_1$-algebra A_1 such that the $container$-domain is a copy of the $list$-domain. Note that the reduct of A, forgetting this copying, yields A_1, meaning that the sort implementation does not change the implementing model. Adding the axioms now, the abstract operations of $Container$ are derived in terms of $List_1$-operations and $copy$. Consequently, the free constructions from the sort implementation to the operation implementation do not change domains. In other words, the implemented operations are fully defined on implementing data. The example $List_0$ **implements** $Container$ in Section 7.2.1 works similarly.

Another typical example is the implementation of queues by arrays with two pointers in Section 7.2.3. Here the implementation of sort $queue$ is given by the product of sorts int, $array$, and int again. Then the $Queue$-operations are defined using this data representation and $Array$-operations. The operation implementation contains the signature of $Queue$ as a subsignature such that each algebra of the implementing level has a $\Sigma(Queue)$-reduct. Choosing an arbitrary $Data$-algebra A, the free construction corresponding to the enrichment of $Array$ $with$ 2 $Pointers$ transforms the free $Array$-algebra over A, together with the initial Int-algebra, into the free $Array$ $with$ 2 $Pointers$ **implements** $Queue$-algebra. As shown in Section 7.2.3, there is a partial abstraction function from the $\Sigma(Queue)$-reduct Q of the latter algebra into the free *Queue*-algebra over A. This corresponds to the abstractor RI which restricts Q to the subalgebra generated by *Queue*-terms and factorizes this subalgebra by the *Queue*-axioms into the free *Queue*-algebra over A. In other words, queues are correctly implemented by arrays with two pointers.

The 2-step implementation corresponds to the notions introduced by Ehrig, Kreowski, Mahr, and Padawitz [EKMP82].

7.5   Composition

It is not usually reasonable to implement a specification as a whole in a single step, because the implementation of large specifications may require many, quite different, design decisions and measures to obtain efficient programs. Hence, stepwise refinement is a most important issue, enabling design decisions to be made one after the other. This reasoning leads to the notion of sequential composition of implementation steps.

There is an easy way to arrive at a sequential composition. As implementations represent a binary relation on specifications, one may just consider its transitive closure (or alternatively its reflexive and transitive closure), which is by definition closed under sequential composition. In this way, the idea of stepwise refinement, as given in Section 7.1, is directly formalized as a sequence of implementation steps. However as the composition is merely a concatenation, all design decisions represented in the implementation steps are kept separately from each other and appear in a particular order. Moreover, if each implementation step makes a certain part of the original specification executable, the evaluation of a specified operation can only be done by a sequence of executions reflecting the sequence of implementation steps. Similarly, the resulting program implementing the specification is scattered over all the intermediate steps.

If one wants to see the program represented by a sequence of implementation steps or if one is interested in a more efficient execution than the sequence of executions provided by the sequence of steps, one needs more sophisticated kinds of sequential compositions. In particular, one may expect that the composition reflects certain properties of the components, for example, it preserves correctness or is an implementation of the same type as its components.

7.5.1   Stepwise implementation

Let $SP_0 \cdots SP_n$ be a sequence of specifications with $n \geq 1$, let $Seq = Impl_1 \cdots Impl_n$ be a sequence of implementations where $Impl_i$ is an implementation of SP_{i-1} by SP_i, for $i = 1, \dots, n$. Then Seq is called a *stepwise implementation* of SP_0 by SP_n. It is called *correct* if each component is correct.

Given stepwise implementations *Seq* of *SP* by *SP'* and *Seq'* of *SP'* by *SP''*, the concatenation *SeqSeq'* obviously yields a stepwise implementation of *SP* by *SP''*, which is correct if *Seq* and *Seq'* are correct.

The same can be done for any particular kind of implementation. As implementation steps are put together in a purely sequential order, without any further interaction, one gets sequential composition on stepwise implementations more or less for free. However if a stepwise implementation consists of two or more steps, it is never of the same kind as its components. In the following paragraphs, we consider alternative notions of sequential compositions.

7.5.2   Renaming implementation

The sequential composition of implementations induced by renamings (see Section 7.4.1) can be done simply by the composition of the renamings. Whether this composition preserves correctness depends on the choice of the abstractor which is not automatically fixed by the renaming.

Let $h: \Sigma(SP_0) \to \Sigma(SP_1)$ and $h': \Sigma(SP_1) \to \Sigma(SP_2)$ be renamings. Then the composition $h' \circ h: \Sigma(SP_0) \to \Sigma(SP_2)$ is again a renaming and induces the renaming implementation $Impl(h' \circ h) = (SP_2, id_{SP_2}, h' \circ h, id_{Mod(SP_2)}, abstr)$ which can be considered as the *sequential composition* of $Impl(h)$ and $Impl(h')$.

It is easy to see that this composition preserves correctness.

7.5.3   Functorial implementation

As the composition of mappings (or functors) $F_1: Mod(SP_1) \longrightarrow Mod(SP_0)$ and $F_2: Mod(SP_2) \longrightarrow Mod(SP_1)$ always yields a mapping (or functor resp.) $F_1 \circ F_2: Mod(SP_2) \longrightarrow Mod(SP_0)$, the corresponding functorial implementations (see Section 7.3.5), $Impl(F_1)$ and $Impl(F_2)$, can be composed into the functorial implementation $Impl(F_1) \circ Impl(F_2) = Impl(F_1 \circ F_2)$. It is easy to see that this composition preserves correctness.

7.5.4   Constructor implementation

Let $Impl = (SP, i, h, \kappa, abstr_{iso})$ be a constructor implementation of SP_0 by SP_1 and $Impl = (SP', i', h', \kappa', abstr_{iso})$ a constructor implementation of SP_1 by SP_2. Then the two enrichments i and i' can be merged into an enrichment $\overline{SP}$ of SP' using the signature morphism h' and the inclusion of signatures into specifications. Formally, the specification $\overline{SP}$ and the two specification morphisms $\hat{\imath}$ and $\hat{h}$ are constructed as the pushout of SP, SP', and the respective specification morphisms, as the following diagram indicates. This means in the case of algebraic specifications that $\overline{SP}$ is the union of SP and SP' after all sorts and operations of SP_1 in SP are renamed according to h'.

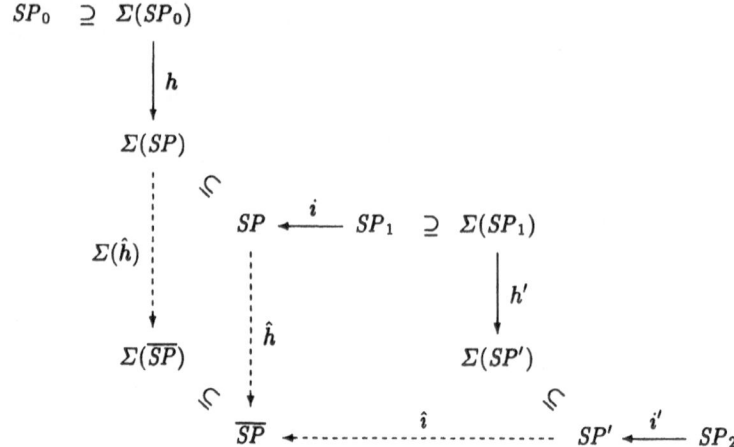

The composition $\bar{\imath} = \hat{\imath} \circ i'$ turns $\overline{SP}$ into an enrichment of SP_2, and the composition $\bar{h} = \Sigma(\hat{h}) \circ h$, where $\Sigma(\hat{h})$ is the signature morphism part of $\hat{h}$, provides a signature morphism from the abstract signature $\Sigma(SP_0)$ into the signature of $\overline{SP}$. The enrichment $\overline{SP}$, together with $\bar{\imath}$ and $\bar{h}$, can be considered as the syntactic part of the sequential composition of $Impl$ and $Impl'$. For the semantic part, one must compose the constructors κ and κ' in a suitable way. They cannnot be composed directly because κ' yields SP'-algebras while κ applies to SP_1-algebras. However the forgetful functor $U_{h'}$ associated with the given signature morphism h' allows the transformation of SP'-algebras into $\Sigma(SP_1)$-algebras. Hence one gets a composition of the constructors if κ can be extended to $\Sigma(SP_1)$-algebras in some natural way. Assuming this, the constructor implementation $Impl \circ Impl' = (\overline{SP}, \bar{\imath}, \bar{h}, \kappa \circ U_{h'} \circ \kappa', abstr_{iso})$ defines the sequential composition of $Impl$ and $Impl'$.

7.5.5   2-step implementation

Consider 2-step implementations (see Section 7.4.3) of SP_0 by SP_1 and SP_1 by SP_2, where the first implementation is given by

$$SP_1 \subseteq SORTIMPL_1 \subseteq OPIMPL_1 \supseteq \Sigma(SP_0)$$

and the second by

$$SP_2 \subseteq SORTIMPL_2 \subseteq OPIMPL_2 \supseteq \Sigma(SP_1).$$

Without going into the technical details, a composition can be constructed using the union of the sort implementations and the union of the operation implementations, yielding a 2-step implementation of SP_0 by SP_2. The composition preserves OP-completeness, but not correctness in general, if the abstractor RI (see Section 7.3.4) is considered.

The problem is that restriction and identification do not interact properly with each other. Consider, for example, the integers with constant 0 and successor *succ* and predecessor *pred* as operations. If one restricts this to the part generated by 0 and *succ*, one gets the non-negative numbers. If one now identifies the latter, due to the equation $succ(x) = succ(succ(x))$, one obtains two equivalence classes $\{0\}$ and $\{1, 2, 3, \ldots\}$, because all numbers except 0 are successors, and as such become identified with their successors. Hence restriction followed by identification yields a domain with two elements. The other way round, the resulting domain has only one element because all integers are equivalent with respect to the equation and the restriction afterwards does not change the domain.

This means that one has to pay a price if one deals with the abstractor *RI*. The compositionality in this case is not for free. While the composition of restrictions is a restriction and the composition of identifications is an identification, the composition of a restriction and an identification does not commute, in general, as the example above shows. Hence the abstractor *RI* is not compositional:

$$RI\,RI \neq RR I\!I = RI.$$

This is not surprising because the effect of equations interferes with the presence of data. To prevent this unfortunate situation and to guarantee the correctness of the composition even in this case, one requires extra sufficient conditions as discussed in [EKMP82]. Those who want to avoid the trouble may stick to a composition of 2-step implementations in the sense of Section 7.5.1.

7.6   Refinement of modular specifications

In the previous sections we have studied different kinds of implementations for specifications and the problem of vertical composition. In the next two sections we will extend this discussion to modular specifications. It is well-known that modularity is an important issue for the design and implementation of large software systems. There is a need to divide large systems into blocks so that system development becomes more manageable, clear, modifiable, and re-usable. These blocks, known as modules, are self-contained entities with individual semantics which can be interconnected in different ways to build up modular systems. This leads to a horizontal structuring for the specification of the modular system. On the other hand we have to consider vertical refinement of modular specifications, leading from requirement definitions, via several intermediate specifications, to the program of the modular system, as discussed in the introduction of this chapter. The final goal, of course, is the correctness of the final modular system with respect to the requirement definition. For this purpose it is desirable that the correctness of each horizontal and vertical step in the development process implies the correctness of the complete process.

In Goguen and Burstall [GB80], software design was explained in terms of a bidimensional refinement process, i.e., horizontal and vertical steps, and it was proposed that each step should preserve correctness and there should be a bidimensional compatibility between horizontal and vertical steps.

In this section we will consider a refinement concept for parameterized and module specifications which satisfy these requirements. The concepts of parameterized and module specifications in this paper include those in [EM85] and [EM90] as special cases, and also allows the consideration of loose constructors, because it is based on general constructors as introduced in Section 7.3. The refinement concept is based on signature and specification morphisms only. In the next section, we extend the general implementation concept introduced in Section 7.3 to the case of parametrized specifications, where, however, correctness and compatibility of horizontal and vertical steps can only be shown in restricted cases, due to the problems discussed in Section 7.5.

7.6.1   Bidimensional compatibility of refinement

Given any kind of modular specification MSP which allows the definition of a horizontal composition $MSP \circ MSP'$ and refinement $R \colon MSP_1 \to MSP_2$ of MSP_1 by MSP_2. Then it should be possible to define a horizontal composition of refinements

$$R_1 \circ R_1' \colon MSP_1 \circ MSP_1' \to MSP_2 \circ MSP_2'$$

and a vertical composition of refinements

$$R_2 * R_1 \colon MSP_1 \to MSP_2 \to MSP_3$$

such that we have the following bidimensional compatibility of refinement between horizontal and vertical composition:

$$
\begin{array}{ccc}
MSP_1 \circ MSP_1' & \xrightarrow{\;(R_2 * R_1) \circ (R_2' * R_1')\;} & MSP_3 \circ MSP_3' \\[2mm]
& R_1 \circ R_1' \quad = \quad R_2 \circ R_2' & \\
& MSP_2 \circ MSP_2' &
\end{array}
$$

An important problem is to show that correctness of the given refinement implies correctness of horizontal and vertical composition, and hence correctness of bidimensional compatibility.

7.6.2   Refinement of parameterized specifications

1. In the following we consider *parameterized specifications* as triples $PSP =$ (PAR, BOD, s), consisting of a *parameter* specification PAR, a *body* specificaton BOD, and a specification morphism $s: PAR \rightarrow BOD$. The semantics of PSP is given by a constructor $\kappa_s: Mod(PAR) \rightarrow Mod(BOD)$.

2. The constructors κ_s depend on the parameterization concept only. Once the concept is chosen κ_s is completely determined by s. For parameterized specifications with initial algebra semantics (see, e.g., [EM85, EK83, EG94]), we have $\kappa_s = F_s: Mod(PAR) \rightarrow Mod(BOD)$, the free construction from $Mod(PAR)$ to $Mod(BOD)$; in the case of final algebra semantics $\kappa_s = CF_s$, the cofree construction [Gan83, EG94]; and for loose semantics $\kappa_s = translate_s: Mod(PAR) \rightarrow \mathcal{P}(Mod(BOD))$. If parameterized specifications are defined by λ-expressions the parameter may occur several times in the body, such that we have specification morphisms $s_1, \ldots, s_n: PAR \rightarrow BOD$. For simplicity, however, we only consider the case $n = 1$.

3. A *refinement* $R: PSP_1 \rightarrow PSP_2$ of parameterized specifications $PSP_i =$ (PAR_i, BOD_i, s_i), for $i = 1, 2$, is given by a pair $R = (p, b)$ of a specification morphism $p: PAR_1 \rightarrow PAR_2$ between the parameters, and a signature morphism b between the body specifications, written $b: \Sigma(BOD_1) \rightarrow$ $\Sigma(BOD_2)$, or $b: BOD_1 \twoheadrightarrow BOD_2$ such that the following diagram commutes.

$$
\begin{array}{ccc}
PAR_1 & \xrightarrow{\;s_1\;} & BOD_1 \\
\downarrow{\scriptstyle p} & = & \downarrow{\scriptstyle b} \\
PAR_2 & \xrightarrow{\;s_2\;} & BOD_2
\end{array}
$$

We do not require that b is a specification morphism in order to have more flexibility for refinements.

4. A refinement $R: PSP_1 \rightarrow PSP_2$ is called *correct* if we have for all $P_2 \in Mod(PAR_2)$

$$U_b \circ \kappa_{s_2}(P_2) \subseteq \kappa_{s_1} \circ U_p(P_2).$$

This means that the model class obtained by PSP_2 is included in that of PSP_2 after restriction of P_2 to PAR_1, i.e., $U_p(P_2)$, and restriction of BOD_2-models B_2 to BOD_1, i.e., $U_b(B_2)$. The composition $U_b \circ \kappa_{s_2}$ is defined pointwise by

$$U_b \circ \kappa_{s_2}(P_2) = \{U_b(B_2) \mid B_2 \in \kappa_{s_2}(P_2)\}.$$

Note that $U_b(B_2)$ for $B_2 \in Mod(BOD_2)$ in general is only a $\Sigma(BOD_1)$-model, since b is just a signature morphism. But $\kappa_{s_1} \circ U_p(P_2)$ consists of BOD_1-models such that the correctness condition implies that $U_p \circ \kappa_{s_2}(P_2)$ consists of BOD_1-models only.

7.6.3 Examples

1. The specifications *Container* and $List_1$ in Section 7.2 can be considered as parameterized specifications over *Data*, where the specification morphisms s_1 and s_2 are inclusions. Let p be the identity on *Data*, and h: *Container* $\dashrightarrow$ $List_1$ the signature morphism h defined in Section 7.2. Then we have the refinement $R = (id, h)$: *Container*(*Data*) $\rightarrow$ $List_1$(*Data*) as shown in the following diagram.

$$
\begin{array}{ccc}
Data & \xrightarrow{\;s_1\;} & Container \\[2pt]
{\scriptstyle id}\big\downarrow & = & \big\downarrow{\scriptstyle h} \\[2pt]
Data & \xrightarrow{\;s_2\;} & List_1
\end{array}
$$

2. In the case of loose semantics, the constructors κ_{s_1} and κ_{s_2} are defined by translation. It has been already mentioned that each $List_1$-algebra L over a *Data*-algebra D, i.e., $L \in \kappa_{s_2}(D)$, becomes a *Container*-algebra C over D, i.e., $C \in \kappa_{s_1}(D)$, after restriction by h, i.e., application of U_h. This means that we have

$$U_h \circ \kappa_{s_2}(D) \subseteq \kappa_{s_1}(D) = \kappa_{s_1} \circ U_{i_d}(D)$$

and hence correctness of the refinement of R in the case of loose semantics. Note that for the type of refinement considered by Sannella and Tarlecki in [ST88b] we have the special case that p and b are respectively identities as specification and signature morphisms. An example of this case is the refinement as above, but replacing $List_1$ by *List* as given in Section 7.2.1.

3. In the case of initial semantics the constructors κ_{s_1} and κ_{s_2} are defined by free constructions F_{s_1}: *Mod*(*Data*) $\rightarrow$ *Mod*(*Container*) and F_{s_2}: *Mod*(*Data*) $\rightarrow$ *Mod*($List_1$). Correctness in this case means

$$U_h \circ F_{s_2}(D) = F_{s_1}(D)$$

which is true, because $F_{s_1}(D)$ and $F_{s_2}(D)$ are both lists constructed freely over D, where $F_{s_2}(D)$ includes an additional concatenation operation.

4. If we replace $List_1$ by *Set* as given in Section 7.2.1 we obtain a refinement R: *Container*(*Data*) $\rightarrow$ *Set*(*Data*) which is correct in the case of loose semantics, but no longer correct in the case of initial semantics, because $U_h \circ F_{s_2}(D)$ is only a quotient of $F_{s_1}(D)$.

Now we are going to study the vertical composition of refinements.

7.6.4 Vertical composition of refinements

1. Given refinements R_1: $PSP_1 \rightarrow PSP_2$ and R_2: $PSP_2 \rightarrow PSP_3$ with $R_i = (p_i, b_i)$, for $i = 1, 2$, the vertical composition $R_2 * R_1$: $PSP_1 \rightarrow PSP_3$ is defined by $R_2 * R_1 = (p_2 \circ p_1, b_2 \circ b_1)$.

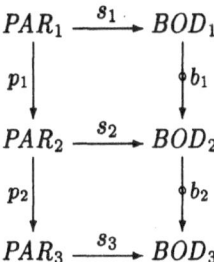

2. Given correct refinements R_1 and R_2 the vertical composition $R_2 * R_1$ is correct (*correctness of vertical refinement*). In fact we have

$$U_{b_2 \circ b_1} \circ \kappa_{s_3}(P_3) \subseteq \kappa_{s_1} \circ U_n \circ P_1(P_3)$$

which can be concluded directly from the correctness of R_1 and R_2 and the equations $U_{b_2 \circ b_1} = U_{b_1} \circ U_{b_2}$, $U_{P_2 \circ P_1} = U_{P_1} \circ U_{P_2}$, which hold for forgetful functors in any institution.

The next step is to consider horizontal composition of refinements, which is based on horizontal composition of parameterized specifications. The simple case is *direct horizontal composition*, where the body BOD_1 of the first is equal to the parameter PAR_2 of the second parameterized specification. In general, however, we only have a parameter passing morphism $g: PAR_2 \rightarrow BOD_1$. This means that we first have to construct the *actualization* of PSP_2 by g, called PSP_{2_g}, before we can define the general horizontal composition $PSP_2 \circ_g PSP_1$ via g as the direct horizontal composition $PSP_{2_g} \circ PSP_1$ of PSP_1 and the actualization PSP_{2_g} of PSP_2. For this reason we also define refinement in two steps: direct horizontal composition and actualization of refinements. We start with actualization.

7.6.5 Actualization of refinement

1. Given a refinement $R_1 = (p_1, b_1): PSP_1 \rightarrow PSP_2$ and a specification morphism $g_1: PAR_1 \rightarrow PAR_3$, called a *parameter passing morphism*, as shown in the diagram below. Then we are able to construct actualizations PSP_3 and PSP_4 of PSP_1 and PSP_2, respectively, and an *actualized refinement* $R_3 = (p_3, b_3): PSP_3 \rightarrow PSP_4$, written $R_3 = R_{1_{g_1}}$.

   In this diagram PAR_4, BOD_3, and BOD_4 are constructed as pushouts in the back, left and front square, respectively. The signature morphism $b_2: BOD_3 \rightarrow BOD_4$ is uniquely defined such that the bottom and right square become commutative diagrams of signature morphisms, using the

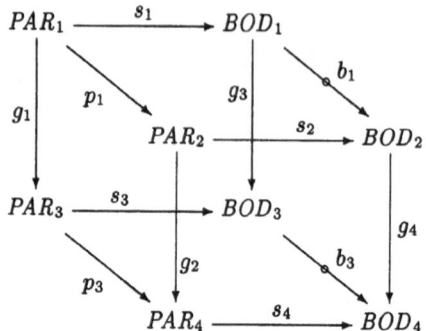

fact that the pushout BOD_3 of specification morphisms is also a pushout of signature morphisms.

Hence we obtain *actualized parameterized specifications* $PSP_3 = (PAR_3, BOD_3, s_3)$ and $PSP_4 = (PAR_4, BOD_4, s_4)$ and an *actualized refinement* $R_3 = (p_3, b_3): PSP_3 \rightarrow PSP_4$.

2. Given a correct refinement $R_1: PSP_1 \rightarrow PSP_2$ between persistent parameterized specifications PSP_1 and PSP_2, i.e., $U_{s_i} \circ \kappa_{s_i}(P_i) = P_i$, for $i = 1, 2$, then also the actualizations PSP_3 and PSP_4 are persistent, and the actualized refinement is correct (*correctness of actualization*) provided that we have (1) an institution with amalgamation and (2) an extension property for persistent constructors.

Given the pushout

$$PAR_1 \xrightarrow{s_1} BOD_1$$
$$g_1 \downarrow \qquad = \qquad \downarrow g_3$$
$$PAR_3 \xrightarrow{s_3} BOD_3$$

this means:

(1) For each $P_3 \in Mod(PAR_3)$, $B_1 \in Mod(BOD_1)$ with $U_{g_1}(P_3) = P_1 = U_{s_1}(B_1)$, there is a unique $B_3 \in Mod(BOD_3)$, written $B_3 = P_3 +_{p_1} B_1$, such that $U_{s_3}(B_3) = P_3$ and $U_h(B_3) = B_1$. B_3 is called the *amalgamation* of P_3 and B_1 via p_1.

(2) Given a persistent constructor κ_{s_1}, i.e., $U_{s_1} \circ \kappa_{s_1}(P_1) = P_1$, then κ_{s_3}, the constructor associated with s_3, is also persistent and satisfies

$$U_{g_3} \circ \kappa_{s_3}(P_3) = \kappa_{s_1} \circ U_{g_1}(P_3).$$

for all $P_3 \in Mod(PAR_3)$. In this case κ_{s_3} is called the *extension* of κ_{s_1} via g_1.

(3) To prove the correctness of actualization, we have to show for all $P_4 \in Mod(PAR_4)$

$$U_{b_3} \circ \kappa_{s_4}(P_4) \subseteq \kappa_{s_3} \circ U_{P_3}(P_4).$$

Due to the uniqueness of amalgamation for $Mod(BOD_3)$, it suffices to show

(a) $U_{s_3} \circ U_{b_3} \circ \kappa_{s_4}(P_4) \subseteq U_{s_3} \circ \kappa_{s_3} \circ U_{P_3}(P_4)$, and
(b) $U_{g_3} \circ U_{b_3} \circ \kappa_{s_4}(P_4) \subseteq U_{g_3} \circ \kappa_{s_3} \circ U_{P_3}(P_4)$.

In fact (a) follows from the persistency of κ_{s_3} and κ_{s_4} (due to the extension property) and (b) follows from correctness of R_1 and the extension property of κ_{s_3} and κ_{s_4} with respect to κ_{s_1} and κ_{s_2}, respectively:

$$
\begin{aligned}
U_{g_3} \circ U_{b_3} \circ \kappa_{s_4}(P_4) &= U_{b_1} \circ U_{g_4} \circ \kappa_{s_4}(P_4) \\
&= U_{b_1} \circ \kappa_{s_2} \circ U_{g_2}(P_4) \\
&\subseteq \kappa_{s_1} \circ U_{p_1} \circ U_{g_2}(P_4) \\
&= \kappa_{s_1} \circ U_{g_1} \circ U_{p_3}(P_4) \\
&= U_{g_3} \circ \kappa_{s_3} \circ U_{p_3}(P_4)
\end{aligned}
$$

7.6.6 Horizontal composition of refinement

1. Given refinements $R_1 = (p_1, b_1)$: $PSP_1 \to PSP_2$ and $R_3 = (p_3, b_3)$: $PSP_3 \to PSP_4$ with $b_1 = p_3$, the *direct horizontal composition* $R_3 \circ R_1$: $PSP_3 \circ PSP_1 \to PSP_4 \circ PSP_2$ is defined by $R_3 \circ R_1 = (p_1, b_3)$.

$$
\begin{array}{ccccc}
PAR_1 & \xrightarrow{s_1} & BOD_1 = PAR_3 & \xrightarrow{s_3} & BOD_3 \\
\downarrow{p_1} & & \downarrow{b_1 = p_3} & & \downarrow{b_3} \\
PAR_2 & \xrightarrow{s_2} & BOD_2 = PAR_4 & \xrightarrow{s_4} & BOD_4
\end{array}
$$

2. Given correct refinements R_1 and R_3 as above, the direct horizontal composition $R_3 \circ R_1$ is correct (*correctness of direct horizontal composition*), provided that the constructors are functorial, i.e., $\kappa_{s_3 \circ s_1} = \kappa_{s_4} \circ \kappa_{s_2}$ and $\kappa_{s_4 \circ s_2} = \kappa_{s_4} \circ \kappa_{s_2}$.

3. Given refinements $R_1 = (p_1, b_1)$: $PSP_1 \to PSP_2$ and $R_3 = (p_3, b_3)$: $PSP_3 \to PSP_4$ with parameter-passing morphism g: $PAR_3 \to BOD_1$, the *horizontal composition* of R_1 and R_3 via g, written $R_3 \circ_g R_1$, is given by the direct horizontal composition of R_1 and the actualized refinement R_{3_g} of R_3 via g, i.e., $R_3 \circ_g R_1 = R_{3_g} \circ R_1$.

4. Given correct refinements R_1 and R_3 as above where all parameterized specifications are persistent, then the horizontal composition of R_1 and R_3 is also correct (*correctness of horizontal composition*), and the induced parameterized specifications are persistent provided that the conditions given in Sections 7.6.5.2 and 7.6.6.2 are satisfied.

7.6.7   Bidimensional compatibility of refinement for parameterized specifications

The bidimensional compatibility formulated for modular specifications in Section 7.6.1 is valid for parameterized specifications with vertical and direct horizontal composition of refinements, and also for vertical and horizontal composition, including compatible actualization steps. This follows directly from the corresponding constructions in Sections 7.6.4–7.6.6 and the composition properties of pushouts. For the special case of monomorphic functorial parameterized specifications, the corresponding result is explicitly shown in [EG94].

7.6.8   Refinement of module specifications

The refinement concepts discussed above for parameterized specifications can be extended to module specifications, similar to [EM90], where in addition to a parameter we have explicit import and export interface specifications:

1. A *module specification MOD* consists of a *parameter* specification *PAR*, an *export* specification *EXP*, an *import* specification *IMP*, a *body* specification *BOD*, and specification morphisms i, e, s, v such that the following diagram commutes.

$$
\begin{array}{ccc}
PAR & \xrightarrow{\ e\ } & EXP \\
{\scriptstyle i}\downarrow & = & \downarrow{\scriptstyle r} \\
IMP & \xrightarrow{\ s\ } & BOD
\end{array}
$$

   The semantics *SEM* constructs, for each import model $I \in Mod(IMP)$, a class of export models $SEM(I) = U_v \circ \kappa_s(I)$, where $\kappa_s \colon Mod(IMP) \to \mathcal{P}(Mod(BOD))$ is a constructor as discussed in Section 7.6.2.

   If the constructors are free constructions, we obtain module specifications in the sense of [EM90], but we may also consider nonmonomorphic constructors like *translate$_s$*

2. A *refinement R*: $MOD_1 \to MOD_2$ of module specifications is given by a triple $R = (r_P, r_E, r_I)$ of specification morphisms such that the following diagram commutes.

   In contrast to parameterized specifications, it is not necessary to require a signature morphism between the body parts, which allows even more flexibility in the design of the bodies.

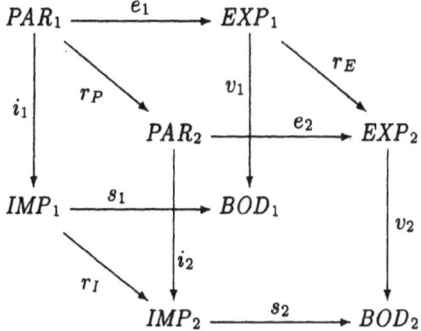

Correctness of a refinement R means that the class of export models constructed by MOD_2 is included in that constructed by MOD_1, taking into account suitable restrictions defined by forgetful functors, i.e., for each $I_2 \in Mod(IMP_2)$ we have

$$U_{r_E} \circ SEM_2(I_2) \subseteq SEM_1 \circ U_{r_I}(I_2).$$

If the constructors are free constructions this corresponds to coherent refinements as studied in [EM90].

3. Similar to parameterized specifications we can define *vertical composition* of refinement for module specifications by

$$R_2 * R_1 = (r_{2_P} \circ r_{1_P}, r_{2_E} \circ r_{1_E}, r_{2_I} \circ r_{1_I})$$

and it is easy to show that correctness of R_1 and R_2 implies the correctness of $R_2 * R_1$.

Regarding *horizontal composition*, different kinds of module interconnections, like composition, union, and actualization, are discussed in [EM90] and can be extended to our more general framework with constructors. This also applies to the compatibility results in [EM90] between horizontal composition and refinement.

4. Finally, let us point out that composition and refinement can already be expressed on the level of *interface specifications INT* $= (PAR, EXP, IMP, e, i)$, which are useful to consider before the design of the body parts. *Realization* of interfaces studied in [EM90] allows the construction of the body parts in a separate step, so that there is full compatibility between horizontal and vertical composition, and realization.

7.7 Implementation of modular systems

We mentioned in the introduction of the previous section the overall goal which is to show correctness of the development process starting from the requirement definition and leading to the final modular system. In the case of refinements, we have shown that the horizontal and vertical refinement steps are corrrectness-preserving and compatible with each other under very mild assumptions concerning the underlying institution and the kind of constructors which are used. The concept of refinement, however, is only of limited use for vertical development in practice. In general it is much more reasonable to consider implementations including constructors and abstractors as discussed in Section 7.3. Hence the central issue of this section is to discuss implementations of parameterized specifications, including horizontal and vertical composition of implementations. Unfortunately, as known from Section 7.5, the vertical composition of implementations in general is not correctness-preserving and for similar reasons this also applies to horizontal composition. This seems to be quite serious for achieving our overall goal using more general notions of implementations. Fortunately, it was shown by Fernando Orejas and his coauthors in [ONS96] that it is not necessary to have correctness of horizontal and vertical composition of implementation on all levels of the development process. In fact, it is sufficient to have correctness of elementary implementation steps at all the specification levels in the development process except the final one, which is usually the programming language level. Only on this final level do we need correctness of horizontal and vertical composition of implementations in order to show correctness of the development process.

This important result will be sketched in this section, using a general concept of implementation including extension, restriction constructors, and abstractors. For this general concept correctness criteria for horizontal and vertical composition are still under development [EKO97]. The situation is much easier in the case of "simple implementations" using extension and constructors, but no restriction and no abstractors. This corresponds roughly to constructor implementations in the sense of Sannella and Tarlecki [ST88b].

We start with the general notion of implementation of parameterized specifications as presented in [ONS96], including extensions, restrictions, constructors, and abstractors, where the refinement concept in the previous section is a special case with identical extensions and abstractors. For simplicity, however, we only consider the case where the parameterized specifications share the same parameter.

7.7.1   Implementation of parameterized specifications

1. Let $PSP_1 = (PAR, BOD_1, s_1)$ and $PSP_2 = (PAR, BOD_2, s_2)$ be parameterized specifications. Then the syntax of an *implementation* $I: PSP_1 \rightarrow PSP_2$ of PSP_1 by PSP_2 is given by $I = (e, b)$, where $e: BOD_2 \rightarrow BOD_3$ is a specification morphism, called *extension*, and $b: BOD_1 \dashrightarrow BOD_3$ is a signature morphism between the body specifications BOD_1 and BOD_3, called *restriction*, such that the following diagram commutes.

$$PAR \xrightarrow{\;s_1\;} BOD_1$$

$$s_2 \downarrow \qquad = \qquad \downarrow b$$

$$BOD_2 \xrightarrow{\;e\;} BOD_3$$

2. The *semantics* of $PSP_i = (PAR, BOD_i, s_i)$ is given by a constructor $\kappa_{s_i} \colon Mod(PAR) \to \mathcal{P}(Mod(BOD_i))$, for $i = 1, 2$, and the *semantics of the implementation* I by a constructor $\kappa_e \colon Mod(BOD_2) \to \mathcal{P}(Mod(BOD_3))$, the forgetful functor $U_b \colon Mod(\Sigma(BOD_3)) \to Mod(\Sigma(BOD_1))$, and an abstractor $\alpha \colon Mod(BOD_1) \to \mathcal{P}(Mod(\Sigma(BOD_1)))$.

   In contrast to the previous section, where the constructor κ_s was completely determined by the specification morphism s and a fixed parameterization concept, we allow in this section *individual constructors* and *abstractors*. These are restricted to the functions between the corresponding model classes, but are not completely determined by the corresponding specifications and specification morphisms. This allows more flexibility, especially if we consider constructions within different institutions. it includes the concept in [ONS96], which allows the implementation of the constructor κ_1 based on s_1 by the constructor κ_2 based on s_2. Moreover, the abstractor α in [ONS96] is determined by the signature morphism b, though it seems far more flexible for α to be based on the specification BOD_1, determined by b or s_1 depending on the abstraction concept.

3. Combining syntax and semantics, an *implementation* $I \colon PSP_1 \to PSP_2$ of $PSP_1 = (PAR, BOD_1, s_1, \kappa_{s_1})$ by $PSP_2 = (PAR, BOD_2, s_2, \kappa_{s_2})$ is given by $I = (e, b, \kappa_e, \alpha)$, where the components are defined as above. The implementation I is called *correct* if for all $P \in Mod(PAR)$ the model class $\kappa_e \circ \kappa_{s_2}(P)$, generated by PSP_2 and the extension, is included in the model class $\kappa_{s_1}(P)$, generated by PSP_1 up to restriction with respect to b and abstraction with respect to α, i.e.,

$$U_b \circ \kappa_e \circ \kappa_{s_2}(P) \subseteq \alpha \circ \kappa_{s_1}(P).$$

All the compositions above are defined "pointwise", e.g.,

$$\kappa_e \circ \kappa_{s_2}(P) = \{B_3 \in Mod(BOD_3) \mid B_3 \in \kappa_e(B_2) \text{ for some } B_2 \in \kappa_{s_2}(P)\}.$$

4. Like the case of refinements in the previous section, it is possible to define *horizontal and vertical composition of implementations*. This is done for vertical and a restricted kind of horizontal composition in [ONS96]. In [EKO95] we discuss the general case including correctness conditions. Unfortunately, several important notions do not satisfy transitivity (see Section 7.5), and for similar reasons horizontal and vertical compositions are not correct in the general case. However we will discuss a simpler notion of implementations in Section 7.7.3 where horizontal and vertical compositions are correct under mild assumptions.

7.7.2   Examples

1. In Example 7.6.3.1 we showed a refinement $R = (id, h)$: $Container(Data)$ $\to List_1(Data)$. According to Section 7.2 the parameterized specification $List(Data)$ is generated by the parameterized specification $List_0(Data)$, the usual elementary specification of lists, and an extension $e: List_0 \to List_1$. This allows the consideration of the same example as an implementation $I = (e, h)$: $Container(Data) \to List_0(Data)$ of $Container(Data)$ by $List_0(Data)$.

$$
\begin{array}{ccc}
Data & \xrightarrow{\ s_1\ } & Container \\[2pt]
\Big\downarrow {\scriptstyle s_2} & = & \Big\downarrow {\scriptstyle h} \\[2pt]
List_0 & \xrightarrow{\ e\ } & List_1
\end{array}
$$

2. Like Example 7.6.3, we can consider the case of loose semantics with constructors $\kappa_{s_1}, \kappa_{s_2}$, and κ_e defined by translation, or the case of initial semantics where the constructors are defined by free constructions. In both cases the abstractor α can be chosen to be the identity.

3. In general, nontrivial choices for the abstractor α are abstraction up to behavioral equivalence and abstraction up to reachability. In the first case, we have

$$\alpha(B_1) = \{B \in Mod(\Sigma(BOD_1)) \mid B \equiv B_1\}$$

where behavioral equivalence $B \equiv B_1$ is defined using all parameter sorts as observable and the remaining ones in BOD_1 as nonobservable.
In the second case, we have

$$\alpha(B_1) = \{B \in Mod(\Sigma(BOD_1)) \mid REACH_{B_1}(B) = B_1\}$$

with $REACH_{s_1}(B) = \bigcap\{B' \subseteq B \mid U_{s_1}(B') = U_{s_1}(B)\}$.

Before we discuss horizontal and vertical composition in the general case of implementations, where we cannot expect good properties concerning the preservation of correctness, let us first consider simple implementations, where one obtains correctness results similar to those of refinements in the previous section.

7.7.3   Simple implementations of parameterized specifications

1. A *simple implementation* $I: PSP_1 \to PSP_2$ of a parameterized specification PSP_1 by PSP_2 with $PSP_i = (PAR, BOD_i, s_i, \kappa_{s_i})$, for $i = 1, 2$, is given by $I = (e, \kappa_e)$ with the specification morphism $e: BOD_2 \to BOD_1$ such that $e \circ s_2 = s_1$, and the constructor $\kappa_e: Mod(BOD_2) \to \mathcal{P}(Mod(BOD_1))$. The simple implementation I is called *correct* if we have for all $P \in Mod(PAR)$

$$\kappa_e \circ \kappa_{s_2}(P) \subseteq \kappa_{s_1}(P).$$

This means that a simple implementation $I = (e, \kappa_e)$ can be considered as a general implementation $I = (e, b, \kappa_e, \alpha)$, where $b = id_{BOD_1}$ and α is the identical abstractor. In fact, simple implementations correspond to constructor implementations in the sense of Sannella and Tarlecki [ST88b].

2. The *vertical composition* of simple implementations $I_1 = (e_1, \kappa_{e_1}) : PSP_1 \to PSP_2$ and $I_2 = (e_2, \kappa_{e_2}) : PSP_2 \to PSP_3$ is given by $I_2 * I_1 = (e_1 \circ e_2, \kappa_{e_1} \circ \kappa_{e_2}) : PSP_1 \to PSP_3$, and correctness of I_1 and I_2 implies that of $I_2 * I_1$.

3. The *direct horizontal composition* of simple implementation $I_1 = (e_1, \kappa_{e_1}) : PSP_1 \to PSP'_1$ and $I_2 = (e_2, \kappa_{e_2}) : PSP_2 \to PSP'_2$ with $BOD_1 = PAR_2$

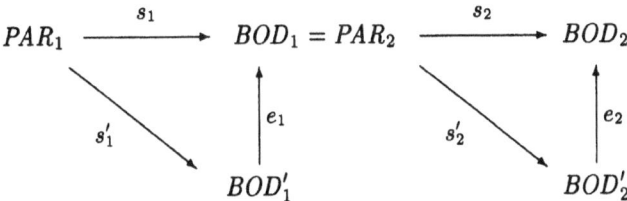

is given by

$$I_2 \circ I_1 = (e_2, \kappa_{e_2}) : PSP_2 \circ PSP_1 \to PSP'_2 \circ_{e_1} PSP'_1$$

where the composite parameterized specifications are given by $PSP_2 \circ PSP_1 = (PAR_1, BOD_2, s_2 \circ s_1, \kappa_{s_2} \circ \kappa_{s_1})$ and $PSP_{2'} \circ_{e_1} PSP_{1'} = (PAR_1, BOD'_2, s'_2 \circ e_1 \circ s'_1, \kappa_{s'_2} \circ \kappa_{e_1} \circ \kappa_{s'_1})$. It is easy to prove that correctness of I_1 and I_2 implies correctness of $I_2 \circ I_1$.

4. *Actualization* and (general) *horizontal composition* of simple implementations can be defined similarly to the case of refinements in the previous section, where horizontal composition $I_2 \circ_g I_1$ via a specification morphism $g : PAR_2 \to BOD_1$ is defined as direct horizontal composition $I_{2_g} \circ I_1$ of I_1 and the actualization I_{2_g} of I_2. In order to prove correctness of I_{2_g}, and hence $I_2 \circ_g I_1$, we need, similarly to the proof in Section 7.6.5.3, persistency and extension properties of the corresponding constructors and the amalgamation property for the institution. The extension property in the case of the individual constructor κ_{s_1} in the situation of Section 7.6.5.2 means that the extension κ_{s_3} of a persistent constructor κ_{s_1} can be constructed by:

$$\kappa_{s_3}(P_3) = \{P_3 +_{p_1} B_1 \mid U_{g_1}(P_3) = P_1, B_1 \in \kappa_{s_1}(P_1)\}.$$

Finally we also obtain *bidimensional compatibility* of simple implementation for parameterized specifications in the sense of Section 7.6.1 using direct or general horizontal composition.

In order to define modular stepwise software design in the sense of [ONS96], leading from specifications in an institution $\mathcal{L}$ to programs in another institution $\mathcal{L}'$, we need to formulate the concepts of translation and realization between parameterized specifications in $\mathcal{L}$ and $\mathcal{L}'$.

7.7.4   Translation and realization

Given two parameterized specifications $PSP = (PAR, BOD, s, \kappa_s)$ in $\mathcal{L}$ and $PSP' = (PAR', BOD', s', \kappa_{s'})$ in $\mathcal{L}'$, we have:

1. A *translation* $T\colon PSP \to PSP'$ if (for simplicity) the syntax of PSP and PSP' concerning the signature of parameter and body are equal. The translation is called *correct* if the semantics is also equal, more precisely $\kappa_s(P) = \kappa_{s'}(P)$, for all $P \in Mod(PAR)$, and $Mod(PAR) \subseteq Mod(PAR')$.
2. A *realization* $R = (b, \alpha)\colon PSP \to PSP'$ if (for simplicity) the signatures PAR and PAR' are equal and $b\colon BOD \rightarrow BOD'$ is a signature morphism. The realization is called *correct* if we have $Mod(PAR) \subseteq Mod(PAR')$ and, for all $P \in Mod(PAR)$,

$$U_b \circ \kappa_{s'}(P) \subseteq \alpha \circ \kappa_s(P)$$

with abstractor $\alpha\colon Mod(BOD) \to \mathcal{P}(Mod(\Sigma(BOD)))$.

It must be noted that not every parameterized specification in $\mathcal{L}$ has a translation in $\mathcal{L}'$ or can be realized in $\mathcal{L}'$. The notion of realization is similar to that of refinement in the previous section, but it allows in addition an abstractor and different institutions.

7.7.5   Modular stepwise software design

According to [ONS96] the process of modular stepwise software design can be defined by a "software design tree" where all the nodes are labeled with parameterized specifications PSP in an institution $\mathcal{L}$ (specification level), except for some of the leaves with labels in another institution $\mathcal{L}'$ (programming level).

Given a parameterized specification PSP_0 in institution $\mathcal{L}$ as the root, we obtain a tree, called *software design tree*, if we allow the following design steps.

(a) *Implementation*: We pick one of the nodes, say, a node with label PSP in $\mathcal{L}$, and implement (within $\mathcal{L}$) PSP in terms of another PSP', i.e., we provide an implementation $I = (e, b, \kappa_e, \alpha)\colon PSP \to PSP'$, where e corresponds to parameterized specification PSP'' with semantics κ_e. Moreover, we allow PSP' and PSP'' to be composed (via direct horizontal composition) of other parameterized specifications, say,

$$PSP' = PSP_n \circ \cdots \circ PSP_1 \quad \text{and} \quad PSP'' = PSP_m \circ \cdots \circ PSP_{n+1}.$$

Then we create m sons of our node with label PSP which are labeled by $PSP_1, \ldots, PSP_m$.

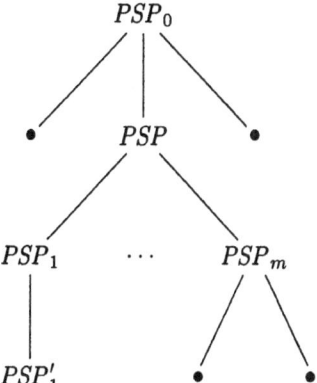

(b) *Translation*: We pick one of the nodes, say, a node with label PSP_1 in $\mathcal{L}$ and provide a translation $T\colon PSP_1 \to PSP'_1$ from PSP_1 to PSP'_1 in $\mathcal{L}'$. In this case we create a single son of our node with label PSP_1, where the son is labeled by PSP'_1.

The software design tree is *complete* if all the leaves are labeled by parameterized specifications $PSP'_1, \ldots, PSP'_n$ in $\mathcal{L}'$. In this case the given parameterized specification PSP_0 in $\mathcal{L}$ is completely realized by $PSP'_1, \ldots, PSP'_n$ in $\mathcal{L}'$. The main problem is to show that this realization is correct. Of course, we require that all the software design steps, implementations in $\mathcal{L}$, and translation from $\mathcal{L}$ to $\mathcal{L}'$, are correct. But what kind of correctness requirements for horizontal and vertical composition are required? One would expect that we need correctness of horizontal and vertical compostion of implementations in $\mathcal{L}$, but this in general is not satisfied. Fortunately, such properties are only required for the programming level institution $\mathcal{L}'$, not for $\mathcal{L}$. In the following, we will call such an institution $\mathcal{L}'$ *well-behaved* and give a short summary of the required properties.

7.7.6   Correctness of stepwise software design

Let us assume that we have a complete software design tree labeled by a parameterized specification PSP_0 in the root and $PSP'_1, \ldots, PSP'_n$ in the leaves where PSP_0 belongs to an institution $\mathcal{L}$ for a specification language, and $PSP'_1, \ldots, PSP'_n$ to a well-behaved institution $\mathcal{L}'$ for a programming language. Moreover, let us assume that all implementation and translation steps used to build up the software design tree are correct, and PSP' is the horizontal composition of $PSP'_1, \ldots, PSP'_n$ via some suitable parameter-passing morphisms $g_1, \ldots, g_n$ which are defined by the software design process. Then PSP' in $\mathcal{L}'$ is a correct realization of PSP_0 in $\mathcal{L}$. This means that the stepwise software design given by the complete software design tree is correct. The proof for this important theorem is given in [ONS96]. It is important to note that this theorem does not require any kind of correctness for horizontal or vertical composition in $\mathcal{L}$, but only for $\mathcal{L}'$. In fact, it is required that $\mathcal{L}'$ is well-behaved in the following sense:

1. $\mathcal{L}'$ has correct vertical and horizontal composition, where the horizontal composition defined in [ONS96] is a mixture of direct and general horizontal compositions using actualizations.
2. The parameterized specifications $PSP' = (PAR', BOD', s', \kappa_{s'})$ in $\mathcal{L}'$ have no axioms in PAR', and $\kappa_{s'}$ is monomorphic and persistent, i.e., $\kappa_{s'}(P')$ for $P' \in Mod(PAR')$ consists of only one model up to isomorphism and $U_{s'} \circ \kappa_{s'}(P') = P'$.
3. The institution $\mathcal{L}'$ has amalgamation and extension of perstistent constructors (see Section 7.7.3.4).

In [ONS96] it is argued that these assumptions for $\mathcal{L}'$ are satisfied for "most" programming languages, especially for high-level programming languages like MODULA. Unfortunately, it is not easy to define explicitly an institution for such languages, so that no explicit verification of well-behavedness of such institutions $\mathcal{L}'$ is so far known. However, the important point is that we do not require that the specification language institution $\mathcal{L}$ is well-behaved in order to conclude correctness of stepwise software design.

Acknowledgements. We are grateful to Frank Drewes and Renate Klempien-Hinrichs for many helpful comments.

8 Specification Languages

Donald Sannella[1] and Martin Wirsing[2]

[1] Laboratory for Foundations of Computer Science, University of Edinburgh,
Edinburgh, Scotland; `dts@dcs.ed.ac.uk`, `http://www.dcs.ed.ac.uk/~dts/`
[2] Institut für Informatik, Ludwig-Maximilians-Universität München, München,
Germany; `wirsing@informatik.uni-muenchen.de`,
`http://www.pst.informatik.uni-muenchen.de/personen/wirsing/`

8.1 Introduction

The basic components of any specification language include: constructs for
specifying the properties of individual program components such as types
and functions; structuring mechanisms for building large specifications in a
modular fashion; a description of the semantics of the language; mechanisms
for performing proofs of properties of specifications; a notion of refinement of
specifications; and a way of relating specifications to programs written in a
programming language. These topics are discussed in other chapters of this
book, and in each case there are various alternatives to choose from. To take a
simple example, the properties of functions may be specified using equations
or using first-order (or even higher-order) formulas as axioms.

A specification language is a commitment to a compatible combination
of these choices. It is a commitment because the syntax of a specification
language determines once and for all what can be expressed and what cannot be expressed. For example, the syntax of formulas determines how the
properties of functions are specified; similarly, a construct for hiding types
and/or functions may or may not be included. The choices must be compatible in various respects. On one hand, certain combinations of choices simply
make no sense; for instance, initial semantics requires the use of axioms no
more complex than conditional equations. More subtly, the inclusion of certain structuring mechanisms complicates the process of proving properties of
specifications and so these might be omitted or restricted in a language in
which proofs are of primary importance.

There is no single best combination of choices because this depends on
many factors including the intended context of use and the range of applications to be addressed. Consequently, this chapter will consider the rationale
for choosing between the various alternatives available. The technical details
of these alternatives are given in other chapters for the most part and will
not be repeated here.

The next section is devoted to a brief overview of some existing specification languages. Section 8.3, which forms the bulk of the chapter, discusses the

various design decisions that would confront the designer of a new specification language. The final short section speculates on some likely future trends. As in other chapters of this book, we focus on property-oriented specifications, although many of the issues discussed relate to specification languages in general.

8.2   Existing specification languages

In this section we give a brief summary of the features of a range of existing specification languages, with a small example of a specification in each language. Further small examples of specifications written in these languages will appear in the next section. Note that these examples are chosen to illustrate language features rather than for their intrinsic interest. See [Wir95] for a more comprehensive survey with historical remarks.

Clear and OBJ3. Clear [BG77, BG80], the first algebraic specification language, had considerable influence on the design of many other languages. Features of Clear include: the use of specification-building operations for constructing specifications in a modular way, with care taken to respect shared sub-specifications, and most of the particular operations used in other languages (see Chapter 6); explicit pushout-based parameterisation (Chapter 6); free generating constraints (Chapter 5); institution independent theory-level semantics (Chapters 4 and 5); a structured proof system [SB83] (Chapter 11); and a notion of implementation of parameterised specifications [SW82] (Chapter 7). OBJ3 [GWM+92] is an executable specification language, also known as an "ultra-high-level" programming language, that can be seen as an implementation of Clear (using associative-commutative rewriting, see Chapter 9) for the institution of order-sorted conditional equational logic. Order-sorted logic is used in order to cope smoothly with partial functions, errors and subsort polymorphism; see Chapter 3. OBJ3 has several variants of Clear's enrich specification-building operation, including so-called "protecting importation" which ensures that the enrichment is persistent. It supports a very flexible "mixfix" notation with operators such as *if__then__else__fi*: *expr* × *stmt* × *stmt* → *stmt*. Two new object-oriented specification languages are based on OBJ3: Maude [Mes93] extends OBJ3 by a notion of objects and by concurrent rewriting for describing the dynamic behaviour of objects; CafeOBJ [DF98] is similar but focuses additionally on behavioural specifications (see Chapter 5).

The following simple example in OBJ3 demonstrates the use of subsorts. The operation `tail` can only be applied to values of sort `NeList` (non-empty lists), which is specified as a subsort of `List` and a supersort of the sort `Elt` of list elements (from the "interface theory" TRIV).

```
obj LIST[X :: TRIV] is
  sorts List NeList .
  op nil : -> List .
```

```
    subsorts Elt < NeList < List .
    op __ : List List -> List [assoc id: nil] .
    op __ : NeList List -> NeList .
    op __ : NeList NeList -> NeList .
    protecting NAT .
    op |_| : List -> Nat .
    eq | nil | = 0 .
    var E : Elt .   var L : List .
    eq | E L | = 1 + | L | .
    op tail_ : NeList -> List .
    var E : Elt .   var L : List .
    eq tail E L = L .
endo
```

ACT ONE and ACT TWO. ACT [CEW93] is an approach to formal software development that includes a language called ACT ONE [Cla89] for writing parameterised specifications, called "types", with conditional equational axioms and initial constraints, and an extension called ACT TWO [Fey88] for writing specified modules. ACT ONE has a pure initial algebra semantics (Chapter 5) where every type is parameterised (non-parameterised types are considered as a degenerate case) and denotes a free functor. Thus it provides only simple specification-building operators (free extension, union, renaming and pushout-based instantiation) but no operation for hiding. A module in ACT TWO consists of four specifications:

Parameter: This describes parameters that are common to the entire module or modular system in which the module appears, e.g., the underlying character set.

Import interface: This describes the sorts and operations that the module requires to be supplied by other modules.

Export interface: This describes the sorts and operations that the module supplies for use by other modules.

Body: This defines the construction of the exported components in terms of the imported components. This construction may involve auxiliary operations that are not exported.

These four specifications are written in ACT ONE extended by permitting first-order axioms (except in the body part). Module-building operations (composition, union, instantiation and renaming) are module-level analogues to specification-building operations. Both ACT ONE and ACT TWO have both a presentation-level and a model-level semantics. For each language these are described separately and are then shown to be compatible.

The following example is an ACT ONE parameterised type of lists. The type ELEM serves as the formal parameter for the type LIST, meaning that the body of LIST is interpreted as an extension of ELEM. List(D) is a so-called "sort with structured name", expressing the dependency of the sort List on the parameter sort D.

```
parameter ELEM is
  sorts D
endpar

type LIST[ELEM] is
  extend
    union NAT, BOOL endunion
  by
    sorts
      List(D)
    constructors
      nil: -> List(D);
      cons: D, List(D) -> List(D);
    functions
      vars x:D, l:List(D);
      func length: List(D) -> Nat;
        length(nil) = zero;
        length(cons(x,l)) = succ(length(l));
      func isempty: List(D) -> Bool;
        isempty(nil) = true;
        isempty(cons(x,l)) = false;
  endext
endtype
```

The operation of instantiating LIST with a specification NAT of natural numbers is called "actualization", and is written as follows:

```
type NATLIST is
  actualize
    LIST by NAT using sortnames Nat for D
  endact
endtype
```

The instantiation of the sort D causes the resulting type NATLIST to contain a sort with the name List(Nat).

ASL. ASL [SW83, Wir86, ST88a] is a kernel language that is intended to provide a sound semantic basis for defining more user-friendly high-level specification languages (e.g., Pluss [BGM89, Bid89] and the meta-language used in [ABB+86]). It consists of a small number of simple but powerful specification-building operations with an institution-independent loose model-class semantics (Chapters 4 and 5). Generating constraints are incorporated via a specification-building operation, called reachable. A specification-building operation called abstract (and a special case of abstract called behaviour) provides observational abstraction (Chapter 5). The absence of initiality constraints means that there is no reason to restrict to the use of conditional equational axioms, so more expressive logics – typically first-order logic with equality – are used. ASL provides explicit λ-calculus-based higher-order parameterisation. [SST92] describes an extension of ASL called ASL+ that also supports specification of (possibly higher-order) parameterised algebras, and distinguishes between these and parameterised specifications, with a formal system for proving satisfaction, cf. [Asp97]. The expressive power of the language makes it possible to model refinement of specifications (Chapter 7) as model class inclusion; more elaborate notions of implementation of ASL specifications are studied in [ST88b]. A system for performing structured proofs of properties of specifications and of correctness of implementation steps (Chapter 11) is described in [Far92], and complete proof systems for subsets of ASL are given in [Wir93, Cen94, HWB97, Hen97].

The following is an ASL specification of bags of natural numbers with a counting function based on a predefined specification of natural numbers (NAT) that is assumed to be monomorphic. The use of reachable ensures that only term-generated models are admissible. The axioms ensure that the data structure of finite multisets of natural numbers is (up to isomorphism) the only model of this specification.

```
BAGcount =
  reachable
    enrich NAT by
      sorts   Bag
      opns    empty : Bag
              cons: Nat, Bag -> Bag
              count: Nat, Bag -> Nat
      axioms  cons(x, cons(y, b)) = cons(y, cons(x, b))
              count(x,empty) = zero
              count(x, cons(x,b)) = succ(count(x, b))
              x≠y ⇒ count(x, cons(y,b)) = count(x, b)
  on {Bag}
```

Using the behaviour operator, the class of models is enlarged to include, e.g., the data structure of finite lists of natural numbers, which is the initial model of BEHAVIOURAL_SET; the models of BAGcount are its final models.

```
BEHAVIOURAL_SET = behaviour BAGcount wrt {Nat}
```

Larch. Larch [GH86, GH93] is a family of specification languages. Each Larch specification has components written in two languages: one designed for a specific programming language, the so-called Larch interface language, and another common to all programming languages, the so-called Larch shared language LSL. LSL is a loose algebraic specification language with equational axioms, generating constraints, and simple structuring operators (rename, combine, and implicit parameterisation) but no operation for hiding, and a theory semantics based on first-order logic with equality. There is a construct for observational abstraction called partitioned by which has the status of a constraint, and specifications, called "traits", may contain assertions about intended theorems using the keyword implies. The LP theorem prover supports proof of properties of LSL traits. Larch interface languages have been

designed for CLU, Ada, C, Modula-3, Smalltalk, and C++. Each of these
is an extension of a programming language with annotations that describe
program properties expressed via abstract concepts from the problem domain
that are defined in LSL traits.

The following is an LSL trait defining bags based on a predefined trait of
natural numbers (NAT). Sorts are declared implicitly by their appearance in
the signature, and the parameters in the declaration of a trait may contain
names of sorts or function symbols. Parameters are used merely to facilitate renaming, so the appearance or non-appearance of parameters has no
semantic consequences. Axioms are given following the keyword asserts. The
generated by assertion amounts to a reachability constraint for Bag, while
the partitioned by assertion says that values of sort Bag that cannot be distinguished using count are equal. The implies clause states intended consequences of the trait, where converts count says that the definition of count
is sufficiently complete.

```
BAGcount (E, Bag): trait
  includes NAT
  introduces
    empty: → Bag
    cons: E, Bag → Bag
    count: E, Bag → Nat
  asserts
    Bag generated by empty, cons
    Bag partitioned by count
    ∀ b: Bag, x,y: E
      count(x,empty) == zero;
      count(x,cons(y,b)) == if x=y then succ(count(x,b))
                                    else count(x,b)
  implies
    converts count
    ∀ x: E, b: Bag
      cons(x,b) ≠ empty
```

A partial instantiation of BAGcount with natural numbers for the element
sort E is achieved by defining a new trait that includes NAT and BAGcount
appropriately renamed:

```
NatBAGcount: trait
  includes
    NAT, BAGcount (Nat, Bag)
```

Note that BAGcount (Nat, Bag) is just an abbreviation for the renaming
BAGcount (Nat for E).

Extended ML. Extended ML (EML) [San91, ST91, KS98] is a framework for
the formal development of modular programs in the Standard ML (SML)

functional programming language [Pau96] from specifications of their required behaviour. The EML language is a "wide-spectrum" language that is based on a large subset of SML, excluding mainly references and input/output. EML thus inherits SML's higher-order polymorphic type system and its facilities for building hierarchically-structured and parameterised program modules with explicit interfaces. EML extends this subset of SML by permitting axioms in module interfaces, for specifying required properties of module components, and in place of code in module bodies, for describing functions in a non-algorithmic way prior to their implementation as SML code. Axioms are written in a language of higher-order logic with equality, with additional features to deal with run-time exceptions and non-terminating functions. Correctness of module bodies with respect to their interfaces is required only up to behavioural equivalence. The semantics of EML [KST94, KST97] is based directly on the static and dynamic semantics of SML [MTH90], extended with a loose model-class semantics to give meaning to the specification constructs. The relationship between EML and SML is thus a formal one based on semantics in contrast to the rather informal relationship between specification language and programming language in the case of Larch. Spectral [KS91] can be seen as an extension of EML with higher-order parameterisation at the module level, dependent types, and logically-constrained subtypes.

The following is an EML specification of a generic sorting module called Sort. The first axiom in its input interface P0 requires the function le to terminate for any choice of arguments. The first axiom in its output interface SORT requires the function sort to terminate with a "proper" value (i.e., not an exception) for any argument. In the second axiom of SORT, the function sorted is a local function whose scope is that axiom. The third axiom in SORT assumes that a polymorphic function count for counting the number of occurrences of a value in a list has been defined elsewhere.

```
signature P0 =
  sig
    eqtype elem
    val le : elem * elem -> bool
    axiom forall(x,y) => le(x,y) terminates
    axiom forall x => le(x,x)
    axiom forall(x,y) => le(x,y) andalso le(y,x) implies x==y
    axiom forall (x,y,z) => le(x,y) andalso le(y,z) implies le(x,z)
  end

signature SORT =
  sig
    structure Elem:P0
    val sort : Elem.elem list -> Elem.elem list
    axiom forall l => (sort l) proper
    axiom let
            fun sorted l =
```

```
            forall (l1,x,l2,y,l3) =>
                l1@[x]@l2@[y]@l3==l implies Elem.le(x,y)
        in
          forall l => sorted(sort l)
        end
    axiom forall l => forall x => count(x,l)==count(x,sort l)
  end

functor Sort(X : PO) : sig include SORT sharing Elem=X end = ?
```

SPECTRUM. SPECTRUM [BFG+93] is a language for developing executable specifications from requirements specifications. It contains explicit support for partial functions, and axioms are not restricted to equational or conditional equational formulas since executability is not the primary aim.

The whole specification style is oriented towards functional programming and the specification of concurrent systems via dataflow nets of (infinite) streams. As a consequence a number of functional language concepts have been integrated into the specification language. The sort system provides parametric polymorphism and sort classes in the style of the functional language Haskell [HPW92]. Furthermore, the language supports the definition of infinite objects by recursion equations, the definition of functions by typed λ-abstractions, and higher-order functions. Functions can be partial, strict, non-strict, continuous. Non-continuous functions are called "mappings" and are only used as auxiliary functions for specification purposes. The constructs for specification in the large and for parameterisation are inspired by ASL.

The semantics of SPECTRUM is loose. Its definition and proof system were influenced by languages and systems of the LCF family [GMW79], cf. [Reg94]. The logic of SPECTRUM is similar to PPλ, the logic of computable functions underlying LCF. In particular, all carrier sets are complete partial orders. The main difference to PPλ is due to the notion of sort class which semantically is treated like the notion of kind in type theory.

The following is a SPECTRUM specification of a sort class T0 that defines the class of totally ordered sorts. The attributes strict and total of $\leq$ ensure that $\leq$ is a total function and that applying $\leq$ to a bottom (i.e., undefined) element yields the bottom element (of Bool) as result.

```
Torder = {
  class T0;
  .≤. : a :: T0 ⇒ a × a → Bool;
  ≤ strict total;
  axioms a :: T0 ⇒ ∀x, y, z: a in
  x ≤ x;                              -- reflexivity
  x ≤ y ∧ y ≤ x ⇒ x = y;             -- antisymmetry
  x ≤ y ∧ y ≤ z ⇒ x ≤ z;             -- transitivity
  x ≤ y ∨ y ≤ x;                      -- linearity
  endaxioms;
}
```

The specification MinLIST extends Torder and a polymorphic specification LIST1 of lists by a partial function min for computing the minimum of a list (cf. POLY_LIST in Section 8.3.5; LIST1 is an extension of POLY_LIST by the function ∈ that checks whether an element is a member of a list). The first axiom ensures that min is undefined for the empty list nil.

```
MinLIST = {
    enriches Torder + LIST1;
    min : a :: TO ⇒ List a → a;
    min strict;
    axioms a :: TO ⇒ ∀x: a, l : List a in
    ¬(δ(min nil));
    ¬(l = nil) ⇒ min(l) ∈ l ∧ (x ∈ l ⇒ min(l) ≤ x);
    endaxioms;
}
```

CASL. CASL [CoF98, Mos97] is a language for specifying requirements and designs. It was designed by combining the most successful features of existing languages to give a common basis for future joint work on algebraic specifications. CASL is intended to be the central member of a coherent family of languages, with simpler languages (e.g., for use with existing term rewriting tools) obtained by restriction, and more advanced languages (e.g., for specifying reactive systems) obtained by extension.

CASL supports both partial and total functions as well as predicates, with axioms of first-order logic with equality and definedness assertions (Chapter 3). Datatype declarations allow concise specification of sorts together with constructors and (optionally) selectors. Subsorts are provided, including logically-constrained subsorts as in Spectral, and OBJ-style mixfix operator syntax is supported. Structuring operators (translation, reduction, union, loose extension, free extension) and pushout-style generic specifications are available for building large specifications. Architectural specifications describe how to build the specified software from separately-developed generic units with specified interfaces. Specification libraries allow the distributed storage and retrieval of named specifications.

The following CASL specification defines ordered lists over a partially-ordered type of elements as a subsort of ordinary lists.

```
spec PO =
    type Elem
    pred __ ≤ __ : Elem * Elem
    vars x,y,z : Elem
    axioms x≤x ;
            x≤y ∧ y≤x ⇒ x=y ;
            x≤y ∧ y≤z ⇒ x≤z
    end
```

```
spec ORDLIST[PO] =
  free type List[Elem] ::= nil | __::__ (hd:?Elem; tl:?List[Elem])
  pred ordered : List[Elem]
  vars a,b : Elem; l : List[Elem]
  axioms ordered(nil) ;
         ordered(a::nil) ;
         a≤b ∧ ordered(b::l) ⇒ ordered(a::(b::l))
  type OrdList[Elem] = { l : List[Elem] . ordered(l) }
  end
```

The constructor $\_\_::\_\_$ is a total function of type $\mathtt{Elem} \times \mathtt{List[Elem]} \rightarrow \mathtt{List[Elem]}$. Suppose that $a\!:\!\mathtt{Elem}$ and $l\!:\!\mathtt{OrdList[Elem]}$. Since $\mathtt{OrdList[Elem]}$ is a subsort of $\mathtt{List[Elem]}$, the term $a\!::\!l$ is well-formed and is of type $\mathtt{List[Elem]}$. We can "cast" this term to type $\mathtt{OrdList[Elem]}$ by writing $a\!::\!l$ as $\mathtt{OrdList[Elem]}$. This term is well-formed and has the indicated type but will have a defined value if and only if $\mathrm{ordered}(a\!::\!l)$.

ORDLIST is instantiated by supplying an actual parameter that fits the formal parameter PO. One way to do this is to supply a specification together with a fitting morphism, for example:

```
ORDLIST[NAT fit Elem |-> Nat, __≤__]
```

which, assuming that NAT is a specification of natural numbers with sort Nat and predicate $\leq$, gives ordered lists over the natural numbers with sort names List[Nat] and OrdList[Nat]. The same effect may be achieved by using a named "view":

```
view PO-NAT : PO to NAT = Elem |-> Nat, __≤__

ORDLIST[view PO-NAT]
```

A view may itself have parameters, which have to be instantiated when it is used.

8.3   Design decisions

8.3.1   Context of the design

Design decisions are influenced by the general context of the design as well as by the taste of the designer and technical feasibility. The context includes the range of uses to which the language will be put, the environment in which the language will be used, and the tools available to support its use. All of these factors dictate to some extent the choices discussed in the following sections.

Range of applications. The intended range of applications is a significant factor in the design of a specification language. A special-purpose language may be able to provide extra support for a given application domain with additional features to cope with special aspects of the domain. Examples of application domains are: safety-critical systems, where mechanisms to support fault analysis might be required; real-time systems, where notations to indicate timing constraints would be provided; and information systems, e.g., supported via the entity-relationship model.

Level of abstraction. Different specification languages aim to describe problems and/or systems at different levels of abstraction. Possible levels include: that of the problem to be solved (requirements specifications); that of the module structure of the system (design specifications); that of the final solution itself (programs). Some languages focus on a single level of abstraction, while others aim to cope with two or even more of these levels. This influences the form of axioms, the semantic approach, and the type of parameterisation that the language supports.

Programming paradigm. The question of which programming paradigm the specification language aims to support (e.g., imperative, functional, object-oriented, concurrent) exerts a strong influence on all aspects of the design of the language. The state of the art in algebraic specification is such that the most obvious context to which it is appropriate is the development of functional programs with non-trivial data structures, so this is where most experience has been accumulated and where the discussion below will be most directly relevant. There are extensions to handle other paradigms, e.g., Chapter 12 discusses object orientation and Chapter 13 discusses concurrency.

Software development process. The extent to which a specification language will support the software development process is a factor in the design of the language. The rôle of the language in the software development process depends on the particular development model adopted. More than one specification language may be required, with different languages to support different phases of development.

Tools. A pervasive influence on the design of a specification language is the desired degree of tool support. Tools might include parsers, type checkers, theorem provers, browsers, library search tools, rapid prototyping tools, graphical visualization tools, version control mechanisms, analysis tools, etc. If extensive tool support is an important factor in the design, then various design decisions may be influenced by the ease with which the resulting language can be mechanically processed. Some tools or meta-tools (e.g., theorem provers for a logical framework like LF [HHP93] or Isabelle [Pau94b]) may already be available; then the language design may be influenced by the constraints built into such tools.

8.3.2   General design principles

Other general design decisions are influenced by the intended mode of use of the specification language and by its underlying semantic concepts.

Readability, writeability, etc. For the intended mode of use one may have different aims such as easy readability, writeability, executability, verifiability, modifiability. These choices often conflict: easy readability and writeability may require high-level, easy-to-use constructs, whereas easy verifiability excludes certain high-level constructs. For example SPECTRUM, EML, and CASL are languages with many high-level features which make them suitable for use in practical applications, but proof is more difficult than, e.g., in LSL or RAP [HG86]. Executability rules out the use of constructs such as unrestricted existential quantifiers; this makes it difficult to express certain abstract properties and therefore inhibits easy writeability.

OBJ3 and CASL support readability by the use of mixfix notation, while OBJ3 and a planned future sublanguage of CASL support executability and verifiability by restricting to conditional equational formulas. ACT ONE was originally designed to be an executable kernel language with a close relation to the underlying mathematical concepts of initial algebra semantics, but after experience with use of the language, a more sophisticated syntax was added to increase readability and useability. LSL sacrifices executability in favour of brevity, clarity, and abstractness. To compensate it provides ways to check and to verify specifications with the help of the LP theorem prover. The absence of a hiding operation makes such support easier to provide.

ASL was designed as a kernel language with high expressive power and orthogonal language constructs. As a consequence it has a well-developed set of proof rules. Because of the expressiveness of observational abstraction, verifiability of specifications containing the observational abstraction operator is still a topic of current research although very considerable progress has been made in the past few years [BH96].

Up to now modifiability has not been taken in account by most languages. Pluss offers syntactic constructs called draft and sketch that are designed to support this feature. Maude supports inheritance and a redefinition mechanism inspired by object-oriented languages. These concepts are semantically well-founded in Maude: inheritance is based on subsorting, and redefinition can be modeled with the structuring operators for renaming and hiding. Through use of a modular natural deduction calculus [Pet94], ASL supports the reuse of proofs when specifications are modified. In formal development using EML, the module system insulates the developer from the need to re-prove correctness from scratch when interfaces change in certain ways in the course of development [Var92].

General semantic decisions. The semantic concepts underlying a language are often influenced by its intended use: e.g., executability induces the existence of initial or free models, while verifiability requires the existence of proof rules that are sound with respect to the semantics.

Typical general semantic decisions are: Should it be possible to write inconsistent specifications, or specifications that can only be realized using non-computable functions? Should every specification have an initial or a free model (e.g., so that term rewriting makes sense)?

- Inconsistent specifications are generally possible with loose semantics, but not with initial semantics where there is always at least one model, which might be trivial. Another way to force consistency is to restrict the form of the axioms to explicit definitions on top of a fixed set of given consistent specifications, as in model-based approaches to specification.

- Models can be forced to be non-computable when full first-order logic is used or when "local" generating or initial constraints are present [Wir90]. ACT ONE protects against these "errors"; most other languages do not. Exceptions are a subset of OBJ3 and other executable specification languages including the executable subsets of SPECTRUM and EML.

- Initial/free models exist for the executable subset of a language where the operational semantics coincides with the mathematical semantics. Typically, this requires conditional equational axioms or more generally Horn or hereditary Harrop formulas as in λProlog [NM88], sufficient completeness with respect to the constructors, and only simple equations (such as associativity, commutativity, and idempotence) between constructors.

The decision depends on the intended use of specifications. If one aims at abstractness for describing requirements of a program or a software system, then the requirement of consistency, existence of initial and free models, and computability of models should be sacrificed. On the other hand, for the design of algorithms and operational descriptions of dynamic behaviour, most of these properties should hold. See [ST96] for an analysis of the relationship between models of specifications and models of programs.

Structure of language definition. A separate question about the semantics concerns the structure of the semantic description of a complex specification language. A convenient option is to define the semantics for a subset of the language or for a more primitive language (the so-called "kernel", which is chosen to be simple, orthogonal, and low-level) and then to translate all syntactic constructs to this kernel (compare the semantics of CIP-L [BBB+85], COLD [FJ92], and RSL [Mil90b]). The problem is to choose the right kernel: if it is too small then the translation may be too complex; if it is too large then the semantics of the kernel language may be too complicated. ASL is a kernel language, and EML, Pluss, and SPECTRUM are all built (in principle) on top of this kernel.

8.3.3   Specification in the small

For specification "in the small", the first problem is to decide on a kind of signature, a logical system for writing axioms, and a notion of algebra for defining the meaning of axioms via the definition of satisfaction. Of course, these choices are not independent since signatures provide names for sorts, operators, etc., that appear in axioms and are interpreted by algebras.

Institutions. In most cases, the choices made will satisfy additional compatibility conditions and so will give rise to an institution (see Chapter 4). This has certain advantages for the design of features for specification in the large; see the next subsection. It is possible to postpone most decisions pertaining to specification in the small by deciding on an institution-independent approach, as with Clear and ASL; this avoids premature commitment, and leads to a family of compatible languages. Still, to actually make use of the language it is necessary to choose a particular institution, and then these decisions must be made. If the particular choices that seem most desirable turn out not to give rise to an institution, then either the use of an institution-independent approach or the features chosen for specification in the small will need to be reconsidered. More likely, as in EML and CASL, the "in the large" features depend to a minor extent on the details of the "in the small" features, and then one can employ an appropriately enriched version of institutions to specify the interface between the two language layers. See [ST86] for the case of EML and [Mos98b] for the case of CASL.

Signatures and type structure. The main point that needs to be decided for the choice of signatures is that of type structure. Should signatures be one-sorted (i.e., untyped) or many-sorted (i.e., typed) and, if many-sorted, should there be simple sorts, polymorphism, subsorts, higher-order types, or sort classes? The standard choice for specification languages is to have typed signatures. Defining types, which imposes conceptual structure on the problem domain, seems to be a natural part of writing specifications. Although untyped calculi such as the λ-calculus have enormous expressive power, they are difficult to use effectively because the power is in a raw form which makes errors easy to commit and difficult to detect. Type systems are used to tame this raw power by distinguishing well-formed expressions from ill-formed ones; the latter are regarded as meaningless and are therefore rejected. The expressive power of the type system determines how many useful expressions are regarded as well-formed, but there is a trade-off between tractability of typing and expressiveness of types. Simple type systems allow static type checking or even static type inference, and automatic disambiguation of expressions. In complex type systems, type checking amounts to a form of theorem proving which means that error detection cannot be fully automatic. LSL has simple sorts and ACT ONE has simple sorts with structured names (see the example in Section 8.2). In OBJ3, ordered sorts provide increased expressive power but as a result some typechecking must be deferred to run-time using the mechanism of "retracts"; the same holds for Spectral's system of dependent types, except that typechecking of logically-constrained subsorts requires theorem proving in general. EML inherits its higher-order polymorphic type system from SML, and the same algorithm can be used to infer types automatically. SPECTRUM has higher-order types and sort class polymorphism but no subsorts, and types can be inferred automatically as in Haskell. CASL has ordered sorts including logically-constrained subsorts as in Spectral. A planned future extension of CASL will include polymorphic types and higher-order functions [HKM98].

The difference between these alternatives can be seen by comparing how lists or bags over an ordered domain of values would be specified in each language. The specification of Sort in Section 8.2 indicates how this is done in EML; lists over an arbitrary type of values are handled via a (built-in) polymorphic type but the order on the values is dealt with via a module parameter. In SPECTRUM, sort classes would be used to handle both kinds of parameterisation – see the specification MinLIST in Section 8.2. Here, polymorphism is a degenerate case of the use of sort classes: List α is a shorthand notation for the declaration of a sort constructor List::(CPO)CPO where CPO is the default class of complete partial orders. Sort constructors can be overloaded, e.g., the additional declaration List::(EQ)EQ asserts that if the argument is of sort class EQ then the result sort is of sort class EQ as well. ACT ONE would use a parameterised type very much like LIST in Section 8.2, but with a parameter specification that requires an order on the parameter type. The use of structured sort names in ACT ONE resembles the use of polymorphism in EML, but the resemblance is merely on a syntactic level. The CASL version of this example given in Section 8.2 is similar except for the use of predicates in place of boolean-valued functions and the definition of ordered lists as a logically-constrained subtype of ordinary lists. The treatment in CafeOBJ would be similar to the CASL version except that the subtype of ordered lists would be axiomatised using a sort membership predicate.

LSL would handle this example by using an **assumes** clause which has to be discharged when the parameter is instantiated, as follows:

```
TO(E): trait
   introduces _≤_: E, E → Bool
   asserts
     ∀ x, y, z: E
       x ≤ x;
       (x ≤ y ∧ y ≤ z) ⇒ x ≤ z;
       (x ≤ y ∧ y ≤ x) == (x = y);
       (x ≤ y ∨ y ≤ x)

BAG_TO (E, Bag): trait
   assumes TO (E)
   includes BAGcount (E, Bag)
```

Let NAT1 be a trait that extends natural numbers by the usual order relation. Then in

```
NatBAG_TO: trait
  includes
    NAT1, BAG_TO (Nat)
```

the assumption TO (Nat) has to be discharged by proving that the ordering on natural numbers is a total order.

Finally, in ASL the way that this example is expressed would depend on the institution in use: in an institution with signatures containing only simple sorts, parameterisation would be used as in ACT ONE; in an institution with ordered sorts, the treatment would be like that in OBJ3; and so on.

Expressiveness of logic. The simplest logic used in algebraic specification is equational logic. This has a simple proof system that is complete and can be easily mechanised, but it only permits simple properties to be conveniently expressed. Therefore additional structuring operators, such as hiding, and additional constraints are necessary to adequately axiomatise some data structures. LSL uses equational logic while OBJ3 and ACT ONE use conditional equational logic. On the other hand a complex logic like higher-order logic is extremely expressive, allowing for example the expression of hiding via higher-order quantification and of generating constraints as ordinary formulas, but no complete proof system can exist and proof search is much less tractable than in simpler logics. EML and SPECTRUM use higher-order logic with equality. A reasonable compromise is to use first-order logic with equality, as in CASL, see Chapter 3. See [Wir90] for results on expressiveness of different first-order specification frameworks. Although there is a trade-off between expressiveness and tractability which is analogous to that in the case of type structure, the choice here also depends on the phase of the development process that is addressed. Ease of expression is crucial for requirements specifications, while equations or conditional equations are appropriate for programs (see, e.g., [GHM88]). Although equational logic is in principle very expressive (see, e.g., [Wir90]), the simplest and most natural way of writing a specification often requires the use of a more complex logic. For example, the function *prime*: *nat* → *bool* for testing primality is easily specified in first-order logic with equality using existential quantification:

$$\forall p : nat. \, (\exists n : nat. \, p > n > 1 \wedge divides\,(n,p)) \Leftrightarrow prime\,(p) = false$$

Specifying this function in equational logic or conditional equational logic is possible (with hidden functions), but the specification is much longer and much less straightforward.

Computational phenomena. Another dimension of complexity concerns the way that the logic deals with computational phenomena such as partial functions, non-termination and exceptions. To a large extent this depends on the context of application, see Section 8.3.1. For example, EML was designed for specifying SML functions, so it needs to cope with exceptions and non-termination. Similarly, one main application area of SPECTRUM is specification of operations on streams (employing non-terminating functions), used to model distributed systems [BDD$^+$92]. Some languages, including ACT ONE, do not attempt to deal with such phenomena and thereby exclude many important examples while avoiding extra complexity. A related choice is that taken by LSL, where all functions are total but the semantics is loose; thus *pop(empty)* would be a stack, but one about which nothing is known. Another option is to attempt to avoid partial functions by viewing them as total functions on a specified subsort whenever possible; this is the approach taken in OBJ3, Spectral, and CASL. For specification languages that attempt to deal with these phenomena directly, there are many choices. This is a very delicate matter since these choices interact with each other and also with the choice of the notion of algebra. For example, one may choose to deal with non-terminating functions but require all functions to be strict, or one may choose to admit non-strict or even non-monotonic functions; the primitives provided by the logic for making assertions about functions will vary accordingly. One needs a way of expressing definedness in any case, but in the case of non-strict functions one also needs a way of quantifying over both defined and undefined values. The choice of the meaning of equality is a subtle issue, with the choices including *strong equality* $t \overset{s}{=} t'$ (the values of t and t' are defined and equal or else are both undefined) and *existential equality* $t \overset{e}{=} t'$ (the values of t and t' are defined and equal). See Chapter 3 for the details of many of these options. CASL has a definedness predicate and both strong and existential equalities. If equality is available on function types, there is a further choice between extensional and intensional equality. EML and SPECTRUM both have a definedness predicate and strong extensional equality. EML also has a predicate for distinguishing exceptional values. Another difference between these two languages is quantification: in EML this ranges over SML-expressible values only, while quantification over functions in SPECTRUM can range either over all functions or alternatively over only the continuous functions.

Attributes. It is sometimes convenient to record certain standardized properties of an operation in a specification as attributes attached to its name in the signature rather than by writing an equivalent formula as an axiom. Examples are associativity and commutativity attributes in OBJ3 and CASL and strictness and totality attributes in SPECTRUM. Although both ways of expressing such a property are semantically equivalent, expressing it in the form of an attribute can enable "syntactic" use to be made of this "logical" information. For example, proofs about associative and commutative operations can use AC-unification in place of proof steps appealing to the corresponding axioms. This simplifies proof search, especially since the unrestricted use of permutative properties like commutativity leads to infinite search paths. Rules of inference for reasoning about strict or total operations are simpler than in the unrestricted case; this again leads to shorter proofs. Such attributes can also be used to restrict the space of possibilities when searching in a library of specifications.

Another kind of attribute that may be recorded with a specification is its intended and/or verified consequences. Examples of such attributes are the property of sufficient completeness in LSL (called **converts**) and RAP as well as the set of derived or required theorems in LSL (using the **implies** clause).

Algebras. As mentioned above, the choice of algebras is closely related to the choice of logic. The usual choice is to take the simplest model that correctly captures the aspects of computation that are of interest. For example, since ACT ONE does not attempt to deal with partial functions, etc., ordinary many-sorted algebras suffice for its purposes. If initial semantics is chosen as in ACT ONE, there is another potential point of interaction, since then homomorphisms have to be defined and initial models must exist in order for specifications to be meaningful. For specifying concurrent or reactive systems, one requires more complex models: transition systems, event structures, Petri nets, etc. But then a more general structural view seems to be necessary: see, e.g., [Rut96].

Semantic approach. The last item is the semantic approach (see Chapter 5). The choice of loose or initial semantics depends mainly on the phase of the software development process that specifications are intended to address; see Section 8.3.1. Requirements specifications are naturally loose, while programs are naturally initial. Design specifications might be either loose if they describe only part of the structure of a program, or initial if they give a complete input/output definition. Another point is that initial semantics requires the use of axioms no more complex than conditional equations. Orthogonal to the choice between initial and loose semantics is the choice between a semantics in terms of model classes or a semantics in terms of theories or presentations. Model-class semantics is the more flexible choice but theories and presentations relate more directly to proof. ASL is a requirements specification language (although the extension of ASL described in [SST92] can be used for design specifications as well) and it therefore has a loose semantics, which is expressed in terms of model classes. ACT ONE and ACT TWO are design specification languages; ACT ONE uses initial semantics (thus the restriction to conditional equational axioms) while ACT TWO uses loose semantics for interface specifications and initial semantics for module body specifications. Both are given in terms of both presentations and models. LSL defines concepts for use in Larch interface language annotations in finished programs or program modules; both languages have loose semantics, expressed in terms of theories. Ordinary structured specifications in CASL are for specifying requirements while architectural specifications are for specifying designs, and the semantics of both are loose and are expressed in terms of model classes.

OBJ3, EML, and SPECTRUM all allow the definition of executable programs as well as the specification of their interfaces. In each case the semantics of interface specifications is loose, while for programs the computation mechanism can be seen as choosing a particular algebra in the class of models described by the program.

8.3.4 Specification in the large

It is generally agreed that specification languages need to supply mechanisms for building large specifications in a modular way from small units.

In the small = in the large? The first question to ask is whether these mechanisms need to be separated from the mechanisms for specification in the small. All algebraic specification languages impose such a separation, and this is natural from the point of view of modular programming where the mechanisms provided for building and composing modules are different from the mechanisms provided for coding algorithms and data structures. However, if the type system used in signatures is sophisticated enough, these levels can be merged. An example is ECC [Luo90], where, e.g., dependent sums are used for tuples of all kinds: tuples of values, tuples of module components, pairs of a signature and a list of axioms, etc. Here there is not even a distinction between signatures and axioms because propositions are regarded as types having proofs as values. In spite of the elegance of a compact framework, it can be argued that the separation between levels provides conceptual clarity and contributes to ease of understanding.

Institution independence. If an institution-independent approach is taken, the features for specification in the large must be defined completely independently of the features for specification in the small, since these will vary from one institution to another. This is related to the principle of orthogonality in programming language design, and seems to be a useful way of decomposing the design of a specification language. Variants of most standard specification-building operations, such as those in Clear and ASL, are available for use in an arbitrary institution [ST88a], so this approach does not impose strong restrictions on the structuring features of the language. These operations are relatively well-understood, and come equipped with proof rules for reasoning about specifications built using them. Although OBJ3 and EML are not institution-independent, much of their underlying foundation has been developed in this framework via Clear and ASL respectively. For example, the theory that underlies EML's methodology for formal development is entirely institution-independent. CASL is not institution-independent but its features for specification in the large are defined for an arbitrary institution enriched with elements for handling compound identifiers and presentations of signature morphisms [Mos98b].

Choice of specification-building operations. Some of the tradeoffs involved in the choice of particular specification-building operations are similar to those in the choice of logic. Decreased expressive power generally leads to increased tractability. For example, proving properties of specifications written in a language with only simple specification-building operations such as union and/or enrich, and rename (e.g., OBJ3, LSL or ACT ONE) is easier than when operations such as observational abstraction are used. In the former case it is possible to translate specifications into "flat" presentations, reducing the problem to ordinary theorem proving; in the latter case separate mechanisms are required for performing proofs in structured specifications. In other words, the semantics of a language with simple specification-building operations can be given in terms of presentations, while the semantics of a language like ASL requires the use of model classes. On the other hand, the structure of a specification conveys information about the conceptual structure of the problem domain, and it seems useful to take advantage of this structure in understanding and analyzing specifications rather than destroying it via flattening. However, a language with simple operations typically requires the use of a complex notion of implementation (Chapter 7) which complicates the problem of proving correctness of implementation steps. The use of observational abstraction makes it possible to identify the class of models of a specification with its class of realizations and to use model class inclusion for refinement. In fact, this does not truly simplify the task of proving correctness of implementations; it just shifts the problem to that of proving properties of specifications. Another possibility is to shift the complexity to the logic used for writing axioms [Wir93], which makes ordinary theorem proving more difficult.

Enrichment. All specification languages contain some form of enrichment, extension or importation construct for building a new specification SP' by adding sorts, operations, and/or axioms to what is already contained in an existing specification SP. But this denotes subtly different things in different languages. In the initial algebra approach embodied in ACT ONE (compare data enrichment in Clear and including/using in OBJ3) it denotes the free construction. In the loose approach used in ASL, LSL, SPECTRUM, EML, and CASL (compare "ordinary" enrichment in Clear), it denotes an extension where taking the reduct of a model of SP' to the signature of SP gives a model of SP. Loose and free extension do not protect the meaning of the original specification SP. The free construction may add new values to the existing sorts of SP (if it is not "sufficiently complete"), or cause existing values in such sorts to become equated (if it is not "hierarchy consistent"). In the loose approach neither of these can arise, but it may happen that some models of SP cannot be extended to models of SP'. In an attempt to avoid such problems, OBJ3 introduces two special ways to import modules: extending is supposed to guarantee that new values are not added to existing sorts, and protecting is also supposed to ensure that existing values are not equated, i.e., that the extension is "persistent". However, the implementation of OBJ3 does not check that these properties actually hold. Similarly, LSL traits may contain assertions that given operations are defined in a sufficiently complete way, and these can be checked using LP. In general, proving persistency is difficult and needs model-theoretic methods unless the new axioms in SP' take the form of explicit definitions. The consequent difficulty in providing tool support for persistency checking might constitute a reason for excluding such assertions from a specification language.

Sharing. When a specification SP is built by combining specifications $SP1$ and $SP2$, there is a potential problem if both $SP1$ and $SP2$ contain a sub-specification SP'. It is not desirable to obtain two "copies" of the common sorts and operations; this is not only untidy, but also makes it difficult to express certain properties that involve operators from both $SP1$ and $SP2$. For example, if SP' is a specification of natural numbers, $SP1$ contains an operation $f1: s1 \rightarrow nat$, $SP2$ contains an operation $f2: nat \rightarrow s2$, and SP contains two copies of nat, then an axiom containing a subexpression of the form $f2(f1(\ldots))$ will not typecheck. There are three basic approaches to avoiding this situation. The first is to take the *disjoint union* of $SP1$ and $SP2$, but with some way of explicitly equating sorts and/or operations that should be shared. This is the approach taken in EML module interface spec-ifications, following SML, using so-called "sharing constraints". The second approach is to take the *set-theoretic union*, identifying a symbol in $SP1$ with any symbol in $SP2$ having the same name (and same type for operations, if overloading is permitted). This may lead to unintended sharing, so mech-anisms may be provided for detecting when the sharing is accidental. For example, "origin consistency" [Jon89] would reject an attempt to combine $SP1$ with $SP2$ when this would identify two symbols that were originally declared in different specifications. For requirements specification this is too restrictive since one may want, e.g., to build a specification for an equivalence relation from specifications of reflexivity, symmetry, and transitivity where the same relation symbol occurs in all three subspecifications. Set-theoretic union is used in ASL, SPECTRUM, LSL, and CASL. Finally, one can take an *amalgamated sum* of $SP1$ and $SP2$ with respect to a list of shared subspeci-fications. This is the approach taken in Clear, OBJ3, and ACT ONE, where the list of shared subspecifications is determined implicitly (compare module body specifications in EML); an alternative is to supply this list explicitly.

Hiding. A specification language requires some means of hiding auxiliary parts of signatures in order to provide adequate expressive power, at least when a logic weaker than second-order logic is used for writing axioms. Hiding can be provided as a specification-building operation, as in the case of **derive** in Clear and ASL, **local** in EML and CASL, **export/hide** in SPECTRUM, and **reveal/hide** in CASL. Alternatively (or additionally), hiding can be achieved via an interface attached to a specification, as with hiding via a signature in

EML or "unit reductions" in CASL architectural specifications. ACT ONE has no hiding mechanism but this facility is provided by the "export part" of an ACT TWO module. Similarly, LSL contains no hiding mechanism because an LSL trait is not intended to be implemented itself but only to provide concepts for use in Larch interface language specifications. Hiding complicates modular proofs, and completeness of a modular proof system for a language with hiding requires an interpolation property for the logic in use [BHK90, Cen94]. The use of an interface for hiding takes explicit account of the fact that reasoning about a specification requires access to hidden components.

Constraints. Constraints bridge the realms of specification in the small and specification in the large. Constraints are sometimes viewed as (complicated) axioms, as in SPECTRUM and LSL, and sometimes as specification-building operations, as in ASL. They are sometimes viewed in both ways, i.e., as specification-building operations that correspond semantically to axioms, as in Clear and CASL, where the axioms obtained may not be expressible in the logic used for writing ordinary axioms. In EML, constraints are part of the type system (via datatype declarations). In Pluss and OBJ3, the type of a specification determines how a constraint is to be interpreted; for example, in a Pluss sketch a constrained sort is required to contain *at least* the values generated by the operations, while in a spec it is required to contain *exactly* these values.

Multiple institutions. It is often convenient to use specialized logical tools to reason about particular aspects of a system. For example, automata theory is useful for reasoning about control and communication subsystems, and numerical analysis is useful for reasoning about scientific calculations; if both things co-exist within a single system then there may be a need for integrating properties of one with properties of the other. Related to this is the problem of describing a multi-paradigm system built from heterogeneous components, such as mixed hardware/software systems. It may also be appropriate to use different institutions at different stages of development of a system to express properties of a system viewed at different levels of abstraction. The simultaneous use of multiple institutions requires some way of relating the institutions to each other, e.g., via the concept of "institution semi-morphism" that maps the models of one institution (the more "concrete" one) to those of another (the more "abstract" one) [ST], see also [MM95]. None of the languages mentioned earlier have mechanisms for the use of multiple institutions, although [ST88b] describes an extension to ASL that permits specifications in one institution to be translated to another institution via an institution semi-morphism; cf. [Tar98] and [ST].

8.3.5   Parameterisation

The intention of parameterisation in a specification language is to abstract away from a part of the specification in order to be able to instantiate it with different data types or specifications. This allows specifications to be defined in a generic fashion so that they may be applied in a variety of contexts that share some common characteristics. See Chapter 6.

Parameterisation is normally regarded as an important specification-structuring mechanism and therefore it really belongs in the previous section. It is treated separately here only in order to keep it from dominating the other topics in that section.

Implicit versus explicit parameterisation. Syntactically, one can distinguish between explicit parameterisation where the parameter part is fixed by means of some particular syntactic notation, and implicit parameterisation where only an instantiation mechanism is provided without any particular notation for abstraction.

Implicit parameterisation is often used in languages for requirements specification in order to re-use previously defined specifications. Implicit parameterisation can be seen as just another use of the renaming mechanism (cf. Section 8.3.4) in languages without explicit parameterisation, such as early versions of LSL.

In explicit parameterisation there is a syntactic construct for abstraction consisting of a formal parameter and a body. The formal parameter is usually a specification that is constrained by a signature and a set of axioms. In (the new version of) LSL it is a list of sort names and operation names. The syntax is similar in all specification languages. Compare:

```
proc LIST(X : ELEM) =
                enrich X by data sorts List ... enden    in Clear
LIST = param X = ELEM; body {enriches X; ... }           in SPECTRUM
obj LIST[X :: ELEM] is sorts List ... endo               in OBJ3
type LIST[D] is sorts List(D) ... endtype                in ACT ONE
LIST = λX:ELEM. enrich X by ...                          in ASL
spec LIST[ELEM] = free type List[Elem] ::= ... end       in CASL
```

where ELEM is a specification containing just the single sort Elem. Explicit parameterisation mechanisms are used for the design of generic components (such as in ACT TWO, COLD, and CASL); often it is influenced by the parameterisation concepts of the underlying programming language (as, e.g., in EML).

Pushout-style versus λ-calculus style parameterisation. In pushout-style parameterisation, which is used in Clear, OBJ3, ACT ONE, and CASL, a parameterised specification consists of a "requirement" specification R together with a "body" specification B that extends R. Instantiation requires an actual parameter specification A that "fits" R, witnessed by a specification morphism from R to A, and the result is obtained by a pushout construction.

On the other hand, in view of the fact that a parameterised specification is a (specification-valued) function, a version of the typed λ-calculus may be used, with λ-abstraction for defining parameterised specifications and instantiation corresponding to β-reduction. This is the approach that is used in ASL, COLD, Spectral, and SPECTRUM. Then many features of the classical λ-calculus carry over to parameterised specifications; in particular, higher-order parameterisation is possible (see [Wir86, SST92, Cen94]) and (by using uncurrying) specifications with several parameters can be defined. The difference with respect to the classical λ-calculus concerns mainly the type system used. The class of admissible parameters may be restricted to specifications that fit a requirement specification (as in ASL) or may be dependent on former parameters (as in Spectral).

For a large subclass of the first-order case, both styles of parameterisation coincide. In particular, if the body of the parameterised specification is an extension of the requirement specification, the combination of renaming with β-reduction yields the same textual result as the pushout construction. But note that in general specification expressions may be more complicated than simple extension. In these cases the λ-calculus approach is more general than the pushout approach.

Parameterised specifications versus specification of parameterised program modules. Parameterisation can be used both at the level of program modules as in Ada and SML and at the level of specifications, and specifications can describe both ordinary (non-parameterised) program modules and parameterised program modules. Thus it is possible to give a parameterised specification of a non-parameterised program module, or a (non-parameterised) specification of a parameterised program module [SST92]. Most specification languages provide only one of these possibilities (Clear, OBJ3, EML, ASL, LSL, SPECTRUM) but some provide both. In those languages that provide both, the two concepts may be identified (ACT ONE, Pluss) or viewed as distinct (Spectral, ASL+, CASL). Higher-order parameterisation of either or both kinds may be available (ASL, ASL+, Spectral, SPECTRUM) or unavailable (Clear, OBJ3, EML, LSL, ACT ONE, Pluss, CASL).

Parameterised specifications provide a flexible mechanism for structuring specifications. These are most appropriate at the level of requirements specifications. Specifications of parameterised program modules define the module's input and output interfaces. These are most appropriate for use in writing design specifications [BST99]. Moreover, this concept ensures a number of compositionality and consistency properties. Languages that identify both concepts support the view (either implicitly or explicitly) that the structure of a requirements specification should match the structure of the eventual implementation. Forcing such a match seems to be inappropriate since structure at the two levels may arise for quite different reasons, see [FJ90].

Polymorphism versus parameterisation. An important design decision is whether both polymorphism and parameterisation should be supported by a specification language. Polymorphism (and even more, sort classes) models the concepts of generic functions and generic sorts "in the small", whereas parameterisation of specifications models the same concepts "in the large".

A typical example is polymorphic lists. Consider the specification POLY_LIST in SPECTRUM:

```
POLY_LIST = {
  sort List a;
  nil : → List a;
  cons : a × List a → List a;
  cons strict total;
  List a generated by nil, cons;
}


POLY_LIST_NB = POLY_LIST + NAT + BOOL
```

and the parameterised specification LIST in Clear:

```
proc LIST (X : ELEM) =
  enrich X by
    data sorts List
         opns  nil: List
               cons: Elem, List -> List
  enden

LIST_NAT = LIST(NAT[Elem is Nat])
LIST_BOOL = LIST(NAT[Elem is Bool])
```

The main difference is in the way that instantiations are made. Using polymorphism, several instantiations can be made within the same specification, whereas using parameterisation the same thing requires the construction of separate (and sometimes many separate) instantiations of the parameterised specification. Polymorphic languages like SPECTRUM and EML offer both concepts, where parameterisation is mainly included for interface descriptions in design specifications.

The following example shows a combination of polymorphism with parameterisation in SPECTRUM where POLY_LIST is used as a parameter:

```
MAP =
  param
    X = POLY_LIST;
  body {
    enriches X;
    map : (a → b) → List a → List b;
    axioms ... ;
    endaxioms;
  }
```

Subsequently, MAP can be instantiated, e.g., with polymorphic specifications of stacks and queues.

Higher-order functions versus parameterisation. Higher-order functions in specifications can sometimes be expressed using (first-order) parameterised specifications. This is used in languages such as OBJ3 that rely on first-order functions. The restriction to first-order functions is regarded by some as an advantage rather than a limitation [Gog90], but this is a controversial issue. For example, the higher-order function map above can be expressed in OBJ3 as follows (this is for the special case where a = b; the case where a ≠ b is similar but notationally more cumbersome, see [GM]):

```
obj UNARY is
  sort Elem .
  op f : Elem -> Elem .
endo

obj MAP [X :: UNARY] is
  protecting LIST[X] .
  op map: List -> List .
  var Y : Elem . var L : List .
  eq map(nil) = nil .
  eq map(cons(Y, L)) = cons(f(Y), map(L)) .
endo
```

The main difference is once more in the way that parameter passing is done. With higher-order functions, several higher-order function calls can be made within the same specification whereas the parameterised approach requires the construction of separate instantiations of the parameterised specification. With higher-order functions, the use of higher-order logic is natural (which makes reasoning more difficult) although it is not necessarily forced, while in the parameterised approach first-order logic suffices.

8.3.6   Syntax

Decisions involved in designing the syntax of a specification language are similar to those in programming language design.

Expression syntax. Some specification languages (e.g., EML) support infix operators while others (e.g., OBJ3, Pluss, and CASL) support a very flexible mixfix notation. The trade-off is between a simple syntax that is easy to parse but difficult to read, and a flexible syntax requiring a powerful and extensible parser. In the context of algebraic specifications the use of at least infix operators seems unavoidable, and greater flexibility seems worth the effort: most specifications are written in order to be read by humans, in contrast to most programs. A proviso is that unrestricted support for mixfix syntax requires the use of a parser for a general context-free language, introducing performance penalties that would probably be regarded as prohibitive in a compiler for a programming language. Haskell's "offside rule", where indentation is used in place of parentheses, would probably be a worthwhile additional feature of specification language syntax.

Name structure. A basic design decision is the name structure to be used in large specifications. Most specification languages have a flat name structure (e.g., OBJ3, SPECTRUM, ASL, LSL, CASL) whereas EML supports SML's "dot" notation, where reference to a name n in a specification SP requires the qualified name $SP.n$. When specifications can be nested as in EML, this leads to names such as $A.B.C.n$. A flat name structure, in which every name is global, is impractical for large specifications: it requires the user to keep track of all names in order to avoid name clashes. A possible solution is via the use of origin consistency to detect accidental name clashes, see Section 8.3.4; this guarantees that names can be disambiguated even when the same name is used for different things in different specifications. A related concept is that of overloaded operation names, where types are used instead of origins for disambiguation.

8.3.7 Software development process

The integration of a specification language in a software process model and the choice of the intended programming paradigm are important decisions for the use of the language in software development.

Software process model. In classical software process models such as the waterfall model or its refinements, as well as in software developments methods based on structured analysis or object-orientation, formal specification does not play an important rôle (see, e.g., [Som92]). This situation could change, however, if formal methods can be integrated with pragmatic techniques and if practically usable tool support for formal specification becomes widely available. Formal specifications are useful as a meta-tool for the formal foundation of pragmatic development models. For example, in [Huß94b, Huß94a] the pragmatic notations (including diagrams) of SSADM have been axiomatised in SPECTRUM which has led to several proposals for improvements of SSADM.

There are several software development methods based on formal approaches including the transformational approach of the CIP [BBD+81] and PROSPECTRA [HK93] projects (see Chapter 14), the B-method [ALN+91], and the methods used in the projects RAISE [NHWG89], KORSO [PW95], and PROCOS [BLH92]. One of the principal design decisions is whether there should be just one language supporting the whole software development process. In this case the language has to incorporate features for describing requirements, designs, and programs. This leads to wide-spectrum languages such as CIP-L, COLD, RSL, and EML with complex semantics and proof rules. An alternative is to decouple the specification language from the programming language, as in LSL (via the use of a separate interface language) and CASL. On the other hand, if several languages and notations are used, as in the case of most pragmatic approaches, then one has to fix the semantic relationships between these notations, derive correctness criteria and provide translations between the different languages. For example, one might want to use SPECTRUM for the requirements specification of a system and then ACT ONE for the design specification and a functional language for coding [WDC+94]. An important issue for the practical acceptance of formal techniques in the design and requirements phase is the graphical representation of formal descriptions (see, e.g., [CEW93]) and the integration of formal texts with usual semi-formal descriptions and diagrammatic notations (see [Wol94]).

For the support of the development and maintenance of specifications, some languages include syntactic features for describing the evolution of a specification and the modification of existing specifications for new purposes. For example, Pluss offers different syntactic constructs for specifications that are still under development and for complete specifications (draft/sketch versus spec). Renaming, hiding, and parameterisation provide means for controlled modification. Moreover some languages support refinement as an expressible relation between specifications (e.g., COLD) whereas others consider it as an external meta-relation (e.g., Clear, ASL).

Programming paradigm. "Pure" algebraic specification languages support the algebraic programming style. Roughly, an algebraic program can be characterized as an executable specification that has first-order positive conditional equations as axioms. In this sense all languages that admit only such axioms (e.g., OBJ3, RAP, ACT, ASF [BHK89], and appropriate executable subsets of other languages) can be considered as algebraic programming languages.

In order to support classical programming paradigms too, one has to extend a pure specification language with appropriate language constructs or to design a specification language that directly integrates the algebraic specification style with programming concepts. This is true even in the case of functional programs, which are closely related to the algebraic specification style. For example, higher-order function types are introduced in EML, Spectral, and SPECTRUM. Object-oriented specification languages like Maude and OS [Bre91] employ subsorting and structuring facilities for (class) inheritance. COLD introduces operators of dynamic logic for writing pre- and postconditions of imperative programs, whereas SMoLCS [AR87] and Maude use labeled transition systems (SMoLCS) and term rewriting (Maude) for specifying sequential and concurrent state transitions. Moreover, languages that support imperative, object-oriented, or concurrent programming include linguistic features for sequential or concurrent control flow.

Some specification languages such as EML and Larch relate directly with a specific programming language rather than merely with a programming paradigm. In this case the semantics of both languages have to be integrated. This may pose difficult semantical problems [KS98]. For EML, a verification semantics had to be developed that extends the structural operational semantics of SML by the semantic foundations of ASL-like specifications.

Support for refinement. When constructing a software system it is not sufficient to construct just one specification: one has to develop different specifications ranging from abstract requirement specifications and design specifications to executable specifications or programs. A basic design decision is whether the notion of refinement should be an expressible relation or an external meta-relation. If refinement is expressible, then the difference between specification-building operations and the refinement relation is a matter of one's point of view.

In pure specification languages, such as ASL, Clear, or ACT ONE, a specification is intended to describe only one level of abstraction of a software system at a time. Notions of refinement and implementation have been studied for these languages (see, e.g., [SW82, SW83, Far92, Cen94]), but they are used on the meta-level in the specification development methodology. For example, the development graph of SPECTRUM (see [PW95]) relies on the notion of refinement as model class inclusion, where a specification SP' is a refinement of a specification SP if any model of SP' is also a model of SP.

Some languages, including COLD, ACT TWO, OBJ3, EML, ASL+, Spectral, and CASL, offer features for representing two different levels of abstraction. COLD and ACT TWO introduce notions of "component" (in COLD) and "module" (in ACT TWO) that relate internal and external descriptions of a software system. In COLD one writes COMP $x:K$:= L for describing a so-called "design component" with name x, interface specification K, and specification L acting as implementation of K; the requirement is that L is a refinement of K. This is similar in EML and Spectral, except that parameterised components are supported (including higher-order parameterisation in the case of Spectral) and L is only required to refine K "up to behavioural equivalence". In ACT TWO a module consists of four parts: export, import, parameter, and body. As in COLD, the body is required to be a refinement of the export part. OBJ3 has the concept of "view" for defining fitting morphisms used for parameter passing. This can also be used for representing a refinement between an OBJ3 theory and an OBJ3 object.

8.4 Future trends

Specification languages that have been under development during the last few years show a number of tendencies in the development of algebraic specification languages. There is a tendency to aim at covering the whole software development process starting from requirements up through executable specifications. Thus they tend to include elements of both loose specifications (for requirements) and initial algebra specifications (in an executable sublanguage). All of them support partial functions in some form, and there is a tendency to favour simple specification-building operations. All of them contain concepts for explicit parameterisation and there is a growing tendency to provide support for specification of parameterised programs. There is a trend to

include features oriented towards particular programming paradigms, mainly
for functional and object-oriented programming.

Though currently studied in the literature, certain aspects of programming are not addressed by present algebraic specification languages:

1. The integration of the diagrammatic features of current software engineering notations such as UML [UML97].
2. The integration of application-dependent features such as description
   techniques for information systems [Tse91, Huß94b, Huß94a] or for real-
   time systems [ZHR92].
3. The combination of different logics in a single specification language for
   the description of heterogeneous systems [AC93, CM93, Tar98].
4. The specification of dynamic systems and the description of dynamic
   properties of systems such as safety and liveness; see Chapter 13.

In each case, more work is needed both on the foundations and on experiments
with applications.

Inventing languages is a seductive activity. The history of programming
languages is littered with languages that were never used by anybody other
than their inventors, and the shorter history of specification languages threat-
ens to show similar tendencies. In particular, the strong focus of attention
in the past has been on the foundations of specification languages and there
has therefore been comparatively little work on putting them into practice.
This is an issue that needs to be addressed in the future for algebraic specification in general. One effort in this direction is the Common Framework
Initiative (CoFI) which developed CASL as the basis of an attempt to create
a focal point for future joint work on algebraic specifications and a platform
for exploitation of past and present work on methodology, support tools, etc.
[Mos97]. For specification languages in particular, acceptance by software en-
gineers requires much more work on finding an appropriate compromise between purity and practicality, between expressiveness and simplicity of learning and use, and on the development of tools that provide truly useful support
to the users of such languages. Today formal specification is practicable for
medium-sized applications (see, e.g., [GM94]) but it not suitable for writing,
e.g., a first sketch of the structure of a system or for describing standardized
architectures as used in large application systems. One possible direction for
future development to address this point is graphical representation of static
and dynamic aspects of the system under development.

Acknowledgements. Thanks to Didier Bert, Răzvan Diaconescu, Chris
George, Bernd Krieg-Brückner, and Horst Reichel for helpful comments.

9 Term Rewriting

Hélène Kirchner

LORIA-CNRS
BP 239 F-54506 Vandœuvre-lès-Nancy Cedex, France
Helene.Kirchner@loria.fr
http://www.loria.fr/~hkirchne/

In the area of system specifications, rewriting techniques have been developed for two main applications: prototyping algebraic specifications of user-defined data types and theorem proving related to program verification. A formulation of term rewriting was proposed first by Evans [Eva51] and later by Knuth and Bendix [KB70]. Originally its main purpose was for generating *canonical term rewriting systems*, which can be used as decision procedures for proving the validity of equalities in some equational theories. Since then, in addition to the validity problem of equational logic, term rewriting has also been applied to inductive theorem proving, checking consistency and completeness of equational or conditional specifications, first-order theorem proving, unification theory, geometry theorem proving, etc. Through various implementations and experiments in automated theorem proving, it has been demonstrated that simplification of formulas by rewriting is indeed an effective way of pruning the search space. On the other hand, with the emergence of equationally specified abstract data types in the late 1970's, term rewriting has also gained considerable popularity as a bridge between programming language theory and program verification. Several specification languages or programming environments, such as LARCH, OBJ, ASF, RAP, MAUDE, ELAN, to cite a few, are using rewriting as their basic evaluation mechanism. Rewriting techniques are also used in functional programming languages such as ML.

This chapter introduces term rewriting and some of its applications from different points of view. Rewriting is first presented in Section 9.1 as an abstract relation on a set, and properties of such relations, mainly confluence and well-foundedness, are introduced. Confluence is the key concept to ensure the determinism of a rewriting relation seen as a computation process. It means that two computations starting from the same term will eventually give the same result. Well-foundedness ensures that any computation terminates. In Section 9.2, a more concrete notion of rewriting is defined on first-order terms. Several examples of rewriting relations on terms or equivalence classes of terms are then given, including rewriting modulo axioms and conditional rewriting. In this part, emphasis is put on the Church–Rosser properties of these different relations, and their application in deciding equality in equational or conditional theories. In Section 9.3, the rewriting logic is defined and is shown to be especially suited to describing concurrent computations.

It also provides a logical framework in which other logics can be represented, and a semantic framework for the specification of languages and systems. Experiments with applications, such as the specification of constraint solving systems or theorem provers, reveal the importance of a strategy to restrict the computation space or to select relevant computations.

The rest of the chapter is concerned with applications of rewriting to proofs of various program properties. Section 9.4 addresses the issue of automatically building, for an equational theory, an equivalent convergent rewrite system, when one exists. This is solved via a completion process that turns equalities into rewrite rules, computes new derived equalities and reduces them. When this process terminates, the result is a confluent and terminating set of rewrite rules. Several generalisations of the initial completion mechanism, based on superposition and simplification, are also surveyed in this section. In Section 9.5, the completion technique is generalised in the context of order-sorted equational specifications. There, rewriting is used not only to evaluate but also to compute the types of terms during evaluation. Eventually, in the context of many-sorted conditional specifications, in Section 9.6, rewrite techniques are used to prove inductive properties in the initial model and completeness of function definitions.

9.1 Rewriting as an abstract reduction system

Most of basic definitions and properties of rewrite systems can be stated abstractly, by considering rewriting as a binary relation on sets. This point of view, taken from [Klo92], has the great advantage of defining these properties once for all for different rewriting relations.

Definition 9.1. An *abstract reduction system* is a structure $\langle \mathcal{T}, \longrightarrow \rangle$ consisting of a set $\mathcal{T}$ and a binary relation $\longrightarrow$ on $\mathcal{T}$. ◇

For each binary relation $\longrightarrow$, $\xrightarrow{+}$ and $\xrightarrow{*}$ respectively denote its transitive and its reflexive transitive closure. On a set $\mathcal{T}$ whose elements are denoted by $t, t', \ldots$, we also define:

$$t \longleftarrow t' \text{ if } t' \longrightarrow t$$
$$t \longleftrightarrow t' \text{ if } t \longrightarrow t' \text{ or } t \longleftarrow t'$$
$$t \xleftarrow{*} t' \text{ if } t' \xrightarrow{*} t$$
$$t \xleftarrow{+} t' \text{ if } t' \xrightarrow{+} t.$$

For any natural number n, the composition of n steps of $\longrightarrow$ or $\longleftrightarrow$ is denoted respectively by $\xrightarrow{n}$ or $\xleftrightarrow{n}$ By convention, $\xrightarrow{0}$ and $\xleftrightarrow{0}$ are simply syntactic equality. As usual, $\xleftrightarrow{+}$ and $\xleftrightarrow{*}$ denote respectively the transitive and the reflexive transitive closure of $\longleftrightarrow$.

Composition of binary relations $\longrightarrow_1$ followed by $\longrightarrow_2$ will be denoted by $\longrightarrow_1 \circ \longrightarrow_2$.

Termination and confluence are two properties often required for a reduction system. They are defined here together with their relationship with other properties, especially local confluence, the Church–Rosser property and the existence of normal forms.

Definition 9.2. Let $\langle \mathcal{T}, \longrightarrow \rangle$ be an abstract reduction system.

- An element $t \in \mathcal{T}$ is a $\longrightarrow$-*normal form* if there exists no $t' \in \mathcal{T}$ such that $t \longrightarrow t'$. Furthermore t' is called a normal form of $t \in \mathcal{T}$ if $t \stackrel{*}{\longrightarrow} t'$ and t' is a $\longrightarrow$-normal form. This will be denoted later by $t \stackrel{!}{\longrightarrow} t'$.
- The relation $\longrightarrow$ is *terminating* (or *strongly normalising*, or *well-founded*) if there is no infinite sequence of related elements $t_1 \longrightarrow t_2 \longrightarrow \ldots \longrightarrow t_i \longrightarrow \ldots$.
- The relation $\longrightarrow$ is *weakly terminating* (or weakly normalising) if every element $t \in \mathcal{T}$ has a normal form.
- The relation $\longrightarrow$ has the *unique normal form property* if, for any $t, t' \in \mathcal{T}$, $t \stackrel{*}{\longleftrightarrow} t'$ and t, t' are normal forms imply $t = t'$. The unique normal form of t, when it exists, is denoted $t \downarrow$. $\diamond$

Example 9.3. Let a, b, c, d be distinct elements of $\mathcal{T}$.
The relation $a \longrightarrow b$ is terminating, but $a \longrightarrow a$ is not.
The relation defined by $a \longrightarrow b, b \longrightarrow a, a \longrightarrow c, b \longrightarrow d$ is weakly terminating but not terminating. It does not have the unique normal form property. ∎

The termination property of an abstract relation associated with an abstract reduction system $\langle \mathcal{T}, \longrightarrow \rangle$ is studied in the general framework of well-founded orderings on the set $\mathcal{T}$.

Proposition 9.4. [MN70] *Let $\langle \mathcal{T}, \longrightarrow \rangle$ be an abstract reduction system. The relation $\longrightarrow$ is terminating on $\mathcal{T}$ if and only if there exists a well-founded ordering $>$ over $\mathcal{T}$ such that $s \longrightarrow t$ implies $s > t$.*

For abstract reduction systems, the Church–Rosser property is expressed as follows:

Definition 9.5. The relation $\longrightarrow$ is *Church–Rosser* on a set $\mathcal{T}$ if $\langle \mathcal{T}, \longrightarrow \rangle$ satisfies

$$\stackrel{*}{\longleftrightarrow} \subseteq \stackrel{*}{\longrightarrow} \circ \stackrel{*}{\longleftarrow}.$$ $\diamond$

A relation is confluent if it satisfies the diamond property defined as follows:

Definition 9.6. The relation $\longrightarrow$ is *confluent* on $\mathcal{T}$ if $\langle \mathcal{T}, \longrightarrow \rangle$ satisfies

$$\stackrel{*}{\longleftarrow} \circ \stackrel{*}{\longrightarrow} \subseteq \stackrel{*}{\longrightarrow} \circ \stackrel{*}{\longleftarrow}.$$ $\diamond$

Since $\xleftarrow{*} \circ \xrightarrow{*} \subseteq \xleftrightarrow{*}$, the Church–Rosser property obviously implies the confluence. The converse is shown by induction on the number of $\longleftrightarrow$-steps that appear in $\xleftrightarrow{*}$.

Theorem 9.7. *The relation* $\longrightarrow$ *is confluent if and only if it is Church–Rosser.*

In the following, $t \downarrow t'$ means that both t and t' have a common descendant, i.e., an element t'' such that $t \xrightarrow{*} t'' \xleftarrow{*} t'$. In this case, the pair (t, t') is said to be *convergent*.

If an element t has two normal forms t_1 and t_2, then by confluence $t_1 \downarrow t_2$ and thus $t_1 = t_2$, since both are normal forms. So we obtain the next result:

Proposition 9.8. *If* $\longrightarrow$ *is confluent, the normal form of any element is unique, provided it exists.*

The confluence property has a local version that only considers two different applications of the relation.

Definition 9.9. The relation $\longrightarrow$ is *locally confluent* on $\mathcal{T}$ if $\langle \mathcal{T}, \longrightarrow \rangle$ satisfies:

$$\longleftarrow \circ \longrightarrow \; \subseteq \; \xrightarrow{*} \circ \xleftarrow{*}.$$

◇

Confluence clearly implies local confluence but the converse is not true, as shown by the next example:

Example 9.10. Consider four distinct elements a, b, c, d of $\mathcal{T}$ and the relation defined by $a \longrightarrow b, b \longrightarrow a, a \longrightarrow c, b \longrightarrow d$. The relation, though locally confluent (consider any possible case of ambiguity) and weakly terminating, is not confluent since $c \xleftarrow{*} a \xrightarrow{*} d$, but neither c nor d can be rewritten.

∎

In this example, of course, the relation is not terminating since there exists a cycle $a \longrightarrow b \longrightarrow a$. Provided the relation terminates, confluence and local confluence are equivalent.

Lemma 9.11. [New42] *If* $\longrightarrow$ *is terminating, then* $\longrightarrow$ *is confluent if and only if* $\longrightarrow$ *is locally confluent.*

Then the Church–Rosser property is equivalent to the local confluence property, under the termination assumption. The results are summarised in the following theorem.

Theorem 9.12. *If the relation* $\longrightarrow$ *is terminating, the following properties are equivalent:*

1. $\longrightarrow$ *is Church–Rosser.*
2. $\longrightarrow$ *is confluent.*

3. $\longrightarrow$ is locally confluent.

4. for any $t, t' \in \mathcal{T}, t \overset{*}{\longleftrightarrow} t'$ if and only if $t \downarrow = t' \downarrow$.

(1) and (2) are equivalent by Theorem 9.7, (2) and (3) are equivalent by Lemma 9.11. (4) clearly implies (1). Finally (1) implies (4) by applying the Church–Rosser property to $t \downarrow$ and $t' \downarrow$.

Definition 9.13. A relation $\longrightarrow$ is *convergent* if it is confluent and terminating. ⋄

Other notions of confluence appear in the literature. Let $\overset{0,1}{\longrightarrow}$ denote at most one application of a rewriting step, and $\overset{0,1}{\longleftarrow}$ the symmetric relation.

Definition 9.14. Let $\langle \mathcal{T}, \longrightarrow \rangle$ an abstract reduction system. The relation $\longrightarrow$ is *strongly confluent* on $\mathcal{T}$ if:

$$\longleftarrow \circ \longrightarrow \; \subseteq \; \overset{0,1}{\longrightarrow} \circ \overset{0,1}{\longleftarrow} .$$ ⋄

Strong confluence implies confluence [New42]. Confluence of $\longrightarrow$ coincides with strong confluence of $\overset{*}{\longrightarrow}$. This property is used in classical proofs of the Church–Rosser property for the lambda-calculus [Bar84]. In general, this may be of some help in establishing the confluence of non-terminating systems. Concerning systems that are only weakly normalising, a method for establishing confluence has been proposed in [CG91, Har89]. The main idea of the following result is to establish a relation between two abstract reduction systems in such a way that one will inherit the confluence property of the other.

Theorem 9.15. [CG91] *Let $\langle \mathcal{T}, \longrightarrow \rangle$ be a weakly terminating abstract reduction system and $\langle \mathcal{T}', \longrightarrow' \rangle$ be a confluent abstract reduction system such that there exists a mapping π from $\mathcal{T}$ to $\mathcal{T}'$ satisfying:*

1. *for all elements s and t in $\mathcal{T}$, if $s \longrightarrow t$ then $\pi(s) \overset{*}{\longleftrightarrow}' \pi(t)$,*
2. *the image with π of a normal form in $\langle \mathcal{T}, \longrightarrow \rangle$ is a normal form in $\langle \mathcal{T}', \longrightarrow' \rangle$, and*
3. *π is injective on the normal forms in $\langle \mathcal{T}, \longrightarrow \rangle$.*

Under these conditions, $\langle \mathcal{T}, \longrightarrow \rangle$ is confluent.

This result can be extended to several special cases, in particular when $\mathcal{T} = \mathcal{T}'$ and $\pi = Id$. One can also see that if $\langle \mathcal{T}', \longrightarrow' \rangle$ has the unique normal form property, under the previous hypothesis on $\langle \mathcal{T}, \longrightarrow \rangle$ and π, $\langle \mathcal{T}, \longrightarrow \rangle$ also has the unique normal form property. This has been applied, for instance, to prove the confluence of the $\lambda\beta$-calculus, as developed in [CG91].

9.2   Rewriting as a relation on terms

Abstract reduction systems are now specialised to terms and equivalence classes of terms. In this context, the Church–Rosser property is especially interesting since it provides a decision procedure for equality.

For the usual notions of signatures, terms and substitutions, the reader should refer to Chapter 2, or to [DJ90]. In this chapter, a signature Σ is composed of a set of sort symbols $\mathcal{S}$ and a set of function symbols $\mathcal{F}$. $\mathcal{S}$ may be omitted when reduced to one (universal) sort. $\mathcal{T}(\Sigma, \mathcal{X})$ denotes the set of terms built on the signature Σ and the set of variables $\mathcal{X}$. Variables are often denoted by letters x, y, z. $\mathcal{T}(\Sigma)$ is the set of ground terms, i.e., without variables. $\mathcal{V}ar(t)$ denotes the set of variables of a term t, $t_{|\omega}$ the subterm of t at position ω, and $t[u]_\omega$ the term t that contains the subterm u at position ω (ω may be omitted). Greek letters $\sigma, \theta, \ldots$ are used to denote substitutions that are completely defined by a mapping of variables to terms, written $\sigma = (x_1 \mapsto t_1) \ldots (x_n \mapsto t_n)$, when $\sigma(x_i) = t_i$, for $i = 1, \ldots, n$. The *domain* of the substitution σ is the finite set of variables $\mathcal{D}om(\sigma) = \{x \mid x \in \mathcal{X} \text{ and } \sigma(x) \neq x\}$, the *range* of σ is the set of terms $\mathcal{R}an(\sigma) = \cup_{x \in \mathcal{D}om(\sigma)}\sigma(x)$. A substitution σ is *ground* if $\mathcal{R}an(\sigma) \subseteq \mathcal{T}(\Sigma)$. Composition of σ and ρ is denoted by $\rho\sigma$.

9.2.1   Term rewriting

The notion of abstract reduction system can be instantiated by considering $\mathcal{T}$ as the set of terms and $\longrightarrow$ as the rewriting relation generated by a set of rewrite rules.

Rewrite systems.

Definition 9.16. A *rewrite rule* is an ordered pair of terms denoted $l \to r$. The terms l and r are respectively called the *left-hand side* and the *right-hand side* of the rule. In a rewrite rule, every variable occurring in the right-hand side of a rule also occurs in the left-hand side. All variables are implicitly universally quantified. A *rewrite system* or *term rewriting system* is a (finite or infinite) set of rewrite rules.   ◇

Example 9.17. Combinatory Logic, originally devised by Schönfinkel (1924) and then by Curry, is an example of a theory with a binary function symbol $\cdot$ (application), constant symbols S, K, I (combinators) and variables. The theory is defined by three axioms that can be applied as rewrite rules, using the following rewrite system CL:

$$((S \cdot x) \cdot y) \cdot z \longrightarrow (x \cdot z) \cdot (y \cdot z)$$
$$(K \cdot x) \cdot y \longrightarrow x$$
$$(I \cdot x) \longrightarrow x.$$

The symbol $\cdot$ is often omitted for improved readability, as are the parentheses associating terms from left to right. For instance xyz stands for $(x \cdot y) \cdot z$.   ∎

A rule is applied by replacing an instance of the left-hand side by the same instance of its right-hand side, but never the converse, contrary to equalities. Note that two rules are considered to be the same if they only differ by a renaming of their variables.

Definition 9.18. Given a rewrite system R, a term t *rewrites* to a term t', which is denoted by $t \longrightarrow_R t'$ if there exist a rule $l \to r$ of R, a position ω in t, a substitution σ, satisfying $t_{|\omega} = \sigma(l)$ and called a *match* from l to $t_{|\omega}$, such that $t' = t[\sigma(r)]_\omega$.   ◇

When t rewrites to t' with a rule $l \to r$ and a substitution σ, it will always be assumed that variables of l and t are disjoint. This is not a restriction, since variables of rules can be renamed without loss of generality. When any of the rewrite system, the rule, the substitution and/or the position require precise definition, a rewriting step is denoted by $t \longrightarrow_R^{\omega, \sigma, l \to r} t'$. The subterm $t_{|\omega}$, where the rewriting step is applied, is called the *redex*. A term that has no redex is said to be *R-irreducible* or in *R-normal form*. An R-irreducible form of t is denoted by $t{\downarrow}_R$. A *rewriting derivation* is any sequence of rewriting steps $t_1 \longrightarrow_R t_2 \longrightarrow_R \ldots$. A rewrite system R induces a reflexive transitive binary relation on terms, called the *rewriting relation* or the *derivability relation*, denoted by $\overset{*}{\longrightarrow}_R$.

Example 9.19. In Combinatory Logic (see Example 9.17), we can derive

$$SKKx \longrightarrow_{CL} Kx(Kx) \longrightarrow_{CL} x.$$

We also have

$$S(KS)Kxyz \longrightarrow_{CL} KSx(Kx)yz \longrightarrow_{CL} S(Kx)yz$$
$$\longrightarrow_{CL} Kxz(yz) \longrightarrow_{CL} x(yz).$$

Abbreviating $S(KS)K$ by B, this establishes that $Bxyz \overset{*}{\longrightarrow}_{CL} x(yz)$, and defines Bxy as the composition $x \circ y$, where x and y are variables representing functions.

Another interesting combinator is SII, often abbreviated as Ω, and called self-applicator. This name is justified by the following derivation:

$$\Omega x = SIIx \longrightarrow_{CL} Ix(Ix) \longrightarrow_{CL} Ixx \longrightarrow_{CL} xx.$$

But the term $\Omega\Omega$ admits a cyclic derivation, since: $\Omega\Omega \overset{+}{\longrightarrow}_{CL} \Omega\Omega$.

It can also be proved that any term F in Combinatory Logic has a fixed point P defined as $\Omega(BF\Omega)$. Indeed:

$$P = \Omega(BF\Omega) \overset{*}{\longrightarrow}_{CL} (BF\Omega)(BF\Omega) \overset{*}{\longrightarrow}_{CL} F(\Omega(BF\Omega)) = FP.$$ ∎

Termination. Working now on terms, instead of an abstract set $\mathcal{T}$, yields methods for proving termination of term rewriting. The rewrite relation $\longrightarrow_R$ is terminating if it is included in a suitable well-founded ordering. A quasi-ordering is a reflexive transitive, binary relation and is well-founded if its strict part is well-founded. A *reduction (quasi-)ordering* is a (quasi-)ordering $>$ closed under context and substitution, i.e., such that for any term t and any substitution σ, if $u > s$ then $t[u] > t[s]$ and $\sigma(u) > \sigma(s)$. If there exists a well-founded reduction ordering $>$ on terms such that $l > r$ for any rewrite rule $(l \rightarrow r) \in R$, then $\longrightarrow_R$ is terminating.

There are essentially two types of well-founded reduction orderings on terms, surveyed in [Der87]. The first one, called *syntactical*, provides the ordering via an analysis of the term structure. Among these orderings are the *recursive path ordering* [Der82], the *lexicographic path ordering* [KL82] and the *recursive decomposition ordering* [JLR82, Les90]. These orderings are convenient since they are based on a concept of precedence which is a partial order on the signature. In addition, they enjoy a property called *incrementality* that enables one to handle an incremental set of ordering problems. This is particularly useful when a new equality must be ordered and added to a set of rewrite rules without modifying their previous orientations. Another family consists of *semantical orderings*, that interpret the terms in another structure where a well-founded ordering is known. For such a purpose, two common ordered sets are the natural numbers and the terms ordered by a syntactical ordering. The first choice enables one to consider functions over natural numbers which are stable under substitutions. The most frequently used are polynomials [Lan79, CL87, Ste90, CL92, Gie95]. The set of terms ordered by a syntactical ordering may also be chosen as a target for the "semantics" [BD86, BL87]. Both methods can be combined as in [KB70] which uses a semantical ordering (polynomial interpretation of degree 1) and a syntactical ordering. A general path ordering, dealing with semantical as well as syntactical aspects, has been proposed in [DH95] and generalises several previous approaches.

Syntactical orderings contain the important family of simplification orderings. A *simplification ordering* is a reduction ordering such that a term is greater than any of its subterms. The last subterm condition suffices for ensuring well-foundedness of simplification orderings, provided that the signature is finite. Simplification orderings can be built from a well-founded ordering on the function symbols $\mathcal{F}$ called a *precedence*. A simple example of path ordering is the *multiset path ordering*. To understand the next definition, the reader should be aware that a multiset over a set $\mathcal{T}$ is a function M from $\mathcal{T}$ to natural numbers. Intuitively, for any element x in $\mathcal{T}$, $M(x)$ specifies the number of occurrences of x in M. A multiset is finite if $M(x) = 0$ for all but finitely many x.

Let $>$ be a partial ordering on $\mathcal{T}$. Its multiset extension $>^{mult}$ is the transitive closure of the following relation on finite multisets over $\mathcal{T}$:

$\{s_1, \ldots, s_{i-1}, s_i, s_{i+1}, \ldots, s_n\} >^{mult} \{s_1, \ldots, s_{i-1}, t_1, \ldots, t_k, s_{i+1}, \ldots, s_n\}$
if for some element s_i, there exist k elements, $k \geq 0$, such that $s_i > t_j$ for
any $j = 1, \ldots, k$. If $>$ is well-founded on $\mathcal{T}$, its multiset extension $>^{mult}$ is
well-founded on finite multisets over $\mathcal{T}$.

Definition 9.20. Let $>_{\mathcal{F}}$ be a precedence on $\mathcal{F}$. The *multiset path ordering*
$>_{mpo}$ is defined on ground terms by $s = f(s_1, .., s_n) >_{mpo} t = g(t_1, \ldots, t_m)$
if at least one of the following conditions holds:

1. $f = g$ and $\{s_1, \ldots, s_n\} >^{mult}_{mpo} \{t_1, \ldots, t_m\}$
2. $f >_{\mathcal{F}} g$ and $\forall j \in \{1, \ldots, m\}, s >_{mpo} t_j$
3. $\exists i \in \{1, \ldots, n\}$ such that either $s_i >_{mpo} t$ or $s_i \sim t$,
   where $\sim$ means equivalent up to permutation of subterms. ◇

Note that the precedence on $\mathcal{F}$ need not be total, and that the condition
$f = g$ could be replaced by $f \sim_{\mathcal{F}} g$ where $\sim_{\mathcal{F}}$ is the equivalence induced on
$\mathcal{F}$ by $>_{\mathcal{F}}$: $f \sim_{\mathcal{F}} g$ if $f >_{\mathcal{F}} g$ and $g >_{\mathcal{F}} f$.

Example 9.21. Let h, k, f be respectively $1, 2, 2$-ary function symbols and x, y
be variables. By the subterm relation, $h(f(x, y)) >_{mpo} f(x, y) >_{mpo} y$.
Assuming a precedence where $k >_{\mathcal{F}} f$, $k(f(x, y), z) >_{mpo} f(x, k(z, y)) >_{mpo}$
y, since both $k(f(x, y), z) >_{mpo} x$ and $k(f(x, y), z) >_{mpo} k(z, y)$. The latter
holds since $f(x, y) >_{mpo} y$. ∎

Another direction for proving termination of rewrite systems has been
opened up more recently by [AG96, AG97]. Looking at rewrite systems as
programs, the idea is that each rule whose left-hand side begins with f is
defining this function f. Let f be called a *defined symbol*. Such a program is
terminating if any recursive call is terminating. This motivates the idea that
only subterms of the right-hand sides that have a defined symbol at top, have
to be considered for the examination of termination behaviour. Clearly a term
without defined symbol is in R-normal form. If no defined symbol occurs in
any right-hand side of rules, any term can be reduced only finitely many times
by R. Thus infinite reductions originate from the fact that defined symbols
are introduced by the right-hand sides of rules. By tracing the introduction of
these defined symbols, information can be obtained about termination of R.
This is the idea behind the dependency pairs criterion. In the next definition,
to avoid handling tuples, an auxiliary symbol F is introduced for any defined
symbol f.

Definition 9.22. If $f(t_1, \ldots, t_n) \to t[g(s_1, \ldots, s_m)]$ is a rule in R, where
f and g are defined symbols, and t is some context term, then the pair
$(F(t_1, \ldots, t_n), G(s_1, \ldots, s_m))$ is a *dependency pair* of R.
    An *R-chain* is a sequence of renamings of dependency pairs such that there
exists a substitution σ with $\sigma(t_i) \xrightarrow{*}_R \sigma(s_{i+1})$, for every two consecutive
pairs (s_i, t_i) and (s_{i+1}, t_{i+1}) in the sequence. ◇

Theorem 9.23. *A rewrite system R is terminating if and only if no infinite R-chain exists.*

Since termination is undecidable, the criterion given by Theorem 9.23 is undecidable too. The technique proposed in [AG96, AG97] is the following: a dependency graph is built, whose nodes are labelled with the dependency pairs. There is an arc from (s,t) to (u,v) if there exists a substitution σ such that $\sigma(t) \xrightarrow{*}_R \sigma(u)$. Since this last property is in general undecidable, the dependency graph has to be approximated by a super-graph with the same cycles. Dependency pairs are then used to generate a set of inequalities. If these equalities can be satisfied by a suitable well-founded ordering, R is terminating.

Theorem 9.24. *If there exists a reduction quasi-ordering $\geq$ such that $l \geq r$ for each rule $l \to r$ in R, $s \geq t$ for each dependency pair (s,t) on a cycle of the dependency graph, and $s > t$ for at least one dependency pair (s,t) on every cycle of the dependency graph, then R is terminating.*

Example 9.25. In the following rewrite system, the defined symbols are *minus* and *quot*.

$$minus(x,0) \to x$$
$$minus(s(x), s(y)) \to minus(x,y)$$
$$quot(0, s(y)) \to 0$$
$$quot(s(x), s(y)) \to s(quot(minus(x,y), s(y)))$$

The associated dependency pairs are

$$(MINUS(s(x), s(y)) \,,\, MINUS(x,y))$$
$$(QUOT(s(x), s(y)) \,\,,\, MINUS(x,y))$$
$$(QUOT(s(x), s(y)) \,\,,\, QUOT(minus(x,y), s(y)))$$

The dependency graph has two cycles corresponding to the first and third dependency pairs. To prove termination, it is sufficient to find a reduction quasi-ordering that satisfies the following inequalities.

$$minus(x,0) \geq x$$
$$minus(s(x), s(y)) \geq minus(x,y)$$
$$quot(0, s(y)) \geq 0$$
$$quot(s(x), s(y)) \geq s(quot(minus(x,y), s(y)))$$
$$MINUS(s(x), s(y)) > MINUS(x,y)$$
$$QUOT(s(x), s(y)) > QUOT(minus(x,y), s(y))$$

A polynomial interpretation in which 0 is mapped to 0, $s(x)$ to $x + 1$, $minus(x,y), quot(x,y), MINUS(x,y), QUOT(x,y)$ all mapped to x, provides the solution.   ∎

Church–Rosser property. Results valid for any abstract reduction system specialise to the set of terms $\mathcal{T}(\Sigma, \mathcal{X})$ and the rewriting relation $\longrightarrow_R$. In this context, the Church–Rosser property states the relation between replacement of equals by equals and rewriting. Reformulating Definition 9.5, we obtain the following:

The relation $\longrightarrow_R$ is Church–Rosser on $\mathcal{T}(\Sigma, \mathcal{X})$ if

$$\forall t, t', \ t \xleftrightarrow{*}_R t' \text{ if and only if } \exists t'', \ t \xrightarrow{*}_R t'' \xleftarrow{*}_R t'.$$

Given termination, this is equivalent to confluence:

$$\forall t, t_1, t_2, \ t_1 \xleftarrow{*}_R t \xrightarrow{*}_R t_2 \text{ implies } \exists t'', \ t_1 \xrightarrow{*}_R t'' \xleftarrow{*}_R t_2.$$

A rewrite system R is Church–Rosser (respectively confluent, locally confluent, terminating, convergent) on $\mathcal{T}(\Sigma, \mathcal{X})$ if $\longrightarrow_R$ is Church–Rosser (respectively confluent, locally confluent, terminating, convergent) on $\mathcal{T}(\Sigma, \mathcal{X})$.

A rewrite system R is *ground Church–Rosser* (respectively *ground confluent, ground locally confluent, ground terminating, ground convergent*) if $\longrightarrow_R$ is Church–Rosser (respectively confluent, locally confluent, terminating, convergent) on the set of ground terms $\mathcal{T}(\Sigma)$.

Example 9.26. Right distributivity of an operator $+$ on an operator $*$, expressed by the rule $(x + y) * z \to (x * z) + (y * z)$, or left distributivity $x * (y + z) \to (x * y) + (x * z)$, taken separately, are confluent rewrite systems. Put together, the rewrite system composed of these two distributivity rules is terminating (using, for instance, the precedence $* >_{\mathcal{F}} +$ in Definition 9.20), but not confluent since

$$(x + y) * (x + y) \xrightarrow{+} ((x * x) + (y * x)) + ((x * y) + (y * y))$$
$$\xrightarrow{+} ((x * x) + (x * y)) + ((y * x) + (y * y)) \qquad \blacksquare$$

The following example is chosen from the area of hardware verification, and illustrates how to use rewrite systems to perform a simple proof of equivalence of two circuits.

Example 9.27. [Kri94] A ripple-carry adder with n bits is composed of n full adders. Inputs of the i-th full adder are the i-th bits $p(i)$ and $q(i)$ of the two binary numbers to be added, and the carry $r(i)$ from the $(i - 1)$-th adder. Outputs of the i-th full adder are the carry $r(i + 1)$ and the i-th bit of the sum $sum(i)$. Inputs of the whole circuit are $p(0), q(0), r(0)$ and outputs are $sum(0), \dots, sum(n)$ and $r(n)$. A ripple-carry adder is described by the following term rewriting system, where, for readability, parentheses for associative Boolean operators $\oplus$ and $\vee$ are omitted.

$$\Sigma = \begin{cases} sorts & Nat, \ Bool \\ \oplus & : Bool \times Bool \mapsto Bool \\ \wedge & : Bool \times Bool \mapsto Bool \\ \vee & : Bool \times Bool \mapsto Bool \\ p & : Nat \qquad\qquad \mapsto Bool \\ q & : Nat \qquad\qquad \mapsto Bool \\ r & : Nat \qquad\qquad \mapsto Bool \\ sum & : Nat \qquad\qquad \mapsto Bool \\ s & : Nat \qquad\qquad \mapsto Nat \end{cases}$$

$$R = \begin{cases} \forall i : Nat, \\ sum(i) \rightarrow p(i) \oplus q(i) \oplus r(i) \\ r(s(i)) \rightarrow (p(i) \wedge q(i)) \vee (p(i) \wedge r(i)) \vee (q(i) \wedge r(i)) \end{cases}$$

Termination of the rewrite system can be proved, for instance, by using the multiset path ordering (see Definition 9.20) induced by the following precedence on function symbols: $sum >_{\mathcal{F}} \oplus, r$ and $r >_{\mathcal{F}} \vee, \wedge, p, q$.

Another kind of adder, a carry-lookahead adder, is described, with auxiliary functions g and h to compute the carry, in the rewrite system below.

$$\Sigma = \begin{cases} sort & Nat, \ Bool \\ g & : Nat \qquad \mapsto Bool \\ h & : Nat \qquad \mapsto Bool \end{cases}$$

$$R = \begin{cases} \forall i : Nat, \\ sum(i) \rightarrow g(i) \oplus p(i) \oplus r(i) \\ g(i) \rightarrow p(i) \wedge q(i) \\ h(i) \rightarrow p(i) \vee q(i) \\ r(s(i)) \rightarrow g(i) \vee (h(i) \wedge r(i)) \end{cases}$$

A proof of equivalence between these two adders is achieved by the fact that normalising the rewrite rules of the second system with the rules for boolean expressions (namely $(x \vee y \rightarrow (x \wedge y) \oplus x \oplus y)$ and $(x \oplus x \rightarrow 0)$), yields the rules of the first system.

$$\begin{aligned} r(s(i)) \longrightarrow\ & g(i) \vee (h(i) \wedge r(i)) \\ \longrightarrow\ & (p(i) \wedge q(i)) \vee ((p(i) \vee q(i)) \wedge r(i)) \\ \longrightarrow\ & (p(i) \wedge q(i)) \vee (p(i) \wedge r(i)) \vee (q(i) \wedge r(i)). \\ sum(i) \longrightarrow\ & g(i) \oplus h(i) \oplus r(i) \\ \longrightarrow\ & (p(i) \wedge q(i)) \oplus (p(i) \vee q(i)) \oplus r(i) \\ \longrightarrow\ & (p(i) \wedge q(i)) \oplus (p(i) \wedge q(i) \oplus p(i) \oplus q(i)) \oplus r(i) \\ \longrightarrow\ & p(i) \oplus q(i) \oplus r(i). \end{aligned}$$

■

Orthogonal systems. For non-terminating systems, confluence can nevertheless be ensured by adding strong syntactic conditions on the left-hand sides. This is the case, for instance, for the class of orthogonal systems.

Definition 9.28. A non-variable term t' *overlaps* a term t if there exists a non-variable position ω in t such that $t_{|\omega}$ and t' are unifiable.

A rewrite system is non-overlapping if there is no overlap between the left-hand sides of the rules. A term is linear if each variable occurs only once. A rewrite system is left-linear if the left-hand side of each rule is a linear term. A rewrite system that is both left-linear and non-overlapping is called *orthogonal*.   ◇

Example 9.29. The system from Combinatory Logic given in Example 9.17 is orthogonal.   ■

Theorem 9.30. *If a rewrite system R is orthogonal, then it is confluent.*

This is proved in [Hue80], using the fact that concurrent reduction (i.e., application in one step of the rewrite of rules of R at disjoint positions) is strongly confluent.

Orthogonal rewrite systems have been extensively studied in the context of the operational semantics of recursive programming languages. Of particular interest is the existence of normalising reduction strategies. In [HL91a, HL91b], it is shown that in an orthogonal rewrite system, each term contains a needed redex, i.e., a redex that has to be rewritten in order to reach a normal form. A call-by-need strategy is obtained by repeatedly rewriting needed redexes, and always succeeds in finding a normal form when it exists. Although it is in general undecidable whether a redex is needed, for the class of strongly sequential systems, needed redexes can be found effectively. Moreover, it is decidable whether an orthogonal rewrite system is strongly sequential.

Decidability results. Most properties of rewrite systems are undecidable. For a rewrite system built on a finite signature with finitely many rewrite rules, it is undecidable whether confluence holds and whether termination holds. However, for ground rewrite systems (i.e., with no variable), confluence is decidable [DHLT87] and termination is decidable [HL78a]. But ground confluence is undecidable as shown in [KNO90].

A rewrite system R provides a decision procedure for an equational theory E if R is finite, $\longrightarrow_R$ is convergent and $\overset{*}{\longleftrightarrow}_R$ coincides with $=_E$. The word problem is the problem to decide whether an equality $s = t$ between two ground terms follows from E. This is a particular case of provability in E for arbitrary terms. If R is ground convergent, the word problem is decidable by reducing both terms s and t to their normal forms, and by testing the syntactic equality of the results. Of course, not every equational theory can be decided by rewriting. Some decidable theories are not finitely presented. Moreover, some finitely presented theories are undecidable (combinatorial logic, for instance). Even some finitely presented and decidable theories are not decidable by rewriting.

Example 9.31. Kapur and Narendran [KN85] found the following finite Thue system, consisting of the only axiom: $h(k(h(x))) = k(h(k(x)))$, with a decidable word problem and without an equivalent finite convergent system.   ∎

9.2.2   Rewriting modulo a set of equalities

A second kind of abstract reduction system is obtained by considering $\mathcal{T}$ as the set of equivalence classes of terms modulo the congruence generated by a set of axioms A, and $\longrightarrow$ as the rewriting relation induced on equivalence classes by a set of rewrite rules. This is especially important for dealing with the problem of commutative, or associative and commutative (AC for short), theories. This requires the elaboration of new abstract concepts, namely the notion of *class rewrite systems*, also called equational term rewriting systems. Let A be any set of axioms with decidable unification, matching and word problems and $\overset{*}{\longleftrightarrow}_A$ be the generated congruence relation on $\mathcal{T}(\Sigma, \mathcal{X})$.

Definition 9.32. A *class rewrite system*, denoted (R/A), is defined by a set of axioms A and a set of rewrite rules R assumed disjoint.   ◇

The class rewrite relation applies to a term if there exists a term in the same equivalence class modulo A that is reducible with R.

Definition 9.33. A term t *rewrites* to t' *with a class rewrite system* (R/A), denoted $t \longrightarrow_{R/A} t'$, if there exist a rewrite rule $(l \to r) \in R$, a term u, a position ω in u and a substitution σ such that $t \overset{*}{\longleftrightarrow}_A u[\sigma(l)]_\omega$, and $t' \overset{*}{\longleftrightarrow}_A u[\sigma(r)]_\omega$.   ◇

A term irreducible for $\longrightarrow_{R/A}$ is said to be in (R/A)-normal form. An (R/A)-normal form of a term t is denoted $t \downarrow_{R/A}$. Let us consider in more detail the termination and confluence properties for this new relation.

A reduction ordering $>$ is A-compatible if $\overset{*}{\longleftrightarrow}_A \circ > \circ \overset{*}{\longleftrightarrow}_A \subseteq >$. Well-founded A-compatible reduction orderings do not exist when R is non-empty and A contains an axiom like idempotency ($x + x = x$), where a variable occurs alone on one side and several times on the other. From an instance σ of a rewrite rule $(l \to r) \in R$, a contradiction to the well-foundedness of $>$ may be built, provided $\sigma(l) > \sigma(r)$:

$$\sigma(l) \overset{*}{\longleftrightarrow}_A \sigma(l) + \sigma(l) > \sigma(r) + \sigma(l) \overset{*}{\longleftrightarrow}_A \sigma(r) + (\sigma(l) + \sigma(l))$$
$$> \sigma(r) + (\sigma(r) + \sigma(l)) \ldots$$

Other axioms that prevent the existence of a well-founded A-compatible reduction ordering are equalities like ($x * 0 = 0$), where a variable occurs on one side and not on the other. Then $0 \longleftrightarrow_A \sigma(l) * 0 > \sigma(r) * 0 \longleftrightarrow_A 0$ provides a contradiction to well-foundedness. Indeed, if such axioms are present, they must be considered as rewrite rules.

The rewrite relation $\longrightarrow_{R/A}$ is not completely satisfactory from an operational point of view: even if R is finite and $\overset{*}{\longleftrightarrow}_A$ decidable, $\longrightarrow_{R/A}$ may not be computable since equivalence classes modulo A may be infinite or not computable. For instance, the axiom $(-x = x)$ generates infinite equivalence classes. To avoid searching through equivalence classes, the idea is to use a weaker relation on terms, called *rewriting modulo A*, which incorporates A in the matching process.

Definition 9.34. Given a class rewrite system (R/A), the term t (R, A)-*rewrites* to t', denoted by $t \longrightarrow_{R,A}^{(l \to r), \omega, \sigma} t'$, if there exist a rewrite rule $(l \to r) \in R$, a position ω in t, a substitution σ, such that $t_{|\omega} \overset{*}{\longleftrightarrow}_A \sigma(l)$, and $t' = t[\sigma(r)]_\omega$. ◇

In order to obtain the same effect as rewriting in equivalence classes, some properties called *confluence and coherence modulo* a set of axioms are needed [Hue80, Jou83, JK86a].

Definition 9.35. The class rewrite system (R/A) is *Church–Rosser* on a set of terms $\mathcal{T}$ if

$$\overset{*}{\longleftrightarrow}_{R \cup A} \subseteq \overset{*}{\longrightarrow}_{R/A} \circ \overset{*}{\longleftrightarrow}_A \circ \overset{*}{\longleftarrow}_{R/A} .$$

The rewriting relation $\longrightarrow_{R,A}$ $((R, A)$ for short) defined on $\mathcal{T}$ is:

- *Church–Rosser modulo A* if
$$\overset{*}{\longleftrightarrow}_{R \cup A} \subseteq \overset{*}{\longrightarrow}_{R,A} \circ \overset{*}{\longleftrightarrow}_A \circ \overset{*}{\longleftarrow}_{R,A}$$

- *confluent modulo A* if
$$\overset{*}{\longleftarrow}_{R,A} \circ \overset{*}{\longrightarrow}_{R,A} \subseteq \overset{*}{\longrightarrow}_{R,A} \circ \overset{*}{\longleftrightarrow}_A \circ \overset{*}{\longleftarrow}_{R,A}$$

- *coherent modulo A* if
$$\overset{*}{\longleftarrow}_{R,A} \circ \overset{*}{\longleftrightarrow}_A \subseteq \overset{*}{\longrightarrow}_{R,A} \circ \overset{*}{\longleftrightarrow}_A \circ \overset{*}{\longleftarrow}_{R,A}$$

- *locally confluent modulo A* if
$$\longleftarrow_{R,A} \circ \longrightarrow_R \subseteq \overset{*}{\longrightarrow}_{R,A} \circ \overset{*}{\longleftrightarrow}_A \circ \overset{*}{\longleftarrow}_{R,A}$$

- *locally coherent modulo A* if
$$\longleftarrow_{R,A} \circ \longleftrightarrow_A \subseteq \overset{*}{\longrightarrow}_{R,A} \circ \overset{*}{\longleftrightarrow}_A \circ \overset{*}{\longleftarrow}_{R,A}$$ ◇

The next theorem relates the properties above to each other.

Theorem 9.36. [JK86a] *The following properties of a class rewrite system (R/A), are equivalent on $\mathcal{T}$* :

1. (R, A) *is Church–Rosser modulo A.*
2. (R, A) *is confluent modulo A and coherent modulo A.*

3. (R, A) is locally confluent and locally coherent modulo A.

4. $\forall t, t'$, $t \xleftrightarrow{*}_{R \cup A} t'$ if and only if $t \downarrow_{R,A} \xleftrightarrow{*}_A t' \downarrow_{R,A}$.

For the specific case of associative and commutative theories, there exists a systematic way to fulfil local coherence by adding extended rules.

Definition 9.37. Let R be a rewrite system and A be the axioms of commutativity and associativity of the symbol f. Any rule $l \to r \in R$, such that the top symbol of l is f, has an *extended rule* $f(l, x) \to f(r, x)$, where x is a variable that does not occur in l (thus nor in r).

Let R^{ext} be R plus all extended rules $f(l, x) \to f(r, x)$ of rules $l \to r \in R$ for which there exist no rule $l' \to r'$ in R and no substitution σ, such that $f(l, x) \xleftrightarrow{*}_{AC} \sigma(l')$ and $f(r, x) \xrightarrow{*}_{AC \cup R/AC} \sigma(r')$.   ◇

Indeed if there exist a rule $l' \to r'$ in R and a substitution σ, such that $f(l, x) \xleftrightarrow{*}_{AC} \sigma(l')$ and $f(r, x) \xrightarrow{*}_{AC \cup R/AC} \sigma(r')$, the extended rule is not needed for achieving local coherence.

Example 9.38. This Church–Rosser axiomatisation of free distributive lattices is taken from [PS81]. Let $\cup$ and $\cap$ be two associative and commutative symbols.

$$AC = \left\{ \begin{array}{ll} x \cup y = y \cup x & \qquad x \cap y = y \cap x \\ (x \cup y) \cup z = x \cup (y \cup z) & \qquad (x \cap y) \cap z = x \cap (y \cap z) \end{array} \right.$$

The following set of rules R

$$R = \left\{ \begin{array}{ll} x \cup (x \cap y) \to x \\ x \cap (y \cup z) \to (x \cap y) \cup (x \cap z) \\ x \cap x \qquad \to x \\ x \cup x \qquad \to x \end{array} \right.$$

is such that (R^{ext}, AC) is Church–Rosser modulo AC and R/AC is terminating.

The term $(b \cup (a \cup b))$ where a, b are constants, can be reduced by the extended rule $(x \cup x) \cup z \to x \cup z$ to $a \cup b$, since $(b \cup (a \cup b))$ is AC-equivalent to $(b \cup b) \cup a$.   ∎

Many mathematical structures (abelian groups, commutative rings for instance) involve associativity and commutativity laws.

Example 9.39. Boolean ring. Since the 1950s it has been known that any term of the Boolean algebra (i.e., any Boolean formula) admits a set of *prime implicants* that can be computed algorithmically [Qui59, SCL70]. However, attempts to find a Church–Rosser class rewrite system for Boolean algebra using a completion procedure failed to terminate in all experiments reported by [Hul80b, PS81]. Only in 1991 was a formal proof of the non-existence of a convergent system for Boolean algebra presented [Soc91]. Previously in 1985,

J. Hsiang observed that there exists nevertheless a convergent class rewrite system in a signature $\mathcal{F} = \{0, 1, \wedge, \vee, \neg, \oplus\}$, where $\oplus$ and $\wedge$ are operators from a Boolean ring. A class rewrite system for Boolean rings, where the conjunction $\wedge$ and the exclusive-or $\oplus$ are associative and commutative, is given by the following sets of rewrite rules and equalities, denoted BR/AC:

$$BR = \begin{cases} x \oplus 0 & \to x \\ x \oplus x & \to 0 \\ x \wedge 0 & \to 0 \\ x \wedge 1 & \to x \\ x \wedge x & \to x \\ x \wedge (y \oplus z) \to (x \wedge y) \oplus (x \wedge z) \end{cases}$$

$$AC = \begin{cases} x \wedge y = y \wedge x & \qquad x \oplus y = y \oplus x \\ (x \wedge y) \wedge z = x \wedge (y \wedge z) & \qquad (x \oplus y) \oplus z = x \oplus (y \oplus z) \end{cases}$$

Using this class rewrite system, we get the following derivation where p, q, r are predicate symbols:

$$(p(x) \wedge p(x)) \oplus p(x) \oplus (q(y) \wedge r(x, y) \wedge 1) \oplus (p(x) \wedge 0) \oplus 1$$
$$\xrightarrow{*}_{BR/AC} q(y) \wedge r(x, y) \oplus 1.$$

In order to deal with Boolean algebra's negation $\neg$ and disjunction $\vee$ [HD83], one may add rules:

$$x \vee y \to (x \wedge y) \oplus x \oplus y$$
$$\neg x \to x \oplus 1$$ ∎

A refinement of rewriting modulo A called normalised rewriting was introduced in [Mar94]. Considering A as a Church–Rosser class rewrite system (S/AC), it consists of normalising modulo AC with S before applying a rule in R. It allows complex theories for A, such as abelian groups, commutative rings or finite fields.

9.2.3 Conditional rewriting

Conditional equalities arise naturally in algebraic specifications of data types, where they provide a way to handle a large class of partial functions and case analysis. The formulas being considered are written "$\Gamma \Rightarrow l = r$", where Γ is a conjunction of equalities $(s_1 = t_1 \wedge \ldots \wedge s_n = t_n)$. Γ and $l = r$ are called respectively the *condition* and the *conclusion* of the conditional equality. In a *conditional rewrite rule*, the conclusion is oriented and this denoted as "$\Gamma \Rightarrow t \to s$". The orientation is performed with respect to an ordering on terms. There may be variables occurring in the condition but not in the conclusion, called extra-variables. Different abstract reduction systems defined on terms, but using different relations, have been proposed for these formulas. First, using a conditional rewrite system R, a given rule may be applied to a term if its condition, instantiated by the matching substitution, is satisfied.

Definition 9.40. Given a conditional rewrite system R, a term t *rewrites* to a term t' if there exist a conditional rewrite rule $\Gamma \Rightarrow l \rightarrow r$ of R, a position ω in t, a substitution σ, satisfying $t_{|\omega} = \sigma(l)$, and a substitution τ for the extra-variables such that $\tau(\sigma(\Gamma))$ holds. Then $t' = t[\omega \leftarrow \tau(\sigma(r))]$.   ◊

When conditional rewriting is used for modelling logic and functional programming, this definition which allows extra-variables in the conditions and right-hand sides of rules, is quite convenient. A conditional rewrite rule is applicable if there exists a substitution for the extra-variables that makes the condition hold.

Example 9.41. Consider the set of conditional rewrite rules with variables x, y, z, x', y':

$$append(nil, y) \rightarrow y$$
$$(x = cons(x', y') \wedge append(y', y) = z) \Rightarrow append(x, y) \rightarrow cons(x', z)$$

With the second rule, $append(cons(a, nil), nil)$ rewrites to $cons(a, nil)$ since the substitution $(x' \mapsto a)(y' \mapsto nil)(z \mapsto nil)$ is a solution of $(cons(a, nil) = cons(u, v) \wedge append(y', nil) = z)$. In this example, there are variables in the right-hand side of the second rule that do not occur in the left-hand side.

■

Definition 9.40 is too general to be efficient, due to the complexity of verifying the condition. This leads us to distinguish more restrictive notions of conditional rewriting relations, according to the kind of evaluation chosen for the conditions. Definition 9.40 is modified accordingly.

First, it is often required that every variable occurring either in the condition or in the right-hand side of a rule also occurs in the left-hand side, which is denoted by $Var(\Gamma) \cup Var(r) \subseteq Var(l)$. This requirement rules out Example 9.41.

Thus in the condition (C) "there exists a substitution τ for the extra-variables such that $\tau(\sigma(\Gamma))$ holds" of Definition 9.40, τ is identity and we are left to check that $\sigma(\Gamma)$ holds.

This is done in different ways, according to different kinds of rewriting relations introduced below. Only the replacements of condition (C) in Definition 9.40 are mentioned:

1. for *natural* conditional rewriting: $t \longrightarrow_{R^{nat}} t'$ if
   (C^{nat}) there exists a proof $\sigma(s_i) \stackrel{*}{\longleftrightarrow}_{R^{nat}} \sigma(t_i)$ for each instantiated component of the condition.
2. for *join* conditional rewriting: $t \longrightarrow_{R^{join}} t'$ if
   (C^{join}) $\sigma(s_i) \downarrow_{R^{join}} = \sigma(t_i) \downarrow_{R^{join}}$ for each instantiated component of the condition.
3. for *normal* conditional rewriting: $t \longrightarrow_{R^{norm}} t'$ if
   (C^{norm}) $\sigma(t_i)$ is a normal form of $\sigma(s_i)$, denoted by $\sigma(s_i) \stackrel{!}{\longrightarrow}_{R^{norm}} \sigma(t_i)$, for each instantiated component of the condition.

(C^{nat}) is very close to equational reasoning with the underlying conditional equalities. However, from a rewriting point of view, it is not quite satisfactory due to the bidirectional use of rewrite rules. In practice, (C^{join}) has been studied more and is often implemented. (C^{norm}) is interesting in relation to logic and functional programming, and when conditions are expressed with Boolean functions.

Example 9.42. [BK86a] Consider the join conditional rewriting relation for the system:

$$f(a) \to a$$
$$x = f(x) \Rightarrow f(x) \to c$$

Since $a \downarrow = f(a) \downarrow$, $f(a) \longrightarrow c$ by applying the second rule. Then $f(f(a)) \longrightarrow c$. But neither c nor $f(c)$ are reducible. ∎

Example 9.43. Consider the normal conditional rewriting relation for the system:

$$even(0) \quad \to true$$
$$even(s(x)) \to odd(x)$$
$$even(x) = false \Rightarrow odd(x) \quad \to true$$
$$even(x) = true \Rightarrow odd(x) \quad \to false$$

Then $even(s(0)) \longrightarrow odd(0) \longrightarrow false$ by applying first the second, then the fourth rule. ∎

In any case, the rewrite relation $\longrightarrow_R$ associated with a conditional rewrite system has an inductive definition, which is fundamental for establishing the properties of conditional rewrite systems.

Definition 9.44. Let R be a conditional rewrite system and R_i the rewrite system defined for $i \geq 0$ as follows:

$$R_0 = \{l \to r \mid l \to r \in R\}$$
$$R_{i+1} = R_i \cup \{\sigma(l) \to \sigma(r) \mid (s_1 = t_1 \wedge \ldots \wedge s_n = t_n) \Rightarrow l \to r \in R,$$
$$\text{and } \forall j = 1, \ldots, n, \ \sigma(s_j) \equiv_i \sigma(t_j)\}$$

where $\equiv_i$ denotes $\overset{*}{\longleftrightarrow}_{R_i}$, $\downarrow_{R_i}$ or $\overset{!}{\longrightarrow}_{R_i}$. Then $t \longrightarrow_R t'$ if and only if $t \longrightarrow_{R_i} t'$ for some $i \geq 0$. ◇

With this definition, an abstract reduction system can be associated with any conditional rewrite system, according to the evaluation chosen for the conditions. All definitions and properties of abstract relations are thus available for these abstract reduction systems.

Obviously, if R^{join} is confluent, then so is the relation R^{nat}, but the converse does not hold, because not all proofs in the condition of a natural rewriting proof can be transformed into joinability proofs.

Example 9.45. [Kap84] Consider the following system, where a, b, a', b', a'' are constants and g, d are any terms.

$$a'' \rightarrow a$$
$$a' \rightarrow b$$
$$b' \rightarrow a$$
$$b' \rightarrow b$$
$$a' = a'' \Rightarrow g \rightarrow d$$

Then $a'' \longleftrightarrow_{Rnat} a \longleftrightarrow_{Rnat} b' \longleftrightarrow_{Rnat} b \longleftrightarrow_{Rnat} a'$, and so $g \longrightarrow_{Rnat} d$. But $a'' \downarrow_{Rjoin} a'$ is false and thus $g \longrightarrow_{Rjoin} d$ is false, as is $g \overset{*}{\longleftrightarrow}_{Rjoin} d$. ∎

In [Kap83, Kap84], an example of a conditional rewrite system is constructed such that the relation $\longrightarrow_R$ is not decidable and the normal form of a term is not computable. Additional conditions have to be added to regain decidable properties.

Definition 9.46. A conditional rewriting relation is *decreasing* if there exists a well-founded extension $>$ of the rewriting relation $\longrightarrow$ which satisfies two additional properties:

- $>$ contains the proper subterm relation.
- for each rule $(s_1 = t_1 \wedge \ldots \wedge s_n = t_n) \Rightarrow l \rightarrow r$, $\sigma(l) > \sigma(s_i)$ and $\sigma(l) > \sigma(t_i)$, for all substitutions σ and all indices i, $1 \leq i \leq n$. ◇

The next result is proved by induction on the ordering that makes the conditional rewriting relation decreasing.

Theorem 9.47. [DO90] *Whenever the relation $\longrightarrow_{Rjoin}$ is decreasing, the relations $\longrightarrow_{Rjoin}$, $\overset{*}{\longrightarrow}_{Rjoin}$ and $\downarrow_{Rjoin}$ are all decidable.*

The notion of decreasing-ness of a rewriting relation is not usable in practice. A property that can be checked on the conditional rewrite system is preferable. This leads to the introduction the notion of a *reductive* system [JW86] that generalises the definition of a *simplifying* system in [Kap87].

Definition 9.48. [JW86] A conditional rewrite system is *reductive* if there exists a well-founded reduction ordering $>$ such that, for each conditional rule $(s_1 = t_1 \wedge \ldots \wedge s_n = t_n) \Rightarrow l \rightarrow r$, and substitution σ ,
(i) $\sigma(l) > \sigma(r)$ and
(ii) $\sigma(l) > \sigma(s_i)$ and $\sigma(l) > \sigma(t_i)$, for all indices i, $1 \leq i \leq n$. ◇

Actually reductive systems capture the finiteness of the evaluation of terms, as explained in [DO90], where it is also proved that if R is a reductive conditional rewrite system, R^{join} is decreasing.

9.3   Rewriting as a logic for concurrent computations

While the emphasis in the previous section was on the Church–Rosser property and the use of rewriting as a decision procedure for equality in equational theories, another point of view is now adopted which strengthens the computational logic aspect of rewriting. Rewriting logic is presented, for instance in [Mes92], as a logic of actions in which logical deduction is identified with concurrent rewriting, i.e., application in one step of rewrite rules at disjoint positions in a term. Rewriting logic provides a general framework for unifying a variety of models of concurrency as argued in [Mes92], but also a logical framework in which other logics can be encoded as proposed in [MM93].

In rewriting logic, the sentences are of the form $t \longrightarrow t'$ where t and t' are terms of $\mathcal{T}(\Sigma, \mathcal{X})$. More generally, some structural axioms may be taken into account and sequents are then defined as sentences of the form $\langle t \rangle_A \longrightarrow \langle t' \rangle_A$, where $\langle t \rangle_A$ denotes the equivalence class of t modulo A. Then the rewrite relation considered is rewriting modulo A. To support intuition, the reader may regard A as being the empty set of axioms, or the associativity and commutativity axioms for some binary symbol. But in theory, A can encode more complex data structures.

Moreover, in rewriting logic, proofs are first-order objects and are represented by proof terms. In order to build proof terms and, later, strategies, rules are labelled with a set $\mathcal{L}$ of ranked label symbols, i.e, symbols with a fixed arity. In order to compose proofs, the infix binary operator ";" is introduced. A *proof term* is by definition a term built on function symbols in $\mathcal{F}$, label symbols in $\mathcal{L}$ and the concatenation operator ";". Let $\mathcal{PT}$ denote the set of proof terms. Taking proof terms into account in the sentences leads one to consider sequents of the form $\pi : \langle t \rangle_A \longrightarrow \langle t' \rangle_A$, where $t, t' \in \mathcal{T}(\Sigma, \mathcal{X})$ and $\pi \in \mathcal{PT}$. The informal meaning of such sentences is that the proof π allows the derivation of t' from t. The labels are crucial in both the later definition of strategies and the construction of models for the rewriting logic.

We are now ready to formalise deduction in rewriting logic. For simplicity, we restrict ourselves here to unconditional rewriting, but everything can be extended to conditional rewriting at the cost of more technical definitions. The reader is referred to [Mes92] for a full treatment of this case.

A *labelled rewrite theory* on the signature $\Sigma = (\mathcal{S}, \mathcal{F})$ is a 5-tuple $\mathcal{R} = (\Sigma, \mathcal{X}, A, \mathcal{L}, R)$ where $\mathcal{X}$ is a countably infinite set of variables, A a set of $\mathcal{T}(\Sigma, \mathcal{X})$-equalities, $\mathcal{L}$ is a set of ranked labels, and R a set of labelled rewrite rules of the form $\ell : l \rightarrow r$, where $\ell \in \mathcal{L}$ and $l, r \in \mathcal{T}(\Sigma, \mathcal{X})$. Each rule $\ell : l \rightarrow r$ has a finite set of variables $Var(l) \cup Var(r) = \{x_1, \ldots, x_n\}$, and the arity of ℓ is exactly the number of distinct variables in this set. This is recorded in the notation $\ell(x_1, \ldots, x_n) : l \rightarrow r$.

A labelled rewrite theory $\mathcal{R}$ entails the sequent $\pi : \langle t \rangle_A \longrightarrow \langle t' \rangle_A$, if the sequent is obtained by the finite application of the deduction rules **REW** presented in Figure 9.1.

Reflexivity For any $t \in \mathcal{T}(\Sigma)$:

$$\overline{t : \langle t \rangle_A \longrightarrow \langle t \rangle_A}$$

Congruence For any $f \in \mathcal{F}$ with $arity(f) = n$:

$$\frac{\pi_1 : \langle t_1 \rangle_A \longrightarrow \langle t_1' \rangle_A \quad \ldots \quad \pi_n : \langle t_n \rangle_A \longrightarrow \langle t_n' \rangle_A}{f(\pi_1, \ldots, \pi_n) : \langle f(t_1, \ldots, t_n) \rangle_A \longrightarrow \langle f(t_1', \ldots, t_n') \rangle_A}$$

Replacement For any rewrite rule $\ell(x_1, \ldots, x_n) : l \to r$ of R,

$$\frac{\pi_1 : \langle t_1 \rangle_A \longrightarrow \langle t_1' \rangle_A \quad \ldots \quad \pi_n : \langle t_n \rangle_A \longrightarrow \langle t_n' \rangle_A}{\ell(\pi_1, \ldots, \pi_n) : \langle l(t_1, \ldots, t_n) \rangle_A \longrightarrow \langle r(t_1', \ldots, t_n') \rangle_A}$$

Transitivity
$$\frac{\pi_1 : \langle t_1 \rangle_A \longrightarrow \langle t_2 \rangle_A \quad \pi_2 : \langle t_2 \rangle_A \longrightarrow \langle t_3 \rangle_A}{\pi_1; \pi_2 : \langle t_1 \rangle_A \longrightarrow \langle t_3 \rangle_A}$$

Fig. 9.1. REW: The rules of rewrite deduction

Example 9.49. Consider as an example the following specification:

$$\Sigma = \begin{cases} sorts & Nat \\ 0 & : \qquad\qquad \mapsto Nat \\ s & : Nat \qquad\quad \mapsto Nat \\ + & : Nat \times Nat \mapsto Nat \end{cases}$$

A contains the commutativity axiom for $+$:

$$A = \{ \forall x, y : Nat, \ x + y = y + x \}$$

with the following set of labelled rewrite rules R:

$$R = \begin{cases} \ell_0 & : 0 + 0 \quad \to 0 \\ \ell_1(x, y) : s(x) + y \to s(x + y) \end{cases}$$

In this rewrite theory, we can prove:

$$\langle 0 + s(0) \rangle_A \longrightarrow \langle s(0) \rangle_A$$

We first use **Reflexivity** to obtain

$$0 : \langle 0 \rangle_A \longrightarrow \langle 0 \rangle_A$$

With **Replacement** using the rewrite rule ℓ_1 and the commutativity of $+$:

$$\frac{0 : \langle 0 \rangle_A \longrightarrow \langle 0 \rangle_A \qquad 0 : \langle 0 \rangle_A \longrightarrow \langle 0 \rangle_A}{\ell_1(0,0) : \langle 0 + s(0) \rangle_A \longrightarrow \langle s(0 + 0) \rangle_A}$$

With **Replacement** using the rewrite rule ℓ_0

$$\ell_0 : \langle 0 + 0 \rangle_A \longrightarrow \langle 0 \rangle_A$$

With **Congruence** for the symbol s:

$$\frac{\ell_0 : \langle 0 + 0 \rangle_A \longrightarrow \langle 0 \rangle_A}{s(\ell_0) : \langle s(0+0) \rangle_A \longrightarrow \langle s(0) \rangle_A}$$

and with **Transitivity**

$$\frac{\ell_1(0,0) : \langle 0 + s(0) \rangle_A \longrightarrow \langle s(0+0) \rangle_A \quad s(\ell_0) : \langle s(0+0) \rangle_A \longrightarrow \langle s(0) \rangle_A}{\ell_1(0,0); s(\ell_0) : \langle 0 + s(0) \rangle_A \longrightarrow \langle s(0) \rangle_A}$$

This concludes the proof and constructs the associated proof term. ∎

Although the rules of rewriting logic are very close to those of equational logic, the difference between equational logic and rewriting logic actually relies upon the requirement for symmetry. Let us now make the link with the more operational notion of rewriting, previously introduced in Section 9.2. For simplicity, let us consider an empty set of structural axioms A and ignore proof terms. Thus sequents are simply written $t \longrightarrow t'$. Concurrent rewriting coincides exactly with deduction in rewriting logic, and sequential rewriting appears as a special case.

For a given rewrite system R, a sequent $t \longrightarrow t'$ is:

- a *zero-step R-rewrite*, if it can be derived from R by application of the rules **Reflexivity** and **Congruence**. In this case t and t' coincide.
- a *one-step concurrent R-rewrite*, if it can be derived from R by application of the rules **Reflexivity**, **Congruence** and at least one application of **Replacement**. When **Replacement** is applied exactly once, then the sequent is called a one-step *sequential R-rewriting*.
- a *concurrent R-rewrite* if it can be derived from R by finite application of the rules in **REW**.

The relation between one-step sequential R-rewriting and the rewrite relation $\longrightarrow_R$ defined on terms as in Section 9.2.1 is defined in the following lemma, proved by induction on the form of the proof of the sequent $t \longrightarrow t'$.

Lemma 9.50. *A sequent $t \longrightarrow t'$ is a one-step sequential R-rewriting if and only if there exist a rule $l \to r$ in R, a substitution σ and a position ω of t such that $t \longrightarrow_R^{\omega,\sigma,l \to r} t'$.*

This result extends to any concurrent rewriting derivation, and is also a consequence of the form of the proof of the sequent.

Lemma 9.51. *For each sequent $t \longrightarrow t'$ computed using the rewriting logic relative to a set of rules R:*

- *either $t = t'$,*
- *or there exists a chain of one-step concurrent R-rewrites:*

$$t = t_0 \longrightarrow t_1 \longrightarrow \ldots \longrightarrow t_{n-1} \longrightarrow t_n = t'.$$

Moreover, in this chain, each step $t_i \longrightarrow t_{i+1}$ can be chosen sequentially.

This last result shows the equivalence between the operational definition of rewriting and the sequential rewriting relation obtained from rewriting logic. The main advantage in rewriting logic is precisely to eliminate an operational definition of rewriting that needs to handle the notion of positions, matching and replacement.

Interesting investigations for rewriting logic include its use for the specification and design of software systems, as well as for semantic integration of different programming paradigms based on a common logic and model theory. In [MM93], it is shown how various logics can be represented in rewriting logic, for example, equational logic, Horn logic, linear logic and the natural semantics [Kah87]. In [KKV95, BKK+96b], the authors experiment rewriting logic to design constraint solvers, theorem provers, the kernel of programming languages, and also to combine computational paradigms in a single logical framework. This is applied, for instance, to the combination of two constraint solvers, of constraint solving and theorem proving, and of constraint solving with logic programming.

However this kind of application reveals the need for the definition of strategies. In practice, a strategy is a way to describe which computations are of interest to the user, and specifies where a given rule should be applied in the term to be reduced. From a theoretical point of view, a strategy as defined in [KKV95, BKK+96b] characterises a subset of all proof terms that can be built in the current rewrite theory. The application of a strategy to a term results in the (possibly empty) set of all terms derivable from the starting term, using this strategy. A strategy fails when it returns an empty set of terms. To illustrate these ideas, the following example illustrates how strategies are defined and implemented in the language ELAN.

Example 9.52. The application of a rewrite rule in ELAN yields, in general, several results. This is due first to equational matching (for instance AC-matching computing several possible substitutions) and, second, to the use of strategies that recursively return several possible results on subterms. So application of a rule to a term produces a set of results that may be empty, or a singleton, or a finite (or infinite) set of reduced terms. In order to take into account sets of results, and to control rule application, the following strategy constructors are provided:

- A labelled rule is a primal strategy. The result of applying a rule labelled ℓ on a term t returns a set of terms. This primal strategy fails if the set of resulting terms is empty.
- Two strategies can be concatenated by the symbol ";", i.e. the second strategy is applied on all results of the first one. $S_1; S_2$ denotes the sequential composition of the two strategies. It fails if either S_1 fails or S_2 fails. Its results are all results of S_1 on which S_2 is applied and gives some results.

- first$(S_1, \ldots, S_n)$ chooses the first strategy S_i in the list that does not fail, and returns all its results. This strategy may return more than one result, or fails if all sub-strategies S_i fail.
- first one$(S_1, \ldots, S_n)$ selects the first result of the first strategy S_i in the list that does not fail. This strategy returns at most one result and fails if all sub-strategies fail.
- dk$(S_1, \ldots, S_n)$ chooses all strategies given in the list of arguments and for each of them returns all its results. This set of results may be empty, in which case the strategy fails.
- The strategy id is the identity that does nothing but never fails.
- fail is the strategy that always fails and never gives any result.
- repeat*(S) applies repeatedly the strategy S until it fails and returns the results of the last unfailing application. This strategy can never fail (zero application of S is possible) and may return more than one result.
- The strategy iterate*(S) is similar to repeat*(S) but returns all intermediate results of repeated applications.

A short illustration of the use of these strategies is provided by the specification of the extraction of elements from a list.

$$\Sigma = \begin{cases} sort \quad Elem, NeList \\ \quad \_ : Elem \quad\quad\quad\quad \mapsto NeList \\ \quad \cdot : Elem, NeList \mapsto NeList \\ element : NeList \quad\quad \mapsto Elem \end{cases}$$

The declaration $\_ : Elem \mapsto NeList$ of the invisible unary operator forces an element of sort $Elem$ to be a non-empty list of sort $NeList$. Extraction is defined by three labelled rules:

$$R = \begin{cases} \forall e : Elem, l : NeList, \\ \quad\quad extract1(e) : element(e) \quad \rightarrow e \\ \quad\quad extract2(e, l) : element(e \cdot l) \rightarrow e \\ \quad\quad extract3(e, l) : element(e \cdot l) \rightarrow element(l) \end{cases}$$

Let a, b, c be three constants of sort $Elem$. Then

$$\text{repeat*}(\text{dk}(extract1, extract2, extract3))(a \cdot b \cdot c)$$

yields the set $\{a, b, c\}$.

$$\text{repeat*}(\text{first}(extract1, extract2, extract3))(a \cdot b \cdot c)$$

yields the set $\{a\}$.

$$\text{iterate*}(\text{dk}(extract1, extract2, extract3))(a \cdot b \cdot c)$$

yields the set $\{element(a \cdot b \cdot c), a, element(b \cdot c), b, element(c), c\}$. ∎

In order to improve further this first language of elementary strategies, a natural idea is to provide the users with the capability to program their own strategies or tactics, as already done in many logical frameworks. The idea of controlling rewriting by rewriting, more precisely by a strategy language based on rewriting, is investigated in [CELM96, GSHH92, BKK96a, BKK98]. Broadly speaking, there are two levels of rewriting: the object (or first-order term) level, and the meta-level that controls the object-level. This approach may be related to transformation rules and tactics presented in the context of Chapter 14.

Using conditional rewriting as a computational paradigm is simple, elegant, reasonably expressive and efficient, as witnessed, for instance, by performance of the ELAN compiler [Vit96, MK98]. Another advantage is that several properties of rewrite programs can be automatically proved. This is the topic of the second part of this chapter, where we focus on techniques for verifying some properties like confluence or well-typedness, and for proving some inductive properties of programs.

9.4   Superposition and simplification

The concept of critical pairs and the superposition process between two formulas introduced in this section appear in many places in automated deduction, and other examples of their use will be given in the following sections. Their role in the local confluence property is explained in the first part of this section, presenting the basic completion procedure of rewrite systems. Completion is based on superposition and simplification by rewriting. In the second part, extensions and applications of these two basic mechanisms are surveyed.

9.4.1   Completion procedure

Results presented in Section 9.1 have shown that the Church–Rosser property of a terminating term rewriting system is equivalent to the local confluence property. This property is important in the context of programming in rewriting logic, for ensuring the unicity of the result computed by a rewrite program. Local confluence, in turn, can be checked on special patterns computed from pairs of rewrite rules, called *critical pairs*. When a set of rewrite rules fails this test because some critical pair is not joinable, a special rule tailored for this case is added to the system. This is the basic idea of the so-called *completion procedure*, designed by Knuth and Bendix [KB70], building on the ideas of Evans [Eva51]. Of course, adding a new rule implies computing new critical pairs and the process is recursively applied. If this process of completion stops, it results in a locally confluent term rewriting system and if, in addition, termination has been incrementally checked, the system is also confluent. Two other possibilities can arise: the process can find an equality that cannot be oriented and then terminates in failure. It can also indefinitely discover non-joinable critical pairs, and hence endlessly add new rewrite rules. Critical pairs are produced by overlaps of two redexes in the same term.

Definition 9.53. [Hue80] Let $(g \to d)$ and $(l \to r)$ be two rules with disjoint sets of variables, such that l overlaps g at a non-variable position ω of g with the most general unifier σ. The *overlapped term* $\sigma(g)$ produces the *critical pair* (p, q) defined by $p = \sigma(g[\omega \leftarrow r])$ and $q = \sigma(d)$. The new equality $(p = q)$ is obtained by *superposition* of $(l \to r)$ on $(g \to d)$ at position ω. $CP(R)$ will denote the set of all critical pairs between rules in R.   ◇

In this definition $(g \to d)$ and $(l \to r)$ may be the same rule up to variable renaming, that is, a rewrite rule can overlap itself. Note that a rule always overlaps itself at the outermost position ϵ, producing a trivial critical pair.

Lemma 9.54. Critical pairs lemma [KB70, Hue80]
Let t, t', t'' be terms in $\mathcal{T}(\Sigma, \mathcal{X})$ such that

$$t' \xleftarrow{\omega, \alpha, l \to r}_R t \xrightarrow{\epsilon, \beta, g \to d}_R t''$$

with ω being a non-variable position in g. Such a proof is called a peak. Then, there exists a critical pair

$$(p, q) = (\sigma(g[\omega \leftarrow r]), \sigma(d))$$

of the rule $(l \to r)$ on the rule $(g \to d)$ at position ω such that σ is the most general unifier of l and $g_{|\omega}$, $\beta\alpha(x) = \tau\sigma(x)$ for all $x \in Var(g) \cup Var(l)$ and for some substitution τ. So $t' = \tau(p)$ and $t'' = \tau(q)$.

Definition 9.55. A critical pair (p, q) is said to be *joinable* (or convergent), denoted by $p \downarrow_R q$, if there exists a rewrite proof $p \xrightarrow{*}_R w \xleftarrow{*}_R q$.   ◇

The next theorem is also called Newman's lemma in the literature.

Theorem 9.56. [KB70, Hue80] *A rewrite system R is locally confluent if and only if any critical pair of R is joinable.*

Since finite systems have a finite number of critical pairs, their local confluence is decidable. Moreover, according to Theorem 9.56, a terminating rewrite system R is confluent if and only if all its critical pairs are joinable. This was the idea of the *superposition test* proposed by Knuth and Bendix [KB70]. The completion process was first studied from an operational point of view and several pioneer implementations were surveyed in [HKK91]. A completion process is composed of elementary tasks, such as rewriting a term in a rule or an equality, orienting an equality, and computing a critical pair. The efficiency of the general process is related to the strategies chosen for combining these tasks. Each task can be formalised as a transition rule (a rewrite rule in some adequate rewrite theory), that transforms sets of equalities and rules. Strategies express the control for application of these transition rules. Separating control from transition rules makes the proofs of correctness and completeness independent of strategies. As in traditional proof theory, these transition rules are studied by looking at their effect on equational proofs. The completion process then appears as a way to modify the axioms in order to reduce a proof to some normal form, called a rewrite proof. This abstract point of view is adopted here. It is based on the relation between the transition rule system that describes the relation between successive pairs of sets R and P computed by the process, and a proof transformation that reduces proofs to rewrite proofs. Following [BD87], the completion process is described by a set of transition rules, which actually provides a rewrite theory for completion. Each transition rule transforms a computation state composed of pairs (P, R), where P is a set of equalities and R a set of rewrite rules. These rules are usually followed by a condition that specifies in which case the rule applies. Each step computes (P_{i+1}, R_{i+1}) from (P_i, R_i) using a rewrite relation $\longmapsto$.

Let $R_* = \bigcup_{i \geq 0} R_i$ be the set of all generated rules and $P_* = \bigcup_{i \geq 0} P_i$ the set of all generated equalities. Let R_∞ and P_∞ be respectively the set of persisting rules and pairs, i.e., the sets effectively generated by completion starting from a set of axioms P_0. Formally:

$$P_\infty = \bigcup_{i \geq 0} \bigcap_{j > i} P_j \text{ and } R_\infty = \bigcup_{i \geq 0} \bigcap_{j > i} R_j.$$

Completion is parameterised by a given reduction ordering, denoted $>$, used to order equalities in a decreasing way. Moreover, rewrite rules are compared by the following ordering: $l \rightarrow r \gg g \rightarrow d$ if

- a subterm of l is an instance of g and not conversely, or
- l and g are equal up to variable renaming, and $r > d$ in the given reduction ordering.

The completion procedure is expressed by the set of rules presented in Figure 9.2.

Orient turns an equality $p = q$, such that $p > q$, into a rewrite rule. **Deduce** adds equational consequences derived from overlaps between rules. **Simplify** uses rules to simplify both sides of equalities and could be written more precisely as two rules **Left-Simplify** and **Right-Simplify** where $q \longrightarrow_R q'$. **Delete** removes any trivial equality. **Compose** simplifies right-hand sides of rules with respect to other rules. **Collapse** simplifies the left-hand side of a rule and turns the result into a new equality, but only when the simplifying rule $g \rightarrow d$ is smaller than the disappearing rule $l \rightarrow r$ in the ordering $\gg$. Note that if any equality is simplified by existing rules before being oriented, the condition of the **Collapse** rule is always satisfied. **Simplify**, **Compose** and **Collapse** are three aspects of *simplification* of formulas by rewriting with current rules in R.

Orient $P \cup \{p = q\}, R \mapsto\!\!\!\!\mapsto P, R \cup \{p \to q\}$
$\qquad\qquad\qquad$ if $p > q$
Deduce $P, R \qquad\qquad\quad \mapsto\!\!\!\!\mapsto P \cup \{p = q\}, R$
$\qquad\qquad\qquad$ if $(p, q) \in CP(R)$
Simplify $P \cup \{p = q\}, R \mapsto\!\!\!\!\mapsto P \cup \{p' = q\}, R$
$\qquad\qquad\qquad$ if $p \longrightarrow_R p'$
Delete $P \cup \{p = p\}, R \mapsto\!\!\!\!\mapsto P, R$
Compose $P, R \cup \{l \to r\} \mapsto\!\!\!\!\mapsto P, R \cup \{l \to r'\}$
$\qquad\qquad\qquad$ if $r \longrightarrow_R r'$
Collapse $P, R \cup \{l \to r\} \mapsto\!\!\!\!\mapsto P \cup \{l' = r\}, R$
$\qquad\qquad\qquad$ if $l \longrightarrow_R^{g \to d} l'$ and $l \to r \gg g \to d$

Fig. 9.2. Standard completion rules

These rules are sound in the following sense: they do not change the equational theory.

Lemma 9.57. [Bac91] *If* $(P, R) \mapsto\!\!\!\!\mapsto (P', R')$, *then* $\stackrel{*}{\longleftrightarrow}_{P \cup R}$ *and* $\stackrel{*}{\longleftrightarrow}_{P' \cup R'}$ *are the same.*

A completion procedure is aimed at transforming an initial set of equalities P_0 and an initial set of rules R_0, usually assumed empty, into a rewrite rule system R_∞ that is convergent and inter-reduced.

With a proof-theoretical point of view, let us now consider the set of all provable equalities $(t = t')$. To each (P_i, R_i) is associated a proof of $(t = t')$ using rules in R_i and equalities in P_i. Following [Bac91], with each rule is associated a proof-transformation rule on some kind of proofs, using rules in R_* and equalities in P_*. The proof-transformation relation is denoted by $\Longrightarrow$. Establishing the correctness of a completion procedure involves the following steps:

- Each sequence of proof transformation using the relation $\Longrightarrow$ terminates and decreases the complexity measure of proofs. For the standard completion rules given in Figure 9.2, the complexity measure is defined as follows, in [Bac91]. The complexity measure of elementary proof steps is given by:

$$c(s \longleftrightarrow_P t) \;\; = (\{s, t\})$$
$$c(s \longleftrightarrow_R^{l \to r} t) = (\{s\}, l \to r)$$

By convention, the complexity of the empty proof Λ is $c(\Lambda) = (\emptyset)$. Complexities of elementary proof steps are compared using the lexicographic combination, denoted $>_{ec}$, of the multiset extension $>^{mult}$ of the reduction ordering for the first component, and the ordering $\gg$ for the second component. Since both $>$ and $\gg$ are well-founded, so is $>_{ec}$. The complexity $c(\mathcal{P})$ of a non-elementary proof $\mathcal{P}$ is the multiset of the complexities of its elementary proof steps. Complexities of non-elementary proofs are compared using the multiset extension $>_c$ of $>_{ec}$, which is also well-founded. Then if $\mathcal{P} \Longrightarrow \mathcal{P}'$, one can check that $c(\mathcal{P}) >_c c(\mathcal{P}')$.

- Every minimal proof is in the desired normal form, in this case a rewrite proof. For that, a completion procedure must satisfy a fairness requirement, which states that all necessary proof transformations are performed.

The fairness hypothesis states that any proof reducible by $\Longrightarrow$ will eventually be reduced. In other words, no reducible proof is forgotten.

Definition 9.58. A derivation $(P_0, R_0) \longmapsto\!\!\!\to (P_1, R_1) \longmapsto\!\!\!\to \ldots$ is *fair* if, whenever $\mathcal{P}$ is a proof in $(P_i \cup R_i)$ reducible by $\Longrightarrow$, there is a proof $\mathcal{P}'$ in $(P_j \cup R_j)$ at some step $j \geq i$ such that $\mathcal{P} \overset{+}{\Longrightarrow} \mathcal{P}'$.   ◇

A sufficient condition to satisfy the fairness hypothesis can be given:

Proposition 9.59. *A derivation* $(P_0, R_0) \longmapsto\!\!\!\to (P_1, R_1) \longmapsto\!\!\!\to \ldots$ *is fair if* $CP(R_\infty)$ *is a subset of* P_*, R_∞ *is reduced and* P_∞ *is empty.*

In practice, the implementation of this abstract fairness condition is often performed through a marking process of rewrite rules, the critical pairs of which have been computed with all other rules, or which have been normalised. When fairness is ensured, proofs have normal forms with respect to $\Longrightarrow$. By induction on $\Longrightarrow$, if a derivation $(P_0, R_0) \longmapsto\!\!\!\to (P_1, R_1) \longmapsto\!\!\!\to \ldots$ is fair, any proof $t \overset{*}{\longleftrightarrow}_{P_i \cup R_i} t'$, for $i \geq 0$, has a rewrite proof $t \overset{*}{\longrightarrow}_{R_\infty} s \overset{*}{\longleftarrow}_{R_\infty} t'$.

Theorem 9.60. *If the derivation* $(P_0, R_0) \longmapsto\!\!\!\to (P_1, R_1) \longmapsto\!\!\!\to \ldots$ *is fair, then* R_∞ *is Church–Rosser and terminating. Moreover* $\overset{*}{\longleftrightarrow}_{P_0 \cup R_0}$ *and* $\overset{*}{\longleftrightarrow}_{R_\infty}$ *coincide on terms.*

The termination property of R_∞ is obvious since the test is incrementally processed for each rule added in R_*, thus in R_∞. The Church–Rosser property results from the fact that any proof has a normal form for $\Longrightarrow$ which is a rewrite proof. Eventually, the fact that $\overset{*}{\longleftrightarrow}_{P_0 \cup R_0} = \overset{*}{\longleftrightarrow}_{R_\infty}$ is a consequence of Lemma 9.57.

The completion may not terminate and may generate an infinite set of rewrite rules.

Example 9.61. [Der89] The theory of idempotent semi-groups (sometimes called bands) is defined by the following set of two axioms:

$$(x * y) * z = x * (y * z)$$
$$x * x = x$$

The completion diverges and generates an infinite set of rewrite rules:

$$(x * y) * z \rightarrow x * (y * z)$$

$$x * x \rightarrow x$$

$$x * (x * z) \rightarrow x * z$$

$$x * (y * (x * y)) \rightarrow x * y$$

$$x * (y * (x * (y * z))) \rightarrow x * (y * z)$$

$$\cdots$$

$$x * (y * (z * (y * (x * (y * (z * x)))))) \rightarrow x * (y * (z * x))$$

$$\cdots$$ ∎

Example 9.62. Rewriting and completion techniques have been applied to the study of several calculi defined for handling explicit substitutions in λ-calculus. The λ-calculus was defined by Church in the 1930s in order to develop a general theory for computable functions. The λ-calculus contains a rule β, called β-reduction, which consists of replacing some variable occurrences by a term. This substitution is described in the meta-language of λ-calculus and introduces the need for variable-renaming operations to take into account the binding-rules environment. This was the motivation for the notation introduced by de Bruijn in the 1970s for counting the binding depth of a variable. Another improvement has been introduced in the 1980s (although the idea was proposed by de Bruijn in 1978) to deal with substitutions: instead of defining them in the meta-language, a new operator is introduced to denote explicitly the substitution to perform. Thus, in addition to a rule (*Beta*) which starts the substitution, other rules are added to propagate it down to variables. Based on this idea, several calculi, regrouped under the name $\lambda\sigma$-calculi, have been studied and progressively refined in order to obtain suitable properties like termination and confluence.

The $\lambda\sigma$-calculi of [ACCL91, CHL96] are first-order rewriting systems. They contain the λ-calculus, written in de Bruijn notation, as a proper subsystem, but they differ by their treatment of substitution, which leads to slightly different confluence properties. The chosen example here is the $\lambda\sigma$-calculus of [ACCL91].

A substitution is represented as a list of terms with an operator *cons* (written ·) and an operator for the empty list (written *id* as it represents the identity substitution). The substitution $u_1 \cdot u_2 \cdot \ldots \cdot u_n \cdot id$ replaces 1 by u_1, ... , n by u_n, and decrements by n all the other (free) indices in the term. The operator ↑ can be seen as the infinite substitution $2 \cdot 3 \cdot 4 \cdot \ldots$ The term $u[s]$ denotes the application of substitution s to u. The composition operator is denoted as ∘. For more details, see [ACCL91, CHL96].

Let $\mathcal{X}$ be a set of term meta-variables, and $\mathcal{Y}$ be a set of substitution meta-variables. The set of *terms* and *explicit substitutions* is inductively defined as:

$$u = 1 \mid X \mid (u\ u) \mid \lambda u \mid u[s]$$
$$s = Y \mid id \mid \uparrow \mid u \cdot s \mid s \circ s$$

(Beta)	$(\lambda u)v \rightarrow u[v \cdot id]$
(App)	$(u\ v)[s] \rightarrow (u[s]\ v[s])$
(VarCons)	$1[u \cdot s] \rightarrow u$
(Id)	$u[id] \rightarrow u$
(Abs)	$(\lambda u)[s] \rightarrow \lambda(u[1 \cdot (s \circ \uparrow)])$
(Clos)	$(u[s])[t] \rightarrow u[s \circ t]$
(IdL)	$id \circ s \rightarrow s$
(ShiftCons)	$\uparrow \circ (u \cdot s) \rightarrow s$
(AssEnv)	$(s_1 \circ s_2) \circ s_3 \rightarrow s_1 \circ (s_2 \circ s_3)$
(MapEnv)	$(u \cdot s) \circ t \rightarrow u[t] \cdot (s \circ t)$
(IdR)	$s \circ id \rightarrow s$
(VarShift)	$1 \cdot \uparrow \rightarrow id$
(SCons)	$1[s] \cdot (\uparrow \circ s) \rightarrow s$

Fig. 9.3. A $\lambda\sigma$ term rewriting system

with $X \in \mathcal{X}$ and $Y \in \mathcal{Y}$. This is a first-order, many-sorted algebra built on the signature:

1	:	$\mapsto$ term
(_ _)	: term term	$\mapsto$ term
(λ_-)	: term	$\mapsto$ term
_[_]	: term substitution	$\mapsto$ term
id	:	$\mapsto$ substitution
$\uparrow$	:	$\mapsto$ substitution
_ · _	: term substitution	$\mapsto$ substitution
_ $\circ$ _	: substitution substitution	$\mapsto$ substitution

The $\lambda\sigma$-calculus is defined as the term rewriting system defined in Figure 9.3. If we drop the rule (Beta), we obtain the rewrite system that performs the application of substitutions.

The main properties of this $\lambda\sigma$ term rewriting system are:

1. The term rewriting system $\lambda\sigma$ is locally confluent on any term [ACCL91].
2. $\lambda\sigma$ is confluent on terms without a substitution variable [Río93].
3. $\lambda\sigma$ is *not* confluent on terms with term and substitution variables [CHL96].

∎

9.4.2 Extensions and applications

A large amount of work has been done on extending the completion techniques to handle rewriting modulo an equational theory or conditional rewriting, and to apply superposition and simplification in other contexts of automated deduction. Let us sketch below a few directions of research. Further references can also be found in [HKLR92, Kir95b].

Unfailing completion. Completion procedures can abort on an equality that cannot be oriented into a terminating rule. The idea of extending completion by computing equational consequences of non-orientable equalities can be traced back to [Bro75, Lan75]. In 1985, the notion of an *unfailing completion* was proposed: in this framework, orientable instances of an equality are used to perform reductions, even if the equality itself is not orientable. For instance, the commutativity axiom $(x * y = y * x)$ may have an instance $(f(x) * x = x * f(x))$ that can be oriented with a lexicographic path ordering. Such an instance reduces the term $(f(a) * a)$ to $(a * f(a))$. This method requires a reduction ordering $>$ which can be extended to an ordering total on ground terms. Deduction of new equalities requires the definition of *ordered critical pairs* between two equalities of E, obtained by unifying one term of an equality with a non-variable subterm of the other. The unfailing completion procedure, also called ordered completion, has only two possible outcomes: either it generates a finite set of equalities, or it diverges. In the first case, it provides a decision procedure for the validity of any equational theorem. In the second, it provides a semi-decision procedure [HR87, Rus87, Bac91, BDP89]. Another use of unfailing completion is a familiar method for mathematical proof, namely proof by contradiction: assume the negation of the formula to be proved and derive a contradiction. To refute this negation, the unfailing completion computes ordered critical pairs, simplifies equalities and eliminates trivial equalities, until it generates a contradiction [BDH86, BDP89]. The unfailing completion has been proved refutationally complete [HR87, Rus87, BDP89, Bac91].

Narrowing and goal solving. Narrowing is a relation on terms which generalises rewriting by using unification instead of matching, in order to find a rule whose application results in an instantiation followed by one reduction step of the term. This relation was first introduced in [Fay79, Hul80a, JKK83], in order to perform unification in equational theories presented by a confluent and terminating rewrite system R. Applying a narrowing step on an equation to be solved (the goal) in order to deduce new equations (or sub-goals) is again a process of superposing a rule of R onto the goal equation. This process is iterated until an equation is found both sides of which are syntactically unifiable. Then the composition of their most general unifier with all the substitutions computed by narrowing yields a unifier modulo R. The narrowing process consists of building all the possible narrowing derivations starting from the equation to be solved, and computes in this way a complete set of unifiers modulo the equational theory defined by R. However, the drawback of such a general method is that it very often diverges and several attempts have been made to restrict the size of the narrowing derivation tree [Hul80a, RKKL85, Rét87, DS88, You89, NRS89, KB91, Chr92, WBK94, LR96]. Another application of narrowing is its use as an operational semantics of logic and functional programming languages like BABEL

[MR92], EQLOG [GM86b] and SLOG [Fri85b], among many others. A survey of the integration of functions into logic programming based on narrowing, presenting both theoretical and practical points of view, can be found in [Han93]. More developments on narrowing can be found in Chapter 10.

Deduction with constraints. Constraints have been introduced in automated deduction since about 1990, although one could find similar ideas in theory resolution [Sti85] and higher-order resolution [Hue72]. A first motivation for introducing constraints in equational deduction processes is provided by completion modulo a set of axioms A [BD89, JK86a]. A serious drawback of this class of completion procedures is an inherent inefficiency, due to the computation of matchers and unifiers modulo A. A natural idea is to use constraints to record unification problems in the theory A and to avoid solving them immediately. Constraints are only checked for satisfiability, which is in general much simpler than finding a complete set of solutions or a solved form, especially in equational theories. Following ideas proposed in [KKR90, MN90, Pet90, BPW89, JM90, Kir95a] on ordering and equality constraints, further investigations have been carried out, in particular, to study completeness of deduction with constraints and the impact of constraints on simplification in [BGLS92, NR92a, NR92b, LS93, Lyn94, NR94, Vig94, Lyn95, BGLS95].

First-order theorem proving. It has been observed since 1975 [Bro75, Lan75, Sla74] that superposition is a restricted form of paramodulation [RW69], while simplification by rewriting is similar to demodulation [WRCS67], used in clausal theorem proving. However, until the 1980s, very few theoretical results were known about the effect of simplifying during deduction. An extension of the classical *semantic trees* method was designed in [Pet83, HR86, Rus89] to show the refutational completeness of several refinements of the resolution and paramodulation rules, that allow free interleaving of deduction steps with simplification steps, without loss of completeness. Another approach with a different proof technique is developed in [BG94a], where various refutationally complete calculi for first-order logic with equality, allowing elimination of redundant formulas, are presented. Ordered completion and completion of equational Horn clauses are derived as special cases.

9.5   Order sorted rewriting

In this section, rewriting and completion techniques are applied to order-sorted theories, whose expressive power covers a large subclass of partial functions and provides some way of handling errors.

Recent decades have seen the emergence of many frameworks integrating the notions of sort, subsort, equality and polymorphism [Obe62, Sch87a,

SNGM89, GM92]. These works assume statements of the form $A \leq B$ (sort inclusion) and $f : A, B \mapsto C$ (operator declaration) to be assertions for a static parsing process. They are used to define well-formed terms and are not directly encompassed in the deduction process. This results in various problems for obtaining a Birkhoff-like completeness theorem or deciding local confluence [JKKM92, Wal92], which leads to additional assumptions on the presentation. Another strange behaviour of the approach, considering sort inclusions and operator declarations as syntactic assertions, is that the standard rewriting tools do not behave as expected.

Example 9.63. The following specification, written here in an OBJ-like syntax and inspired by [SNGM89], illustrates two problems.

$$
\begin{array}{llll}
S = A, B & \mathcal{F} = f : A \mapsto A & R = & a \to b \\
\quad A \leq B & \quad a : \ \mapsto A & & \forall x : A, \\
& \quad b : \ \mapsto B & & \quad f(x) \to x
\end{array}
$$

First, since $a \xrightarrow{*} b$, one usually expects $f(a) \xrightarrow{*} f(b)$. However, while $f(a)$ is a syntactically well-formed term, $f(b)$ is not. Indeed, $f(b)$ is semantically well-formed since b, being equal to a, also gets the sort A. Second, even if, in order to avoid the previous problem, another declaration $f : B \mapsto B$ is added, the rewrite system of this specification presents another pathological feature. Although there is no (standard) critical pair between the two rewrite rules, the system is not locally confluent: the term $f(a)$ can be rewritten into a using the second rule, then to b, or into $f(b)$ using the first rule. But both b and $f(b)$ are then irreducible. Again, in order-sorted logic, since $a \to b$ and $a : A$, then b is also of sort A. Once this sort information is deduced, the peak is convergent owing to the application of $\forall x : A, f(x) \to x$.   ∎

A simple way to overcome these problems is to restrict to sort-decreasing rules [KKM88], i.e., rewrite rules that always decrease the sort of the rewritten term. Then a well-formed term always rewrites to another well-formed term. Completion procedures for order-sorted algebraic specifications have been designed [GKK90, Gan89, Wal92, Com92, Wer93, HKK94a, HKK98, Com98a, Com98b, Wer98], several of them using the restriction to sort-decreasing rules. Another solution is to deduce new sorts from equalities or rewrite rules, and to deal with the semantic sorts informally introduced in Example 9.63. This approach is followed in particular in [HKK98], where an algebraic framework based on rewriting techniques is proposed to perform order-sorted deductions. It is sketched below to illustrate how typing information can be handled and integrated in the rewriting process.

Assume a signature $\Sigma = (\mathcal{S}, \mathcal{F})$, with a universal sort U, and let $\mathcal{X}$ be an $\mathcal{S}$-sorted set of variables with declarations $(x :: A)$. Σ-Formulas are either existential formulas $(Ex\ t)$, or membership formulas $(t : A)$, or equalities $(t = t')$, and a presentation $\mathcal{P}$ is a set of Σ-formulas and variable declarations. A model $\mathcal{M}$ of a presentation $\mathcal{P}$ is a Σ-algebra such that $\forall A \in \mathcal{S}, A^{\mathcal{M}} \neq \emptyset$,

ExSubterm	$\dfrac{Ex\ t[u]}{Ex\ u}$	**ExMember**	$\dfrac{t:A}{Ex\ t}$
ExEqual	$\dfrac{t=t'}{Ex\ t}$	**Reflexive**	$\dfrac{Ex\ t}{t=t}$
Globality	$\dfrac{Ex\ t}{t:U}$	**VarMember**	$\dfrac{x::A}{x:A}$
ExReplace	$\dfrac{Ex\ t[u]\quad u=v}{Ex\ t[v]}$	**MeReplace**	$\dfrac{t[u]:A\quad u=v}{t[v]:A}$
EqReplace	$\dfrac{t[u]=w\quad u=v}{t[v]=w}$	**ExSubst**	$\dfrac{Ex\ t}{Ex\ \sigma(t)}$
MeSubst	$\dfrac{t:A}{\sigma(t):A}$	**EqSubst**	$\dfrac{t=t'}{\sigma(t)=\sigma(t')}$

Fig. 9.4. Order-sorted deduction

$U^{\mathcal{M}} = M$, the carrier of $\mathcal{M}$, and $\forall f \in \mathcal{F}, f^{\mathcal{M}}$ is a partial function from $M^{arity(f)}$ to M. A sound and complete deduction system for the class of such models is given in Figure 9.4.

The most important point to notice is the interaction between equality and membership formula expressed in the rule **MeReplace**. This leads to undecidability of typing and therefore of matching and unification. However, using an extended term structure that includes sort information in each term node, specific unification and matching algorithms can be designed and proved decidable. This structure is called a decorated term.

Definition 9.64. A *decoration* S is an expression of the form S' or $s \cup S'$, where S' is a set of sorts and s is a set-variable. A *decorated term* is either $x^{:S}$ with x a variable, or of the form $f(t_1^{:S_1}, \ldots, t_n^{:S_n})^{:S}$, where $t_1^{:S_1}, \ldots, t_n^{:S_n}$ are decorated terms, $f \in \mathcal{F}$ with arity n. $t^{:\downarrow\emptyset}$ represents t with empty decorations everywhere, except at positions with variables, where $x^{:S}$ becomes $x^{:\{A\}}$ if $(x :: A) \in \mathcal{P}$. ◇

When a term t is decorated by several sorts, say A, B, C, this means that in any model, t denotes an element common to the three subsets interpreting A, B, C, which then have a non-empty intersection. We are thus implicitly working with an extended sort structure that contains sets of sorts corresponding to sort intersections. In order to keep the same model construction and avoid empty sorts, any sort intersection introduced must be non-empty. For that, it is sufficient to require sort-inheritance. Informally this property states that the sort relation given in the presentation contains enough sorts to have the following property: if a term t can be proven in the presentation to be of sorts A and B, then there must exist a sort C with $C \leq A$ and $C \leq B$.

Definition 9.65. A presentation $\mathcal{P}$ is *sort inheriting* if for any term t, and sorts $A, \ldots, B$ in $\mathcal{S}$

$$\mathcal{P} \models t : A \wedge \ldots \wedge t : B \;\Rightarrow\; \exists C \in \mathcal{S}, \; C \leq A \;\wedge \ldots \wedge\; C \leq B \qquad \diamond$$

Sort inheritance is close to a semantic notion of regularity and is in general undecidable, just as membership formulas. It is implied by syntactic regularity and, in particular, any regular OBJ specification is sort inheriting. Sort inheritance can be understood as the property of the sort lattice to be closed by semantic sort intersection. A part of this closure operation can be performed on the initial subsort relation, but in general more intersections may be discovered during a completion process.

Matchers and unifiers are computed using the term structure as usual, but also using the sort information given in decorations and the sort structure. Since matching and unification only use the sort information available in the decorated term at unification or matching time, they are correct, but generally not complete, with respect to order-sorted deduction. Assuming sort inheritance, decorated unification is unitary.

Decorated rewrite systems have two kinds of rules. A *decorated rewrite rule* is an ordered pair of decorated terms, written $(p^{:S} \to q^{:S'})$, satisfying the usual property that variables of the right-hand side always appear in the left-hand side. Such rules implement the equational part of the presentation. For instance, the decorated rule $f(x^{:\{A\}})^{:\{A\}} \to x^{:\{A\}}$ corresponds to the rewrite rule $(\forall x :: A, \; f(x) \to x)$.

Another kind of rule is introduced, whose purpose is to increase decorations. A *decoration rewrite rule* is of the form $l^{:s} \to l^{:s \cup S_l}$ and only applies if the decoration matched by s is not included in S. Such rules implement the parsing part of the presentation. For instance, the decoration rule $f(x^{:\{A\}})^{:s} \to f(x^{:\{A\}})^{:s \cup \{A\}}$ corresponds to the rank declaration $f : A \mapsto A$, i.e., to the formulas $(\forall x :: A, \; f(x) : A)$.

The completion process proposed on decorated terms is based on the hypothesis that the axiomatisation is modularised in three parts: (i) equalities $(t = t')$, (ii) membership formulas $(t : A)$ and (iii) sort inclusions $(A \leq B)$. (i) and (ii) are handled via rewriting rules, and are thus modified and enriched during completion. On the contrary, (iii) is stable during the *whole* completion.

With each presentation $\mathcal{P}$ is associated a triple (D, E, R), computed as follows:

$$
\begin{array}{lll}
\text{With } Ex\ t & \text{is associated } (t^{:\downarrow\emptyset})^{:s} \to (t^{:\downarrow\emptyset})^{:s \cup \{U\}} & \text{in } D \\
\text{With } t : A & \text{is associated } (t^{:\downarrow\emptyset})^{:s} \to (t^{:\downarrow\emptyset})^{:s \cup \{A\}} & \text{in } D \\
\text{With } t = t' & \text{is associated } s^{:\downarrow\emptyset} = t^{:\downarrow\emptyset} & \text{in } E \\
\text{With } t \to t' & \text{is associated } t^{:\downarrow\emptyset} \to t'^{:\downarrow\emptyset} & \text{in } R
\end{array}
$$

The goal of the completion process is to find a presentation $\mathcal{P}_\infty$, defined by $(D_\infty, E_\infty, R_\infty)$, such that E_∞ is the empty set and any formula valid in $\mathcal{P}$ has a rewrite proof using $R_\infty \cup D_\infty$. More precisely:

$$\mathcal{P} \models Ex\ t \text{ if and only if } \exists A, S, t_0\ :\ t \xrightarrow{*} t_0^{:S \cup \{A\}}$$
$$(\mathcal{P}_\infty \text{ is existentially complete})$$
$$\mathcal{P} \models t : A \text{ if and only if } \exists S, t_0, A' \leq A\ :\ t \xrightarrow{*} t_0^{:S \cup \{A'\}}$$
$$(\mathcal{P}_\infty \text{ is type complete})$$
$$\mathcal{P} \models s = t \text{ if and only if } \exists A, S, t_0\ :\ s \xrightarrow{*} t_0^{:S \cup \{A\}} \xleftarrow{*} t$$
$$(\mathcal{P}_\infty \text{ is Church–Rosser})$$

The completion process is based on critical pairs computation, and simplification of rules and equalities using existing rewrite rules. A precise description can be found in [HKK94a, HKK98]. Assuming sort inheritance, when a fair completion does not fail, the resulting set of rewrite rules provides a way to prove not only equational theorems of the form $(t = t')$, but also typing theorems $(t : A)$ or existential formulas $(Ex\ t)$.

Theorem 9.66. *Assume $\mathcal{P}_0$ is a sort inheriting presentation. If the completion process constructs a derivation $(D_0, E_0, R_0) \mapsto\mapsto (D_1, E_1, R_1) \mapsto\mapsto \ldots$, such that E_∞ is empty and $CP(D_\infty \cup R_\infty)$ is a subset of $E_* \cup D_*$, $\mathcal{P}_\infty$ is terminating, Church–Rosser, type complete and existentially complete on all valid terms. Moreover $\mathcal{P}_0$ and $\mathcal{P}_\infty$ have the same equational theorems.*

The completion process is then adapted to detect the need for a sort structure extension and the failure of sort inheritance. If E contains $(p^{:S \cup \{B\}} = x^{:S' \cup \{A\}})$, with $(x :: A) \in \mathcal{P}$ and not $A \leq B$, the process fails, but detects that a sort inclusion is missing and proposes to add $A \leq B$. If D contains $(p^{:s} \to p^{:s \cup \{A\} \cup S})$ and $(p'^{:s} \to p'^{:s \cup \{B\} \cup S'})$, with $\sigma(p^{:\emptyset}) = \sigma(p'^{:\emptyset})$ for some unifying substitution σ, and if there exists no $C \leq A, B$, the process fails. It also detects intersecting sorts without a common subsort and proposes to add a new sort $[A, B]$, intuitively corresponding to the intersection of A and B, and a new membership formula $(\sigma(p) : [A, B])$, witnessing the fact that this intersection is non-empty.

These are conservative extensions of $\mathcal{P}_0$ but completion has to be restarted. Further work on how to ensure the sort inheritance hypothesis is found in [HKK94b].

Example 9.67. Applying this approach to Example 9.63 leads to the following steps of computation. The presentation is first translated into membership and equality formulas (first column below). Then, the formulas of the first column are translated into decoration rules and decorated equalities in the second column:

$$f(x) : A \qquad\qquad f(x^{:\{A\}})^{:s} \to f(x^{:\{A\}})^{:s \cup \{A\}}$$
$$a : A \qquad\qquad\qquad\quad a^{:s} \to a^{:s \cup \{A\}}$$
$$b : B \qquad\qquad\qquad\quad b^{:s} \to b^{:s \cup \{B\}}.$$
$$a = b \qquad\qquad\qquad\quad a^{:\emptyset} = b^{:\emptyset}$$
$$f(x) = x \qquad\qquad f(x^{:\{A\}})^{:\emptyset} = x^{:\{A\}}$$

The presentation is saturated, using a completion process, into the following:

$$f(x^{:\{A\}})^{:s} \rightarrow f(x^{:\{A\}})^{:s\cup\{A\}}$$
$$a^{:s} \rightarrow a^{:s\cup\{A\}}$$
$$b^{:s} \rightarrow b^{:s\cup\{A,B\}}$$
$$a^{:\{A\}} \rightarrow b^{:\{A,B\}}$$
$$f(x^{:\{A\}})^{:\{A\}} \rightarrow x^{:\{A\}}$$

In this saturated presentation, the term $f(b^{:\emptyset})^{:\emptyset}$ is first decorated (using the decoration rules) into $f(b^{:\{A,B\}})^{:\{A\}}$. Then it is rewritten using the last decorated rewrite rule above into $b^{:\{A,B\}}$. ∎

In other order-sorted frameworks, there are terms that may be syntactically ill-formed, although they rewrite to well-formed terms. The classical order-sorted specification of stacks of natural numbers provides as an example the term $top(pop(push(2, push(1, nil))))$. An operational solution, proposed in [GJM85, GD92, JKKM92], is to add some retracts at execution time, but this does not solve the problem in full generality. The decorated terms approach does not need retracts, or anything else, to handle these terms.

Example 9.68. Let us consider the classical example of stacks (of sort St) and non-empty stacks (of sort $NeSt$) of natural numbers (of sort Nat), with the sort inclusion $NeSt \leq St$. Let $x :: Nat, y :: NeSt, z :: St$ be sorted variables. We define the following sets D of decoration rewrite rules and R of decorated rewrite rules.

$$D = \begin{cases} nil^{:s} & \rightarrow nil^{:s\cup\{St\}} \\ push(x, z)^{:s} & \rightarrow push(x, z)^{:s\cup\{NeSt\}} \\ pop(y)^{:s} & \rightarrow pop(y)^{:s\cup\{St\}} \\ top(y)^{:s} & \rightarrow top(y)^{:s\cup\{Nat\}} \\ 1^{:s} & \rightarrow 1^{:s\cup\{Nat\}} \\ 2^{:s} & \rightarrow 2^{:s\cup\{Nat\}} \end{cases}$$

$$R = \begin{cases} top(push(x, z)^{:\{NeSt\}}) & \rightarrow x \\ pop(push(x, z)^{:\{NeSt\}}) & \rightarrow z \end{cases}$$

Reducing $top(pop(push(2, push(1, nil))))$ with D yields:

$$top(pop(push(2^{:\{Nat\}}, push(1^{:\{Nat\}}, nil^{:\{St\}})^{:\{NeSt\}})^{:\{NeSt\}}))$$

and then (with R and D):

$$1^{:\{Nat\}}$$

However, reducing $top(pop(push(1, nil)))$ gives $top(nil^{:\{St\}})$. This term is not valid, since top decoration in its normal form is empty. No retract is needed in this context. ∎

Another approach to order-sorted specifications, described in [BJM97], is based on membership equational logic, which has strong similarities with the inference rules proposed in Figure 9.4. Atomic formulas are equalities and sort membership assertions, while sentences are Horn clauses. Instead of dealing with sort information with rewrite rules as above, this approach uses tree automata to perform sort computations.

9.6 Inductive proofs

Rewriting techniques are also used to prove properties that hold in the initial algebra of an equational or conditional theory. The proof of such theorems may need an induction principle. A surprising feature of the completion procedure is that it is powerful enough to be used as an inductive theorem prover by applying the *proof by consistency method* [KM87]. The methods presented here do not require the explicit expression of an induction principle. This is why they are called *implicit induction methods*, in contrast to *explicit induction* mechanised, for instance, in NQTHM, LP or PVS.

The general principle is as follows: Let R be a ground convergent rewrite system that presents the theory E. An equational theorem $(t = t')$ holds in the initial algebra of E if and only if the completion of $R \cup \{t = t'\}$ does not derive an inconsistency. Broadly speaking, the discovery of an inconsistency witnesses the fact that the two sets of equalities R and $R \cup \{(t = t')\}$ do not define the same initial algebra. Different methods for detecting inconsistency have been proposed. In 1980, in the works of Musser [Mus80] and Goguen [Gog80], an inconsistency is the equality between Boolean terms $(true = false)$. This assumes an axiomatisation of Booleans and an equality predicate eq_s for each sort s. This also requires that any expression $eq_s(t, t')$ reduces to either *true* or *false*. In 1982, Huet and Hullot showed how the notion of constructors allows dropping the requirements about the equality predicate [HH82], provided there is no relation between constructors. An inconsistency is then just an equality between terms only built with different constructors. In 1986, Jouannaud and Kounalis, following previous ideas of Dershowitz [Der83] and Rémy [Rém82], introduced the concept of *ground reducibility* [JK86b], which allows handling relations between constructors. In their approach, an inconsistency is an equality between two irreducible ground terms. Given a ground convergent rewrite system R, a term t is *ground reducible* with R if all its ground instances are R-reducible. The key of their approach is that if $(l \rightarrow r)$ is a rewrite rule such that l is ground reducible with R, and if $R \cup \{l \rightarrow r\}$ is ground convergent, then $(l = r)$ is an inductive theorem of R. In 1988, Bachmair adapted unfailing completion to produce a procedure for proof by consistency that is refutationally complete [Bac88]. His method avoids two drawbacks of the previous methods: first, it does not fail with a non-orientable equality, which was a case for which nothing could be concluded in other methods. Second, inductive proofs are not concerned with the production of a (ground) Church–Rosser set of rules, which was obtained as a side effect in previous approaches. Following Fribourg [Fri86], Bachmair proposed a linear proof method that considerably reduces the number of equalities to be deduced.

In 1990, another major step was taken by formalising the rewriting induction method that does not require confluence, or ground confluence, of the rewrite system [KR90a, Red90]. The method gives a sufficient condition for proving inductive theorems. In the case where confluence is retained, the method is refutationally complete. The essential idea underlying rewriting induction is that, when a (conditional or unconditional) rewrite system R is terminating, the corresponding rewrite relation is a well-founded ordering on terms and can be used to support inductive reasoning. The method combines simplification by rewriting with the selection of significant induction schemas. The simplification strategy uses axioms, previously proved conjectures and instances of the conjecture itself as soon as they are smaller than the current conjecture, with respect to some well-founded relation [Bou94, BR95a, BKR95].

Despite their differences, the various techniques mentioned above have something in common: the computation of critical pairs in proofs by coherence corresponds to the instantiation of the conjecture to be proved in explicit induction, and also to the generation of induction schemas in rewriting induction.

9.6.1 Rewriting induction

This section is aimed at providing an introduction to the rewriting induction technique. In full generality, formulas to be proved are clauses of the form

$$(\bigwedge_{i=1,\dots,n} s_i = t_i \;\Rightarrow\; \bigvee_{j=1,\dots,m} u_j = v_j),$$

$>$ is a well-founded ordering on terms and R is a reductive conditional rewrite system with rules $(s_1 = t_1 \wedge \dots \wedge s_n = t_n) \Rightarrow l \to r$ such that the set of variables occurring in the right-hand side and in the condition of the rule, precisely $Var(r) \bigcup_{i=1,\dots,n} Var(s_i) \cup Var(t_i)$, is a subset of $Var(l)$.

Inductively valid clauses are implications that hold in the initial algebra defined by R. For a general discussion of inductive validity of first-order equational clauses, the reader should refer to [WG94].

Definition 9.69. A clause $(\bigwedge_{i=1,\dots,n} s_i = t_i \Rightarrow \bigvee_{j=1,\dots,m} u_j = v_j)$ is an *inductive theorem of* R if, for any ground substitution σ, whenever for all i, $i = 1,\dots,n$, $\sigma(s_i) \overset{*}{\longleftrightarrow}_R \sigma(t_i)$, there exists $j \in [1,\dots,m]$ such that $\sigma(u_j) \overset{*}{\longleftrightarrow}_R \sigma(v_j)$. The clause c is called inductively valid in R, which is denoted by $R \models_{ind} c$. ◇

To perform a proof by induction, it is necessary to provide some induction schemas. In the approach described below, these schemas are provided first by selecting *induction variables* on which induction has to be applied, and second by a special set of terms called the *test set*.

Definition 9.70. Let R be a conditional rewrite system and c be a clause or a term. The *set of induction variables* of c, denoted $indvar(c)$, is defined as follows: $x \in indvar(c)$ if for any ground substitution σ such that $Dom(\sigma) = Var(c) - \{x\}$, there exists a model of R in which $\sigma(c)$ is false.   ◇

To compute induction variables of c effectively, one can check that x occurs in a non-variable subterm u of c at position ω and there exists a conditional rewrite rule $(s_1 = t_1 \wedge \ldots \wedge s_n = t_n) \Rightarrow l \rightarrow r$ in R such that u and l are unifiable and either $l_{|\omega}$ is not a variable, or $l_{|\omega}$ is a non-linear variable of l, or $l_{|\omega}$ belongs to $\bigcup_{i=1,\ldots,n} indvar(s_i) \cup indvar(t_i)$.

Example 9.71. Consider the following system that defines the two predicates *odd* and *even* on natural numbers.

$$even(0) \rightarrow true$$
$$even(s(0)) \rightarrow false$$
$$even(s(s(x))) \rightarrow even(x)$$
$$even(x) = true \qquad\qquad odd(x) \rightarrow false$$
$$even(x) = false \Rightarrow \qquad odd(x) \rightarrow true$$

The variable x is an induction variable of both the term $even(x)$ and the term $odd(x)$.   ∎

The notion of cover sets introduced below provides a finite description of the set of irreducible ground terms for a set of conditional rewrite rules R: any ground R-irreducible term is an instance of some element in the cover set. The notion was introduced in [Red90, ZKK88].

Definition 9.72. A *cover set* for a set of conditional rewrite rules R is a finite set $CS(R)$ of R-irreducible terms that satisfies: for any R-irreducible ground term u, there exists a term t in $CS(R)$ and a ground substitution σ such that $\sigma(t) = u$.   ◇

Test sets have an additional property, which needs the notion of weak irreducibility. For a set of conditional rewrite rules R, a term t is weakly R-irreducible if for any subterm u, such that there exists a rule $\Gamma \Rightarrow l \rightarrow r$ and a substitution σ with $\sigma(l) = u$, $\sigma(\Gamma)$ is unsatisfiable in the theory defined by R.

Definition 9.73. A set of terms $TS(R)$ is a *test set* for a set of conditional rewrite rules R if $TS(R)$ is a cover set for R, such that for any term t and any instantiation σ of induction variables of t by terms of $TS(R)$, if $\sigma(t)$ is weakly R-irreducible, then there exists a ground substitution τ such that $\tau(\sigma(t))$ is ground and R-irreducible.   ◇

Test sets are used to build substitutions σ that instantiate variables by terms of $TS(R)$, appropriately renamed. $TSS(R)$ denotes this set of test set substitutions σ that associates with an induction variable, a term of the test set $TS(R)$.

Example 9.74. Consider the specification (Σ, R):

$$\Sigma = \begin{cases} sort & Int \\ 0 : & \mapsto Int \\ s : Int \mapsto Int \\ p : Int \mapsto Int \end{cases}$$

$$R = \begin{cases} \forall x, y : Int, & \\ \quad x + 0 \to x & p(x) + y \to x + p(y) \\ \quad 0 + x \to x & x + p(y) \to p(x + y) \\ \quad s(x) + y \to x + s(y) & s(p(x)) \to x \\ \quad x + s(y) \to s(x + y) & p(s(x)) \to x. \end{cases}$$

Here a test set is $TS(R) = \{0, s(z), p(z)\}$. For an induction variable x, $TSS(R) = \{(x \mapsto 0), (x \mapsto s(z)), (x \mapsto p(z))\}$. ∎

The construction of a test set for a rewrite system R is decidable, and algorithms are given for the equational case in [Kou90], and for the conditional case with free constructors in [KR90a, BR93]. The role of a test set for refuting conjecture is explained by the following criterion of provable inconsistency and its relation to non-inductive consequence.

Definition 9.75. Let R be a conditional rewrite system and $>$ a well-founded ordering that contains R. A clause $(\bigwedge_{i=1,\dots,n} s_i = t_i \Rightarrow \bigvee_{j=1,\dots,m} u_j = v_j)$ is *provably inconsistent* if there exists a test set substitution σ in $TSS(R)$ such that, for any $j = 1, \dots, m$, $\sigma(u_j)$ and $\sigma(v_j)$ are distinct and the maximal elements of $\{\sigma(u_j), \sigma(v_j)\}$ with respect to $>$ are weakly R-irreducible. ◇

This definition is simpler when R is a rewrite system. An equality $(g = d)$ is then provably inconsistent with respect to R if there exists a substitution σ in $TSS(R)$, such that $\sigma(g)$ and $\sigma(d)$ are distinct and the maximal elements of $\{\sigma(g), \sigma(d)\}$ with respect to $>$ are R-irreducible.

This notion of provable inconsistency provides the basis for a proof by refutation.

Theorem 9.76. [Bou94] *Let R be a ground confluent conditional rewrite system. If a clause C is provably inconsistent, then C is not an inductive consequence of R.*

Example 9.77. Consider the "union" operator defined on lists built from constructors $\{cons, nil\}$:

$$union(nil, nil) \to nil$$
$$union(cons(x, nil), nil) \to cons(x, nil)$$
$$union(nil, cons(x, l)) \to cons(x, l)$$
$$union(cons(x, l), cons(y, l')) \to cons(x, cons(y, union(l, l')))$$

R is ground confluent and terminating. $S(R) = \{nil, cons(x, l)\}$. Consider the conjecture

$$union(l, l') = union(l', l).$$

Among the test set instances, we have:

$$union(cons(x, l), cons(y, l')) = union(cons(y, l'), cons(x, l))$$

which simplifies to

$$cons(x, cons(y, union(l, l'))) = cons(y, cons(x, union(l', l))).$$

Among the test set instances of this last conjecture, we have:

$$cons(x, cons(y, union(nil, nil))) = cons(y, cons(x, union(nil, nil)))$$

which simplifies to

$$cons(x, cons(y, nil)) = cons(y, cons(x, nil))$$

which is provably inconsistent. So $union(l, l') = union(l', l)$ is not an inductive consequence of R. ■

The rewriting induction process is described below in the equational case, and the first-order clause case is handled in Chapter 10. The procedure is based on the expansion of an equality (or a clause) to be proved, using test sets and rewriting. Intuitively, an expansion is a reduced instance of an equality. Rewriting is performed on the greatest term of the instantiated equality or on both terms if they are incomparable in the ordering.

Definition 9.78. The *expansion set* of an equality $(p = q)$ such that $p \not< q$, with respect to a test set $TS(R)$, is the set of equalities $Exp(p = q)$ defined as
$Exp(p = q) = \{p' = \sigma(q) \mid \sigma \in TSS(R) \text{ and } \sigma(p) \longrightarrow_R p'\}$ if $p > q$,
$Exp(p = q) = \{p' = q' \mid \sigma \in TSS(R) \text{ and } \sigma(p) \longrightarrow_R p', \sigma(q) \longrightarrow_R q'\}$ if p and q are incomparable. ◇

Redundant formulas may be eliminated. For instance, an equality $(p = p)$ trivially holds and thus may be deleted. Simplification of the original equalities to be proved and their expansions, is performed.

Given a set R of rewrite rules, the rewriting induction process is parameterised by a reduction ordering $>$ and transforms two sets of equalities: C

Expand $C \cup \{p = q\}, H \mapsto\!\!\!\to C \cup Exp(p = q), H \cup \{p = q\}$
Delete $C \cup \{p = p\}, H \mapsto\!\!\!\to C, H$
Simplify $C \cup \{p = q\}, H \mapsto\!\!\!\to C \cup \{p' = q\}, H$
$\qquad\qquad\qquad$ if $p \longrightarrow_{RUHUC} p'$

Fig. 9.5. Rewriting Induction rules

contains the set of equalities to be proved and H registers equalities from C, that can be used as inductive hypotheses, after being expanded. The process is described by the set of rules in Figure 9.5.

These rules are instances of more complex rules for conditional theories given in [Bou94].

Theorem 9.79. *If there is a derivation* $(C, \emptyset) \mapsto\!\!\!\to (C_1, H_1) \mapsto\!\!\!\to \ldots \mapsto\!\!\!\to (\emptyset, H)$*, for some set* H*, then all equalities in* C *are inductive consequences of* R*.*

The proof of this result can be found in [Red90] for the equational case, and in [Bou94] for the conditional case.

Example 9.80. An easy illustration of the method is provided by the proof of the commutativity of $+$ on natural numbers. Consider the specification (Σ, R)

$$\Sigma = sort\ Int \qquad\qquad R = \forall x, y : Int,$$
$$\quad 0 : \qquad \mapsto Int \qquad\qquad\quad x + 0 \qquad \to x$$
$$\quad s : \quad Int \mapsto Int \qquad\qquad x + s(y) \quad \to s(x + y)$$

The chosen ordering is the multiset path ordering induced by $+ > s > 0$. Note that $TS(R) = \{0, s(z)\}$. Let $C_0 = \{u + v = v + u\}$. The equality is first expanded into $C_1 = \{u = 0 + u, s(u + z) = s(z) + u\}$. Then $H_1 = \{u + v = v + u\}$. Again the first equality in C_1 is expanded.

$$C_2 = \{0 = 0, s(y) = 0 + s(y), s(u + z) = s(z) + u\} \text{ and}$$
$$H_2 = \{u + v = v + u, u = 0 + u\}.$$

The first equality in C_2 is deleted and the second is simplified using the orientable equality $0 + u \to 0$ in H_2 to $s(y) = s(y)$, which can be deleted too. We are left with

$$C_3 = \{s(u + z) = s(z) + u\} \text{ and}$$
$$H_3 = \{u + v = v + u, u = 0 + u\}.$$

Again the first equality in C_3 is expanded.

$$C_4 = \{s(0 + z) = s(z), s(s(y) + z) = s(s(z) + y)\}$$
$$H_4 = \{u + v = v + u, u = 0 + u, s(u + z) = s(z) + u\}.$$

Refute $C \cup \{c\}, H \mapsto\!\!\!\!\rightarrow refute$
$\qquad$ if c is provably inconsistent

Fig. 9.6. Refutation rule

Then $s(0 + z) = s(z)$ is simplified (by $0 + u \to u$) and deleted.
$s(s(y) + z) = s(s(z) + y)$ is simplified twice by $s(z) + u \to s(z + u)$ and yields
$s(s(y + z)) = s(s(z + y))$.
Using then $u + v = v + u$ to simplify again, we eventually obtain a trivial
equality which is deleted. We end up with

$\qquad C_4 = \emptyset$ and
$\qquad H_4 = \{u + v = v + u, u = 0 + u, s(u + z) = s(z) + u\}.$ ∎

Returning to the assumption of confluence for the conditional rewrite
system R, the rules for Rewriting Induction may be used in a refutationally
complete method for inductive proofs. In order to obtain this refutation completeness result, a new rule, the **Refute** rule given in Figure 9.6, must be
added.

Theorem 9.81. [Bou94] *Let R be a ground convergent conditional rewrite
system. Let $(C_0, \emptyset) \mapsto\!\!\!\!\rightarrow (C_1, H_1) \mapsto\!\!\!\!\rightarrow ....$ be a derivation using rules for Rewriting
Induction and* **Refute**. *If there exists j such that* **Refute** *applies to (C_j, H_j),
then C_0 is not an inductive consequence of R.*

Example 9.82. Consider the specification

$\Sigma = sort \ Int$ $\qquad\qquad\qquad R = \forall x, y : Int,$
$\quad 0 \quad : \qquad\qquad \mapsto Int \qquad\qquad x + 0 \qquad \to x$
$\quad s \quad : Int \qquad \mapsto Int \qquad\qquad x + s(y) \quad \to s(x + y)$
$\quad + \quad : Int, Int \mapsto Int \qquad\qquad x - 0 \qquad \to x$
$\quad - \quad : Int, Int \mapsto Int \qquad\qquad 0 - x \qquad \to 0$
$\qquad\qquad\qquad\qquad\qquad\qquad\qquad s(x) - s(y) \to x - y.$

The following conjecture $\forall x, y, z : Int, \ (x - y) + y = x$ is not valid in the initial
algebra of the specification, since the process derives the equality $s(0) = 0$
and this contradiction refutes the conjecture. ∎

More experiments with this technique, including those with conditional
rewrite systems, can be found in [BR95a, Bou97]. In the particular case of
Boolean specifications where conditions are Boolean expressions, some of the
concepts defined above have been efficiently implemented in the system Spike
[BR95b].

9.6.2 Sufficient completeness

An important application of proofs by induction is checking sufficient completeness of specifications. Informally, this property ensures that the axioms of the specification defining a function f (here, the rewrite rules whose left-hand side is headed by the symbol f) cover all possible cases of arguments for the defined function f. For this purpose, the signature is usually divided into a set C of *constructors* and a set D of *defined function* symbols. The specification is said to be *sufficiently complete* with respect to constructors if for each sort $s \in \Sigma$, for any ground term t of sort s, there exists a constructor term t' of sort s in $\mathcal{T}(C)$ such that $t \xleftrightarrow{*}_R t'$. Sufficient completeness is undecidable in general. For non-conditional specifications, several criteria have been proposed [GH78, Bid81, HH82, Thi84, Kou85, Com86, JK89, LLT90] and are all based on the use of rewrite systems. In the conditional case, the class of considered specifications is restricted in [BR90] to systems with Boolean conditions. The technique presented below is described in [Bou96] in the more general framework of parameterised specifications.

Definition 9.83. The specification is said to be *operationally sufficiently complete* with respect to constructors if, for each sort $s \in \Sigma$, for any ground term t of sort s, there exists a constructor term t' of sort s in $\mathcal{T}(C)$ such that $t \xrightarrow{*}_R t'$. ◇

To check this property, for each defined function symbol, a pattern tree is built. The root of the tree is the term $f(x_1, \dots, x_n)$ if f has arity n, and the leaves give a partition of all possible arguments for f.

Definition 9.84. A term t is *reducible by case* with R if there exist a non-empty sequence of conditional rewrite rules $(\Gamma_1 \Rightarrow l_1 \to r_1, \dots, \Gamma_k \Rightarrow l_k \to r_k)$, a set of positions $\omega_1, \dots, \omega_k$ in t and a set of substitutions $\sigma_1, \dots, \sigma_k$ such that $t_{|\omega_1} = \sigma_1(l_1), \dots, t_{|\omega_k} = \sigma_k(l_k)$ and $\sigma_1(\Gamma_1) \vee \dots \vee \sigma_k(\Gamma_k)$ is inductively valid in R. ◇

Example 9.85. In Example 9.71, the term $odd(x)$ is reducible by case since there exist two rules with left-hand side $odd(x)$ and for which the disjunction of conditions $even(x) = true \vee even(s(x)) = true$ is inductively valid. On the contrary, the term $even(x)$ is not reducible by case. ∎

The pattern tree for f is built as follows: starting from the root, if at some node the term t with variable $x_1, \dots, x_n$ is R-irreducible by case, variables occurring at some specific positions are substituted by constructor terms of the form $c(y_1, \dots, y_m)$ where c is a constructor of the corresponding sort. The construction of the tree is achieved when all leaves are R-reducible by case, in which case f is completely defined, or not extensible any further. Then any R-irreducible leaf provides a useful pattern for introducing additional rules. A precise description of the algorithm is given in [Bou96]. In [BJM97, BJ97] the techniques presented here are generalised to the framework of membership equational logic.

9.7 Further reading

This chapter is by no means an exhaustive description of rewriting theory and applications. Excellent surveys have been written by G. Huet and D. Oppen [HO80], J.-W. Klop [Klo92], J. Avenhaus and K. Madlener [AM90], N. Dershowitz and J.-P. Jouannaud [DJ90, DJ91], D. Plaisted [Pla93], and the term rewriting approach to automated theorem proving is surveyed, for instance, in [HKLR92]. The book [BN98] offers a good introduction to the discipline of term rewriting. Recent advances can be found by looking at proceedings of the Conference on Rewriting Techniques and Applications, in Springer Lecture Notes in Computer Science volumes 202, 256, 355, 488, 690, 914, 1103, 1232 and 1379. Volume 909 in the same series, entitled Term Rewriting, is devoted to selected presentations at the French Spring School of Theoretical Computer Science in 1993.

Acknowledgements. This work has been partially supported by the Esprit Basic Research Working Group 22457 - Construction of Computational Logics II and by the Esprit Working Group 29432 - Common Framework Initiative Working Group COFI_WG.

10   Proof in Flat Specifications

Peter Padawitz

Informatik 5, Universität Dortmund, D-44221 Dortmund, Germany
padawitz@cs.uni-dortmund.de
http://ls5.cs.uni-dortmund.de/padawitz.html

10.1   Introduction

This chapter deals with the verification of data types. We put particular
emphasis on

- a uniform syntax for constructor-based specifications of both visible and
  hidden data types,

- Gentzen clauses, rules and proofs as a uniform schema for presenting
  (proofs of) conjectures in a natural, flexible, structured and implemen-
  table way that keeps the gap between informal reasoning and formal
  deduction as small as possible,

- a simple model- and proof-theoretical basis to which all more or less
  advanced rules and methods can be reduced for showing their correctness,

- providing the reader with syntactical criteria for the main conditions on
  a specification that shall be amenable to efficient proof and prototyping
  methods.

Deductive aspects of specifications are also treated in other chapters of
this book. Sections 2.6 through 2.9 provide basic notions and results for equational reasoning, i.e. for proving equational conjectures about specifications
with unconditional equational axioms. Section 4.5 deals with entailment (=
inference) relations in general. Chapter 9 handles axiomatization and proof in
terms of rewrite systems. Chapter 11 presents proof systems for specifications
that are given as expressions consisting of specification-building operations
(see Sections 4.6 and 6.3). Chapter 14 provides design rules for specifications
similarly to those that have been developed for functional and logic programs
(see [MPS93]).

In [Pad88a] we pointed out that the theoretical background of formal
specification originates in four mathematical research areas: recursion theory
(REC), automated theorem proving (ATP), abstract data types (ADT) and
term rewriting systems (TRS). The boundaries between these areas are about
to vanish: functional and logic programming and the development of specification languages fused REC and ADT to type theory; automated reasoning
systems involve rewrite methods. However, the results of REC, ATP, ADT
and TRS still have little impact on software design. This situation might be
remedied if the users of REC, ATP, ADT and TRS, respectively, would also
look a little more beyond the special theory they are employing and take

into account other foundations and results as well. So far ATP is mainly concerned with *relational* proof methods and their use in knowledge bases and classical AI applications. Its origin in algebra makes TRS view programs and data types as purely *functional* rewrite systems. REC deals with operational and denotational models of *computing and complexity*, while ADT focuses on the mere formal *description* of higher levels in software design, such as modularization, parameterization and refinement.

Only tailor-made syntactical, model- and proof-theoretic frameworks will meet the actual goals of formal software development. Although deductive and operational concepts still attract less attention in the ADT community than the model-theoretic issues, we face a great number of rule-based systems that are claimed to support the design of programs and data types. Less than thirty years ago there were only *assertion logics* for verifying imperative programs and the *fixpoint theory* of recursive functions. Nowadays the approaches concerned with deduction range from rewrite systems, type theory and frameworks dealing with the interfaces between several logics (see Chapter 4) to a large number of implemented provers and proof checkers. [ARS96] lists more than sixty currently available automated reasoning systems.

Although this chapter is supposed to give an overview of the state of the art in its topic, we think that the non-specialized reader would not gain much from a pure list of current deduction-oriented approaches to formal specification. Instead, we intend to integrate the main concepts, notions and definitions of several approaches into a uniform framework, which is based on **many-sorted predicate logic with equality** and admits functional, relational, state-oriented ("transitional") and all kinds of mixed specification and reasoning. Moreover, the uniform framework allows us to handle **visible** structures, which are representable by unique **normal forms**, along with **hidden** structures whose identity is defined in terms of transition relations, attributes and observations. Many-sorted predicate logic with equality is also sufficient for incorporating functional sorts, local definitions, polymorphism, parameterization and refinements. Instead of employing *order-sorted logic* (see Sections 2.10.4 and 9.5) we simulate subtype relations by unary predicates and include definedness predicates for expressing partiality into our basic notion of a **standard specification** (Definition 10.5).

For any expression (term or formula) e, $\mathbf{var}(e)$ denotes the set of variables occurring in e. e is **linear** if each variable occurs in e at most once. e is **ground** if $var(e)$ is empty. Both $\mathbf{e}(\mathbf{t})$ and $\mathbf{e}[\mathbf{t/u}]$ stand for expressions that include the subexpression t. $e[t/u]$ indicates that the subexpression u of e has been replaced by t. Set brackets { and } in a formula enclose either quantified variables or optional subexpressions. For instance, $t\{\equiv u\}$ stands for t or $t \equiv u$.

We start with a well-known data type, not for illustrating the particular features of standard specifications, but for fixing the syntactic environment these specifications are based upon. Precise definitions are given afterwards.

Example 10.1. The following specification consists of a **parameter specifi-cation** ENTRY and an extension LIST of ENTRY for presenting finite lists of elements taken from a model of ENTRY. While the axioms of the param-eter are arbitrary *Gentzen clauses* (Definition 10.4), the extension adds only Horn clauses to ENTRY.

ENTRY
    sorts   *entry entry′ bool*
    funs   *true, false*: → *bool*
               not: *bool* → *bool*
               eq: *entry* * *entry* → *bool*
    preds $\_ \neq \_, \_ \leq \_, \_ > \_$: *entry* * *entry*
    vars   *b*: *bool*   *x, y*: *entry*
    axioms $b \equiv true \lor b \equiv false$
               $true \equiv false \Rightarrow FALSE$
               $not(true) \equiv false$
               $not(false) \equiv true$
               $eq(x, y) \equiv true \Longleftrightarrow x \equiv y$
               $eq(x, y) \equiv false \Longleftrightarrow x \not\equiv y$
               $x \leq y \lor x > y$
               $x \leq y \land x > y \Rightarrow FALSE$

LIST = **signature of** ENTRY +
    vissorts *list* = *list(entry)*
    consts *nil*: → *list*
                $\_ :: \_$: *entry* * *list* → *list*
    defuns $[\_]$: *entry* → *list*
                $\_@\_$: *list* * *list* → *list*
                map: (*entry* → *entry′*) * *list* → *list(entry′)*
                filter: (*entry* → *bool*) * *list* → *list*
                remove: *entry* * *list* → *list*
    preds $\_ \in \_$: *entry* * *list*
               $\_ \notin \_$: *entry* * *list*
               sorted: *list*
               exists, forall: (*entry* → *bool*) * *list*
    vars   *x, y*: *entry* *L, L′*: *list* *f*: *entry* → *entry′* *g*: *entry* → *bool*
    Horn axioms
               $[x] \equiv x :: nil$
               $nil@L \equiv L$
               $(x :: L)@L′ \equiv x :: (L@L′)$
               $map(f, nil) \equiv nil$
               $map(f, x :: L) \equiv f(x) :: map(f, L)$
               $filter(g, nil) \equiv nil$
               $filter(g, x :: L) \equiv x :: filter(g, L) \Leftarrow g(x) \equiv true$
               $filter(g, x :: L) \equiv filter(g, L) \Leftarrow g(x) \equiv false$
  (A) $remove(x, L) \equiv filter(\lambda y.not(eq(x, y)), L)$
               $x \in y :: L \Leftarrow x \equiv y \lor x \in L$
               $x \notin nil$
               $x \notin y :: L \Leftarrow x \not\equiv y \land x \notin L$

$$sorted(nil)$$
$$sorted([x])$$
$$sorted(x :: y :: L) \Leftarrow x \leq y \wedge sorted(y :: L)$$
$$exists(g, x :: L) \Leftarrow g(x) \equiv true \vee exists(g, L)$$
$$forall(f, nil)$$
$$forall(g, x :: L) \Leftarrow g(x) \equiv true \wedge forall(g, L)$$

LIST takes the signature of ENTRY and adds *visible sorts* (**vissorts**), *constructors* (**consts**), *defined functions* (**defuns**) and *predicates* (**preds**). Structured sort symbols such as *list(entry)* stand for polymorphic types in the sense of functional programming languages. Sorts of a parameter specification such as *entry* correspond to type variables. Predicates specified by Horn clauses will be interpreted as least relations satisfying their axioms. Given a term $t(x)$ with variable(s) x, the λ-abstraction $\lambda x.t(x)$ is a short notation for an additional constructor, while the application $t(u)$ stands for the term $apply(t, u)$ where *apply* is an additional defined function. For instance, Axiom A actually reads as follows:

$$remove(x, L) \equiv filter(c(x), L)$$
$$apply(c(x), y) \equiv eq(x, y)$$

where $c : entry \rightarrow (entry \rightarrow bool)$ is an implicit constructor and $apply : ((entry \rightarrow bool) \times entry) \rightarrow bool$ is an implicit defined function.   ∎

Note the difference between predicates on the one hand and Boolean (defined) functions on the other hand. For obtaining efficient proof methods we will assume that all defined functions are total(izable), in other words: they have decidable domains. But Boolean functions with decidable domains do not have the expressive power of predicates because the latter can be regarded as Boolean functions with *semi*-decidable domains.

A pair $PSP = (PAR, SP)$ such as (ENTRY,LIST) is called a **parameterized specification** (see Section 6.4). Its semantics is given by the class of **actualizations of PSP by PAR-models**. An actualization of PSP is an unparameterized specification **SP(A)** that combines SP with a PAR-model A in the following way. The carrier elements of A are regarded as constructors of $SP(A)$. The interpretation F^A of a function or predicate $F \in PAR$ in A becomes the set $\{F(a) \equiv F^A(a) \mid a \in A\}$ resp. $\{F(a) \mid a \in F^A\}$ of axioms of $SP(A)$. A single actualization of PSP may also be built upon several PAR-models if SP involves several instances of a structured sort symbol such as $list(entry)$ and $list(entry')$.

If PAR is empty, then the semantics of PSP reduces to a single model, namely the initial SP-model (Theorem 10.11). If PAR is non-empty, all proofs of conjectures about PSP must presuppose an *arbitrary* actualization of PSP. Then the axioms of PAR are the only properties of the parameter such proofs will refer to.

Hence even in the parametric case we reason about a flat specification (see Section 2.7) because we have a single – though arbitrary – actualization in mind. This differs from the approach taken in Chapter 11 where the proof systems yield, in the first place, a *structured operational semantics* for specification-building operations (see Sections 4.6 and 6.3). The specifications we deal with here may also *originate* in expressions over specification-building operations. But these are somewhat "flattened away" in our presentations.

Section 10.2 introduces *standard specifications* as a uniform syntactical schema for presenting visible as well as hidden types. Loose, reachable and initial semantics as well as deductive and inductive theorems of standard specifications are defined, discussed and related to each other in various ways also in Section 10.2. Section 10.3 is devoted to the special model- and proof-theoretical issues that come up with hidden sorts, infinity and behavioural equivalence. Coinductivity as a syntactical criterion for the congruence property of behavioural equivalence and the derivation of axioms for complements deserve particular attention in Section 10.3. Moreover, Sections 10.2 and 10.3 provide the model- and proof-theoretical basis mentioned above to which examples and results presented in the sequel refer: the *cut calculus*, the *initial model* and, as its natural extension to hidden types, the *final model*.

Section 10.4 introduces the most natural and flexible schema of a *Gentzen proof* and shows how goal solving, program synthesis and other verification tasks fit into this schema. Section 10.5 deals with the main properties of a standard specifications: *completeness, confluence, termination* and *consistency* with respect to base models. Syntactical criteria that ensure these properties rely on the use of the (Horn) axioms of a standard specifications as rewrite rules (see Chapter 9). Section 10.5 provides both the general notions and results that justify this use and criteria for confluence and consistency that cover more or less special situations. *Implicit induction* denotes a class of proof methods, which are based on notions discussed in Section 10.5. Criteria for proving theorems about standard specifications by implicit induction are developed in Section 10.6. Section 10.7 puts all results presented so far together and infers the main rules for building up Gentzen proofs. The rules can roughly be classified into *unfold rules* that "execute" function and predicate definitions, lemma applications and fixpoint rules. Finally, Section 10.8 presents narrowing as the extension of rewriting to a basic rule for solving goals and embeds narrowing into Gentzen proofs.

10.2   Standard specifications

For the basic notions of many-sorted logic we refer the reader to Chapter 2. Here we set up the definition of a standard specification *SP*, which refines that of a many-sorted one insofar as the constructors of the specified domains are separated from the functions to be defined by *SP* and insofar as visible domains are distinguished from hidden ones whose objects are identified by their behaviour, i.e. their reaction to *actions* performed on them.

Given a set S of sorts, $s_1, \ldots, s_n \in S$ and an S-sorted set A, $A_{s_1 \ldots s_n}$ stands for the product $A_{s_1} \times \cdots \times A_{s_n}$. An S-**sorted relation** R **on** A is an S-sorted set such that for all $s \in S$, $R_s \subseteq A_s \times A_s$. R extends to an S^+-sorted relation as follows: $(a_1, \ldots, a_n) R(b_1, \ldots, b_n)$ iff for all $1 \le i \le n$, $a_i R_{s_i} b_i$.

We introduce special many-sorted signatures that separate visible from hidden sorts and data constructors from defined functions and include "undefined elements" as well as particular predicates for specifying definedness and equality relations.

Definition 10.2. A standard signature $\Sigma = (S, CO, DF, PR)$ consists of a set $S = visS \uplus hidS$ of **sorts** and disjoint[1] S^+-sorted sets CO of **constructors**, DF of **defined functions** and PR of **predicates**. The sorts of $visS$ resp. $hidS$ are called **visible** resp. **hidden**. Given $w \in S^+$, a symbol $f \in (CO \cup DF \cup PR)_w$ is **visible** resp. **hidden** if $w \in visS^+$ resp. $w \notin visS^+$. Both constructors and defined functions are called **function symbols**.

As usual, a function symbol $f \in (CO \cup DF)_{ws}$ is written as $f \colon w \to s$. If $w = \varepsilon$, f is called **a constant**. A predicate $r \in PR_w$ is written as $r \colon w$. For all $s \in S$, Σ implicitly contains the **equality predicate** $\equiv \colon ss$ and the **bisimulation predicate** $\sim \colon ss$. For all $s \in visS$, $\equiv \colon ss$ agrees with $\sim \colon ss$, and Σ implicitly contains the **bottom constant** $\bot \colon \to s$ and the **definedness predicate** $Def \colon s$.

Non-equality predicates are called **logical predicates**.

Given an S-sorted set X of variables, $T_\Sigma(X)$ denotes the S-sorted set of Σ-terms over X. Let $r \colon w \in PR$ and $t \in T_\Sigma(X)_w$. t is **visible** if $w \in visS^+$. t is **hidden** if $w \notin visS^+$. $r(t)$ is a Σ-**atom**. If r is a logical predicate resp. an equality predicate, then $r(t)$ is a **logical atom** resp. an **equation**. An equation $t \equiv t'$ is **visible** resp. **hidden** if t and t' are visible resp. hidden.

For all hidden constructors $f \colon w \to s$, s is hidden. Intuitively, hidden constructors do not construct visible data. The S-sorted set $NF_\Sigma(X)$ of Σ-**normal forms** is inductively defined as follows:

- $X \cup \{\bot \colon \to s \mid s \in visS\} \subseteq NF_\Sigma(X)$.
- For all visible constructors $f \colon w \to s$ and $t \in (NF_\Sigma(X) \setminus \{\bot\})_w$, $f(t) \in NF_\Sigma(X)_s$.
- For all hidden constructors $f \colon w \to s$ and $t \in NF_\Sigma(X)_w$, $f(t) \in NF_\Sigma(X)_s$.

T_Σ and NF_Σ denote the sets of all ground Σ-terms and Σ-normal forms, respectively. $s \in S$ is an **empty sort** if $T_{\Sigma,s}$ is empty. Σ is **inhabited** if all sorts of Σ are non-empty. Σ is **visible** if $hidS$ is empty. The greatest visible subsignature of Σ is denoted by $vis\Sigma$.   ◇

Note that all visible Σ-normal forms are $vis\Sigma$-terms. The purpose of ground normal forms is to represent data. A specification is *functional* if

---

[1] This does not exclude syntactic overloading for the sake of simpler notations.

all visible data have unique normal form representations (Definition 10.18). Hence visible data are identified by the *structure* of their normal forms and thus functions and predicates can be specified in terms of this structure and conjectures about them can be proved by induction on the structure. Hidden data also have normal forms, but here the representation need not be unique. Instead, the identity of hidden objects is defined in terms of the bisimulation predicates of Σ. The same conceptual difference between visible and hidden data also applies to definedness: visible sorts have bottom constants in order to make undefined values visible; hidden normal forms, however, are mere names for objects, so that hidden bottom constants would not make sense.

Let Σ be a standard signature with sort set S and X be an S-sorted set of variables.

Given a further signature $\Sigma' = (S', CO', DF', PR')$, a **signature morphism** $\sigma: \Sigma' \to \Sigma$ consists of a function $\sigma_{sorts}: S' \to S$ and $(S')^+$-sorted sets of functions $\sigma_{funs} = \{\sigma_w: (CO'_w \cup DF'_w) \to (CO_{\sigma(w)} \cup DF_{\sigma(w)})\}$ and $\sigma_{preds} = \{\sigma_w: PR'_w \to PR_{\sigma(w)}\}$ such that for all $f: w \to s \in CO' \cup DF'$, $\sigma(f): \sigma(w) \to \sigma(s)$ and for all $r: w \in PR'$, $\sigma(r): \sigma(w)$ (see Section 2.5).

Definition 10.3 (substitutions). An S-sorted function $\sigma: X \to T_\Sigma(X)$ is called a **substitution** over Σ.[2] The **domain of** σ, $dom(\sigma)$, is the set of all variables x with $x\sigma \neq x$. Given $Y \subseteq X$, σ_Y denotes the restriction of σ with $dom(\sigma) \subseteq Y$. The substitution with the empty domain is called the **identity substitution** and denoted by id. σ is **ground** resp. **ground normal** if $dom(\sigma)\sigma \subseteq T_\Sigma$ resp. $dom(\sigma)\sigma \subseteq NF_\Sigma$. T_Σ^X denotes the set of ground substitutions over Σ. The **instance** $t\sigma$ of a term or atom t **by** σ is obtained from t by replacing each variable x by $x\sigma$. A term or atom u **subsumes** t, written $u \leq t$, if t is an instance of u. t is **unifiable** with u and σ is a **unifier** of t and u if $t\sigma = u\sigma$. σ **subsumes** a substitution τ, written $\sigma \leq \tau$, if for all $x \in X$, $x\sigma \leq x\tau$. σ is **minimal** or **most general** with respect to a property P of substitutions if σ fulfils P and subsumes all substitutions fulfilling P.

Let $Y, Z \subseteq X$ and σ, τ be substitutions. The composition $\sigma\tau$ is defined by $x(\sigma\tau) = (x\sigma)\tau$ for all $x \in X$. σ is a **renaming of Y away from Z** if $dom(\sigma)\sigma \subseteq X$, $|Y\sigma| = |Y|$ and $Y\sigma \cap Z = \emptyset$. σ is a **skolemization** if for all $x \in dom(\rho)$, $x\sigma$ is a "new" constant. Skolemizations prevent variables of an expression e from being replaced when a further substitution is applied to $e\rho$ (Definition 10.12). ◇

Definition 10.4. A Σ-**goal** is a finite conjunction G of Σ-atoms. The empty conjunction is called **empty goal** and denoted by $\emptyset$ or $TRUE$. Given $Y \subseteq X$ and two goals G, H, the formula $\exists Y G$ is an **existential goal**. A **goal set** is a finite disjunction of existential goals. The empty disjunction is denoted by $FALSE$. If GS and HS are goal sets, then $GS \Leftarrow HS$ is a **Gentzen clause**.

---

[2] Both substitutions and signature morphisms are usually denoted by small Greek letters.

Let $r(t)$ be a logical atom, $f(t) \equiv u$ be an equation, H be a goal and ϕ be an existential goal. The formula $r(t) \Leftarrow H$ is a **Horn clause for** r. $f(t) \equiv u \Leftarrow H$ is a **Horn clause for** f. $r(t) \Rightarrow (H \Rightarrow \phi)$ is a **co-Horn clause for** r. Several Horn clauses $p \Leftarrow H_1, \ldots, p \Leftarrow H_n$ with the same conclusion are sometimes combined to the equivalent formula $p \Leftarrow (H_1 \vee \cdots \vee H_n)$.   ◇

Gentzen clauses are as expressive as arbitrary first-order formulas.[3] Moreover, functional-logic programs are merely Horn clauses and most conjectures about them, such as pre/post-conditions, invariants, etc., are given as Gentzen clauses. As being implications Gentzen clauses can be passed over directly to tableau (top-down, backward) proof calculi. They need not be transformed into space-consuming normal forms before a theorem prover can deal with them. Skolemization and other normalizations often destroy the original structure (and easy-to-understand meaning) of conjectures so that user interaction with a prover that processes normalized formulas is mostly impossible.

While conjectures may be presented as arbitrary Gentzen clauses, the axioms of a standard specification are restricted to Horn and co-Horn clauses. This complies with usual syntactic schemas adopted by functional, logic and even state- or object-oriented programs. Semantically, the restriction to Horn and co-Horn clauses guarantees the existence of initial and final models and thus of concrete implementations. These models also entail a number of "meta-theorems" that lead to proof rules, which the theorem prover can apply automatically (see Section 10.3).

Definition 10.5. A **standard specification** $SP = (\Sigma, AX)$ consists of a standard signature Σ and a set AX of Horn and co-Horn clauses, the **axioms** of SP, such that each predicate of Σ is either a **safety predicate** or a **liveness predicate** and the following conditions hold true.

- All axioms for functions and safety predicates are Horn clauses.
- All axioms for liveness predicates are co-Horn clauses.
- Bisimulation predicates for hidden sorts are liveness predicates.
- AX implicitly contains the following **partiality axioms** for each visible constructor $c: s_1 \ldots s_n \to s$ of Σ and all $1 \le i \le n$:

$$c(x_1, \ldots, x_{i-1}, \bot, x_{i+1}, \ldots, x_n) \equiv \bot$$
$$Def(c(x_1, \ldots, x_n)) \Leftarrow Def(x_1) \wedge \cdots \wedge Def(x_n).$$

- For all other Horn axioms $C = (f(t_1, \ldots, t_n)\{\equiv u\} \Leftarrow H)$,

(1) f is a defined function or a logical predicate $\ne Def$, $t = (t_1, \ldots, t_n)$ is a tuple of normal forms, each variable of u occurs in t or H,

---

[3] The lack of explicit negation is compensated by the fact that predicates often induce axioms for their complements by simply negating the axioms (see Sections 10.3 and 10.4).

(2)   for all $1 \leq i \leq n$, $t_i \in X \cup \{\bot\}$ or $Def(t_i) \in H$,[4]

(3) C consists of visible symbols if f is visible,

(4) H is sequential (see below) and does not contain liveness predicates.

(5)   For all co-Horn axioms $r(x) \Rightarrow (H \Rightarrow GS)$, r is a liveness predicate, $r(x) \wedge H$ and GS are sequential (see below) and H does not contain liveness predicates.

(6) AX implicitly contains the axiom $x \sim y \Rightarrow y \sim x$ and each other axiom for $\sim$ has one of the following forms:

$$x \sim y \;\Rightarrow\; f(x,z) \sim f(y,z)$$
$$x \sim y \;\Rightarrow\; (r(x,z) \Rightarrow r(y,z))$$
$$x \sim y \;\Rightarrow\; (\delta(x,z,x') \Rightarrow \exists y'(\delta(y,z,y') \wedge x' \sim y'))$$

where $f: sv \to s'$ is a defined function and $r : sv$ and $\delta : svs''$ are safety predicates with $s, s'' \in hidS$ and $v \in visS^*$. f, r and δ are called **functional**, **relational** and **transitional** s-actions, respectively.

TA, **SPR** and **LPR** denote the sets of transitional actions, logical safety predicates resp. liveness predicates of SP. Given a set Y of hidden variables, the set of **sequential goals with target** Y is inductively defined as follows:

- The empty goal is sequential with empty target.

- If G is sequential with target Y and $r(t)$ is an atom such that $r \notin TA$ is a logical or visible equality predicate, then $G \wedge r(t)$ is sequential with target Y.

- If G is sequential with target Y and $\delta(t, u, x)$ is an atom such that $\delta \in TA$ and x is a variable that does not occur in G, t or u, then $G \wedge \delta(t, u, x)$ is sequential with target $Y \cup \{x\}$.

A goal set $\exists X_1 G_1 \vee \ldots \exists X_n G_n$ is **sequential** if for all $1 \leq i \leq n$, G_i is sequential with some target $Y_i \subseteq X_i$.

SP is **visible** if Σ is visible (see Definition 10.2). By (3), SP has a visible subspecification **visSP** that consists of $vis\Sigma$ and the (Horn) axioms for $vis\Sigma$.

◇

Conditions (4) and (5) are crucial for the class of SP-models to be closed under quotients by bisimulation predicates (see Theorem 10.8), in other words, for SP to be **behaviourally consistent** in the sense of [BHW95]). In particular, (4) implies that Horn-axiom premises do not contain hidden equations. This condition is already present in previous approaches to behavioural specifications (see, e.g., [BW82b], Corollary 4; [Wir90], Theorem 5.4.5; [BHW95], Example 3.24). Since we will reason about SP with respect to the final model, which is the quotient by bisimilarity of the initial model (see Section 10.3), behavioural consistency is crucial.

---

[4] For the sake of brevity, Def-atoms in the premise of an axiom are usually omitted.

(4) and (5) also ensure that liveness predicates can be defined in terms of safety predicates, but not vice versa. This allows us to construct the final model in a hierarchical way: the initial model of the Horn axioms is extended by greatest solutions of the co-Horn axioms and then factored through bisimilarity. (5) guarantees that the corresponding *consequence operator* Ψ is monotonic and thus greatest solutions do exist (see Theorem 10.19). By the way, the hierarchy condition involved in (5) resembles the stratification condition employed for constructing minimal models of logic programs (see [ABW88]).

Conditions (1) and (2) allow us to specify partial functions that are continuous in the sense of recursive function theory. The partiality axioms imply that visible *constructors* are strict, i.e. preserve bottom constants. Condition (2) admits non-strict defined functions that map $\bot$ to "defined values".

Example 10.6. Let NAT and LIST be visible specifications of natural number arithmetic and finite lists of elements of a sort *entry*. We extend LIST + NAT by a specification of **infinite streams**:

INFSEQ = LIST + NAT +
    hidsorts *stream* = *stream*(*entry*)
    consts   _&_ : *entry* * *stream* → *stream*
              01: → *stream*(*nat*)
              nats: *nat* → *stream*(*nat*)
              odds: *stream* → *stream*
              zip: *stream* * *stream* → *stream*
              map: (*entry* → *entry'*) * *stream* → *stream*(*entry'*)
    defuns   *head*: *stream* → *entry*
              tail: *stream* → *stream*
              10: → *stream*(*nat*)
              _#_ : *list* * *stream* → *stream*
              evens: *stream* → *stream*
    preds   _ ~ _ : *stream* * *stream*
              exists, *forall*, *forallExists*: (*entry* → *bool*) * *stream*
              fair: *stream*
    vars   *n*: *nat*   *x*: *entry*   *L*: *list*   *s*, *s'*, *s1*, *s2*: *stream*
              f: *entry* → *entry'*   *g*: *entry* → *bool*
    Horn axioms
              $head(x\&s) \equiv x$ $tail(x\&s) \equiv s$
              $head(01) \equiv 0$ $tail(01) \equiv 10$
              $head(nats(n)) \equiv n$ $tail(nats(n)) \equiv nats(n+1)$
              $head(odds(s))) \equiv head(s)$ $tail(odds(s)) \equiv odds(tail(tail(s)))$
              $head(zip(s,s')) \equiv head(s)$ $tail(zip(s,s')) \equiv zip(s', tail(s))$
              $head(map(f,s)) \equiv f(head(s))$
                                        $tail(map(f,s)) \equiv map(f, tail(s))$

              $10 \equiv 1\&01$
              $nil\#s \equiv s$
              $(x :: L)\#s \equiv x\&t \ \Leftarrow \ L\#s \equiv t$
              $evens(s) \equiv odds(tail(s))$
              $exists(g,s) \ \Leftarrow \ g(head(s)) \equiv true \lor exists(g, tail(s))$

co-Horn axioms

$$s \sim s' \Rightarrow head(s) \equiv head(s')$$
$$s \sim s' \Rightarrow tail(s) \sim tail(s')$$
$$forall(g, s) \Rightarrow g(head(s)) \equiv true \land forall(g, tail(s))$$
$$forallExists(g, s) \Rightarrow exists(g, s) \land forallExists(g, tail(s))$$

(A) $fair(s) \Rightarrow forallExists(\lambda x.eq(x, 0), s)$

(B) $fair(s) \Rightarrow \exists\{L, s'\}(0 \notin L \land s \equiv L\#(0\&s') \land fair(s'))$

We expect a model that interprets INFSEQ as follows. & appends an entry to a stream. 01 and 10 denote the streams whose elements alternate between zeros and ones. $nats(n)$ generates the stream of all numbers starting from n. $odds(s)$ returns the stream of all elements of s that have odd-numbered positions in s. *zip* merges two streams into a single stream by alternatively appending an element of one stream to an element of the other stream. # concatenates a list and a stream into a stream. *head*, *tail*, *map*, *exists* and *forall* have the same meaning as stream functions as they have as list functions. $\sim$ is stream equality. $forallExists(g, s)$ is valid iff $g(x) = true$ for infinitely many $x \in s$. Hence $fair(s)$ holds true iff s contains infinitely many zeros. This results from Axiom A as well as from Axiom B.   ∎

As to liveness and safety predicates, modal logic usually adopts a different view. The former are called safety properties insofar as they are *invariance* conditions. The latter are called liveness properties because they describe *reachability* conditions (see, e.g., [Lar88]). This view is reasonable if liveness and safety should tell us how the validity of predicates referring to *individual* states changes in time. But here a (hidden) predicate always refers to the *set* of states that are reachable from a single one. Roughly said, the validity of a safety predicate for a state set S can be deduced *inductively* from considering finite subsets of S, while the validity of a liveness predicate for S can only be "decided" if the entire – maybe infinite – set is taken into account. Nevertheless, despite the different liveness-versus-safety view both modal logic and (co-)Horn logic formalize invariance and reachability conditions as least and greatest fixpoints, respectively.

Definition 10.7 (semantics). Let S be the set of sorts of Σ. A Σ-**structure** A consists of an S-sorted set, the **carrier** of A, also denoted by A, a function $f^A\colon A_w \to A_s$ for each function symbol $f\colon w \to s$ and a relation $r^A \subseteq A_w$ for each predicate $r\colon w$ of Σ.[5] Given two Σ-structures A and B, a Σ-**homomorphism** $h\colon A \to B$ is a homomorphism in the sense of Section 2.3 such that $h(r^A) \subseteq r^B$ for all predicates r of Σ. A and B are Σ-**isomorphic**, written: $A \cong B$, if $g \circ h = id_A$ and $h \circ g = id_B$ hold true for some Σ-homomorphism $g\colon B \to A$.

Let $\approx$ be an S-sorted equivalence relation on A. $\approx$ is A-**compatible** with a function symbol $f\colon w \to s \in \Sigma$ if for all $a, b \in A_w$, $a \approx b$ implies $f^A(a) \approx f^A(b)$. $\approx$ is a Σ-**congruence** if $\approx$ is A-compatible with all function

---

[5] If Σ has no predicates, a Σ-structure is a Σ-algebra (see Section 2.2).

symbols of Σ. $\approx$ is A-**compatible** with a predicate $r \in \Sigma$ if for all $a \in r^A$, $a \approx b$ implies $b \in r^A$.

If $\approx$ is a Σ-congruence, then the **quotient** $A/\approx$ **of** A **by** $\approx$ is the Σ-structure that interprets sorts and function symbols as in Section 2.3. The equivalence class of all $b \in A$ with $a \approx b$ for some $a \in A$ is denoted by $[a]$. This notation extends to tuples: $[(a_1, \ldots, a_n)]$ is an abbreviation of $([a_1], \ldots, [a_n])$. $A/\approx$ interprets each predicate $r \in \Sigma$ as the set of equivalence classes $[a]$ with $b \in r^A$ for some $b \approx a$.

Given a Σ-structure A and a signature morphism $\sigma: \Sigma' \to \Sigma$, the definition of the σ-**reduct** $A|_\sigma$ of A (see Section 2.5) is extended to predicates: for all $r \in \Sigma'$, $r^{A|_\sigma}$ is defined as $\sigma(r)^A$. If σ is an inclusion, i.e. $\Sigma' \subseteq \Sigma$, we write $A_{\Sigma'}$ instead of $A|_\sigma$ and call $A_{\Sigma'}$ the Σ'-**reduct of** A.

The interpretation of Σ-terms in a Σ-structure A depends on a **valuation** of variables in A, i.e. an S-sorted function $b: X \to A$. The unique Σ-homomorphism extending b to a function from $T_\Sigma(X)$ to A is denoted by b^*. If t is a ground term, then for all $b: X \to A$, $b^*(t)$ has the same value and we write t^A instead of $b^*(t)$. A is **reachable** if for all $a \in A$ there is a ground term t with $t^A = a$. Each Σ-structure A has a unique reachable substructure, denoted by **gen(A)** (see Section 2.2).

A valuation $b: X \to A$ **solves** an equation $t \equiv t'$ **in** A if $b^*(t) = b^*(t')$. b **solves** a logical atom $r(t)$ in A if $b^*(t) \in r^A$. The notion of a solution is extended from atoms and equations to complex first-order formulas as usual. If b solves a formula F in A, we write $A \models_b F$. A **satisfies** F or F is **valid** in A, written $A \models F$, if all valuations in A solve F in A.

Let $SP = (\Sigma, AX)$ be a standard specification. A Σ-structure A is an SP-**model** if A satisfies all axioms of SP. The classes of all resp. all reachable SP-models are denoted by **Mod(SP)** resp. **Gen(SP)** and called the **loose** resp. **reachable semantics** of SP.   ◇

The main problem caused by the requirement that SP is behaviourally consistent (see above) comes with the set TA of transitional actions of SP (Definition 10.5). Given a Σ-structure A, let's call $\sim^A$ a **bisimulation** if A satisfies the axioms for $\sim$. The point is that a bisimulation is usually not compatible with transitional actions, at least not with respect to their target (Definition 10.5). The congruence property (wrt TA) of $\sim^A$ and the bisimulation property of $\sim^A$ coincide if and only if TA consists of the input-output relations of *functional* actions.[6] Figure 10.13 illustrates both properties schematically. On the one hand, a congruence $\approx$ that is not A-compatible with a predicate r may violate the monotonicity of the step from A to $A/\approx$ in the sense that all goals satisfied by A remain valid in the quotient. On the other hand, the standard model of a standard specification SP will be a quotient (Definition 10.20). Hence we have to make sure that the axioms

---

[6] This coincidence is probably the reason why most behavioural-semantics approaches only consider functional actions.

of SP are preserved by the quotient constructions. With respect to TA, this is accomplished by Conditions 10.5(4) and (5) on the axioms of SP – provided that $\sim^A$ is A-compatible with all other functions and predicates of SP (Theorem 10.8). If A is the initial SP-model, then A-compatibility with functions and *safety* predicates will be guaranteed by the additional condition of coinductivity (Definition 10.24).

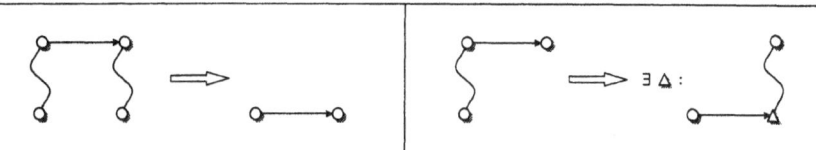

Fig. 10.1. Congruence versus bisimulation

Theorem 10.8. [Pad98] *Let A be an SP-model such that $\sim^A$ is compatible with $\Sigma \setminus TA$. Then $B =_{def} A/\sim^A$ is an SP-model, which interprets the logical predicates of Σ as follows:*

- *For all $r \in SPR \setminus TA \cup LPR$, $[a] \in r^B \iff_{def} a \in r^A$.*
- *For all $\delta \in TA$, $([a],[b],[c]) \in \delta^B \iff_{def} \exists\, c' \sim^A c : (a,b,c') \in \delta^A$.*

Let us now turn to the particular models induced by a standard specification.

Definition 10.9. Given a class $\mathcal{C}$ of Σ-structures, $I \in \mathcal{C}$ is **initial** in $\mathcal{C}$ if for all $A \in \mathcal{C}$ there is a unique Σ-homomorphism $ini^A : I \to A$. $T \in \mathcal{C}$ is **final** in $\mathcal{C}$ if for all $A \in \mathcal{C}$ there is a unique Σ-homomorphism $fin^A : A \to T$.   ◇

Each two initial (resp. final) Σ-structures are Σ-isomorphic. The initial model of SP is usually presented as the quotient of T_Σ by the equivalence relation consisting of all ground equations that are derivable from the axioms of SP via the *cut calculus:*[7]

Definition 10.10. Let $SP = (\Sigma, AX)$ be a standard specification. The **cut calculus for** SP consists of all Horn (!) axioms of SP, the following **congruence axioms** stating that $\equiv$ is a Σ-congruence and compatible with all predicates of Σ:

---

[7] See Theorem 2.23 for the case of specifications whose axioms are unconditional equations.

$$x \equiv x$$

$$y \equiv x \ \Leftarrow \ x \equiv y$$

$$f(x_1, \ldots, x_n) \equiv f(x_1, \ldots, x_{i-1}, y, x_{i+1}, \ldots, x_n) \ \Leftarrow \ x_i \equiv y$$

for all function symbols $f: s_1 \ldots s_n \to s \in \Sigma$ and $1 \le i \le n$

$$r(x_1, \ldots, x_n) \ \Leftarrow \ r(x_1, \ldots, x_{i-1}, y, x_{i+1}, \ldots, x_n) \wedge x_i \equiv y$$

for all predicates $r : s_1 \ldots s_n \in \Sigma$ and $1 \le i \le n$.

and the following inference rules. Let p be an atom and G, H be goals.[8]

instantiation $\dfrac{p \Leftarrow H}{p\sigma \Leftarrow H\sigma} \ \Downarrow$ for all substitutions σ over Σ

cut $\dfrac{p \Leftarrow G \wedge H, \ H}{p \Leftarrow G} \ \Downarrow$

$\wedge$-generation $\dfrac{G, \ H}{G \wedge H} \ \Downarrow$

Given a formula F, we write $SP \vdash_{cut} F$ if there is a derivation of F employing only axioms and rules of the cut calculus for SP. The set of derivable equations $t \equiv t'$ induces the **SP-equivalence** relation $\equiv_{SP}$:

$$t \equiv_{SP} t' \ \Longleftrightarrow_{def} \ SP \vdash_{cut} t \equiv t'.$$

A normal form u is a **normal form of** a term t if t and u are SP-equivalent.

SP-equivalence yields an interpretation of $\equiv$ in T_Σ. If one interprets all logical safety predicates r of SP accordingly, i.e.

$$r_{SP}(t) \ \Longleftrightarrow_{def} \ SP \vdash_{cut} r(t),$$

then T_Σ becomes a $(\Sigma \setminus LPR)$-structure and SP-equivalence becomes a Σ-congruence. The quotient of T_Σ by $\equiv_{SP}$ is called the **initial SP-model** and denoted by **Ini(SP)**. A first-order formula satisfied by $Ini(SP)$ is called an **initial theorem** of SP.   ⋄

Theorem 10.11. *Let $SP = (\Sigma, AX)$ be a standard specification. Each SP-model A with $A_{\Sigma \setminus LPR} \cong Ini(SP)$ is an initial Σ-structure in $Mod(SP)$.*

The choice of $Ini(SP)$ as the standard model of a (Horn) specification has been motivated thoroughly in the literature (see, e.g., [MG85, EM85, Pad88a]). Initial models reduce the reasoning about visible data types to inductive theorem proving. Initial semantics complies with function sorts, polymorphism and parameter specifications (see Examples 10.1 and 10.6).

---

[8] Arrows attached to rules indicate the direction(s) in which the rules provide implications. Hence rules with $\Downarrow$ are used for forward reasoning, while rules with $\Uparrow$ build up top-down proofs.

Function sorts such as $s \rightarrow s'$ are associated with an implicit apply-operator that extends the inductive definition of $T_\Sigma(X)$ as follows:

$$t \in T_\Sigma(X)_{s \rightarrow s'} \wedge u \in T_\Sigma(X)_s \Rightarrow t(u) \in T_\Sigma(X)_{s'}.$$

As, for each Σ-structure A, $A_{s_1 \ldots s_n}$ is defined as the product $A_{s_1} \times \ldots \times A_{s_n}$, $A_{s \rightarrow s'}$ is defined as the function space $[A_s \rightarrow A_{s'}]$. Theorem 10.11 remains valid, although $Mod(SP)$ must be replaced by $Gen(SP)$ [MTW88b].

Definition 10.12 (Gentzen clause theorems). Let $SP = (\Sigma, AX)$ be a standard specification. A Horn clause $p \Leftarrow H$ is a **deductive theorem** of SP if there is a skolemization ρ with $dom(\rho) = var(H)$ (Definition 10.3) such that $(\Sigma, AX \cup H\rho) \vdash_{cut} p\rho$.

A Gentzen clause $GS \Leftarrow HS$ is an **inductive theorem** of SP if for all goals H of HS and $\sigma \in T_\Sigma^X$, $SP \vdash_{cut} H\sigma$ implies $SP \vdash_{cut} G\tau$ for some goal G of GS and $\tau \in T_\Sigma^X$ such that $\tau_V = \sigma_V$ where V is the set of free variables of $GS \Leftarrow HS$.[9]

DTh(SP) and **ITh(SP)** denote the sets of deductive and inductive theorems of SP, respectively.   ◇

The initial SP-model interprets safety predicates as the **least relations** satisfying their axioms. This follows from *Kleene's fixpoint theorem*, which is also used for defining the semantics of pure logic programs. *Tarski's fixpoint theorem* forbids the use of negation, while Kleene's theorem enforces more restrictive conditions on the axioms of a specification. Let us recapitulate (set-theoretical versions) of both theorems because some axioms of a standard specifications may be – in contrast to pure logic programs – Horn clauses with (universal) quantifiers in premises and even non-Horn clauses.

Theorem 10.13 (fixpoint theorems). (see, e.g., [LNS82]) *Let A be a set and $\Phi: \wp(A) \rightarrow \wp(A)$ be a monotonic function (with respect to set inclusion). $B \subseteq A$ is a **fixpoint** of Φ if $\Phi(B) = B$. Φ is **upward continuous** if for all increasing chains $B_1 \subseteq B_2 \subseteq B_3 \subseteq \ldots$ of subsets of A, $\Phi(\cup_{i \in \mathbb{N}} B_i)$ is a subset of $\cup_{i \in \mathbb{N}} \Phi(B_i)$. Φ is **downward continuous** if for all decreasing chains $B_1 \supseteq B_2 \supseteq B_3 \supseteq \ldots$ of subsets of A, $\cap_{i \in \mathbb{N}} \Phi(B_i)$ is a subset of $\Phi(\cap_{i \in \mathbb{N}} B_i)$. The sets*

$$\Phi^\infty =_{def} \cup_{i \in \mathbb{N}} \Phi^i(\emptyset) \quad and \quad \Phi_\infty =_{def} \cap_{i \in \mathbb{N}} \Phi^i(A)$$

*are called the **Kleene closures** of Φ.*

Tarski's Theorem
   $\mu\Phi =_{def} \cap\{B \subseteq A \mid \Phi(B) \subseteq B\}$ is the least fixpoint of Φ. $\Phi^\infty \subseteq \mu\Phi$.
   $\nu\Phi =_{def} \cup\{B \subseteq A \mid B \subseteq \Phi(B)\}$ is the greatest fixpoint of Φ. $\nu\Phi \subseteq \Phi_\infty$.

---

[9] For equational Gentzen clauses without quantifiers, this characterization of inductive theorems agrees with Definition 9.27.

Kleene's Theorem

Φ *is upward continuous* $\quad\Rightarrow\quad \Phi(\Phi^\infty) \subseteq \Phi^\infty \quad\Rightarrow\quad \mu\Phi = \Phi^\infty.$

Φ *is downward continuous* $\quad\Rightarrow\quad \Phi_\infty \subseteq \Phi(\Phi_\infty) \quad\Rightarrow\quad \nu\Phi = \Phi_\infty.$

When applying fixpoint theorems, we sometimes refer to the *relational version* of *SP*:

Definition 10.14 (relational version). Let $SP = (\Sigma, AX)$ be a standard specification and $f\colon w \to s$ be a defined function of *SP*. The **graph** or **input-output relation** $r_f : ws$ **of** f is the implicit predicate $r_f : w$ of *SP* with the Horn axiom

$$r_f(x, y) \quad\Leftarrow\quad f(x) \equiv y.$$

An equation $f(t) \equiv u$ with defined function f and $t, u \in NF_\Sigma(X)$ is called **flat**. *SP* is **flat** if defined functions do only occur in flat equations of *AX*. If *SP* is flat, then the **relational version** of *SP* is the specification $rel(SP)$ obtained from *SP* by replacing all defined functions of Σ by their graphs and each flat equation $f(t) \equiv u$ occurring in *AX* by the logical atom $r_f(t, u)$.

$\diamond$

Theorem 10.15. *Let SP be a functional specification (Definition 10.18) with signature Σ and rel(SP) be the relational version of SP. For all defined functions $f\colon w \to s \in \Sigma$, logical safety predicates $r : w \in \Sigma$, $t \in NF_{\Sigma,w}$ and $u \in NF_{\Sigma,s}$,*

$$f(t) \equiv_{SP} u \quad\Longleftrightarrow\quad rel(SP) \vdash_{cut} r_f(t, u),$$

$$SP \vdash_{cut} r(t) \quad\Longleftrightarrow\quad rel(SP) \vdash_{cut} r(t).$$

A proof sketch for this theorem is given in Section 10.5.

The carrier elements of $Ini(rel(SP))$ are equivalence classes consisting of normal forms. Since *SP* is functional, each equivalence class $[t]$ is either a singleton or all terms of $[t]$ contain $\bot$ (see Definition 10.5).

Theorem 10.16 (initial = least fixpoint semantics). *Let $SP = (\Sigma, AX)$ be a functional specification,*

$$rel(SP) = (rel(\Sigma), rel(AX))$$

be the relational version of SP and C be the class of $rel(\Sigma)$-structures that interpret the function symbols of $rel(\Sigma)$ the same as $Ini(rel(SP))$ does. The **consequence operator** $\Phi\colon C \to C$ *is defined as follows: For all logical safety predicates $r \in \Sigma$, defined functions $f \in \Sigma$, $b\colon X \to A$ and $A \in C$,*

$$b^*(nf(t)) \in r^{\Phi(A)} \Longleftrightarrow_{def} \exists (r(t) \Leftarrow H) \in AX : A \models_b H,$$

$$b^*(nf(t, u)) \in r_f^{\Phi(A)} \Longleftrightarrow_{def} \exists (f(t) \equiv u \Leftarrow H) \in AX : A \models_b H.$$

Φ *is monotonic with respect to set inclusion on C where*

$$A \subseteq B \quad\Longleftrightarrow_{def}\quad \forall r \in PR : r^A \subseteq r^B.$$

Hence by Tarski's Theorem, Φ has the least fixpoint $\mu\Phi = \cap\{A \in \mathcal{C} \mid \Phi(A) \subseteq A\}$.

Φ is upward continuous. Hence by Kleene's Theorem, $\mu\Phi$ agrees with $\Phi^{\infty} = \cup_{i \in \mathbb{N}}\Phi^{i}(\bot)$ where $\bot$ is the least element of $\mathcal{C}$.

Moreover, $\mu\Phi = Ini(rel(SP))$ because $Ini(rel(SP))$ interprets all logical predicates as the least relations satisfying their axioms.[10]

Theorems 10.15 and 10.16 allow us to identify a functional specification *SP* with its relational version and, consequently, the initial *SP*-model with the least fixpoint of Φ. It remains to interpret the liveness predicates of *SP*. This will be done in Section 10.3.

Theorem 10.17 (validity and provability). *Let $SP = (\Sigma, AX)$ be an inhabited standard specification, $\sigma : \Sigma' \to \Sigma$ be a signature morphism and $\approx$ be a Σ-congruence on $Ini(SP)$. For all $t, t' \in T_{\Sigma'}$,*

$$[t] \approx_{\sigma} [t'] \iff_{def} [\sigma(t)] \approx [\sigma(t')]$$

defines a Σ'-congruence on $Ini(SP)|_{\sigma}$ such that $gen(Ini(SP)|_{\sigma})/\approx_{\sigma}$ agrees with $gen(Ini(SP)/\approx|_{\sigma})$. Given a predicate $r \in \Sigma'$, $\approx_{\sigma}$ is compatible with r if $\approx$ is compatible $\sigma(r)$ (see Definition 10.7). Let Σ_{σ} consist of Σ and a **representation predicate** *$rep_s : \sigma(s)$ for each sort (tuple) $s \in \Sigma'$. Let AX_{σ} consist of AX and a Horn axiom*

$$rep(\sigma(f)(x_1, \ldots, x_n)) \Leftarrow rep(x_1) \wedge \cdots \wedge rep(x_n)$$

for each function symbol $f \in \Sigma'$. Let $SP_{\sigma} = (\Sigma_{\sigma}, AX_{\sigma})$.

(1) For all Horn clauses $C = (p \Leftarrow H)$ over Σ,

$$Mod(SP) \models C \iff C \in DTh(SP),$$
$$Gen(SP) \models C \iff \forall \sigma \in T_{\Sigma}^{X} : (\Sigma, AX \cup H\sigma) \vdash_{cut} p\sigma.$$

(2) For all Gentzen clauses C over Σ,

$$Ini(SP) \models C \iff C \in ITh(SP).$$

(3) For all Gentzen clauses C over Σ',

$$Ini(SP)|_{\sigma} \models C \iff \sigma(C) \in ITh(SP),$$
$$gen(Ini(SP)|_{\sigma}) \models C(x) \iff Ini(SP_{\sigma}) \models \sigma(C)(x) \Leftarrow rep(x),$$
$$gen(Ini(SP)/\approx|_{\sigma}) \models C(x)$$
$$\iff Ini(SP_{\sigma})/\approx \models \sigma(C)(x) \Leftarrow rep(x)$$

where x is a tuple consisting of all variables of C.

From now on, we assume that the signature of a non-parameter specification is inhabited. This is crucial for the $\Leftarrow$-directions of (1), which establish soundness properties of the cut calculus. Without the assumption of inhab-

---

[10] This follows almost directly from the soundness of $\vdash_{cut}$ with respect to $\mathcal{C}$.

itedness the free variables of axioms and derived goals had to be universally quantified as it in the equational calculus of Section 2.8. Otherwise one may infer invalid goals from valid ones whenever empty sorts are involved. For instance, a supposedly invalid equation $a \equiv b$ were derivable from the axioms $F(x) \equiv a$ and $F(x) \equiv b$ if $sort(x)$ is empty. Apart from the assumption of inhabitedness, *equational* deductive theorems of a standard specification with only *equational* axioms are exactly those formulas that are derivable via the equational calculus of Section 2.8.

(1) and (2) imply $Ini(SP) \models g$ iff $Gen(SP) \models g$ for all goals g and $Ini(SP) \models g$ iff $Mod(SP) \models g$ for all *ground* goals g. For *conditional* Horn clauses C, however, the hierarchy is strict: C may hold in $Ini(SP)$, but not in $Gen(SP)$, and C may hold in $Gen(SP)$, but not in $Mod(SP)$. Invariance properties of the form $r(f(x)) \Leftarrow r(x)$, stating that f preserves r, are usually valid only in $Ini(SP)$, in particular if r is defined inductively on the structure of normal forms (see, e.g., [Pad88a], Example 4.3.4). This is one of the reasons for choosing initial semantics as *the* semantics and the inductive theory as *the* theory of a (visible) data type. Still some basic conditions must be added to the syntactical ones of Definition 10.5 in order to justify the separation of constructors from defined functions:

Definition 10.18 (functional specification). Let SP be a standard specification with signature Σ. SP is **complete** if each ground Σ-term has a normal form (Definitions 10.2, 10.10). SP is **consistent** if each two SP-equivalent ground normal forms are equal. SP is **functional** if SP is complete and consistent. If SP is functional, we denote the unique normal form of a ground Σ-term (tuple) t by $\mathbf{nf(t)}$.   ◇

If SP is functional, then the defined functions of SP are totalized in the sense that for all ground terms t, $SP \not\vdash_{cut} Def(t)$ implies $t \equiv_{SP} \bot$. The converse holds true because all axioms for Def are partiality axioms for Σ (Definition 10.5) and thus $SP \not\vdash_{cut} Def(\bot)$.

Functionality is defined for any standard specification SP, but actually concerns only its visible subspecification because our standard model of SP will be a final one where $\equiv_{SP}$ defines only the equality of visible data. Completeness seems to forbid the specification of partial functions. For instance, the classical theory of partial-recursive functions (REC; see Section 10.1) admits *divergence*, while completeness demands convergence into normal forms, at least into bottom constants. However, when it comes to proof methods, REC hides divergence by modelling a partial function as the supremum of its *total* approximations. But approximations and their suprema may be specified as *total* defined functions, i.e. within a complete specification (see, e.g., [Pad91, Pad92]). Alternatively, certain partial functions may be modelled as hidden constructors along with a liveness predicate expressing divergence (see [Gor95]).

10.3   Final semantics and bisimilarity

Final semantics was introduced for modelling **permutative types** such as
finite sets, finite bags (multisets) and functions with a finite domain (stores,
arrays, indexed lists) (see, e.g., [GGM76, Wan79, Kam83]). These types are
still constructor-based, but *constructor equations* are needed to axiomatize
data equality. Hence specifications of permutative types are complete, but
not consistent. From a model-theoretic viewpoint, initial semantics is sufficient for handling permutative types. Constructor equations are Horn axioms
and thus we obtain an initial model as in the case of standard specifications
where the only constructor equations are the partiality axioms (see Definition 10.5). From a proof-theoretic viewpoint, however, initial semantics is less
appropriate. Efficient resolution- and rewriting-oriented proof methods treat
constructor equations CE on a lower level than other axioms (see, e.g., [Plo72,
Sti85, JK86a]). Normal forms are replaced by equivalence classes of normal
forms modulo the equivalence relation $\equiv_{CE}$ induced by CE. Resolution and
rewriting modulo $\equiv_{CE}$ work well if CE is restricted to particular axioms such
as associativity, commutativity, idempotence, etc. Otherwise corresponding
proof rules are difficult to handle.

With respect to final semantics, constructor equations become *theorems*
that can be derived, e.g., by **context induction** (see [Hen91a]). Studies
in category theory and modal logic dealing with coalgebras, coinduction and
greatest fixpoints suggested both subsuming permutative types under hidden
types and adopting final semantics as well for **object types** and **infinite
types** such as streams and processes (see, e.g., [AW89, GD94b, Rei95, Rut96,
GM97, JR97]). In order to keep the proof-theoretic benefit of initial semantics
we present even infinite types as *functional* specifications (Definition 10.18).
Consistency is not restrictive here because hidden data equality is defined
in terms of bisimulation and not equality predicates. Completeness, however,
seems to be restrictive. How can uncountably many streams be represented by
countably many normal forms? The answer is that not all elements of a carrier
set need to be covered by normal forms as long as the theory of a specification
SP is preserved whenever further normal forms are added to SP. Since a
theorem about a hidden carrier cannot be proved by structural induction
(because the structure is hidden), it will mostly be valid in extensions of SP
by further normal forms. For instance, many streams are not represented as
normal forms of INFSEQ, while a single stream may have several normal
forms (see Example 10.6). Whereas *visible* data are considered to be equal if
they have the same normal forms, bisimulation predicates take over the rôle
of identifying terms representing the same *hidden* objects.

Besides separating visible from hidden sorts we have distinguished safety
from liveness predicates. By Theorem 10.16, the initial model interprets safety
predicates as least relations satisfying their (Horn) axioms. Dually, liveness
predicates are interpreted as greatest relations satisfying their (co-Horn) axioms. Horn axioms for a visible symbol f usually entail an inductive definition

of f in the initial model. Category theory set up the duality between inductively defined algebras and coinductively defined coalgebras (see, e.g., [AM82, Gor95, Rei95, JR97]). Initial algebras and final coalgebras are isomorphisms built up from constructors in one direction and actions ("destructors") in the other. For instance, the final coalgebra of infinite streams (see Example 10.6) splits into the constructor & and the actions *head* and *tail*. We keep the duality in mind, but follow more algebraic lines by constructing final models as quotients of initial ones. Let us first "dualize" Theorem 10.16 for characterizing the initial model not only as a least, but also as a greatest fixpoint:

Theorem 10.19 (greatest fixpoint semantics, bisimilarity). *Let* $SP = (\Sigma, AX)$ *be a functional specification and* C *be the class of* Σ-*structures that interpret* $\Sigma \setminus LPR$ *the same as* $Ini(SP)$ *does. The* **consequence operator** $\Psi : C \to C$ *is defined as follows: For all* $r \in LPR$, $b : X \to A$ *and* $A \in C$,

$$b(x) \in r^{\Psi(A)} \quad\Longleftrightarrow_{def}\quad \forall(r(x) \Rightarrow C) \in AX : A \models_b C.$$

By Condition 10.5(4), Ψ *is monotonic with respect to set inclusion on* C *(see Theorem 10.16). Hence by Tarski's Theorem,* Ψ *has the greatest fixpoint* $\nu\Psi = \cup\{A \in C \mid A \subseteq \Psi(A)\}$.

If Ψ *is downward continuous, then by Kleene's Theorem,* $\nu\Psi$ *agrees with* $\Psi_\infty = \cap_{i\in\mathbb{N}}\Psi^i(\top)$ *where* $\top$ *is the greatest element of* C.

The interpretation of $\sim$ *in* $\nu\Psi$ *is called* SP-**bisimilarity** *and denoted by* $\sim_{SP}$. SP *is* **behaviourally congruent** *if* $\sim_{SP}$ *is* $\nu\Psi$-*compatible with* $\Sigma \setminus TA$.

By Condition 10.5(6), SP-bisimilarity is an equivalence relation. For all visible sorts s, $\sim_{SP,s}$ coincides with the equality on $Ini(SP)_s$.

Ψ is downward continuous if for all co-Horn axioms $r(x) \Rightarrow (H \Rightarrow GS)$ of SP, the goal set GS does not involve existential quantifiers. Otherwise the downward continuity of Ψ is nor guaranteed. If, for instance, AX includes a co-Horn clause of the form $r(x) \Rightarrow \exists y q(x, y)$, then Ψ is downward continuous only if $r^{\cap_i \Psi(B_i)}$ is a subset of $r^{\Psi(\cap_i B_i)}$. But this means that $\forall i \exists y : q^{B_i}(x, y)$ implies $\exists y \forall i : q^{B_i}(x, y)$, which does *not* always hold true.

The μ-calculus of modal logic (see [Sti92]) provides the idea for a condition on co-Horn axioms that guarantees the continuity of Ψ. Given a predicate r and a modal formula $F(r)$, $\mu r.F(r)$ and $\nu r.F(r)$ denote the least resp. greatest interpretation of r that satisfies $r(x) \Leftrightarrow F(r)$. First a certain *polarity condition* on the structure of F ensures that the consequence operator induced by F is monotonic and thus Tarski's Theorem implies that there least and greatest interpretations do exist. F is even (upward and downward) continuous if the transition system that underlies any modal-formula interpretation is *finitely branching*. This property can be generalized to a criterion for the downward continuity of Ψ as defined in Theorem 10.19 and thus for the iterative constructability of $\nu\Psi$ even if co-Horn axioms involve existential quantifiers. We do not elaborate this criterion here, but note that it holds true for all sample specifications presented in this chapter.

Definition 10.20 (final semantics). Let SP be a functional and behaviourally congruent specification and $\nu\Psi$ be defined as in Theorem 10.19. The **final SP-model $Fin(SP)$** is the quotient of $\nu\Psi$ by $\sim_{SP}$ that interprets the predicates of Σ as follows:

- For all $r \in SPR \setminus TA \cup LPR$, $[a] \in r^{Fin} \iff_{def} a \in r^{\nu\Psi}$.
- For all $\delta \in TA$, $([a], [b], [c]) \in \delta^{Fin} \iff_{def} \exists\, c' \sim_{SP} c : (a, b, c') \in \delta^{\nu\Psi}$.

A first-order formula satisfied by $Fin(SP)$ is called a **final theorem** of SP.

$\diamond$

By Theorem 10.8, the final SP-model is an SP-model. $Fin(SP)$ coincides with $Ini(SP)$ if SP has neither hidden sorts nor liveness predicates.

Example 10.21. Condition 10.5(6) implies that $\sim_{SP,s}$ is the entire product $Ini(SP)_s^2$ if all s-actions are functions into hidden sorts. To illustrate this consider the following standard specification of integer numbers:

```
INT
    hidsorts int
    consts   0: → int
             _ + 1: int → int
             _ − 1: int → int
             _ + _ : int * int → int
             _ − _ : int * int → int
    defuns   succ, pred: int → int
    preds    is0: int
             _ ~ _ : int * int
    vars     x, y: int
    Horn axioms
             succ(0) ≡ 0 + 1
             pred(0) ≡ 0 − 1
             is0(0)
             succ(x + 1) ≡ succ(x) + 1
             pred(x + 1) ≡ x
             succ(x − 1) ≡ x
             pred(x − 1) ≡ pred(x) − 1
             succ(x + y) ≡ (x + y) + 1
             pred(x + y) ≡ (x + y) − 1
             succ(x − y) ≡ (x − y) + 1
             pred(x − y) ≡ (x − y) − 1
    co-Horn axioms
             x ~ y ⇒ succ(x) ~ succ(y)
             x ~ y ⇒ pred(x) ~ pred(y)
             x ~ y ⇒ (is0(x) ⇒ is0(y))
```

The final INT-model $Fin(INT)$ (see below) is isomorphic to $\mathbb{Z}$. For instance, the normal form equations $(x + 1) - 1 = x$ and $(x - 1) + 1 = x$ are final theorems of INT. INT-bisimilarity would identify *all* ground INT-terms if the last axiom for $\sim$ were omitted. ∎

If all actions of a standard specification SP are functional or relational, then SP-bisimilarity can be characterized as *contextual* equivalence. Given a hidden sort s, an s-**context** is a visible term or atom consisting of visible symbols, functional or relational actions and a single s-sorted variable. The **contextual SP-equivalence** $\approx_{SP}$ is defined as follows:

$$[t] \approx_{SP} [t'] \iff_{def} \forall\ s\text{-contexts } c : \begin{cases} c(t) \equiv_{SP} c(t') & \text{if } c \text{ is a term} \\ SP \vdash_{cut} c(t) \Leftrightarrow c(t') & \text{otherwise} \end{cases}$$

(see Section 5.5). For instance, it is easy to see that contextual INFSEQ-equivalence coincides with INFSEQ-bisimilarity (Example 10.6).

Definition 10.22 (hierarchical models). Let Σ and Σ' be standard signatures with $\Sigma' \subseteq \Sigma$. Let A and B be reachable Σ'- resp. Σ-structures (see Definition 10.7).

B is **complete wrt** A if $B_{\Sigma'} \cong gen(B_{\Sigma'})$ or, equivalently, for all sorts $s \in \Sigma'$ and $t \in T_{\Sigma,s}$ there is $t \in T_{\Sigma'}$ such that B satisfies $t \equiv t'$. B is **consistent wrt** A if $gen(B_{\Sigma'}) \cong A$ or, equivalently, for all ground Σ'-atoms p, $B \models p$ implies $A \models p$. Given a standard specification SP with signature Σ, an SP-model B is **hierarchical over** A if B is complete and consistent wrt A (and thus $B_{\Sigma'} \cong A$) and if for all logical safety predicates $r \in \Sigma \setminus \Sigma'$ and $t \in T_\Sigma$, $B \models r(t)$ implies $Ini(SP) \models r(t)$.   ◇

Note that SP is functional (see Definition 10.18) iff $Ini(SP)$ is hierarchical over $Ini(SP')$ where SP' consists of the constructors, bottom constants, definedness predicates and partiality axioms of SP (see Definition 10.5). Since visible Σ-normal forms are $vis\Sigma$-terms (see Definition 10.2), all reachable SP-models are complete wrt $Ini(visSP)$ (see Definition 10.5) whenever SP is complete. We summarize the main properties of the final SP-model:

Theorem 10.23. *Let SP be a functional and behaviourally congruent specification.*

(1) *For all safety predicates $r : w$ of SP, $r^{Fin(SP)}$ is the least subset of $Fin(SP)_w$ that satisfies the Horn axioms of SP.*

(2) *For all defined functions $f : w \to s$ of SP, the graph $r_f : ws$ of f is the least subset of $Fin(SP)_{ws}$ that satisfies the Horn axioms of the relational version of SP.*

(3) *For all liveness predicates $r : w$ of SP, $r^{Fin(SP)}$ is the greatest subset of $Fin(SP)_w$ that satisfies the co-Horn axioms of SP.*

(4) *$Fin(SP)$ is final in the class of hierarchical SP-models over $Ini(visSP)$ [Pad98].*

(1)-(3) imply that the fixpoint rules presented in Section 10.7 are sound. [Pad98] provides syntactical conditions on Horn axioms for hidden symbols of SP that guarantee behavioural congruence:

Definition 10.24 (coinductivity). Let Σ be a standard signature. Given hidden (empty or singleton) term tuples t, u and visible term tuples a, b, an atom $p = p(t, a, b, u)$ is **oriented** if t and a are normal forms and p has one of the following forms:

(1) $\delta(t, ab, u)$ for a transitional action δ,

(2) $r(t, a)$ for another hidden safety predicate r,

(3) $f(t, a) \equiv b$ for a hidden defined function f with visible range sort,

(4) $f(t, a) \equiv u$ for a hidden defined function f with hidden range sort.

p is **basic** if f resp. f is an action. A Horn clause

$$p_0(t_0, a_0, b_0, u_0) \;\Leftarrow\; G_0 \wedge \bigwedge_{i=1}^{n} (p_i(t_i, a_i, b_i, u_i) \wedge G_i)$$

is **coinductive** if

(5) $t_0, a_0 \in NF_{vis\Sigma}(X)$ and $b_0, u_0 \in NF_{\Sigma}(X)$ **or**

(6) p_0 is basic, $a_0 \in NF_{vis\Sigma}(X)$ and there are $t \in NF_{vis\Sigma}(X)$, $c, d \in NF_{\Sigma}(X)$ and a term v over $vis\Sigma \cup DF$ such that $t_0 = c(t)$, $c \notin X$ and $(b_0, u_0) = d(v(t))$

and for all $0 \le i \le n$, G_i is a goal over $vis\Sigma$, $p_i(t_i, a_i, b_i, u_i)$ is oriented, $b_i, u_i \in NF_{vis\Sigma}(X)$,[11] $var(t_i, a_i) \subseteq V_{i-1}$ and $var(b_0, u_0) \subseteq V_n$ where $V_0 = var(t_0, a_0, G_0)$ and $V_i = V_{i-1} \uplus var(b_i, u_i) \cup var(G_i)$.

SP is **coinductive** if all axioms for hidden defined functions or safety predicates of SP are coinductive. $\diamond$

Coinductive clauses have sequential premises (see Definition 10.5). With respect to functional actions, the notion of coinductivity was inspired by *coinductive coalgebra definitions* (see [Rei95, JR97]). With respect to transitional actions, it generalizes typical formats of transition system specifications used in process logics (see [Plo83, GV92]).

a and b comprise all visible arguments of an oriented atom $p(t, a, b, u)$. a collects the "input", b the "output" arguments. In Cases (2)-(4), the separation of a from b is fixed. In Case (1), it is usually determined by the requirement that the axioms for δ are coinductive. Intuitively, coinductivity enforce a data flow through the above Horn clause; starting out from t_0, a_0 to t_i, $i > 0$; from b_i, u_i to (t_j, a_j), $j > i$; and from b_i, u_i "back" to u_0.

Theorem 10.25. [Pad98] *A functional and coinductive specification SP is behaviourally congruent if the consequence operator Ψ of Theorem 10.19 is downward continuous.*

Example 10.26. INFSEQ (see Example 10.6) is coinductive. Suppose we want to define stream comprehension analogously to list comprehension by the following axioms (see *filter* in Example 10.1):

---

[11] Note that the only hidden normal forms of $NF_{vis\Sigma}(X)$ are variables.

$$head(filter(g,s)) \equiv head(s) \Leftarrow g(head(s)) \equiv true, \qquad (10.1)$$
$$head(filter(g,s)) \equiv head(filter(g,tail(s))) \Leftarrow g(head(s)) \equiv false. \qquad (10.2)$$

(10.2) is not coinductive! In terms of the clause in Definition 10.24, $t_0 = filter(g,s)$, $a_0 = u_0 = \varepsilon$ and $b_0 = head(filter(g,tail(s)))$. Condition 10.24(5) is violated because t_0 is not a variable. Condition 10.24(6) does not hold because the occurrence of the defined function $head$ in b_0 precedes the constructor $filter$. In fact, INFSEQ-bisimilarity cannot be compatible with $filter$ already for intuitive reasons: $filter$ may return finite streams, while the actions $head$ and $tail$ only allow us to specify infinite streams. $filter$ can only be part of a specification that comprises both finite and infinite streams. This specification replaces the functional actions $head$ and $tail$ of INFSEQ by a transitional action $\longrightarrow$: $stream \times nat \times stream$ (see [Pad98]).   ∎

Similarly, CCS-like process types can be specified coinductively in a quite straightforward way.

For the rest of this section suppose that $SP = (\Sigma, AX)$ is functional and behaviourally congruent. Hence SP has all the – proof-theoretically significant – properties stated in Theorem 10.23. As a first consequence we obtain a couple of initial or final "meta-theorems" representing background knowledge about SP.

Lemma 10.27. *Let r be a logical safety predicate and f be a defined function of SP, $AX_r = \{r(t_i) \Leftarrow H_i \mid 1 \le i \le n\}$ and $AX_f = \{f(t_i) \equiv u_i \Leftarrow H_i \mid 1 \le i \le n\}$ be the sets of axioms of SP for r resp. f, $Y_i = var(r(t_i) \Leftarrow H_i)$ resp. $Y_i = var(f(t_i) \equiv u_i \Leftarrow H_i)$ and x,y be (tuples of distinct) variables that do not occur in Y_i. The formulas*

$$r(x) \iff \bigvee_{i=1}^n \exists Y_i(x \equiv t_i \wedge H_i),$$
$$f(x) \equiv y \iff \bigvee_{i=1}^n \exists Y_i(x \equiv t_i \wedge y \equiv u_i \wedge H_i),$$

*called the **iff-completions** of r and f, respectively, are both initial and final theorems of SP.*

For instance,

$$sorted(L) \iff L \equiv nil$$
$$\vee\ \exists\{x\}(L \equiv [x])$$
$$\vee\ \exists\{x,y,L'\}(L \equiv x :: y :: L' \wedge x \le y \wedge sorted(y :: L'))$$

is the iff-completion of $sorted : list$ (see Example 10.1).

Analogously, liveness predicates are completed with respect to the greatest fixpoint $\nu\Psi$ of the consequence operator Ψ (see Theorem 10.19):

Lemma 10.28. *Let r be a liveness predicate of SP and $AX_r = \{r(t_i) \Rightarrow \phi_i \mid 1 \leq i \leq n\}$ be the set of axioms of SP for r, $Y_i = var(r(t_i) \Rightarrow \phi_i)$ and x be a tuple of distinct variables that do not occur in Y_i. The formula*

$$r(x) \iff \bigwedge_{i=1}^{n} \forall Y_i(x \equiv t_i \Rightarrow \phi_i),$$

called the **iff-completion** *of r, is valid in $\nu\Psi$ and a final theorem of SP.*

For instance,

$$s \sim s' \iff head(s) \equiv head(s') \wedge tail(s) \sim tail(s') \tag{10.3}$$

is the iff-completion of $\sim$: *stream* $\times$ *stream* (see Example 10.6).

By negating the iff-completion of a predicate r one often obtains axioms for the complement of r:

Definition 10.29. Given predicates $r : w$ and $\bar{r} : w$ of SP, $\bar{r}$ is called the *SP-* **complement** of r if $\bar{r}^{Fin(SP)}$ is the complement of $r^{Fin(SP)}$ i.e. $\bar{r}^{Fin(SP)} = Fin(SP)_w \setminus r^{Fin(SP)}$. ◇

For instance, clause (10.3) entails that the *least* interpretation of $\not\sim$: *stream* $\times$ *stream*, specified by the Horn clause

$$s \not\sim s' \Leftarrow head(s) \not\equiv head(s') \vee tail(s) \not\sim tail(s'), \tag{10.4}$$

is the INFSEQ-complement of $\sim$. Similarly, the iff-completion of the *safety* predicate *sorted* (see above) leads to the following co-Horn axioms for the *liveness* predicate *unsorted*:

$$unsorted(L) \Rightarrow (L \equiv nil \Rightarrow FALSE)$$
$$unsorted(L) \Rightarrow (L \equiv [x] \Rightarrow FALSE)$$
$$unsorted(L) \Rightarrow ((L \equiv x :: y :: L' \wedge x \leq y) \Rightarrow unsorted(y :: L')).$$

Both – alternative – axioms for the liveness predicate *fair* of INFSEQ (Example 10.6) yield generalized[12] Horn axioms for *unfair*:

$$unfair(s) \Leftarrow existsForall(\lambda x.not(eq(x,0)), s)$$
$$unfair(s) \Leftarrow \forall\{L, s'\}((0 \notin L \wedge s \sim L\#(0\&s')) \Rightarrow unfair(s')).$$

Lemmas 10.27, 10.28 and 10.30 (see below) are crucial for the correctness of inference rules presented in Section 10.7. Let us now turn to the general proof schema for reasoning about standard specifications.

---

[12] See above.

10.4   Gentzen proofs and program synthesis

In the previous section we have defined the basic theories of a standard specification that capture its loose, reachable, initial and final semantics, respectively (see Definitions 10.12, 10.20). We discussed the dominant rôle of initial semantics because it reflects the intuition behind *free*-constructor-based types and entails efficient proof rules for reasoning about such *visible* types. A data type DT that involves non-free constructors, actions operating on hidden sorts or non-inductive liveness predicates, however, is better given a final semantics, which comes as a quotient of the initial model by behavioural equivalence, provided that DT can be presented as a coinductive specification SP (see Definition 10.24). As proofs of safety conditions and the visible part of SP are mainly based on induction and least fixpoint properties, so can proofs of liveness conditions and the hidden part of SP be based on greatest fixpoint properties. Moreover, "meta-theorems" such as Lemmas 10.27 and 10.28, which hold true under general requirements to SP, provide for **simplification rules** that speed up proofs considerably because they can be applied – in contrast to fixpoint rules – almost automatically (see Section 10.7).

Previous examples should have justified our assertion that Gentzen clauses are sufficient for presenting not only a data type DT itself, but also all interesting conjectures about DT, such as invariants and other conditions on functions or predicates of DT that one might demand in order to call them "correct". In contrast to the formulas used in classical ATP approaches (see Section 10.1) Gentzen clauses are not the result of a normalization process that often hides the meaning of the original formulas considerably and thus makes proofs difficult to follow, let alone to control interactively. In particular, inductive proofs *require* interaction for guiding them towards induction hypotheses or setting up generalizations. Hence we focus on **Gentzen rules**, i.e. proof rules that relate (finite) sets (= conjunctions) of Gentzen clauses to each other. A derivation consisting of Gentzen rule applications is called a **Gentzen proof**. A Gentzen proof is **successful** if it terminates with the Gentzen clause $TRUE$. In contrast to a derivation with the cut calculus (Definition 10.10), a Gentzen proof proceeds

- **top-down** or **backward** because it starts with the conjecture to be proved and argues backwards by establishing inverse implications between successively derived clause sets;

- **linear** since a Gentzen rule is always applied to the clause set derived at latest, in other words, the inference relation built up of a set of Gentzen rules is the transitive closure of the set of rule instances.

A Gentzen clause $GS \Leftarrow HS$ is an implication (first level) such that both the conclusion GS and the premise HS are goal sets (second level), which are disjunctions of goals (third level), which are conjunctions of atoms (fourth level). Most steps of a Gentzen proof work on the second level, i.e. modify GS

or HS separately. On the first level, this means applying the following two Gentzen rules: Let SP be a standard specification with final model $Fin(SP)$.

$$\textbf{head expansion} \qquad \frac{GS_0 \ \Leftarrow \ HS}{GS_1 \ \Leftarrow \ HS} \Uparrow \quad \begin{cases} \text{if } GS_0 \textbf{ expands to } GS_1, \\ \text{i.e. } Fin(SP) \models GS_0 \Leftarrow GS_1 \end{cases}$$

$$\textbf{body contraction} \qquad \frac{GS \ \Leftarrow \ HS_0}{GS \ \Leftarrow \ HS_1} \Uparrow \quad \begin{cases} \text{if } HS_0 \textbf{ contracts to } HS_1, \\ \text{i.e. } Fin(SP) \models HS_0 \Rightarrow HS_1 \end{cases}$$

Rules for expanding resp. contracting formulas are called **expansion** resp. **contraction rules**. Contractions establish implications in the direction in which the proof proceeds. Expansions yield implications in the opposite direction.[13] A rule is **correct** if the implication is valid in the final model of any specification for which such a model exists (see Theorem 10.23). Since Gentzen proofs always proceed backwards, all Gentzen rules are expansions. The arrow attached to a rule indicates whether it is an expansion ($\Uparrow$), a contraction ($\Downarrow$) or both ($\Updownarrow$). In the last case the rule establishes an equivalence between two formulas and is called a **simplification**, which appeals to the fact that it often simplifies a formula in some respect. Most Gentzen rules are presented in Section 10.7.

Besides proving the correctness of a specification SP by showing the final validity of selected conjectures, SP may be tested by **solving goals**, in particular equations, in (existentially quantified) variables. In functional languages, solving is restricted to ground term evaluation, i.e. to solving equations of the form $t \equiv x$ with $x \in X \setminus var(t)$. In logic languages, a goal to be solved is called a query and provides the input to a logic program. Most queries can be answered by applying Gentzen rules. A goal H is a **solved goal** if H is a conjunction $x_1 \equiv t_1 \wedge \cdots \wedge x_n \equiv t_n$ where $x_1, \ldots, x_n$ are variables and $t_1, \ldots, t_n$ are normal forms. Then a clause $G \Leftarrow H$ holds true iff $G[t_1/x_1, \ldots, t_n/x_n]$ holds true. Hence G is solved by expanding G to a solved goal. In pure logic programming, solved goals are the only presentations of solutions. CLP (= constraint logic programming) accepts more general forms of solutions, like the goal $x > 5$, which covers an infinite set of solved goals.

Conversely, G is unsolvable if G contracts to $FALSE$. In Section 10.8, we deal with particular rules for deriving solved goals and discuss their completeness: does a normal form substitution σ solve a goal G only if the rules expand G to the solved goal $\bigwedge_{x \in var(t)} x \equiv x\sigma$?

For obtaining solved goals one needs to stepwise remove logical predicates and defined functions from the goal to be solved. These rules, which are called **unfold rules**, are direct consequences of Lemmas 10.27 and 10.28 and will be presented in Section 10.7. If all logical predicates and defined functions have been removed, we might still be left with goals consisting of equations

---

[13] *Inductive logic programming* calls expansions *inductive* rules and contractions *deductive* rules [MdR94].

between normal forms. The following lemma allows us to reduce such goals to solved ones.

Lemma 10.30. *Let SP be a functional specification, c, d be different visible constructors of SP, $x = (x_1, \ldots, x_n)$ and $y = (y_1, \ldots, y_n)$. Then the following equivalences are final theorems of SP:*

$$c(x) \equiv c(y) \quad \Longleftrightarrow \quad (x_1 \equiv y_1 \wedge \cdots \wedge x_n \equiv y_n) \vee (c(x) \equiv c(y) \equiv \bot)$$
$$c(x) \equiv d(y) \quad \Longleftrightarrow \quad c(x) \equiv d(y) \equiv \bot$$
$$c(x) \equiv \bot \quad \Longleftrightarrow \quad x_1 \equiv \bot \vee \cdots \vee x_n \equiv \bot$$

$$c \equiv \bot \quad \Longleftrightarrow \quad FALSE$$
$$x \equiv \bot \wedge Def(t) \quad \Longleftrightarrow \quad FALSE$$

Moreover, if for all different visible constructors c, d of SP, SP is extended by the Horn axioms

$$c(x) \not\equiv c(y) \quad \Leftarrow \quad (x_i \not\equiv y_i \wedge c(x) \not\equiv \bot) \vee (x_i \not\equiv y_i \wedge c(y) \not\equiv \bot)$$
$$\qquad \qquad \qquad \text{for all } 1 \le i \le n$$
$$c(x) \not\equiv d(y) \quad \Leftarrow \quad c(x) \not\equiv \bot \vee d(y) \not\equiv \bot$$
$$c(x) \not\equiv \bot \quad \Leftarrow \quad x_1 \not\equiv \bot \wedge \cdots \wedge x_n \not\equiv \bot$$
$$c \not\equiv \bot,$$

then for all visible sorts s of SP, $\not\equiv^{Ini(SP)}: ss$ is the complement of SP-equivalence (Definition 10.29).

As Lemma 10.28 leads to inference rules for handling hidden equations and inequations, Lemma 10.30 provides us with rules for visible equations and inequations. If they are applied as soon as possible, they usually reduce the search space of a Gentzen proof significantly. Therefore, many theorem provers realize them as part of a built-in simplifier that works in the background and applies rules automatically (see Section 10.7).

Given an arbitrary subspecification SP' of SP, $Ini(SP)$ is consistent wrt $Ini(SP')$ iff for all ground Σ'-atoms p, $SP \vdash_{cut} p$ implies $SP' \vdash_{cut} p$, intuitively: iff the step from SP' to SP does not lead to "confusion", i.e. properties of SP'-data, for example equivalences, that did not already hold in SP'. Consistency is a crucial proof obligation when specifications are developed hierarchically. But it also occurs when one designs *in-the-small*, namely when a defined function or safety predicate is redefined by replacing its axioms by equivalent ones. Then consistency is accompanied by a change of signatures and thus replaced the more general property of inductive equivalence along signature morphisms:

Definition 10.31 (inductive equivalence). Given two specifications $SP_i = (\Sigma_i, AX_i)$, $i = 1, 2$, and signature morphisms $\sigma \colon \Sigma' \to \Sigma_1$ and

$\tau\colon \Sigma' \to \Sigma_2$, SP_1 and SP_2 are **inductively equivalent along** (σ, τ) if $gen(Ini(SP_1)|_\sigma)$ and $gen(Ini(SP_2)|_\tau)$ are isomorphic or, equivalently, for all ground Σ'-atoms p,

$$SP_1 \vdash_{cut} \sigma(p) \quad \Longleftrightarrow \quad SP_2 \vdash_{cut} \tau(p). \qquad\qquad \diamond$$

Σ' comprises the functions and predicates to be redefined together with those that remain the same, while $\Sigma_1 \setminus \Sigma'$ resp. $\Sigma_2 \setminus \Sigma'$ are the auxiliary functions and predicates used in SP_1- resp. SP_2-axioms for Σ'. The proof that SP_1 and SP_2 are inductively equivalent splits into three steps.

(1) For all functions f and logical predicates r of Σ', $\sigma(f)(x) \equiv \tau(f)(x)$ and $\sigma(r)(x) \Leftrightarrow \tau(r)(x)$ are inductive theorems of $SP =_{def} SP_1 \cup SP_2$.

(2) $Ini(SP_1 \cup SP_2)$ is consistent wrt $Ini(SP_1)$.

(3) $Ini(SP_1 \cup SP_2)$ is consistent wrt $Ini(SP_2)$.

Indeed, (1)-(3) imply

$$SP_1 \vdash_{cut} \sigma(p) \overset{(2)}{\Longleftrightarrow} SP \vdash_{cut} \sigma(p) \overset{(1)}{\Longleftrightarrow} SP \vdash_{cut} \tau(p) \overset{(3)}{\Longleftrightarrow} SP_2 \vdash_{cut} \tau(p)$$

for all ground Σ'-atoms p. Roughly said, Condition (1) states that certain functions and predicates occurring in SP_1 and SP_2 are equivalent with respect to the inductive theory of the *union* of these specifications. Conditions (2) and (3) ensure that the equivalences are preserved when SP_1 (resp. SP_2) is *replaced* by SP_2 (resp. SP_2).

Consistency is also a proof obligation when a "concrete" specification CSP is supposed to implement an "abstract" specification ASP. The implementation itself is presented as a further specification ISP and related to CSP and ASP via signature morphisms $\alpha\colon ASP \to ISP$ and $\kappa\colon CSP \to ISP$ (see Chapter 7). In terms of Section 10.2, one may call ISP a **correct implementation of** ASP **by** CSP if $gen(Fin(ISP)|_\alpha)$ is a hierarchical ASP-model over $Ini(visASP)$ and if $Fin(ISP)|_\kappa$ is consistent wrt $Fin(CSP)$ (Definition 10.22). Both conditions involve consistency requirements, which can often be satisfied with the help of Theorem 10.48.

We have defined the semantics of a parameterized specification $PSP = (PAR, SP)$ as the class of actualizations of PSP by PAR-models where an actualization $SP(A)$ combines a PAR-model A with SP in the way described in Section 10.1. A may be given as a structure whose signature, say, Σ, is different from the signature of PAR, say, Σ'. Then we need a signature morphism $\sigma\colon \Sigma' \to \Sigma$ and it is not A, but the σ-reduct $A|_\sigma$ of A, that is combined with SP. The actualizations of PSP are often described as values of a **functor** $F_{PSP}\colon Mod(PAR) \to Mod(SP)$ (see [EM85], Chapter 7). Here F_{PSP} maps A to $Fin(SP(A))$. The question whether, for a single PAR-model A, $Fin(SP(A))$ is consistent wrt A is generalized to the question on which subclass $\mathcal{C}$ of $Mod(PAR)$ F_{PSP} is **persistent**, i.e. maps *each* $A \in \mathcal{C}$ to an SP-model that is consistent wrt A.

Deductive program synthesis and **transformation** deals with rules that relate inductively equivalent specifications to each other (see, e.g., Chapter 14 and [MPS93]). The rules were mostly derived from proofs of program equivalences. For instance, let axioms for two functions f, g be given and suppose the task is to replace the axiom

$$(f; g)(x) \equiv g(f(x)) \tag{10.1}$$

for their composition by new axioms that do not refer to f and g. We may start an expansion of (1) using the axioms for f, g as well as functions and predicates f or g refers to. At some stage of the expansion we may obtain a clause that includes a term of the form $g(f(t))$. We rewrite $g(f(t))$ into $(f; g)(t)$, i.e. perform a **fold step**, which is actually an induction step (see Section 10.7). We proceed with the expansion until – hopefully – we are left with Horn clauses that satisfy the syntactic conditions of a standard specification on Horn axioms for $(f; g)$. We exchange them for (10.1) and conclude that (10.1) is a final *theorem* of the modified specification. The expansion can also be regarded as the derivation of a solution of (10.1) in the function variable $(f; g)$. In the functional programming and term rewriting communities, rule-based systems for solving goals in function variables are called **higher-order unification** procedures. The solutions come as λ-expressions, though never involve fixpoint operators, which means that these procedures do not (yet?) synthesize *recursive* programs. The restriction is due to the purely functional treatment of formulas and proofs in those communities. Traditional approaches to program synthesis got further here because they are more logic-oriented. In particular, fold steps realize induction steps (see above) and these are indispensible for deriving recursive programs.

In order to show how classical methods of program synthesis fit into the functional-logic approach let us reformulate the well-known assertion method for verifying iterative programs in terms of a Gentzen rule, called **Hoare induction**. Given the input-output relation $r : A \times C$ of a function $f : A \to C$ to be defined, the task is to construct an iterative program for f. An **iterative program for** f is a set of axioms of the following form:

$$f(x) \equiv iter(in(x))$$
$$iter(y) \equiv iter(loop_1(y)) \ \Leftarrow \ G_1(y)$$

. . .

$$iter(y) \equiv iter(loop_n(y)) \ \Leftarrow \ G_n(y)$$
$$iter(y) \equiv out(y) \ \Leftarrow \ G_{n+1}(y)$$

where $iter : B \to C$, $in : A \to B$, $out : B \to C$, $loop_1, \ldots, loop_n : B \to B$ and the *guards* $G_1, \ldots, G_{n+1}$ are auxiliary functions and goals, respectively. The program for f and the axioms for r are supposed to be part of a specification SP such that $r(x, z) \Leftarrow f(x) \equiv z$ is valid in the $Fin(SP)$. This clause is the premise of Hoare induction:

Hoare induction

$$\frac{r(x,z) \;\Leftarrow\; f(x) \equiv z}{}$$ ⇑

$$inv(x, in(x))$$

$$\wedge \quad inv(x, loop_1(y)) \;\Leftarrow\; inv(x,y) \wedge G_1(y)$$

$$\cdots$$

$$\wedge \quad inv(x, loop_n(y)) \;\Leftarrow\; inv(x,y) \wedge G_n(y)$$

$$\wedge \quad r(x, out(y)) \;\Leftarrow\; inv(x,y) \wedge G_{n+1}(y)$$

The conclusion of the rule "generates" the predicate $inv : A \times B$, called the **Hoare invariant**. inv may be regarded as a generalization of r. While r relates the input domain A of f to the output domain C of f, inv relates A to the intermediate domain B. A symmetric rule captures *subgoal induction* [MW77]. Instead of the Hoare invariant, this rule generates a **subgoal invariant** $sgi : B \times C$, which relates the intermediate domain B to the output domain of f and actually coincides with the input-output relation of *iter*. sug can be defined in terms of inv and vice versa:

$$sgi(y,z) \;\Longleftrightarrow\; \forall x : r(x,z) \Leftarrow inv(x,y),$$

$$inv(x,y) \;\Longleftrightarrow\; \forall z : r(x,z) \Leftarrow sgi(y,z).$$

Example 10.32. The following specification includes an iterative program for a function *split* that, given $n \in \mathbb{N}$ and a list L, decomposes L into sublists $L_1, \ldots, L_k$ such that $L_1, \ldots, L_{k-1}$ have length n.

PARTITION = LIST + NAT +
  defuns   *split*: *nat* * *list* → *list(list)*
      iter: *nat* * *list* * *nat* * *list(list)* → *list(list)*
      flatten: *list(list)* → *list*
      length: *list* → *nat*
  preds   *lengthOK*: *nat* * *list(list)*
  vars   *x*: *entry*   *n, k*: *nat*   *L, L'*: *list*   *P*: *list(list)*
  Horn axioms
      $split(n, L) \equiv iter(n, L, 0, nil, nil)$
      $iter(n, x :: L, k, L', P) \equiv iter(n, L, k+1, L'@[x], P) \Leftarrow k < n$
      $iter(n, x :: L, k, L', P) \equiv iter(n, x :: L, 0, nil, P@[L']) \Leftarrow k \equiv n$
      $iter(n, nil, k, L', P) \equiv P@[L']$
      $flatten(nil) \equiv nil$
      $flatten(L :: P) \equiv L@flatten(P)$
      $lengthOK(n, nil)$
      $lengthOK(n, [L]) \Leftarrow length(L) \leq n$
      $lengthOK(n, L :: P) \Leftarrow length(L) \equiv n \wedge lengthOK(n, P)$
      $length(nil) \equiv 0$
      $length(x :: L) \equiv length(L) + 1$

Let $A = nat \times list$ and $C = list(list)$. *split* is supposed to satisfy the following input-output relation $r : A \times C$:

$$r(n, L, P) \;\Longleftrightarrow_{def}\; L \equiv flatten(P) \wedge lengthOK(n, P).$$

To prove $r(n, L, P) \Leftarrow split(n, L) \equiv P$, we set $B = A \times A \times C$, apply Hoare induction and obtain the following requirements to a Hoare invariant inv : $A \times B$:

$inv(n, L, n, L, 0, nil, nil)$

$inv(n, L, n', L', k + 1, L''@[x], P)$

$\qquad \Leftarrow \quad inv(n, L, n', x :: L', k, L'', P) \wedge k < n'$

$inv(n, L, n', x :: L', 0, nil, P@[L''])$

$\qquad \Leftarrow \quad inv(n, L, n', x :: L', k, L'', P) \wedge k \equiv n'$

$r(n, L, flatten(P@[L''])) \quad \Leftarrow \quad inv(n, L, n', nil, k, L'', P).$

In the course of expanding these clauses to $TRUE$ (using rules presented in Section 10.7) one "derives" the following definition of inv:

$$inv(n, L, n', L', k, L'', P) \Longleftrightarrow_{def} L \equiv flatten(P@[L''@L']) \wedge n \equiv n'$$
$$\wedge \; k \equiv length(L'')$$
$$\wedge \; lengthOK(n, P@[L'']). \qquad\blacksquare$$

10.5   Confluence and consistency

In this section we deal with confluence, which is the crucial condition on a standard specification SP that allows us to switch back and forth between SP and its relational version $rel(SP)$ (Definition 10.14). Confluence ensures that SP and $rel(SP)$ are equivalent in the sense of Theorem 10.15. Together with Theorem 10.16 this implies that $Ini(SP)$ interprets not only the logical predicates, but also the graphs of all defined functions of SP as the least relations satisfying their axioms (Theorem 10.23(2)). In turn, this guarantees the correctness of most important proof rules that involve the functional part of SP, namely unfolding, normal form handling, fixpoint induction (Section 10.7) and its derivatives such as Hoare induction (see above). Moreover, confluence provides us with simple criteria for the consistency of SP (Definition 10.18) and the consistency of $Ini(SP)$ with respect to the initial models of subspecifications of SP (Definition 10.22). Recall that the consistency of SP is also needed for showing that $Fin(SP)$ is final in the class of hierarchical SP-models over $Ini(visSP)$ (Theorem 10.23(4)). Morever, we have seen in the previous section how the consistency of the initial model of $SP_1 \cup SP_2$ is used for proving that SP_1 and SP_2 are inductively equivalent.

The notion of confluence is based on the **rewrite relation** $\longrightarrow_{SP}$ that is derived from the axioms of SP. For dealing with predicates as well as with functions we first define the *(goal) reduction calculus* for SP that yields the basic rules $\longrightarrow_{SP}$ is built upon. Both the cut calculus (Definition 10.10) and the reduction calculus are used for proving goals. However, the former produces bottom-up proofs starting out from axioms of SP, while the latter proves a goal top-down by transforming it into the empty goal. Confluence is characterized as the equivalence of both calculi with respect to ground goals (Theorem 10.36).

A **fresh** or **extra variable** of a Horn clause $C = (t\{\equiv u\} \Leftarrow H)^{14}$ is a variable that occurs in u or H, but not in t. **fresh**(C) denotes the set of fresh variables of the clause. Note that Condition 10.5(1) on a standard specification implies that all fresh variables of C occur in the premise H.

Definition 10.33. Let $SP = (\Sigma, AX)$ be a standard specification. The **reduction calculus for** SP consists of the following rules for reducing goals. Let G be a goal and σ be a substitution over Σ.

rewriting upon C $\quad \dfrac{G(t\sigma)}{G(u\sigma) \wedge H\sigma}\Uparrow \quad$ if $C = (t \equiv u \Leftarrow H) \in AX$
$\qquad\qquad\qquad\qquad\qquad\qquad\qquad$ and $\textit{fresh}(C)\sigma \subseteq NF_\Sigma(X)$

resolution upon C $\quad \dfrac{r(t\sigma) \wedge G}{H\sigma \wedge G}\Uparrow \quad$ if $C = (r(t) \Leftarrow H) \in AX,\ r \neq \equiv$
$\qquad\qquad\qquad\qquad\qquad\qquad\qquad$ and $\textit{fresh}(C)\sigma \subseteq NF_\Sigma(X)$

reflection $\qquad\qquad \dfrac{t \equiv t \wedge G}{G}\Uparrow$

A sequence $G_1, \ldots, G_n$ of goals such that for all $1 \leq i < n$, G_{i+1} is obtained from G_i by applying one of the above rules, is called an SP-**reduction of** G_1 into G_n and we write $G_1 \vdash_{SP} G_n$. A goal G is SP-**convergent** if $G \vdash_{SP} \emptyset$.

The SP-**rewrite relation** is a binary relation both on terms and goals and defined as follows:

$$t \longrightarrow_{SP} u \quad\Longleftrightarrow_{def}\quad \begin{cases} \text{there is a rewrite step } r(t) \vdash_{SP} r(u) \wedge H \\ \text{such that } H \vdash_{SP} \emptyset, \end{cases}$$

$$G(t) \longrightarrow_{SP} G(t') \quad\Longleftrightarrow_{def}\quad t \longrightarrow_{SP} t'.$$

A term or goal u is an SP-**reduct** of a term or goal t if $t \xrightarrow{*}_{SP} u$. t is SP-**reduced** if $t \xrightarrow{*}_{SP} u$ implies $t = u$. A substitution τ is an SP-**reduct** of a substitution σ if $x\sigma \xrightarrow{*}_{SP} x\tau$ for all $x \in X$. A goal G is **strongly** SP-**convergent** if all SP-reducts of G are SP-convergent.

Two terms are SP-**joinable** if they have a common SP-reduct. SP is (ground) **confluent** if for all ground terms t, each two SP-reducts of t are SP-joinable. $\qquad\qquad\qquad\qquad\qquad\qquad\qquad\qquad\qquad\qquad\qquad\quad \diamond$

If SP has no logical predicates, then $\xrightarrow{*}_{SP}$ coincides with *join conditional rewriting* in the sense of Section 9.2.3. By Condition 10.5(1), all Σ-normal forms are SP-reduced. $\vdash_{SP}$ and $\longrightarrow_{SP}$ are correct with respect to $Ini(SP)$, i.e. for all goals G, H and terms t, u,

$$G \vdash_{SP} H \quad\Longrightarrow\quad Ini(SP) \models G \Leftarrow H, \tag{10.1}$$

$$t \xrightarrow{*}_{SP} u \quad\Longrightarrow\quad t \equiv u \vdash_{SP} \emptyset. \tag{10.2}$$

---

[14] Recall that set brackets in a formula enclose optional subexpressions. $t\{\equiv u\}$ stands for t or $t \equiv u$.

Moreover,

$$G(t) \xrightarrow{*}_{SP} G(u) \vdash_{SP} \emptyset \implies G(t) \vdash_{SP} \emptyset, \qquad (10.3)$$

$$t \text{ and } u \text{ are } SP\text{-joinable} \iff t \equiv u \vdash_{SP} \emptyset. \qquad (10.4)$$

Lemma 10.34. *SP is confluent iff all SP-convergent ground goals are strongly SP-convergent.*

Proof. "⇐": Let u, u' be reducts of a ground term t. Since $t \xrightarrow{*}_{SP} u$ and $u \equiv u$ is convergent, (10.3) implies $t \equiv u \vdash_{SP} \emptyset$. Since $t \xrightarrow{*}_{SP} u'$ and convergent ground goals are strongly convergent, we conclude $u' \equiv u \vdash_{SP} \emptyset$. Hence by (10.4), u' and u are joinable.

"⇒": Let $r(t)$ be a ground atom and u be a ground term such that $r(t) \vdash_{SP} \emptyset$ and $t \xrightarrow{*}_{SP} u$. We show $r(u) \vdash_{SP} \emptyset$ by induction on a shortest reduction $r(t), G_1, \ldots, G_n, \emptyset$. Let the step from $r(t)$ to G_1 be a rewrite step. Then $G_1 = (r(t') \wedge H)$ and $t \longrightarrow_{SP} t'$ for some v, H. Since SP is confluent, t' and u have a common reduct v. By induction hypothesis, $r(t') \vdash_{SP} \emptyset$ and $t' \xrightarrow{*}_{SP} v$ imply $r(v) \vdash_{SP} \emptyset$. Since $u \xrightarrow{*}_{SP} v$, (10.3) implies $r(u) \vdash_{SP} \emptyset$.

Let the step from $r(t)$ to G_1 be a resolution step. Then there is an axiom $r(v) \Leftarrow H$ in SP such that $t = v\sigma$ and $G_1 = H\sigma$ for some σ. By Conditions 10.5(1) and (2), v is a tuple $(v_1, \ldots, v_n)$ of normal forms such that for all $1 \leq i \leq n$, $v_i \in X \cup \{\bot\}$ or $Def(v_i) \in H$. Moreover, there are $1 \leq k \leq n$ and a term u_k such that $v_k\sigma \longrightarrow_{SP} u_k$ and $u = (v_1\sigma, \ldots, u_k, \ldots, v_n\sigma)$. Hence $v_k \in X$ or $Def(v_k) \in H$.

Suppose that there is τ with $\sigma \xrightarrow{*}_{SP} \tau$ and $u \xrightarrow{*}_{SP} v\tau$. By induction hypothesis, $H\sigma \vdash_{SP} \emptyset$ implies $H\tau \vdash_{SP} \emptyset$. Hence $r(v\tau)$ and thus $r(u)$ are convergent. If $v_k \in X$, we define τ by $v_k\tau = u_k$ and $x\tau = x\sigma$ for all $x \in X \setminus \{v_k\}$ and indeed obtain $\sigma \xrightarrow{*}_{SP} \tau$ and $u = (v_1\sigma, \ldots, u_k, \ldots, v_n\sigma) \xrightarrow{*}_{SP} (v_1\tau, \ldots, u_k, \ldots, v_n\tau) = v\tau$.

Let $Def(v_k) \in H$. Since v_k is a normal form and $Def(v_k\sigma)$ is a subgoal of $H\sigma$ and thus convergent, the axioms for Def (see Definition 10.5) imply that there is σ' with $\sigma \xrightarrow{*}_{SP} \sigma'$ and $Def(x\sigma') \vdash_{SP} \emptyset$ for all $x \in var(v_k)$. Hence $Def(x\sigma) \vdash_{SP} \emptyset$ and thus $v_k\sigma \xrightarrow{+}_{SP} u_k$ implies $u_k = v_k'\tau'$ and $v_k'\rho = v_k$ for a term v_k', a substitution τ' and a variable renaming ρ such that v_k' is linear and for all $x, y \in var(v_k')$, $x\rho = y\rho$ implies $x\tau' \xleftarrow{*}_{SP} x\rho\sigma = y\rho\sigma \xrightarrow{*}_{SP} y\tau'$. Since SP is confluent, we conclude $v_k'\tau' \xrightarrow{*}_{SP} v_k\tau$ for some τ with $x\tau' \xrightarrow{*}_{SP} x\rho\tau$ for all $x \in var(v_k')$. Hence $x\rho\sigma \xrightarrow{*}_{SP} x\tau' \xrightarrow{*}_{SP} x\rho\tau$ for all $x \in var(v_k')$ and thus $x\sigma \xrightarrow{*}_{SP} x\tau$ for all $x \in var(v_k)$. Moreover, $u_k = v_k'\tau' \xrightarrow{*}_{SP} v_k\tau$. W.l.o.g. we define $x\tau = x\sigma$ for all $x \in X \setminus var(v_k)$ and obtain again $\sigma \xrightarrow{*}_{SP} \tau$ and $u = (v_1\sigma, \ldots, u_k, \ldots, v_n\sigma) \xrightarrow{*}_{SP} v\tau$.

Let the step from $r(t)$ to G_1 be a reflection step. Then $r(t) = (t' \equiv t')$, $t' \xrightarrow{*}_{SP} u'$ and $r(u) = (u' \equiv t')$ for some t', u'. Since $u' \equiv u'$ is convergent, (10.3) implies $r(u) \vdash_{SP} \emptyset$. □

Lemma 10.35. *Let SP be complete and confluent. Then for all congruence axioms and Horn axioms $C = (p \Leftarrow H)$ of SP and $\sigma \colon X \to T_\Sigma$, $H\sigma \vdash_{SP} \emptyset$ implies $p\sigma \vdash_{SP} \emptyset$ (see Definition 10.10).*

Proof. Since SP is complete, there is $\tau \colon X \to NF_\Sigma$ such that for all $x \in X$, $x\sigma \equiv_{SP} x\tau$. Since SP is confluent and normal forms are SP-reduced, we conclude $x\sigma \xrightarrow{*}_{SP} x\tau$ for all $x \in X$. Hence by Lemma 10.34, $H\sigma \vdash_{SP} \emptyset$ implies $H\tau \vdash_{SP} \emptyset$. We show $p\tau \vdash_{SP} \emptyset$.

Let C be an axiom of SP. If $p = (t \equiv u)$, then rewriting upon C transforms $p\tau$ into the goal $(u \equiv u \wedge H)\tau$, and the reflection rule leads to $H\tau$. Since $H\tau \vdash_{SP} \emptyset$, we conclude $p\tau \vdash_{SP} \emptyset$. If $p = r(t)$ for some $r \neq\equiv$, then resolution upon C transforms $p\tau$ into $H\tau$. Since $H\tau \vdash_{SP} \emptyset$, we conclude $p\tau \vdash_{SP} \emptyset$.

Let $C = (x \equiv x)$. Then the reflection rule implies $p\tau \vdash_{SP} \emptyset$. Let $C = (y \equiv x \Leftarrow x \equiv y)$. Then by (10.4), $H\tau \vdash_{SP} \emptyset$ implies that $x\tau$ and $y\tau$ are SP-joinable. Hence, again by (10.4), $p\tau \vdash_{SP} \emptyset$.

Let $C = (f(x_1,\ldots,x_n) \equiv f(x_1,\ldots,x_{i-1},y,x_{i+1},\ldots,x_n) \Leftarrow x_i \equiv y)$. Then by (10.4), $H\tau \vdash_{SP} \emptyset$ implies that $x_i\tau$ and $y\tau$ have a common SP-reduct, say t. Hence $f(x_1\tau,\ldots,x_{i-1}\tau,t,x_{i+1}\tau,\ldots,x_n\tau)$ is a common SP-reduct of $f(x_1,\ldots,x_n)\tau$ and $f(x_1,\ldots,x_{i-1},y,x_{i+1},\ldots,x_n)\tau$. Again by (10.4), we conclude $p\tau \vdash_{SP} \emptyset$.

Let $C = (r(x_1,\ldots,x_n) \Leftarrow r(x_1,\ldots,x_{i-1},y,x_{i+1},\ldots,x_n) \wedge x_i \equiv y)$. Then $H\tau \vdash_{SP} \emptyset$ implies $r(x_1,\ldots,x_{i-1},y,x_{i+1},\ldots,x_n)\tau \vdash_{SP} \emptyset$ and by (10.4), $x_i\tau$ and $y\tau$ have a common SP-reduct, say t. Hence by Lemma 10.34, the atom $r(x_1\tau,\ldots,x_{i-1}\tau,t,x_{i+1}\tau,\ldots,x_n\tau)$ is SP-convergent. By (10.3), we conclude $p\tau \vdash_{SP} \emptyset$.

By (10.3), $p\tau \vdash_{SP} \emptyset$ implies $p\sigma \vdash_{SP} \emptyset$.   □

By (10.1), $\vdash_{SP}$ is correct with respect to $\vdash_{cut}$. The completeness of the reduction calculus almost coincides with the confluence of SP:

Theorem 10.36 (Church–Rosser Theorem).[15] *A complete specification SP is confluent iff for all ground goals G, $SP \vdash_{cut} G$ implies $G \vdash_{SP} \emptyset$.*

Proof. "$\Leftarrow$": By Lemma 10.34, it is sufficient to show that convergent goals are strongly convergent. So let $G \vdash_{SP} \emptyset$ and $G \xrightarrow{*}_{SP} G'$. By (10.1) and (10.2), $SP \vdash_{cut} G'$. Hence by assumption, $G' \vdash_{SP} \emptyset$.

Using Lemma 10.35, the $\Rightarrow$-part can be shown easily by induction on the length of a shortest cut calculus proof of G.   □

Corollary 10.37 (consistency criterion I). *A complete specification SP is consistent and thus functional (Definition 10.18) iff it is confluent.*

Proof. "$\Leftarrow$": Let t, t' be two SP-equivalent ground normal forms. By the Church–Rosser Theorem, $t \equiv t' \vdash_{SP} \emptyset$. By (10.4), t and t' are joinable. By Condition 10.5(1), normal forms are joinable iff they are equal.

---

[15] This generalizes the classical Church–Rosser Theorem 9.1 for unconditional equational rewrite systems to specifications whose axioms are arbitrary Horn clauses.

"⇒": Let u, u' be reducts of a ground term t. Since SP is complete, there are normal forms v, v' of u and u', respectively. By (10.2), u and u' and thus v and v' are SP-equivalent. Since SP is consistent, v and v' are equal and thus a common reduct of t.   □

One of the main consequences of Theorem 10.36 is Theorem 10.15, which is used, for instance, in the proofs of Theorem 10.23(2) and Lemma 10.27.

We sketch the proof of Theorem 10.15. Without loss of generality suppose that SP is flat (Definition 10.14). Since SP is functional, $rel(SP) = (rel(\Sigma), rel(AX))$ is also functional. Hence by Theorem 10.36, the two logical equivalences in the statement of Theorem 10.15 are equivalent to the following ones: for all defined functions $f \in \Sigma$, logical predicates $r \in \Sigma$ and $t, u \in T_{rel(\Sigma)}$,

$$f(t) \equiv u \vdash_{SP} \emptyset \iff r_f(t, u) \vdash_{rel(SP)} \emptyset, \tag{10.5}$$

$$r(t) \vdash_{SP} \emptyset \iff r(t) \vdash_{rel(SP)} \emptyset. \tag{10.6}$$

Since SP is flat, (10.5) and (10.6) can be shown easily by induction on the length of SP- resp. $rel(SP)$-reductions.

For semantical as well as deductive reasons the confluence of a standard specification is as important as its behavioural congruence. For ensuring the latter property we have the criterion of coinductivity (Definition 10.24). The development of criteria for confluence has a long tradition in term rewriting (see Chapter 9). The general approach is to localize the property insofar as joinability needs to be shown only for *critical pairs*, i.e. for reducts that stem from axiom superpositions, and then to lift the joinability of critical pairs to full confluence by arguing inductively along a *reduction ordering*.

Definition 10.38. A standard specification SP with signature Σ is **terminating** if all SP-reduced terms are normal forms and there is a **reduction ordering** $>$ **for** SP, i.e. a well-founded and transitive relation on terms and goals[16] such that the following two conditions hold true:

axiom compatibility	$r(t\sigma) > H\sigma$ resp. $v(t\sigma) > (v(u\sigma) \wedge H\sigma)$ for all Horn axioms $C = (r(t) \Leftarrow H)$ resp. $C = (t \equiv u \Leftarrow H)$ of SP, terms and atoms $v(x)$ and $\sigma \in T_\Sigma^X$ such that $fresh(C)\sigma \subseteq NF_\Sigma$ and $H\sigma$ is SP-convergent.
subterm compatibility	$t > u$ for all $t \in T_\Sigma$ and all proper subterms u of t.

Let $>$ be a reduction ordering for SP. An SP-reduction $G_1, \ldots, G_n$ is **decreasing** if $G_i > G_{i+1}$ for all $1 \le i < n$. Given a goal G, **safe(G)** and **live(G)** denote the sets of atoms $r(t)$ of G such that r is a safety resp. liveness

---

[16] Here a goal is regarded as the multiset of its atoms and an equation as the multiset of its two sides.

predicate. Let R be a set of liveness predicates of SP and HC be a set of Horn clauses $r(t) \Leftarrow H$ such that $r \in R$, $live(H) \subseteq R$ and $\nu\Psi \models HC$ (see Theorem 10.19). HC is $>$-**reductive** if for all $r \in R$, $(r(t) \Leftarrow H) \in HC$ and $\sigma \in T_\Sigma^X$,

$$fresh(r(t) \Leftarrow H)\sigma \subseteq NF_\Sigma \wedge safe(H)\sigma \vdash_{SP} \emptyset \quad \Rightarrow \quad r(t\sigma) > H\sigma. \qquad \diamond$$

The axiom compatibility of $>$ implies that all SP-reductions into the empty goal are decreasing. Moreover, rewrite steps decrease ground terms, i.e. $t \longrightarrow_{SP} t'$ implies $t > t'$. Moreover, the assumptions that $>$ is well-founded and all SP-reduced terms are normal forms imply that each ground term has a normal SP-reduct (Definition 10.33). Hence a terminating specification is always complete (Definition 10.18). The subterm compatibility of $>$ is needed for the above-mentioned induction step from critical pairs to the contexts in which they may occur.

The assumption "$H\sigma$ is SP-convergent" in axiom compatibility makes termination a rather weak requirement. The assumption says that a rewrite or resolution step must decrease a ground goal only if the premise of the applied axiom instance is convergent.

If HC is $>$-reductive, then the co-Horn axioms for R can be replaced by Horn axioms such that the resulting specification HSP is terminating in a slightly stronger sense: in axiom compatibility, it reads $safe(H)\sigma$ instead of $H\sigma$. Let us show that the final model of HSP and the final model of SP are isomorphic:

Theorem 10.39 (Horn axioms for liveness predicates). *Suppose that* $SP = (\Sigma, AX)$ *is terminating and confluent, the consequence operator* Ψ *of Theorem 10.19 is downward continuous and* R *and* HC *are as in Definition 10.38. Construct HSP from SP by replacing the co-Horn axioms for* R *with HC. If HC is* $>$-*reductive, then* $Fin(HSP)$ *and* $Fin(SP)$ *are isomorphic.*

Proof. By Theorem 10.19, it is sufficient to show $r^{Ini(HSP)} = r^{\nu\Psi}$ for all $r : w \in R$. Since $\nu\Psi$ satisfies HC and $r^{Ini(HSP)}$ is the least interpretation of r that satisfies HC, $r^{Ini(HSP)}$ is a subset of $r^{\nu\Psi}$ and thus it remains to show $r^{\nu\Psi} \subseteq r^{Ini(HSP)}$ or, equivalently,

$$\forall a \in Ini(HSP)_w \setminus r^{Ini(HSP)} : a \notin r^{\nu\Psi}. \tag{10.7}$$

So let $a \in Ini(HSP)_w \setminus r^{Ini(HSP)}$. Then there is $t \in T_\Sigma$ such that $[t] = a$ and $HSP \nvdash_{cut} r(t)$. Hence by (10.1), $r(t)$ is not HSP-convergent. We say that a goal G **fails after** i **steps** iff each decreasing HSP-reduction of G consists of at most i goals and the last one of these is not the empty goal. Since $>$ is well-founded, $r(t) \nvdash_{HSP} \emptyset$ implies that there is $i \in \mathbb{N}$ such that $r(t)$ fails after i steps. Hence (10.7) reduces to:

$$\exists i \in \mathbb{N} : r(t) \text{ fails after } i \text{ steps} \quad \Rightarrow \quad [t] \notin r^{\nu\Psi}. \tag{10.8}$$

Since Ψ is downward continuous, $\nu\Psi = \Psi_\infty = \cap_{i\in\mathbb{N}}\Psi^i(\top)$ by Theorem 10.19. Hence (10.8) follows from:

$$r(t) \text{ fails after } i \text{ steps} \quad \Rightarrow \quad [t] \notin r^{\Psi^{i+1}(\top)}. \tag{10.9}$$

We show (10.9) by induction on i.

Suppose that $r(t)$ fails after 0 steps. Then $r(t) \not\succ H\sigma$ for all $(r(u) \Leftarrow H) \in HC$ and $\sigma \in T_\Sigma^X$ such that $u\sigma = t$ and $fresh(r(u) \Leftarrow H)\sigma \subseteq NF_\Sigma$. Since HC is $\succ$-reductive, $safe(H)\sigma$ is not SP-convergent. Since SP is terminating, SP is complete. Since SP is confluent, Theorem 10.36 implies $SP \not\vdash_{cut} safe(H)\sigma$. Hence $Ini(SP)$ does not satisfy $safe(H)\sigma$ and thus, by the definition of $\top$ (see Theorem 10.19), $\top$ does not satisfy $H\sigma$. Hence $[t] = [u\sigma] \notin r^{\Psi(\top)}$ by the definition of Ψ (see Theorem 10.19).

Suppose that $r(t)$ fails after $i > 0$ steps. Let $(r(u) \Leftarrow H) \in HC$ and $\sigma \in T_\Sigma^X$ such that $u\sigma = t$ and $fresh(r(u) \Leftarrow H)\sigma \subseteq NF_\Sigma$. If $safe(H)\sigma$ is not SP-convergent, we can conclude as above that $\Psi^i(\top)$ does not satisfy $H\sigma$ and thus $[t] = [u\sigma] \notin r^{\Psi^{i+1}(\top)}$ by the definition of Ψ.

Let $safe(H)\sigma$ be SP-convergent. Then $r(t) > H\sigma$ because HC is $\succ$-reductive. Hence $H\sigma$ fails after $i - 1$ steps. By induction hypothesis, $[v\sigma] \notin q^{\Psi^i(\top)}$ for all $q(v) \in live(H)\sigma$. Hence, if $live(H)$ is not empty, we conclude $[t] = [u\sigma] \notin r^{\Psi^{i+1}(\top)}$ from the definition of Ψ. If $live(H)$ is empty, then $H = safe(H)$ and thus $H\sigma$ is SP-convergent. Hence there is a decreasing SP-reduction of $H\sigma$ into the empty goal. Since all SP-reductions are HSP-reductions, we conclude that $H\sigma$ does *not* fail, which contradicts the above assertion that $H\sigma$ fails after $i - 1$ steps.   □

If the assumptions of Theorem 10.39 hold true, even the complement of a safety predicate can be specified by Horn axioms and thus both predicates may be used in Horn axioms for further safety predicates or defined functions. For instance, we turn the co-Horn axioms for *unsorted* (see Section 10.3) into the following (generalized) Horn axiom HC by inverting an outer implication sign:

$$unsorted(L)$$
$$\Leftarrow \forall\{x, y, L'\}((L \equiv x :: y :: L' \wedge x \le y) \Rightarrow unsorted(y :: L')).$$

Since $\nu\Psi$ satisfies HC and there is a suitable reduction ordering $>$ such that HC is $\succ$-reductive, we conclude from Theorem 10.39 that HC provides an equivalent specification of *unsorted*. Some ideas involved in the proof of Theorem 10.39 (especially Implication (10.9)) are quite similar to arguments that justify **negation as failure** in logic programming (see [Apt90], Section 5).

Termination is assumed in the Superposition Theorem given below, which characterizes confluence as *subjoinability* of *critical clauses*:

Definition 10.40 (critical clause).[17] Let $SP = (\Sigma, AX)$ be a standard specification, v be a term or an atom with a single occurrence of a variable x,

$$C = t\{\equiv u\} \Leftarrow H \quad \text{and} \quad C' = t' \equiv u' \Leftarrow H'$$

be Horn axioms of SP and σ, τ be minimal substitutions (see Definition 10.3) such that $t\sigma = v[t'\tau/x]$ and the position of x in v agrees with the position of a function symbol in t. Then the Horn clause

$$CC = v[u'\tau/x]\{\equiv u\sigma\} \Leftarrow H\sigma \wedge H'\tau$$

is the SP-**critical clause induced by** C **on** C' **at** $t\sigma$. CC is an SP-**overlay** if $v = x$.

SP is **weakly orthogonal**[18] if for each SP-critical clause $t\{\equiv t'\} \Leftarrow H$, $t = t'$ or $Ini(SP) \models \neg H$. ◇

Due to Condition 10.5(1) on SP, each SP-critical clause is either an overlay or induced by a non-partiality axiom on a partiality axiom. Moreover, each overlay is induced by two (conditional) equations. If SP has no Horn axioms with fresh variables and no critical clause is induced by two different clauses C, C', then SP is weakly orthogonal. If, however, C has fresh variables, then C may induce a critical clause $t \equiv t' \Leftarrow H$ on C *itself* such that t is different from t' (see Example 10.44).

Definition 10.41 (subjoinability).[19] Let $SP = (\Sigma, AX)$ be a standard specification. Given a binary relation $>$ on terms and goals, a Horn clause $p \Leftarrow H$ is **subjoinable below** a term or atom t if

$$H\sigma \vdash_{SP} \emptyset \quad \Rightarrow \quad p\sigma \vdash_{SP} \emptyset$$

for all $\sigma \in T_{\Sigma}^X$ such that all SP-convergent ground goals $G < t\sigma$ are strongly SP-convergent (Definition 10.33). ◇

Subjoinability is a variant of inductive validity (Definition 10.12) where the cut calculus is replaced by the reduction calculus.

Theorem 10.42 (Superposition Theorem I).[20] *A terminating specification is confluent iff for all terms and atoms t, each SP-critical clause induced at t is subjoinable below t.*

---

[17] See Definition 9.22 for the equational case.

[18] See Definition 9.12 for the equational case.

[19] See Definition 9.23 for the equational case.

[20] In the equational case this theorem combines the *Buchberger–Newman Lemma* (see Lemma 9.1) with the *Knuth–Bendix Theorem* (see Theorem 9.9). In the ground equational case it agrees with Lemmas 7 and 8 of [Küc89]. In the case of conditional-equational axioms without fresh variables it agrees with [DOS88a], Theorem 2, which, however, assumes termination and joinability in a stronger sense.

Proof. "only-if": Let SP be confluent, $>$ be a reduction ordering for SP, C, C', CC be as in Definition 10.40 and $\rho \in T_\Sigma^X$ such that $H\sigma\rho \wedge H'\tau\rho$ is SP-convergent and all SP-convergent ground goals $G < t\sigma\rho$ are strongly SP-convergent. Since $>$ is axiom compatible, $t\sigma\rho > H\sigma\rho \wedge H'\tau\rho$. Hence $H\sigma\rho$ and $H'\tau\rho$ are strongly SP-convergent. Since $>$ is well-founded, there is a normal SP-reduct ρ' of ρ. Hence $H\sigma\rho \xrightarrow{*}_{SP} H\sigma\rho'$, $H'\tau\rho \xrightarrow{*}_{SP} H'\tau\rho'$ and thus $H\sigma\rho'$ and $H'\tau\rho'$ are SP-convergent. Since C and C' are axioms of SP, $t\sigma\rho'\{\equiv u\sigma\rho'\}$ is SP-convergent and $t\sigma\rho' = v[t'\tau/x]\rho' \longrightarrow_{SP} v[u'\tau/x]\rho'$. Since SP is confluent, $v[u'\tau/x]\rho'\{\equiv u\sigma\rho'\}$ and thus $v[u'\tau/x]\rho\{\equiv u\sigma\rho\}$ are SP-convergent.

The "if"-part of the proof is carried out in [Pad92] (Theorem 6.10). It is shown by induction on p along $>$ that all SP-convergent ground atoms p are strongly SP-convergent. Moreover, the proof depends on the fact that all Horn axioms of SP are subjoinable below their left-hand sides. □

Note that the axiom compatibility of $>$ is needed in this proof only for axioms with fresh variables. If, for instance, *fresh*(C) were empty, we could choose $x\rho'$ as $x\rho$ for all $x \in var(C)$ and would get the same result.

One immediately obtains

Corollary 10.43 (confluence criterion I).[21] *A terminating and weakly orthogonal specification is confluent.*

For instance, this criterion implies that the constructor subspecification of any standard specification is confluent.

In contrast to pure equational logic and rewriting we employ here rather complicated notions. This is necessary because standard specifications should cover non-trivial data types and functional-logic programs built upon them. Consequently, we must deal with *conditional* axioms and take into account *fresh variables* from the very begininning. We would cheat the user of the theory if we would restrict ourselves to the simple equational case and consider the incorporation of axiom premises and fresh variables a "straightforward" extension. Take a look at research papers on conditional rewriting with or without extra variables and you will face quite various notions of reduction, termination, critical pair, joinability, confluence and normal form (see, e.g., [BK86a, JW86, Kap87, DOS88b, DOS88a, BG89, GW93, AL94, SMI95]). The variety is due to the number of application areas for rewriting, which range from theorem proving and prototyping to programming and compiler construction.

Example 10.44. The following visible specification comprises typical functional-logic programs and thus should be functional in order to be amenable to efficient proof methods.

---

[21] See Theorem 9.6 for the equational case.

DIV&REP = ENTRY + NAT +
   vissorts *tree*
   consts   *leaf* : *entry* → *tree*
            _#_ : *tree* ∗ *tree* → *tree*
   defuns *div* : *nat* ∗ *nat* → (*nat* ∗ *nat*)
            rep&min : *tree* ∗ *nat* → *tree* ∗ *nat*
            min : *entry* ∗ *entry* → *entry*
   vars m, n, q, r : *nat* x, y, z : *entry* $T1, T2, U1, U2$: *tree*
   Horn axioms
            $div(m, n) \equiv (0, m) \Leftarrow m < n$
            $div(m, n) \equiv (q + 1, r) \Leftarrow m \geq n > 0 \wedge div(m - n, n) \equiv (q, r)$
            $div(m, 0) \equiv bottom$
            $rep\&min(leaf(x), y) \equiv (leaf(y), x)$
            $rep\&min(T1\#T2, x) \equiv (U1\#U2, min(y, z))$
                $\Leftarrow rep\&min(T1, x) \equiv (U1, y) \wedge rep\&min(T2, x) \equiv (U2, z)$
            $min(x, y) \equiv x \Leftarrow x \leq y$
            $min(x, y) \equiv y \Leftarrow y \leq x$

EXT = DIV&REP +
   defuns *repByMin* : *tree* → *tree*
   vars x : *entry* $T1, T2$: *tree*
   Horn axioms
            $repByMin(T1) \equiv T2 \Leftarrow rep\&min(T1, x) \equiv (T2, x)$

$div(m, n)$ returns the quotient of m and n and the remainder of the division. $repByMin(T)$ replaces all entries of the tree T by their minimum. DIV&REP can be implemented directly in a functional language such as *Standard ML* [Pau96]:

```
exception bottom
fun div(m,n) = if m < n then (0,m)
                  else if m >= n > 0
                        then let val (q,r) = div(m-n,n)
                              in (q+1,r) end
                        else raise bottom
infix #
datatype tree = leaf of nat | # of tree*tree
fun rep&min(leaf(x),y) = (leaf(y),x)
|   rep&min(T1#T2,x) = let val (U1,y) = rep&min(T1,x)
                            val (U2,z) = rep&min(T2,y)
                        in (U1#U2,min(y,z)) end
```

repByMin can be implemented directly only in a logic language that provides a mechanism for solving equations. In order to apply the axiom the premise equation $rep\&min(T_1, x) \equiv (T_2, x)$ must be solved in x and T_2, but no solution is obtained by simply rewriting $rep\&min(T_1, x)$. Pure logic programming would enforce us to replace the defined functions *rep&min* and *repByMin* by their graphs, say r resp. q:

$r(leaf(x), y, leaf(y), x)$

$r(T_1 \# T_2, x, U_1 \# U_2, min(y, z)) \Leftarrow r(T_1, x, U_1, y) \wedge r(T_2, x, U_2, z)$

$q(T_1, T_2) \Leftarrow r(T_1, x, T_2, x).$

Then we could compute the solution. However, this means throwing out the baby with the bathwater. Reasoning about *repByMin* on the basis of *these* axioms would prevent us from exploiting the otherwise obvious functional dependency between arguments of $r_{repByMin}$. The axioms for $r_{rep\&min}$ are a purely relational *implementation* of *repByMin* as Pettorossi's program using λ-abstraction is a purely functional *implementation* of *repByMin* [PS87].

Given a ground term t, the equation $rep\&min(t, x) \equiv (T_2, x)$ has unique solution in the initial DIV&REP-model. This fact should be sufficient for concluding that EXT is functional if DIV&REP is so.   ∎

The axioms for *div* and *rep&min* illustrate a syntactical criterion for confluence, which is similar to coinductivity (Definition 10.24). Axiom variables may "flow" only from the conclusion's left-hand side to left-hand sides of premise equations and from right-hand sides of premise equations to *subsequent* premise equations or "back" to the conclusion's right-hand side:

Definition 10.45. A Horn clause

$f(t)\{\equiv u\} \quad \Leftarrow \quad t_1 \equiv u_1 \wedge \cdots \wedge t_n \equiv u_n \wedge H$

is **deterministic up to** $Y \subseteq X$ if for $V_0 = var(t)$ and all $1 \leq i \leq n$, u_i is a normal form,

- $Def(u_i) \in H$ or $u_i \in X \cup \{\bot\}$,
- $var(t_i) \subseteq V_{i-1}, var(u) \subseteq V_n$ and $Y = var(H) \backslash V_n$ for $V_i = V_{i-1} \uplus var(u_i)$.
                                                                        ◇

A Horn clause without fresh variables is deterministic up to the empty set of variables: set n=0. Hence determinism generalizes the usual, but too restrictive (see above), assumption in conditional rewriting theory. In contrast to similar notions in [BG89], [GW93], [AL94] and [SMI95], our notion of a deterministic clause $f(t) \equiv u \Leftarrow G$ takes into consideration only those fresh variables that occur in u. A non-empty Y indicates that $f(t)$ rewrites to u if H is solvable in Y, but the solution is not part of the "result" u. This case must be admitted, in particular if H includes an atom $r(v, w)$ and, given an instance v' of v, the rewrite step should only depend on the *existence* of an instance w' of w such that $r(v', w')$ holds true.

The axioms for *div*, *rep&min* and *min* (see Example 10.44) are deterministic up to $\emptyset$, but

$repByMin(T_1) \equiv T_2 \quad \Leftarrow \quad rep\&min(T_1, x) \equiv (T_2, x)$ (10.10)

is not deterministic, not even up to $\{x\}$, as one might think at first sight. Although T_2 is "used", but x is not, both variables are "generated" simultaneously and thus the premise of (10.10) can be regarded – in terms of Definition 10.45 – neither as $t_1 \equiv u_1$ nor as H. Clause (10.10) may be regarded as a network specification.[22] While T_1 denotes an "input port" and T_2 denotes an "output port", x is "fed back" from the output to the input port. Fortunately, the function $repByMin$, which is to be defined by solutions of the premise of (10.10), does not occur there, too. Hence we can employ the fact that unique solutions are obtained already with respect to the *base* specification DIV&REP. In other words, the overlay induced by Clause (10.10) on itself and given by:

$$T_2 \equiv T_3 \ \Leftarrow \ rep\&min(T_1, x) \equiv (T_2, x) \wedge rep\&min(T_1, y) \equiv (T_3, y)$$
$$(10.11)$$

is an inductive DIV&REP-theorem.

Theorem 10.46 (confluence criterion II). [Pad97] *Let SP be terminating, AX be the set of Horn axioms of SP and SP′ be a confluent subspecification of SP with signature Σ' and Horn axiom set AX′ such that $NF_{\Sigma,s} = NF_{\Sigma',s}$ for all sorts $s \in \Sigma'$. SP is confluent if*

(a) for all $(f(t)\{\equiv u\} \Leftarrow H) \in AX \setminus AX'$, $f \in \Sigma \setminus \Sigma'$,

(b) for all $C \in AX \setminus AX'$, C is deterministic or the overlay induced by C on itself is an inductive theorem of SP′,

(c) all SP-overlays induced by two different clauses of $AX \setminus AX'$ are inductive theorems of SP′.

To apply the criterion to Example 10.44 we first assume that NAT and (the instances of) ENTRY are confluent and set $SP' = $ ENTRY+NAT and $SP = $ DIV&REP. Then (a) holds true trivially. (b) is valid because the axioms for for div, $rep\&min$ and min are deterministic. The only two different axioms of $SP \setminus SP'$ that induce an overlay are the two axioms for min. It reads as follows:

$$x \equiv y \ \Leftarrow \ x \leq y \wedge y \leq x.$$

Since this is an axiom of ENTRY, we conclude (c). Hence Theorem 10.46 implies that DIV&REP is confluent. Secondly, we set $SP' = $ DIV&REP and $SP = $ EXT. Again, (a) holds true trivially. (b) is valid because (10.11) is an inductive SP'-theorem, which can be shown easily, e.g. by inductive expansion (Section 10.7). (c) holds true trivially because no overlay is induced by two different axioms of $SP \setminus SP'$. Hence confluence criterion II implies that EXT is confluent.

Given a standard specification SP, the subspecification SP' consisting of the constructors, bottom constants, definedness predicates and partiality

---

[22] See [Pad92], Example 8.7, for a more complex example.

axioms of SP is always confluent. Hence this may always be the first choice for SP' when applying Theorem 10.46.

The Superposition Theorem and its corollaries 10.43 and 10.46 are the main reasons for requiring that SP is terminating. Though termination also implies completeness, it is mostly easier to show this condition directly. As a side-effect one may already use completeness when defining a reduction ordering, which is usually done inductively on the structure of terms and goals (see, e.g., Definition 9.10). If the definition splits into several cases, the transitivity of $>$ is often harder to prove than the well-foundedness. If SP is complete, then the following combination of an ordering on Σ with a term resp. proof ordering yields a reduction ordering for most standard specifications.

Theorem 10.47 (termination criterion). (see [Pad92], Section 6.2) *Suppose that SP is complete and all SP-reduced terms are normal forms. A binary relation on the set of defined functions and logical safety predicates of Σ is defined as follows:*

$$f >_\Sigma g \iff_{def} \exists : (f(t)\{\equiv t\} \Leftarrow H) \in AX \wedge g \text{ occurs in } u \text{ or } H.$$

SP is terminating if for all $C = (f(t)\{\equiv t\} \Leftarrow H) \in AX$, atoms or subterms $g(v)$ of u or H, $\sigma \in T_\Sigma^X$ and normal forms t', v' of $t\sigma$ resp. $v\sigma$ such that $g >_\Sigma^* f$, $H\sigma$ is convergent and $fresh(C)\sigma$ consists of normal forms, one of the following conditions holds true:

- *f and g are function symbols and v' is smaller than t' with respect to a fixed well-founded ordering on NF_Σ.*
- *f and g are logical predicates and $g(v')$ has a shorter cut calculus proof than $f(t')$.*

We leave it to the reader to show that the termination criterion applies to the specification EXT of Example 10.44. The reduction ordering $>$ built up from the signature ordering $>_\Sigma$ satisfies $f(t) > u$ for all defined functions and logical safety predicates f and all ground normal forms u. Note that this is necessary for $>$ to be compatible with axioms that involve fresh variables (see Definition 10.38).

To sum up, Corollary 10.43 and Theorems 10.46 and 10.47 provide us with criteria for confluence resp. termination and thus – via the Church–Rosser Theorem – for consistency (Corollary 10.37) and the equivalence of SP and $rel(SP)$ (Theorem 10.15).

Confluence criteria are also useful for proving the consistency of $Ini(SP)$ wrt $Ini(SP')$ where SP' is an arbitrary subspecification of SP (Definition 10.22). In Section 10.4, we pointed out that the more general consistency property comes up as proof obligation when two specifications are shown to be inductively equivalent or when implementations are shown to be correct. Hence we close this section by presenting four criteria for the consistency of $Ini(SP)$ wrt $Ini(SP')$, which impose increasingly weaker assumptions on SP.

Theorem 10.48 (consistency criteria II). [Pad97] *Let AX be the set of Horn axioms of SP and SP′ be a subspecification of SP with signature Σ′ and Horn axiom set AX′.*

(1) **proof by consistency** *Let Σ = Σ′. Ini(SP) is consistent wrt Ini(SP′) iff AX \ AX′ consists of inductive theorems of SP′.*

Ini(SP) is consistent wrt Ini(SP′) if

(2) for all function symbols $f: w \to s \in \Sigma \setminus \Sigma'$, $s \in \Sigma \setminus \Sigma'$, and for all $(r(t) \Leftarrow H) \in AX \setminus AX'$, $r \in \Sigma \setminus \Sigma'$.

Ini(SP) is consistent wrt Ini(SP′) if SP is functional, for all $(f(t)\{\equiv u\} \Leftarrow H) \in AX \setminus AX'$, $f \in \Sigma \setminus \Sigma'$, and

(3) for all constructors $c: w \to s \in \Sigma \setminus \Sigma'$, $s \in \Sigma \setminus \Sigma'$, or

(4) SP′ is complete and all axioms of SP′ are deterministic up to $\emptyset$.

10.6 Implicit induction

Theorem 10.48(1) forbids signature extensions ($\Sigma = \Sigma'$). Hence it is less useful for showing consistency than for proving inductive theorems. But even in Case $\Sigma = \Sigma'$ consistency can be reduced to functionality – if $AX \setminus AX'$ is *SP′-reducible*:

Definition 10.49. A set CS of Horn clauses is (ground) **SP-reducible** if for all $(t\{\equiv u\} \Leftarrow H) \in CS$ and $\sigma \in T_\Sigma^X$ such that $H\sigma$ is $(SP \cup CS)$-convergent, $t\sigma$ is not SP-reduced.   ◇

Lemma 10.50.[23] *Let CS be a set of Horn clauses such that SP ∪ CS is terminating.*

(1) If $SP \cup CS$ is confluent and CS is SP-reducible, then $Ini(SP \cup CS)$ is consistent wrt Ini(SP).

(2) If SP is confluent and $Ini(SP \cup CS)$ is consistent wrt Ini(SP), then CS is SP-reducible.

Theorem 10.42 reduces confluence to the subjoinability of critical clauses. Lemma 10.50 suggests a variant that characterizes confluence and *reductive* consistency simultaneously (Theorem 10.54). Given a subspecification $SP′$ of SP, consistency of $Ini(SP)$ wrt $Ini(SP′)$ becomes reductive consistency if $\vdash_{cut}$ is replaced by convergence:

Definition 10.51. Given a subspecification $SP′$ of SP, SP is **reductively consistent wrt** $SP′$ if for all ground atoms p over $SP′$, $p \vdash_{SP} \emptyset$ implies $p \vdash_{SP'} \emptyset$.   ◇

---

[23] This generalizes [JK89], Theorem 1, to conditional axioms.

Proposition 10.52 (confluence criterion III). *Given a confluent sub-specification SP' of SP, SP is confluent if SP is reductively consistent wrt SP'.*

Proof. Let u, u' be SP-reducts of a ground term t. Hence $t \equiv u$ and $t \equiv u'$ are SP-convergent. Since SP is reductively consistent wrt SP', both equations are SP'-convergent and thus SP'-joinable (Definition 10.33), i.e. $u \xrightarrow{*}_{SP'} v \xleftarrow{*}_{SP'} t \xrightarrow{*}_{SP'} v' \xleftarrow{*}_{SP'} u'$ for some v, v'. Since SP' is confluent, v and v' and thus u and u' are SP'-joinable. □

Secondly, we modify the definition of subjoinability (Definition 10.41):

Definition 10.53 (SP'-subjoinability). Given a subspecification SP' of SP and a binary relation $>$ on goals, a Horn clause $p \Leftarrow H$ is SP'-**subjoinable below** a term or atom t if

$$H\sigma \vdash_{SP} \emptyset \quad \Rightarrow \quad p\sigma \vdash_{SP} \emptyset$$

for all $\sigma \in T_\Sigma^X$ such that all SP-convergent ground goals $G < t\sigma$ are strongly SP-convergent and SP'-convergent.

Given two subsets A, B of AX, (A, B) is (SP', SP)-**convergent** if for all terms and atoms t, each SP-critical clause $u\{\equiv v\} \Leftarrow H$ induced by some $C \in A$ on some $C' \in B$ at t is SP'-subjoinable below $t\{\equiv v\}$.   ◇

Theorem 10.54 (Superposition Theorem II). *Let SP be terminating with reduction ordering $>$ and CS be a set of Horn clauses over Σ.*

(a) $SP \cup CS$ is confluent and reductively consistent wrt SP if $>$ is compatible with CS and $(AX \cup CS, AX \cup CS)$ is $(SP, SP \cup CS)$-convergent.

(b) Let $SP \cup CS$ be confluent. If the $>$ is compatible with CS or CS has no fresh variables[24], then $(AX \cup CS, AX \cup CS)$ is $(SP, SP \cup CS)$-convergent.

Proof. Quite similar to the proof of Theorem 10.42. For (a), one shows by induction on p along $>$ that all $(SP \cup CS)$-convergent ground atoms p are strongly $(SP \cup CS)$-convergent and SP-convergent.   □

The criterion for the confluence of $SP \cup CS$ in (a) can be weakened if SP is already confluent:

Lemma 10.55. [Pad96b] *Let SP be confluent and CS be an SP-reducible set of Horn clauses such that $SP \cup CS$ is terminating. $(AX \cup CS, AX \cup CS)$ is $(SP, SP \cup CS)$-convergent if (CS, AX) is $(SP, SP \cup CS)$-convergent.*

These results lead to a method for proving inductive theorems without explicit induction steps, provided that the conjectures are Horn clauses and there is a reduction ordering that is compatible with AX and CS (Definition 10.38):

---

[24] See the remark following Theorem 10.42.

Theorem 10.56 (implicit induction I). *Let SP be confluent and terminating and CS be a set of Horn clauses over Σ.*

(a) If CS is SP-reducible, (CS, AX) is $(SP, SP \cup CS)$-convergent and the reduction ordering for SP is compatible with CS, then CS consists of inductive theorems of SP.

(b) Let CS consist of inductive theorems of SP. If the reduction ordering for SP is compatible with CS or CS has no fresh variables, then (CS, AX) is $(SP, SP \cup CS)$-convergent.

Proof. (a) By Lemma 10.55, $(AX \cup CS, AX \cup CS)$ is $(SP, SP \cup CS)$-convergent. Hence by Theorem 10.54(a), $SP \cup CS$ is confluent and reductively consistent wrt SP and thus by the Church–Rosser Theorem, $Ini(SP \cup CS)$ is consistent wrt $Ini(SP)$. Hence Theorem 10.48(1) implies $CS \subseteq ITh(SP)$.

(b) By Theorem 10.48(1), $Ini(SP \cup CS)$ is consistent wrt SP. Since SP is confluent, the Church–Rosser Theorem implies that $SP \cup CS$ is reductively consistent wrt SP. Hence by Proposition 10.52, $SP \cup CS$ is confluent and thus by Theorem 10.54(b), (CS, AX) is $(SP, SP \cup CS)$-convergent. □

By (b), the previous theorem also allows us to disprove inductive conjectures. In contrast to forbidding fresh variables in AX, it might be reasonable to exclude them from CS. (b) says that in this case the reduction ordering need not be compatible with the conjectures.

Critical clauses of conjectures on axioms correspond to cases proofs by explicit induction split into. Consequently, the implicit-induction approaches of [Red90], [KR90b] and [BKR95] replace the notion of a critical clause of CS on AX by that of a cover or test set for CS, which, roughly said, subsumes all solutions of CS. The conjectures are proved by induction on the test set along a reduction ordering for SP (see, e.g., Section 9.6). One may construct the test set from the minimal substitutions that generate critical clauses of CS on AX. In contrast to Theorem 10.56(a), the corresponding validity criterion (Theorem 10.58) does not require the reduction ordering for SP to be compatible with CS. Instead, CS and all lemmas used in proofs that critical clauses are SP-subjoinable must be *bounded*:

Definition 10.57. Given a reduction ordering $>$ for SP, a clause $t\{\equiv u\} \Leftarrow H$ is **bounded** by an atom p if for all goals $G(x)$ and $\sigma, \tau \in T_\Sigma^X$, $p\sigma \geq G(t\tau)$ implies $p\sigma > (G(u\tau) \wedge H\tau)$. ◇

Theorem 10.58 (implicit induction II). [Pad96b] *Suppose that SP is confluent and terminating and CS is a set of Horn clauses over Σ. For any Horn clause C, let $crit_C$ be the set of $(SP \cup CS)$-critical clauses of C on AX and $Y_C = var(crit_C) \setminus var(C)$. If (CS, AX) is $(SP, SP \cup CS)$-convergent and for all $C = (t\{\equiv u\} \Leftarrow H) \in CS$,*

$$\exists Y_C : \bigvee \{G \wedge \bigwedge_{x \in dom(\sigma)} x \equiv x\sigma \mid (v\{\equiv u\sigma\} \Leftarrow H\sigma \wedge G) \in crit_C\}$$

is an inductive theorem of SP and C is bounded by all conclusions of clauses of CS, then CS consists of inductive theorems of SP.

Let us finally present two criteria for the *SP*-subjoinability of a critical clause *CC* (Definition 10.53). The first one simply requires the inductive validity of *CC*. The second one demands that the conclusion of *CC* is reducible into the premise by applying inductive theorems of *SP* as well as bounded clauses of *CS*, which are thus used as implicit induction hypotheses!

Lemma 10.59 (criteria for *SP*-subjoinability). *Let SP be confluent and terminating, CS be a set of Horn clauses over Σ and $CC = (u\{\equiv v\} \Leftarrow H)$ be an $(SP \cup CS)$-critical clause induced by some C on some C' at some t. CC is SP-subjoinable below $t\{\equiv v\}$ if C and C' are bounded by $t\{\equiv v\}$ and*

(1) CC is an inductive theorem of SP or

(2) there is a subset DS of $ITh(SP) \cup CS$ consisting of clauses bounded by $t\{\equiv v\}$ such that for all $\sigma \in T_\Sigma^X$ there is a DS-reduction of $u\sigma\{\equiv v\sigma\}$ into a subset of $H\sigma$.

Proof. Let $\sigma \in T_\Sigma^X$ such that $H\sigma$ is $(SP \cup CS)$-convergent and all $(SP \cup CS)$-convergent ground goals $G < t\sigma\{\equiv v\sigma\}$ are *SP*-convergent and strongly $(SP \cup CS)$-convergent. Since C, C' are bounded by $t\{\equiv v\}$, we have $t\sigma\{\equiv v\sigma\} > (u\sigma\{\equiv v\sigma\} \wedge H\sigma)$. Hence $H\sigma$ is *SP*-convergent. In Case (1), the Church–Rosser Theorem implies that $u\sigma\{\equiv v\sigma\}$ is *SP*-convergent. In Case (2) there is a subset H' of σ such that $t\sigma\{\equiv v\sigma\} > u\sigma\{\equiv v\sigma\} \vdash_{DS} H' \vdash_{SP} \emptyset$. Suppose that for all ground goals G,

$$t\sigma\{\equiv v\sigma\} > G \vdash_{DS} \emptyset \text{ implies } G \vdash_{SP} \emptyset. \tag{10.12}$$

Then $u\sigma\{\equiv v\sigma\}$ is *SP*-convergent as well.

We show (10.12) by induction on the length of a shortest *DS*-reduction R of G into $\emptyset$. There are $C = (l\{\equiv r\} \Leftarrow H'') \in DS$, a goal $G'(x)$ and $\tau \in T_\Sigma^X$ such that $G = G'(l\tau)$ and the shortest *DS*-reduction of $G'(r\tau) \wedge H''\tau$ into $\emptyset$ is shorter than R. Since C is bounded by $t\{\equiv v\}$, we have $t\sigma\{\equiv v\sigma\} > G'(r\tau) \wedge H''\tau$. Hence by induction hypothesis, $G'(r\tau) \wedge H''\tau$ is *SP*-convergent and thus *SP*-convergent and strongly $(SP \cup CS)$-convergent. If $C \in CS$, the convergence implies $G \vdash_{SP \cup CS} \emptyset$ and thus $G \vdash_{SP} \emptyset$ because $t\sigma\{\equiv v\sigma\} > G$. If C is an inductive theorem of *SP*, then $H''\tau \vdash_{SP} \emptyset$ implies $SP \vdash_{cut} l\tau\{\equiv r\tau\}$ and thus by the Church–Rosser Theorem, $l\tau\{\equiv r\tau\}$ is *SP*-convergent. Hence $G = G'(l\tau)$ is $(SP \cup CS)$-convergent because $G'(r\tau)$ is strongly $(SP \cup CS)$-convergent. Since $t\sigma\{\equiv v\sigma\} > G$, we conclude $G \vdash_{SP} \emptyset$.   □

Theorems 10.56 and 10.58 provide – together with the above criteria for *SP*-subjoinability – two methods for proving inductive theorems implicitly. Similar procedures have been established by [Red90], [KR90b] and [BKR95] (see Section 9.6). In all these cases reduction orderings on formulas take over the rôle of induction orderings, which are typical for *explicit* induction. Any implicit-induction method is based on the equivalence of inductive validity and consistency (Theorem 10.48(1)), originally stated in [HH82] and [KM87]. Current research on explicit induction develops strategies for guiding proofs towards induction hypotheses by recognizing conjecture *skeletons* within subgoals and applying *wave rules* (see, e.g., [Hut90, BSvH$^+$93]), while implicit induction tries to make such strategies superfluous. However, the range of applications for implicit induction is quite limited because of a number of restrictions and assumptions such as the Horn clause form of conjectures and the involvement of reduction orderings in proofs. Comparative analyses of both approaches can be found in [Pad92] and [BBH94].

10.7   Unfolding and explicit induction

Let us now continue the presentation of Gentzen rules for building up Gentzen proofs, which we started in Section 10.4. First a couple of useful expansions of (conjunctions of) Gentzen clauses:

body splitting
$$\frac{GS \;\Leftarrow\; HS \lor HS'}{(GS \;\Leftarrow\; HS) \land (GS \;\Leftarrow\; HS')} \;\Updownarrow$$

body reduction
$$\frac{GS(t) \;\Leftarrow\; H(t) \land t \equiv u}{GS(u) \;\Leftarrow\; H(u)} \;\Updownarrow$$

subsumption
$$\frac{\exists Y (G_1 \lor \cdots \lor G_n) \;\Leftarrow\; \exists Z (H_1 \lor \cdots \lor H_k)}{TRUE} \;\Updownarrow$$

> if the premise $\exists Z(H_1 \lor \cdots \lor H_k)$ **subsumes** the conclusion $\exists Y(G_1 \lor \cdots \lor G_n)$, i.e. for all $1 \le i \le k$ there are $1 \le j \le n$ and a substitution σ such that $G_j \sigma_Y \subseteq H_i$ (see Definition 10.3)

Intuitively, subsumption is a "syntactic implication" that generalizes the θ-subsumption of universally quantified disjunctions [Plo70]. For instance, the goal $x \ge 0 \land \exists y(x < s(y))$ subsumes $\exists y(x < y)$.

Many non-inductive Gentzen proofs consist of head expansions, body contractions (Section 10.4) and a final subsumption of the head (= conclusion) by the body (= premise) of the Gentzen clause that was derived at latest. Head expansions and body contractions are applications of **goal set rules** each of which transforms either the premise or the conclusion of a Gentzen clause. Roughly said, a proof of a Gentzen clause $GS \Leftarrow HS$ that consists of head expansions and body contractions brings GS and HS "closer and closer together". Of course, it may happen that the expanded conclusion fails to meet the contracted premise, i.e. the latter will never subsume the former. Such failures can often be avoided if simplifications, which turn formulas into equivalent ones, are preferred to pure expansions or contractions (see Section 10.4).

The term "simplification" indicates that these rules often simplify the structure of their redices, i.e. the (sub)formulas they are applied to. Automated theorem provers usually apply simplification rules automatically. The proofs get shorter and more readable if simplifications are executed in the background and these steps are not recorded explicitly. As they are equivalence transformations, simplifications do not change the semantics (= set of solutions) of a formula, but they will modify the syntactic structure. This may lead to failures in cases where a rule was applicable before the modification, but not afterwards. For instance, an induction step can be performed only if an induction hypothesis has been generated. But an induction hypothesis can be recognized automatically only from its syntactic structure. Hence intermediate simplifications may block induction steps. Nevertheless, theorem provers can work reasonably fast only if they are equipped with **built-in simplifiers** that apply simplification rules automatically.

Simplifiers "optimize" theorem provers. They "massage" formulas for the purpose of speeding up the proof process. Since they perform in the background, the verifier may concentrate on those parts of a proof that need his guidance. In Section 10.4, we have discussed the analogy between program synthesis and proof construction. Both activities: designing a program and building a proof, are successful only if a reasonable balance between automatic support and user interaction is maintained. On the one hand, the set of inductive theorems of a specification SP is not recursively enumerable. On the other hand, restrictions of the formal setting that aim at enumerable or even decidable theories cuts down the range of applications for deductive methods considerably.

Rewriting-oriented theorem provers and goal solvers simplify formulas usually by **normalization**, i.e. by reducing terms and goals into reduced ones. Normalization also increases the efficiency of functional-logic language interpreters significantly [Han94]. General-purpose provers like Isabelle [Pau94b] employ rewriting for simplifying formulas, for instance for removing propositional operators. The simplifier of our program verifier *Expander* [Pad94] applies some of the simplification rules given below automatically and normalizes subgoals constructed from standard functions and predicates on natural numbers, lists, bags, sets and maps. Here normalization is actually (partial) evaluation with respect to fixed parameter models (see Section 10.1). Inference rules with built-in standard type evaluating/solving mechanisms are presented in, e.g., [Sti85], [JK86a] and [AB92]. CLP (= constraint logic programming) integrates goal solvers tailored to standard types into logic-language interpreters or compilers. The efficiency of a theorem prover that uses partial evaluators is increased considerably if the formulas to be evaluated are compiled into, say, ML expressions, normalized by ML functions and then retranslated into the prover language. In principle, any simplification rule could be implemented in this way as long as the evaluating programs preserve the (initial/final) semantics of their input. The separation of simplification from higher proof issues such as induction steps makes theorem provers designable in a modular way. The simplifier can be developed and extended more or less independently of other parts of the prover.

In the sequel we suppose that SP is functional and behaviourally congruent. Fundamental simplification rules for Gentzen proofs over SP can be derived directly from "meta-theorems" stated and discussed in previous sections. They are concerned with different parts of the signature of SP: Lemma 10.27 with safety predicates and defined functions, Lemma 10.28 with liveness predicates, Lemma 10.30 with visible equalities and inequalities, and Lemma 10.39 with liveness predicates turned into safety predicates. All these results contribute to the correctness of the following Gentzen rules.

total atom unfolding

$$\frac{\exists Y\,(r(t) \wedge G)}{\exists Z_1(t \equiv t_1 \wedge H_1 \wedge G) \vee \cdots \vee \exists Z_n(t \equiv t_n \wedge H_n \wedge G)} \quad \Updownarrow$$

partial atom unfolding

$$\frac{\exists Y\,(r(t) \wedge G)}{\exists Z_{i_1}(t \equiv t_{i_1} \wedge H_{i_1} \wedge G) \vee \cdots \vee \exists Z_{i_k}(t \equiv t_{i_k} \wedge H_{i_k} \wedge G)} \quad \Uparrow$$

if r is a logical safety predicate of SP, $\{r(t_i) \Leftarrow H_i \mid 1 \le i \le n\}$ is the set of axioms for r, $Z_i = Y \uplus var(r(t_i) \Leftarrow H_i)$ and $\{i_1, \ldots, i_k\} \subseteq \{1, \ldots, n\}$

total term unfolding

$$\frac{\exists Y\, G(f(t))}{\exists Z_1(t \equiv t_1 \wedge H_1 \wedge G(u_1)) \vee \cdots \vee \exists Z_n(t \equiv t_n \wedge H_n \wedge G(u_n))} \quad \Updownarrow$$

partial term unfolding

$$\frac{\exists Y\, G(f(t))}{\exists Z_{i_1}(t \equiv t_{i_1} \wedge H_1 \wedge G(u_{i_1})) \vee \cdots \vee \exists Z_{i_k}(t \equiv t_{i_k} \wedge H_n \wedge G(u_{i_k}))} \quad \Uparrow$$

if f is a defined function of SP, $\{f(t_i) \equiv u_i \Leftarrow H_i \mid 1 \le i \le n\}$ is the set of axioms for f, $Z_i = Y \uplus var(f(t_i) \equiv u_i \Leftarrow H_i)$ and $\{i_1, \ldots, i_k\} \subseteq \{1, \ldots, n\}$

normal form splitting

$$\frac{c(t) \equiv c(u) \wedge G}{(t_1 \equiv u_1 \wedge \cdots \wedge t_n \equiv u_n \wedge G) \vee (c(t) \equiv c(u) \equiv \bot \wedge G)} \quad \Updownarrow$$

$$\frac{c(t) \equiv \bot \wedge G}{(t_1 \equiv \bot \wedge G) \vee \cdots \vee (t_n \equiv \bot \wedge G)} \quad \Updownarrow$$

if c is a visible constructor, $t = (t_1, \ldots, t_n)$ and $u = (u_1, \ldots, u_n)$

normal form clash

$$\frac{c(t) \equiv d(u) \wedge G}{c(t) \equiv d(u) \equiv \bot \wedge G} \quad \Updownarrow \quad \text{if } c \text{ and } d \text{ are different visible constructors}$$

bottom removal

$$\frac{c \equiv \bot \wedge G}{FALSE} \updownarrow \quad \text{if } c \text{ is a visible constructor} \qquad \frac{t \equiv \bot \wedge Def(t)}{FALSE} \updownarrow$$

If there are many axioms for r or f, then total unfolding produces many goals. A clever implementation of these rules combines them with normal form splitting and clash and does not generate goals that would be removed by the latter rules almost immediately. In effect, these rule combinations are *simplication* instances of partial unfolding.

variable removal $\qquad \dfrac{\exists Y (x \equiv t \wedge G(x))}{\exists Y \setminus \{x\} G(t)} \updownarrow \quad \text{if } x \in Y \setminus var(t)$

If x occurs more than once in $G(x)$ and t is not a normal form, then variable removal leads to multiple occurrences of t in $G(t)$ and thus to multiple redices for defined function unfolding. Hence it is reasonable to restrict the applications of this rule to normal forms t. Note that without this case of variable removal we could never eliminate valid, but non-reflective equations. The following rule generalizes variable removal from x to a term u, but, since u may be an arbitrary term, it is only an expansion rule.

matching rule $\qquad \dfrac{\exists Y \uplus dom(\sigma)(u \equiv u\sigma \wedge G)}{\exists Y G\sigma} \upuparrows \quad \text{if } var(u) \cap var(u\sigma) = \emptyset$

A special case is

reflection $\qquad \dfrac{t \equiv t \wedge G}{G} \updownarrow$

The following rule can be regarded both as a kind of atom unfolding and as the *hidden*-term counterpart of normal form splitting.

hidden equation unfolding

$$\frac{\exists Y (t \equiv u \wedge G)}{\exists Y \forall \{z_1, \ldots, z_n\} (f_1(t, z_1) \equiv f_1(u, z_1) \wedge \ldots \wedge f_n(t, z_n) \equiv f_n(u, z_n) \wedge G)} \updownarrow$$

if $s = sort(t)$ is hidden, $f_1, \ldots, f_n$ are the s-actions of SP and all s-actions are functional (Definition 10.5)

The main steps of a proof by *context induction* [Hen91a], i.e. by induction on the size of contexts (see Section 10.3), are applications of hidden equation unfolding. If there are also relational or transitional s-actions, equations between s-sorted terms must be submitted to coinduction (see below).

Unfolding rules apply only axioms. The following rules allow us to use an arbitrary Gentzen clause

$$C \quad = \quad \exists Y (\{u_0 \equiv\} u_1 \vee \cdots \vee \{u_0 \equiv\} u_n) \Leftarrow \exists Z H$$

with $Y \cap var(u_0) = \emptyset$ as a lemma within proofs of other theorems.

lemma contraction

$$\frac{\exists Z (G\{t_0\} \wedge H\sigma)}{\exists Y ((G(t_1) \wedge \{t_0 \equiv\} t_1) \vee \cdots \vee (G(t_n) \wedge \{t_0 \equiv\} t_n))} \Downarrow$$

if for all $0 \leq i \leq n$, $t_i = u_i \sigma$ and $(Y \cup Z) \cap dom(\sigma) = \emptyset$

lemma expansion

$$\frac{\exists Y (G_1(t_1) \vee \cdots \vee G_n(t_n))}{\exists Y \uplus Z (H\sigma \wedge G_1\{u_0\}\sigma \wedge \cdots \wedge G_n\{u_0\}\sigma \wedge \bigwedge_{x \in V} x \equiv x\sigma)} \Uparrow$$

if V is the set of free variables of the rule premise and for all $1 \leq i \leq n$, $t_i \sigma = u_i \sigma$ and $(Y \cup Z) \cap dom(\sigma) = Y \cap var(t_i) \cap var(G_i) = \emptyset$

If C is a Horn clause, say, $r(t) \Leftarrow H$, and r is a logical predicate, then lemma expansion agrees with (input) **resolution** as known from logic programming and classical theorem proving. If C is a conditional equation, say, $t \equiv u \Leftarrow H$, then lemma expansion coincides with **paramodulation**, which, in turn, generalizes rewriting (Definition 10.33) and narrowing (Definition 10.62). In contrast to unfolding, lemma expansion can be applied only if certain parts of C are *unifiable* with corresponding parts of the rule premise (see Definition 10.3). Lemma contraction is applicable only if parts of C *subsume* corresponding parts of the rule premise.

Lemma contraction and expansion can be turned easily into rules that capture various forms of explicit induction. The idea is to use a conjecture such as C above as a lemma in the proof of C. However, the actual instances of C applied in the proof must represent induction hypotheses. This is ensured as follows.

Let x be a list of the free variables of C and $\gg: sort(x) \times sort(x)$ be a safety predicate of SP such that $var(u_0)$ is a subset of x and the final model of SP interprets $\gg$ as a well-founded relation. $\gg^{Fin(SP)}$ is supposed to be the induction ordering along which C shall be proved. It need not be known at the beginning of a proof, but should be supplied at latest when induction steps are to be executed. Here are the variants of lemma contraction and expansion that apply C as an induction hypothesis and perform explicit induction steps: Let $\rho: x \to X$ be a renaming of x away from x (Definition 10.3).

inductive contraction

$$\frac{\exists Z(\boldsymbol{x} \gg \boldsymbol{x}\rho\sigma \wedge H\rho\sigma \wedge G\{t_0\})}{\exists Y((G(t_1) \wedge \{t_0 \equiv\}t_1) \vee \cdots \vee (G(t_n) \wedge \{t_0 \equiv\}t_n))} \quad \Downarrow$$

if for all $0 \leq i \leq n$, $t_i = u_i\rho\sigma$ and $(Y \cup Z) \cap dom(\sigma) = \emptyset$

inductive expansion

$$\frac{\exists Y(G_1(t_1) \vee \cdots \vee G_n(t_n))}{\exists Y \uplus Z(\boldsymbol{x} \gg \boldsymbol{x}\rho\sigma \wedge H\rho\sigma \wedge G_1\{u_0\rho\}\sigma \wedge \ldots \wedge G_n\{u_0\rho\}\sigma \wedge \bigwedge_{x \in V} x \equiv x\sigma)} \quad \Uparrow$$

if V is the set of free variables of the rule premise and for all $1 \leq i \leq n$, $t_i\sigma = u_i\rho\sigma$ and $(Y \cup Z) \cap dom(\sigma) = Y \cap var(t_i) \cap var(G_i) = \emptyset$

Inductive contraction assumes, inductive expansion produces the **descent condition** $\boldsymbol{x} \gg \boldsymbol{x}\rho\sigma$, which indicates that the instance $C\sigma$ of the conjecture C is regarded as an induction hypothesis. Since descent conditions do not occur in C itself, they can only be generated in the course of proving C. Symmetrically, they will be removed eventually before the proof of C can be finished successfully. This means they are valid and hence the instances of C applied as lemmas have really been induction hypotheses. In other words, the above rules are correct *in the context of successful proofs of C*.[25] The rules also admit simultaneous induction where the conjecture consists of several Gentzen clauses, but each induction step uses only one of them.

Term or *structural* induction[26] actually covers all induction schemas that are based upon well-founded data orderings. Theorem provers employing term induction are, for instance, the Boyer–Moore prover [BM79], INKA [Hut90], CLAM [BSvH+93], the Larch Prover [GH93] and Isabelle [Pau94b]. If data are presented as normal forms, term induction often means induction on the size of a normal form. The only conceptual alternative to structural induction is **computational induction** where one induces on the length of computations of a program or derivations of a formula. A classical example of computational induction is Park's **fixpoint induction** [Par69] for reasoning about partial-recursive functions. To prove a property of f one induces on the indices of functions f_i, $i \in \mathbb{N}$, that approximate f.

Fixpoint induction is based on a variant of Kleene's fixpoint theorem (Theorem 10.13) where the powerset CPO (complete partial order) $\wp(A)$ is replaced by a CPO of partial functions. Park just employed the fixpoint property, while Morris [Mor71] used the approximating functions f_i explicitly. The essence of his method – for which the notion "computational induction" was introduced – is, roughly said, to prove a goal $G(f(t))$ by showing $G(f_0(t))$ and $G(f_i(t)) \Rightarrow G(f_{i+1}(t))$. Suppose that f is specified by the axiom

---

[25] For proofs of this conclusion, see [Pad91], Theorem 5.2; [Pad92], Theorem 5.5; or [Pad96a], Theorems 3.6 and 3.11.

[26] This was originally formulated in [Bur69].

$f(x) \equiv y \Leftarrow H(f, x, y)$. Then $f_{i+1}(x) \equiv y \Leftarrow H(f_i, x, y)$ is the axiom for f_{i+1}. Moreover, a proof of $G(f_{i+1}(t))$ consists of a proof of $\exists y : (H(f_i, t, y) \wedge G(y))$ and a final application of the axiom for f_{i+1}. Since i correlates with the length of a proof of $G(f_i(t))$, computational induction is indeed induction on proof lengths. But it can also be regarded as induction on *formulas* (here: $G(f(t))$) and simulated by inductive expansion/contraction if the ordering $\gg$ takes into account the entire conjecture C and not only sub*terms* of C.

Instead of working out a corresponding generalization of inductive expansion/contraction we derive from Theorem 10.23(1)-(3) the following rules whose soundness is a direct consequence of the interpretation of predicates as the least resp. greatest fixpoints satisfying their axioms. Let $r_1, \ldots, r_n$ be logical predicates of SP or $rel(SP)$ (Definition 10.14) and AX be the set of axioms for $r_1, \ldots, r_n$.

fixpoint induction

$$\frac{\bigwedge_{i=1}^{n} GS_i(x) \Leftarrow r_i(x)}{\exists q_1, \ldots, q_n : (\bigwedge_{C \in AX} C[q_1/r_1, \ldots, q_n/r_n] \wedge \bigwedge_{i=1}^{n} GS_i(x) \Leftarrow q_i(x))} \Updownarrow$$

if $r_1, \ldots, r_n$ are safety predicates

coinduction

$$\frac{\bigwedge_{i=1}^{n} GS_i(x) \Rightarrow r_i(x)}{\exists q_1, \ldots, q_n : (\bigwedge_{C \in AX} C[q_1/r_1, \ldots, q_n/r_n] \wedge \bigwedge_{i=1}^{n} GS_i(x) \Rightarrow q_i(x))} \Updownarrow$$

if $r_1, \ldots, r_n$ are liveness predicates

When applying these rules one finds out that fixpoint induction is tailored to other kinds of conjectures than coinduction. The former handles conjectures $GS_i(x) \Leftarrow r_i(x)$ stating that GS_i holds true for all data satisfying r_i. In many cases, the safety predicate r_i is the graph of a defined function f and GS_i describes the *desired* input-output relation of f. Coinduction, however, deals with inverse conjectures $GS_i(x) \Rightarrow r_i(x)$, which state that r_i holds true for all data specified by GS_i. Here GS_i is often solved (see Section 10.4), i.e. consists of equations $y \equiv t$ where t is a normal form (and y is a component of x). Then the conjecture says that all data that can be represented as instances of these normal forms satisfy the liveness predicate r_i. For bisimulation predicates r_i, coinduction becomes a rule for proving (even conditional) hidden equations. In order to match the rule premise, a hidden equation $H \Rightarrow t \equiv t'$ must first be turned into the clause $(H \wedge x \equiv t \wedge y \equiv t') \Rightarrow x \equiv y$ before the rule can be applied.

Both rules are simplifications. The downward implication $\Downarrow$ holds true because $Fin(SP)$ satisfies AX and thus we may define q_i as r_i. The upward implication $\Uparrow$ is valid because all solutions of AX in $r_1, \ldots, r_n$ are super- resp. subsets of the interpretations of $r_1, \ldots, r_n$ in $Fin(SP)$ because these are the least resp. greatest solutions of AX (Theorem 10.23). In other words, if the final model satisfies $C[q_1/r_1, \ldots, q_n/r_n]$ for all $C \in AX$, then it also satisfies $q_i(x) \Leftarrow r_i(x)$ resp. $q_i(x) \Rightarrow r_i(x)$ for all $1 \leq i \leq n$. Hence the respective rule premise follows from the rule conclusion $GS_i(x) \Leftarrow q_i(x)$ resp. $GS_i(x) \Rightarrow q_i(x)$.

q_i is an existentially quantified predicate variable whose possible value ranges between GS_i and r_i (with respect to set inclusion). Choosing q_i strictly smaller resp. greater than GS_i means to **generalize** resp. **co-generalize** GS_i. On the one hand, the term "generalize" complies with the intuition that a smaller relation expresses a stronger condition. On the other hand, co-generalizing often means to weaken a given solved goal set by augmenting it (disjunctively) with further equations $y \equiv t$ (see above).

Example 10.60. (see Example 10.32)

> PART2 = PARTITION +
>> **preds** $part : list * list(list)$
>> **vars** $x, y : entry$ $L, L' : list$ $P : list(list)$
>> **Horn axioms**
>>> $part([x], [[x]])$ (1)
>>> $part(x :: y :: L, [x] :: P) \Leftarrow part(y :: L, P)$ (2)
>>> $part(x :: y :: L, (x :: L') :: P) \Leftarrow part(y :: L, L' :: P)$ (3)
>> **conjects** $L \equiv flatten(P) \Leftarrow part(L, P)$ (1)

$part(L, P)$ holds true iff P is a partition of L into successive sublists of L. If L is a singleton consisting of x, then $[[x]]$ is the only partition of L (Axiom 1). If L starts with x and y, then there are two ways of obtaining a partition of L from a partition P of $y :: L$: either one adds the singleton $[x]$ to the elements of P (Axiom 2) or one appends x to the first element L' of P and adds $x\&L'$ to P (Axiom 3). We show the correctness of *part* by expanding Conjecture 1:[27]

$L \equiv flatten(P) \Leftarrow part(L, P)$
fixpoint induction without generalization $(q = GS = (L \equiv flatten(P)))$
$\vdash [x] \equiv flatten([[x]]),$
   $x :: y :: L \equiv flatten([x] :: P) \Leftarrow y :: L \equiv flatten(P),$
   $x :: y :: L \equiv flatten((x :: L') :: P) \Leftarrow y :: L \equiv flatten(L' :: P)$
body reduction
$\vdash [x] \equiv flatten([[x]]),$
   $x :: flatten(P) \equiv flatten([x] :: P),$
   $x :: flatten(L' :: P) \equiv flatten((x :: L') :: P)$
partial *flatten*-term unfolding
$\vdash \exists\{L, P\}([[x]] \equiv L :: P \wedge [x] \equiv L@flatten(P)),$
   $\exists\{L', P'\}([x] :: P \equiv L' :: P' \wedge x :: flatten(P) \equiv L'@flatten(P')),$
   $\exists\{L'', P'\}((x :: L') :: P \equiv L'' :: P' \wedge x :: flatten(L' :: P) \equiv L''@flatten(P'))$
total [ ]-term unfolding

---

[27] Colons connect Gentzen clauses conjunctively.

$\vdash \exists\{L, P\}((x :: nil) :: nil \equiv L :: P \wedge x :: nil \equiv L@flatten(P)),$
  $\exists\{L', P'\}((x :: nil) :: P \equiv L' :: P' \wedge x :: flatten(P) \equiv L'@flatten(P')),$
  $\exists\{L'', P'\}((x :: L') :: P \equiv L'' :: P' \wedge x :: flatten(L' :: P) \equiv L''@flatten(P'))$
normal form splitting, clash and variable removal
$\vdash x :: nil \equiv (x :: nil)@flatten(nil),$
  $x :: flatten(P) \equiv (x :: nil)@flatten(P),$
  $x :: flatten(L' :: P) \equiv (x :: L')@flatten(P)$
partial @-term unfolding
$\vdash x :: nil \equiv x :: flatten(nil),$
  $x :: flatten(P) \equiv x :: flatten(P),$
  $x :: flatten(L' :: P) \equiv x :: (L'@flatten(P))$
partial $flatten$-term unfolding
$\vdash x :: nil \equiv x :: nil,$
  $x :: flatten(P) \equiv x :: flatten(P),$
  $x :: (L'@flatten(P)) \equiv x :: (L'@flatten(P))$
reflection
$\vdash TRUE$ ∎

On the one hand, many proofs involving inductive contraction/expansion can be simplified considerably if fixpoint induction is used instead. Since the fixpoint induction generates them automatically, using this rule relieves us from interactively guiding the proof towards induction hypotheses, which is required if inductive contraction/expansion shall be applied eventually. However, the fixpoint rules are not applicable to every Gentzen clause conjecture. For instance, if it has the form $GS(x) \Leftarrow (r(x) \wedge H(x))$, i.e. GS need not hold for all data satisfying r, but only for those that satisfy H, then we must either turn the conjecture into the "Curried form" $(GS(x) \Leftarrow H(x)) \Leftarrow r(x)$ or find axioms for a subpredicate r' of r such that $r'(x)$ is equivalent to $r(x) \wedge H(x)$. The first solution is simple, but may leave us with a quite complex conclusion of fixpoint induction because an implication has to be substituted for each occurrence of r in AX. Hence the second solution is more convenient, especially since program transformation rules may help us constructing axioms for r' from those for r. Moreover, the combination of such transformation rules with fixpoint rules would lead to proof rules for other conjecture patterns than those handled by fixpoint induction or coinduction.

The rule for subgoal induction mentioned in Section 10.4 is a special case of fixpoint induction for two predicates r_1, r_2, namely the graphs of f and $iter$, and where no generalization is involved ($q_i = GS_i$; see above). Hoare induction can be derived from subgoal induction by transforming subgoal invariants into Hoare invariants (see Section 10.4).

Example 10.61. (see Example 10.6). We show that the infinite stream 01 of alternating zeros and ones is fair by expanding the clause $fair(01)$:

  $fair(01)$
inverse body reduction

$\vdash s \equiv 01 \implies \mathit{fair}(s)$

coinduction with co-generalization $(q \neq GS = (s \equiv 01))$

$\vdash q(s) \implies \exists\{L, s'\}(0 \notin L \land s \equiv L\#(0\&s') \land q(s')), \qquad s \equiv 01 \implies q(s)$

body reduction

$\vdash q(s) \implies \exists\{L, s'\}(0 \notin L \land s \equiv L\#(0\&s') \land q(s')), \qquad q(01)$

total q-atom unfolding by applying the axiom $q(s) \Leftarrow (s \equiv 01 \lor s \equiv 10)$

$\vdash (s \equiv 01 \lor s \equiv 10) \implies \exists\{L, s'\}(0 \notin L \land s \equiv L\#(0\&s') \land (s' \equiv 01 \lor s' \equiv 10)),$
$\quad 01 \equiv 01 \lor 01 \equiv 10$

body splitting, reflection and head success

$\vdash s \equiv 01 \implies \exists\{L, s'\}(0 \notin L \land s \equiv L\#(0\&s') \land (s' \equiv 01 \lor s' \equiv 10)),$
$\quad s \equiv 10 \implies \exists\{L, s'\}(0 \notin L \land s \equiv L\#(0\&s') \land (s' \equiv 01 \lor s' \equiv 10))$

body reduction

$\vdash \exists\{L, s'\}(0 \notin L \land 01 \equiv L\#(0\&s') \land (s' \equiv 01 \lor s' \equiv 10)),$
$\quad \exists\{L, s'\}(0 \notin L \land 10 \equiv L\#(0\&s') \land (s' \equiv 01 \lor s' \equiv 10))$

a Boolean transformation

$\vdash \exists\{L, s'\}((0 \notin L \land 01 \equiv L\#(0\&s') \land s' \equiv 01) \lor$
$\qquad\qquad (0 \notin L \land 01 \equiv L\#(0\&s') \land s' \equiv 10)),$
$\quad \exists\{L, s'\}((0 \notin L \land 10 \equiv L\#(0\&s') \land s' \equiv 01) \lor$
$\qquad\qquad (0 \notin L \land 10 \equiv L\#(0\&s') \land s' \equiv 10))$

variable removal

$\vdash \exists\{L\}((0 \notin L \land 01 \equiv L\#(0\&01)) \lor (0 \notin L \land 01 \equiv L\#(0\&10))),$
$\quad \exists\{L\}((0 \notin L \land 10 \equiv L\#(0\&01)) \lor (0 \notin L \land 10 \equiv L\#(0\&10)))$

partial #-term unfolding and variable removal

$\vdash (0 \notin \mathit{nil} \land 01 \equiv 0\&01) \lor \exists\{L\}(0 \notin L \land 01 \equiv L\#(0\&10)),$
$\quad \exists\{L\}(0 \notin L \land 10 \equiv L\#(0\&01)) \lor \exists\{x, L\}(0 \notin x\&L \land 10 \equiv x\&(L\#(0\&10)))$

goal elimination

$\vdash 0 \notin \mathit{nil} \land 01 \equiv 0\&01,$
$\quad \exists\{x, L\}(0 \notin x\&L \land 10 \equiv x\&(L\#(0\&10)))$

partial $\notin$-atom and total 10-term unfolding

$\vdash 01 \equiv 0\&01,$
$\quad \exists\{x, L\}(0 \neq x \land 0 \notin L \land 1\&01 \equiv x\&(L\#(0\&10)))$

hidden equation unfolding

$\vdash \mathit{head}(01) \equiv \mathit{head}(0\&10) \land \mathit{tail}(01) \equiv \mathit{tail}(0\&10),$
$\quad \exists\{x, L\}(0 \neq x \land 0 \notin L \land \mathit{head}(1\&01) \equiv \mathit{head}(x\&(L\#(0\&10))) \land$
$\qquad\qquad \mathit{tail}(1\&01) \equiv \mathit{tail}(x\&(L\#(0\&10))))$

partial head-term and tail-term unfolding

$\vdash 0 \equiv 0 \land 10 \equiv 10,$
$\quad \exists\{x, L\}(0 \neq x \land 0 \notin L \land 1 \equiv x \land 01 \equiv L\#(0\&10))$

reflection, variable removal and head success

$\vdash \exists\{L\}(0 \neq 1 \land 0 \notin L \land 01 \equiv L\#(0\&10))$

partial $\neq$-atom and #-term unfolding and variable removal

$\vdash 0 \notin \mathit{nil} \land 01 \equiv 0\&10$

partial $\notin$-atom unfolding

$\vdash 01 \equiv 0\&10$

hidden equation unfolding
$\vdash head(01) \equiv head(0\&10) \wedge tail(01) \equiv tail(0\&10)$
partial *head*-term and *tail*-term unfolding
$\vdash 0 \equiv 0 \wedge 10 \equiv 10$
reflection
$\vdash TRUE$ ∎

We were lucky here because INFSEQ-bisimilarity is defined only in terms
of the *functional* actions *head* and *tail* and thus we could reduce all hidden
equations by hidden equation unfolding. In the general case one may use the
following rule that is derived from coinduction. Let s be a hidden sort, AX
be the set of axioms of SP for $\sim$: ss and $t, t' \in T_\Sigma(X)_s$.

$$\textbf{hidden equation expansion} \quad \frac{H \Rightarrow t \equiv t'}{\bigwedge_{C \in AX} C[q/\sim]} \Uparrow \quad \text{if} \quad \left\{ \begin{array}{l} q(x,x) \\ q(t,t') \Leftarrow H \\ q(t',t) \Leftarrow H \end{array} \right\}$$

are the axioms of SP for q

For instance, based on a specification of finite and infinite streams, [Pad97]
presents a Gentzen proof of the hidden-type benchmark equation

$$filter(f, map(g, s)) \equiv map(g, filter(f \circ g, s)) \tag{10.1}$$

(see Example 10.26 and [Gor95]). The proof starts with hidden equation
expansion and proceeds with inductive contraction upon subconjectures. Note
that – in contrast to Example 10.61 – (10.1) is a statement about *infinitely
many* streams.

10.8   Goal solving

Goal solving as a particular verification problem werde already mentioned
in Section 10.4. Rules and methods for solving equations and other goals
with respect to a functional specification are usually based on *narrowing*
procedures, originally invented by [Sla74] and [Hul80a]. Narrowing generalizes rewriting and resolution (Definition 10.33) insofar as narrowing – besides
transforming a term or atom – generates a substitution that may solve a given
goal. Usually only the generalization of rewriting is called narrowing, while
the corresponding generalization of resolution as defined in Definition 10.33
yields the input resolution in the sense of automated theorem proving and
logic programming. Resolution is the basic rule for solving goals that consist of logical predicates and normal forms. Narrowing adapts the resolution
principle to goals that involve defined functions, in particular to sets of equations. Hence interpreters of functional-logic programming languages actually
implement narrowing procedures [Han94].

Unfolding and lemma expansion also combine rewriting with substitution. But these rules do not separate the rewritten goal from the generated substitution. Instead, the latter is made part of the former: the substitution is presented as an (equational) subgoal. For the sake of clarity (and efficiency), solution procedures should better keep substitutions different from goals. Pairs of a goal and a substitution whose domain does not include variables of the goal are called constraints:

Definition 10.62. Let $SP = (\Sigma, AX)$ be a standard specification. A Σ-**constraint** $\langle G, \tau \rangle$ consists of a goal G and a substitution $\tau: X \to NF_\Sigma(X)$ such that $var(G) \cap dom(\sigma) = \emptyset$. The **narrowing calculus for** SP consists of the following rules for reducing Σ-constraints. Let G be a goal and σ be a be a minimal unifier of two terms t and t' and $(p \Leftarrow H) \in AX$ such that $H\sigma$ is SP-narrowable and $fresh(p \Leftarrow H)\sigma$ consists of normal forms.

narrowing
$$\frac{\langle G(t), \tau \rangle}{\langle G(u)\sigma \wedge H\sigma, \tau\sigma_{var(G(t))} \rangle} \Uparrow \quad \text{if } t \notin X \text{ and } p = (t' \equiv u)$$

resolution
$$\frac{\langle r(t) \wedge G, \tau \rangle}{\langle H\sigma \wedge G\sigma, \tau\sigma_{var(r(t)\wedge G)} \rangle} \Uparrow \quad \text{if } r \neq\equiv \text{ and } p = r(t')$$

unification
$$\frac{\langle t \equiv t' \wedge G, \tau \rangle}{\langle G\sigma, \tau\sigma \rangle} \Uparrow$$

If one starts with the identity substitution id, Condition 10.5(1) ensures that all generated substitutions map into $NF_\Sigma(X)$. Suitable variable renamings must guarantee that the variable condition on constraints is preserved by rule applications.

A sequence $cs_1, \ldots, cs_n$ of Σ-constraints such that for all $1 \leq i < n$, cs_{i+1} is obtained from cs_i by applying one of the above rules, is called an SP-**narrowing reduction** of cs_1 into cs_n and we write $cs_1 \vdash\!\!\sqrt{}_{SP} cs_n$. Given a goal G, $\tau: X \to NF_\Sigma(X)$ is a **narrowing solution** of G if $\langle G, id \rangle \vdash\!\!\sqrt{}_{SP} \langle \emptyset, \tau \rangle$. G is SP-**narrowable** if $\langle G, id \rangle \vdash\!\!\sqrt{}_{SP} cs$ for some cs. Otherwise G is SP-**narrowed**.   ◇

The arrow $\Uparrow$ indicates that the rules of the narrowing calculus are – like those of the reduction calculus – expansions. Since rule premises and conclusions are constraints, the expansion property here means:

$$\langle G, \tau \rangle \vdash\!\!\sqrt{}_{SP} \langle G', \tau' \rangle \quad \Rightarrow \quad Ini(SP) \models G\tau' \Leftarrow G'. \tag{10.1}$$

Hence, in particular,

$$\langle G, id \rangle \vdash\!\!\sqrt{}_{SP} \langle G', \tau \rangle \quad \Rightarrow \quad SP \vdash_{cut} G\tau \tag{10.2}$$

(see (10.1) in Section 10.5).

The relationship between $\vdash_{SP}$ (Definition 10.33) and $\vdash\!\!\sqrt{}_{SP}$ is as follows: If $\langle G, \tau \rangle \vdash\!\!\sqrt{}_{SP} \langle G', \tau' \rangle$, then $G\sigma \vdash_{SP} G'$ for σ with $\tau\sigma = \tau'$. Conversely, given $\tau \colon X \to NF_\Sigma(X)$, if $G\tau$ is SP-convergent, then τ is a narrowing solution of G.

Let us again suppose that $SP = (\Sigma, AX)$ is functional and behaviourally congruent. The Church–Rosser Theorem 10.36 tells us that functionality implies not only that the reduction calculus is complete, but also that the narrowing calculus is **solution complete**, i.e. for all goals G and ground normal substitutions τ,

$$SP \vdash_{cut} G\tau \quad \Rightarrow \quad \langle G, id \rangle \vdash\!\!\sqrt{}_{SP} \langle \emptyset, \tau \rangle. \tag{10.3}$$

Hence goals can be disproved with narrowing and resolution: if each narrowing reduction of $\langle G, id \rangle$ is either infinite or ends with a constraint $\langle H, \tau \rangle$ such that H is SP-narrowed, then G is unsolvable in $Ini(SP)$:

$$\forall \tau \colon \langle G, id \rangle \not\!\!\!\vdash\!\!\sqrt{}_{SP} \langle \emptyset, \tau \rangle \quad \Rightarrow \quad \forall \tau \colon SP \not\vdash_{cut} G\tau.$$

Since narrowing reductions are expansions, a Gentzen clause $GS \Leftarrow HS$ may sometimes be proved by establishing narrowing reductions of all constraints $\langle G, id \rangle$ with $G \in GS$ into, say, $\langle G_1, \tau_1 \rangle, \ldots, \langle G_n, \tau_n \rangle$ and then expanding the goal set

$$(G_1 \wedge \bigwedge_{x \in dom(\tau_1)} x \equiv x\tau_1) \vee \cdots \vee (G_n \wedge \bigwedge_{x \in dom(\tau_n)} x \equiv x\tau_n) \tag{10.4}$$

into HS or contracting HS into (10.4) by applying Gentzen rules. This also shows the similarity between narrowing and unfolding (Section 10.7). On the one hand, narrowing improves over unfolding because the latter might generate a number of invalid equations that are rejected by narrowing if their respective sides are not unifiable. On the other hand, unfolding applies all axioms for a function or predicate in parallel, while narrowing considers only one axiom in each step. Applying all axioms simultaneously is crucial for the contraction property of unfolding. A sequence of unfolding steps yields the levels of a *proof tree* whose nodes are actually constraints and whose paths are narrowing reductions. Given a goal G, a narrowing procedure shall derive a *complete* set Sol of solutions, which means that each ground solution of G is subsumed by some $\sigma \in Sol$. In general, the outdegree of the generated *narrowing tree* increases both with the number of narrowable parts of a goal and with the number of axioms that are applicable to the same redex. However, the latter agrees with the number of possible overlays at a single redex (Definition 10.40), and this number is usually small, if not one.

Consequently, the efficiency of a narrowing procedure depends on the number of redices (subterms or atoms) of the goal to be solved. A *redex selector* assigns to each narrowable goal a single redex. The intention is to figure out *complete* redex selectors, which reduce narrowing trees as much as possible, but nevertheless return complete sets of narrowing solutions.

Definition 10.63. Given an atom or a term $t \notin X$ and a goal G with a unique occurrence of the variable x, the expression $G \bullet t$ is a **position** of the goal $G[t/x]$. For a substitution σ, $(G \bullet t)\sigma$ stands for the position $G\sigma_{X \setminus \{x\}} \bullet t\sigma$ (Definition 10.3).

Let σ be a minimal unifier of two terms t and t' and $(p \Leftarrow H) \in AX$ such that $H\sigma$ is SP-narrowable and $fresh(p \Leftarrow H)\sigma$ consists of normal forms.

- If $t \notin X$ and $p = (t' \equiv u)$, then $\langle G(u)\sigma \wedge H\sigma, \sigma_{var(G(t))} \rangle$ is a **narrowing reduct** of $G \bullet t$.

- If $p = r(t')$ and $r \neq \equiv$, then $\langle H\sigma \wedge G\sigma, \sigma_{var(r(t) \wedge G)} \rangle$ is a **narrowing reduct** of $(x \wedge G) \bullet r(t)$.

- $\langle G\sigma, \sigma \rangle$ is a **narrowing reduct** of $(x \wedge G) \bullet (t \equiv t')$.

A position $G \bullet t$ with a narrowing reduct is a **narrowing redex** of $G[t/x]$. A function Sel that assigns to each SP-narrowable goal G a redex of G is a **redex selector**. Sel is **uniform** if $Sel(G)\sigma = Sel(G\sigma)$ for all goals G and ground normal substitutions σ.   ◇

A redex selector prunes narrowing trees with respect to their breadth, but does not reduce their depth. The latter is accomplished if narrowing reducts are *simplified* before they are committed to further narrowing steps (see Section 10.7). Here simplification means **normalization** (i.e. reduction into reduced goals), normal form splitting, clash and variable removal (see, e.g., [Han94]). Since any simplification rule turns a formula into an equivalent one, simplifications can always be used in order to shorten narrowing reductions. However, simplifiers may also remove successful narrowing reductions whose resulting solutions are not achieved on other paths of the narrowing tree. This effect can be avoided if simplifications are restricted to *reductive* ones, i.e. those that comply with a reduction ordering for SP.

Definition 10.64. A function that assigns to each goal G a goal $Sim(G)$ such that $G \Leftrightarrow Sim(G)$ is an inductive theorem of SP is a **simplifier**. Given a reduction ordering $>$ for SP (Definition 10.38), Sim is $>$-**reductive** if $G\sigma \geq Sim(G)\sigma$ for all goals G and ground normal substitutions σ.

Given a redex selector Sel and a simplifier Sim, let $succs$ be the function from constraints to sets of constraints, defined by:

$$succs^{Sel}_{Sim}\langle G, \tau \rangle$$
$$= \{\langle Sim(G'), \tau\sigma \rangle \mid \langle G', \sigma \rangle \text{ is a narrowing reduct of } Sel(G)\}.$$

A narrowing reduction $cs_1, \ldots, cs_n$ such that for all $1 \leq i < n$, $cs_{i+1} \in succs^{Sel}_{Sim}(cs_i)$, is called an (Sel, Sim)-**narrowing reduction** of cs_1 into cs_n and we write $cs_1 \vdash^{Sel}_{Sim} cs_n$.   ◇

Theorem 10.65. ([Pad96a], Lemmas 10.4 and 10.9) *Let Sel be a redex selector and Sim be a simplifier. For all constraints $\langle G, id \rangle$ and $\sigma : X \to NF_\Sigma(X)$, $G\sigma$ is SP-convergent if $\langle G, id \rangle \vdash^{Sel}_{Sim} \langle \emptyset, \tau \rangle$. Moreover, if Sel is uniform, $>$ is* *a reduction ordering for SP and Sim is $>$-reductive, then the converse holds true for all ground normal substitutions τ, more precisely:*

$$G\tau \vdash_{SP} \emptyset \quad \Rightarrow \quad \exists \sigma \leq \tau : \langle G, id \rangle \vdash_{Sim}^{Sel} \langle \emptyset, \sigma \rangle$$

where $\leq$ denotes the subsumption relation on substitutions (Definition 10.3).

From this and the Church–Rosser Theorem 10.36 we conclude that the set of (Sel, Sim)-narrowing reductions is solution complete, i.e. for all goals G and ground normal substitutions τ,

$$SP \vdash_{cut} G\tau \quad \Rightarrow \quad \exists \sigma \leq \tau : \langle G, id \rangle \vdash_{Sim}^{Sel} \langle \emptyset, \sigma \rangle \tag{10.5}$$

(see (10.3)), provided that the assumptions of Theorem 10.65 hold true. By combining (10.2) with (10.5) we obtain the following refutation rule:

(Sel, Sim)**-refutation** $\dfrac{G}{FALSE} \updownarrow$

> if *Sel* is a uniform redex selector, *Sim* is a $>$-reductive simplifier and for all $\sigma : X \to NF_\Sigma(X)$, $\langle G, id \rangle \nvdash_{Sim}^{Sel} \langle \emptyset, \sigma \rangle$.

Narrowing procedures select either innermost or outermost redex positions (see, e.g., [Fri85a, AEH94]). The completeness of *SP* implies that innermost redex selectors are uniform (see [Pad96a], Lemma 9.3). Choosing innermost redices becomes inefficient if many of them occur in unsolvable subgoals, which are not reducible into the empty goal and thus do not contribute to the derivation of solutions. With outermost redex selection many of these subgoals will have been removed before the redex selector meets them. Unfortunately, outermost redex selectors are not uniform unless additional constraints are imposed on the axioms, e.g. *non-sub-unifiability* (see [Ech88] and [Pad88a], Section 8.4). This is a rather restrictive condition that forbids axiom premises. *Needed narrowing* (see [AEH94]) overcomes the condition by replacing narrowing with pure substitution whenever an outermost narrowing step fails because a normal form t' in the axiom to be applied is not unifiable with a non-normal form t in the goal to be narrowed. The substitution step ensures that the want-to-be redex is at least instantiated. Then a new redex is selected, t will eventually be reduced to a normal form and again compared with t'. If both terms t and t' are normal forms, then either they are unifiable and the narrowing step is carried out or we have a normal form clash and thus may reject the application of the given axiom to the given goal position without even instantiating the goal (see [Pad94]; [Pad96a], Section 9).

Whether the redex selector works inner- or outermost, a smart simplifier, which detects and removes unsolvable goals as early as possible, decreases the number of redices occurring in the course of a reduction significantly. Narrowing and simplification are also combined with *residuation* and *lazy evaluation* where unification steps are delayed until certain variables of the terms to be unified have been instantiated "sufficiently" (see [LFA93, Han95]). Term and atom unfolding perform lazy unification in the sense that these rules generate an equation E as a subgoal, while narrowing would try to unify the two sides of E. With respect to (Sel, Sim)-narrowing, a complementary unfold rule performing eager unification reads as follows. Its correctness follows from [Pad96a], Lemmas 10.5 and 10.10.

(Sel, Sim)-**unfolding**

$$\frac{G}{\exists Y((G_1 \wedge \bigwedge_{x \in dom(\tau_1)} x \equiv x\tau_1) \vee \cdots \vee (G_n \wedge \bigwedge_{x \in dom(\tau_n)} x \equiv x\tau_n))} \quad \Updownarrow$$

if Sel is a uniform redex selector, Sim is a $>$-reductive simplifier, $succ_{Sim}^{Sel}\langle G, id \rangle = \{\langle G_1, \tau_1 \rangle, \ldots, \langle G_n, \tau_n \rangle\}$, $Y = \biguplus_{i=1}^n var(dom(\tau_i)\tau_i)$ and $Y \cap (var(G) \cup dom(\tau_i)) = \emptyset$

Atom and term unfolding as defined Section 10.7 applies all axioms for a function or predicate f to an arbitrary redex of the form $G \bullet f(t)$. (Sel, Sim)-unfolding applies from these axioms only those whose left-hand sides are unifiable with the redex selected by Sel and discards the others. Some of the discarded axioms can only be rejected by atom and term unfolding if these rules are combined with normal form splitting and clash.

11 Proof Systems for Structured Specifications and Their Refinements

Michel Bidoit[1], María Victoria Cengarle[2], and Rolf Hennicker[2]

[1] LSV, CNRS and ENS de Cachan, 61 avenue du Président Wilson,
F-94235 Cachan Cedex, France
[2] Institut für Informatik, Ludwig-Maximilians-Universität München,
Oettingenstr. 67, D-80538 München, Germany

11.1 Introduction

Reasoning about specifications is one of the fundamental activities in the process of formal program development. This ranges from proving the consequences of a specification, during the prototyping or testing phase for a requirements specification, to proving the correctness of refinements (or implementations) of specifications. The main proof techniques for algebraic specifications have their origin in equational Horn logic and term rewriting. These proof methods have been well studied in the case of nonstructured specifications (see Chapters 9 and 10). For large systems of specifications built using the structuring operators of specification languages, relatively few proof techniques have been developed yet; for such proof systems, see [SB83, HST94, Wir93, Far92, Cen94, HWB97].

In this chapter we focus on proof systems designed particularly for modular specifications. The aim is to concentrate on the structuring concepts, while abstracting as much as possible from the particular logic and proof methods used for specification "in the small". Technically this is achieved by considering a kernel language of structured specifications defined over an arbitrary institution (see Chapter 4). Hence the proof methods presented can be instantiated by any kind of proof system chosen for a particular logic. The structuring operators of the kernel language are representative of a wide spectrum of specification languages (see Chapter 8). They include operators for the combination of specifications, and for hiding (export) and renaming specifications. The semantics of a specification expression is determined by the loose model-class semantics described in Chapter 5.

On this basis we study proof systems for verifying two kinds of statements. On the one hand, we are interested in proving that a sentence φ is a consequence (i.e., theorem) of a specification SP, denoted by SP $\models \varphi$, which means that φ is valid in all models of SP. On the other hand, we are interested in proving refinement relations, written $SP_1 \rightsquigarrow SP_2$, between a more "abstract" specification SP_1 and a more "concrete" specification SP_2. The refinement relation is based on model class inclusion, i.e., it is a simplified version of the implementation definition given in Chapter 7.

After having introduced the syntax and semantics of the kernel specification language in Section 11.2, we first consider different kinds of proof systems for proving theorems of specifications (over the kernel language). In Section 11.3, noncompositional proof systems are studied, where the underlying idea is to compute, for a given specification, a flat, unstructured set of axioms and rules which (combined with some standard proof system for the chosen logic) may be used for deriving theorems of the specification. Here the most prominent representative is the normal-form approach based on the laws of the module algebra [BHK90]. The drawback of this method is that it does not respect the modular structure of specifications. Therefore we present in Section 11.4 a structured proof system whose derivations are performed in accordance with the modular structure of a specification. In this case the idea is to provide for each specification-building operator, Op, proof rules of the form

$$\frac{SP_1 \vdash \varphi_1, \ldots, SP_n \vdash \varphi_n}{Op(SP_1, \ldots, SP_n) \vdash \varphi}$$

where the intended meaning is: if $\varphi_1, \ldots, \varphi_n$ are respectively theorems of $SP_1, \ldots, SP_n$, then φ is a theorem of $Op(SP_1, \ldots, SP_n)$.

An important issue in program development is the proof of refinements. In Section 11.5 we present a proof system for deriving refinement relations $SP_1 \rightsquigarrow SP_2$, following the modular structure of the specification SP_1 to be implemented. The proof system for refinements can be built on top of any proof system for theorems of a specification.

In many specification languages, some parameterization concept is available in order to define specifications in a generic way. Based on the λ-calculus style of parameterization (see Chapter 8), we present in Section 11.6 an extension of the kernel language to include parameterization. We also extend the structured proof systems of the kernel language for proving theorems and refinements in the context of parameterized specifications.

For all proof systems considered so far, we provide soundness and (relative) completeness results. Up to Section 11.6 the proof systems are given for specifications over an arbitrary institution. In the remaining sections we consider the particular institution of many-sorted first-order logic with equality, in which we can define two further kinds of specification-building operators, namely the reachability and observability operators. The reachability operator increases the expressive power of the specification language by imposing reachability constraints (i.e., generation principles) on the models of a specification. Two observability operators are considered (see also Chapter 5) which are particularly useful in the context of refinements (see below). In Section 11.7 we first present noncompositional proof systems for specifications with reachability and observability operators. In both cases the idea is to introduce particular semiformal rules (the "infinitary induction" and "infinitary context induction" rules) which allow one to characterize the reachability and the observational equality of elements. In this way proof systems consisting of a set of (nonlogical) axioms and rules, adapted to the particular form of a specification expression, can be constructed. Again soundness and (relative) completeness results for the proof systems of the extended language are provided. We then show that a sound and complete structured proof system can also be defined for the extended specification language, provided we allow infinitary sentences (hence working in the institution of many-sorted infinitary first-order logic with equality).

Finally we show how the proof system for refinement studied in Section 11.5 can be extended to take into account the new specification-building operators, and we also consider observational refinements of specifications.

11.2   The kernel specification language

11.2.1   Basic concepts and notations

The basis of the kernel specification language to be introduced in the next subsection is the concept of an *institution* as defined in Chapter 4. Recall that an institution is a tuple of the form[1]

$$\text{INS} = \langle \text{Sign}, \text{Sen}, \text{Str}, (\models_\Sigma)_{\Sigma \in |\text{Sign}|} \rangle$$

where Sign is a category of signatures, Sen: Sign $\rightarrow$ Set and Str: Sign^{op} $\rightarrow$ Cat are functors, and, for each signature Σ, $\models_\Sigma$ is a satisfaction relation such that the satisfaction condition holds. The elements of Sen(Σ) are called Σ-sentences and the objects of Str(Σ) are called Σ-structures. As in Chapter 4 we use the following notation.

Notation. For any signature morphism $\sigma\colon \Sigma \rightarrow \Sigma'$, the function Sen($\sigma$): Sen($\Sigma$) $\rightarrow$ Sen(Σ') (between sets of sentences) is, for simplicity, also denoted by σ, and the functor Str(σ): Str(Σ') $\rightarrow$ Str(Σ) is denoted by $\_|_\sigma$. Hence the satisfaction condition can be rephrased as $A' \models_{\Sigma'} \sigma(\varphi)$ if and only if $A'|_\sigma \models_\Sigma \varphi$ for any $\sigma\colon \Sigma \rightarrow \Sigma'$, $\varphi \in$ Sen(Σ), and $A' \in |\text{Str}(\Sigma')|$. The satisfaction relation is extended as usual to sets of Σ-sentences and classes of Σ-structures. For instance, if C is a class of Σ-structures and φ is a Σ-sentence, then $C \models \varphi$ if and only if $A \models \varphi$ for each $A \in C$.

Here and in the following sections, we assume a fixed, arbitrary institution INS is given.

Institutions provide a model-theoretic view of logical systems. However we still require the notion of a proof system which allows us to extract (soundly) the theorems of a theory presented over an institution. More precisely, given a presentation $\langle \Sigma, E \rangle$ (called a *basic specification* later on) with $\Sigma \in |\text{Sign}|$ and $E \subseteq$ Sen(Σ), we want to compute its associated theory, consisting of the

---

[1] In Chapter 4, the functor Str is called Mod. We reserve the use of this name for the collection of models associated with a specification as defined in Section 11.2.2.

sentences $\varphi \in \text{Sen}(\Sigma)$ which are implied by the presentation, i.e., those φ for which, for all $A \in |\text{Str}(\Sigma)|$, if $A \models_\Sigma E$ then $A \models_\Sigma \varphi$. By abuse of notation, we then also write $E \models_\Sigma \varphi$ where $\models_\Sigma$ denotes the consequence relation (see Chapter 4).

A proof-theoretic view of institutions is captured by the notions of *entailment relation* and *entailment system* as defined in Chapter 4. An entailment relation on $\text{Sen}(\Sigma)$ is a relation $\vdash \subseteq \wp(\text{Sen}(\Sigma)) \times \text{Sen}(\Sigma)$ satisfying the reflexivity, transitivity, and weakening conditions. Moreover, an entailment system (w.r.t. the given institution INS) is a tuple of the form

$$\text{ES} = \langle \text{Sign}, \text{Sen}, (\vdash_\Sigma)_{\Sigma \in |\text{Sign}|} \rangle$$

such that, for each $\Sigma \in |\text{Sign}|$, $\vdash_\Sigma$ is an entailment relation on $\text{Sen}(\Sigma)$, preserved under translation of sentences by signature morphisms. (Intuitively this means that the entailment relations are constructed in a uniform way.) An entailment system is *sound (complete)* if $\vdash_\Sigma$ is sound (complete) w.r.t. the semantic consequence relation $\models_\Sigma$ for each $\Sigma \in |\text{Sign}|$.

For dealing with proofs, we consider entailment systems which are generated by the axioms and rules of a proof system. A proof system for INS consists of a (uniformly defined) family of sets $\Pi(\Sigma)$ (for $\Sigma \in |\text{Sign}|$), where the elements of $\Pi(\Sigma)$ are pairs (Φ, φ), consisting of a set $\Phi \subseteq \text{Sen}(\Sigma)$ and a Σ-sentence φ. The pair (Φ, φ) can be considered as an (instantiated) proof rule or as an axiom if $\Phi = \emptyset$.

Definition 11.1 (Proof system). A *proof system* for INS is a family $\Pi = (\Pi(\Sigma))_{\Sigma \in |\text{Sign}|}$ such that

for each $\Sigma \in |\text{Sign}|$, $\Pi(\Sigma) \subseteq \wp(\text{Sen}(\Sigma)) \times \text{Sen}(\Sigma)$, and
for each signature morphism $\sigma: \Sigma \to \Sigma'$, if $(\Phi, \varphi) \in \Pi(\Sigma)$ then $(\sigma(\Phi), \sigma(\varphi)) \in \Pi(\Sigma')$.

Π is *sound* if $\Pi(\Sigma) \subseteq \models_\Sigma$ for each $\Sigma \in |\text{Sign}|$. Π is called *formal* (or *finitary*) if $\Pi(\Sigma) \subseteq \wp_f(\text{Sen}(\Sigma)) \times \text{Sen}(\Sigma)$ for each $\Sigma \in |\text{Sign}|$, where $\wp_f(\text{Sen}(\Sigma))$ is the set of the *finite* subsets of $\text{Sen}(\Sigma)$; otherwise Π is called *semiformal* (or *infinitary*). ◇

Usually a proof system is presented by a finite set of axiom schemes and inference rules containing metavariables for the sentences [Men87]. Then $\Pi(\Sigma)$ just consists of all instantiations (of the axiom schemes and inference rules) where the metavariables are replaced by sentences; see also the proof system presented for the first-order logic with equality **FOLEq** in the example below. Formally, any relation R between sets of Σ-sentences and Σ-sentences generates an entailment relation in the following way.

Definition 11.2. Given $R \subseteq \wp(\text{Sen}(\Sigma)) \times \text{Sen}(\Sigma)$, where $\Sigma \in |\text{Sign}|$, $\vdash_R$ is the least entailment relation on $\text{Sen}(\Sigma)$ containing R. ◇

Notation. $\Phi \vdash_R \varphi$ stands for $(\Phi, \varphi) \in \vdash_R$, $\vdash_R \varphi$ stands for $\emptyset \vdash_R \varphi$.

Given $R \subseteq \wp(\text{Sen}(\Sigma)) \times \text{Sen}(\Sigma)$, we call φ an *axiom* of R if $(\emptyset, \varphi) \in R$. Obviously, for each set Φ of Σ-sentences, the set $\{\varphi \in \text{Sen}(\Sigma) \mid \Phi \vdash_R \varphi\}$ is the least set of Σ-sentences which contains all sentences of Φ and all axioms of R, and is closed under the "rules of inference" of R [Kei71].

If R is finitary, i.e., $R \subseteq \wp_f(\text{Sen}(\Sigma)) \times \text{Sen}(\Sigma)$, then a necessary and sufficient condition for $\Phi \vdash \varphi$ is that there exists a finite sequence $\varphi_1, \dots, \varphi_k$ of Σ-sentences, called *deduction* [Men87], such that $\varphi_k = \varphi$, and for each $i \in \{1, \dots, k\}$

1. φ_i is an axiom (i.e., $(\emptyset, \varphi) \in R$), or
2. φ_i is a hypothesis (i.e., $\varphi_i \in \Phi$), or
3. there is a set of indices $J \subseteq \{1, \dots, i-1\}$ such that $(\{\varphi_j : j \in J\}, \varphi_i) \in R$.[2]

The advantage of the inductive definition of $\vdash_R$ is that it provides a procedure for testing theoremhood, which is semidecidable if the proof system is formal.

It is straightforward to show that any proof system for INS generates an entailment system, i.e., we have the following proposition.

Proposition 11.3. *Let Π be a proof system for* INS. *Then*

$$\text{ES}_\Pi \overset{\text{def}}{=} \langle \text{Sign}, \text{Sen}, (\vdash_{\Pi(\Sigma)})_{\Sigma \in |\text{Sign}|} \rangle$$

is an entailment system. If Π is sound then ES_Π is sound.

Remark 11.4. If the proof system Π is sound, then INS together with ES_Π constitute a general logic (see Chapter 4).

Definition 11.5. A proof system Π for INS is called *complete* if ES_Π is complete, i.e., for each $\Sigma \in |\text{Sign}|$, $\Phi \subseteq \text{Sen}(\Sigma)$, and Σ-sentence φ, if $\Phi \models_\Sigma \varphi$ then $\Phi \vdash_{\Pi(\Sigma)} \varphi$.   ◇

Example 11.6 (First-order logic with equality). An important example is the institution **FOEq** of first-order logic with equality as defined in Chapter 4, Section 4.2. We use here a slightly different institution, called **FOLEq**, where for each first-order signature Σ the sentences over Σ are arbitrary first-order Σ-formulas (possibly containing free variables)[3] and the Σ-structures have nonempty carrier sets.[4] We will use the following notations and definitions:

- **FOLEq** $= \langle$**FOLSig, FOLWff, FOLStr**, $\models \rangle$

---

[2] In fact, the first case is subsumed by the third with $J = \emptyset$; traditionally, however, these two cases are distinguished as one distinguishes axioms from inference rules.

[3] This is necessary since formulas occurring in the inference rules of proof systems for **FOLEq** may contain free variables.

[4] Hence no pathological situation concerning the satisfaction relation can arise [KK67].

- A (first-order) signature $\Sigma \in |\mathbf{FOLSig}|$ is a triple $\langle S, F, P \rangle$ of *sorts* S, *function symbols* F, and *predicate symbols* P. With each $f \in F$, a *profile* $s_1, \ldots, s_n \to s$ is associated and with each $p \in P$, a *profile* $s_1, \ldots, s_n$ is associated, where $s_1, \ldots, s_n, s \in S$. We often write $\mathrm{Sorts}(\Sigma)$ for S, $\mathrm{Opns}(\Sigma)$ for F, and $\mathrm{Preds}(\Sigma)$ for P.

- A *signature morphism* $\sigma \colon \Sigma \to \Sigma'$ from a signature $\Sigma = \langle S, F, P \rangle$ to a signature $\Sigma' = \langle S', F', P' \rangle$ is a triple $\langle \sigma_{\mathrm{Sorts}}, \sigma_{\mathrm{Opns}}, \sigma_{\mathrm{Preds}} \rangle$ of mappings $\sigma_{\mathrm{Sorts}} \colon S \to S'$, $\sigma_{\mathrm{Opns}} \colon F \to F'$, $\sigma_{\mathrm{Preds}} \colon P \to P'$, such that the profiles are respected.

- For each signature $\Sigma \in |\mathbf{FOLSig}|$, we assume that we have a $\mathrm{Sorts}(\Sigma)$-sorted family $X = (X_s)_{s \in \mathrm{Sorts}(\Sigma)}$ of denumerable sets X_s of variables (of sort s). The set $T_\Sigma(X)$ of Σ-terms with variables in X and the set $\mathbf{FOLWff}(\Sigma)$ of first-order Σ-formulas with equality and with (possibly free) variables in X are defined as usual. Substitution of a term t of sort s for a variable $x \in X_s$ in a formula φ is denoted by $\varphi[t/x]$.

- For each signature $\Sigma \in |\mathbf{FOLSig}|$, the category $\mathbf{FOLStr}(\Sigma)$ of (total) Σ-structures (whose carrier sets are not empty) and Σ-homomorphisms is defined as usual. If Σ contains no predicate symbols, i.e., is of the form $\langle S, F, \emptyset \rangle$ (abbreviated to $\langle S, F \rangle$), then Σ-structures are called Σ-*algebras*. For each signature morphism $\sigma \colon \Sigma \to \Sigma'$, $\mathbf{FOLStr}(\sigma)$ is the usual forgetful functor and is denoted by $\_|_\sigma$.

- The satisfaction relation $\models_\Sigma$ between Σ-structures and Σ-formulas is defined as usual. Sometimes we drop the index Σ if it is obvious from the context.

Similarly, the institution $\mathbf{IFOLEq}$ of infinitary first-order logic with denumerable conjunctions and disjunctions is defined.

An example of a (formal) proof system for the institution $\mathbf{FOLEq}$ is the following Hilbert-like system $\Pi(\mathbf{FOLEq})$, which is sound and complete [Men87]. For each signature Σ,

(A1) $\varphi_1 \Rightarrow (\varphi_2 \Rightarrow \varphi_1)$ is an axiom of $\Pi(\mathbf{FOLEq})$, i.e.,
  if $\varphi_1, \varphi_2 \in \mathbf{FOLWff}(\Sigma)$ then $(\emptyset, \varphi_1 \Rightarrow (\varphi_2 \Rightarrow \varphi_1)) \in \Pi(\mathbf{FOLEq})(\Sigma)$;

(A2) $(\varphi_1 \Rightarrow (\varphi_2 \Rightarrow \varphi_3)) \Rightarrow ((\varphi_1 \Rightarrow \varphi_2) \Rightarrow (\varphi_1 \Rightarrow \varphi_3))$ is an axiom of $\Pi(\mathbf{FOLEq})$;

(A3) $(\neg\varphi_2 \Rightarrow \neg\varphi_1) \Rightarrow ((\neg\varphi_2 \Rightarrow \varphi_1) \Rightarrow \varphi_2)$ is an axiom of $\Pi(\mathbf{FOLEq})$;

(A4) if $s \in \mathrm{Sorts}(\Sigma)$ and φ_1 contains no free occurrences of $x \in X_s$,[5] then $(\forall x : s. \; \varphi_1 \Rightarrow \varphi_2) \Rightarrow (\varphi_1 \Rightarrow (\forall x : s. \; \varphi_2))$ is an axiom of $\Pi(\mathbf{FOLEq})$;

(A5) if t is a Σ-term of sort s free for $x \in X_s$ in φ,[6] then $(\forall x : s. \; \varphi) \Rightarrow \varphi[t/x]$ is an axiom of $\Pi(\mathbf{FOLEq})$;

---

[5] The notion of free occurrence of a variable in a formula is defined as usual.

[6] A term t is free for a variable x in a formula φ if no occurrence of x in the formula φ lies within the scope of any quantifier $(\forall x')$, where x' is a variable in the term t. Thus, the substitution of the term for the variable in the formula causes no variable trapping in the term.

(A6) if $s \in \text{Sorts}(\Sigma)$ and $x \in X_s$,
   then $\forall x : s. \ x = x$ is an axiom of $\Pi(\textbf{FOLEq})$;

(A7) if $s \in \text{Sorts}(\Sigma)$ and $x, x' \in X_s$,
   then $x = x' \Rightarrow (\varphi(x, x) \Rightarrow \varphi(x, x'))^7$ is an axiom of $\Pi(\textbf{FOLEq})$;

(MP) $\dfrac{\varphi_1 \Rightarrow \varphi_2, \ \varphi_1}{\varphi_2}$ is an inference rule of $\Pi(\textbf{FOLEq})$, i.e.,

   if $\varphi_1, \varphi_2 \in \textbf{FOLWff}(\Sigma)$ then $(\{\varphi_1 \Rightarrow \varphi_2, \varphi_1\}, \varphi_2) \in \Pi(\textbf{FOLEq})(\Sigma)$;

(Gen) $\dfrac{\varphi}{\forall x : s. \ \varphi}$ is an inference rule of $\Pi(\textbf{FOLEq})$.

Note that φ, φ_1, φ_2, and φ_3 may be considered as metavariables which can be instantiated by well-formed formulas. Hence $\Pi(\textbf{FOLEq})$ can be presented just by giving seven axiom schemes and two inference rules.

A sound and complete proof system $\Pi(\textbf{IFOLEq})$ for infinitary first-order logic can be found in [Kei71]. It contains, for any denumerable set Φ of formulas, infinitary axioms of the form $\bigwedge \Phi \Rightarrow \varphi$ with $\varphi \in \Phi$, and infinitary inference rules of the form $\dfrac{\psi \Rightarrow \varphi \ \text{for all} \ \varphi \in \Phi}{\psi \Rightarrow \bigwedge \Phi}$. ∎

The next proposition is important for proving the equivalence of the two noncompositional proof systems for specification expressions considered in Section 11.3.

Proposition 11.7. *Let Π be a proof system for* INS. *For each signature $\Sigma \in |\text{Sign}|$ and for all $\Phi, \Phi' \subseteq \text{Sen}(\Sigma)$ and $\varphi \in \text{Sen}(\Sigma)$:*

$$\Phi' \vdash_{\Pi(\Sigma) \cup \Phi} \varphi \ \text{if and only if} \ \Phi' \cup \Phi \vdash_{\Pi(\Sigma)} \varphi$$

where $\Pi(\Sigma) \cup \Phi$ stands for $\Pi(\Sigma) \cup \{(\emptyset, \phi) \mid \phi \in \Phi\}$.

11.2.2   Structured specifications: the kernel language

In this section we define a kernel language for specifications in the institution INS, consisting of presentations over INS (which we call basic specifications) and three specification-building operators. The semantics of the language is defined by two functions Sig and Mod, returning respectively the signature and the model class associated with each specification expression, see Chapter 4, Section 4.6.

Assumptions. We require that the given institution INS has semicomposable signatures, see Chapter 4.[8] Moreover, we assume that Sign is an inclusive

---

[7] $\varphi(x, x')$ arises from $\varphi(x, x)$ by replacing some, but not necessarily all, free occurrences of x by x', with the proviso that x' is free for x in $\varphi(x, x)$. Thus, $\varphi(x, x')$ may or may not contain free occurrences of x.

[8] An institution with semicomposable signatures is called an *exact* institution in [Mes89] and *semiexact* in [DGS93]. This prerequisite is equivalent to the amalgamation lemma, see Chapter 4.

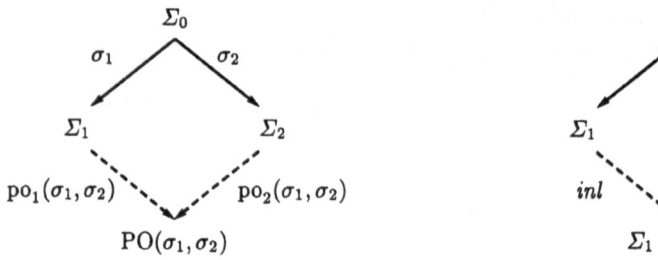

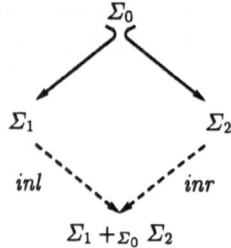

Fig. 11.1. Pushout diagram   **Fig. 11.2.** Amalgamated sum

category with initial object $\emptyset$ and inclusion-preserving pushouts [DGS93].[9] If $\sigma: \Sigma \to \Sigma'$ is an inclusion (also called embedding) morphism, then we write $\sigma: \Sigma \hookrightarrow \Sigma'$, or simply $\Sigma \subseteq \Sigma'$.[10] In this case, for any Σ'-structure A we write $A|_\Sigma$ for $A|_\sigma$. Similarly, if C is a class of Σ-structures, then $C|_\Sigma$ denotes the class of all Σ-structures $A|_\Sigma$ with $A \in C$. Pushouts are depicted as in Figure 11.1. If σ_1 and σ_2 are inclusions, the respective pushout morphisms are denoted by *inl* and *inr*, and the pushout signature is denoted by $\Sigma_1 +_{\Sigma_0} \Sigma_2$ and is called the *amalgamated sum* of Σ_1 and Σ_2 over Σ_0, see Figure 11.2.[11] An isomorphism between signatures is also called a *renaming*.

Example 11.8 (FOLEq). If Σ_1, Σ_2 are first-order signatures such that Σ_1 is included componentwise in Σ_2 (i.e., Sorts$(\Sigma_1) \subseteq$ Sorts(Σ_2), Opns$(\Sigma_1) \subseteq$ Opns(Σ_2), and Preds$(\Sigma_1) \subseteq$ Preds(Σ_2)), then Σ_1 is called a *subsignature* of Σ_2, denoted by $\Sigma_1 \subseteq \Sigma_2$. Inclusion morphisms are defined as subsignature relations $\Sigma_1 \subseteq \Sigma_2$.   ∎

Definition 11.9 (Structured specifications). The collection Spec of *specification expressions* (or specifications for short) in INS, together with the functions Sig: Spec $\to$ |Sign| and Mod: Spec $\to$ $\{C \subseteq \text{Str}(\Sigma) \mid \Sigma \in \text{Sign}\}$ (returning the signature and the class of models of a given specification, respectively), are inductively defined as follows.

1. A *basic specification* is a pair $\langle \Sigma, E \rangle$ where Σ is a signature and E is a set of Σ-sentences.
   $$\text{Sig}(\langle \Sigma, E \rangle) \overset{\text{def}}{=} \Sigma$$
   $$\text{Mod}(\langle \Sigma, E \rangle) \overset{\text{def}}{=} \{A \in \text{Str}(\Sigma) \mid A \models_\Sigma E\}$$

---

[9] Inclusive categories capture the notion of image subobjects. In other words, if $\sigma: \Sigma \to \Sigma'$ is a signature morphism, then it makes sense to talk about $\sigma(\Sigma) \in$ |Sign| (which is intuitively the image of Σ under σ).

[10] Note that, by definition of inclusive categories, there is at most one inclusion morphism between two signatures, see [DGS93].

[11] The pushout can be chosen in such a way that either *inl* or *inr* is an inclusion, since Sign is assumed to be a category with inclusion-preserving pushouts, see [DGS93].

2. If SP_1 and SP_2 are specification expressions, and Σ is a signature such that $\Sigma \subseteq \text{Sig}(SP_1)$ and $\Sigma \subseteq \text{Sig}(SP_2)$, then the *amalgamated sum (or union)* $SP_1 +_\Sigma SP_2$ of SP_1 and SP_2 w.r.t. Σ is a specification expression.

$\text{Sig}(SP_1 +_\Sigma SP_2) \stackrel{\text{def}}{=} \text{Sig}(SP_1) +_\Sigma \text{Sig}(SP_2)$ (see Figure 11.2)

$\text{Mod}(SP_1 +_\Sigma SP_2) \stackrel{\text{def}}{=} \{A \in \text{Str}(\text{Sig}(SP_1 +_\Sigma SP_2)) \mid A|_{inl} \in \text{Mod}(SP_1)$
$\text{and } A|_{inr} \in \text{Mod}(SP_2)\}$

3. If SP is a specification expression and Σ is a signature such that $\Sigma \subseteq \text{Sig}(SP)$, then $SP|_\Sigma$ is a specification expression, called an *export* specification.

$\text{Sig}(SP|_\Sigma) \stackrel{\text{def}}{=} \Sigma$

$\text{Mod}(SP|_\Sigma) \stackrel{\text{def}}{=} \text{Mod}(SP)|_\Sigma$

The symbols of SP that do not occur in Σ are called the *hidden symbols* of $SP|_\Sigma$, whereas the symbols of Σ are its *exported symbols*.

4. If SP is a specification expression and $r\colon \text{Sig}(SP) \to \Sigma$ is a renaming, then **rename** SP **by** r is a specification expression called the *renaming of* SP *by* r.

$\text{Sig}(\textbf{rename } SP \textbf{ by } r) \stackrel{\text{def}}{=} \Sigma$

$\text{Mod}(\textbf{rename } SP \textbf{ by } r) \stackrel{\text{def}}{=} \{A \in \text{Str}(\Sigma) \mid A|_r \in \text{Mod}(SP)\}$ ◇

Amalgamated sum, export, and renaming are called *specification-building operators*. Further operators can be derived, for instance, the *plain sum (or union)* defined as follows:

$SP_1 + SP_2 \stackrel{\text{def}}{=} SP_1 +_\Sigma SP_2$ where $\Sigma = \text{Sig}(SP_1) \cap \text{Sig}(SP_2)$.[12]

The morphisms *inl* and *inr* associated with a plain sum are the usual embeddings since no undesired name conflicts can arise.

If Σ is the signature of a specification expression SP, then we write $\text{Sorts}(SP)$ to denote $\text{Sorts}(\Sigma)$, $\text{Opns}(SP)$ to denote $\text{Opns}(\Sigma)$, and $\text{Preds}(SP)$ to denote $\text{Preds}(\Sigma)$. If $SP = \langle \Sigma, E\rangle$ is a basic specification, then the axioms E of SP are also denoted by $\text{Axs}(SP)$.

Example 11.10 (CONT). The following basic specification CONT (in the institution **FOLEq**) describes properties of containers which can store arbitrary elements. It has a sort Cont for the containers and a sort Elem for the elements of containers. Elements can be stored in a container with the operation add and the membership test is denoted by the function symbol member.

```
spec CONT
    sorts Bool, Elem, Cont
    functions true, false :  → Bool
              empty       :  → Cont
              add         : Elem, Cont → Cont
              member      : Elem, Cont → Bool
```

---

[12] For this definition to make sense, the category of signatures must have pullbacks, which would mean that INS is a *reasonable* institution in the terminology of [DGS93].

axioms ∀x, y: Elem, s: Cont.
       true ≠ false ∧
       member(x, empty) = false ∧
       member(x, add(x, s)) = true ∧
       (x ≠ y ⇒ member(x, add(y, s)) = member(x, s))
endspec

Note that CONT has only one axiom, which is a universally quantified conjunction of formulas. Of course, the axioms of CONT could equally be presented by a *set* of universally quantified formulas. For technical convenience, in the following, we often disregard which kind of presentation of the axioms of a (basic) specification is chosen. ∎

Example 11.11 (CONT-I). The specification CONT-I constructed below can be considered as an implementation of containers by stacks. It is based on a common specification STACK of stacks which is combined with a specification MEMBER, that implements the membership test over stacks. In order to adapt the sum of these specifications (which is called STACK-EXT) to the signature of the container specification, an appropriate renaming is performed and the signature of CONT is exported from the renamed specification (i.e., the stack operations top and pop used for the implementation are hidden).

spec STACK
    sorts Elem, Stack
    functions emptyst : → Stack
             push   : Elem, Stack → Stack
             pop   : Stack → Stack
             top   : Stack → Elem
    axioms ∀x: Elem, s: Stack.
          push(x, s) ≠ emptyst ∧
          pop(push(x, s)) = s ∧
          top(push(x, s)) = x
endspec

spec MEMBER
    sorts Bool, Elem, Stack
    functions true, false : → Bool
             emptyst   : → Stack
             push   : Elem, Stack → Stack
             pop   : Stack → Stack
             top   : Stack → Elem
             member   : Elem, Stack → Bool
    axioms ∀x: Elem, s: Stack.
          true ≠ false ∧
          member(x, emptyst) = false ∧
          [s ≠ emptyst ⇒
            ((top(s) = x ⇒ member(x, s) = true) ∧
            (top(s) ≠ x ⇒ member(x, s) = member(x, pop(s))))]
endspec

spec STACK-EXT = STACK + MEMBER

spec CONT-I = (**rename** STACK-EXT **by** r)$|_{\text{Sig(CONT)}}$

where r renames Stack into Cont, emptyst into empty, and push into add.

∎

For any specification expression SP, a Sig(SP)-sentence is also called an SP-sentence. Given an SP-sentence φ, we write SP $\models_\Sigma \varphi$ if Mod(SP) $\models_\Sigma \varphi$, i.e., if φ is satisfied by all models of SP. In this case φ is called a *theorem* of SP. It is a central aim of this chapter to study proof systems that allow us to prove that a given sentence is a theorem of a specification. In the following, we consider two different possibilities: noncompositional proof systems and structured proof systems which allow us to perform proofs according to the structure of a given specification.

11.3   Noncompositional proof systems

Let INS = $\langle$Sign, Sen, Str, $(\models_\Sigma)_{\Sigma \in |\text{Sign}|}\rangle$ be an arbitrary institution which satisfies the assumptions of Section 11.2.2 and, moreover, let Π be a sound proof system for INS. There are two approaches to noncompositional proof systems: either a normal form nf(SP) equivalent to a given specification SP is computed whose theorems are exactly those of SP, or a proof system is designed for each particular specification.

In both cases, hidden symbols play an important role in the derivation of theorems. Consider, for example, a first-order signature with just one sort and three constants a, b, and c of that sort. Define the basic specification consisting of that signature and two axioms $a = c$ and $b = c$. Now hide the symbol c. In the resulting specification, there are just two (visible) constants, and no "visible axiom". Nevertheless, the equality $a = b$ is satisfied by each of its models.

11.3.1   Proof system associated with a specification

In this subsection we construct an individual proof system for each specification SP that allows us to prove theorems of SP. Technically this is done by providing, for each structured specification SP, a set Π_{SP} of particular "nonlogical" axioms and rules according to the form of SP. We show that when combining Π_{SP} with the given proof system Π for the underlying institution INS, the result is a sound proof system for SP (which is also complete if Π is complete).

The proof systems Π_{SP} are inductively defined according to the structure of the specification SP. In the simplest case SP is a basic specification and Π_{SP} just consists of the axioms of SP. In all other cases the proof system for SP is constructed by using the (appropriately modified) proof systems for the specifications from which SP was built by the specification-building operators.

As already sketched above, the hidden symbols cannot be simply forgotten when trying to derive the theorems that are valid in a specification. However special care is then needed because of possible name conflicts with hidden symbols, as shown in the following examples.

Example 11.12. There may be name conflicts caused by the combination of "export" and "rename". Indeed, assume that SP is a basic specification in **FOLEq** with one sort s, three constants a, b, c of sort s, and the axiom $a = c$. Now hide the symbol c, i.e., construct the specification $SP' = SP|_\Sigma$ with $\Sigma = \langle \{s\}, \{a, b\}, \emptyset \rangle$. Since we cannot forget axioms concerning hidden symbols, the equation $a = c$ is also an axiom of (the proof system for) SP'. In a next step we "re-introduce" the constant c by renaming b to c, i.e., we consider the specification **rename** SP' **by** r where $r_{\text{Sorts}}(s) = s$, $r_{\text{Opns}}(a) = a$, and $r_{\text{Opns}}(b) = c$. The renamed specification now has two constants a and c. If we construct the proof system for **rename** SP' **by** r by renaming only the symbols of Σ according to r, then $a = c$ is also an axiom of (the proof system for) **rename** SP' **by** r. Hence we can derive $a = c$ which, nevertheless, is not valid in all models of **rename** SP' **by** r. ∎

The problem in this example is that a previously hidden symbol is re-introduced by the renaming. This difficulty is solved by using a pushout construction which renames not only the (visible) symbols of SP', but also all its hidden symbols in a consistent way when constructing the proof system for **rename** SP' **by** r.

Example 11.13. There may also be name conflicts caused by the combination of "export" and "amalgamated sum". Indeed, let SP be as in the previous example and let SP_2 be a basic specification with the same signature as SP, but with the axiom $b = c$. Now hide c in SP, i.e., construct the specification $SP_1 = SP|_\Sigma$ where $\Sigma = \langle \{s\}, \{a, b\}, \emptyset \rangle$. Again, since we cannot forget axioms concerning hidden symbols, the equation $a = c$ is also an axiom of (the proof system for) SP_1. Now construct the amalgamated sum $SP_1 +_\Sigma SP_2$ (which is just the plain sum $SP_1 + SP_2$). Clearly, if we assume that the proof system for $SP_1 +_\Sigma SP_2$ is simply the union of the proof systems for the single specifications, then both equations $a = c$ and $b = c$ belong (as axioms) to the proof system for $SP_1 +_\Sigma SP_2$. This means that we can derive $a = b$, which is not valid in all models of $SP_1 +_\Sigma SP_2$. ∎

The problem in the last example is that the constant c which is hidden in SP_1 is identified with the "visible" constant c of SP_2. This confusion can again be avoided using a pushout construction which consistently renames the hidden symbols of SP_1 (and, symmetrically, of SP_2) when combining the proof systems for SP_1 and SP_2. In summary, these considerations lead to the following definition of the symbols of a specification.

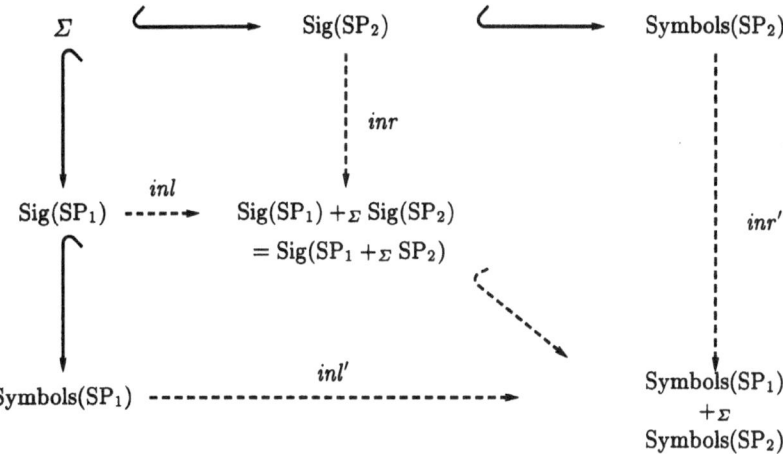

Fig. 11.3. Symbols of an amalgamated union

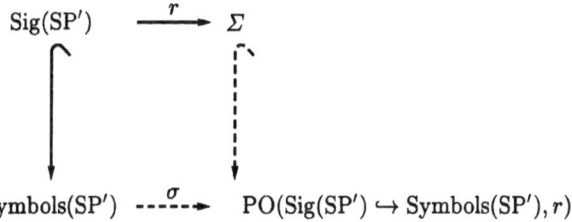

Fig. 11.4. Symbols of a renamed specification

Definition 11.14 (Symbols). For any specification expression SP, the signature Symbols(SP) of the symbols of SP is inductively defined as follows.

1. If SP is a specification expression of the form $\langle \Sigma, E \rangle$,
   then Symbols(SP) $\overset{\text{def}}{=}$ Σ.
2. If SP is a specification expression of the form $SP_1 +_\Sigma SP_2$,
   then Symbols(SP) $\overset{\text{def}}{=}$ Symbols(SP_1) $+_\Sigma$ Symbols(SP_2) (see Figure 11.3).
3. If SP is a specification expression of the form $SP'|_\Sigma$,
   then Symbols(SP) $\overset{\text{def}}{=}$ Symbols(SP').
4. If SP is a specification expression of the form **rename** SP' by r,
   then Symbols(SP) $\overset{\text{def}}{=}$ PO(Sig(SP') $\hookrightarrow$ Symbols(SP'), r) (see Figure 11.4).

$\diamond$

Note that Symbols(SP) is always well defined since pushout signatures are defined up to isomorphism and, by assumption, preserve inclusions. Hence we can always choose Symbols(SP) such that Sig(SP) $\subseteq$ Symbols(SP).

Intuitively, the symbols in Symbols(SP) that are not included in Sig(SP) are the hidden symbols of SP. The pushout technique ensures that hidden symbols are not confused with other symbols when computing the proof system for an amalgamated union or for a specification obtained by renaming.

Example 11.15 (Symbols(CONT-I)). Let CONT-I be the specification of Example 11.11. Obviously, the signature of the symbols of the (sub)specifications STACK, MEMBER, STACK-EXT, and **rename** STACK-EXT by r is just the signature of each of the given specifications, since no export operator is applied in these cases. Then, by definition of the symbols of an export specification, Symbols(CONT-I) = Sig(**rename** STACK-EXT by r) which is the signature

$$\langle\{\text{Bool}, \text{Elem}, \text{Cont}\},$$
$$\{\text{true}, \text{false} : \to \text{Bool},$$
$$\text{empty} : \to \text{Cont},$$
$$\text{add} : \text{Elem}, \text{Cont} \to \text{Cont},$$
$$\text{pop} : \text{Cont} \to \text{Cont},$$
$$\text{top} : \text{Cont} \to \text{Elem},$$
$$\text{member} : \text{Elem}, \text{Cont} \to \text{Bool}\}\rangle .$$

$\blacksquare$

Definition 11.16 (Π_{SP}). For any specification expression SP, the corresponding proof system Π_{SP} is inductively defined as follows.

1. If SP is a basic specification $\langle \Sigma, E \rangle$,
   then $\Pi_{SP} \stackrel{\text{def}}{=} E$.
2. If SP is a specification expression of the form $SP_1 +_\Sigma SP_2$,
   then $\Pi_{SP} \stackrel{\text{def}}{=} inl'(\Pi_{SP_1}) \cup inr'(\Pi_{SP_2})$
   where $inl' = \text{po}_1(\Sigma \hookrightarrow \text{Symbols}(SP_1), \Sigma \hookrightarrow \text{Symbols}(SP_2))$,
   $\qquad inr' = \text{po}_2(\Sigma \hookrightarrow \text{Symbols}(SP_1), \Sigma \hookrightarrow \text{Symbols}(SP_2))$
   (see Figure 11.3), and $inl'(\Pi_{SP_1})$ and $inr'(\Pi_{SP_2})$ are respectively the straightforward extensions of *inl'* and *inr'* to proof systems.
3. If SP is a specification expression of the form $SP'|_\Sigma$,
   then $\Pi_{SP} \stackrel{\text{def}}{=} \Pi_{SP'}$.
4. If SP is a specification expression of the form **rename** SP' by r,
   then $\Pi_{SP} \stackrel{\text{def}}{=} \sigma(\Pi_{SP'})$,
   where $\sigma = \text{po}_1(\text{Sig}(SP') \hookrightarrow \text{Symbols}(SP'), r)$ (see Figures 11.1 and 11.4).

$\diamond$

For any specification SP, the proof system Π_{SP} can be combined with the given proof system Π for deriving SP-sentences. We write $\vdash_{\Pi_{SP}} \varphi$ if φ is an SP-sentence such that $\vdash_{\Pi(\Sigma)\cup\Pi_{SP}} \varphi$, where $\Sigma = \text{Symbols}(SP)$.[13] The next theorem states that in this way we obtain a sound proof system for SP. Moreover, if Π is complete, we even obtain a complete proof system for SP.

---

[13] By $\Pi(\Sigma) \cup \Pi_{SP}$ we simply mean, in the same way as in Proposition 11.7, the proof system defined by $\Pi(\Sigma) \cup \{(\emptyset, \phi) \mid \phi \in \Pi_{SP}\}$. In Section 11.7 inference rules, as well as axioms, will be included in Π_{SP}.

Theorem 11.17 (Soundness and completeness of $\Pi_{\text{SP}} \cup \Pi$). *Let* SP *be a specification expression and let φ be an SP-sentence. If $\vdash_{\Pi_{\text{SP}}} \varphi$, then* SP $\models_{\text{Sig(SP)}} \varphi$. *The converse holds if Π is complete.*

Proof. The theorem is a consequence of the equivalence of the given proof system and the normal form approach studied in the following subsections (see Corollary 11.21 and Proposition 11.24).   □

Example 11.18 ($\Pi_{\text{CONT-I}}$). The proof system of the specification CONT-I of Example 11.11 is constructed by first, combining the axioms of the (sub)specifications STACK and MEMBER, and then, renaming the resulting set of sentences according to r. Hence $\Pi_{\text{CONT-I}}$ consists of the following sentences over the signature Symbols(CONT-I):

$\Pi_{\text{CONT-I}} =$
$\{\ \forall x : \text{Elem}, s : \text{Cont}.$
      $\text{add}(x, s) \neq \text{empty} \wedge \text{pop}(\text{add}(x, s)) = s \wedge \text{top}(\text{add}(x, s)) = x,$
   $\forall x : \text{Elem}, s : \text{Cont}.$
      $\text{true} \neq \text{false} \wedge \text{member}(x, \text{empty}) = \text{false} \wedge$
      $[s \neq \text{empty} \Rightarrow ((\text{top}(s) = x \Rightarrow \text{member}(x, s) = \text{true}) \wedge$
                $(\text{top}(s) \neq x \Rightarrow \text{member}(x, s) = \text{member}(x, \text{pop}(s))))] \ \}$

Proofs of theorems of CONT-I can now be performed by combining the "non-logical" axioms of the proof system $\Pi_{\text{CONT-I}}$ with the axioms and rules of $\Pi(\textbf{FOLEq})(\text{Symbols(CONT-I)})$ (which is just the first-order proof system $\Pi(\textbf{FOLEq})$ used for formulas over the signature Symbols(CONT-I)). For instance, one can derive $\vdash_{\Pi_{\text{CONT-I}}} \varphi$ where φ is the Sig(CONT-I)-sentence $x \neq y \Rightarrow \text{member}(x, \text{add}(y, s)) = \text{member}(x, s)$. We will not give a proof here as an explicit proof of this theorem, using the structured proof system to be introduced in Section 11.4, will be presented in Example 11.27.   ∎

11.3.2   Normal form of a given specification

The idea of the normal form approach is to compute, for a given specification SP, an equivalent but "simpler" specification nf(SP), called the *normal form* of SP, such that one can use just the axioms of this simpler specification (together with the proof system Π associated with the underlying institution) in order to derive the theorems of the given specification SP.

Two specification expressions are said to be *equivalent* if they have the same signature and the same model class.

Since we have to take into account the hidden symbols of a specification expression SP, a normal form equivalent to SP cannot generally be just a basic specification, it has to be a basic specification restricted to an export signature (at least, if we restrict ourselves to finite presentations; for examples and proofs of this fact, see [BM84, BT87, Maj77, Maj79, TWW82, Wir90]).

More precisely, given a specification expression SP, there exists a *normal form* specification expression nf(SP) of the form $\langle \text{Symbols(SP)}, E \rangle|_{\text{Sig(SP)}}$

equivalent to SP. For computing the normal form of SP it is enough to compute the symbols of SP (as described in the previous subsection) and an appropriate set E of axioms. This can be done inductively on the structure of the given specification.

Definition 11.19 (Normal form). The normal form nf(SP) associated with a specification expression SP is defined by induction on the structure of SP as follows.

1. If SP is a specification expression of the form $\langle \Sigma, E \rangle$,
   then nf(SP) $\stackrel{\text{def}}{=} \langle \Sigma, E \rangle|_{\Sigma}$.
2. If SP is a specification expression of the form $SP_1 +_{\Sigma} SP_2$
   with nf(SP$_i$) $= \langle \text{Symbols}(SP_i), E_i \rangle|_{\text{Sig}(SP_i)}$ for $i = 1, 2$,
   then nf(SP) $\stackrel{\text{def}}{=} \langle \text{Symbols}(SP), inl'(E_1) \cup inr'(E_2) \rangle|_{\text{Sig}(SP)}$,
   where inl' and inr' are the signature morphisms shown in Figure 11.3.
3. If SP is a specification expression of the form $SP'|_{\Sigma}$
   with nf(SP$'$) $= \langle \text{Symbols}(SP'), E \rangle|_{\text{Sig}(SP')}$,
   then nf(SP) $\stackrel{\text{def}}{=} \langle \text{Symbols}(SP'), E \rangle|_{\Sigma}$.
4. If SP is a specification expression of the form **rename** SP$'$ **by** r
   with nf(SP$'$) $= \langle \text{Symbols}(SP'), E \rangle|_{\text{Sig}(SP')}$,
   then nf(SP) $\stackrel{\text{def}}{=} \langle \text{Symbols}(SP), \sigma(E) \rangle|_{\text{Sig}(SP)}$,
   where σ is the signature morphism shown in Figure 11.4. ◇

Theorem 11.20. *For any specification expression SP, nf(SP) is well defined and, moreover, equivalent to SP.*

Proof. The proof is straightforward by induction on the structure of SP, using the amalgamation lemma (which holds since, by assumption, INS has semicomposable signatures) for the induction step when SP is an amalgamated union or a renaming. □

As a direct consequence of the last theorem, we obtain that the normal form of a specification SP provides a means for proving the theorems of SP. This proof method is sound, and also complete provided Π is complete.

Notation. We write SP $\vdash_{nf} \varphi$ if φ is an SP-sentence such that $E \vdash_{\Pi(\Sigma)} \varphi$, where nf(SP) $= \langle \Sigma, E \rangle|_{\Sigma'}$.

Corollary 11.21 (Soundness and completeness of the normal form approach). *Let SP be a specification expression and let φ be an SP-sentence. If SP $\vdash_{nf} \varphi$, then SP $\models_{\text{Sig}(SP)} \varphi$. The converse holds if Π is complete.*

Proof. Let nf(SP) $= \langle \Sigma, E \rangle|_{\Sigma'}$.
Then   SP $\vdash_{nf} \varphi$
iff $E \vdash_{\Pi(\Sigma)} \varphi$ (by definition)
implies $\langle \Sigma, E \rangle \models \varphi$ (since Π is sound) (∗)
iff   nf(SP) $\models_{\text{Sig}(SP)} \varphi$ (by satisfaction condition)
iff   SP $\models_{\text{Sig}(SP)} \varphi$ (by Theorem 11.20).
The converse of (∗) holds if Π is complete. □

*Example 11.22 (***nf**(CONT-I)*)*. The normal form of the specification CONT-I (see Example 11.11) is given by the specification

$$\text{nf}(\text{CONT-I}) = \langle \text{Symbols}(\text{CONT-I}), \Pi_{\text{CONT-I}} \rangle|_{\text{Sig}(\text{CONT})}$$

where Symbols(CONT-I) is the signature of the symbols of CONT-I shown in Example 11.15, and $\Pi_{\text{CONT-I}}$ is the set of sentences presented in Example 11.18. For proving theorems using the normal form approach, i.e., for proving CONT-I $\vdash_{\text{nf}} \varphi$, we can use the first-order proof system $\Pi(\textbf{FOLEq})$ and prove $\Pi_{\text{CONT-I}} \vdash_{\Pi(\textbf{FOLEq})(\text{Symbols}(\text{CONT-I}))} \varphi$ from the hypotheses $\Pi_{\text{CONT-I}}$. We will see in the next section that such proofs are equivalent to proofs with the proof system $\Pi_{\text{CONT-I}}$ considered in Section 11.3.1. ∎

11.3.3 Equivalence of $\vdash_{\Pi_{\text{SP}}}$ and $\vdash_{\text{nf}}$

Let us now compare the two proposals for noncompositional proof systems. As a first step we see that if $\text{nf}(\text{SP}) = \langle \Sigma, E \rangle|_{\Sigma'}$, then $\Pi_{\text{SP}} = E$, and vice versa (in particular, $\Sigma = \text{Symbols}(\text{SP})$ and $\Sigma' = \text{Sig}(\text{SP})$).

Lemma 11.23. *For any specification expression* SP, $\text{nf}(\text{SP}) = \langle \Sigma, E \rangle|_{\Sigma'}$ *if and only if* $\Pi_{\text{SP}} = E$.

Proof. The thesis can be shown by induction on the structure of SP. □

This lemma shows that computing the proof system Π_{SP} for a specification SP yields the same set of axioms as when computing the axioms of the normal form of SP. Hence for proving theorems of SP, there should be no difference between using the proof system Π_{SP} and using the normal form approach. That our intuition is right can be formally stated using the result of Proposition 11.7.

Proposition 11.24. *Let* SP *be a specification expression and let* φ *be an* SP*-sentence.* $\vdash_{\Pi_{\text{SP}}} \varphi$ *if and only if* SP $\vdash_{\text{nf}} \varphi$.

Proof. Let $\text{nf}(\text{SP}) = \langle \Sigma, E \rangle|_{\Sigma'}$.

$\vdash_{\Pi_{\text{SP}}} \varphi$

iff $\vdash_{\Pi(\Sigma) \cup \Pi_{\text{SP}}} \varphi$ (by definition)

iff $\vdash_{\Pi(\Sigma) \cup E} \varphi$ (by Lemma 11.23)

iff $E \vdash_{\Pi(\Sigma)} \varphi$ (by Proposition 11.7)

iff SP $\vdash_{\text{nf}} \varphi$ (by definition).   □

In summary, the two noncompositional proof systems are equivalent for specifications of the kernel language. However, they differ in the philosophy by which they were inspired. This becomes evident when the kernel language is extended by further operators (for reachability and observability) in the framework of the **FOLEq** institution in Section 11.7. In this case, the normal form approach needs infinitary axioms while proof systems for specifications use semiformal rules (for instance, infinitary induction). That is, a proof system Π_{SP} consists not only of axioms (as above) but may also contain rules designed according to the form of SP.

11.4   Structured proof systems

The disadvantage of the approaches detailed in Section 11.3 is not only the technical complexity of computing a normal form or a proof system, which can be of importance for larger specifications, but also the loss of structure. The structured nature of the specification is ignored, and too many axioms may cause the proof to become hard to handle. In this section we give a modular alternative, in the spirit of the proof system of Chapter 4, which takes the structure of specifications into account.

The structured proof system, called Π_s, is presented, as usual, by a set of axioms and rules which now define a relation $\vdash_{\Pi_s}$ between specification expressions and sentences. This proof system was inspired by the structured proof system of [Wir93]. It is based on the given proof system Π for the institution INS.

Definition 11.25 (Π_s). The relation $\vdash_{\Pi_s}$ between specifications SP and SP-sentences φ is inductively defined as follows:

(basic) $\langle \Sigma, E \rangle \vdash_{\Pi_s} \varphi$ for all $\varphi \in E$

(pi) $\dfrac{\text{SP} \vdash_{\Pi_s} \varphi_i \ (i \in I)}{\text{SP} \vdash_{\Pi_s} \varphi}$ if $\{\varphi_i : i \in I\} \vdash_{\Pi(\mathrm{Sig(SP)})} \varphi$

(sum-l) $\dfrac{\text{SP}_1 \vdash_{\Pi_s} \varphi}{\text{SP}_1 +_\Sigma \text{SP}_2 \vdash_{\Pi_s} inl(\varphi)}$

where $inl = po_1(\Sigma \hookrightarrow \mathrm{Sig(SP_1)}, \Sigma \hookrightarrow \mathrm{Sig(SP_2)})$

(sum-r) $\dfrac{\text{SP}_2 \vdash_{\Pi_s} \varphi}{\text{SP}_1 +_\Sigma \text{SP}_2 \vdash_{\Pi_s} inr(\varphi)}$

where $inr = po_2(\Sigma \hookrightarrow \mathrm{Sig(SP_1)}, \Sigma \hookrightarrow \mathrm{Sig(SP_2)})$

(exp) $\dfrac{\text{SP} \vdash_{\Pi_s} \varphi}{\text{SP}|_\Sigma \vdash_{\Pi_s} \varphi}$ if $\varphi \in \mathrm{Sen}(\Sigma)$

(ren) $\dfrac{\text{SP} \vdash_{\Pi_s} r^{-1}(\varphi)}{\textbf{rename SP by } r \vdash_{\Pi_s} \varphi}$ ◇

The proof system Π_s is sound as shown in the next theorem.

Theorem 11.26 (Soundness of Π_s). *Let* SP *be a specification expression and let* φ *be an* SP-sentence. *If* SP $\vdash_{\Pi_s} \varphi$, *then* SP $\models_{\mathrm{Sig(SP)}} \varphi$.

Proof. The theorem is proved by induction on the length of the derivation for SP $\vdash_{\Pi_s} \varphi$.   □

Example 11.27. Consider the specification CONT-I of Example 11.11. Let us prove CONT-I $\vdash_{\Pi_s} \varphi$, where φ is the sentence $x \neq y \Rightarrow \mathrm{member}(x, \mathrm{add}(y, s)) = \mathrm{member}(x, s)$ (used as an axiom in the specification CONT of Example 11.10):

1. MEMBER $\vdash_{\Pi_s} s \neq$ emptyst $\Rightarrow$
   $((\text{top}(s) = x \Rightarrow \text{member}(x, s) = \text{true}) \land$
   $(\text{top}(s) \neq x \Rightarrow \text{member}(x, s) = \text{member}(x, \text{pop}(s))))$
                                                                          (basic), (pi)
2. STACK $\vdash_{\Pi_s}$ push$(y, s) \neq$ emptyst   (basic), (pi)
3. STACK-EXT $\vdash_{\Pi_s} s \neq$ emptyst $\Rightarrow$
   $((\text{top}(s) = x \Rightarrow \text{member}(x, s) = \text{true}) \land$
   $(\text{top}(s) \neq x \Rightarrow \text{member}(x, s) = \text{member}(x, \text{pop}(s))))$
                                                                          (sum-r), (1)
4. STACK-EXT $\vdash_{\Pi_s}$ push$(y, s) \neq$ emptyst   (sum-l), (2)
5. STACK-EXT $\vdash_{\Pi_s}$ push$(y, s) \neq$ emptyst $\Rightarrow$
   $((\text{top}(\text{push}(y, s)) = x \Rightarrow \text{member}(x, \text{push}(y, s)) = \text{true}) \land$
   $(\text{top}(\text{push}(y, s)) \neq x \Rightarrow \text{member}(x, \text{push}(y, s)) =$
   $\text{member}(x, \text{pop}(\text{push}(y, s)))))$
                                                                          (pi), (3)
6. STACK-EXT $\vdash_{\Pi_s} (\text{top}(\text{push}(y, s)) = x \Rightarrow \text{member}(x, \text{push}(y, s)) = \text{true}) \land$
   $(\text{top}(\text{push}(y, s)) \neq x \Rightarrow \text{member}(x, \text{push}(y, s)) =$
   $\text{member}(x, \text{pop}(\text{push}(y, s))))$
                                                                          (pi), (4,5)
7. STACK $\vdash_{\Pi_s}$ top$(\text{push}(y, s)) = y$ (basic), (pi)
8. STACK-EXT $\vdash_{\Pi_s}$ top$(\text{push}(y, s)) = y$ (sum-l), (7)
9. STACK-EXT $\vdash_{\Pi_s} (y = x \Rightarrow \text{member}(x, \text{push}(y, s)) = \text{true}) \land$
   $(y \neq x \Rightarrow \text{member}(x, \text{push}(y, s)) =$
   $\text{member}(x, \text{pop}(\text{push}(y, s))))$
                                                                          (pi), (6,8)
10. STACK $\vdash_{\Pi_s}$ pop$(\text{push}(y, s)) = s$ (basic), (pi)
11. STACK-EXT $\vdash_{\Pi_s}$ pop$(\text{push}(y, s)) = s$ (sum-l), (10)
12. STACK-EXT $\vdash_{\Pi_s} (y = x \Rightarrow \text{member}(x, \text{push}(y, s)) = \text{true}) \land$
    $(y \neq x \Rightarrow \text{member}(x, \text{push}(y, s)) = \text{member}(x, s))$
                                                                          (pi), (9,11)
13. STACK-EXT $\vdash_{\Pi_s} y \neq x \Rightarrow \text{member}(x, \text{push}(y, s)) = \text{member}(x, s)$ (pi), (12)
14. rename STACK-EXT by $r \vdash_{\Pi_s}$
    $y \neq x \Rightarrow \text{member}(x, \text{add}(y, s)) = \text{member}(x, s)$ (ren), (13)
15. CONT-I $\vdash_{\Pi_s} y \neq x \Rightarrow \text{member}(x, \text{add}(y, s)) = \text{member}(x, s)$ (exp), (14)
16. CONT-I $\vdash_{\Pi_s} x \neq y \Rightarrow \text{member}(x, \text{add}(y, s)) = \text{member}(x, s)$ (pi), (15)

                                                                          ∎

In the rest of this section we address the completeness of Π_s. The issue is converse to that of soundness, that is, whether SP $\models_{\text{Sig(SP)}} \varphi$ implies SP $\vdash_{\Pi_s} \varphi$. To begin, we consider the institution **FOLEq** and the (sound and complete) proof system $\Pi(\textbf{FOLEq})$ of Example 11.6. We show that in this case the structured proof system Π_s (based on $\Pi(\textbf{FOLEq})$) is complete if the axioms of the basic specifications are closed sentences.

Theorem 11.28. *Let* SP *be a specification expression in* **FOLEq** *such that the axioms of the basic specifications are closed. Let* φ *be an* SP*-sentence. If* SP $\models_{\text{Sig(SP)}} \varphi$, *then* SP $\vdash_{\Pi_s} \varphi$.

Proof. The proof is by induction on the structure of SP. The critical case is when SP is of the form $SP_1 +_\Sigma SP_2$, for which Theorem 11.20, as well as the interpolation property, are needed [Cen94].   □

In the general case, the completeness of Π_s is ensured if, in addition to the assumptions of Section 11.2.2, we have:[14]

- INS has conjunctions and negations, that is, for each $\Sigma \in |Sign|$, $Sen(\Sigma)$ is closed under finite conjunctions and negations (with the obvious meaning);
- INS is compact, that is, whenever $\Phi \models_\Sigma \varphi$, then there exists a finite set $\Psi \subseteq \Phi$ such that $\Psi \models_\Sigma \varphi$.
- Π is finitary, complete, and satisfies both the deduction theorem (see, e.g., [Men87]) and the interpolation property.

Thereby we say that Π satisfies the interpolation property if the following holds: if $\varphi_1 \vdash_{\Pi(\Sigma)} \varphi_2$ where φ_i is a Σ_i-sentence with $\Sigma_i \subseteq \Sigma$ ($i = 1, 2$), then there exists a $(\Sigma_1 \cap \Sigma_2)$-sentence[15] φ such that $\varphi_1 \vdash_{\Pi(\Sigma)} \varphi$ and $\varphi \vdash_{\Pi(\Sigma)} \varphi_2$ [Bar77, BM86].[16] The sentence φ is called interpolant. In equational logic, for example, interpolation does not hold [BHK90].[17] Classical and intuitionistic first-order predicate calculi [BM86], as well as the logic of partial terms [Bee85], satisfy interpolation.

If INS has denumerable conjunctions and negations and Π satisfies the conditions stated above, apart from being finitary, i.e., Π has a semiformal rule of the form

$$\frac{\varphi_i : i \in I}{\varphi}$$

with denumerable I, then Π_s is complete. In particular, Theorem 11.28 can be extended to the institution **IFOLEq** of infinitary first-order logic (see Example 11.6).

The structured proof system Π_s has the advantage, mentioned above, that it handles the modular structure of the specification of interest. Nevertheless, in practice, one does not work purely in this way: sometimes we *do* compute a halfway normal form, in particular, when dealing with the combination of two specifications by means of the sum. That is, the techniques presented in Section 11.3 and in the present subsection deal with the extremes, but a system designer normally works somewhere in between. An interesting discussion about this subject can be found in [HST94].

---

[14] See also [Bor98].

[15] Here we need to assume, moreover, that the category of signatures has pullbacks, i.e., that INS is reasonable [DGS93].

[16] A different notion of interpolation (defined on the semantic level) is given in [DGS93], where sets of sentences rather than single sentences are allowed as hypotheses.

[17] Note, however, that equational logic satisfies the interpolation property as defined in [DGS93].

11.5   Proof systems for refinement

A system specified in an abstract manner is implemented by any structure in its class of models. This structure is usually specified by adding design decisions to the original specification in a stepwise way. Refinement is a partial order relation that holds between a more "abstract" and a more "concrete" specification of that chain. That is, two specifications are in this relation if they have the same signature and if the model class of the "concrete" specification is included in the model class of the "abstract" specification, which is expressed by the following definition; see Chapter 7 and also [ST88b, SW83, Wir93].

Definition 11.29. A specification SPI *refines* (or *implements*) a specification SP, written $SP \rightsquigarrow SPI$, if $Sig(SPI) = Sig(SP)$ and $Mod(SPI) \subseteq Mod(SP)$.
◊

   The refinement relation has some obvious properties: given a signature Σ and sets of Σ-sentences E and E', if $E' \vdash_{\Pi(\Sigma)} \varphi$ for every $\varphi \in E$, then $\langle \Sigma, E \rangle \rightsquigarrow \langle \Sigma, E' \rangle$. Furthermore, all specification-building operators considered here are monotonic with respect to $\rightsquigarrow$, e.g., if $SP \rightsquigarrow SPI$, then $SP +_\Sigma SP' \rightsquigarrow SPI +_\Sigma SP'$ (provided the specification expressions are well formed).
   In the following we present a proof system, denoted by $\Pi_{\rightsquigarrow_s}$, which allows us to derive a binary relation $SP \rightsquigarrow_s SPI$ inductively according to the modular structure of the specification SP, such that whenever $SP \rightsquigarrow_s SPI$, then SPI refines SP. Thereby we assume given an arbitrary sound proof system Π_{spec} for the derivation of the theorems φ of a specification expression SP (see Sections 11.3 and 11.4). We write $SP \vdash \varphi$ if φ can be proved by the axioms and rules of Π_{spec}. In order to introduce $\Pi_{\rightsquigarrow_s}$ we need the notion of extension.

Definition 11.30. A specification SP_2 is a *(syntactic) extension* of a specification SP_1 if $Sig(SP_1) \subseteq Sig(SP_2)$. SP_2 is a *conservative* extension of SP_1 if it is a syntactic extension of SP_1 such that $Mod(SP_1) = Mod(SP_2)|_{Sig(SP_1)}$.
◊

Definition 11.31 (Proof system $\Pi_{\rightsquigarrow_s}$ for refinement). The proof system $\Pi_{\rightsquigarrow_s}$ for the refinement relation consists of the following axiom scheme and inference rules, that inductively define a relation $\rightsquigarrow_s$ between specifications:

(ref-basic) $\langle \Sigma, E \rangle \rightsquigarrow_s SPI$ if $Sig(SPI) = \Sigma$ and $SPI \vdash \varphi$ for every $\varphi \in E$

(ref-sum) $\dfrac{SP_1 \rightsquigarrow_s SPI|_{inl}, \ SP_2 \rightsquigarrow_s SPI|_{inr}}{SP_1 +_\Sigma SP_2 \rightsquigarrow_s SPI}$ if $Sig(SPI) = Sig(SP_1 +_\Sigma SP_2)$

where $inl = po_1(\Sigma \hookrightarrow Sig(SP_1), \Sigma \hookrightarrow Sig(SP_2))$
and $inr = po_2(\Sigma \hookrightarrow Sig(SP_1), \Sigma \hookrightarrow Sig(SP_2))$

(ref-exp) $\dfrac{SP \leadsto_{\!s} SPI'}{SP|_{\Sigma} \leadsto_{\!s} SPI}$ if $\Sigma = Sig(SPI) \subseteq Sig(SP)$, and
if SPI' is a conservative extension of SPI

(ref-ren) $\dfrac{SP \leadsto_{\!s} \text{rename } SPI \text{ by } r^{-1}}{\text{rename } SP \text{ by } r \leadsto_{\!s} SPI}$ ◊

Theorem 11.32 (Soundness and completeness of $\Pi_{\leadsto_{\!s}}$). *Given specifications SP and SPI, if SP $\leadsto_{\!s}$ SPI, then SP $\leadsto$ SPI. The converse holds if Π_{spec} is complete.*

Proof. Soundness can be proved by induction on the length of the derivation, completeness by induction on the structure of SP.   □

Remark 11.33. The proof system $\Pi_{\leadsto_{\!s}}$ has the following difficulty. Rule (ref-exp) requires an appropriate conservative extension of SPI to be found. That means, completeness is achieved at the cost of a proof obligation which is of semantic nature. Assume, for the moment, that we are working in the institution **FOLEq**. Then, to prove conservativeness, we have to show that each model of SPI can be extended by interpretations of the symbols $Sig(SPI') \setminus Sig(SPI)$, in such a way that the extension is a model of SPI'. In general, this has to be done by a semantical construction. However, if the axioms for the function symbols in $Sig(SPI') \setminus Sig(SPI)$ are in a constructive form, such as explicit or implicit definition, then it is well known that conservativeness can be checked by syntactic conditions (under mild assumptions).

Example 11.34. Let CONT and CONT-I be the specifications of Examples 11.10 and 11.11, respectively. In order to prove CONT $\leadsto_{\!s}$ CONT-I, we only have to show that CONT-I $\vdash \varphi$ for every $\varphi \in Axs(\text{CONT})$ (since CONT is a basic specification). This can be done using the structured proof system of Section 11.4, as illustrated in Example 11.27. More elaborate structured proofs of implementation relations will be shown in Example 11.48 in the context of parameterized specifications, and in Example 11.70 in the context of the reachability and observability operators.   ∎

11.6   Parameterization

In this section we define parameterized specifications, together with associated proof systems. There are many practical examples that motivate parameterization, for instance, storing structures like stacks, lists, and queues, which are parameterized with respect to the elements to be stored.

In the literature one can find two major approaches to parameterization, the so-called pushout approach (e.g., [BG80, EM85]), and that based on the λ-calculus (e.g., [SST92, Wir90]; see also Chapter 8 for a discussion of both approaches). The drawback of the former approach is that it does not easily generalize to abstraction over arbitrary higher-order variables. The need for higher-order parameterization arises naturally: an interpreter for a language with block structure may need a stack as an actual parameter (in order to save the return address each time a new block is entered), whereby the stack is also a parameterized specification. In this chapter, we discuss only the λ-calculus parameterization mechanism. For proof systems in the framework of pushout parameterization, the reader may consult [Pad84, Pad87, Pad88b].

We define a language of simply typed λ-terms whose words are the specification expressions of Section 11.2.2, together with variables, λ-abstraction, and application. This language allows the definition of parameterized specification expressions but not the specification of parameterized program modules (see Chapter 8 and [SST92]). There may be a parameter restriction (also called *semantic type*) associated with the formal parameter variable of a λ-abstraction. An application is well defined if the actual parameter refines (or implements) the parameter restriction. That is, we can require an argument to "behave" like a specification expression intended, in particular, to satisfy a number of axioms. This implies that a parameterized specification applied to an argument may yield an erroneous expression in two ways, namely, either by violating the parameter restriction, or by disregarding the formation laws (also called *syntactic typing*) of specification expressions (see Definition 11.9). In order to keep the presentation as simple as possible, we do not take syntactic typing of specification expressions into account here, and assume always that specifications are well formed. A proof system that allows the derivation of syntactic well-definedness is given in [Cen94, CW95].

In the following we define a proof system that allows us to derive the theorems of a (parameterized and instantiated) specification λ-term without considering parameter restrictions. In the same way as in Section 11.5, this proof system is used for the definition of an inference system for refinement relations between specification λ-terms. In particular, this mechanism can then be applied to the verification of parameter restrictions. By adding parameter restrictions and combining the two inference systems, we finally obtain a system for deriving theorems and refinements in the context of parameterized specifications.

11.6.1   A λ-calculus for parameterized specifications

We enhance the kernel language defined over the institution INS (see Section 11.2.2) with parameterization based on the simply typed λ-calculus. The set of types of the λ-terms is the closure under arrow application of the type sp, which is the base type of specification terms. As a notational convention, we choose X, Y, Z to denote variables of the λ-calculus of parameterized specifications.

Pure λ-terms are defined just as a set of typed variables, closed under λ-abstraction and application. Additionally in our case, we have as constants the *basic* specifications over the chosen institution and, moreover, the specification-building operators as further term constructors. In particular, any specification of the kernel language is a λ-term of type sp without variables. Types and terms are formally defined as follows.

Definition 11.35 (Types). The set T of *types* is inductively defined by

$$T ::= sp \mid (T \to T) \qquad \diamond$$

Definition 11.36 (λ-Terms). Let $Var = (Var_\tau)_{\tau \in T}$ be a family of (denumerable) sets Var_τ of variables of type τ. The family $\Lambda = (\Lambda_\tau)_{\tau \in T}$ of sets Λ_τ of λ-terms of type τ over Var and INS is inductively defined by

$$
\begin{aligned}
\Lambda_{sp} ::= &\ \langle \Sigma, E \rangle && \text{for each } \Sigma \in |\text{Sign}| \text{ and } E \subseteq \text{Sen}(\Sigma) \\
\mid &\ \Lambda_{sp} +_\Sigma \Lambda_{sp} && \text{for each } \Sigma \in |\text{Sign}| \\
\mid &\ \Lambda_{sp}|_\Sigma && \text{for each } \Sigma \in |\text{Sign}| \\
\mid &\ \textbf{rename } \Lambda_{sp} \textbf{ by } r && \text{for each renaming } r \\
\Lambda_{(\tau_1 \to \tau_2)} ::= &\ (\lambda Var_{\tau_1}.\Lambda_{\tau_2}) \\
\Lambda_{\tau_2} ::= &\ (\Lambda_{(\tau_1 \to \tau_2)}\ \Lambda_{\tau_1}) \\
\mid &\ Var_{\tau_2} && \diamond
\end{aligned}
$$

The semantics of the λ-terms is straightforward. The semantics $[\![M]\!]$ of a term M is a function on environments where an environment assigns values to variables. If e is an environment and M is of type sp, then $[\![M]\!]_e$ is a specification in the collection Spec of specification expressions (see Section 11.2.2), whereby we assume that the formation rules of specification expressions are fulfilled.[18] If M is of arrow type, then $[\![M]\!]_e$ is a function. More precisely, the semantics of types and λ-terms is defined as follows.

Definition 11.37 (Semantics of types and terms). The semantics $[\![\tau]\!]$ of a type τ is defined by induction on the structure of τ:

$[\![sp]\!] \overset{\text{def}}{=} \text{Spec}$

$[\![(\tau_1 \to \tau_2)]\!] \overset{\text{def}}{=} ([\![\tau_1]\!] \to [\![\tau_2]\!])$, the collection of all mappings from $[\![\tau_1]\!]$ to $[\![\tau_2]\!]$.

An *environment* is a family $e = (e_\tau)_{\tau \in T}$ of functions $e_\tau : Var_\tau \to [\![\tau]\!]$ for each $\tau \in T$. The index τ is usually omitted and we simply write e. We let $e_{[X:=S]}$ denote the environment defined in terms of an environment e, a variable $X \in Var_\tau$, and a value $S \in [\![\tau]\!]$, by

$$(e_{[X:=S]})(Y) \overset{\text{def}}{=} \begin{cases} S & \text{if } Y = X, \\ e(Y) & \text{if } Y \neq X. \end{cases}$$

The semantics $[\![M]\!]$ of a λ-term M is a function on environments defined by induction on the structure of M:

$[\![\langle \Sigma, E \rangle]\!]_e \overset{\text{def}}{=} \langle \Sigma, E \rangle$;

$[\![M +_\Sigma N]\!]_e \overset{\text{def}}{=} [\![M]\!]_e +_\Sigma [\![N]\!]_e$;

$[\![M|_\Sigma]\!]_e \overset{\text{def}}{=} [\![M]\!]_e|_\Sigma$;

---

[18] Otherwise, $[\![M]\!]_e$ should be a distinct error value [Cen94].

$[\![\text{rename } M \text{ by } r]\!]_e \overset{\text{def}}{=} \text{rename } [\![M]\!]_e \text{ by } r$;

$[\![(\lambda X.M)]\!]_e : [\![\tau_1]\!] \rightarrow [\![\tau_2]\!]$ (where τ_1 is the type of X and τ_2 that of M) is the mapping defined by $[\![(\lambda X.M)]\!]_e (S) \overset{\text{def}}{=} [\![M]\!]_{e_{[X:=S]}}$ for all $S \in [\![\tau_1]\!]$;

$[\![(M \ N)]\!]_e \overset{\text{def}}{=} [\![M]\!]_e ([\![N]\!]_e)$;

$[\![X]\!]_e \overset{\text{def}}{=} e(X)$.   ◇

The notions of free variables of a term (denoted by $\mathrm{FV}(M)$, if M is a λ-term) and of substitution of terms for (free) variables (denoted by $M[X := N]$, if X is a variable and M, N are λ-terms with X and N of the same type) are defined as usual. In particular, M is a ground λ-term if M contains no free variable.

Definition 11.38 (β-equality). The binary relation of β-reduction is the least set of pairs of λ-terms containing $((\lambda X.M) \ N) \rightarrow_\beta M[X := N]$[19] for every M, N, X with X and N of the same type, that commutes with the specification-building operators, λ-abstraction, and application. The binary relation of β-equality is the equivalence relation generated by the reflexive, symmetric, and transitive closure of β-reduction.   ◇

Lemma 11.39. *For any λ-terms M, N of the same type, $M =_\beta N$ implies $[\![M]\!] = [\![N]\!]$.*

This lemma can be proved by induction on the definition of β-equality. As a consequence of being a simply typed λ-calculus, for any term M there is a unique β-normal form N (i.e., N is such that $N \not\rightarrow_\beta$) with $M =_\beta N$.

The notion of β-reduction only uses substitution, that is, no name adjustment, typically through a parameter-passing morphism, takes place. A proposal in this direction in the framework of the λ-calculus approach to parameterization can be found in [Bau93].

Example 11.40 (CONT$_{par}$). A parameterized specification of containers in the institution **FOLEq** can be defined by

$$\mathrm{CONT}_{par} \overset{\text{def}}{=} (\lambda X.X +_{\text{ELEM}} \mathrm{CONT}),$$

where ELEM is the signature containing only the sort Elem, and CONT is the specification of containers as defined in Example 11.10. The amalgamated union ensures that no name conflict can arise by applying CONT$_{par}$ to an argument that contains, besides Elem, further symbols of CONT.   ■

Example 11.41 (CONT-I$_{par}$). In the same way we can use the specification CONT-I, that implements CONT, in order to define (at a more concrete level)

---

[19] We adopt the *variable convention* of [Bar81] and require that free variables and bound variables have different names. Because of the variable convention, for $M[X := N]$ we do not have to be concerned about variable capturing in the case where M is an abstraction term.

a parameterized specification of containers. That is, we define

$$\text{CONT-I}_{par} \overset{\text{def}}{=} (\lambda X.X +_{\text{ELEM}} \text{CONT-I}),$$

where ELEM is the signature as in the example above.   ∎

11.6.2   Proving theorems of specification λ-terms

In this subsection we present a proof system Π_{ps} which allows us to derive
theorems of specification λ-terms of type sp. This proof system extends the
structured proof system Π_s of Section 11.4, and contains additional rules
according to the λ-term constructors. Its judgements are of the form

$M \vdash_{\Pi_{ps}} \varphi$ where M is a ground term of type sp and φ is a $[\![M]\!]$-
sentence.

In the same way as Π_s, Π_{ps} is also based on a given proof system Π for the
underlying institution (see rule (prf-pi) below) and, moreover, on the axioms
and rules of β-equality[20] (see rule (prf-refl) below).

Definition 11.42 (Π_{ps}).   The relation $\vdash_{\Pi_{ps}}$ between ground λ-terms
$M \in \Lambda_{sp}$ and $[\![M]\!]$-sentences φ is inductively defined as follows:

(prf-ax) $\langle \Sigma, E \rangle \vdash_{\Pi_{ps}} \varphi$ for all $\varphi \in E$

(prf-pi) $\dfrac{M \vdash_{\Pi_{ps}} \varphi_i \ (i \in I)}{M \vdash_{\Pi_{ps}} \varphi}$ if $\{\varphi_i : i \in I\} \vdash_{\Pi(\text{Sig}([\![M]\!]))} \varphi$

(prf-sum-l) $\dfrac{M_1 \vdash_{\Pi_{ps}} \varphi}{M_1 +_\Sigma M_2 \vdash_{\Pi_{ps}} inl(\varphi)}$

(prf-sum-r) $\dfrac{M_2 \vdash_{\Pi_{ps}} \varphi}{M_1 +_\Sigma M_2 \vdash_{\Pi_{ps}} inr(\varphi)}$

(prf-exp) $\dfrac{M \vdash_{\Pi_{ps}} \varphi}{M|_\Sigma \vdash_{\Pi_{ps}} \varphi}$ if φ is a Σ-sentence

(prf-ren) $\dfrac{M \vdash_{\Pi_{ps}} r^{-1}(\varphi)}{\textbf{rename } M \textbf{ by } r \vdash_{\Pi_{ps}} \varphi}$

(prf-refl) $\dfrac{M_1 \vdash_{\Pi_{ps}} \varphi, M_1 =_\beta M_2}{M_2 \vdash_{\Pi_{ps}} \varphi}$ ◇

Note that if $M \vdash_{\Pi_{ps}} \varphi$, then M is necessarily of basic type, that is, it
cannot be a λ-abstraction. Nevertheless it may contain occurrences of the
"λ" symbol, e.g., if M is of the form $((\lambda X.N_1) \ N_2)$, where N_1 is of type sp.
In this case, theorems of M can be proved using the rule (prf-refl) in the
following way. Suppose that we have derived $N_1[X := N_2] \vdash_{\Pi_{ps}} \varphi$. Then, by
rule (prf-refl), we can infer $M \vdash_{\Pi_{ps}} \varphi$.

---

[20] Defined according to Definition 11.38.

Remark 11.43. If M is a λ-abstraction, then any sentence φ which is valid in all applications $(M\ N)$ of M to an arbitrary argument N may be considered as a theorem of M. Hence, according to the following soundness theorem, for proving a theorem φ of M, it is sufficient to prove $(M\ N) \vdash_{\Pi_{ps}} \varphi$ for all arguments N of M. For instance, let M be a parameterized specification of the form $(\lambda X.X +_{\Sigma} SP)$, where SP is a specification of the kernel language. If we can prove $SP \vdash_{\Pi_s} \varphi$, then we can also prove $SP \vdash_{\Pi_{ps}} \varphi$ (since $\Pi_s \subseteq \Pi_{ps}$). Therefore by (prf-sum-r), $N +_{\Sigma} SP \vdash_{\Pi_{ps}} inr(\varphi)$ for any N of type *sp* and hence, by rule (prf-refl), $(M\ N) \vdash_{\Pi_{ps}} inr(\varphi)$ for all arguments N of M. This means that we can deduce $inr(\varphi)$ for any application of M and, therefore, $inr(\varphi)$ may be considered as a theorem of M.

Theorem 11.44 (Soundness). *Let M be a ground term of type sp and let φ be a $[\![M]\!]$-sentence. If $M \vdash_{\Pi_{ps}} \varphi$, then $[\![M]\!] \models_{\mathrm{Sig}([\![M]\!])} \varphi$.*

The theorem can be proved by induction on the length of the derivation of $M \vdash_{\Pi_{ps}} \varphi$ [Cen94]. Thereby one shows that $M \vdash_{\Pi_{ps}} \varphi$ implies $[\![M]\!] \vdash_{\Pi_s} \varphi$. This is fairly obvious if M is a β-normal form; if M is not a β-normal form, one computes the unique β-normal form N with $M =_{\beta} N$, thus the result applies to N and, by (prf-refl), also to M. The theorem is then a consequence of the correctness of Π_s (see Theorem 11.26).

Furthermore, given a ground term M of type *sp*, one can show that $[\![M]\!] \vdash_{\Pi_s} \varphi$ implies $M \vdash_{\Pi_{ps}} \varphi$. Indeed, this implication is evident since $\Pi_s \subseteq \Pi_{ps}$, and because of (prf-refl). Completeness therefore depends on the same conditions that determine the completeness of Π_s. That is, under certain assumptions on INS and Π (see Theorem 11.28 and the discussion of completeness of Π_s in Section 11.4), the proof system Π_{ps} is also complete.

Example 11.45. Let us specify containers of natural numbers by instantiating the parameterized specification CONT-I$_{par}$ of Example 11.41 with a specification NAT of the natural numbers as follows:
NATCONT-I = (CONT-I$_{par}$ **rename** NAT **by** r'), where r' renames **Nat** into **Elem**. Let φ be the sentence $x \neq y \Rightarrow member(x, add(y, s)) = member(x, s)$. Now we show that NATCONT-I $\vdash_{\Pi_{ps}} \varphi$:

1. CONT-I $\vdash_{\Pi_{ps}} \varphi$ analogously to Example 11.27
2. (**rename** NAT **by** r') $+_{\text{ELEM}}$ CONT-I $\vdash_{\Pi_{ps}} \varphi$ (prf-sum-r), (1)
3. NATCONT-I $\vdash_{\Pi_{ps}} \varphi$ (prf-refl), (2)

Note that in the last step, rule (prf-refl) can be applied since

$$
\begin{aligned}
\text{NATCONT-I} &= (\text{CONT-I}_{par}\ \textbf{rename NAT by } r') \\
&=_{\beta} (X +_{\text{ELEM}} \text{CONT-I})[X := \textbf{rename NAT by } r'] \\
&= (\textbf{rename NAT by } r') +_{\text{ELEM}} \text{CONT-I} .
\end{aligned}
$$

Note also that φ is a CONT-axiom (see Example 11.10) and, moreover, that NATCONT-I$|_{inr} \vdash_{\Pi_{ps}} \psi$ for every $\psi \in \text{Axs(CONT)}$.   ∎

11.6.3   Refinement of parameterized specifications

The refinement relation between specifications can be extended to values in $[\![\tau]\!]$, for arbitrary τ, by a standard pointwise definition. If $\tau = sp$, and if S_1 and S_2 are values in $[\![sp]\!]$, then $S_1 \rightsquigarrow S_2$ if S_2 refines S_1 according to Definition 11.29. If $\tau = (\tau_1 \to \tau_2)$ and F_1 and F_2 are values in $[\![(\tau_1 \to \tau_2)]\!]$, then $F_1 \rightsquigarrow F_2$ if $F_1(S) \rightsquigarrow F_2(S)$ for all $S \in [\![\tau_1]\!]$.

Since all the specification-building operators considered here are monotonic with respect to $\rightsquigarrow$, it can be proved that monotonicity extends to denotable values, i.e., if N_1 and N_2 are ground terms such that $[\![N_1]\!] \rightsquigarrow [\![N_2]\!]$, then $[\![(M\ N_1)]\!] \rightsquigarrow [\![(M\ N_2)]\!]$ for any ground term M of an appropriate type. As a consequence, given denotable values F_1, F_2, S_1, and S_2, if $F_1 \rightsquigarrow F_2$ and $S_1 \rightsquigarrow S_2$, then $F_1(S_1) \rightsquigarrow F_2(S_2)$ [Cen94, Cen95].

The refinement relation between specification terms of type sp can be derived by the following proof system, which extends the proof system for refinements presented in Section 11.5 by additional rules that handle λ-abstraction and application. Its judgements are of the form $M \rightsquigarrow_{ps} N$, where M and N are ground terms of type sp.

Considering the axiom scheme (ref-basic) below, one can see that this system, analogously to that introduced in Section 11.5, is based on a proof system for proving theorems of a specification. In our case, we have chosen the proof system Π_{ps} of the previous subsection. In principle, however, any other proof system could be used.

Definition 11.46 (Proof system $\Pi_{\rightsquigarrow_{ps}}$ for refinement). The proof system $\Pi_{\rightsquigarrow_{ps}}$ for the refinement relation consists of the following axiom scheme and inference rules that inductively define a relation $\rightsquigarrow_{ps}$ between ground λ-terms of type sp:

(ref-basic) $\langle \Sigma, E \rangle \rightsquigarrow_{ps} M$ if $M \vdash_{\Pi_{ps}} \varphi$ for all $\varphi \in E$

(ref-sum) $\dfrac{M_1 \rightsquigarrow_{ps} N|_{inl},\ M_2 \rightsquigarrow_{ps} N|_{inr}}{M_1 +_{\Sigma} M_2 \rightsquigarrow_{ps} N}$

(ref-exp) $\dfrac{M \rightsquigarrow_{ps} M_2}{M|_{\Sigma} \rightsquigarrow_{ps} M_1}$ $\begin{array}{l}\text{if } \Sigma = \mathrm{Sig}([\![M_1]\!]) \subseteq \mathrm{Sig}([\![M]\!]), \text{ and}\\ \text{if } [\![M_2]\!] \text{ is a conservative extension of } [\![M_1]\!]\end{array}$

(ref-ren) $\dfrac{M \rightsquigarrow_{ps} \textbf{rename } N \textbf{ by } r^{-1}}{\textbf{rename } M \textbf{ by } r \rightsquigarrow_{ps} N}$

(ref-refl) $\dfrac{M_1 \rightsquigarrow_{ps} N,\ M_1 =_{\beta} M_2}{M_2 \rightsquigarrow_{ps} N}$ ◇

Theorem 11.47 (Soundness and completeness of $\Pi_{\rightsquigarrow_{ps}}$). *Let M and N be ground terms of type sp. If $M \rightsquigarrow_{ps} N$, then $[\![M]\!] \rightsquigarrow [\![N]\!]$. The converse holds if Π_{ps} is complete.*

Example 11.48. Consider the following (abstract) specification of containers of natural numbers obtained by instantiating the parameterized specification CONT_{par} of Example 11.40 with the specification NAT of the natural numbers: $\text{NATCONT} = (\text{CONT}_{par} \text{ rename NAT by } r')$, where r' renames Nat into Elem. Let us prove that NATCONT-I of Example 11.45 is an implementation of NATCONT, i.e., $\text{NATCONT} \leadsto_{ps} \text{NATCONT-I}$:

1. $\text{CONT} \leadsto_{ps} \text{NATCONT-I}|_{inr}$
   (ref-basic), since $\text{NATCONT-I}|_{inr} \vdash_{\Pi_{ps}} \varphi$ for every $\varphi \in \text{Axs(CONT)}$; see Example 11.45
2. $\text{NAT} \leadsto_{ps} \text{rename } (\text{NATCONT-I}|_{inl}) \text{ by } (r')^{-1}$
   (ref-basic), since one can prove $\text{rename } (\text{NATCONT-I}|_{inl}) \text{ by } (r')^{-1} \vdash_{\Pi_{ps}} \varphi$ for every $\varphi \in \text{Axs(NAT)}$
3. $\text{rename NAT by } r' \leadsto_{ps} \text{NATCONT-I}|_{inl}$ (ref-ren), (2)
4. $(\text{rename NAT by } r') +_{\text{ELEM}} \text{CONT} \leadsto_{ps} \text{NATCONT-I}$ (ref-sum), (1,3)
5. $\text{NATCONT} \leadsto_{ps} \text{NATCONT-I}$ (ref-refl), (4)

                                                                   ■

11.6.4 Semantic typing

We can demand that an actual parameter exhibits a particular behavior in order to be a valid argument of a parameterized specification. Then the derivation of the refinement relation between specification λ-terms provides a means for verifying that an argument satisfies the semantic type imposed on the parameters of a parameterized specification.

First, we enhance the calculus of parameterized specifications by allowing an additional parameter restriction, accompanying the formal parameter variable of a λ-abstraction. Formally, we define

$$\Lambda_{(\tau_1 \to \tau_2)} ::= (\lambda \, Var_{\tau_1} : \Lambda_{\tau_1} . \Lambda_{\tau_2})$$

With respect to the semantics of this generalized λ-calculus, we have to take into account that the parameter restriction may be violated by an actual parameter of a λ-abstraction. Therefore, it is necessary to introduce a distinguished semantic value $\bot$ to denote erroneous λ-applications. Hence, the semantics of the base type sp is defined by $[\![sp]\!] = \text{Spec} \cup \{\bot\}$, and the semantics of arbitrary arrow types τ is inductively defined in the same way as in Section 11.6.1 over this new base case. The semantics of the term constructors that correspond to the specification-building operators (see Section 11.6.1) are extended in the obvious (strict) way. The semantics of a term $(\lambda X : R.M)$ of type $(\tau_1 \to \tau_2)$ in an environment e is a function from $[\![\tau_1]\!]$ to $[\![\tau_2]\!]$, which is now defined by

$$[\![(\lambda X : R.M)]\!]_e (S) \stackrel{\text{def}}{=} \begin{cases} [\![M]\!]_{e[X:=S]} & \text{if } [\![R]\!]_e \leadsto S \\ \bot_{\tau_2} & \text{otherwise,} \end{cases}$$

where $\bot_\tau$ is a "totally undefined" constant for each interpretation of types in T, inductively defined on the structure of τ by

$\perp_{sp} \in [\![sp]\!] \overset{\text{def}}{=} \perp$

$\perp_{(\tau_1 \to \tau_2)} \in [\![(\tau_1 \to \tau_2)]\!]$ is the constant function defined by $\perp_{(\tau_1 \to \tau_2)}(S) \overset{\text{def}}{=} \perp_{\tau_2}$ for each $S \in [\![\tau_1]\!]$.

The semantics of basic specification terms, variables, and application have the same (inductive) definition as in Section 11.6.1.

Example 11.49. (a) A parameterized specification of sorted lists in **FOLEq** can be defined by $\text{SORT}_{par} \overset{\text{def}}{=} (\lambda X : \text{ORD}.X +_{\text{RELEM}} \text{SORT})$, where RELEM is the signature containing the sort **Elem** and a binary relation R over **Elem**, SORT is the usual specification of lists over elements of sort **Elem** supplied with a sorting function, and ORD is the specification with signature RELEM and the axioms of partial order for R. The parameter restriction ensures that any actual parameter will be equipped with a partial order, so that it makes sense to compute the (topological) sort of a list of elements of sort **Elem**.

(b) It is possible to require that one parameter refines another, by writing, for instance, $(\lambda X.(\lambda Y : X.M))$. ∎

The introduction of parameter restrictions requires the redefinition of the notions of β-reduction and β-equality of Definition 11.38; we have to state as base case the following rule of *conditional* β-relation:

(cond-beta) $\dfrac{R \leadsto_{ps} N}{((\lambda X : R.M)\ N) \to_\beta M[X := N]}$

This means that, in the same way as in [Fei89], a step of β-reduction is only possible if the parameter restriction associated with the λ-abstraction can be derived for the actual parameter. Note that Lemma 11.39 remains valid with this new definition of β-equality.

In order to take into account parameter restrictions (i.e., semantic typing) in the proof systems, we replace the axiom of β-reduction (see Definition 11.38) with the rule (cond-beta), and generalize the systems Π_{ps} and $\Pi_{\leadsto_{ps}}$ to terms of the extended language (whereby the axioms and rules remain unchanged). Note that the systems are now mutually dependent, since as before $\Pi_{\leadsto_{ps}}$ uses proofs of Π_{ps} (see rule (ref-basic)) and Π_{ps} uses proofs of β-equality (see rule (prf-refl)), for which proofs of refinements are now necessary (see rule (cond-beta)). These derivations of refinement relations may be considered as parameter-passing verification.

The soundness results of Theorems 11.44 and 11.47 carry over to the new system. Moreover, if the conditions for completeness of Π_s (see the discussions in Sections 11.4 and 11.6.2) are satisfied, then the new system is also complete (for ground terms of base type).

Remark 11.50. Similarly to Remark 11.43, if M is a λ-abstraction of the form $(\lambda X : R.M')$, then for proving a theorem φ over M it is now sufficient to prove

$(M \ N) \vdash_{\Pi_{ps}} \varphi$ for all correct arguments N of M. For this it is enough to prove $M'[X := R] \vdash_{\Pi_{ps}} \varphi$, because $R \rightsquigarrow_{ps} N$ holds for any correct parameter N of M and hence, by monotonicity, $M'[X := N] \vdash_{\Pi_{ps}} \varphi$. Thus, by (prf-refl), $((\lambda X : R.M') \ N) \vdash_{\Pi_{ps}} \varphi$. This means φ is a theorem of any (valid) application of $(\lambda X : R.M)$.

Further Extensions. The drawback of the parameter restrictions in their present form is that they require an actual parameter to match *exactly* the signature of the parameter restriction. This can be relaxed by requiring an argument to implement the parameter restriction R when restricted to the signature of R, i.e., if $[\![R]\!]_e \rightsquigarrow S|_{\mathrm{Sig}([\![R]\!]_e)}$ then $[\![(\lambda X : R.M)]\!]_e(S) \stackrel{\mathrm{def}}{=} [\![M]\!]_{e_{[X:=S]}}$. Therefore, in order to be able to derive that an actual parameter conforms to a parameter restriction, we have to compute the signature of parameter restrictions and thus further terms denoting signatures are necessary.

If we have a signature base type, say sg, and terms denoting signatures, then we could use them in those places where a signature is necessary, for instance, $M|_N$ if M is of type sp and N of type sg. Soon we discover that specification-building operators also need to become signature-building operators, and the function Sig has to play the role of a term constructor as well. For a detailed study, see [Cen94].

In summary, the defined λ-calculus, together with the extensions discussed, supplies a powerful typing mechanism whose syntactic angle is relaxed with respect to previous approaches and its semantic angle is given more expressiveness. By permitting the derivation of syntactic and semantic typing, parameter verification is supported.

An objective of further research is an extension of Π_{ps} and $\Pi_{\rightsquigarrow_{ps}}$ such that it will be possible to derive $M \vdash_{\Pi_{ps}} \varphi$, and $M \rightsquigarrow_{ps} N$, for M and N not necessarily of type sp but of the same arbitrary type.

11.7 Reachability and observability operators

The kernel specification language considered in Sections 11.2–11.5 will now be extended in two directions. On the one hand, we will consider a *reachability* specification-building operator, able to express *reachability constraints* (see Chapter 5, Section 4). On the other hand, we will consider two new operators, *observational behavior* and *observational quotient*, that are necessary to express properties related to the observable behavior of software systems (see Chapter 5, Section 4). The motivations for considering observability operators are strongly related to the refinement relation between specifications as discussed in Section 11.5. Indeed, in contrast with the idea that requirements should describe those properties which have to be satisfied by implementations, many classical refinements of, for instance, set-like data structures do not preserve all requirements axioms. A careful analysis shows that only the observable consequences of the axioms are preserved. This means that instead of attempting to prove that a specification SPI is a refinement of a given requirements specification SP (an attempt that will often fail in practice), it is much more satisfactory to attempt to prove that the specification SPI is a refinement of the specification "behavior of SP", where the latter is assumed to capture the observable consequences of the requirements provided by SP. For a more detailed discussion of the refinement relation and the related problems, see Chapter 7.

11.7.1   Extension of the kernel specification language

In order to define the new specification-building operators, we will assume that specifications are built in the framework of the **FOLEq** institution (see Example 11.6, Section 11.2.1); unfortunately it is not convenient to define these operators in the framework of an arbitrary institution, since we will rely heavily on the (internal) structure of the signatures considered. However, note that even if for technical reasons we work within the framework of the **FOLEq** institution, nevertheless we assume, in addition, that all specification signatures are countable and contain no predicate symbols, i.e., they are of the form (S, F), with S and F countable.[21] To define the new specification-building operators, we first need some preliminary definitions.

Definition 11.51 (Preliminary technical definitions). Let $\Sigma = (S, F)$ be a signature.

1. The *reachability constraint* over Σ induced by a subset $\Omega \subseteq F$ of function symbols is the pair $\mathcal{R}_\Omega = (S_\Omega, \Omega)$, with $S_\Omega = \{s \in S \mid$ there exist $f \in \Omega$ with range$(f) = s\}$. A sort $s \in S_\Omega$ is called a *constrained sort* and a function symbol $f \in \Omega$ is called a *constructor symbol* (or briefly constructor).

2. A *constructor term* of sort s is a term $t \in T_{\Sigma'}(X')_s$, where $\Sigma' = (S, \Omega)$, $X' = (X'_s)_{s \in S}$ with $X'_s = X_s$, if $s \in S \backslash S_\Omega$, and $X'_s = \emptyset$, if $s \in S_\Omega$ (thereby $X = (X_s)_{s \in S}$ is the generally assumed family of countably infinite sets X_s of variables of sort $s \in S$). The set of constructor terms is denoted by T_Σ^Ω.

3. A Σ-algebra A satisfies the reachability constraint induced by Ω, denoted by $A \models \mathcal{R}_\Omega$, if, for all $s \in S$ and for all $a \in A_s$, there exists a constructor term $t \in T_{\Sigma'}(X')_s$ of sort s and a valuation $\alpha \colon X' \to A$, such that $I_\alpha(t) = a$. (Thereby $I_\alpha \colon T_{\Sigma'}(X') \to A$ denotes the unique Σ'-morphism

---

[21] Indeed the reachability specification-building operator could be easily defined for specifications containing predicate symbols. The assumption that specification signatures do not contain predicate symbols is necessary for the observability operators, since otherwise the definition of the observational equality (see Chapter 5, Section 4) would have to be extended to take into account predicate symbols. However, note that predicate symbols, though not used in specification signatures, will be used later on when defining the hidden symbols of a specification.

that extends the valuation α. Note that the definition is independent of X since X_s is countably infinite for all $s \in S$.)

4. Let In $\subseteq S$ be a set of sorts. The signature Σ is *sensible* w.r.t. In if for all $s \in S \setminus$ In (and hence for all $s \in S$), there exists a term t of sort s built with function symbols in Σ and variables of sort $s \in$ In.

5. Let Obs and In be two sets of sorts such that Obs $\subseteq S$ and In $\subseteq S$, and such that Σ is sensible w.r.t. In. Then for any Σ-algebra A, $\approx_{\mathrm{Obs,In},A}$ denotes the *(partial) observational equality* on A induced by Obs and In (see Chapter 5, Section 4).

Beh$_{\approx_{\mathrm{Obs,In}}}$ is the corresponding *observational behavior operator* defined by, for any class C of Σ-algebras:

Beh$_{\approx_{\mathrm{Obs,In}}}(C) \stackrel{\mathrm{def}}{=} \{A \in \mathrm{Alg}(\Sigma) \mid A/\approx_{\mathrm{Obs,In},A} \in C\}$.

$./\approx_{\mathrm{Obs,In}}$ is the corresponding *observational quotient operator* defined by, for any class C of Σ-algebras: $C/\approx_{\mathrm{Obs,In}} \stackrel{\mathrm{def}}{=} \{A/\approx_{\mathrm{Obs,In},A} \mid A \in C\}$. ◇

We now extend the kernel specification language defined in Section 11.2.2 by new specification-building primitives:

Definition 11.52 (Structured specifications). Definition 11.9 (instantiated by the specific institution **FOLEq**) is extended by the following specification-building primitives:

5. If SP is a specification expression, and if Ω is a set of function symbols such that $\Omega \subseteq \mathrm{Opns}(\mathrm{SP})$, then **reach** SP **w.r.t.** Ω is a specification expression.

   Sig(**reach** SP **w.r.t.** Ω) $\stackrel{\mathrm{def}}{=}$ Sig(SP)

   Mod(**reach** SP **w.r.t.** Ω) $\stackrel{\mathrm{def}}{=} \{A \in \mathrm{Mod}(\mathrm{SP}) \mid A \models \mathcal{R}_\Omega\}$

   where $\mathcal{R}_\Omega$ is the reachability constraint over Sig(SP) induced by Ω.

6. If SP is a specification expression, if Obs and In are two sets of sorts such that Obs $\subseteq$ Sorts(SP) and In $\subseteq$ Sorts(SP), and if, moreover, Sig(SP) is sensible w.r.t. In, then **behavior** SP **w.r.t.** Obs, In is a specification expression.

   Sig(**behavior** SP **w.r.t.** Obs, In) $\stackrel{\mathrm{def}}{=}$ Sig(SP)

   Mod(**behavior** SP **w.r.t.** Obs, In) $\stackrel{\mathrm{def}}{=}$ Beh$_{\approx_{\mathrm{Obs,In}}}$(Mod(SP))

   where Beh$_{\approx_{\mathrm{Obs,In}}}$ is the observational behavior operator defined above.

7. Under the same assumptions as in 6 above, SP$/\approx_{\mathrm{Obs,In}}$ is a specification expression.

   Sig(SP$/\approx_{\mathrm{Obs,In}}$) $\stackrel{\mathrm{def}}{=}$ Sig(SP)

   Mod(SP$/\approx_{\mathrm{Obs,In}}$) $\stackrel{\mathrm{def}}{=}$ Iso(Mod(SP)$/\approx_{\mathrm{Obs,In}}$)

   where Iso closes the given class under isomorphism, and $./\approx_{\mathrm{Obs,In}}$ is the observational quotient operator defined above.

An observational abstraction specification-building primitive [SW83, ST88b] can be defined as a derived operator, using the relationships between observational behavior and observational abstraction (see Chapter 5):

   abstract SP **w.r.t.** Obs, In $\stackrel{\mathrm{def}}{=}$ **behavior** (SP$/\approx_{\mathrm{Obs,In}}$) **w.r.t.** Obs, In. ◇

Unfortunately it is impossible to define sound and complete *finitary* proof systems for specifications built with reachability and observability operators in the framework of the **FOLEq** institution. We will therefore study below two alternatives: the first is to use infinitary proof rules, hence defining semi-formal proof systems. The second is to use infinitary sentences as nonlogical axioms, hence defining the proof systems in the framework of the **IFOLEq** institution (see Example 11.6). In both cases we will use the fact that there is a natural "embedding" of the **FOLEq** institution into the **IFOLEq** institution. As a consequence, any specification SP of our specification language (defined in the framework of the **FOLEq** institution) can also be considered as a specification of the corresponding specification language defined in the framework of the **IFOLEq** institution. The important point is that, while the reachability and observability operators are proper new operators when working in the framework of the **FOLEq** institution (since they increase the expressive power of the kernel specification language), this is no longer the case when we work in the framework of the **IFOLEq** institution: then the three new operators can indeed be considered as derived operators of the kernel specification language, as explained in the next subsection.

11.7.2   Reachability and observability operators as derived operators

In this subsection we show that, in the framework of the **IFOLEq** institution, the reachability, observational behavior, and observational quotient operators can be considered as derived operators of the specification language.

The reachability operator.

Proposition 11.53. *Let* SPR *be a specification expression of the form* **reach** SP *w.r.t.* Ω, *and let* $\Sigma = (S, F) = \text{Sig}(\text{SP})$ *(hence* $\Omega \subseteq F$*). Then* SPR *is equivalent to the specification* $\text{SP} + \langle \Sigma, \{\text{GEN}_s^{\Omega} \mid s \in S_{\Omega}\}\rangle$, *where* S_{Ω} *is defined as in Definition 11.51(1) and where* GEN_s^{Ω} *is the infinitary sentence defined by* $\text{GEN}_s^{\Omega} \overset{\text{def}}{=} \forall x\!:\!s.\ \bigvee_{t \in (T_{\Sigma}^{\Omega})_s} \exists \text{Var}(t).\ x = t$. *Thereby* $\exists \text{Var}(t)$ *is an abbreviation for* $\exists x_1 : s_1 \dots \exists x_n : s_n$ *where* $x_1, \dots, x_n$ *are the variables (of sort* $s_1, \dots, s_n$*) of the term* t.

Proof. A Σ-algebra A is a model of SPR if and only if A is a model of SP and, moreover, $A \models \mathcal{R}_{\Omega}$. But $A \models \mathcal{R}_{\Omega}$ if and only if $A \models \text{GEN}_s^{\Omega}$, for each $s \in S_{\Omega}$. □

For each sort $s \in S_{\Omega}$, the infinitary disjunction GEN_s^{Ω} expresses exactly that all elements of sort s are denotable by some constructor term $t \in (T_{\Sigma}^{\Omega})_s$.[22]

---
[22] Infinitary formulas of this kind were used in [MVS85].

The observational behavior operator. The observational behavior operator can also be considered as a derived operator of the kernel specification language, provided we allow predicate symbols in the signatures of specifications and, again, infinitary axioms. The main idea is to introduce on the one hand unary predicate symbols T_s, together with an appropriate axiomatization $\mathrm{Ax}[T_s]$, such that, for each algebra A satisfying $\mathrm{Ax}[T_s]$, the interpretation of T_s in A is just the domain of $\approx_{\mathrm{Obs,In},A}$ (w.r.t. the sort s). On the other hand, we introduce binary predicate symbols $\sim_s$ and an axiomatization $\mathrm{Ax}[\sim_s]$, such that $\sim_s$ denotes the observational equality $\approx_{\mathrm{Obs,In},A}$ for each algebra A.[23] Since in the definition of the observational behavior operator, Σ-algebras A and their observational quotient $A/\approx_{\mathrm{Obs,In},A}$ are related, the next idea is to encode both algebras within a single structure. For describing the connection between any Σ-algebra A and its quotient $A/\approx_{\mathrm{Obs,In},A}$, we first consider them as algebras over different (isomorphic) signatures Σ and $\mathrm{Copy}(\Sigma)$, and we specify an appropriate "pseudo" epimorphism c which connects sorts of Σ with their copy in $\mathrm{Copy}(\Sigma)$.

Proposition 11.54. *Let* SPB *be a specification expression of the form* behavior SP *w.r.t.* Obs, In, *and let* $\Sigma = (S, F)$ *be the signature of* SP *(hence also of* SPB*). We have* Obs $\subseteq S$, In $\subseteq S$, *and the signature* Σ *is sensible w.r.t.* In. SPB *is equivalent to the specification* $\mathrm{SP}'|_\Sigma$, *where* SP' *is defined by:*

$\mathrm{SP}' = (\text{rename SP by Copy}) +$
$\quad \langle\, (S \cup \mathrm{Copy}(S),\ F \cup \mathrm{Copy}(F) \cup \{T_s \mid s \in S \setminus \mathrm{In}\} \cup \{\sim_s, c_s \mid s \in S\}),$
$\quad \{\mathrm{Ax}[T_s] \mid s \in S \setminus \mathrm{In}\} \cup \{\mathrm{Ax}[\sim_s] \mid s \in S\} \cup \mathrm{Ax}[\{c_s \mid s \in S\}]\,\rangle$

where:

1. $\mathrm{Copy}(\Sigma)$ *is a disjoint copy of the signature* Σ *(i.e., we have* $\Sigma \cap \mathrm{Copy}(\Sigma) = \emptyset$*), and* $\mathrm{Copy}\colon \Sigma \to \mathrm{Copy}(\Sigma)$ *is the corresponding bijective signature morphism.*
2. *For each sort* $s \in S \setminus \mathrm{In}$, T_s *is a unary predicate symbol of domain* s, *intended to denote the values of sort* s *for which the partial observational equality is defined.*
3. *For each sort* $s \in S$, $\sim_s$ *is an infix binary predicate symbol of domain* s, s, *intended to denote the values of sort* s *which are observationally equal.*
4. *For each sort* $s \in S$, $c_s\colon s \to \mathrm{Copy}(s)$ *is a new function symbol (i.e.,* $c_s \notin \Sigma \cup \mathrm{Copy}(\Sigma)$*) connecting the sort* s *and its copy* $\mathrm{Copy}(s)$.
5. *For each sort* $s \in S \setminus \mathrm{In}$, $\mathrm{Ax}[T_s]$ *is the infinitary sentence defined by:*

$$\mathrm{Ax}[T_s] \stackrel{\mathrm{def}}{=} \forall x\colon s.\ \Big[T_s(x) \Leftrightarrow \bigvee_{t \in T_\Sigma(X_{\mathrm{In}})_s} \exists \mathrm{Var}(t).\, x = t\Big],$$

   where $(X_{\mathrm{In}})_s \stackrel{\mathrm{def}}{=} X_s$, *if* $s \in \mathrm{In}$, *and* $(X_{\mathrm{In}})_s \stackrel{\mathrm{def}}{=} \emptyset$, *if* $s \in S \setminus \mathrm{In}$.
6. *For each sort* $s \in S$, $\mathrm{Ax}[\sim_s]$ *is the infinitary sentence defined by:*

$$\mathrm{Ax}[\sim_s] \stackrel{\mathrm{def}}{=} \forall x, y\colon s.\ \Big[x \sim_s y \Leftrightarrow \bigwedge_{c \in T_\Sigma(X_{\mathrm{In}} \cup \{z_s\})_\bullet} \forall \mathrm{Var}(c).\, c[x] = c[y]\Big]$$

   if $s \in \mathrm{In}$,

---

[23] Another approach which axiomatizes the observational equality in higher-order logic is presented in [HS96].

$\text{Ax}[\sim_s] \stackrel{\text{def}}{=} \forall x, y : s.$
$$\left[x \sim_s y \Leftrightarrow T_s(x) \wedge T_s(y) \wedge \bigwedge_{c \in T_\Sigma(X_{\text{In}} \cup \{z_s\})} \forall \text{Var}(c). c[x] = c[y] \right]$$
if $s \in S \setminus \text{In}$,
where $T_\Sigma(X_{\text{In}} \cup \{z_s\})$ is the set of the observable contexts, with context variable z_s of sort s (see Chapter 5, Section 4), $\text{Var}(c)$ denotes the set of the variables of the context c, apart from the context variable z_s, and $c[x]$ and $c[y]$ denote respectively the terms obtained by substituting the variables x and y for the context variable z_s in c.

7. $\text{Ax}[\{c_s \mid s \in S\}] \stackrel{\text{def}}{=} [\text{Hom}] \cup [\text{Epi}] \cup [\sim=]$, where:
   [Hom] is the union, for each function symbol f
      with profile $s_1, \ldots, s_n \to s$ in F, of the sentence:
      $\forall x_1 : s_1. \ldots .\forall x_n : s_n. T_{s_{r_1}}(x_{r_1}) \wedge \ldots \wedge T_{s_{r_k}}(x_{r_k}) \Rightarrow$
      $\qquad c_s(f(x_1, \ldots, x_n)) = \text{Copy}(f)(c_{s_1}(x_1), \ldots, c_{s_n}(x_n))$,
      where $s_{r_1}, \ldots, s_{r_k}$ are the non input sorts among the argument sorts of f,
   [Epi] is the union of:
   - for each sort $s \in \text{In}$, the sentence:
     $\forall x' : \text{Copy}(s). \exists x : s. x' = c_s(x)$,
   - for each sort $s \in S \setminus \text{In}$, the sentence:
     $\forall x' : \text{Copy}(s). \exists x : s. T_s(x) \wedge x' = c_s(x)$,
   [$\sim=$] is the union of:
   - for each sort $s \in \text{In}$, the sentence:
     $\forall x, y : s. x \sim_s y \Leftrightarrow c_s(x) = c_s(y)$,
   - for each sort $s \in S \setminus \text{In}$, the sentence:
     $\forall x, y : s. [T_s(x) \wedge T_s(y) \Rightarrow (x \sim_s y \Leftrightarrow c_s(x) = c_s(y))]$.

Proof. Let B be a model of SP', and let $A \stackrel{\text{def}}{=} B|_\Sigma$ (hence A is a model of SP'$|_\Sigma$). The axioms $\text{Ax}[T_s]$ ensure that the interpretations of the predicate symbols T_s in B coincide with $\text{Dom}(\approx_{\text{Obs,In},A})$. The axioms $\text{Ax}[\sim_s]$ ensure that the interpretations of the predicate symbols $\sim_s$ in B coincide with $\approx_{\text{Obs,In},A}$. The axioms $\text{Ax}[\{c_s \mid s \in S\}]$ ensure that the interpretations of the function symbols c_s in B can be considered as defining a surjective Σ-morphism from $\text{Dom}(\approx_{\text{Obs,In},A})$ to $(B|_{\text{Copy}(\Sigma)})|_{\text{Copy}}$, the kernel of which coincides with the kernel of the quotient projection from $\text{Dom}(\approx_{\text{Obs,In},A})$ to $A/\approx_{\text{Obs,In},A}$. Hence $(B|_{\text{Copy}(\Sigma)})|_{\text{Copy}}$ (which, by definition of SP', is a model of SP) is isomorphic to $A/\approx_{\text{Obs,In},A}$. Hence models of SP' are exactly those algebras comprising a Σ-part A and a $\text{Copy}(\Sigma)$-part which, when restricted along Copy to Σ, is both a model of SP and isomorphic to $A/\approx_{\text{Obs,In},A}$. This is enough to show that SPB and SP'$|_\Sigma$ are equivalent specifications (for more details, see [Hen97]). $\qquad \square$

The observational quotient operator. To show that the observational quotient operator is a derived operator of the kernel specification language, we use exactly the same ideas as for the observational behavior operator, but now in the "reverse" direction.

Proposition 11.55. *Let* SPQ *be a specification expression of the form* SP/$\approx_{\text{Obs,In}}$, *and let* $\Sigma = (S, F)$ *be the signature of* SP *(hence also of* SPQ*). We have* Obs $\subseteq S$, In $\subseteq S$, *and the signature* Σ *is sensible w.r.t.* In. SPQ *is equivalent to the specification* SP$'|_{\Sigma}$, *where* SP$'$ *is defined by:*
SP$'$ = (**rename** SP **by** Copy) +
$\quad\quad \langle \, (\; S \cup \text{Copy}(S), \;\; F \cup \text{Copy}(F) \cup \{T_{\text{Copy}(s)} \mid s \in S \setminus \text{In}\} \cup$
$\quad\quad\quad \{\sim_{\text{Copy}(s)}, c_{\text{Copy}(s)} \mid s \in S\} \,),$
$\quad\quad\quad \{ \; \text{Ax}[T_{\text{Copy}(s)}] \mid s \in S \setminus \text{In}\} \cup \{\text{Ax}[\sim_{\text{Copy}(s)}] \mid s \in S\} \cup$
$\quad\quad\quad \text{Ax}[\{c_{\text{Copy}(s)} \mid s \in S\}] \, \rangle$

where:

1. *As for the observational behavior operator,* $\text{Copy}(\Sigma)$ *is a disjoint copy of the signature* Σ, *and* Copy: $\Sigma \to \text{Copy}(\Sigma)$ *is the corresponding bijective signature morphism.*
2. *For each sort* $s \in S \setminus \text{In}$, $T_{\text{Copy}(s)}$ *is a unary predicate symbol of domain* $\text{Copy}(s)$ *intended to denote the values of sort* $\text{Copy}(s)$ *for which the partial observational equality (w.r.t.* $\text{Copy}(\text{Obs})$ *and* $\text{Copy}(\text{In})$*) is defined.*
3. *For each sort* $s \in S$, $\sim_{\text{Copy}(s)}$ *is an infix binary predicate symbol of domain* $\text{Copy}(s), \text{Copy}(s)$ *intended to denote the values of sort* $\text{Copy}(s)$ *which are observationally equal (w.r.t.* $\text{Copy}(\text{Obs})$ *and* $\text{Copy}(\text{In})$*).*
4. *For each sort* $s \in S$, $c_{\text{Copy}(s)} : \text{Copy}(s) \to s$ *is a new function symbol (i.e.,* $c_{\text{Copy}(s)} \notin \Sigma \cup \text{Copy}(\Sigma)$*) connecting the sort* $\text{Copy}(s)$ *to* s.
5. *For each sort* $s \in S \setminus \text{In}$, $\text{Ax}[T_{\text{Copy}(s)}]$ *is the same infinitary sentence as* $\text{Ax}[T_s]$, *but now with* $\text{Copy}(\Sigma)$, $\text{Copy}(\text{In})$, *and* $\text{Copy}(s)$ *instead of* Σ, In, *and* s *respectively.*
6. *For each sort* $s \in S$, $\text{Ax}[\sim_{\text{Copy}(s)}]$ *is the same infinitary sentence as* $\text{Ax}[\sim_s]$, *but now with* $\text{Copy}(\Sigma)$, $\text{Copy}(\text{In})$, *and* $\text{Copy}(s)$ *instead of* Σ, In, *and* s *respectively.*
7. $\text{Ax}[\{c_{\text{Copy}(s)} \mid s \in S\}]$ *is defined similarly to* $\text{Ax}[\{c_s \mid s \in S\}]$, *but w.r.t. the function symbols* $c_{\text{Copy}(s)}$, *instead of the function symbols* c_s.

Proof. Similar to that of Proposition 11.54. It is easy to show that the models of SP$'$ comprise a $\text{Copy}(\Sigma)$-part A^{Copy}, which is simply a renamed version of a model A of SP, and a Σ-part isomorphic to $A/\approx_{\text{Obs,In},A}$. This is sufficient to show that SPQ and SP$'|_{\Sigma}$ are equivalent specifications.   □

11.7.3   Noncompositional proof systems

Propositions 11.53–11.55 show that the reachability and observability operators can be considered as derived operators of the kernel specification language, in the framework of the **IFOLEq** institution. Using these results, it is easy to define, for each specification SP of our specification language (defined in the framework of the **FOLEq** institution), an equivalent normal form nf(SP) which, however, is then a specification with infinitary axioms, and hence a specification in the framework of the **IFOLEq** institution. Then, using a sound and complete proof system $\Pi(\textbf{IFOLEq})$ for the institution **IFOLEq** (for instance, that presented in [Kei71]), we obtain a sound and complete proof method for proving theorems of SP (see Section 11.3.2).

We will now show how the noncompositional approach of Section 11.3.1, where an individual proof system is constructed for each specification expression, can be extended to take into account the reachability and observability operators, by introducing adequate additional (infinitary) proof rules. Let us stress that, for this noncompositional approach, we will stay strictly within the framework of the **FOLEq** institution (and hence we will just rely on the standard proof system $\Pi(\textbf{FOLEq})$ for first-order logic). In that case, the results of Propositions 11.53–11.55 cannot be used directly. However, they provide us with the necessary intuition about which additional rules are necessary: the idea we shall use is to "reflect" the infinitary axioms of the **IFOLEq** equivalent specifications by adequate infinitary proof rules. In the following, we study each new specification-building operator independently, and then we define the noncompositional proof system associated with each specification expression.

Rules for the reachability operator. We will define the proof system associated with **reach** SP w.r.t. Ω as being the proof system for SP suitably enriched by some (infinitary) rules which, intuitively speaking, reflect the axioms GEN_s^Ω. To achieve this, we define the infinitary induction rules associated with a signature Σ and a set of constructors Ω:

Definition 11.56 (Infinitary induction rules). Let $\Sigma = (S, F)$ be a signature and $\Omega \subseteq F$ be a set of constructors. The infinitary induction rules associated with Σ and Ω, denoted by $(\text{iI}_{\Sigma,\Omega})$, are given by the countable set $(\text{iI}_{\Sigma,\Omega}) \stackrel{\text{def}}{=} \{(\text{iI}_{\Sigma,\Omega,s}) \mid s \in S_\Omega\}$ of infinitary proof rules:

$$(\text{iI}_{\Sigma,\Omega,s}) \quad \frac{\varphi[t/x] \text{ for all } t \in (T_\Sigma^\Omega)_s}{\forall x : s. \, \varphi}$$

where φ is an arbitrary first-order formula built with arbitrary symbols and with at least a free variable x of sort s.   ◇

The infinitary induction rules $(\text{iI}_{\Sigma,\Omega})$ express that a formula $\forall x : s. \, \varphi$ is derivable if all instantiations $\varphi[t/x]$, where the variable x is replaced by the constructor terms t, are derivable. The soundness of the infinitary induction rules is obvious (compare with the infinitary sentences GEN_s^Ω).

Rules for the observability operators. We can use Proposition 11.54 to understand how to derive a proof system for **behavior** SP w.r.t. Obs, In from a proof system for SP: we will add to the proof system for **rename** SP **by** Copy (obtained from the proof system for SP, as explained in Section 11.3.1) new hidden predicate symbols T_s, $\sim_s$, new hidden function symbols c_s, the finitary axioms $\text{Ax}[\{c_s \mid s \in S\}]$, and new finitary axioms and infinitary rules that will reflect the infinitary axioms $\text{Ax}[T_s]$ and $\text{Ax}[\sim_s]$.

Definition 11.57 (Infinitary rules for the observational equality).
Let $\Sigma = (S, F)$ be a signature and Obs, In $\subseteq S$ (such that Σ is sensible w.r.t. In). Let $\{\sim_s : s\ s \mid s \in S\}$ be a set of binary predicate symbols and let $\{T_s : s \mid s \in S \setminus \text{In}\}$ be a set of unary predicate symbols.

1. $(T_{\Sigma,\text{In}})$ is the following countable set of axioms:
   $$\{T_s(t) \mid t \in T_\Sigma(X_{\text{In}})_s,\ s \in S \setminus \text{In}\}$$

2. $(iT_{\Sigma,\text{In}})$ is the countable set $(iT_{\Sigma,\text{In}}) \stackrel{\text{def}}{=} \{(iT_{\Sigma,\text{In},s}) \mid s \in S \setminus \text{In}\}$ of infinitary proof rules for relativized induction:
   $$(iT_{\Sigma,\text{In},s}) \quad \frac{\varphi[t/x] \text{ for all } t \in T_\Sigma(X_{\text{In}})_s}{\forall x : s.\, T_s(x) \Rightarrow \varphi}$$

   where φ is an arbitrary first-order formula built with arbitrary symbols and with at least a free variable x of sort s.

3. $(iCI_{\sim,\Sigma,\text{Obs},\text{In}})$ is the countable set $(iCI_{\sim,\Sigma,\text{Obs},\text{In}}) \stackrel{\text{def}}{=} \{(iCI_{\sim,\Sigma,\text{Obs},\text{In},s}) \mid s \in S\}$ of infinitary context induction proof rules, defined as follows.
   If $s \in \text{In}$ then:
   $$(iCI_{\sim,\Sigma,\text{Obs},\text{In},s}) \quad \frac{\varphi \vee \forall\text{Var}(c).\, c[x] = c[y] \text{ for all } c \in T_\Sigma(X_{\text{In}} \cup \{z_s\})}{\varphi \vee x \sim_s y}$$
   If $s \in S \setminus \text{In}$ then $(iCI_{\sim,\Sigma,\text{Obs},\text{In},s})$ is the rule:
   $$\frac{\varphi \vee (T_s(x) \wedge T_s(y) \wedge \forall\text{Var}(c).\, c[x] = c[y]) \text{ for all } c \in T_\Sigma(X_{\text{In}} \cup \{z_s\})}{\varphi \vee x \sim_s y}$$

   In the above rules, φ denotes an arbitrary first-order formula.

4. $(\text{Comp}_{\sim,\Sigma,\text{Obs},\text{In}})$ is the following countable set of compatibility axioms:
   $$\{\forall x, y : s.\, x \sim_s y \Rightarrow c[x] = c[y] \mid s \in S,\ c \in T_\Sigma(X_{\text{In}} \cup \{z_s\})\} \cup$$
   $$\{\forall x, y : s.\, x \sim_s y \Rightarrow T_s(x) \wedge T_s(y) \mid s \in S \setminus \text{In}\} \qquad \diamond$$

The axioms $(T_{\Sigma,\text{In}})$, together with the rules $(iT_{\Sigma,\text{In}})$, reflect the infinitary axioms $\text{Ax}[T_s]$, and the rules $(iCI_{\sim,\Sigma,\text{Obs},\text{In}})$, together with the axioms $(\text{Comp}_{\sim,\Sigma,\text{Obs},\text{In}})$, reflect the infinitary axioms $\text{Ax}[\sim_s]$.

Similarly, the proof system for $SP/\approx_{\text{Obs},\text{In}}$ will be derived from the proof system for **rename** SP **by** Copy by adding new hidden predicate symbols $T_{\text{Copy}(s)}$, $\sim_{\text{Copy}(s)}$, new hidden function symbols $c_{\text{Copy}(s)}$, the finitary axioms $\text{Ax}[\{c_{\text{Copy}(s)} \mid s \in S\}]$, and new finitary axioms and infinitary rules that will reflect the infinitary axioms $\text{Ax}[T_{\text{Copy}(s)}]$ and $\text{Ax}[\sim_{\text{Copy}(s)}]$. For this purpose, we again use the infinitary rules for the observational equality defined in Definition 11.57 (but now w.r.t. Copy(Σ), Copy(Obs), and Copy(In)).

Definition of the proof systems. We now have all the necessary ingredients to define the proof system associated with a structured specification that may contain reachability or observability operators. First, we define, as in Section 11.3.1, the set of all visible and hidden symbols of such specifications.

Definition 11.58 (Symbols of a structured specification). The inductive definition of the symbols of a specification (see Definition 11.14) is extended as follows:

5. If SP is a specification expression of the form **reach** SP' w.r.t. Ω,
   then Symbols(SP) $\overset{\text{def}}{=}$ Symbols(SP').
6. If SP is a specification expression of the form **behavior** SP' w.r.t. Obs, In,
   then Symbols(SP) $\overset{\text{def}}{=}$ PO(Copy, Sig(SP') $\hookrightarrow$ Symbols(SP')) $\cup$ Sig(SP')
   $\cup \{T_s \mid s \in S \setminus \text{In}\} \cup \{\sim_s, c_s \mid s \in S\}$.
7. If SP is a specification expression of the form SP'/$\approx_{\text{Obs,In}}$,
   then Symbols(SP) $\overset{\text{def}}{=}$ PO(Copy, Sig(SP') $\hookrightarrow$ Symbols(SP')) $\cup$ Sig(SP')
   $\cup \{T_{\text{Copy}(s)} \mid s \in S \setminus \text{In}\} \cup \{\sim_{\text{Copy}(s)}, c_{\text{Copy}(s)} \mid s \in S\}$.

Since pushout signatures are defined up to isomorphism, we can always choose the pushout signature used above in (6) and (7) to be a disjoint copy of Symbols(SP') which contains Copy(Sig(SP')). Note also that, although specification signatures do not contain predicate symbols, symbol signatures may now contain (hidden) predicate symbols. Without loss of generality, we assume that all newly introduced symbols are fresh.   ◇

The proof system associated with a specification expression can now be defined as follows:

Definition 11.59 (Extension of Π_{SP}). Definition 11.16 of the proof system Π_{SP} associated with a specification expression SP is extended as follows:

5. If SP is a specification expression of the form **reach** SP' w.r.t. Ω,
   then $\Pi_{\text{SP}} \overset{\text{def}}{=} \Pi_{\text{SP'}} \cup (\text{iI}_{\text{Sig(SP)},\Omega})$.
6. If SP is a specification expression of the form **behavior** SP' w.r.t. Obs, In,
   then $\Pi_{\text{SP}} \overset{\text{def}}{=}$ Copy$^\star(\Pi_{\text{SP'}}) \cup (\text{T}_{\text{Sig(SP)},\text{In}}) \cup (\text{iT}_{\text{Sig(SP)},\text{In}}) \cup$
   $(\text{iCI}_{\sim,\text{Sig(SP)},\text{Obs},\text{In}}) \cup (\text{Comp}_{\sim,\text{Sig(SP)},\text{Obs},\text{In}}) \cup \text{Ax}[\{c_s \mid s \in S\}]$.
7. If SP is a specification expression of the form SP'/$\approx_{\text{Obs,In}}$,
   then $\Pi_{\text{SP}} \overset{\text{def}}{=}$ Copy$^\star(\Pi_{\text{SP'}}) \cup (\text{T}_{\text{Copy(Sig(SP))},\text{Copy(In)}}) \cup$
   $\qquad (\text{iT}_{\text{Copy(Sig(SP))},\text{Copy(In)}}) \cup$
   $\qquad (\text{iCI}_{\sim,\text{Copy(Sig(SP))},\text{Copy(Obs)},\text{Copy(In)}}) \cup$
   $\qquad (\text{Comp}_{\sim,\text{Copy(Sig(SP))},\text{Copy(Obs)},\text{Copy(In)}}) \cup$
   $\qquad \text{Ax}[\{c_{\text{Copy}(s)} \mid s \in S\}]$.

where in (6) and (7), Copy$^\star \overset{\text{def}}{=}$ po$_2$(Copy, Sig(SP') $\hookrightarrow$ Symbols(SP')), i.e., Copy* produces a disjoint copy of Symbols(SP').   ◇

As in Section 11.3.1, for any specification SP, the proof system Π_{SP} is combined with the proof system $\Pi(\textbf{FOLEq})$ for deriving SP-sentences. We write $\vdash_{\Pi_{\text{SP}}} \varphi$ if φ is an SP-sentence such that $\vdash_{\Pi(\textbf{FOLEq})(\Sigma) \cup \Pi_{\text{SP}}} \varphi$, where $\Sigma = $ Symbols(SP). The next theorem states that by this method we obtain a sound and complete proof system for SP (since $\Pi(\textbf{FOLEq})$ is complete).

Theorem 11.60 (Soundness and completeness of the proof systems).
Let SP *be a specification expression and let* φ *be an* SP*-sentence.*
SP $\models_{\mathrm{Sig(SP)}} \varphi$ *if and only if* $\vdash_{\Pi_{\mathrm{SP}}} \varphi$.

For a proof (which relies on the extended omitting-types theorem for showing the completeness of the infinitary rules), see [HWB97].

11.7.4   Structured proof systems

Let us now discuss how to obtain a structured proof system, in the sense of Section 11.4, for the reachability and observability operators. Unfortunately, it is not yet known how to define a sound and complete structured proof system (even with infinitary rules) for our extended specification language in the framework of the **FOLEq** institution. Therefore we will again use the results established in Section 11.7.2 and work within the **IFOLEq** institution. In that case, we could derive directly from Propositions 11.53–11.55, proof rules for the reachability, observational behavior, and observational quotient operators. It seems therefore rather easy to obtain a sound and complete structured proof system for the extended specification language in the framework of the **IFOLEq** institution. However, let us stress that the structured proof system obtained in that way is perhaps not so useful from a practical point of view, since the proof rules for the reachability and observability operators are quite complex.[24] The aim of this subsection is to show that we can define "practically better" proof rules. For this we again study each specification-building operator separately.

The reachability operator. Let us consider a specification of the form **reach** SP **w.r.t.** Ω, and let $\Sigma = (S, F)$ be the signature of SP (hence $\Omega \subseteq F$). A direct application of Proposition 11.53 leads to the following proof rule:

$$(\mathrm{R}) \quad \frac{\mathrm{SP} + \langle \Sigma, \{\mathrm{GEN}_s^{\Omega} \mid s \in S_{\Omega}\}\rangle \vdash_{\Pi_s} \varphi}{\mathbf{reach}\ \mathrm{SP}\ \mathbf{w.r.t.}\ \Omega \vdash_{\Pi_s} \varphi}$$

Using the rule (sum-l), we derive from (R) the following rule:

$$(\mathrm{reach}) \quad \frac{\mathrm{SP} \vdash_{\Pi_s} \varphi}{\mathbf{reach}\ \mathrm{SP}\ \mathbf{w.r.t.}\ \Omega \vdash_{\Pi_s} \varphi}$$

Using the rules (sum-r) and (basic), we derive from (R) the following rules:

$$(\mathrm{reach\text{-}GEN}_s) \quad \mathbf{reach}\ \mathrm{SP}\ \mathbf{w.r.t.}\ \Omega \vdash_{\Pi_s} \mathrm{GEN}_s^{\Omega} \quad (\text{for each sort } s \in S_{\Omega})$$

---

[24] This criticism also applies to the noncompositional proof systems defined in the previous subsection.

It is intuitively clear that the set of rules $\{(\text{reach})\} \cup \{(\text{reach-GEN}_s) \mid s \in S_\Omega\}$ is equivalent to rule (R) above, and it will provide a set of sound and complete structured proof rules for the reachability operator. Now, to understand how this set of rules will allow us to prove theorems of specifications with reachability constraints, let us stress that the following rule is a derived rule of $\Pi(\textbf{IFOLEq})$, for each sort $s \in S_\Omega$:

$$(\text{iI}_s) \quad \frac{\text{GEN}_s^\Omega, \ \varphi[t/x] \text{ for all } t \in (T_\Sigma^\Omega)_s}{\forall x : s. \ \varphi}$$

where φ is an arbitrary formula composed of arbitrary symbols and at least a free variable x of sort s.

Hence, whenever we consider a specification expression SP' containing a subspecification of the form **reach** SP **w.r.t.** Ω, the sentences GEN_s^Ω can first be propagated to the larger specification SP' using the rules of the structured proof system of Definition 11.25, and then we can apply the infinitary induction rules (iI_s) to SP'. In practice, for proving the infinitely many hypotheses $\varphi[t/x]$ for all $t \in (T_\Sigma^\Omega)_s$, one would use an induction scheme like structural induction with respect to $(T_\Sigma^\Omega)_s$.

Example 11.61. As an example let us consider the usual specification NAT of the natural numbers with constructors $\Omega = \{0, \text{succ}\}$ and the following extension NAT1 of NAT:

    spec NAT1 = NAT + NATf

where

    spec NATf
        sorts Nat
        functions 0   : $\to$ **Nat**
                  succ, f : **Nat** $\to$ **Nat**
        axioms $\forall n : \text{Nat}. \ \text{f}(0) = 0 \wedge \text{f}(\text{succ}(n)) = \text{f}(n)$
    endspec

Let us prove that $\text{NAT1} \vdash_\Pi, \forall n : \text{Nat}. \ \text{f}(n) = 0$:

1. $\text{NATf} \vdash_\Pi, \forall n : \text{Nat}. \ \text{f}(0) = 0 \wedge \text{f}(\text{succ}(n)) = \text{f}(n)$ (basic)
2. $\text{NAT1} \vdash_\Pi, \forall n : \text{Nat}. \ \text{f}(0) = 0 \wedge \text{f}(\text{succ}(n)) = \text{f}(n)$ (sum-r), (1)
3. $\text{NAT1} \vdash_\Pi, \text{f}(t) = 0$ for all $t \in (T_{\text{Sig(NAT)}}^\Omega)_{\text{Nat}}$
                            by induction on the structure of the terms t, using (2)
4. $\text{NAT} \vdash_\Pi, \text{GEN}_{\text{Nat}}^\Omega$ (reach-GEN$_{\text{Nat}}$)
5. $\text{NAT1} \vdash_\Pi, \text{GEN}_{\text{Nat}}^\Omega$ (sum-l), (4)
6. $\text{NAT1} \vdash_\Pi, \forall n : \text{Nat}. \ \text{f}(n) = 0$ (pi) with (iI_{Nat}), using (5) and (3)

                                                                          ■

Remark 11.62. Let us stress that a more naive approach would lead to an incomplete set of structured proof rules. Indeed, instead of the rules (reach-GEN$_s$) above (which require the use of infinitary sentences), one could be tempted to try to avoid infinitary sentences by introducing just the following infinitary proof rules (for each $s \in S_\Omega$):

$$(\text{reach-iI}_s) \quad \frac{\text{reach SP w.r.t. } \Omega \vdash_{\Pi_s} \varphi[t/x] \text{ for all } t \in (T_\Sigma^\Omega)_s}{\text{reach SP w.r.t. } \Omega \vdash_{\Pi_s} \forall x:s.\varphi}$$

These rules are obviously sound, but unfortunately not complete. For instance, in the example above, it would not be possible to derive the desired theorem using just the rules (reach), (reach-iI$_s$), and the structured rules of Definition 11.25. The problem is that the rules (reach-iI$_{\text{Nat}}$) can only be applied to NAT, but not to the larger specification NAT1. This clearly demonstrates why it is unfortunately not possible to define the structured proof system for specifications with reachability constraints in the framework of the institution **FOLEq** (at least not in this way).

The observability operators. In Section 11.7.2 we defined an infinitary axiomatization of the observational equality $\approx_{\text{Obs,In}}$, using auxiliary predicate symbols T_s and $\sim_s$ and infinitary axioms $\text{Ax}[T_s]$ and $\text{Ax}[\sim_s]$ (i.e., again formulas of the institution **IFOLEq**), which characterize the observational equality $\approx_{\text{Obs,In},A}$ for each Σ-algebra A. We will again use this infinitary axiomatization, but in a different way compared to Proposition 11.55:

Proposition 11.63. *Let us consider a specification expression of the form* $SP/\approx_{\text{Obs,In}}$, *and let* $\Sigma = (S,F)$ *be the signature of SP. We have* $\text{Obs} \subseteq S$, $\text{In} \subseteq S$, *and the signature* Σ *is sensible w.r.t. In. Let* φ *be a closed* Σ-*formula. Let* $\mathcal{L}(\Sigma) \stackrel{\text{def}}{=} \Sigma \cup \{T_s \mid s \in S \setminus \text{In}\} \cup \{\sim_s \mid s \in S\}$ *and let* $\text{AX}[\mathcal{L}] \stackrel{\text{def}}{=} \{\text{Ax}[T_s] \mid s \in S \setminus \text{In}\} \cup \{\text{Ax}[\sim_s] \mid s \in S\}$ *(see Proposition 11.54).* $SP/\approx_{\text{Obs,In}} \models \varphi$ *if and only if* $SP + \langle \mathcal{L}(\Sigma), \text{AX}[\mathcal{L}] \rangle \models \mathcal{L}(\varphi)$, *where* $\mathcal{L}(\varphi)$ *is a syntactic translation of* φ *inductively defined as follows:*[25]

1. *If* φ *is an equation* $l = r$ *between two terms* l, r *of sort* s,

   then $\mathcal{L}(\varphi) \stackrel{\text{def}}{=} l \sim_s r$.

2. $\mathcal{L}(\neg\varphi) \stackrel{\text{def}}{=} \neg(\mathcal{L}(\varphi))$, $\mathcal{L}(\varphi \wedge \varphi') \stackrel{\text{def}}{=} \mathcal{L}(\varphi) \wedge \mathcal{L}(\varphi')$.

3. *If* $s \in \text{In}$, *then* $\mathcal{L}(\forall x:s.\varphi) \stackrel{\text{def}}{=} \forall x:s.\mathcal{L}(\varphi)$,

   and if $s \in S \setminus \text{In}$, *then* $\mathcal{L}(\forall x:s.\varphi) \stackrel{\text{def}}{=} \forall x:s.(T_s(x) \Rightarrow \mathcal{L}(\varphi))$.

For a proof see [BH96].

According to the above proposition, the proof of a theorem φ of a specification $SP/\approx_{\text{Obs,In}}$ can be reduced to the proof of a corresponding theorem, called "lifting" of φ and denoted by $\mathcal{L}(\varphi)$, with respect to an axiomatic extension of SP obtained by adding the predicate symbols T_s and $\sim_s$ and the axioms $\text{AX}[\mathcal{L}]$ to SP. (Hence this extension of SP is defined in the institution **IFOLEq**.) This leads to the following proof rule:

$$(\text{quot}) \quad \frac{SP + \langle \mathcal{L}(\Sigma), \text{AX}[\mathcal{L}] \rangle \vdash_{\Pi_s} \mathcal{L}(\varphi)}{SP/\approx_{\text{Obs,In}} \vdash_{\Pi_s} \varphi}$$

---

[25] Similar constructions, called relativizations, are used in [Wir93, HS96].

where $\Sigma = \mathrm{Sig}(SP)$ and $\mathcal{L}(\Sigma), AX[\mathcal{L}], \mathcal{L}(\varphi)$ are defined in Proposition 11.63.

For the observational behavior operator we use the fact that $\mathrm{Beh}_{\approx_{Obs,In}}$ and $./\approx_{Obs,In}$ form a Galois connection. Hence, for any specification SP, we know that $\mathrm{Mod}((\mathbf{behavior}\ SP\ \mathbf{w.r.t.}\ \mathrm{Obs,In})/\approx_{Obs,In}) \subseteq \mathrm{Mod}(SP)$. Therefore $SP \models \varphi$ implies $(\mathbf{behavior}\ SP\ \mathbf{w.r.t.}\ \mathrm{Obs,In})/\approx_{Obs,In} \models \varphi$. Using Proposition 11.63, this leads to the rule:

(beh-1)
$$\frac{SP \vdash_{\varPi_s} \varphi}{\mathbf{behavior}\ SP\ \mathbf{w.r.t.}\ \mathrm{Obs,In} + \langle \mathcal{L}(\Sigma), AX[\mathcal{L}]\rangle \vdash_{\varPi_s} \mathcal{L}(\varphi)}$$

where $\Sigma = \mathrm{Sig}(SP)$ and $\mathcal{L}(\Sigma), AX[\mathcal{L}], \mathcal{L}(\varphi)$ are defined in Proposition 11.63.

Now we remark that $(\mathbf{behavior}\ SP\ \mathbf{w.r.t.}\ \mathrm{Obs,In} + \langle \mathcal{L}(\Sigma), AX[\mathcal{L}]\rangle)|_{\Sigma}$ and $\mathbf{behavior}\ SP\ \mathbf{w.r.t.}\ \mathrm{Obs,In}$ are equivalent specifications. Hence, for Σ-formulas φ, we obtain the rule:

(beh-2)
$$\frac{\mathbf{behavior}\ SP\ \mathbf{w.r.t.}\ \mathrm{Obs,In} + \langle \mathcal{L}(\Sigma), AX[\mathcal{L}]\rangle \vdash_{\varPi_s} \varphi}{\mathbf{behavior}\ SP\ \mathbf{w.r.t.}\ \mathrm{Obs,In} \vdash_{\varPi_s} \varphi}$$

where $\Sigma = \mathrm{Sig}(SP)$, $\mathcal{L}(\Sigma), AX[\mathcal{L}]$ are defined in Proposition 11.63, and φ is a Σ-formula.

Intuitively, the idea behind the two proof rules (beh-1) and (beh-2) above is to derive theorems of $\mathbf{behavior}\ SP\ \mathbf{w.r.t.}\ \mathrm{Obs,In}$ by using lifted versions of the theorems of SP, and then the axioms $AX[\mathcal{L}]$ and the proof rules of the underlying proof system to eliminate the hidden symbols.

Definition of the structured proof system. The rules introduced in the previous subsection lead to a sound and complete proof system for specifications with reachability and observability operators. In the following, we assume that $\varPi_s$ denotes the structured proof system of Definition 11.25 with the underlying proof system $\varPi(\mathbf{IFOLEq})$ of infinitary first-order logic (see Example 11.6), extended by the proof rules (reach), (reach-GEN$_s$), (quot), (beh-1), and (beh-2).

Theorem 11.64 (Soundness and completeness of $\varPi_s$). *Let SP be a specification expression (with reachability and/or observability operators) such that the axioms of the basic specifications are closed formulas. Let φ be an SP-sentence. SP $\models_{\mathrm{Sig}(SP)} \varphi$ if and only if SP $\vdash_{\varPi_s} \varphi$.*

For a proof (which relies on a normal form construction in the institution **IFOLEq** and on the interpolation property), see [Hen97].

The above proof system $\varPi_s$ is based on infinitary first-order logic because not only the formulas GEN_s^{Ω} of (reach-GEN$_s$) but also the axioms $AX[\mathcal{L}]$, which describe the observational equality, are infinitary sentences. In order to simplify the rules (quot), (beh-1), and (beh-2), one can try to replace the infinitary axiomatization of the observational equality by a finitary one. Unfortunately, it is generally not possible to find a finitary axiomatization of the observational equality in first-order logic (see, e.g., [Wol87, Sch92]). However, it is shown in [BH96] that the observational equality can always be axiomatized by finitary axioms if we use auxiliary (hidden) sorts and functions symbols. As a consequence, the basic specification $\langle \mathcal{L}(\Sigma), AX[\mathcal{L}] \rangle$ occurring in the rules (quot), (beh-1), and (beh-2) can always be replaced by a specification of the form HID $+ \langle \mathcal{L}(\Sigma) \cup \mathrm{Sig}(\mathrm{HID}), \mathrm{FINAX}[\mathcal{L}] \rangle$, where HID is a specification of the **FOLEq** institution (possibly containing a reachability operator) and FINAX[$\mathcal{L}$] is a set of finitary first-order axioms which specify the observational equality with the help of the hidden sorts and function symbols of HID. Hence the use of the infinitary logic can be restricted to the formulas GEN_s^{Ω} which are themselves only necessary if reachability constraints occur in the specification of interest.

A further improvement can be obtained by noticing that we do not need a finitary axiomatization of the observational equality which is valid for arbitrary algebras, but only for the models of the underlying specification (see again [BH96]). This last remark allows us in particular to obtain finitary axiomatizations without a hidden part in many cases, so that we can replace the rule (quot) above by the rule

$$(\text{quot'}) \quad \frac{SP + \langle \mathcal{L}(\Sigma), \mathrm{FINAX}[\mathcal{L}] \rangle \vdash_{\Pi_s} \mathcal{L}(\varphi)}{SP/\approx_{\mathrm{Obs,In}} \vdash_{\Pi_s} \varphi}$$

provided FINAX[$\mathcal{L}$] is a correct finitary axiomatization of the observational equality with respect to the models of SP.[26]

Example 11.65. The following specification STATE-I can be considered as an observational implementation of environments (also called states) which assign natural numbers to identifiers (see Example 11.70). In the specification STATE-I a state is represented by a sequence of pairs consisting of an identifier and its associated value. The reachability constraint requires that states are constructed by the constant init, the operation $\langle .,. \rangle$ (forming the pair of an identifier and a natural number) and the operation .&. (concatenating two sequences of pairs). The specification STATE-I is built on top of the usual specification NAT of the natural numbers and a specification ID of identifiers. It exports the signature Σ-STATE which contains only the typical state operations init, update, and lookup.

---

[26] To achieve the same simplifications for the rules (beh-1) and (beh-2) would require a finitary axiomatization without a hidden part of the observational equality with respect to the models of **behavior** SP w.r.t. Obs, In which, in practice, is not easy to find.

```
spec STATE-SEQ0
    sorts Nat, Id, State
    functions init   :  → State
              ⟨.,.⟩   : Id, Nat → State
              .&.    : State, State → State
              update : Id, Nat, State → State
              lookup : Id, State → Nat
    axioms ∀x, y : Id,  n : Nat,  s, t, u : State.
           s&init = s ∧ init&s = s ∧
           (s&t)&u = s&(t&u) ∧
           update(x, n, s) = ⟨x, n⟩&s ∧
           lookup(x, ⟨x, n⟩&s) = n ∧
           (x ≠ y ⟹ lookup(x, ⟨y, n⟩&s) = lookup(x, s))
endspec

spec STATE-SEQ = reach NAT + ID + STATE-SEQ0 w.r.t. {init, ⟨.,.⟩, .&.}

spec STATE-I = STATE-SEQ|_{Σ-STATE}
where  Σ-STATE = Sig(STATE-SEQ) \ {⟨.,.⟩, .&.}
```

Consider the sorts Obs = {Nat, Id} to be observable and let all sorts be input sorts, i.e., In = Sorts(STATE-I). One can show, using the method described in [BH96], that it is "enough" to consider the "crucial" context $\text{lookup}(x, z_{State})$ to obtain a finitary axiomatization of the observational equality, valid for all models of STATE-I, with FINAX[$\mathcal{L}$] defined as follows:[27]

$$\text{FINAX}[\mathcal{L}] \overset{\text{def}}{=} \forall S_1, S_2 : \text{State}.$$
$$[S_1 \sim_{State} S_2 \Leftrightarrow (\forall x : \text{Id. lookup}(x, S_1) = \text{lookup}(x, S_2))].$$

Now consider the observational quotient specification STATE-I/$\approx_{\text{Obs,In}}$. We want to prove that the (implicitly universally quantified) equation $\text{update}(x, n, \text{update}(x, m, s)) = \text{update}(x, n, s)$ is valid in the observational quotient of STATE-I (although this equation is obviously not valid in STATE-I). For this we can first prove:

$$\text{STATE-I} + \langle \mathcal{L}(\text{Sig}(\text{STATE-I})), \text{FINAX}[\mathcal{L}] \rangle \vdash \text{update}(x, n, \text{update}(x, m, s))$$
$$\sim_{State} \text{update}(x, n, s)$$

using the proof rules of the structured proof system (which is not difficult). Then, we can apply the simplified rule (quot') and we obtain:

$$\text{STATE-I}/\approx_{\text{Obs,In}} \vdash \text{update}(x, n, \text{update}(x, m, s)) = \text{update}(x, n, s)$$

Similarly, we can prove that the update operation is commutative in the observational quotient of STATE-I, i.e., that:

$$\text{STATE-I}/\approx_{\text{Obs,In}} \vdash x \neq y \Rightarrow \text{update}(x, n, \text{update}(y, m, s)) =$$
$$\text{update}(y, m, \text{update}(x, n, s)). \qquad \blacksquare$$

---

[27] Since the observational equality is total, the predicates T_s are not needed in the lifting here. Moreover, since the predicates $\sim_s$ coincide with the equality on observable sorts, we only need one predicate $\sim_{State}$.

11.7.5 Proof systems for refinement

The refinement relation between specifications introduced in Section 11.5 (Definition 11.29) can be extended in a straightforward way to specification expressions built with the reachability and observability operators. As explained in the introduction of this section, the main motivation for having extended the kernel specification language with observability operators is to make the refinement relation more meaningful with respect to practical applications. Indeed using the observational behavior operator, we can define the following notion of observational refinement which captures the idea that, from the observational point of view, it is sufficient if an implementation satisfies the intended behavior of the given specification (and not necessarily the specification itself).

Definition 11.66 (Observational refinement). A specification SPI is an *observational refinement* of a specification SP w.r.t. a set $Obs \subseteq Sorts(SP)$ of observable sorts and a set $In \subseteq Sorts(SP)$ of input sorts, written $SP \overset{Obs,In}{\leadsto} SPI$, if the observational behavior specification **behavior** SP w.r.t. Obs, In is refined by the specification SPI (in the sense of Definition 11.29), i.e., if **behavior** SP w.r.t. Obs, In $\leadsto$ SPI. ◇

Proving an observational refinement relation $SP \overset{Obs,In}{\leadsto} SPI$ is thus equivalent to proving the refinement relation **behavior** SP w.r.t. Obs, In $\leadsto$ SPI. For this purpose we extend the proof system for refinements defined in Section 11.5 as follows.

Definition 11.67 (Extension of the proof system $\Pi_{\leadsto}$ for refinement). Definition 11.31 of the proof system $\Pi_{\leadsto}$ for the refinement relation between structured specifications is extended as follows (again assuming given a sound proof system Π_{spec} for the derivation of the theorems of a specification expression):

$$\text{(ref-reach)} \quad \frac{SP \leadsto_s SPI}{\textbf{reach } SP \textbf{ w.r.t. } \Omega \leadsto_s SPI} \quad \begin{array}{l} \text{if } Mod(SPI) \models \mathcal{R}_\Omega \\ \text{(cf. Definition 11.51)} \end{array}$$

$$\text{(ref-beh)} \quad \frac{SP \leadsto_s SPI/\approx_{Obs,In}}{\textbf{behavior } SP \textbf{ w.r.t. } Obs, In \leadsto_s SPI} \qquad \qquad ◇$$

Remark 11.68. The proof system $\Pi_{\leadsto}$ as defined above contains two difficulties:

1. To apply rule *(ref-reach)*, one has to prove the validity of a reachability constraint. In practice, this can often be achieved by proving a "sufficient completeness" condition (see, e.g., [Wir90, Hen97]) and/or by applying the techniques for proving reachability constraints studied in [Far92].
2. There is no rule for the observational quotient, i.e., if the implementation relation to be proved is of the form $SP/\approx_{Obs,In} \leadsto SPI$, one has to

perform a direct proof. As a consequence, the proof system for the refinement relation is complete (see below) only if the specification to be implemented does not contain observational quotient operators. This is not a severe restriction, since we will usually not apply an observational quotient operator to a specification to be implemented, but, on the contrary, in this case we will apply an observational behavior operator. Then the observational quotient operator will only appear inside an implementation proof on the right-hand side of the implementation relation when applying rule (ref-beh). Hence, for finally applying (ref-basic) in an implementation proof, it is enough to have sound and complete proof rules for the satisfaction of formulas with respect to specifications containing observational quotient operators (which were considered in the previous section). However, a consequence of the absence of proof rules for the refinement of observational quotient specifications is that we cannot directly apply our proof rules for proving an implementation relation of the form **abstract** SP **w.r.t.** Obs, In $\rightsquigarrow$ SPI as considered, e.g., in [ST88b].[28] In this case we have to add the following rule, which is sound since Mod(**behavior** SP **w.r.t.** Obs, In) $\subseteq$ Mod(**abstract** SP **w.r.t.** Obs, In):

$$\text{(ref-abstract)}\quad \frac{\textbf{behavior SP w.r.t. Obs}, \text{In} \rightsquigarrow_s \text{SPI}}{\textbf{abstract SP w.r.t. Obs}, \text{In} \rightsquigarrow_s \text{SPI}}$$

If the specification SP is behaviorally closed, i.e., if we have Mod(SP) $\subseteq$ Mod(**behavior** SP **w.r.t.** Obs, In), we know from Theorem 5.19 that Mod(**abstract** SP **w.r.t.** Obs, In) = Mod(**behavior** SP **w.r.t.** Obs, In) and the rule above is also complete.

Theorem 11.69 (Soundness and completeness of $\Pi_{\rightsquigarrow}$). *For any specification* SP *containing no observational quotient operator and for any specification* SPI, SP $\rightsquigarrow$ SPI *if and only if* SP $\rightsquigarrow_s$ SPI.

For a proof see [HWB97].

Example 11.70. The following specification STATE provides an abstract description of states (see Example 11.65). The last two axioms of the underlying specification STATE0 express that only the current value of an identifier is stored and that the update operation is commutative (when different identifiers x and y are used). The operations init and update are declared as constructors for states.

```
spec STATE0
    sorts Nat, Id, State
    functions init   :   → State
              update : Id, Nat, State → State
              lookup : Id, State → Nat
```

---

[28] Remember that the observational abstraction operator was defined as a derived operator of the language using the observational behavior and the observational quotient operators (see Section 11.7.1).

axioms $\forall x, y : \text{Id}, \ n, m : \text{Nat}, \ s : \text{State}.$
$\text{lookup}(x, \text{update}(x, n, s)) = n \ \wedge$
$(x \neq y \Rightarrow \text{lookup}(x, \text{update}(y, n, s)) = \text{lookup}(x, s)) \ \wedge$
$\text{update}(x, n, \text{update}(x, m, s)) = \text{update}(x, n, s) \ \wedge$
$(x \neq y \Rightarrow \text{update}(x, n, \text{update}(y, m, s)) =$
$\qquad\qquad \text{update}(y, m, \text{update}(x, n, s)) \ )$
endspec

spec STATE = **reach** NAT + ID + STATE0 **w.r.t.** $\{\text{init}, \text{update}\}$

States can be implemented by sequences of pairs (consisting of an identifier and its associated value), as described in the specification STATE-I of Example 11.65. In this implementation, each sequence stores not only the current value of an identifier x, but also all previous values of x. Hence the last two axioms of STATE0 are not satisfied by the implementation. Nevertheless STATE-I is a correct observational refinement of STATE if we consider only the sorts Nat and Id as observable (and all sorts as input sorts), i.e., with Obs and In as in Example 11.65, we have:

$$\text{STATE} \ ^{\text{Obs}, \text{In}}\rightsquigarrow \text{STATE-I}$$

To prove this, we have to prove that **behavior** STATE **w.r.t.** Obs, In $\rightsquigarrow$ STATE-I, which can be done by applying the rules of the proof system for refinement as illustrated below. Thereby STATE0-Ax denotes the axioms of the basic specification STATE0.

1. STATE-I$/\approx_{\text{Obs}, \text{In}} \vdash$ STATE0-Ax $\qquad\qquad$ (see Example 11.65)
2. STATE0 $\rightsquigarrow_s$ STATE-I$/\approx_{\text{Obs}, \text{In}}$ $\qquad\qquad$ (ref-basic), (1)
3. NAT + ID $\rightsquigarrow_s$ (STATE-I$/\approx_{\text{Obs}, \text{In}})|_{\text{Sig}(\text{NAT}+\text{ID})}$ $\qquad$ (ref-sum)
4. NAT + ID + STATE0 $\rightsquigarrow_s$ STATE-I$/\approx_{\text{Obs}, \text{In}}$ $\qquad$ (ref-sum) with (3) and (2)
5. Mod(STATE-I$/\approx_{\text{Obs}, \text{In}}) \models \mathcal{R}_{\{\text{init}, \text{update}\}}$ $\qquad\qquad$ (see below)
6. STATE $\rightsquigarrow_s$ STATE-I$/\approx_{\text{Obs}, \text{In}}$ $\qquad$ (ref-reach) with (4) and (5)
7. **behavior** STATE **w.r.t.** Obs, In $\rightsquigarrow_s$ STATE-I $\qquad$ (ref-beh) with (6)

For the application of the rule (ref-reach), one has to show that the models of STATE-I$/\approx_{\text{Obs}, \text{In}}$ satisfy the reachability constraint $\mathcal{R}_{\{\text{init}, \text{update}\}}$. This, however, is a simple task since the underlying specification STATE-SEQ already satisfies this reachability constraint, which follows from the reachability constraint of STATE-SEQ and the validity of the equations $\langle x, n \rangle = \langle x, n \rangle \& \text{init}$ and $\langle x, n \rangle \& s = \text{update}(x, n, s)$ in STATE-SEQ. $\qquad\blacksquare$

12 Object Specification

Hans-Dieter Ehrich

Abteilung Datenbanken, Technische Universität,
Postfach 3329, D-38023 Braunschweig, Germany
HD.Ehrich@tu-bs.de

12.1 Introduction

From an object-oriented point of view, software systems are considered to be dynamic collections of autonomous objects that interact with each other. Autonomy means that each object encapsulates all features needed to act as an independent computing agent: individual attributes (data), methods (operations), behavior (process), and communication facilities. Each object also has a unique identity that is immutable throughout its lifetime. Coincidentally, object-orientation comes with an elaborate system of classes and types, facilitating structuring and re-use of software.

The object approach is widely accepted in software technology, and there are object-oriented programming languages, database systems, and software development methods. The basic idea is not new, because the essential features were already present in the programming language SIMULA [DMN67]. Wider acceptance came with SMALLTALK [KG76, GR83].

While the object approach is successful in practice, it finds more scepticism than enthusiasm among theoreticians. Object-orientation is often considered an ugly area that lacks conceptual coherence. It is true that much of the work in the field cannot serve as counterevidence. But matters are changing: there is a growing interest in clean concepts and reliable foundations. In fact, object-orientation badly needs theoretical underpinning, for improving practice and facilitating teaching. Object theory is evolving into its own area of study.

Our approach to object specification combines ideas from algebraic data type specification, conceptual data modeling, behavior modeling, specification of reactive systems, and concurrency theory. Unlike conventional specification approaches, object specification shows its real virtue when dealing with open, reactive, and distributed systems.

We concentrate on fundamental concepts and constructions rather than languages and systems. Our approach is based on experiences with developing the OBLOG family of languages and their semantic foundations [SSE87]. OBLOG is being developed into a commercial product [ESD93]. In the academic world, there are two related developments: TROLL [JSHS96, SJH93, HJ95, EH96, DH97] and GNOME [SR94]. The common semantic basis is linear-time temporal logic.

It is not the purpose of this chapter to give an overview of object specification languages, nor of their underlying models and theories. Models, theories, and languages are diverse enough to make a systematic overview, let alone a uniform treatment, impossible within the limits of this chapter. Rather, the author gives his own view and presentation that is based on experiences with developing TROLL, and on joint work with Amílcar and Cristina Sernadas and others on foundations and concepts [SSE87, SFSE88, EGS90, ES90, SJE92, EDS93, EJDS94, SHJE94]. Hints to related work are given in Section 12.7.

The chapter is organized as follows. In the next section, we briefly explain basic concepts: objects, classes, types, and systems. In Section 12.3, we describe a denotational model of object classes that is tailored towards temporal logic. Section 12.4 explains temporal class specification, i.e., the use of linear-time temporal logic for in-the-small specification of object classes. Section 12.5 discusses in-the-large structuring concepts that are typical for object-orientation: inheritance and composition. Inheritance concepts include is-a and as-a specialization, hiding, overriding, overlaying, and covariance. Composition concepts include generalization and aggregation. For specifying interaction among object components, temporal logic is sufficient. For interaction among concurrent objects, an extension of the logic is necessary. We briefly explain our approach using Distributed Temporal Logic, DTL.

12.2   Basic concepts

12.2.1   Objects

An object is an encapsulated unit of structure and behavior. It has an identity which persists through change. Objects communicate with each other. They are typed, related by inheritance, and composed to form complex objects.

A widely accepted, informal object model has been formulated by Wegner [Weg89]: *An object has a set of operations and a local shared state (data) that remembers the effect of operations. The value that an operation on an object returns can depend on the object's state as well as the operation's arguments. The state of an object serves as a local memory that is shared by operations on it. In particular, other previously executed operations can affect the value that a given operation returns. An object can learn from experience, storing the cumulative effect of its experience – its invocation history – in its state.*

The operations of an object are usually called *methods*. In the object model of object-oriented programming, a method may change state and deliver a value. This model also underlies object specification languages like FOOPS and ETOILE.

The object model of object-oriented databases is more restricted; it separates state-changing "proper" methods from "read" methods that are free of side-effects, called *attributes*. This model also underlies TROLL. For the sake of simplicity, we adopt this model here.

12.2.2   Classes

An object *class* represents the prototypical structure and behavior of objects. We illustrate the concept by means of two examples, state variables and banks. For class specifications, we use an ad hoc notation that is closer to the formal model adopted in this chapter than to any of the specification languages mentioned above. Using linear-time temporal logic, however, the specification style is closest in spirit to TROLL.

Example 12.1. There is a kind of object that is so basic and omnipresent in computing that it has many names: *state variable* in programming, *attribute* in databases and conceptual modeling, *field* in file organization, *slot* or *fluent* in artificial intelligence, and *memory cell, register, word, byte, bit*, and *flip-flop* in hardware.

The formal specification of a class $\mathtt{Var}\,[s]$ of state variables, or variables for short, of sort s is as follows. A variable x has an attribute $x.\mathtt{val}$ of sort s denoting its current value. It has actions $x.\mathtt{c}$ for creation, $x:=s$ for assigning values (i.e., an action $x:=\mathtt{v}$, for every value $\mathtt{v}$ of sort s), and $x.\mathtt{d}$ for deletion. In the axioms, we use the *until* temporal operator $\varphi\,\mathcal{U}^{\circ}\,\psi$ for expressing that φ holds from now on, until ψ holds for the next time (see Section 12.4 for a precise definition). We use the notation $\triangleright a$ for expressing that action a is enabled, i.e., may occur at the current state, and $\circ\,a$ for expressing that action a has just occurred.

> *class* $\mathtt{Var}\,[s]$;
>   *uses* s;
>   *var* $x:\mathtt{Var}\,[s]$;
>   *attributes* $x.\mathtt{val}:s$;
>   *actions* $x.\mathtt{c}$; $x:=s$; $x.\mathtt{d}$;
>   *axioms var* $\mathtt{v,w}:s$;
> $\circ\,x.\mathtt{c} \Rightarrow (\neg\triangleright x.\mathtt{c} \wedge \triangleright x:=\mathtt{v} \wedge \triangleright x.\mathtt{d})\ \mathcal{U}^{\circ}\,\circ\,x.\mathtt{d},$
> $\circ\,x.\mathtt{d} \Rightarrow \neg\triangleright x.\mathtt{c} \wedge \neg\triangleright x:=\mathtt{v} \wedge \neg\triangleright x.\mathtt{d},$
> $\circ\,x:=\mathtt{v} \Rightarrow x.\mathtt{val}=\mathtt{v}\ \mathcal{U}^{\circ}\,(\circ\,x:=\mathtt{w} \vee \circ\,x.\mathtt{d})$
> *end*

The axioms say that (1) after creation and before deletion, another creation is disabled but value assignment and deletion are enabled; (2) after deletion, no action is enabled; (3) the value after assignment is retained until the next assignment or deletion.

Notation. In object-oriented notation, it is common practice to omit the local object variable x. We follow this practice in the following. The example then reads as follows.

> *class* $\mathtt{Var}\,[s]$;
>   *uses* s;
>   *attributes* $\mathtt{val}:s$;

actions c; :=s; d;
axioms var v,w: s;
     ∘c ⇒ (¬▷c ∧ ▷:=v ∧ ▷d) $\mathcal{U}^\circ$∘d,
     ∘d ⇒ ¬▷c ∧ ¬▷:=v ∧ ¬▷d,
     ∘:=v ⇒ val=v $\mathcal{U}^\circ$ (∘:=w ∨∘d)
end   ■

Example 12.2. The objects in this example are banks and accounts.

An account has a holder and a balance as attributes, and it has actions to open and close an account and to credit and debit money. In the formal specification, we use the temporal operators Y for *previous* (yesterday) and P for *sometime in the past.* The axioms say that (1) after an account is opened, it has the holder and balance given as actual parameters; (2) crediting increases the balance by the amount credited; (3) debiting decreases the balance by the amount debited; (4) an amount m may only be debited if the account once had at least balance m (a simple solvency criterion).

class Account;
   uses money, Person;
   attributes holder: Person, balance: money;
   actions   *open(Person,money), credit(money),
                  debit(money), +close;
   axioms var p:Person, m,n:money;
     ∘open(p,m) ⇒ holder=p ∧ balance=m,
     ∘credit(m) ⇒ ∘balance:=balance+m,
     ∘debit(m) ⇒ ∘balance:=balance-m,
     ∘debit(m) ⇒ P balance ≥ m
end

* denotes a birth action that may only occur at the beginning, and + denotes a death action that terminates the life of an object. After a birth action, no birth action is enabled, and after a death action, no action is enabled. This may be specified explicitly but we refrain from doing so.

The second and third axioms are abbreviations for

    ∘credit(m) ∧ Y balance=n ⇒ balance=n+m and
    ∘debit(m) ∧ Y balance=n ⇒ balance=n-m, respectively.

As usual, a frame rule is assumed stating that attributes do not change values unless specified otherwise. Among the consequences of the axioms and this frame rule, we have, e.g.,

    ∘open(p,m) ⇒ G holder=p.

G is the *always* temporal operator.

A bank is a complex object. It has a collection of accounts as components, providing access to their individual attributes and actions. The data type acct# gives the set of permissible account numbers. A bank has an owner as an attribute, and actions for establishing a bank, for transferring money from one account to another, and for liquidating a bank. In the following specification, we use the temporal operators X for *next* (tomorrow), $\mathcal{P}^+$ for *precedes in the future*, and $\mathcal{U}^\circ$ for *until*, as introduced above. The axioms say that (1) in the beginning, the owner is the person who established the bank; (2) transferred amounts are debited immediately after transfer; (3) transferred amounts are eventually credited; however, as long as a transfer is not credited, the same amount may not be transferred between the same accounts again; (4) there must not be two different transfers of the same amount from the same account at the same time; (5) transferred amounts are credited before any other amount is debited from the same account.

The third and fourth axioms are a simple way to ensure correct transaction management.

```
class Bank;
    uses money, acct#, Person, Account;
    attributes owner: Person;
    components Acc(acct#): Account;
    actions    *establish(Person),
               transfer(Account,Account,money),
               +liquidate;
    axioms var p: Person, from, to, to2: acct#, m, n: money;
        ∘establish(p) ⇒ owner=p,
        ∘transfer(from,to,m) ⇒ X ∘Acc(from).debit(m),
        ∘transfer(from,to,m) ⇒
                ¬∘transfer(from,to,m) U° ∘Acc(to).credit(m),
        ∘transfer(from,to,m)∧∘transfer(from,to2,m)⇒to=to2,
        ∘transfer(from,to,m) ⇒
                ∘Acc(to).credit(m) P⁺∘Acc(to).debit(n)
end
```

The following liveness property is a consequence of the third axiom.

> ∘transfer(from,to,m) ⇒ F ∘Acc(to).debit(n).

F is the *sometime in the future* temporal operator. The property says that transfers are eventually credited.

So far, the example is unpractical because it does not describe transfers between different banks. One solution is to introduce a complex object **BankWorld** having all banks as components. Transfers between banks may then be specified in the **BankWorld** class, in the same way as transfers between accounts are specified above. A more appropriate solution, however, uses DTL, an extension of temporal logic towards describing concurrent behavior (see Example 12.49).   ∎

For formal treatment, we abstract from irrelevant details and put those which are relevant into a convenient form. Let $\Sigma = (S, \Omega)$ be a data signature where S is a set of sorts, and Ω is an $S^* \times S$-indexed set family of operation symbols. We assume that bool $\in S$. If $\omega \in \Omega_{s_1 \ldots s_n, s}$, we write $\omega : s_1 \times \ldots \times$

$s_n \to s \in \Omega$ or, briefly, $\omega : x \to s$ if $x = s_1 \ldots s_n$ and the reference to Ω is clear from context. Let U be a Σ-algebra.

The syntax of a class specification is given by its attribute and action symbols. For ease of presentation, we assume that these symbols are not parameterized. Parameterized symbols will occur in practice but may be expanded into their instantiations in U by substituting actual parameters in all possible ways.

For the sake of formal simplicity, we treat action symbols as attribute symbols with sort act, to be interpreted by value set $\{\triangleright, \circ, \triangleright \circ, \varepsilon\}$. These values indicate that an action is enabled ($\triangleright$), or has occurred ($\circ$), or both ($\triangleright \circ$), or none of these (ε). In this way, class states are uniformly characterized by a family of sets of attributes. We write $\triangleright a$, $\circ a$, $\triangleright \circ a$, or εa, if action a is in the corresponding state. Further action states may be introduced, for instance, expressing scoping for dealing with dynamic roles and phases [EJD93].

Definition 12.3. Let Σ be a data signature. A *class signature* over Σ is an S-indexed set family $\Gamma = \{\Gamma_s\}_{s \in S}$ of attribute symbols.   ◇

The intended interpretation of a class signature Γ is a Γ-*class* C (see Definition 12.30) describing the generic structure and behavior of object instances (see Section 12.2.3). If there is no danger of confusion, we omit the Γ-prefix. We adopt the view that *objects are sequential processes* and *systems are concurrent collections of objects*. Our denotational models for classes and systems are based on labeled event structures, see Section 12.3 below.

An in-the-small class specification is a pair $Cspec = (\Gamma, \Phi)$ where Γ is a class signature and Φ is a set of class formulas, see Definition 12.37. Different logics may be applied. Our choices, linear-time temporal logic and an extension towards distribution, are elaborated in Sections 12.4 and 12.6.

For dealing with in-the-large specification, morphisms between class signatures and specifications are essential in order to capture relevant constructions like inheritance, hiding, generalization, aggregation, etc. on the syntactic level. The corresponding semantic relationships are captured by class morphisms, see Definition 12.31.

Definition 12.4. Let Γ_1 and Γ_2 be class signatures over the same data signature. A *class signature morphism* $\gamma : \Gamma_1 \to \Gamma_2$ is an S-indexed family of total functions.

For class specifications $Cspec_1 = (\Gamma_1, \Phi_1)$ and $Cspec_2 = (\Gamma_2, \Phi_2)$, a *class specification morphism* $\gamma : Cspec_1 \to Cspec_2$ is a class signature morphism $\gamma : \Gamma_1 \to \Gamma_2$, preserving properties in the sense that $\Phi_2 \models \gamma(\Phi_1)$ holds.   ◇

As usual, $\gamma(\Phi_1)$ is obtained by translating formulas syntactically via γ, replacing every attribute symbol by its γ image. The class specification morphism condition states that the translated axioms $\gamma(\Phi_1)$ must be entailed by Φ_2. Satisfaction and entailment for our version of temporal logic are given in Definition 12.36.

Remark 12.5. The category **csig** of class signatures and class signature morphisms is complete and cocomplete. Colimits describe syntactic composition of signatures and specifications.

Given a Σ-algebra U, a class signature determines the possible states of a class: a state is a value assignment to attributes.

Definition 12.6. Let Γ be a class signature. A *class!state* over U is an S-indexed set family of total functions $\eta\colon \Gamma \to U$. The set of class states over Γ and U is denoted by $[\Gamma \to U]$, or $[\Gamma]$ for short if U is clear from the context.

Let $\gamma\colon \Gamma_1 \to \Gamma_2$ be a class signature morphism. The *class state morphism* $\gamma^*\colon [\Gamma_2] \to [\Gamma_1]$, defined by γ, is given by $\gamma^*(\eta) = \eta \circ \gamma$ for each $\eta \in [\Gamma_2]$.

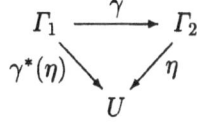

$\diamond$

For an action symbol a, we write $\circ[\eta]a$ for $\eta(a) \in \{\circ, \circ\triangleright\}$, $\triangleright[\eta]a$ for $\eta(a) \in \{\triangleright, \circ\triangleright\}$, $\triangleright\circ[\eta]a$ for $\eta(a) = \circ\triangleright$, and $\varepsilon[\eta]a$ for $\eta(a) = \varepsilon$. Note that $\triangleright\circ[\eta]a$ iff $\triangleright[\eta]a \wedge \circ[\eta]a$, and $\varepsilon[\eta]a$ iff $\neg(\triangleright[\eta]a \vee \circ[\eta]a)$.

12.2.3   Types

An object type is a static domain of object instances of the same class. It consists of a set of object *identities* for representing instances, and a class. Object type morphisms (direct and reverse) describe various kinds of object relationship: inheritance, hiding, generalization, and aggregation.

Definition 12.7. An *object type signature* is a pair $\Theta = (Id, \Gamma)$ where Id is a set of *object identities* and Γ is a class signature. $\diamond$

The *object instance signature* Υ, determined by Θ, provides an individual *object signature* $i.\Gamma$ for each object with identity i. If $a \in \Gamma$, then its instance for object i is denoted by $i.a$. Technically, Υ is the coproduct $\coprod_{i \in Id} \Gamma$ in **csig**.

Object type signatures may be related by morphisms, mapping identities and relating class signatures by class signature morphisms. We refrain from giving a formal definition because the concept is not needed in this chapter. However, it is implicit in object type morphisms, see below. A corresponding remark applies to object instance signature morphisms.

Definition 12.8. An *object type* with signature $\Theta = (Id, \Gamma)$ is a pair $T = (Id, C)$ where C is a Γ-class. $\diamond$

An *instance of* T is a pair $i.C$ where $i \in Id$ is an identity. The object instances of T are given by $\overline{T} = Id.C = \{i.C \mid i \in Id\}$. Each object instance $i.C$ is an individual process that is an isomorphic copy of class C with signature $i.\Gamma$.

Definition 12.9. Let $T_1 = (Id_1, C_1)$ and $T_2 = (Id_2, C_2)$ be object types. An *object type morphism* $\tau: T_1 \rightrightarrows T_2$ consists of a map $\tau': Id_1 \to Id_2$ and a class morphism $\tau'': C_1 \to C_2$.   ◇

Class morphisms are given in Definition 12.31. A class morphism $\tau'': C_1 \to C_2$ has an underlying class signature morphism $\gamma: \Gamma_2 \to \Gamma_1$ going in the opposite direction. The corresponding class state morphism $\gamma^*: [\Gamma_1] \to [\Gamma_2]$ maps C_1 states to C_2 states.

For identities $i \in Id_1$ and $j \in Id_2$, we may define an *object instance morphism* $i\text{-}j.\tau: i.C_1 \to j.C_2$ as the corresponding isomorphic copy of τ.

We give examples of relationships typically expressed as object type morphisms.

Example 12.10. The fact that each customer *is-a* person is expressed by an object type morphism $\tau: Customer \rightrightarrows Person$ such that $\tau': Id_{\texttt{Customer}} \hookrightarrow Id_{\texttt{Person}}$ is an inclusion. The class morphism $\tau'': C_{\texttt{Customer}} \to C_{\texttt{Person}}$ restricts the customer class to its underlying person class, see Example 12.32 for a more detailed treatment. The underlying class signature morphism $\gamma: \Gamma_{\texttt{Person}} \to \Gamma_{\texttt{Customer}}$ embeds person attributes into those of the customer class. Conversely, the class state morphism $\gamma^*: [\Gamma_{\texttt{Customer}}] \to [\Gamma_{\texttt{Person}}]$ reduces customer states to person states, ignoring the additional customer attributes.

The example reflects temporal specialization in the sense that persons may temporarily and repeatedly play the role of customers.   ∎

Example 12.11. The specification $Customer = Person + Company$ expresses that each *Customer* is either a *Person* or a *Company*. This is *generalization* or alternative choice. The customer signature contains the attributes (including actions) that are common to both persons and companies, and that are relevant to their roles as customers. Fixing these is a design decision. For the sake of simplicity, we assume that the *Person* and *Company* classes are already restricted to these roles, i.e., we assume $\Gamma_{\texttt{Customer}} = \Gamma_{\texttt{Person}} = \Gamma_{\texttt{Company}}$.

The object type morphisms $\tau_1: Person \rightrightarrows Customer$ and $\tau_2: Company \rightrightarrows Customer$ embed the constituent types into the generalization (see Sections 12.3.3 and 12.5.2). In fact, $+$ denotes coproduct in the category of object types and object type morphisms over a fixed class signature.

Note that each person *is-a* customer here, the other way round to the previous example. So we have identity inclusions $\tau_1': Id_{\texttt{Person}} \hookrightarrow Id_{\texttt{Customer}}$ and $\tau_2': Id_{\texttt{Company}} \hookrightarrow Id_{\texttt{Customer}}$ as well as class morphisms $\tau_1'': C_{\texttt{Person}} \to C_{\texttt{Customer}}$ and $\tau_2'': C_{\texttt{Company}} \to C_{\texttt{Customer}}$. The class signature morphisms are identities under the assumption mentioned above, and so are the class state morphisms.   ∎

Example 12.12. Components in the sense of Example 12.2 are meant to be interpreted by *object aggregation*: complex objects are built from constituent objects and synchronize the behaviors of these components. For example, we may specify $Car = Motor \bowtie Chassis$ meaning that each Car object is a complex object consisting of a $Motor$ object and a $Chassis$ object. The car signature contains all motor and chassis attributes and distinguishes between them, i.e., $\Gamma_{Car} = \Gamma_{Motor} + \Gamma_{Chassis}$ is the disjoint union with inclusions $\gamma_1 \colon \Gamma_{Motor} \hookrightarrow \Gamma_{Car}$ and $\gamma_2 \colon \Gamma_{Chassis} \hookrightarrow \Gamma_{Car}$.

We have object type morphisms $\tau_1 \colon Car \rightrightarrows Motor$ and $\tau_2 \colon Car \rightrightarrows Chassis$ that give the projections to the components. In fact, $\bowtie$ denotes product in the category of object types and object type morphisms. $id_{Car} = id_{Motor} \times id_{Chassis}$ is the cartesian product with projections $\tau_1' \colon id_{Car} \rightarrow id_{Motor}$ and $\tau_2' \colon id_{Car} \rightarrow id_{Chassis}$. Class morphisms $\tau_1'' \colon C_{Car} \rightarrow C_{Motor}$ and $\tau_2'' \colon C_{Car} \rightarrow C_{Chassis}$ are projections of the product in the category of classes, see Section 12.3.3. ∎

There are relationships between object types that are more naturally expressed by a variant of object type morphism where the class morphism goes the other way. We give examples after the formal definition.

Definition 12.13. Let $T_1 = (Id_1, C_1)$ and $T_2 = (Id_2, C_2)$ be object types. A *reverse object type morphism* $\tau \colon T_1 \rightleftarrows T_2$ consists of a map $\tau' \colon Id_1 \rightarrow Id_2$ and a class morphism $\tau'' \colon C_2 \rightarrow C_1$. ◇

The class morphism $\tau'' \colon C_2 \rightarrow C_1$ has an underlying class signature morphism $\gamma \colon \Gamma_1 \rightarrow \Gamma_2$. Thus, the map between identities goes the same way as the class signature morphism whereas the class state morphisms conform with the class morphisms.

For identities $i \in Id_1$ and $j \in Id_2$, the *reverse object instance morphism* $i\text{-}j.\tau \colon i.C_2 \rightarrow j.C_1$ is the corresponding isomorphic copy of τ.

We give examples of typical relationships expressed as reverse object type morphisms.

Example 12.14. The fact that a special kind of rectangle may be viewed *as-a* square is expressed by a reverse object type morphism $\tau \colon Square \rightleftarrows Rectangle$. Again, each square *is-a* rectangle, so τ' is an inclusion. The class signature morphism embeds square attributes into those of the rectangle class, $\gamma \colon \Gamma_{Square} \rightarrow \Gamma_{Rectangle}$. The class state morphism goes the other way, $\gamma^* \colon [\Gamma_{Rectangle}] \rightarrow [\Gamma_{Square}]$, reducing rectangle states to square states and ignoring the additional rectangle attributes. The class morphism goes the same way, $\tau'' \colon C_{Rectangle} \rightarrow C_{Square}$, projecting rectangle behavior to square behavior. See Example 12.33 for a more detailed treatment. ∎

Like is-a inheritance, as-a inheritance is intended to express a kind of temporal specialization. The difference is that the former means that *all* features are inherited and new ones may be added, whereas the latter means that *some* features are inherited and possibly used in a special way.

Example 12.15. The hiding relationship between databases and their views is expressed by a reverse object type morphism $\tau \colon View \rightleftarrows Database$. Databases and their views are different objects, so we have $Id_{\text{View}} \cap Id_{\text{Database}} = \emptyset$. The class signature morphism maps view attributes into those of the database class. The class morphism in turn projects database behavior onto view behavior, and the class state morphism reduces database states to view states.

∎

Remark 12.16. Given a category **class** of classes and class morphisms, object types and object type morphisms form a category **ot(class)** that inherits limits and colimits from **class**. Comments on **class** are given at the end of Section 12.3.

12.2.4   Systems

An object system is a collection of object instances operating concurrently and interacting by exchanging messages. Our interaction model is synchronous and symmetric "handshake".

Definition 12.17. An *object system signature* is a pair $\Sigma = (S, \Omega)$, where $S = S^d + S^o$ is a set of *sorts*, subdivided into disjoint sets of *data sorts* S^d and *object sorts* S^o. $\Omega = \{\Omega_{x,s}\}_{x \in S^*, s \in S}$ is an $S^* \times S$-indexed set family of *operation symbols*.   ◇

A data signature is the special case where the set S^o of object sorts is empty, so we use the same notation. As for data signatures, interpretation is given in a Σ-algebra U. The intended interpretation of a data sort $d \in S^d$ in U is a set U_d of data elements that is the carrier set of d in U. The data type of sort d is defined by this set and the operations defined on it.

The intended interpretation of an object sort $s \in S^o$ in U is a set $U_s = Id_s.C_s$ of object instances $i.C_s$ with signatures $i.\Gamma_s$, $i \in Id_s$. The class C_s describes the common structure and behavior of its instances. The object type of sort s is defined by the set U_s of instances and the operations defined on it. For simplicity, we ignore object operations in the following, i.e., we consider object types to be sets of instances with the same behavior, as introduced in the previous section.

The intended interpretation of an object system signature is the concurrent composition of its instances, see Section 12.6.

In a sense, a data type is an object type with a trivial class only allowing for static "behavior". On the other hand, the identity parts Id_s of object types are like data types: they may be specified with any of the data specification techniques.

Definition 12.18. The *object instance signature* Υ over Σ and U is given by $\Upsilon = (Id, \Gamma)$ where $Id = \{Id_s\}_{s \in S^o}$ is the set family of object identities, and $\Gamma = \{i.\Gamma_s\}_{i \in Id_s, s \in S^o}$ is the set family of individual object signatures.   ◇

Definition 12.19. An *object system specification* is a pair $Sspec = (\Upsilon, \Phi)$ where Υ is an object instance signature, and $\Phi = \{\Phi_i\}_{i \in Id_s, s \in S^\circ}$ is the set family of object axioms. ◇

There is a problem with using temporal logic for object axioms. For specifying single classes, temporal logic is sufficient. For specifying concurrent object systems, however, temporal logic is not expressive enough: we extend it towards distributed temporal logic in Section 12.6.2.

12.3   A denotational class model

In order to model the sequential behavior of objects and the concurrent behavior of object systems, many models of concurrency may be adopted and integrated in our framework. The question is which model to adopt. According to the classification in [SNW93], we may choose along three dimensions: (1) a behavior or a system model, (2) an interleaving or a noninterleaving model, and (3) a linear or a branching time model.

For denotational semantics, a behavior model is more appropriate than a system model because it is more abstract. For linear-time temporal logic, an interleaving model is sufficient because concurrency cannot be expressed. However, we need a noninterleaving model in order to capture concurrency. According to [SNW93], *labeled event structures* are a fair choice when a non-interleaving denotational model is needed.

We also follow this advice for the following reasons. The relationship between event structures and other models of concurrency is well investigated, and the interpretation structures of temporal logic, linear-time as well as branching-time, can be embedded into event structures, leaving room for extending the logic while staying in the interpretation domain. See Section 12.7 for references to related work.

12.3.1   Event structures

An elegant aspect of event structures is that they capture the behavior of systems, single system runs, objects, and single object runs in one and the same model. We review the basic definitions of prime event structures, concentrating on the case where causality is the reflexive and transitive closure of a base relation of "step causality". We use the notation $\rightarrow$ for the latter, and $\rightarrow^*$ for causality[1]. $\rightarrow^+$ denotes the irreflexive transitive closure of $\rightarrow$.

Definition 12.20. A *prime event structure* – or *event structure* for short – is a triple $E = (Ev, \rightarrow^*, \#)$ where Ev is a set of events, and $\rightarrow^*, \# \subseteq Ev \times Ev$ are binary relations called *causality* and *conflict*, respectively. Causality $\rightarrow^*$

---

[1] The standard notation for causality is $\leq$ but this symbol is also standard for subsorting expressing inheritance (see Section 12.5.1). We use it for the latter.

is a partial ordering, and conflict $\#$ is symmetric and irreflexive. For each event $e \in E$, its *local configuration* $\downarrow e = \{e' \mid e' \rightarrow^* e\}$ is finite. Conflict propagates over causality, i.e., $e \# e' \rightarrow^* e'' \Rightarrow e \# e''$ for all $e, e', e'' \in Ev$.

Two events $e, e' \in Ev$ are *causally related*, $e \sim e'$, iff $(e \rightarrow^* e' \vee e' \rightarrow^* e)$.

Two events $e, e' \in Ev$ are *concurrent*, $e \, co \, e'$, iff $\neg e \sim e' \wedge \neg e \# e'$.

A *configuration* in E is a downward-closed conflict-free set of events $C \subseteq Ev$, i.e., the following conditions hold: (1) $\forall e \in C : \downarrow e \subseteq C$, and (2) $\forall e_1, e_2 \in C : \neg(e_1 \# e_2)$. The set of configurations in E is denoted by $\mathcal{C}(E)$.

A *life cycle* in E is a maximal configuration in E. The set of life cycles in E is denoted by $\mathcal{L}(E)$.   ◇

In the following, order-theoretic notions refer to causality. The finiteness condition for local configurations is a temporal reachability condition: only events $e \in Ev$ that may happen within finite time after the system start are taken into consideration. Obviously, we have $\mathcal{L}(E) \subseteq \mathcal{C}(E)$ for every event structure E. Intuitively, a life cycle represents a possible run of a system, and a configuration represents a (not necessarily proper) prefix of it. This represents a state of a part of the system. Please note that life cycles may be finite or infinite, so the behaviors of terminating programs and nonterminating reactive systems can both be modeled adequately.

Proposition 12.21. *For event structures E_1 and E_2, we have $E_1 = E_2$ iff $\mathcal{C}(E_1) = \mathcal{C}(E_2)$ iff $\mathcal{L}(E_1) = \mathcal{L}(E_2)$.*

Definition 12.22. *For $i = 1, 2$, let $E_i = (Ev_i, \rightarrow_i^*, \#_i)$ be event structures. An* event structure morphism $g : E_1 \rightarrow E_2$ *is a partial function $g : Ev_1 \rightharpoonup Ev_2$ such that, for every configuration $C \in \mathcal{C}(E_1)$, the following conditions hold: (1) $g(C) \in \mathcal{C}(E_2)$, and (2) $\forall e_1, e_2 \in C \cap dom(g) : g(e_1) = g(e_2) \Rightarrow e_1 = e_2$.*   ◇

Event structure morphisms preserve configurations and are injective on them. An event structure morphism $g : E_1 \rightarrow E_2$ is often used to express how behavior in E_1 synchronizes with behavior in E_2: the occurrence of event e_1 in $dom(g)$ implies the simultaneous occurrence of event $e_2 = g(e_1)$ in E_2 and vice versa.

The following lemma gives a characterization of event structure morphisms that only uses local configurations [WN95].

Lemma 12.23. *For $i = 1, 2$, let $E_i = (Ev_i, \rightarrow_i^*, \#_i)$ be event structures. Let $g : Ev_1 \rightharpoonup Ev_2$ be a partial function. Then g is an event structure morphism iff the following conditions hold for all $e, e_1, e_2 \in dom(g)$: (1) $\downarrow g(e) \subseteq g(\downarrow e)$, and (2) $(g(e_1) = g(e_2) \vee g(e_1) \# _2 g(e_2)) \Rightarrow (e_1 = e_2 \vee e_1 \#_1 e_2)$.*

Remark 12.24. The category **ev** of event structures is complete and has coproducts. Products and coproducts are useful for modeling concurrent composition and generalization, respectively [WN95]. Pullbacks may be utilized for modeling composition with sharing, e.g., event sharing for handshake communication.

12.3.2 Event groves

In this subsection, we concentrate on *sequential* event structures because they are appropriate models for classes: they provide an interpretation domain for class specifications using temporal logic.

Since conflict is a derived concept in sequential event structures, we introduce a simpler model that presents precisely the sequential event structures.

Definition 12.25. An *event grove* is an acyclic graph $G = (Ev, \rightarrow)$ such that, for all events $e_1, e_2 \in Ev$, we have $e_1 \sim e_2$ if there is an event $e_3 \in Ev$ for which $e_1 \rightarrow^* e_3$ and $e_2 \rightarrow^* e_3$ holds.   ◇

Remember that $e_1 \sim e_2$ means that $e_1 \rightarrow^* e_2$ or $e_2 \rightarrow^* e_1$. The definition means that, if both events e_1 and e_2 are causal for some future event, then they are causally related. Thus, an event grove is a set of rooted trees. An event grove $G = (Ev, \rightarrow)$ determines a prime event structure in a canonical way by letting all causally unrelated events be in conflict.

Definition 12.26. Let $G = (Ev, \rightarrow)$ be an event grove. The *event structure presented by* G is $E(G) = (Ev, \rightarrow^*, \#)$ where, for all $e_1, e_2 \in Ev$, $e_1 \# e_2$ holds iff $\neg e_1 \sim e_2$.   ◇

It is obvious that this defines a valid event structure. The configurations and life cycles are totally ordered: they are linear traces of events. If G is an event grove, we write $\mathcal{C}(G)$ for $\mathcal{C}(E(G))$ and $\mathcal{L}(G)$ for $\mathcal{L}(E(G))$.

The concurrency relation *co* of an event grove is empty. Event structures without concurrency have been called sequential in [LT87]; they are precisely those that can be presented by event groves. Sequential event structures are the event structures of synchronization trees [Win84] that are an appropriate behavior model of interleaving concurrency. They are elementary in the sense of [NPW81].

Sequential event structures allow for global timing of events along their traces: each initial event is timed 0, and the times of successor events are increased by 1.

Definition 12.27. Let $G = (Ev, \rightarrow)$ be an event grove. The *timing function* $\tau_G \colon Ev \rightarrow \mathbb{N}$ of G is defined by $\tau_G(e) = 0$ iff e is minimal, and $\tau_G(e') = \tau_G(e) + 1$ iff $e \rightarrow e'$ holds.   ◇

Definition 12.28. Let $G_1 = (Ev_1, \rightarrow_1)$ and $G_2 = (Ev_2, \rightarrow_2)$ be event groves. An *event grove morphism* $g \colon G_1 \rightarrow G_2$ is a graph morphism preserving time, i.e., a total map $g \colon Ev_1 \rightarrow Ev_2$ such that, for all events $e, e' \in Ev_1$, we have $e \rightarrow_1 e'$ implies $g(e) \rightarrow_2 g(e')$, and $\tau_{G_1}(e) = \tau_{G_2}(g(e))$.   ◇

A morphism between event groves coincides with a total morphism between the event structures presented by them, i.e., a map $g \colon Ev_1 \rightarrow Ev_2$ is an event grove morphism $g: G_1 \to G_2$ iff it is a total event structure morphism $g: E(G_1) \to E(G_2)$. The rationale for restricting ourselves to total morphisms is that event groves related by morphisms should be fully synchronized: traces are mapped in a bijective way. Different traces, however, may be mapped to the same trace.

Remark 12.29. The category **egr** of event groves with their morphisms is complete and cocomplete. Products in **egr** do not coincide with those in **ev**; while the latter model concurrent object composition, the former model sequential object aggregation. Pullbacks may also be utilized for modeling aggregation with sharing. Coproducts coincide with those in **ev**. Pushouts may be utilized for modeling generalized objects with shared constituents.

12.3.3   Classes

In order to model object classes, events are labeled by class states describing the current values of attributes. Let Γ be a class signature.

Definition 12.30. A Γ-*class* – or *class* for short if Γ is clear from the context – is a triple $C = (G, \lambda, \Gamma)$ where $G = (Ev, \to)$ is an event grove and $\lambda: Ev \to [\Gamma]$ is a map called Γ-*labelling for* G. λ must satisfy the following constraint: for all events $e \in Ev$ and all action symbols $a \in \Gamma_{\mathbf{acs}}$, $\circ[\lambda(e)]a$ implies that there is an event $e' \in Ev$ such that $e' \to e$ and $\triangleright[\lambda(e')]a$.   ◇

Classes are related by class morphisms which are event structure morphisms preserving labels in a canonical way.

Definition 12.31. Let $C_i = (G_i, \lambda_i, \Gamma_i)$, $i = 1, 2$, be classes over the same data signature. A *class morphism* $f: C_1 \to C_2$ is a pair $f = (g, \gamma)$ where $g: G_1 \to G_2$ is an event grove morphism and $\gamma: \Gamma_2 \to \Gamma_1$ is a class signature morphism such that $\gamma^* \circ \lambda_1 = \lambda_2 \circ g$. f is called *strict* iff γ is injective.

$$
\begin{array}{ccccc}
G_1 & \xrightarrow{\ \lambda_1\ } & [\Gamma_1] & & \Gamma_1 \\
{\scriptstyle g}\downarrow & & \downarrow{\scriptstyle \gamma^*} & & \uparrow{\scriptstyle \gamma} \\
G_2 & \xrightarrow{\ \lambda_2\ } & [\Gamma_2] & & \Gamma_2
\end{array}
$$

◇

A helpful intuition for a class morphism $f: C_1 \to C_2$ is that f extracts C_2 as a kind of "view" from C_1, with events projected by g and states reduced via γ^*.

Example 12.32. The class morphism $\tau'': C_{\mathsf{Customer}} \to C_{\mathsf{Person}}$ from Example 12.10 is given by $\tau'' = (g, \gamma)$ where g views customer events as person events, and γ embeds person attributes into those of customers. g is neither injective nor surjective: different customer traces may appear as the same person trace, namely, if the differences are only in the customer specific parts of the states, and there may be person traces that do not appear for customers.   ∎

Example 12.33. The class morphism τ'': $C_{\texttt{Rectangle}} \to C_{\texttt{Square}}$ from Example 12.14 is given by $\tau'' = (g, \gamma)$ where $g = id$ is the identity map: rectangle events *are* square events!

What happens is that each rectangle event represents a rectangle state, together with the trace of events that led to this state. With rectangle state labels reduced to square state labels along this trace, the same events define a square trace ending in a particular square state. For instance, if we assume that the square side length attribute x is inherited as the first of the two rectangle side length attributes (x, y), then the square image of a rectangle in state $(3,7)$ is a square with side length 3. Of course, rectangles with equal side lengths coincide with their square images. Different square traces may carry the same state information along their way, so they are not distinguishable. Also some moves in a square trace may look spontaneous because nothing (visible) happens: the changes (if any) are in the specific rectangle parts of the states.

γ embeds square attributes into those of rectangles. Thus, γ^* forgets about the specific rectangle attributes.   ■

Remark 12.34. Corresponding to event grove category **egr**, we may define a category **class** of classes and class morphisms, and full subcategories Γ-**class** for each class signature Γ. Limits and colimits in **egr** carry over to **class**. Products in **class** model object aggregation. Pullbacks may be utilized for modeling object aggregation with overlapping components. Coproducts in Γ-**class** are useful for modeling generalization or alternative choice, and also for distinguishing instances. Pushouts may be utilized for modeling generalized classes with overlapping constituents.

12.4   Class specification in the small

For specifying classes by explicit axiomatic description, we may choose among many possible logics; see [AGM92] for a comprehensive overview. We choose a temporal logic because it is the simplest logic that can express safety as well as liveness and fairness properties. Since the pioneering work of Pnueli [Pnu77], temporal logic has been widely accepted as a suitable formalism for giving axiomatic descriptions of system dynamics on a high level of abstraction. Temporal logic has been succesfully applied to a number of problems in reactive systems specification. It is the simplest logic that can deal not only with safety properties, but also with liveness properties. See Section 12.7 for references to the literature.

Among the many temporal logics that have been suggested and investigated, we choose a propositional linear-time temporal logic which we call PTL.

In most approaches, a future-oriented temporal logic is adopted. We include past-oriented operators as well. This does not add expressive power

[LS95] but it adds convenience for specification, and it does not complicate the logic too much.

12.4.1 Temporal logic

Let Σ be an object system signature, and let Γ be a class signature (see Definitions 12.17 and 12.3, respectively).

Definition 12.35. The *syntax* of PTL is given by

$$\text{PTL} ::= \text{ATOM} \mid (\text{PTL} \Rightarrow \text{PTL}) \mid (\text{PTL} \, \mathcal{U} \, \text{PTL}) \mid (\text{PTL} \, \mathcal{S} \, \text{PTL})$$
$$\text{ATOM} ::= \text{false} \mid T_\Sigma \, \theta \, T_\Sigma \mid T_\Gamma \, \theta \, T_\Sigma \qquad \diamond$$

In the atoms, θ is a comparison operator like $=, \leq, \dots$ It is used for comparing data terms with data terms, and class terms with data terms. Class terms are just attribute constants, including actions.

Class terms are *flexible*, i.e., we intend to give them time-dependent meanings. The other symbols are *rigid*, i.e., we intend to give them time-independent meanings.

false is the usual logical constant, $\Rightarrow$ denotes logical implication, $\mathcal{U}$ is the *until* temporal operator, and $\mathcal{S}$ is the *since* temporal operator.

$\varphi \, \mathcal{U} \, \psi$ means that φ will always be true from the next moment on until ψ becomes true for the next time; φ need not be true any more as soon as ψ holds; ψ must eventually become true.

$\psi \, \mathcal{S} \, \varphi$ means that, up to the previous moment, ψ was always true since φ was true for the last time; ψ need not have been true as long as φ was; φ must have once been true.

As usual, we introduce derived connectives as follows. $\neg \varphi$ for $\varphi \Rightarrow \text{false}$, *true* for $\neg \text{false}$, $\varphi \vee \psi$ for $\neg \varphi \Rightarrow \psi$, etc.

We also introduce the following future-oriented, derived temporal operators. $\mathsf{X} \varphi$ for $\text{false} \, \mathcal{U} \, \varphi$, $\mathsf{X}^? \varphi$ for $\neg(\mathsf{X}(\neg \varphi))$, $\mathsf{F} \varphi$ for $\varphi \vee \text{true} \, \mathcal{U} \, \varphi$, $\mathsf{G} \varphi$ for $\neg(\mathsf{F}(\neg \varphi))$, $\varphi \, \mathcal{U}^\circ \psi$ for $\varphi \wedge \varphi \, \mathcal{U} \, \psi$, and $\varphi \, \mathcal{P}^+ \psi$ for $\neg((\neg \varphi) \, \mathcal{U} \, \psi)$.

$\mathsf{X} \varphi$ means that φ is true at the next point in time (*tomorrow*) which must exist. $\mathsf{X}^? \varphi$ means that φ is true at the next point in time if it exists. $\mathsf{F} \varphi$ means that φ is eventually true. $\mathsf{G} \varphi$ means that φ is true forever. $\varphi \, \mathcal{U}^\circ \psi$ means that, from *this* moment on, φ will always be true until ψ becomes true for the next time. $\varphi \, \mathcal{P}^+ \psi$ means that φ will precede ψ, i.e., φ will be true before ψ becomes true. The latter need never happen.

Corresponding past-oriented derived temporal operators are introduced as follows. $\mathsf{Y} \varphi$ stands for $\text{false} \, \mathcal{S} \, \varphi$, $\mathsf{Y}^? \varphi$ for $\neg(\mathsf{Y}(\neg \varphi))$, $\mathsf{P} \varphi$ for $\varphi \vee \text{true} \, \mathcal{S} \, \varphi$, $\mathsf{H} \varphi$ for $\neg(\mathsf{P}(\neg \varphi))$, $\varphi \, \mathcal{S}^\circ \psi$ for $\varphi \wedge \varphi \, \mathcal{S} \, \psi$, and $\varphi \, \mathcal{P}^- \psi$ for $\neg((\neg \psi) \, \mathcal{S} \, \varphi)$.

$\mathsf{Y} \varphi$ means that φ was true at the previous point in time (*yesterday*) which must exist. $\mathsf{Y}^? \varphi$ means that φ was true at the previous point in time if it exists. $\mathsf{P} \varphi$ means that φ was once true. $\mathsf{H} \varphi$ means that φ was always true. $\varphi \, \mathcal{S}^\circ \psi$ means that, up to *this* moment, ψ has always been true since φ was true for the last time. $\varphi \; \mathcal{P}^- \; \psi$ means that, if φ ever was true, then φ preceded ψ, i.e., φ was true before ψ was true.

Furthermore, we apply the usual rules for omitting brackets.

Formal interpretation is described in terms of *possible worlds*, cast in event grove terminology. If $G = (Ev, \rightarrow)$ is an event grove, then Ev is the set of possible worlds, $\rightarrow$ is an accessibility relation corresponding to *next*, and $\rightarrow^*$ is an accessibility relation corresponding to *eventually*.

Interpretation is given at events in class life cycles. Let U be a Σ-algebra, $C = (G, \lambda, \Gamma)$ a class, $L \in \mathcal{L}(G)$ a life cycle in G, and $e \in L$ an event. Satisfaction $C, L@e \models \varphi$ means that φ is valid at event e in life cycle L in class C.

Definition 12.36. The satisfaction relation $\models$ is inductively defined as follows.

$C, L@e \models false$ 	does not hold;

$C, L@e \models t_1 \theta t_2$ 	holds iff $e \in L$ and $t_{1U} \theta_U t_{2U}$;

$C, L@e \models a\theta t$ 	holds iff $e \in L$ and $\lambda(e)(a)\theta_U t_U$;

$C, L@e \models (\varphi \Rightarrow \psi)$ holds iff $e \in L$ and $C, L@e \models \varphi$ implies $C, L@e \models \psi$;

$C, L@e \models (\varphi \; \mathcal{U} \; \psi)$ holds iff $e \in L$ and there is a future event $e' \in L$,
$\qquad\qquad\qquad\qquad\quad e \rightarrow^+ e'$, where $C, L@e' \models \psi$ holds, and $C, L@e'' \models \varphi$
$\qquad\qquad\qquad\qquad\quad$ holds for every event $e'' \in L$ such that $e \rightarrow^+ e'' \rightarrow^+ e'$;

$C, L@e \models (\psi \; \mathcal{S} \; \varphi)$ holds iff $e \in L$ and there is a past event $e' \in L$,
$\qquad\qquad\qquad\qquad\quad e' \rightarrow^+ e$, where $C, L@e' \models \varphi$ holds, and $C, L@e'' \models \psi$
$\qquad\qquad\qquad\qquad\quad$ holds for every event $e'' \in L$ such that $e' \rightarrow^+ e'' \rightarrow^+ e$.

$\diamond$

By the abbreviations introduced above, we may derive satisfaction conditions for the other connectives and temporal operators, e.g.,

$C, L@e \models (X \; \varphi)$ 	holds iff $e \in L$ and there is a next event $e' \in L$,
$\qquad\qquad\qquad\qquad\quad e \rightarrow e'$, where $C, L@e' \models \varphi$ holds.

As usual, a formula φ is said to be *satisfiable* in a class C iff $C, L@e \models \varphi$ holds for some life cycle L in C and some event e in L. A formula φ is said to be *valid* in C, in symbols $C \models \varphi$, iff $C, L@e \models \varphi$ holds for all life cycles L in C and all events e in L. A set of formulas Φ *entails* a formula φ in C, in symbols $C, \Phi \models \varphi$, iff φ holds at every event in every life cycle in C at which all formulas in Φ hold.

12.4.2 Temporal specification

Let Γ be a class signature as given in Definition 12.3.

Definition 12.37. A *class specification* is a pair $Cspec = (\Gamma, \Phi)$ where Φ is a set of PTL formulas over Γ. 	$\diamond$

Semantics is defined in the subcategory Γ-**class** $\subseteq$ **class** where the signature Γ is fixed and, for every morphism $f = (g, \gamma)$, γ is the identity morphism.

Γ-**class** has final elements[2]. Such an element is unique up to isomorphism. It represents the most liberal behavior over Γ. Intuitively, its traces are all those that can be formed over Γ.

A corresponding result holds for subcategories of classes satisfying given temporal axioms. Let $Cspec = (\Gamma, \Phi)$ be a specification. Let $Cspec$-**class** be the full subcategory of Γ-**class** consisting of all classes over Γ satisfying Φ.

$Cspec$-**class** has final elements. A final element is obtained as the maximal subclass of a final element in Γ-**class** satisfying Φ.

Corresponding finality results are given in [CSS94].

We may utilize these results for assigning standard abstract semantics to temporal class specifications. In order to increase specification power and manageability, we envisage giving direct axiomatic specification only to basic classes. Derived and complex classes may be specified using in-the-large structuring mechanisms like those described in the next section.

12.5   Class specification in the large

For an effective specification method, in-the-large structuring mechanisms are indispensable. We express structuring by structured object sorts and corresponding object type morphisms.

Object sorts are structured in two ways, by order sorting and by object sort constructors. Inheritance (is-a and as-a) and hiding are expressed by binary ordering relationships on object sorts. Composition is expressed by binary sort constructors for generalization and aggregation. Interaction between components of a complex object can be specified in the framework developed so far.

We note in passing that state variables (see Example 12.1) may be considered as specified by a sort constructor $Var: S^d \rightarrow S^o$. This is one way to express parameterization. Parameterization is a powerful in-the-large concept. Unfortunately, because of space limitations, we cannot go into this issue here.

12.5.1   Inheritance

Inheritance describes how one class re-uses features from another. On a representation level, the inheriting class adopts attributes and possibly adds some. On an implementation level, the inheriting class re-uses code, i.e., it implements inherited attributes (and actions) in the same way as in the original class.

---

[2] For obvious reasons, we avoid the term *object* for the elements of a category although it is standard in category theory.

We look at inheritance from a semantic point of view. This means that the inheriting class provides attributes (and actions) with the same syntax and semantics as the original. Nothing is implied regarding implementation. In fact, even if its semantics does not change, an inherited action may have a separate implementation, for instance a more efficient one. This approach allows for multiple representations and implementations, even within one type. This may be useful, for instance, in a distributed environment with heterogeneous processors.

Let $\Sigma = (S, \Omega)$ be an object system signature where $S = S^d + S^o$, with S^d and S^o being respectively the sets of data and object sorts. Let U be a Σ-algebra.

On the syntactic level, inheritance is expressed by *object sort ordering*, i.e., by a partial ordering $\leq$ on object sorts S^o.

Let $r, s \in S^o$ be object sorts. The intended interpretation of $r \leq s$ is that each object of sort r inherits from an object of sort s. Depending on the kind of inheritance considered, it is formalized in our approach by an object type morphism $\tau \colon U_r \rightrightarrows U_s$ or a reverse object type morphism $\tau \colon U_r \rightleftarrows U_s$.

We distinguish between *is-a inheritance* $r \lesssim s$, *as-a inheritance* $r \precsim s$, and *hiding* $r \sqsubseteq s$. For the first two, we indicate how to generalize the concepts so as to allow for overriding and overlaying. All the forms of inheritance may be multiple.

Is-a inheritance. An is-a inheritance relationship $r \lesssim s$ is intended to express a kind of temporal specialization. Technically, $r \lesssim s$ is to be interpreted by an is-a inheritance morphism defined as follows.

Definition 12.38. An *is-a inheritance morphism* is an object type morphism $\tau \colon U_r \rightrightarrows U_s$ such that $\tau' \colon Id_r \subseteq Id_s$ is an inclusion and $\tau'' \colon C_r \to C_s$ is strict.   ◇

For example, each customer is a person, see Example 12.10. In systems where objects are represented by their identities, is-a inheritance naturally leads to polymorphism because the type of an object is not unique.

The class morphism $\tau'' \colon C_r \to C_s$ that goes with an is-a inheritance morphism reflects the associated class inheritance. Referring to Example 12.32, if a customer is a person, then the class morphism $\tau_1'' \colon C_{\text{Customer}} \to C_{\text{Person}}$ says that every customer life cycle has a valid, underlying person life cycle.

Is-a inheritance may be *multiple*, i.e., we have $r \lesssim s_1, \ldots, r \lesssim s_n$ where the object sorts s_k, $1 \leq k \leq n$, are distinct. On the interpretation level, this leads to object type morphisms $\tau_k \colon U_r \rightrightarrows U_{s_k}$, $1 \leq k \leq n$, expressing from which objects in $U_{s_1}, \ldots, U_{s_n}$ a given object in U_r inherits and that which is being inherited.

The *multiple inheritance problem* does not appear in our context; it has to do with resolving naming conflicts on the syntactic level. An elegant categorial treatment of this issue may be found in [LcP90].

It is obvious from the definition that is-a inheritance $\lesssim$ is a reflexive and transitive relation. Antisymmetry does not hold in a strict sense but $r \lesssim s$ and $s \lesssim r$ imply that we have $Id_r = Id_s$, and C_r and C_s are equivalent in a rather strong sense. Because of space limitations, we cannot elaborate on this point.

As-a inheritance. Like is-a inheritance, as-a inheritance $r \lesssim s$ is intended to express a kind of temporal specialization. The difference is that the former means that *all* features are inherited and new ones may be added, whereas as-a inheritance means that *some* features are inherited and possibly used in a special way.

Technically, $r \lesssim s$ is intended to be interpreted by an as-a inheritance morphism defined as follows.

Definition 12.39. An *as-a inheritance morphism* is a reverse object type morphism $\tau: U_r \rightleftarrows U_s$ such that $\tau': Id_r \subseteq Id_s$ is an inclusion and $\tau'': C_s \rightarrow C_r$ is strict. ◇

For example, we view a rectangle as a square in Example 12.14. On the identity level, there is no difference to is-a inheritance. In fact, each square is a rectangle as well. Consequently, as-a inheritance also leads to polymorphism.

The difference is that on the class level the morphisms go the other way. For example, if we take a rectangle as a square, then there is a class morphism $\tau_2'': C_{\text{Rectangle}} \rightarrow C_{\text{Square}}$ showing how rectangles can be viewed as squares, see Example 12.33.

The class morphism $\tau'': C_s \rightarrow C_r$ that is associated with an as-a inheritance morphism reflects the intended behavior relationship. The rectangle-as-a-square example suggests that it is natural to assume that the underlying event grove morphism is an identity.

We may also have multiple as-a inheritance. In fact, multiple inheritance may involve any combination of is-a and as-a.

It is obvious from the definition that as-a inheritance $\lesssim$ is a reflexive and transitive relation. As for antisymmetry, the situation is similar to the is-a case.

Overriding and overlaying. While pure inheritance, as introduced above, only allows for adding new attributes (and actions) in addition to adopting the old ones, overriding allows the possibility of changing some of the latter, i.e., give them a new meaning. Overlaying provides new versions of attributes while keeping the existing versions accessible.

We describe the situation for is-a inheritance; the case is similar for as-a inheritance. A *weak* is-a relationship allowing for overriding is denoted by $r \leq s$. Let $U_r = (Id_r, C_r)$ and $U_s = (Id_s, C_s)$. Technically, $r \leq s$ is to be interpreted by

- an inclusion $Id_r \subseteq Id_s$,
- strict class morphisms $C_r \to C' \leftarrow C_s$ for some class C', and
- a class signature injection $\gamma\colon \Gamma_s \to \Gamma_r$,

where Γ_s and Γ_r are respectively the signatures of C_r and C_s. Let Γ' be the signature of C'. For ease of notation, we assume that the signature morphisms underlying the strict class morphisms are inclusions. Then we have $\Gamma' \subseteq \Gamma_r \cap \Gamma_s$.

If $a_s \in \Gamma_s$, then $a_r = \gamma(a_s)$ is a_s's new version in class C_r. We may have one of the following cases.

$a_s \in \Gamma'$: a_s is inherited ...
- $a_r = a_s$: ... and coincides with the new version.
  $a_r \neq a_s$: ... but is overlayed by a_r ...
      $a_r \in \Gamma'$: ... that is equal to some other inherited item.
- $a_r \notin \Gamma'$: ... that is new.
$a_s \notin \Gamma'$: a_s is overridden by a_r ...
    $a_r \in \Gamma'$: ... that is equal to some other inherited item.
- $a_r \notin \Gamma'$: ... that is new.

The bullet • indicates cases of practical interest.

Hiding. While is-a and as-a inheritance reflect different kinds of temporal specialization, the intention of hiding is quite different. It may therefore not be justified to subsume it under inheritance, but the formalities on the class level are similar to as-a inheritance.

A hiding relationship $r \sqsubset s$ is intended to express that r-objects are *interfaces* or *views* of s-objects. These are dependent objects like as-a specializations, but they have identities of their own.

Definition 12.40.   A *hiding morphism* is a reverse object type morphism $\tau\colon U_r \rightleftarrows U_s$ such that $Id_r \cap Id_s = \emptyset$ and $\tau''\colon C_s \to C_r$ is strict.   ◇

There is no point in allowing for overriding or overlaying, interfaces are designed to reflect fully what happens in the base object.

In analogy to multiple inheritance, we may have "multiple interfaces", i.e., objects that are interfaces of several base objects. Multiple interfaces synchronize their base objects because each action, attribute, etc., in the interface is shared by all base objects. Categorically, the synchronized behavior is given by a limit in **class**.

Conversely, we may also have several interfaces of the same base object, even of the same sort. The analogous inheritance situation is to have several specializations of the same object – which cannot be of the same sort, though.

There is another difference to as-a specialization. The intended process with the latter is to use only the restricted features of the specialization while the other features are not accessible, because they are out of scope.

Technically, this is enforced by giving the specialization the same identity and relying on the usual object-oriented rule that identities are unique at any time. In contrast, giving interfaces and their base objects different identities means that they may coexist in a state.

It is obvious from the definition that hiding $\sqsubset$ is an irreflexive relation. Transitivity holds if the identity sets involved are pairwise disjoint. As for antisymmetry, the situation is similar to the is-a and as-a cases.

Covariance vs. contravariance. There is one more aspect of inheritance that is worth mentioning, namely, changing sorts of inherited attributes. In functional approaches using subsorting, operators are *contravariant* with respect to range sorts in a natural way: if $\omega: s \to t$ and $s' \leq s$, then we have $\omega: s' \to t'$ for any t' such that $t \leq t'$. This also holds for attributes and other object-sorted functions if we work with inheritance. For instance, if *spouse : Person* is an object-valued attribute for *Person*, each *Male* is a *Person*, and each *Person* is a *LivingBeing*, then we naturally have *spouse : LivingBeing* as an attribute for *Male*.

While this is undoubtedly true, it is not particularly helpful. If we work with inheritance in object-oriented systems, we endeavor to refine the specification with more detail. In our example, if we introduce *Male* and *Female* as special kinds of *Person*, then we take the opportunity to refine the *spouse* attribute as well and specify *spouse : Female* for *Male*, and *spouse : Male* for *Female*.

Therefore, it is natural to have *covariance* with respect to range sorts in object-oriented systems: if $\omega: s \to t$ and $s' \leq s$, then we impose $\omega: s' \to t'$ for a suitable $t' \leq t$.

12.5.2   Composition

Composition describes how constituent objects are put together to form complex objects. On the syntactic level, composition is expressed by object sort constructors. We have two kinds of composition, generalization $r \oplus s$ and aggregation $r \otimes s$. These constructors are associative, so the same constructor may be iterated a finite number of times.

On the semantic level, generalization is expressed by coproducts in a category $\mathbf{ot}(\Gamma\text{-}\mathbf{class})$ of object types over a given class subcategory $\Gamma\text{-}\mathbf{class}$ with a fixed signature Γ, and aggregation is expressed by products in the category $\mathbf{ot}(\mathbf{class})$ of object types over the category **class** of all classes with varying signatures.

Generalization. Composition by generalization $r \oplus s$ is intended to express how objects can be put together to reflect alternative choice among its constituents. We assume that we have $\Gamma_r = \Gamma_s = \Gamma$ for the class signatures of sorts r and s, respectively, see Example 12.11 for the motivation.

Interpretation is given by $U_{r\oplus s} = U_r + U_s$, the coproduct in $\mathbf{ot}(\Gamma\text{-class})$. The coproduct morphisms describe how the constituent object types are embedded into the generalization.

If $U_r = (Id_r, C_r)$ and $U_s = (Id_s, C_s)$, then $U_{r\oplus s} = (Id_{r\oplus s}, C_{r\oplus s})$ where $Id_{r\oplus s} = Id_r + Id_s$ is the disjoint union, and $C_{r\oplus s} = (G_r + G_s, \lambda_r + \lambda_s, \Gamma)$ where $G_r + G_s$ is the coproduct of event groves, and $\lambda_r + \lambda_s : Ev_r + Ev_s \to [\Gamma]$ associates the original label with each event.

If $G_r = (Ev_r, \to_r)$ and $G_s = (Ev_s, \to_s)$ are event groves, their coproduct is $G_r + G_s = (Ev_r + Ev_s, \to_r + \to_s)$. The associated sequential event structure $E(G_r + G_s)$ has conflict relation $\# = \#_r + \#_s + Ev_r \times Ev_s$ provided that Ev_r and Ev_s are disjoint (otherwise, disjoint copies of these sets must be taken). Intuitively, this means that no events from different constituents of a generalization may be in the same life cycle, so there is a choice at the beginning to pursue a life cycle that lies in one of the constituents.

Example 12.41. We refer to Example 12.11 where *Customer = Person + Company*. Given specifications for *Person* and *Company* with object sorts Person and Company, respectively, we may specify Customer = Person $\oplus$ Company in order to achieve the desired interpretation. ∎

For object sorts $r, s \in S^o$, we have $r \lesssim r \oplus s$ and $s \lesssim r \oplus s$. For instance, in the example above, we may view a person as a customer.

Aggregation. Composition by aggregation $r \otimes s$ is intended to express how objects are put together to form complex objects. A life cycle of a complex object is a step-by-step synchronization of life cycles of its components, sharing events at each step.

Interpretation is given by $U_{r\otimes s} = U_r \bowtie U_s$, the product in the category $\mathbf{ot}(\mathbf{class})$. The product morphisms describe projections from a complex object to its components.

If $U_r = (Id_r, C_r)$ and $U_s = (Id_s, C_s)$, then $U_{r\otimes s} = (Id_{r\otimes s}, C_{r\otimes s})$ where $Id_{r\otimes s} = Id_r \times Id_s$ is the cartesian product, and $C_{r\otimes s} = (G_r \times G_s, \lambda_r \times \lambda_s, \Gamma_r + \Gamma_s)$ where $G_r \times G_s$ is the product of event groves, and $\lambda_r \times \lambda_s : Ev_r \times Ev_s \to [\Gamma_r + \Gamma_s]$ associates the labelings of component events with a complex event: if, say, $a \in \Gamma_r$, then $\lambda_r \times \lambda_s(e_r, e_s)(a) = \lambda_r(e_r)(a)$.

If $G_r = (Ev_r, \to_r)$ and $G_s = (Ev_s, \to_s)$ are event groves, their product is given by $G_r \times G_s = (Ev_{r\otimes s}, \to_{r\otimes s})$ where $Ev_{r\otimes s} = \{(e_r, e_s) \mid e_r \in E_r, e_s \in E_s,$ and $\tau_{G_r}(e_r) = \tau_{G_s}(e_s)\}$, and, for all $e_r, e'_r \in Ev_r$ and all $e_s, e'_s \in Ev_s$, $(e_r, e_s) \to (e'_r, e'_s)$ holds iff both $(e_r \to_r e'_r)$ and $(e_s \to_s e'_s)$ hold.

Example 12.42. We refer to Example 12.12 where *Car = Motor $\bowtie$ Chassis*. Given specifications for *Motor* and *Chassis* with object sorts Motor and Chassis, respectively, we may specify Car = Motor $\otimes$ Chassis in order to achieve the desired interpretation. ∎

For object sorts $r, s \in S^o$, we impose $r \sqsubseteq r \otimes s$ and $s \sqsubseteq r \otimes s$. For instance, in the example above, we view both a motor and a chassis as car interfaces, giving a partial view and hiding the rest. As a consequence, in any interpretation, each component belongs to precisely one object. However, note that we may have several hiding relationships on r or s or both so that we may have shared components.

Component interaction. Synchronous interaction between components of a complex object may be described by temporal axioms within the aggregation.

Example 12.43. In the Bank Example 12.2, we may replace the second and fifth axioms by the strict rule

 $\circ\, \texttt{transfer(from,to,m)} \Rightarrow (\circ\, \texttt{A(from).debit(m)} \land \circ\, \texttt{A(to).credit(m)})$

expressing that a transfer action calls a debit action in the `from` account and a credit action in the `to` account. This means that all three actions must occur simultanously.   ■

Interaction axioms are not restricted to this simple kind of *event calling* although these describe many cases of interest. It is conceivable to have liveness axioms like $\circ\, \alpha \Rightarrow \mathsf{F} \circ \beta$, safety axioms like $\circ\, \alpha \Rightarrow \mathsf{G} \neg \triangleright \beta$, exclusion axioms like $\circ\, \alpha \Rightarrow \neg \circ \beta$, etc.

12.6   Object systems

For coping with object systems, it is not practical to work with a sequential model that requires global synchronization of all events. With a concurrent model, however, we must be careful to stay within manageable limits: a model powerful enough to deal with all aspects of concurrency is hard to work with. Our solution is to use linear-time, distributed temporal logic for specification, and locally sequential, labeled event structures as a denotational model. This is expressive enough to reflect the kind of concurrency that often occurs in practice, for instance, in networks of work stations. On the other hand, the model is close enough to sequential classes and temporal logic to allow for a smooth extension of familiar concepts.

Let $\varSigma = (S, \varOmega)$ be an object system signature, interpreted by a data and object algebra U. Let $U_r = (Id_r, C_r)$ and $U_s = (Id_s, C_s)$ be object types in U. Any two instances $i.C_r \in U_r$ and $u.C_s \in U_s$ are assumed to operate concurrently subject to given inheritance constraints. For simplicity, we assume that there are no such constraints so that a system consists of independent objects $i.C_r, u.C_s, \ldots$ The classes are not necessarily distinct.

12.6.1   A denotational system model

In order to show the principle, we will assume that the system consists of just two objects $i.C_r$ and $u.C_s$, operating concurrently. Its behavior is given by $B = C_r \times C_s$, the concurrent product of their classes. We briefly describe the product construction for labeled prime event structures, first treating unlabeled ones and then adding labels. The construction is due to Vaandrager [Vaa89].

Let $E_r = (Ev_r, \rightarrow_r, \#_r)$ and $E_s = (Ev_s, \rightarrow_s, \#_s)$ be event structures. Let $Ev_r^{\perp} = Ev_r + \{\perp_r\}$, $Ev_s^{\perp} = Ev_s + \{\perp_s\}$, and $Z = Ev_r^{\perp} \times Ev_s^{\perp} - \{(\perp_r, \perp_s)\}$. An element (e_r, e_s) where $e_r \in Ev_r$ and $e_s \in Ev_s$ characterizes a shared event where e_r and e_s happen synchronously. An element $(e_r, \perp_s)$ characterizes an event e_r occurring in isolation in E_r, i.e., concurrently to whatever happens in E_s. Correspondingly, an element $(\perp_r, e_s)$ characterizes an event e_s occurring in isolation in E_s.

For $e = (e_r, e_s), e' = (e'_r, e'_s) \in Z$, let $e \rhd e'$ iff $e_r \rightarrow_r^* e'_r$ or $e_s \rightarrow_s^* e'_s$. This auxiliary relation keeps track of local causalities but is not a causality relation itself. For a subset $X \subseteq Z$, let $\rhd_X = \rhd \cap (X \times X)$.

Definition 12.44. A subset $X \subseteq Z$ is a *preconfiguration* iff (1) $pr_r(X) \in C(E_r)$ and $pr_s(X) \in C(E_s)$ and (2) $\rhd_X^+$ is a partial order. X is called a *complete prime* iff it has a unique maximal element with respect to $\rhd_X^+$.

◇

A preconfiguration X characterizes a set of events that has occurred at a given moment, reflected locally by configurations $pr_r(X)$ in E_r and $pr_s(X)$ in E_s. Condition (2) states that the events of the components may occur only once and that both components must agree on the causal relationships between events: causal loops are not allowed.

The following definition applies a standard procedure to define a prime event structure from its finite configurations.

Definition 12.45. The *concurrent product of E_r and E_s* is defined by $E_r \times E_s = (Ev, \rightarrow^*, \#)$ where $Ev = \{X \mid X \text{ is a complete prime}\}$, $X \rightarrow^* Y$ iff $X \subseteq Y$, and $X \# Y$ iff $X \cup Y$ is not a preconfiguration.

◇

A life cycle in $E_r \times E_s$ is a union of life cycles in E_r and E_s, sharing events at synchronization points. If E_r and E_s are sequential, then the local life cycles are linear traces. Thus, a global system life cycle consists of local traces that are glued together at shared events.

In order to make the product construction complete, we have to add labels. We restrict our attention to classes where we have the definitions available. Let $C_r = (G_r, \lambda_r, \Gamma_r)$ and $C_s = (G_s, \lambda_s, \Gamma_s)$ be classes. Of course, the concurrent product of event groves is defined to be that of the prime event structures they present. Let $\lambda': Z \rightarrow [\Gamma_r + \Gamma_s]$ be defined by $\lambda'(e_r, \perp_s) = \lambda_r(e_r)$, $\lambda'(\perp_r, e_s) = \lambda_s(e_s)$, and $\lambda'(e_r, e_s)(a) = \lambda_r(e_r)(a)$ if $a \in \Gamma_r$, and $= \lambda_s(e_s)(a)$ if $a \in \Gamma_s$. Note that Γ_r and Γ_s are disjoint in our context.

Definition 12.46. The *concurrent product of* C_r *and* C_s is defined by $C_r \times C_s = (G_r \times G_s, \lambda, \Gamma_r + \Gamma_s)$ where $\lambda(X) = \lambda'(x)$ such that x is the unique maximal element in X with respect to $\rhd_X^+$.   ◇

Isolated events retain their local labels, and synchronized events obtain their labels in the same way as in aggregated objects, see Section 12.5.2. Note that concurrent events always have distinct labels – there is no "auto-concurrency". A constructive definition of event structure product is given in [LG91].

12.6.2 Distributed temporal specification

An object system specification consists of an object instance signature Υ and a set family of object axioms Φ, see Definition 12.19. Our specification logic is based on a version of the *n-agent logic*, see Section 12.7 for hints to the literature. An agent corresponds to an object that may be thought of as a site in a distributed system. Temporal descriptions are given locally from the viewpoints of agents. Interaction is handled by incorporating statements about other agents. For example, agent i may prescribe that, whenever i sends a message to agent u, the latter will eventually acknowledge receipt. The actions of agent i sending and agent u receiving may be modeled to occur concurrently. Acknowledgement, however, requires interaction; it may be modeled as a synchronized joint event where both talk to each other: i says "I sent something to you" while u says "I received something from you".

We introduce our version of n-agent logic called DTL. Let $Id = \{i, u, \dots\}$ be a given set of agent identities, also called *localities*.

Definition 12.47. The *syntax* of DTL, i.e., its set of well-formed formulas Φ, is given by

$$\text{DTL} ::= \text{ATOM} \mid (\text{DTL} \Rightarrow \text{DTL}) \mid (\text{DTL} \, \mathcal{U}_{Id} \, \text{DTL}) \mid (\text{DTL} \, \mathcal{S}_{Id} \, \text{DTL})$$

where ATOM is defined as in Definition 12.35.   ◇

The only difference with respect to PTL is that we have temporal operators $\mathcal{U}_i$ and $\mathcal{S}_i$ indexed by localities. The intended meaning of $\varphi \, \mathcal{U}_i \, \psi$ is that $\varphi \, \mathcal{U} \, \psi$ holds locally at agent i, and correspondingly for $\mathcal{S}_i$. As for PTL, we may introduce abbreviations X_i, $X_i^?$, F_i, G_i, $\mathcal{U}_i^\circ$, and $\mathcal{P}_i^+$ as well as Y_i, $Y_i^?$, P_i, H_i, $\mathcal{S}_i^\circ$ and $\mathcal{P}_i^-$.

Additionally, we introduce the abbreviation @u for $X_u^?$ *true*, meaning that there is interaction with u at this point in time, i.e., @u holds precisely at events shared with u.

Specification in DTL is bound to localities. That means that the formulas consist of a set family $\Phi = \{\Phi\}_{i \in Id}$ of *local axioms*. The notation $i : \varphi$ means that $\varphi \in \Phi_i$ where $i \in Id$.

Interaction is specified by referring to another locality by using its local temporal operators. We illustrate the idea by examples.

Example 12.48. In order to give the examples a personal touch, we read "I" for i and "you" for u.

$i : \mathsf{P}_u\,\varphi$ I hear that φ was once valid for you

$i : \mathsf{G}_i(@u \Rightarrow \mathsf{X}_i\,\varphi)$ whenever I talk to you, I have φ the next day

$i : \varphi \Rightarrow \mathsf{X}_i\,@u$ if φ holds, then I talk to you the next day

$i : \mathsf{X}_u\,@i$ you tell me that you will contact me tomorrow

$i : \varphi \Rightarrow \mathsf{X}_u\,\psi$ if φ holds, then you tell me that ψ will hold for you tomorrow

$i : \mathsf{Y}_i\,\mathsf{F}_u\,\varphi$ you told me yesterday that φ will hold for you some time

$i : (i.\circ \mathrm{send}(x) \Rightarrow \mathsf{F}_i\ \mathsf{P}_u\,u.\circ \mathrm{receive}(x))$ if I send x, then I will eventually obtain an acknowledgement from you that you received x

∎

Interpretation is given at an event e in a life cycle L in a distributed system $B = \prod_{i \in Id} C_i$. L is a web of local linear traces $L_i, L_u, \ldots$ sharing events. Satisfaction is defined locally. Let $i \in Id$ be a locality, and let $e \in L$. $B, L@e \models_i \varphi$ means that φ holds locally at event e in life cycle L_i in system B.

Here the generality of Definition 12.36 pays off: we may replace C by B, $\models$ by $\models_i$, $\mathcal{U}$ by $\mathcal{U}_i$, $\mathcal{S}$ by $\mathcal{S}_i$, $\rightarrow$ by $\rightarrow_i$, etc. and add the two following rules in order to obtain the definition of local validity.

$B, L@e \models_i (\varphi\,\mathcal{U}_u\,\psi)$iff $e \in L_i, e \in L_u$, and $B, L@e \models_u (\varphi\,\mathcal{U}_u\,\psi)$ hold;

$B, L@e \models_i (\psi\,\mathcal{S}_u\,\varphi)$iff $e \in L_i, e \in L_u$, and $B, L@e \models_u (\psi\,\mathcal{S}_u\,\varphi)$ hold.

The life cycles consist of n local traces that may share events. These shared events are points of interaction. Note that the local state $\downarrow e$ of an event $e \in Ev_i$ may contain events of other objects, for instance, an event e'' that is causal in u for an interaction event e' with u, that is in turn causal for e, i.e., we have $e'' \rightarrow_u e' \rightarrow_i e$. In this sense, agents obtain full historical information about others they have talked to – and about those the others have talked to, etc.

Example 12.49. For illustration, we refer to Example 12.2 and give a concurrent version of the Bank class.

```
class Bank;
    uses money, acct#, Person, Account;
    attributes owner: Person;
    components Acc(acct#): Account;
    actions    *establish(Person),
               transfer(Account,Account,Bank,money);
               receive(Account,Bank,Account,money);
               +liquidate;
    axioms var  p: Person, from, to, to2: acct#,
                b,b2: Bank, m, n: money;
```

```
◦establish(p) ⇒ owner=p,
◦transfer(from,to,b,m) ⇒ X ◦Acc(from).debit(m),
◦transfer(from,to,b,m) ⇒ ¬◦transfer(from,to,b,m)
        𝒰° P_b b.◦receive(from,self,to,m),
◦transfer(from,to,b,m) ∧ ◦transfer(from,to2,b2,m)
        ⇒ to=to2 ∧ b=b2,
◦receive(from,b,to,m) ⇒
        ◦Acc(to).credit(m) 𝒫⁺ ◦Acc(to).debit(n),
◦receive(from,b,to,m) ⇒ F ◦Acc(to).credit(m)
```
end

Comparing with Example 12.2, we have introduced one more parameter in the transfer action, namely, the bank b to which the money is to be transferred. We have also added an action `receive(from,b,to,m)` of receiving money from an account of another bank b. The standard variable `self` denotes "this" bank; it is used when there is a need to refer to the hidden local object variable (see Example 12.1).

By introducing **Bank** variables like b, we can talk locally about transfers to and from another bank. Remember that local variables and localities are omitted. For instance, the full version of the third axiom is

```
self: self.◦transfer(from,to,b,m) ⇒
        ¬self.◦transfer(from,to,b,m)
        𝒰°_self 𝒫⁺_b b.◦receive(from,self,to,m).
```

The axioms should be understandable without further explanation.

The following liveness property for bank i is a consequence of the third axiom.

```
        i.◦transfer(from,to,b,m) ⇒ F_i P_b b.◦receive(from,i,to,m).
```
The property says that transfers are eventually acknowledged as having been received. ∎

Let $Sspec = (\Upsilon, \Phi)$ be an object system specification with a set $Id = \{i, u, \ldots\}$ of identities. Its models are distributed life cycles in systems $B = \prod_{i \in Id} C_i$ satisfying Φ, i.e., $B, L@e \models_i \Phi_i$ for every identity $i \in Id$ and every event $e \in L_i$. The structure of the model category is subject to further study. For a related but slightly different setting, an initiality result has been proved in [ES95].

12.7   Related work

There are a number of methods for object-oriented analysis, modeling, and design. These methods are informal, or semiformal at best. However, they come with methodological guidelines and graphical notations. They help to make formal languages fit for use, so they do have their benefits in the early modeling and design stages. Recently, there has been considerable activity to unify these methods: the Booch [Boo94], OMT [RBP+91], and OOSE methods [Jac92] merged into the universal modeling language UML [FS97] that has been submitted to the Object Management Group to be considered as a standard. Another successful OO analysis method is Fusion [CAB+94].

However, these methods are too inprecise and ambiguous when it comes to animation, verification, and forecasting of system properties, and when it comes to generating test cases or even implementations from specifications.

Among the logic-based formal methods, the work reported here is based on experiences with developing the OBLOG family of languages and their semantic foundations. OBLOG is being developed into a commercial product [ESD93]. In the academic world, there are several related developments: TROLL [SJH93, SHJE94, JSHS96, HJ95, EH96, Har97, DH97], GNOME [SR94], LCM [FW93], and ALBERT [DDPW94].

There are other approaches to formal object specification with a sound theoretical basis. The ones most closely related to ours are FOOPS [GM87b, RS92, GS95a] and MAUDE [Mes93]. FOOPS is based on OBJ3 [GW88], which is in turn based on equational logic. MAUDE is based on rewriting logic that is a uniform model of concurrency [Mes92]. Other language projects working on related ideas are OOZE [AG91] and ETOILE [AB95a]. We also acknowledge inspiration by other work, e.g., [AZ95, Bee95, FM92].

In the OBLOG family, TROLL3 [EH96, Har97] is the first to address problems of concurrency and communication, and to integrate benefits from the informal methods mentioned above. Thus there is a graphical notation for TROLL3 called OMTROLL that adopts elements from OMT. [EH96] gives a brief overview of TROLL3 and OMTROLL and their logic foundations.

The denotational object and system models adopted in this chapter are based on prime event structures. These are abstract and elegant, and powerful enough to model full concurrency. Their relationship to other models of concurrency like labeled transition systems and Petri nets is well investigated, see [NPW81, NRT92, WN95]. They have also been used as a semantic basis for extending the logic to deal with concurrency, see below.

As for the many logics that might be useful for class and system specification, we refer to [AGM92] for a comprehensive overview. The specification logic defined in this paper is based on temporal logic, see [Pnu77] for the pioneering paper and [MP92, MP95] for more recent textbooks. Temporal logic has also been used as the backbone of a large-scale programming system [Tan94]. Our logic is influenced by OSL [SSC95] and the logic of [Aba89] that goes back to [GPSS80] where a sound and complete proof system is also given.

The extension of PTL towards concurrency, introduced in Section 12.6, is based on n-agent logic that was introduced and developed in [LT87, LMRT91, LRT92, MT92b, Thi94, Ram96]. A distinguishing feature is that the assumption of an omnipresent observer of the entire system under consideration is dropped and replaced by a local causal perspective. n-agent logics can explicitly distinguish sequential agents (localities) in the system, refer to the local viewpoint of each agent, and express communication between agents [LT87, LRT92, Ram96]. The last is the major feature of distribution.

These ideas have been addressed in the context of object orientation [ES95] and used to axiomatize a significant subset of the GNOME language [Cal96]. In [ECSD98], two distributed logics are defined, an operational one with basic communication facilities, and a more elaborate one that can talk about communication in an elegant, implicit way. The main result is that there is a sound and complete translation from the latter to the former.

12.8   Concluding remarks

The theory outlined here is taking shape but is not yet complete. It has to be elaborated and refined in several respects.

In particular, the details of class and system composition have to be worked out. For instance, laws for object type construction terms like $T + T \simeq T$, $(T_1 + T_2) \bowtie T_3 \simeq (T_1 \bowtie T_3) + (T_2 \bowtie T_3)$, $(T_1 + T_2) \times T_3 \simeq (T_1 \times T_3) + (T_2 \times T_3)$, etc. have to be set up and proved, based on a suitable equivalence relation $\simeq$. This is the basis for an in-the-large algebraic treatment and optimization of concurrent system construction.

Further research will focus on interaction and modularization concepts in truly concurrent models. DTL describes synchronous and symmetric interaction from a local point of view. For the TROLL language, a richer spectrum of interaction concepts is envisaged, including modes of asynchronous directed interaction. These can be explained on DTL grounds.

As for modularization, we envisage a module concept that reflects generic building blocks of software. In particular, instantiation as well as horizontal and vertical composition of modules must be supported. For the latter, a module must incorporate a reification step between an external interface on a higher level of abstraction, and an internal interface on a lower level of abstraction [Den96b, DE95, Den96a].

Another issue of practical importance is real time. Real-time constraints set limits as to when an action may or must occur or terminate, and how long it may take from a triggering event to the corresponding reaction.

In the longer term, deductive capabilities and default handling must be better understood and eventually incorporated. The role of deduction is to predict the effect of a design before it is implemented. Defaults enhance modularity by allowing assertions to be made in a local object, even when the vocabulary needed to specify their exceptions is unavailable.

Along with these theoretical developments, languages and systems have to be developed, supported by tools and tested in application case and field studies. The TROLL project has proceeded in this direction, and its experiences are encouraging.

Acknowledgements. The author is grateful for help and inspiration from many sides. Amílcar and Cristina Sernadas have been faithful partners in developing the theory. The other partners in the ISCORE, COMPASS, and ASPIRE projects have made so many inputs that it is impossible to acknowledge in detail. Grit Denker and Juliana Küster Filipe have contributed fundamentally to internal discussions on the subjects of the chapter. The practical work on TROLL has provided important input, especially that by Gunter Saake, Thorsten Hartmann, Jan Kusch, and Peter Hartel. Mojgan Kowsari has put TROLL into practice and has given useful feedback. Grit Denker, Juliana Küster Filipe, Michaela Huhn, and Narciso Martí-Oliet have reviewed earlier versions of the paper and given valuable hints. All these contributions are gratefully acknowledged. Naturally, only the author is responsible for all remaining deficiencies.

This work was partly supported by the EU under ESPRIT-III BRA WG 6112 COMPASS, ESPRIT-III BRA WG 3023 ISCORE and ESPRIT-IV WG 22704 ASPIRE.

13   Algebraic Specification of Concurrent Systems

Egidio Astesiano[1], Manfred Broy[2], and Gianna Reggio[1]

[1] DISI – Dipartimento di Informatica e Scienze dell'Informazione
Università di Genova, Via Dodecaneso 35, Genova 16146, Italy
{astes,reggio}@disi.unige.it   http://www.disi.unige.it
[2] Institut für Informatik der Technischen Universität München
D-80333 München, Germany
broy@informatik.tu-muenchen.de
http://wwwbroy.informatik.tu-muenchen.de/

Introduction

A process is a unit with the capacity of performing an activity by which it may interact with other units and/or with the environment. The interactions may involve communicating, synchronizing, cooperating, acting in parallel, competing for resources with other processes and/or with the environment. By "concurrent systems" we mean processes which may consist of other processes (or in turn concurrent systems) operating concurrently.

Most software systems are concerned with concurrent systems and thus it is of paramount importance to provide good formal support to the specification, design, and implementation of concurrent systems. Algebraic/logic methods have also found interesting applications in this field, especially to treat at the right level of abstraction the relevant features of a system, helping to hide the unnecessary details and thus to master system complexity.

Due to the particularly complex nature of concurrent systems, and contrary to the case of classical (static) data structures, there are different ways of exploiting algebraic methods in concurrency. First of all, we do not have a single satisfactory model and view for processes and concurrent systems, like input–output functions for sequential input–output systems. Hence, algebraic methods need to be applied to different models. Moreover, in the literature, we can distinguish at least four kinds of approaches.

A1 The algebraic techniques are used at the metalevel, for instance, in the definition or in the use of specification languages. Then a specification involves defining one or more expressions of the language, representing one or more systems. This is, for example, the case in ACP, CCS, and CSP [BK86b, Mil89, Hoa85].

A2 A particular specification language (technique) for concurrent systems is complemented with the possibility of abstractly specifying the (static) data handled by the systems considered using algebraic specifications.

We can qualify the approaches of this kind by the slogan "plus algebraic specifications of static data types".

A3 These methods use particular algebraic specifications having "dynamic sorts", which are sorts whose elements are/correspond to concurrent systems. In such approaches there is only one "algebraic model" (for instance, a first-order structure or algebra) in which some elements represent concurrent systems.

We can qualify the approaches of this kind as "algebraic specifications of dynamic-data types", which are types of dynamic data (processes/ concurrent systems).

A4 These methods allow us to specify an (abstract) data type, which is dynamically changing with time. In such approaches we have different "algebraic" models corresponding to different states of the system.

We can qualify the approaches of this kind as "algebraic specifications of dynamic data-types"; here the data types are dynamic.

We have organized the paper around the classification above, providing significant illustrative examples for each of the classes. The list of the examples is not exhaustive; moreover, we have given a greater emphasis to the approaches representing an extension to concurrency of algebraic specification techniques. For example, this is why for **A1** we have presented in some detail only CCS, the Calculus of Communicating Systems of R. Milner, as the first and paradigmatic example, though the various versions of CSP, ACP, and the like are of comparable importance as for abundance of literature, theoretical investigations and illustrative applications. Indeed the viewpoint of the process algebra approach is more concerned with formal models of processes via appropriate combinators, in which case the specification problem is handled by adopting a model-oriented approach. The same applies to Petri nets, which represent the earliest attempt (apart from automata) to provide formal models for processes and are as important as CCS, CSP, and the like. Here, within **A2**, we have outlined a formalism concerned with algebraic extensions of Petri nets.

To present a more complete overview, we should also treat another class of approaches, which can be termed "algebraic techniques/tools for dynamics". These are interesting approaches where technical tools developed in the algebraic field are used formally to capture the dynamic nature of processes. Among them, we can recall the use of the hidden sort algebras and specifications, see, for instance, [GD94b], and the use of coalgebras and coalgebraic specifications, see [JR97], also for further references. However we cannot cover these approaches, essentially for lack of space; moreover, coalgebraic methods methods are quite recent and in full development, compared to those covered in this chapter.

Similarly we do not have space to present other methods, where the usual algebraic specifications of static data types are used instead in a particular clever way to specify processes, see, for instance, [BCPR96].

Particular examples of approaches of the four kinds, presented in Sections 13.1 – 13.4 respectively, are neither a complete list, nor have they been chosen because we think they are the best representatives.

Our rationale has been mainly to present representatives. In particular, there is no intention of providing a comparative study of the methods. This is a goal outside the scope of the book.

The general notions about the specification of concurrency needed for understanding the approaches presented are briefly summarized at the beginning of the various sections.

We use a common example for the presentation of all approaches, a very simple concurrent system consisting of a buffer and a user, informally described below.

The Bit example

The system **Bit** (called Bit since it is really very small) consists of two components in parallel: a user and a buffer. The buffer is organized as a queue and contains integers; it may obviously receive and return integer values; it may break down, in which case its content will be 10^{10}, and, moreover, it may happen that the last element of its content is duplicated.

When the system is started by the environment, the buffer is empty and the user puts in sequence 0 and 1 on the buffer; then it gets the first element from the buffer. If this element is the number 0 the user must inform the environment of the correct working of the buffer, otherwise it must signal that there is an error.

Thus **Bit** is an interactive concurrent system with components having both autonomous activities (as the buffer failures) and cooperations (the user writing/reading the buffer), and using some static data (integers); furthermore it also has some relevant static/functional aspects, as the queue organization of the buffer.

Some relevant requirements on **Bit** are:

R0 The buffer must always be able to receive any integer value.

R1 When the user is terminated, it cannot perform an activity again.

R2 In at least one case, the system must behave correctly.

R3 After being started, it will eventually signal OK or ERROR.

R4 OK and ERROR are signaled at most once, and it cannot happen that both are signaled.

R5 The user puts integers on and gets integers from the buffer.

13.1   Process algebras

Process Algebras and Calculi, exemplified by CCS, CSP, ACP, and the like, are the most notable example of the use of algebraic methods in the definition and the use of specification languages (approach **A1**).

Labeled transition systems (abbreviated to lts), as models of processes, underlie CCS and many other variations of process algebras, and are also used in many logical/algebraic specification formalisms. Thus we start this section with the fundamental concepts about lts's and their semantics. Note that the first appearance of lts was in the theory of nondeterministic automata; however, the key idea of using labeled transitions to represent the capabilities of interactions (or participation in events) for describing open systems is generally attributed to Robin Milner in CCS. The related fundamental concept of bisimulation semantics, especially its formalization by maximum fixpoint, is due to David Park.

13.1.1   Modeling processes with labeled transition systems

For the first use of labeled transition systems for the modeling concurrency, see [Mil80, Plo83].

A *labeled transition system* (*lts*) is a triple

$$\langle STATE, LABEL, \rightarrow \rangle,$$

where $STATE$ and $LABEL$ are two sets, the *states* and *labels* of the system, and $\rightarrow \subseteq STATE \times LABEL \times STATE$ is the *transition relation*. A triple $\langle s, l, s' \rangle \in \rightarrow$ is said to be a *transition* and is usually written $s \xrightarrow{l} s'$.

Given an lts we can associate with each $s_0 \in STATE$ the so-called *transition tree*, that is, the tree whose root is s_0, where the order of the branches is not considered, two identically decorated subtrees with the same root are considered as a unique subtree, and if it has a node n decorated with s and $s \xrightarrow{l} s'$, then it has a node n' decorated with s' and an arc decorated with l from n to n'.

A process P is thus modeled by a transition tree determined by an lts $\langle STATE, LABEL, \rightarrow \rangle$ and an initial state $s_0 \in STATE$; the nodes in the tree represent the intermediate (interesting) states of the life of P, and the arcs of the tree the possibilities of P of passing from one state to another. It is important to note here that an arc (a transition) $s \xrightarrow{l} s'$ has the following meaning: P in the state s has the *capability* of passing into the state s' by performing a transition, where the label l represents the interaction with the environment during such a move; thus l contains information on the conditions on the environment for the capability to become effective, and on the transformation of such environment induced by the execution of the action.

Concurrent systems, which are processes having cooperating components that are in turn other processes (or concurrent systems), can be modeled through particular lts obtained by composing other lts describing such components.

By associating with a process P the transition tree having root P we give P an operational semantics: two processes are operationally equivalent whenever the associated transition trees are the same, see [Mil80]. However in most cases such semantics is too fine, since it takes into account all operational details of the process activity. It may happen that two processes which we consider semantically equivalent have associated different transition trees. A simple case is when we consider the trees associated with two deterministic processes interacting with the environment only by returning a final result (e.g., two PASCAL programs) represented by two states p and p': they only perform internal activities except for the last transitions, and thus the associated transition trees reported below are:

$$p \xrightarrow{\tau} p_1 \xrightarrow{\tau} \quad \ldots \quad \xrightarrow{\tau} p_n \xrightarrow{OUT(r)} p_F$$

$$p' \xrightarrow{\tau} p'_1 \xrightarrow{\tau} \quad \ldots \quad \xrightarrow{\tau} p'_m \xrightarrow{OUT(r')} p'_F$$

If we consider an input–output semantics, then the two processes are equivalent iff p, p' are equivalent w.r.t. the input and r, r' are equivalent; the differences concerning other aspects (intermediate states, number of intermediate transitions, etc.) are not considered.

From this simple example, we can also appreciate that we get various interesting semantics on processes modeled by lts depending on what we observe (see, e.g., [Mil80, NH84]). For instance, consider the well–known strong bisimulation semantics of Park [Par81] and Milner [Mil80] and the trace semantics [Hoa85]. In the first case, two processes are equivalent iff they have the same associated transition trees after the states have been forgotten. In the second case, two processes are equivalent iff the corresponding sets of traces (streams of labels), obtained traveling along the maximal paths of the associated transition trees, are the same. In general, the semantics of processes depends on what we are interested in observing.

Now we show how to define precisely strong bisimulation over an lts $\langle STATE, LABEL, \rightarrow \rangle$. A binary relation R on $STATE$ is a *strong bisimulation* iff, for all $s_1, s_2 \in STATE$, if $s_1\ R\ s_2$, then

1. if $s_1 \xrightarrow{l} s'_1$, then there exists s'_2 such that $s'_1\ R\ s'_2$ and $s_2 \xrightarrow{l} s'_2$;

2. if $s_2 \xrightarrow{l} s'_2$, then there exists s'_1 such that $s'_1\ R\ s'_2$ and $s_1 \xrightarrow{l} s'_1$.

The *maximum strong bisimulation* $\sim$ for an lts is defined as the union of all strong bisimulations. We have that $\sim$ is a strong bisimulation and that for all strong bisimulations R, $R \subseteq \sim$.

Similarly we can define weak bisimulation over an lts; in this case the internal transitions, i.e., those corresponding to a null interaction with the environment, are not considered when they have no visible consequence. Technically we use $\tau \in LABEL$ to label internal transitions.[1] We define an auxiliary transition relation

$$\Rightarrow \subseteq STATE \times LABEL \times STATE$$

---

[1] The symbol τ was used for the first time by Milner for CCS internal transitions, see Section 13.1.2.

as follows:

$$s \overset{\tau}{\Rightarrow} s,$$

if $s \overset{l}{\longrightarrow} s'$, then $s \overset{l}{\Rightarrow} s'$,

if $s \overset{\tau}{\longrightarrow} s'$ and $s' \overset{l}{\Rightarrow} s''$, then $s \overset{l}{\Rightarrow} s''$,

if $s \overset{l}{\Rightarrow} s'$ and $s' \overset{\tau}{\longrightarrow} s''$, then $s \overset{l}{\Rightarrow} s''$.

A binary relation R on $STATE$ is a *weak bisimulation* iff, for all $s_1, s_2 \in STATE$, if $s_1 \, R \, s_2$, then

1. if $s_1 \overset{l}{\Rightarrow} s_1'$, then there exists s_2' such that $s_1' \, R \, s_2'$ and $s_2 \overset{l}{\Rightarrow} s_2'$;
2. if $s_2 \overset{l}{\Rightarrow} s_2'$, then there exists s_1' such that $s_1' \, R \, s_2'$ and $s_1 \overset{l}{\Rightarrow} s_1'$.

The *maximum weak bisimulation* $\approx$ is the union of all weak bisimulations. We have that $\approx$ is a weak bisimulation and that for all weak bisimulations R, $R \subseteq \approx$.

Example 13.1. (**Bit using labeled transition systems**) Here we give the lts modeling the two components of **Bit**, the user and the buffer, and **Bit** itself respectively.

$$USER = \langle STATE_U, LABEL_U, \rightarrow_U \rangle$$

$STATE_U =$
$\quad \{Initial, Putting\_0, Putting\_1, Reading, Terminated\} \cup \{Read_i \mid i \in \mathbb{Z}\}$
$LABEL_U = \{START, ERROR, OK\} \cup \{PUT_i, GET_i \mid i \in \mathbb{Z}\}$
$\rightarrow_U$ is graphically represented by depicting the resulting graph in Figure 13.1. Notice that in the state *Reading* the user has infinite action capabilities, one for each possible value that can be obtained from the buffer.

$$BUFFER = \langle \mathbb{Z}^*, LABEL_B, \rightarrow_B \rangle$$

$LABEL_B = \{RECEIVE_i, RETURN_i \mid i \in \mathbb{Z}\} \cup \{\tau\}$
$\rightarrow_B$ contains the following triples, where $i \in \mathbb{Z}, q \in \mathbb{Z}^*$:

$$q \xrightarrow{RECEIVE_i}_B q \cdot i \qquad i \cdot q \xrightarrow{RETURN_i}_B q$$
$$i \cdot q \xrightarrow{\tau}_B i \cdot i \cdot q \qquad q \xrightarrow{\tau}_B 10^{10},$$

$$SYSTEM = \langle STATE_S, LABEL_S, \rightarrow_S \rangle$$

$STATE_S$ consists of pairs of states of the buffer and the user.
$LABEL_S = \{START, \tau, OK, ERROR\}$
$\rightarrow_S$ contains the following triples, where $i \in \mathbb{Z}, u, u' \in STATE_U, b, b' \in STATE_B, \langle \rangle$ is the empty stream:

$$\langle \langle \rangle, u \rangle \xrightarrow{START}_S \langle \langle \rangle, u' \rangle \quad \text{if } u \xrightarrow{START}_U u'$$
$$\langle b, u \rangle \xrightarrow{\tau}_S \langle b', u' \rangle \quad \text{if } b \xrightarrow{RECEIVE_i}_B b' \text{ and } u \xrightarrow{PUT_i}_U u',$$
$$\langle b, u \rangle \xrightarrow{\tau}_S \langle b', u' \rangle \quad \text{if } b \xrightarrow{RETURN_i}_B b' \text{ and } u \xrightarrow{GET_i}_U u',$$
$$\langle b, u \rangle \xrightarrow{OK}_S \langle b, u' \rangle \quad \text{if } u \xrightarrow{OK}_U u',$$

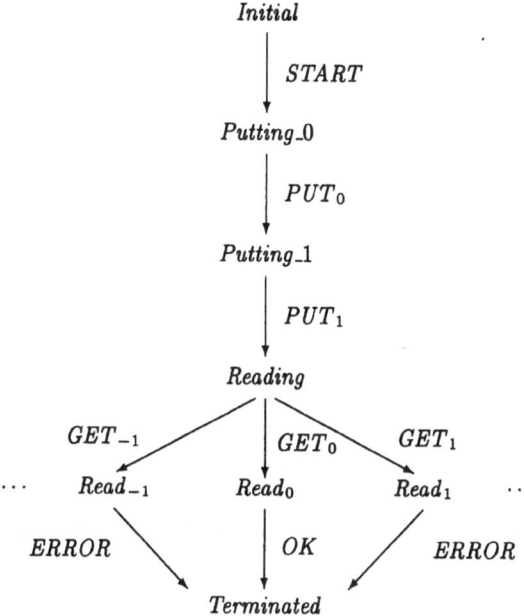

Fig. 13.1. The transitions of the user

$$\langle b, u \rangle \xrightarrow{\;ERROR\;}_S \langle b, u' \rangle \qquad \text{if } u \xrightarrow{\;ERROR\;}_U u',$$
$$\langle b, u \rangle \xrightarrow{\;\tau\;}_S \langle b', u \rangle \qquad \text{if } b \xrightarrow{\;\tau\;}_B b'$$

Notice that $SYSTEM$, defined in a modular way by using $USER$ and $BUFFER$, is an example of how we can compose processes operating in parallel. However, if we forget how it has been defined, then we cannot see its concurrent structure. For example, the fact that the transition

$$\langle 0 \cdot 1, Reading \rangle \xrightarrow{\;\tau\;}_S \langle 1, Read_0 \rangle$$

corresponds to the synchronous execution of the buffer and user action capabilities, labeled by $RETURN_0$ and GET_0 respectively, cannot be deduced by examining $SYSTEM$ alone. ∎

13.1.2   Process calculi and algebras

By process calculi and algebras we mean those approaches which specify processes, whose foremost representatives are the many formalisms known under the collective names CCS, CSP, and ACP. The formalisms, though quite different in some fundamental technical aspects, share some basic underlying ideas:

- as in λ–calculi, processes are represented by terms built over a set of combinators concerning all aspects of process behavior, from flow of control of single processes to operators for composing processes in parallel;

the combinators are different, both as a matter of taste and for technical reasons;

- processes are essentially modeled by transition trees;
- the primitive means of interaction between processes is synchronization, that can be interpreted equivalently as synchronization of exchanging data and simultaneous participation in an event;
- emphasis is laid on algebraic laws stating equivalence of processes;
- a concept of refinement is based on containment of behaviors: refining means reducing the amount of possible behaviors.

We present the basic features of CCS, considered a breakthrough in the field, and will briefly comment on ACP and CSP.

CCS, developed by Robin Milner, basically adopts an operational (transition) semantics, associating with each process a transition tree (graph); on the basis of the transition semantics, some equivalences are defined on the processes (various bisimulations and operational equivalences), and laws are proven stating equivalences on processes; the set of laws is usually a complete axiomatization of the semantics over finite processes.

We refer to [Mil89] as a basic reference.

For explanatory purposes, we can start by looking at CCS as a language for describing possibly infinite transition trees.

If $\mathcal{A}$ denotes a set of basic names, then $\overline{\mathcal{A}} = \{\overline{a} \mid a \in \mathcal{A}\}$ is the set of the conames and $\mathcal{L} = \mathcal{A} \cup \overline{\mathcal{A}}$, with $\overline{\overline{l}} = l$. A special label τ indicates the so-called silent action, i.e., an action not visible outside, since it corresponds to a communication taking place within the process; the set of the actions (or, more acurately, capabilities of action), i.e., the labels, is then $\mathcal{ACT} = \mathcal{L} \cup \{\tau\}$, ranged over by α.

First we have the basic combinators for describing finite depth transition trees:

(1) prefixing $\alpha . E$
(2) summation $\Sigma_{i \in I} E_i$, I an indexing set

where E denotes a generic CCS expression.

Assume that E represents a tree ◭ *t* with root E, then $\alpha . E$ represents the tree

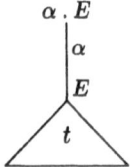

with root $\alpha . E$. This is formalized by an appropriate semantic clause (inductive rule defining $\rightarrow$)

Act $$\dfrac{}{\alpha . E \xrightarrow{\alpha} E} \; .$$

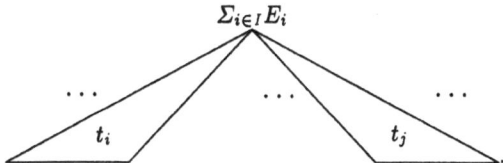

Assume that each E_i represents a tree $\quad$ with root E_i, then $\Sigma_{i \in I} E_i$ represents the tree

$$\Sigma_{i \in I} E_i$$

The related semantic clause is

Sum$_j$ $$\dfrac{E_j \xrightarrow{\alpha} E'_j}{\Sigma_{i \in I} E_i \xrightarrow{\alpha} E'_j} \quad j \in I \; .$$

In other words clause (1) allows us to describe adding an arc and clause (2) to describe branching. Notice that for $I = \emptyset$, clause (2) defines one expression, also written **nil** or **0**, corresponding to a leaf on a tree.

Infinite depth trees are defined as usual by recursion, say $A =_{\text{def}} P$, where the name of the process A (a constant) may appear in P, which denotes an expression. Of course multiple recursion is possible. The semantics is as usual

Con $$\dfrac{P \xrightarrow{\alpha} P'}{A \xrightarrow{\alpha} P'} \quad A =_{\text{def}} P \; .$$

To handle interaction between processes, a basic combinator for parallelism is introduced:

(3) $E_1 \parallel E_2$

There are many ways to define the semantics of the combinator $\parallel$; the original one, which we report here, formalizes a notion of synchronization/ communication by handshaking communication (**Comm$_3$**) and of parallel execution by interleaving (**Comm$_1$, Comm$_2$**).

Comm$_1$ $$\dfrac{E \xrightarrow{\alpha} E'}{E \parallel F \xrightarrow{\alpha} E' \parallel F} \qquad\qquad \textbf{Comm}_2 \quad \dfrac{F \xrightarrow{\alpha} F'}{E \parallel F \xrightarrow{\alpha} E \parallel F'}$$

Comm$_3$ $$\dfrac{E \xrightarrow{l} E' \quad F \xrightarrow{\bar{l}} F'}{E \parallel F \xrightarrow{\tau} E' \parallel F'} \quad l \in \mathcal{L}$$

Rule **Comm$_3$** says that synchronization may take place whenever the capabilities are complementary (l and $\bar{l}$).

In CCS we also have two other operations,

(4) E/L Restriction

(5) $E[f]$ Relabeling,

where $L \subseteq \mathcal{L}$ denotes a set of nonsilent actions, and f is a function from $\mathcal{L}$ to $\mathcal{L}$ such that $\overline{f(l)} = f(\bar{l})$; f can be extended to $\mathcal{ACT}$ by setting $f(\tau) = \tau$.

The semantics of (4) and (5) is given by

$$\textbf{Res} \quad \frac{E \xrightarrow{\alpha} E'}{E/L \xrightarrow{\alpha} E'/L} \quad (\alpha \notin L \cup \overline{L}) \qquad \textbf{Rel} \quad \frac{E \xrightarrow{\alpha} E'}{E[f] \xrightarrow{f(\alpha)} E'[f]} .$$

Notice that relabeling is essentially a user facility for defining processes, with no behavioral meaning, while restriction means hiding from the outside all the action capabilities in L and its complementary set.

The given semantics is operational in nature (though it can also be given in a denotational way) and serves the main purpose of associating a labeled transition system with process expressions.

As we have seen in the preceding section, various semantics can be associated with lts. Depending on the chosen semantics, various laws can be proved about CCS expressions. For example, adopting strong bisimulation semantics, denoted by $\sim$, the following laws hold ($P + Q = \Sigma\{P, Q\}$)

(1) $(\alpha . Q)/L = \begin{cases} \textbf{nil} & \text{if } \alpha \in L \cup \overline{L} \\ \alpha . (Q/L) & \text{otherwise} \end{cases}$

(2) $(\alpha . Q)[f] = f(\alpha) . Q[f]$

(3) $(Q + R)/L = Q/L + R/L$

(4) $(Q + R)[f] = Q[f] + R[f]$

(5) $P + Q = Q + P$

(6) $P + (Q + R) = (P + Q) + R$

(7) $P + P = P$

(8) $P + \textbf{nil} = P$

Also a fundamental law, called the expansion law, can be proved, showing that we can eliminate from a process expression the parallel operator, restriction, and relabeling, thus showing the essence of interleaving semantics as reducing parallel execution to nondeterministic choice. A simpler form concerned only with the parallel combinator is as follows:

(9) $P_1 \parallel \dots \parallel P_n =$
$\Sigma\{\alpha . (P_1 \dots \parallel P_i' \parallel \dots \parallel P_n) \mid P_i \xrightarrow{\alpha} P_i', 1 \le i \le n\} +$
$\Sigma\{\tau . (P_1 \dots \parallel P_i' \parallel \dots \parallel P_j' \parallel \dots \parallel P_n) \mid$

$$P_i \xrightarrow{l} P_i', P_j \xrightarrow{\bar{l}} P_j', 1 \le i < j \le n\}$$

It can be shown that strong bisimulation is a congruence for CCS, since it is substitutive under all combinators and recursive definitions. For example, if $P_1 \sim P_2$, then we have

$\alpha . P_1 \sim \alpha . P_2 \qquad P_1 + Q \sim P_2 + Q$, etc.

Also, under reasonable conditions, recursive definitions uniquely identify a process modulo its bisimulation.

Unfortunately, passing to weak bisimulation (observation equivalence in [Mil89]), we do not get a congruence any longer and thus another equivalence is introduced, called equality (or observation congruence), which implies weak bisimulation. It can be shown that, since CCS is as powerful as Turing machines, no effective axiomatization of equality exists. However the laws provide effective axiomatization for smaller classes of processes, such as the finite processes.

Our presentation shows the role of laws in CCS, as derived theorems from essentially operational semantics. A somewhat different approach has been taken by ACP, mainly developed by Bergstra, Klop, and Baeten (see [BW90] also for references). There the starting point is a complete axiomatization (usually by equations or conditional axioms) of some equivalence (e.g., strong or weak bisimulation) for finite processes; thus two finite processes are equivalent iff their equality can be proven by equational/conditional deduction. Then recursion is added and semantics is again given in terms of graphs, labeled transition systems, or projective limits.

The approach is highly hierarchical, introducing laws for new combinators in a conservative way. Some of the combinators introduced in ACP are due to the technical needs for obtaining complete axiomatizations.

Different again is the CSP approach [Hoa85], where the semantics is denotational and the laws are derived from semantics, and used for reasoning about correctness. The denoted values are different, depending on the richness of the combinators; they range from sets of traces to the so-called refusal sets.

A common problem with the process algebra/calculi approaches is the enormous variety of possible meaningful semantics and thus of the associated derived laws; in one paper [vG90] Van Glabbeek analyzes from a modal logic unifying viewpoint, as many as 155 different semantics.

Example 13.2. (**Bit using** CCS) The process corresponding to the user is defined by
$$USER = START.\overline{PUT_0}.\overline{PUT_1}.(GET_0.\overline{OK}.\textbf{nil} + \Sigma_{i\in\mathbb{Z}-\{0\}}GET_i.\overline{ERROR}.\textbf{nil});$$
the process buffer is defined by the following, mutually recursive definitions
$$BUFFER_{()} = \Sigma_{i\in\mathbb{Z}}PUT_i . BUFFER_i + \tau . BUFFER_{10^{10}}$$
$$BUFFER_{i_1\cdots i_k} =$$
$$\Sigma_{i\in\mathbb{Z}}PUT_i . BUFFER_{i\cdot i_1\cdots i_k} + \overline{GET_{i_k}} . BUFFER_{i_1\cdots i_{k-1}} +$$
$$\tau . BUFFER_{10^{10}} + \tau . BUFFER_{i_1\cdot i_1\cdots i_k}, \qquad i_1\cdot\ldots\cdot i_k \in \mathbb{Z}^+$$
Finally the system is the parallel composition of the two processes above (initially the buffer is empty)
$$SYSTEM = (BUFFER_{()} \parallel USER)/\{PUT_i, GET_i \mid i \in \mathbb{Z}\}. \qquad \blacksquare$$

The example illustrates the use of CCS in the specification phase, which follows a model-oriented approach: with the help of the CCS language a process is described and then a class of models is defined corresponding to the equivalence class of the process (w.r.t. some equivalence).

13.2   Algebraic specification of static data types

In this section we briefly present some specification techniques following the approach **A2** to the algebraic specification of concurrent systems, that is, approaches integrating a formalism for the concurrent aspects with algebraic specifications of the static data types.

13.2.1   Process calculi plus algebraic specification of data types

In this subsection we briefly present two specification formalisms, LOTOS and PSF, designed following approach **A2**, where the processes are defined by a process–algebra style calculus. The differences between LOTOS and PSF are in the formalism for the algebraic specification part (ACT ONE [EFH83] and ASF [BHK89], respectively) and in the combinators of the process calculus chosen (inspired by those of CCS and ACP, respectively, see Section 13.1).

Process specification formalism (PSF). PSF [MV89, MV90] is the process specification formalism developed by Mauw and Veltink as a base for a set of tools to support the process algebras. The main goal in the design of PSF was to provide a specification language with a formal syntax similar to the process algebra ACP [BW90, Section 4] but also with a notion of data type; to this end ASF (the Algebraic Specification Formalism of [BHK89]) has been incorporated.

The basic specification formalism is equational logic with total algebras. The theory and language of ASF are adopted for handling modular and parameterized specifications.

A PSF specification consists of a series of modules, divided into data modules and process modules. Data modules are algebraic specifications with initial semantics. Process modules are ACP specifications of processes. Formally, a process module consists of

- declarations of the operation symbols for actions and processes (which may have the static data as arguments),
- explicit definitions of the synchronization among such actions,
- process definitions of the form $P(x_1, \ldots, x_n) =$ ACP–expression, in which the operators like "+", "$\|$", ";", **"hide"** and **"encaps"**, elementary processes, pure atomic actions, and also P (thus allowing recursive definitions) may appear.

Processes are particular data structures obtained by a given (equational) axiomatization which determines a particular semantics over these structures, embodying ideas of concurrency. This is best understood by looking at the hidden basic concurrent models behind process algebra, which are lts as in CCS and many other approaches; then the axioms provide semantics like strong, trace, or bisimulation semantics and others, see Section 13.1.1. The hidden model is made evident in some presentations of PSF, where ACP processes are described by means of lts. In any case, since ACP essentially provides a language schema for processes, it is irrelevant, other than for building the tools, how its semantics is given, either by equations or by labeled transitions plus semantic equivalences.

It is instead important to note that in PSF:

– the synchronization of actions can be defined explicitly in the communication part; as a consequence, the synchronization mechanism is not fixed and is parameterized;
– the execution mode is interleaving.

The interface between processes and data types is as follows:

– the atomic actions may have as components some values of the specified data types;
– it is possible to define recursively families of processes indexed on the elements of some sort;
– an infinitary nondeterministic choice indexed on the elements of a sort is available.

The semantics of the data part is a classical algebraic semantics by initiality; the semantics of processes is strong bisimulation, which gives a congruence over the term algebra. Thus the semantics identifies an isomorphism class of structures, as for a data type.

The data part is strictly separated from the process part. Thus it is an **A2** approach; but the concurrent structure here is also specified algebraically, though with a fixed set of primitives parameterized on the actions and the synchronization structure. The result is a completely algebraic specification to which all the techniques and results of ASF can be conveniently applied.

Particularly powerful are the modularization mechanisms in PSF, which are borrowed from ASF but truly deal with the integration of data types and processes; the module concept also supports the import and export of processes and actions.

There is a vast literature on the use of process algebras, with a detailed treatment of classical examples and correctness proofs for implementation. However, these examples should not be confused with applications of a specification method like PSF, which have indeed been introduced for supporting industrial applications. Clearly PSF is applicable to a wide range of significant cases in practice, see, for instance, [MV93], but we see a limitation in its strict policy of message passing and no provision for data sharing. In many cases some amount of coding is required which is not in the spirit of abstract specifications. The same remark applies to execution modes other than interleaving, which have to be simulated by appropriate use of synchronization and restriction mechanisms.

PSF has been devised as a basis for the development of a toolset (see, e.g., [MV89, MV91, PSF97]); in particular, a simulator, a term rewriting, and a proof assistant have been implemented.

LOTOS. LOTOS was probably the first internationally known (since 1984), algebraic specification formalism for concurrency [BB87, ISO89]; most importantly, it is an official ISO specification language for open distributed systems, a qualification which alone would rank it high in an ideal value scale of possible important applications. However, LOTOS is interesting also because it represents an early paradigm of which PSF can be considered an improvement. Because of this, we do not go into a detailed discussion of LOTOS; it is enough to compare it with PSF to understand its structure.

LOTOS adds classical algebraic specifications into a language for concurrency like PSF; but it uses ACT ONE [EFH83] instead of ASF and a process description based on an extension of CCS with several derived combinators (e.g., input/output of structured values, sequential composition with possible value passing, enabling/disabling operators) instead of the process algebra ACP. The basic specification formalism (equational logic with total algebras) and process bisimulation semantics are the same.

PSF is an improvement over LOTOS (see a discussion in [MV89]), since it allows more freedom in the definition of synchronization mechanisms and supports import/export of action/processes, thus becoming more flexible for stepwise development.

Throughout these years LOTOS has been used in several practical applications and nowadays tools for helping to write correct LOTOS specifications have been developed (see, for instance, the ESPRIT project LOTOSHERE [vE91]). Recently a new, revised version of LOTOS (E-LOTOS, for Enhancement to LOTOS) has been developed and presented as a standard [LOT97], taking into account the needs that emerged through its application; enhancements concern the data part (built-in, partial operations), the concurrency part (noninterleaving semantics, real time, priorities), and the whole organization of the specifications (introduction of modules).

Example 13.3. (**Bit using** LOTOS) The data part is given by the specification *INT_QUEUE*, shown in Appendix A, and by the following:

> **spec** *MESSAGE* =
>   **sorts** *message*   ** messages exchanged with the environment
>   **opns** *OK, ERROR:* → *message*

Bit is given as the parallel composition of two processes corresponding to the buffer and the user.

The gates of such processes and their connections are graphically represented in Figure 13.2.

Below "?" and "!" distinguish input/output actions, ";", "... → ...", "[]", "|||", and "**i**" denote respectively action prefixing, Boolean guards, nondeterministic choice, parallel combinator and internal action.

In the definition of *BUFFER*, *Put* and *Get* are the gates and *q* is a process parameter of sort *queue*.

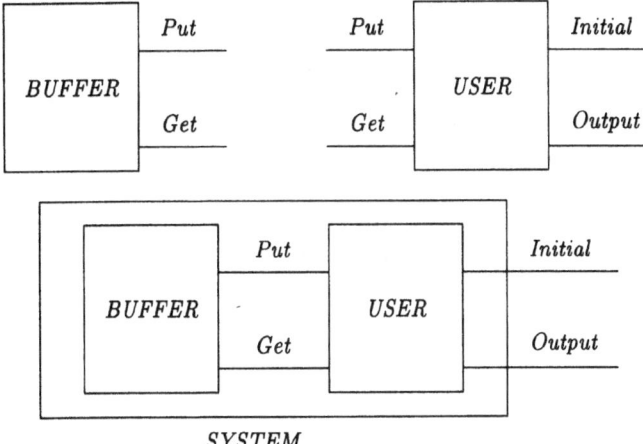

Fig. 13.2. Structure of the LOTOS specification of **Bit**

process $BUFFER[Put, Get](q : queue) :=$
$[Not\_Empty(q)] \rightarrow Get\,!\,First(q); BUFFER[Put, Get](Remove(q))$ []
$\qquad\qquad Put\,?\,x : int; BUFFER[Put, Get](Put(x,q))$ []
$\qquad\qquad \mathbf{i};\, BUFFER[Put, Get](Put(10^{10}, Empty))$ []
$\qquad\qquad \mathbf{i};\, BUFFER[Put, Get](Dup(q))$
end process

process $USER[Initial, Put, Get, Output] :=$
$Initial\,?;\, Put\,!\,0;\, Put\,!\,1;\, Get\,?\,x : int;$
$\qquad ([not(x = 0)] \rightarrow Output\,!\,ERROR$ []
$\qquad [x = 0] \rightarrow \qquad Output\,!\,OK)$
end process

process $System[Initial, Output] :=$
$BUFFER[Put, Get](Empty) \,|||\, USER[Initial, Put, Get, Output]$
end process   ∎

Note the similarity between LOTOS and CCS.

13.2.2 Petri nets

Petri nets are among the earliest and most influential models of concurrency.

Net models. Here we briefly present (elementary) nets, the basic models for the various specification formalisms generally called "Petri nets"; all variants arise either by putting some restrictions on the allowed nets, adding minor features, or offering more compact/simple ways to present the nets (e.g., colored nets, high-level nets, ...); in some cases, a slightly different terminology is used. [Rei85, RT86] give general overviews and further references for net models, while [Rei98] is more concerned with the use of nets for specifying concurrent systems.

A *net* N is a triple $\langle P, T, F \rangle$, where P and T are two sets and F is a subset of $(P \times T) \cup (T \times P)$; P, T, and F are respectively called the sets of the *places*, the *transitions*, and the *arcs* of the net; F is also called the *flow relation*.

Nets are graphically represented by depicting places, transitions, and arcs respectively as circles, boxes, and arrows, see, e.g., Figure 13.3.

Example 13.4. (**Bit using Petri nets**) A Petri net modeling **Bit** is reported in Figure 13.3; note that to make the drawing small enough we have assumed that the buffer may only fail when its content is 01.   ∎

The behavior of a net is defined as follows.

Given a transition $t \in T$, we define $\cdot t = \{p \mid p \in P, \langle p, t \rangle \in F\}$ (the *preconditions* of t) and $t\cdot = \{p \mid p \in P, \langle t, p \rangle \in F\}$ (the *postconditions* of t).

Any function s from P into $\mathbb{N}$ is called a *(global) state* (or *marking*) of N; graphically represented by putting $s(p)$ •'s (called *tokens*) on the place p, for any $p \in P$; the net in Figure 13.3 is in a state characterized just by one token on each of the places *Initial* and *Empty*.

A transition $t \in T$ is *enabled* in a state s (or *may fire*) iff, for any $p \in \cdot t$, $s(p) > 1$. If t is enabled in s, then

$$s' = \lambda p. \begin{cases} s(p) - 1 & \text{if } \in \cdot t - t\cdot \\ s(p) + 1 & \text{if } p \in t\cdot - \cdot t \\ s(p) & \text{otherwise} \end{cases}$$

is called the *successor state of s with t* (or *the state obtained after the firing of t in s*) and $s \xrightarrow{t} s'$ is called a *step* in N.

A *system net* (*s-net* for short) is a net N, together with a state, called the *initial state*.

The single steps of an s-net can be composed in runs: $\langle s_i \xrightarrow{t_i} s_{i+1} \rangle_{i \in I}$ with s_0 the initial state and $I = \{0, \ldots, n\}$ ($I = \mathbb{N}$) is a *finite (infinite) interleaved run*.

Empty: 1; *Initial*: 1 $\xrightarrow{START}$ *Empty*: 1; *Putting_0*: 1 $\xrightarrow{INT1}$
0: 1; *Putting_1*: 1 $\xrightarrow{INT2}$ 01: 1; *Reading*: 1 $\xrightarrow{INT3}$ 01: 1; *Ok*: 1 $\xrightarrow{BREAK}$
10^{10}: 1; *Ok*: 1 $\xrightarrow{OK}$ 10^{10}: 1; *Terminated*: 1

is an example of an interleaved run of the net in Figure 13.3, where a state is represented by listing the places with the number of their tokens, forgetting those without tokens.

Some occurrences of transitions in an interleaved run that are seen as ordered one after the other may be independent. Thus they may also have

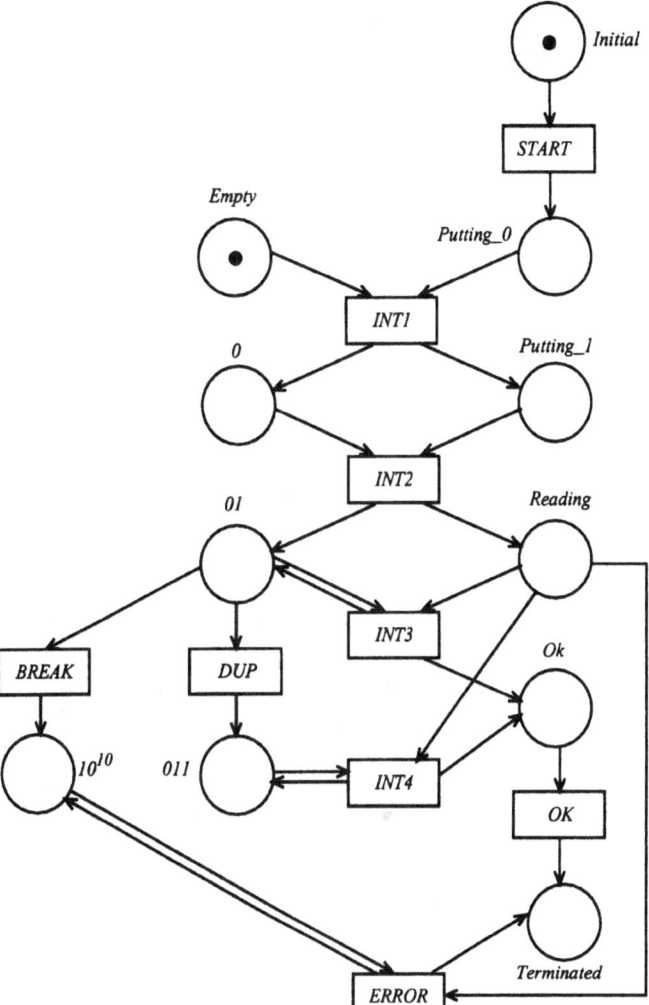

Fig. 13.3. A Petri net modeling **Bit**

happened in the reverse order; in the run depicted above, the two last transitions are independent. We give another definition of a run, making explicit the concurrent aspects.

Concurrent runs are represented by special nets; the underlying idea is that a concurrent run of N is a net, whose places and transitions are labeled with the places and transitions of the original net N and correspond to their occurrences and firings in the run, and where the flow relation corresponds to the causal relationships among them.

Due to lack of space we skip the complete definition of concurrent runs, and just give in Figure 13.4, as an example, the concurrent run corresponding to the run above; here we can see how the two last transitions are not causally related.

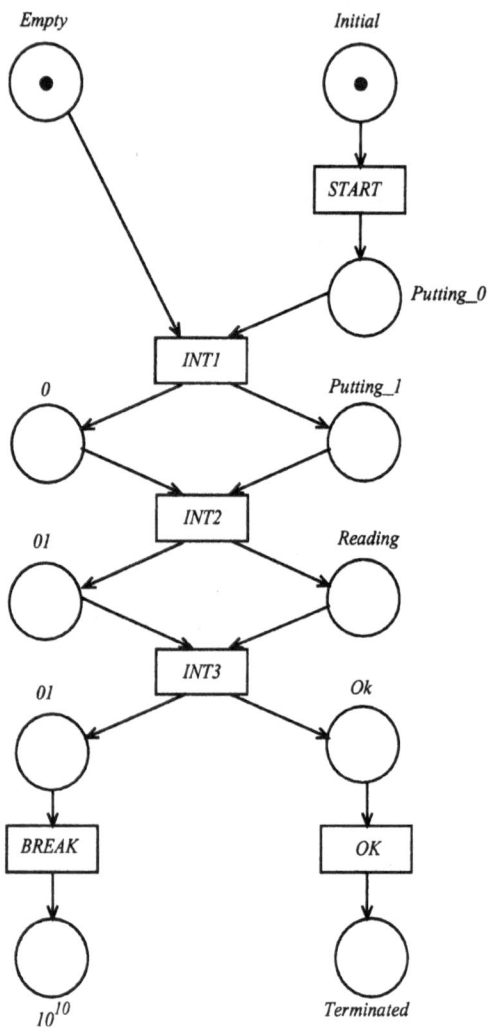

Fig. 13.4. A concurrent run of the net modeling **Bit**

The features of concurrent systems that can be nicely handled by modeling them with nets are as follows:

- local and global states abstractly represented as token distributions over the places;
- atomic actions (the transitions);
- (local) causality and effect between actions and states;
- mutual independence/causality of actions;

while there is no acceptable way to model

- the (distributed) structure of the system;
- the modular decomposition of systems;
- open/interacting/reactive systems (no distinction between internal and external transitions/places: in the net in Figure 13.3, *START*, *OK*, and *ERROR* should be the external transitions), and therefore no way to define modularly the net corresponding to a complex system by putting together the nets corresponding to its components.

Clearly many new formalisms extending nets with other features have been proposed in the literature overcoming some of these problems.

Nets are particularly apt for analyzing the modeled systems, formally checking whether the system has or does not have some properties, including both safety and liveness properties (see Section 13.2.3); several techniques have been provided for that, one of the most relevant is based on "place–invariants".

A *place–invariant* is a linear combination (summation) with (also zero) integer coefficients of the number of tokens contained in the places, which is not changed by the firing of the transitions; all place–invariants of a net may be found by solving a linear system of equations with integer coefficients, thus it is possible to have software tools for finding them.

For example, two invariants of the net in Figure 13.3 are:

$$1 \cdot Initial + 1 \cdot Putting\_0 + 1 \cdot Putting\_1 + 1 \cdot Reading + 1 \cdot Ok + 1 \cdot Terminated = 1$$

$$1 \cdot Empty + 1 \cdot 0 + 1 \cdot 01 + 1 \cdot 011 + 1 \cdot 10^{10} = 1$$

(corresponding to saying that the user and the buffer are always in one and only one state); while

$$1 \cdot Ok - 1 \cdot 01 = 0$$

is not an invariant (the user may be in the Ok state while the buffer content is not 01).

High-level algebraic Petri nets. Nets, as presented in Section 13.2.2, allow us to model several aspects of concurrent systems. However, if we want to use them for significant examples we have to handle very large (if not infinite) nets; e.g., consider the Petri net modeling a system using integer numbers. Moreover in practical applications we have also to handle complex data structures.

To fix the first kind of problem, the basic nets have been extended in several ways: a transition firing can test for the presence of, delete, and add finite sets of tokens (creating Petri nets as originally defined). Later on, the unique black token was replaced by colored tokens, producing the so-called "Colored nets"; here the firing of a transition also depends on the colors of the tokens present in the places, and its firing deletes and adds sets of colored tokens. More generally, tokens can be considered to be data taken in some data structure; this leads to the so-called "algebraic nets" where such a data structure is given by a (many-sorted) algebra.

A richer structure of the tokens allows us to introduce "high-level" nets, where the basic idea is that the arcs are decorated with symbolic expressions, describing in a compact way the sets of tokens causing the firing of a transition, and those deleted and added by such firing. Moreover, now the transitions may also be decorated by some expression corresponding to performing some checks on the tokens present in the places in the premise of the transition, for instance, putting in relation the color of the token in a place with that of the token in another place (the token of type natural in one place should be the length of the token of type queue in another). In the literature it is possible to find several proposals for high-level algebraic nets (see, for instance, [Rei91, BCM88, DH91, Vau87]). The reader interested in a more detailed study of such nets may, for instance, consult [JR91]; here we just briefly present their basic features.

A *high-level algebraic net system* consists of:

- a net $N = \langle P, T, F \rangle$ (*the schema*);
- a signature Σ and a set of sorted variables X;
- an association with each place in P of a sort of Σ (places are typed with the sorts);
- an association with each arc in F of a set of terms built on Σ and X having the sort of the place source or target of the arc (*arc inscriptions*);
- an association with each transition in T of a first-order (conditional, ...) formula built on Σ and X, where only the variables appearing in the inscriptions of the arcs entering in the transition may appear (*transition inscriptions*);
- a Σ–algebra A (the *data part*);
- an association with each place p in P of a set of elements of $|A|_s$, where s is the sort associated with p (*initial state*).

In some approaches, sets are replaced by multisets, and in others the signatures and the algebras are extended to have sorts and operations for handling sets/multisets of elements of the original sorts; thus arcs are inscribed by terms of sort $set(s)/mset(s)$ and the initial marking consists of elements of the carriers of these set/mset sorts. The algebras used in algebraic nets may be of whichever kind; e.g., there are approaches using homogeneous total algebras and others using many-sorted partial algebras.

The abstraction level of high-level algebraic nets may be enhanced by abstractly giving the data part as an algebraic specification with initial semantics.

Example 13.5. (**Bit using algebraic high-level nets**) Here we give a specification of **Bit** which is an improvement on that presented in Figure 13.3; the buffer is now organized as a queue and contains integer numbers, and so it is possible to check whether its first element is 0. Notice that the corresponding non-high-level net is infinite.

The data part is given by the following algebraic specification with initial semantics

```
spec  DATA =
    enrich INT_QUEUE by
    sorts   token
    opns   •: → token
```

The place *Buffer* has sort *queue*, while *Initial*, *Putting_0*, *Putting_1*, *Reading*, *Ok*, and *Terminated* have sort *token*.

The net is shown in Figure 13.5.   ■

Like classical Petri nets, high-level algebraic nets suffer from their lacking modularity.

13.2.3   Temporal logic

In the field of concurrency, specifications following an axiomatic, or, better, property–oriented style, have been widely used, in general to give the formal specification of the requirements on a concurrent system. In these cases, a specification is just a set of formulas of some logic expressing the requirements on the specified system; among the commonest and most relevant requirements, we have:

1. *liveness properties*: (under some condition) something good will happen eventually in the system; for instance, the system will eventually react to the reception of some stimuli/after that some situation has been reached;
2. *safety properties*: (under some condition) something bad will never happen in the system; for instance, after receiving some stimuli/reaching some situation, some (incorrect) output will not be produced/some (incorrect) situation will be not reached;
3. *fairness properties*: the repeated choice between two alternative activities of the system must be fair (i.e., it cannot happen that one of the two alternatives will be chosen forever); e.g., in the case of two processes trying to access a shared resource, it cannot happen that only one will succeed;

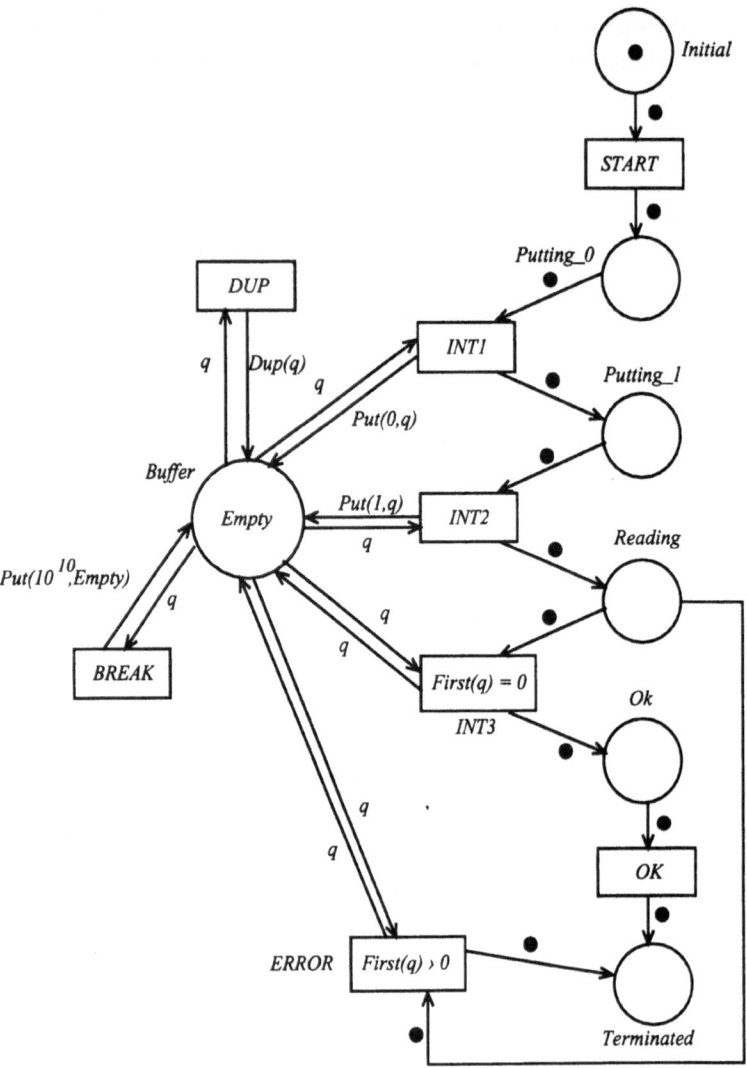

Fig. 13.5. A high-level algebraic net modeling **Bit**

4. *priority/precedence properties*: some activity can be executed iff some other activity cannot be executed; e.g., a process component of a system can write to a buffer iff no other process in the system can read it, a process with priority n can access a resource iff no process with priority higher than n can;

5. ....

First-order logic is not sufficient to express properties such as those above in full generality: some properties are related to the evolution of the system over time (2), others related to the possible activities of a system (1), or of the components of a system. Following the way indicated by Pnueli in a landmark paper [Pnu77], in recent years various modal/temporal logics have been widely and satisfactorily used to express properties like (1) and (2), or special variants particularly tailored for the system model chosen have been developed (see, e.g., [CR97, Mil89]). Concerning properties such as (3, 4), no fully satisfactory proposal has yet been found (some initial attempts are in [Reg91, ES95]).

Here we only briefly introduce the basic (linear and branching) temporal logics, and afterwards we show how first-order temporal logic with equality may be used to give requirement specifications of concurrent systems where the static data are specified by loose algebraic specifications, following an **A2** approach; in Section 13.3, we present an alternative approach based on temporal logic of kind **A3**.

Variations of temporal logics. Here we briefly recall the definition of a linear/branching temporal logic and give some examples of use of its formulas to express requirements on concurrent systems; for references, see [Sti92, Pnu86, Eme90].

A propositional linear temporal logic. TL is a propositional linear temporal logic with a minimal set of combinators.

Assume that Q is a set of propositional symbols; thus the *formulas* of TL are defined by:

$$\phi ::= Q \mid \phi_1 \Rightarrow \phi_2 \mid \neg \phi \mid \phi_1 \, \mathcal{U} \, \phi_2$$

The *models* of TL are sequences of states $\mathcal{M} = \langle s_i \rangle_{i \in \mathbb{N}}$, where a state s is a function from Q into $\{T, F\}$, the set of the truth values.

The *validity* of a formula ϕ over a model $\mathcal{M} = \langle s_i \rangle_{i \in \mathbb{N}}$ is defined as follows:

$$\mathcal{M} \models \phi \quad \text{iff} \quad i, \mathcal{M} \models \phi \text{ for all } i \in \mathbb{N},$$

where

- $i, \mathcal{M} \models Q$ iff $s_i(Q) = T$
- $i, \mathcal{M} \models \phi_1 \Rightarrow \phi_2$ iff $i, \mathcal{M} \models \neg \phi_1$ or $i, \mathcal{M} \models \phi_2$
- $i, \mathcal{M} \models \neg \phi$ iff $i, \mathcal{M} \not\models \phi$
- $i, \mathcal{M} \models \phi_1 \, \mathcal{U} \, \phi_2$ iff there exists $j \geq 0$ such that for all h, $i < h < j$, $h, \mathcal{M} \models \phi_1$ and $j, \mathcal{M} \models \phi_2$.

The term "linear" means that in this case the behavior of a system is modeled by a set of executions represented by linear sequences of states, and thus at a given instant state, there is exactly one successor state.

When TL is used to specify a system, we have that the models and the formulas represent respectively the executions of that system and the properties on such executions. For example, $\phi_1 \, \mathcal{U} \, \phi_2$ corresponds to saying that in any execution the property represented by ϕ_1 holds until the property represented by ϕ_2 holds, and that ϕ_2 surely will hold. Thus a set of TL formulas Π could be used to specify the requirements on a system: Π determines the class of all systems whose possible executions are included in the class of the models of Π.

$\mathcal{U}$ is the basic combinator; many others suitable for expressing further relevant properties can be derived; among them:

- **true** , **false** , $\vee$, $\wedge$, and $\Leftrightarrow$, defined in the usual way
- $\Diamond \, \phi =_{\text{def}}$ **true** $\mathcal{U} \, \phi$ (eventually the property represented by ϕ will hold)
- $\Box \, \phi =_{\text{def}} \neg \Diamond \neg \phi$ (the property represented by ϕ will hold forever)
- $\phi_1 \, \mathcal{WU} \, \phi_2 =_{\text{def}} (\phi_1 \, \mathcal{U} \, \phi_2) \vee \Box \, \phi_1$ (the property represented by ϕ_1 holds until ϕ_2 will hold, but it is not required that ϕ_2 will eventually hold).

A propositional branching temporal logic. BTL is a propositional branching temporal logic with a minimal set of combinators given, following a CTL style. The term "branching" means that in this case the behavior of a system is modeled by a tree whose nodes are decorated by states, and thus at a given instant there may be several different successor states.

As before, assume that Q is a set of propositional symbols; then the *formulas* of BTL are the following, where $\triangle$ is the combinator for "quantifying over paths":

$$\phi ::= Q \mid \triangle \, \phi \mid \phi_1 \Rightarrow \phi_2 \mid \neg \phi \mid \phi_1 \, \mathcal{U} \, \phi_2.$$

The *models* of BTL are transition systems or Kripke structures, where a function from Q into the set of the truth values is associated with each state. Precisely, a model $\mathcal{M}$ is a triple $\langle STATE, \rightarrow, v \rangle$, where $STATE$ is a set, $\rightarrow \subseteq STATE^2$ and v is a function from $STATE$ into the set of the functions from Q into $\{T, F\}$.

The *validity* of a formula ϕ over a model $\mathcal{M}$ is defined as follows.

First we define the set of paths over $\mathcal{M}$:

$PATH(\mathcal{M}) =$
$\{\langle s_i \rangle_{i \in \mathbb{N}} \mid \forall i \in \mathbb{N}. ((s_i \rightarrow s_{i+1} \vee (\forall j. j > i \Rightarrow (s_j = s_i \wedge \not\exists s. s_i \rightarrow s))) \}.$

Given $\sigma = \langle s_i \rangle_{i \in \mathbb{N}} \in PATH(\mathcal{M})$ and $h \geq 0$, $\sigma\lfloor_h$ denotes the path $s_h \, s_{h+1} \, s_{h+2} \, \cdots$.

$$\mathcal{M} \models \phi \quad \text{iff} \quad \sigma, \mathcal{M} \models \phi \quad \text{for all } \sigma \in PATH(\mathcal{M}),$$

where

- $\sigma, \mathcal{M} \models Q$ iff $v(s_0)(Q) = T$
- $\sigma, \mathcal{M} \models \phi_1 \Rightarrow \phi_2$ iff $\sigma, \mathcal{M} \models \neg \phi_1$ or $\sigma, \mathcal{M} \models \phi_2$

- $\sigma, \mathcal{M} \models \neg \phi$ iff $\sigma, \mathcal{M} \not\models \phi$
- $\sigma, \mathcal{M} \models \phi_1 \, \mathcal{U} \, \phi_2$ iff there exists $j \geq 0$ such that for all $h, 0 < h < j$, $\sigma\lfloor_h, \mathcal{M} \models \phi_1$ and $\sigma\lfloor_j, \mathcal{M} \models \phi_2$
- $\sigma, \mathcal{M} \models \triangle \, \phi$ iff $\sigma', \mathcal{M} \models \phi$ for all $\sigma' \in PATH(\mathcal{M})$ such that $s_0 = s_0'$.

When BTL is used to specify a system, we have that a model represents the whole behavior of such a system, i.e., all its possible executions and at which point the nondeterministic choices are made, and then the formulas represent properties on such behavior. For example, $\neg \, \triangle \, (\, \mathbf{true} \, \mathcal{U} \, \phi)$ corresponds to saying that it is not true that in any case the executions of the system will eventually satisfy the property represented by ϕ (if ϕ corresponds to failing, then the formula requires that the system has at least an execution without failures). Thus a set of BTL formulas Φ could be used to specify the requirements on a system: Φ determines the class of all systems whose behavior is described by an element of the class of the models of Φ.

$\triangle$ is the basic branching combinator; many others that are suitable to express further relevant system properties can be derived; among them

$$\nabla \phi =_{\mathrm{def}} \neg \, \triangle \, \neg \, \phi$$

(at least in one case, i.e., the property represented by ϕ holds in at least one path).

The derived combinators for the path formulas, $\diamond$ and $\square$, can be defined as for the linear–time logic.

Further temporal logics. In the previous paragraphs we have briefly sketched two simple logics. In the literature and in the "practice" of specification of concurrent systems, a large number of variants have been proposed; the differences are related to:

anchored version In the model a state (set of states) is singled out to be initial, determined by a special propositional symbol, and the validity of a formula is changed to hold only on such states (paths starting from such states). Formally, for the branching–time case, assume that $s' \in STATE$ is the initial state, thus

$$\mathcal{M} \models \phi \quad \text{iff} \quad \sigma, \mathcal{M} \models \phi \text{ for all } \sigma \in PATH(\mathcal{M}) \text{ such that } s_0 = s'.$$

edge formulas The models, instead of just being sequences (trees) of states, allow the labeling of the transitions from a state to another; thus they are sequences of states and labels or trees where the arcs are labeled (labeled transition systems). Clearly, the formulas are extended to include "edge formulas" expressing conditions on the next label (see, e.g., [Lam83, CR97]).

first-order The basic formulas, instead of being propositional, are first-order. Now there is the problem of the evaluation of the variables appearing in a formula; usually the symbols appearing in a formula (operations, predicates, variables) are classified into: *rigid*, whose interpretation does not depend on the state where the formula is evaluated, and *flexible*, whose interpretation depends on the state where the formula is evaluated. Consequently a model consists of a standard first-order structure, a variable evaluation (for interpreting the rigid symbols) and a sequence (tree) of states, where with each state is associated a first-order structure and a variable evaluation, for interpreting the flexible symbols. Clearly the carriers of such structures must coincide with those of the structure used for rigid symbols (no sensible and usable proposal is available for overcoming this restriction).

Temporal logic and algebraic specifications (A2). First-order temporal logics allow us to give specifications of concurrent systems, where the properties on the dynamic activity are given using the temporal combinators, while the involved (static) data structures are specified by first-order loose algebraic specifications. The sort symbols plus the rigid symbols give the signature of the data structure, while the flexible symbols describe the states of the system (notice a similarity with the dynamic data-type approach in Section 13.4).

Example 13.6. (**Bit using temporal logic (requirements)**) In this case we try to give some requirements on **Bit** (see p. 469), instead of specifying its design, as done in previous examples.

The static data structure is now specified by the following loose algebraic specification, where we do not fix the policy followed by the buffer for storing values (e.g., as a queue or as a stack).[2]

> **spec** *INT_BUNCH* =
> **enrich** *INT* **by**
> **sorts**   *bunch*
> **opns**   *Empty*: $\to$ *bunch*
>   *Put*: *int* $\times$ *bunch* $\to$ *bunch*
>   *First*: *bunch* $\to$ *int*
>   *Remove*: *bunch* $\to$ *bunch*
> **preds** *Not_Empty*: *bunch*
>   *Is_In*: *int* $\times$ *bunch*
> **axioms** *Empty* **and** *Put* are generators for *bunch*
> $\neg\ Is\_In(i, Empty)$
> $Is\_In(i, Put(i', b)) \Leftrightarrow (i = i'\ \lor\ Is\_In(i, b))$
> $Not\_Empty(b) \Leftrightarrow \exists\, i.\ Is\_In(i, b)$
> $Is\_In(First(b), b)$
> $Is\_In(i, Remove(b)) \Rightarrow Is\_In(i, b)$

---

[2] In this chapter, for simplicity, we omit the universal quantifiers when writing the specification axioms; thus, e.g., $\neg\ Is\_In(i, Empty)$ stands for $\forall\, i: int.\ \neg\ Is\_In(i, Empty)$.

The states of the system are then characterized by the following flexible symbols:

opns $Buf\_Cont: \rightarrow bunch$ ******   the buffer content
preds $Putting: int$
     **   is the user putting a given integer in the buffer ?
     $Reading: int$
     **   is the user reading a given integer from the buffer ?
     $Terminated:$
     **   has the user terminated its activity ?
     $Error:$
     **   has something erroneous happened in the system ?

The following formulas express requirements on **Bit**.

$$Putting(i) \Rightarrow \triangle \diamond \ (Is\_In(i, Buf\_Cont) \lor Error)$$

If the user is putting i in the buffer, then in any case, eventually, either i will be in the buffer or something erroneous will happen **(R5)**.

$$\triangle \Box \ (Reading(i) \Rightarrow i = First(Buf\_Cont))$$

or equivalently

$$Reading(i) \Rightarrow i = First(Buf\_Cont)$$

In any case, always, if the user is reading i, then i is the first element of the buffer **(R5)**.

$$Terminated \Rightarrow \Box \ Terminated$$

Once the user has terminated its activity, it cannot restart **(R1)**.

$$\triangledown \Box \neg \ Error$$

There always exists a possible "correct" behavior **(R2)**.
     There is no way to express **(R0)**, since we cannot express the buffer capabilities of interacting with the user within the system.   ∎

13.2.4   Streams and data flow

A model of concurrency where the data structures representing the flow of interaction are made explicit is data flow concepts based on streams. Of course, these temporal formulas do not specify all the interesting properties of a buffer; rather they specify a subset. Temporal logic is not well suited for comprehensive specifications. Therefore it is better to combine it with methods that are more appropriate for specifying safety properties.

Data flow models. For references, see, for instance, [Bro87, Bro97, Bro93, Bro96].

Data flow models of systems are often represented by data flow graphs (also called data flow diagrams). A *data flow graph* is a directed graph, the nodes of which are called data flow nodes and the arcs of which are called data flow arcs. Some arcs may have no sources. These are called *input arcs*. Others may have no target. These are called *output arcs*.

Data flow models are used in many methods in software engineering. They provide a structural view of a system by representing the computing agents by data flow nodes and their communication interconnection by the arrows connecting them. Although data flow diagrams are used in nearly all methods (SA, SADT, SSADM, OMT, SDL, etc.), as well as in many books on operating systems, their meaning is often not well defined and leads to many misinterpretations. The stream model can help to provide a precise meaning for data flow graphics.

There are a number of variations of data flow models. In acyclic data flow models often only one data value is associated with each data flow arc. The data flow nodes are then functions that receive their arguments on their input arcs (one on each arc) and produce one result on each of their output arcs. The data flow diagram accordingly shows a computation tree or a computation graph. This is related to the single assignment languages.

In more sophisticated data flow graphs, we associate a stream of data elements with each data flow arc. This leads to Kahn networks [Kah74]. In the deterministic case each node is associated with a stream processing function that receives its argument streams on its input arcs (one at each arc) and produces one result stream on each of its output arcs. These graphs may be cyclic. This leads to cyclic (recursive) definitions for the streams associated with the arcs. A simple mathematical model for data flow diagrams can be obtained by *stream processing functions*. Nondeterministic data flow diagrams can be handled by associating sets of functions with each node.

The idea of data flow was heavily influenced by the concept of Petri nets. Pioneering papers on data flow were based on the firing rule semantics of Petri nets [Den80]. On the other hand the development of data flow influenced the generaliZation of Petri nets. High-level Petri nets are special cases of such data flow diagrams. Both places and transitions in Petri nets can be seen as data flow nodes.

Functional system specification. In this section we give a brief summary of the basic mathematical concepts of stream-based functional system models. We consider system components with a finite number of input and output channels. Messages are exchanged over the channels. A channel history is mathematically modeled by a stream of messages. The behavior of a (deterministic) component corresponds to a function mapping the streams on its input channels onto streams for its output channels.

A *stream* of messages over a given message set M is a finite or infinite sequence of messages. We define the set of stream by

$$M^\omega =_{\mathrm{def}} M^* \cup M^\infty.$$

By $x \frown y$ we denote the result of concatenating two streams x and y. We assume that $x \frown y = x$, if x is infinite. By $\langle \rangle$ we denote the empty stream. For simplicity we write for $a \in M$, $x \in M^\omega$

$a \frown x$ instead of $\langle a \rangle \frown x$ and $x \frown a$ instead of $x \frown \langle a \rangle$.

If a stream x is a *prefix* of a stream y, we write $x \sqsubseteq y$. The relation $\sqsubseteq$ is called *prefix order*. It is formally specified by

$$x \sqsubseteq y =_{\mathrm{def}} \exists z \in M^\omega : x \frown z = y.$$

The relation $\sqsubseteq$ is a partial order on the set of streams. The empty stream $\langle \rangle$ is the least element.

Given a partially ordered set, a subset is called *directed* if, for any pair of elements in S, there exists an upper bound in S. A partially ordered set is called *complete* if, for every directed set of streams, there exists a least upper bound. The set of streams ordered by the prefix order is complete. The least upper bound of a directed set S is denoted by *lub S*.

The behavior of deterministic interactive systems with n input channels and m output channels is modeled by prefix monotonic functions

$$f : (M^\omega)^n \to (M^\omega)^m$$

called *(m, n)-ary stream processing functions*.

A function f mapping a complete partially ordered set onto a complete partially ordered set is called *continuous*, if, for every directed set S,

$$f(lub\ S) = lub\ \{f(x) \mid x \in S\}$$

The set of all prefix continuous stream processing functions of functionality $(M^\omega)^n \to (M^\omega)^m$ is denoted by

$$SPF^n_m.$$

For simplicity, we do not consider type information here and assume M to be just a set of messages.

The following functions on streams are useful in specifications:

$rt : M^\omega \to M^\omega$	rest of a stream
$ft : M^\omega \to M \cup \{\bot\}$	first element of a stream
$\# : M^\omega \to \mathbb{N} \cup \{\infty\}$	length of a stream
$\_\text{\textcircled{c}}\_ : \mathcal{P}(M) \times M^\omega \to M^\omega$	filter of a stream

Here $\perp$ is used as a dummy to avoid partial functions. These functions are easily specified by the following algebraic equations (let $x \in M^\omega$, $m \in M$, $S \subseteq M$):

$$rt(\langle\rangle) = \langle\rangle, \qquad\qquad rt(m \hat{\ } x) = x,$$
$$ft(\langle\rangle) = \perp, \qquad\qquad ft(m \hat{\ } x) = m,$$
$$\#(\langle\rangle) = 0, \qquad\qquad \#(m \hat{\ } x) = 1 + \#(x),$$

$$S \copyright \langle\rangle = \langle\rangle,$$
$$S \copyright (m \hat{\ } x) = m \hat{\ } (S \copyright x), \quad \text{if } m \in S$$
$$S \copyright (m \hat{\ } x) = S \copyright x, \qquad \text{if } m \notin S$$

These axioms specify the functions completely. They are useful in proofs, too.

Stream processing functions can easily be specified by logical formulas in the style of algebraic equations as is demonstrated for the running example below. Given such functions, we may compose them.

We use two forms of composition: parallel composition and sequential composition. Given functions $f \in SPF^n_k$, $g \in SPF^k_m$ we write

$$f ; g$$

for the *sequential composition* of f and g which yields a function in SPF^n_m, where

$$(f ; g)(x) = g(f(x)).$$

Given functions $f \in SPF^n_m$, $g \in SPF^{n'}_{m'}$ we write

$$f \| g$$

for the *parallel composition* of f and g which yields a function in $SPF^{n+n'}_{m+m'}$, where (let $x \in (M^\omega)^n$, $y \in (M^\omega)^{n'}$):

$$(f \| g)(\langle x, y \rangle) = \langle f(x), g(y) \rangle$$

Finally, given a function

$$f \in SPF^n_m$$

we may construct a function by the feedback operator leading an output line back to an input line. We write

$$\mu^k_j f \in SPF^{n-1}_{m-1}$$

for the function defined by the equation ($1 \le k \le n, 1 \le j \le m$)

$$\mu^k_j f(x_1, \ldots, x_{k-1}, x_{k+1}, \ldots, x_n) = (y_1, \ldots, y_{j-1}, y_{j+1}, \ldots, y_m)$$

where z is the prefix least stream such that the following equation holds

$$f(x_1, \ldots, x_{k-1}, z, x_{k+1}, \ldots, x_n) = (y_1, \ldots, y_{j-1}, z, y_{j+1}, \ldots, y_m)$$

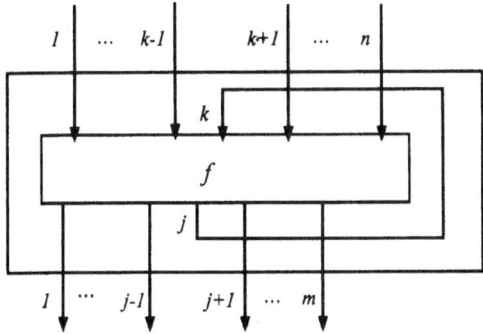

Fig. 13.6. Data flow graph for feedback

See Figure 13.6.

Since f is prefix monotonic, such a stream least solution (least fix point) always exists. Of course, it is unique.

By $SPEC_m^n$ we denote the set of all predicates Q where

$$Q: SPF_m^n \to \{T, F\}$$

The set $SPEC_m^n$ denotes the set of all component specifications for a component with n input channels and m output channels.

We want to compose specifications of components to networks. Each form of composition introduced for functions can be extended to component specifications in a straightforward way. Given component specifications $Q \in SPEC_k^n$, $R \in SPEC_m^k$ we write

$$Q; R$$

for the predicate in $SPEC_m^n$ where

$$(Q; R)(f) \Leftrightarrow \exists\, q, r.\, f = q; r \wedge Q(q) \wedge R(r)$$

Trivially, we have for all specifications $Q \in SPEC_m^n$ the following equations, where I denotes the identity function:

$$Q; I = Q \qquad \text{and} \qquad I; Q = Q.$$

Given specifications $Q \in SPEC_m^n$, $R \in SPEC_{m'}^{n'}$ we write

$$Q \| R$$

for the predicate in $SPEC_{m+m'}^{n+n'}$ where

$$(Q \| R)(f) \Leftrightarrow \exists\, q, r.\, f = q \| r \wedge Q(q) \wedge R(r).$$

Feedback also carries over in a straightforward manner to specifications.

$$(\mu_j^k Q)(f) \equiv \exists\, f'.\, Q(f') \wedge f = (\mu_j^k f')$$

Any data flow graph can be described by parallel composition and feedback. This is easily seen. To build a compositional form for a given data flow diagram where the nodes are described by specifications of stream processing functions, we form a large parallel composition of all data flow nodes. Then we connect the output lines to input lines by feedback as shown by the data flow diagram.

A specification $\overline{Q} \in SPEC_m^n$ is called a *property refinement* of a specification $Q \in SPEC_m^n$ if, for all functions f, we have $\overline{Q}(f) \Rightarrow Q(f)$. We write then

$$\overline{Q} \Rightarrow Q$$

In other words, $\overline{Q}$ is a property refinement of Q if the set of functions described by $\overline{Q}$ is a subset of the set of functions described by Q. More sophisticated notions of refinement are obtained by abstraction and representation specifications as introduced in [Bro97].

A pair of specifications A and R are called abstraction and representation, if

$$R; A = I$$

Let A_1 be an abstraction specification and R_2 be a representation specification. A specification C' is called a refinement of specification C if we have

$$C' \Rightarrow A_1; C; R_2$$

Given the corresponding abstraction specification A_2 and a representation specification R_1, the identities

$$R_1; A_1 = I \qquad R_2; A_2 = I$$

allows us to deduce

$$R_1; C'; A_2 \Rightarrow C.$$

The actual specification of data flow nodes can be done by logical formulas describing the relationship between the input and output streams.

The strong aspect of stream processing concepts is their modularity. They allow for a modular specification, composition, and refinement of interacting systems.

Example 13.7. (**Bit using stream functions**)
First we specify the two components, buffer and user.
For each component we first give its functionality, then we give the specifying axioms.

$$BUFFER\colon (\mathbb{N} \cup \{\, GET\})^\omega \to \mathbb{N}^\omega$$

$\forall h \in \mathbb{N}, z \in \mathbb{N}^*, x \in (\mathbb{N} \cup \{ GET\})^\omega$:
$BUFFER(h \frown z \frown GET \frown x) = h \frown BUFFER(z \frown x) \vee$
$BUFFER(h \frown z \frown GET \frown x) = h \frown BUFFER(h \frown z \frown GET \frown x) \vee$
$BUFFER(h \frown z \frown GET \frown x) = \langle 10^{10}\rangle$

The user is modeled by the function

$USER: \mathbb{N}^\omega \times \{ START\}^\omega \to (\mathbb{N} \cup \{ GET\})^\omega \times \{OK, ERROR\}^\omega$

which is specified by the equation

$USER(x, \langle START\rangle) = \langle 0 \frown 1 \frown GET, INSPECT(x)\rangle$

where the auxiliary function

$INSPECT: \mathbb{N}^\omega \to \{OK, ERROR\}^\omega$

is specified by the equations $(i \in \mathbb{N} \wedge x \in \mathbb{N}^*)$

$INSPECT(0 \frown x) = \langle OK\rangle$
$INSPECT(i \frown x) = \langle ERROR\rangle, \quad i \neq 0$

The system is formed by the parallel composition of *BUFFER* and *USER*, and a feedback of $Output_I$ to *Input* and of *Output* to $Input_I$; see its graphical explanation in Figure 13.7.

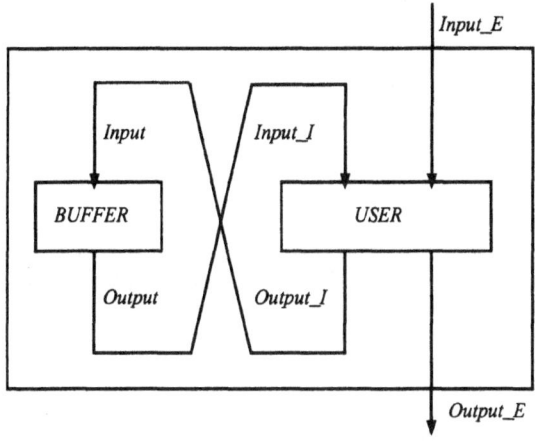

Fig. 13.7. Structure of **Bit** specified using stream-processing functions

The structure of the data flow diagram is captured logically by the equations

$Output = BUFFER(Input)$
$(Output\_I, Output\_E) = USER(Input\_I, Input\_E)$
$Input\_I = Output$
$Input = Output\_I$
$Input\_E = \langle START\rangle$

Given these equations and the specifyig equations above, we may begin a straightforward reasoning about the value of *Output_E*.   ∎

Integration with algebraic data type specifications. The integration of stream processing functions with algebraic specification is quite simple. Streams are nothing other than an abstract data type, just slightly more complex because they may be infinite. The data types forming the messages in a stream are easily specified by algebraic specifications.

Stream processing functions can also be specified by algebraic specifications. Similarly sequential composition, parallel composition, and feedback can be described by algebraic specifications. Here we need higher-order algebraic specifications, of course. The algebraic equations for the composition operators lead to a rich algebra of stream processing components.

13.3   Dynamic-data types (A3)

As shown in the section on **CCS**, labeled transition systems (lts) are an effective way to give an operational semantics to a process algebra. In this section we show how lts can be directly used for the specification of system behavior.

13.3.1   Labeled transition logic (LTL)

The main references for **LTL** are [AR87, AR96b, CR97]; the first appearance is in [AMRW85]. Notice that in the past the terms "algebraic transition systems" (e.g., in [AMRW85, AR87]) and "dynamic specifications" (e.g., in [CR97, Reg93, AR96a]) have also been used for the specifications built using **LTL**.

To model processes **LTL** uses labeled transition systems, see Section 13.1.1, and supplies two different kinds of specifications at different levels of abstraction:

requirement for expressing the requirements on a concurrent system; a requirement specification should determine a class of nonisomorphic models, all those formally and abstractly describing systems having such requirements;

design for expressing the abstract design of a concurrent system, to define abstractly and formally the way in which we intend to design the system; a design specification should determine one model, the one formally and abstractly describing the designed system.

LT-Structures. An lts can be represented by a first-order structure (an algebra with predicates) A on a signature with at least two sorts, *state* and *label*, whose elements correspond respectively to the states and labels of the system, and a predicate $\_ \xrightarrow{} \_: state \times label \times state$ representing the transition relation. The triple $\langle |A|_{state}, |A|_{label}, \rightarrow_A \rangle$ is the corresponding lts. Obviously we can have lts whose states are built by states of other lts (for modeling concurrent systems); in such a case we use structures whose signature has different sorts corresponding to states and labels, and different predicates corresponding to transition relations.

In a formal model for concurrent systems we may need to consider data, too (for example, the data manipulated by a system such as natural numbers); to handle these cases we consider structures which also have sorts that just correspond to data and not to states or labels of lts.

The first-order structures (algebras) corresponding to lts are called *LT-structures* and are formally defined as follows.

- An *LT-signature* $LT\Sigma$ is a pair $\langle \Sigma, DS \rangle$, where:
  * $\Sigma = \langle S, \Omega, \Pi \rangle$ is a first-order signature,
  * $DS \subseteq S$ (the elements in DS are the *dynamic sorts*, which are the sorts corresponding to states of lts),
  * for all $ds \in DS$ there exist a sort $lab\_ds \in S - DS$ (labels of the transitions of the processes of sort ds) and a predicate
    $\_ \xrightarrow{} \_: ds \times lab\_ds \times ds \in \Pi^3$ (transition relation of the dynamic elements of sort ds).
- An *LT-structure* on $LT\Sigma$ (abbreviated to $LT\Sigma$-structure) is a Σ-first-order structure (a Σ-algebra with predicates).

Design LT-specifications. LT-specifications are particular algebraic/logic specifications for LT-structures where conditional formulas are used as axioms; since transitions are described by predicate symbols, such formulas also allow us to express properties on the activity of processes.

An *LT-specification* is a pair $SP = \langle LT\Sigma, AX \rangle$, where $LT\Sigma = \langle \Sigma, DS \rangle$ is an LT-signature and AX a set of conditional formulas on $LT\Sigma$ having form

$$\bigwedge_{i=1,\dots,n} \alpha_i \Rightarrow \alpha_{n+1},$$

where for $i = 1, \dots, n + 1$, α_i is an *atom*, i.e., a formula having the form either $t = t'$ or $p(t_1, \dots, t_m)$.

We can give SP different semantics, as initial and observational, briefly presented below.

The initial semantics of SP determines one (up to isomorphism) LT-structure, precisely $IMod(SP) = T_\Sigma / \equiv_{AX}$, where $\equiv_{AX}$ is the congruence over T_Σ generated by the Birkhoff deductive system for conditional specifications, sound and complete w.r.t. the models of SP and the atomic formulas, see Chapter 3.

---

[3] In this paper, for some of the operation and predicate symbols we use a mixfix notation; for instance, $\_ \xrightarrow{} \_: ds \; lab\_ds \; ds$ means that we shall write $t \xrightarrow{t'} t''$ instead of $\xrightarrow{} (t, t', t'')$.

Since in an LT-structure the transitions correspond to the truth of the ground atoms built by the transition predicates, we have that the transitions in the initial model of an LT-specification are just those whose corresponding atoms can be proved by using the Birkhoff system.

In most cases the initial semantics of an LT-specification is too fine, since it takes into account all details of the activity of the processes (as intermediate states). It may happen that we want to consider as semantically equivalent two processes having different associated transition trees, see Section 13.1.1.

There is a general way to give an observational semantics to LT-specifications introduced in the general case of conditional specifications by some of the authors and A. Giovini (see [AGR92] for a full presentation); this approach is well suited for use in LT-specifications, which specify concurrent systems, since it generalizes and extends the Milner–Park's bisimulation technique to a purely algebraic setting.

Example 13.8. (**Bit using LTL**) We first specify the two components of the system, the buffer and the user, and then how they cooperate.

Below "**dsort** $\ldots : \ldots$" is the construct for declaring dynamic sorts, the second argument is the syntactic form of the transition predicate; thus

dsort $buffer: \ \_ \xrightarrow{} \_$

declares the dynamic sort $buffer$, the associated sort of the labels lab_buffer, and the transition predicate $\_ \xrightarrow{} \_: buffer \times lab\_buffer \times buffer$.

spec $BUFFER =$
    enrich $INT\_QUEUE[buffer/queue]$ **by**
    dsorts $buffer: \ \_ \xrightarrow{} \_$
    opns $\quad \tau: \ \to lab\_buffer$
    $\quad\quad\quad RECEIVE, RETURN: int \to lab\_buffer$
    axioms $Not\_Empty(b) \Rightarrow b \xrightarrow{RETURN(First(b))} Remove(b)$
    $\quad\quad\quad b \xrightarrow{RECEIVE(i)} Put(i,b)$
    $\quad\quad\quad b \xrightarrow{\tau} Put(10^{10}, Empty)$
    $\quad\quad\quad Not\_Empty(b) \Rightarrow b \xrightarrow{\tau} Dup(b)$

spec $USER\_STATUS =$
    enrich INT **by**
    sorts $\quad user\_status$
    opns $\quad Initial, Putting\_0, Putting\_1, Reading, Terminated: \ \to user\_status$
    $\quad\quad\quad Read: int \to user\_status$

spec $USER =$
    enrich $USER\_STATUS[user/user\_status]$ **by**
    dsorts $user: \ \_ \xrightarrow{} \_$
    opns $\quad START, OK, ERROR: \ \to lab\_user$
    $\quad\quad\quad PUT, GET: int \to lab\_user$
    axioms $Initial \xrightarrow{START} Putting\_0$
    $\quad\quad\quad Putting\_0 \xrightarrow{PUT(0)} Putting\_1$
    $\quad\quad\quad Putting\_1 \xrightarrow{PUT(1)} Reading$

$$Reading \xrightarrow{GET(i)} Read(i)$$
$$Read(0) \xrightarrow{OK} Terminated$$
$$NotEq(i,0) \Rightarrow Read(i) \xrightarrow{ERROR} Terminated$$

spec $SYSTEM =$
     enrich $BUFFER + USER$ **by**
     dsorts $system : \_ \xrightarrow{} \_$
     opns $\_ \mid \_ : buffer \times user \to system$
             $START, OK, ERROR, \tau : \; \to lab\_system$
     axioms $u \xrightarrow{START} u' \Rightarrow Empty \mid u \xrightarrow{START} Empty \mid u'$
               $b \xrightarrow{RECEIVE(i)} b' \wedge u \xrightarrow{PUT(i)} u' \Rightarrow b \mid u \xrightarrow{\tau} b' \mid u'$
               $b \xrightarrow{RETURN(i)} b' \wedge u \xrightarrow{GET(i)} u' \Rightarrow b \mid u \xrightarrow{\tau} b' \mid u'$
               $u \xrightarrow{OK} u' \Rightarrow b \mid u \xrightarrow{OK} b \mid u'$
               $u \xrightarrow{ERROR} u' \Rightarrow b \mid u \xrightarrow{ERROR} b \mid u'$
               $b \xrightarrow{\tau} b' \Rightarrow b \mid u \xrightarrow{\tau} b' \mid u$ ∎

Requirement LT-specifications. $SP = \langle LT\Sigma, AX \rangle$ with loose semantics determines the class of its *models*, $Mod(SP)$, i.e., all $LT\Sigma$–structures satisfying all formulas in AX.

LT-specifications with loose semantics can be used to specify the requirements on a concurrent system, thus determining a class of systems (all those satisfying such requirements), instead of abstractly defining one particular system. However, conditional formulas are too limited to express all relevant requirements on concurrent systems, thus various extensions of first-order logic are used, e.g., including combinators of the branching–time temporal logic [CR97], the deontic logic [CR96], using the concept of "abstract event" [AR93b], etc. Below we briefly present the extension of [CR97] with branching-time temporal combinators (see Section 13.2.3).

Let $LT\Sigma = \langle\langle S, \Omega, \Pi \rangle, DS\rangle$ be an LT-signature, L an $LT\Sigma$-structure, and $ds \in DS$. We need the following technical definitions. $PATH(L, ds)$ denotes the set of the *paths* for the elements of sort ds, i.e., all sequences of transitions having the form either (1) or (2) below:

(1) $d_0 \; l_0 \; d_1 \; l_1 \; d_2 \; l_2 \; \ldots$
(2) $d_0 \; l_0 \; d_1 \; l_1 \; d_2 \; l_2 \; \ldots \; d_n$ $n \geq 0$

where for all $i \geq 0$, $d_i \in |L|_{ds}$, $l_i \in |L|_{lab\_ds}$, and $(d_i, l_i, d_{i+1}) \in \to_L$.

$FirstS(\sigma)$ denotes the *first state* of σ; and $FirstL(\sigma)$ denotes the *first label* of σ, if exists, i.e., if σ is not just a state.

$\sigma \in PATH(L, ds)$ is *maximal* iff either it is infinite or there do not exist l, d' such that $\langle d_n, l, d' \rangle \in \to_L$.

Given $\sigma = d_0 \; l_0 \; d_1 \; l_1 \; d_2 \; l_2 \; \ldots$ and $h \geq 0$, if d_h exists, then $\sigma\lfloor_h$ denotes the path $d_h \; l_h \; d_{h+1} \; l_{h+1} \; d_{h+2} \; \ldots$, otherwise undefined.

The set of *formulas*, denoted by $F(LT\Sigma, X)$, and the family of the sets of *path formulas*, denoted by $\langle PF(LT\Sigma, X)_{ds}\rangle_{ds\in DS}$, on $LT\Sigma$, and variables X are defined by multiple induction as follows. For each $s \in S$ and $ds \in DS$:

formulas

- $p(t_1, \ldots, t_n) \in F(LT\Sigma, X)$ $p : s_1 \times \cdots \times s_n \in \Pi,\ t_i \in |T_{LT\Sigma}(X)|_{s_i}$
- $t_1 = t_2 \in F(LT\Sigma, X)$ $t_1, t_2 \in |T_{LT\Sigma}(X)|_s$
- $\neg \phi, \phi \Rightarrow \phi', \forall\, x.\, \phi \in F(LT\Sigma, X)$ $\phi, \phi' \in F(LT\Sigma, X), x \in X$
- $\triangle\, (t, \pi) \in F(LT\Sigma, X)$ $t \in |T_{LT\Sigma}(X)|_{ds},\ \pi \in PF(LT\Sigma, X)_{ds}$

path formulas

- $[\lambda\, x.\, \phi] \in PF(LT\Sigma, X)_{ds}$ $x \in X_{ds},\ \phi \in F(LT\Sigma, X)$
- $\langle \lambda x.\, \phi \rangle \in PF(LT\Sigma, X)_{ds}$ $x \in X_{lab\_ds},\ \phi \in F(LT\Sigma, X)$
- $\pi_1\, \mathcal{U}\, \pi_2 \in PF(LT\Sigma, X)_{ds}$ $\pi_1, \pi_2 \in PF(LT\Sigma, X)_{ds}$
- $\neg\, \pi, \pi \Rightarrow \pi', \forall\, x.\, \pi \in PF(LT\Sigma, X)_{ds}$ $\pi, \pi' \in PF(LT\Sigma, X)_{ds}, x \in X_s$

The formulas of such logic include the usual formulas of first-order logic with equality; if $LT\Sigma$ contains dynamic sorts, they also include formulas built with the transition predicates.

The formula $\triangle\, (t, \pi)$ can be read as "for every path σ starting in the state denoted by t, the path formula π holds on σ". We anchor these formulas to states, following the ideas in [MP89]. The major difference with the classical temporal logic of Section 13.2.3 is that we do not specify a single system but, in general, one or many types of systems, so there is not a single initial state but several, hence the need for an explicit reference to states (through terms) in the formulas built with $\triangle$. The formula $[\lambda\, x.\, \phi]$ holds on the path σ whenever ϕ holds at the first state of σ; while the formula $\langle \lambda x.\, \phi \rangle$ holds on the path σ if σ is not just a single state and ϕ holds at the first label of σ.

Let L be an $LT\Sigma$–structure and v a variable evaluation of X in L; then we define by multiple induction:

- the validity of $\phi \in F(LT\Sigma, X)$ in L w.r.t. v (written L, $v \models \phi$),
- the validity of $\pi \in PF(LT\Sigma, X)$ on a path σ in L w.r.t. v (written L, $v, \sigma \models \pi$),

as follows:

- L, $v \models p(t_1, \ldots, t_n)$ iff $\langle v^{\#}(t_1), \ldots, v^{\#}(t_n) \rangle \in p_{\mathrm{L}}$
- L, $v \models t_1 = t_2$ iff $v^{\#}(t_1) = v^{\#}(t_2)$
- L, $v \models \triangle\, (t, \pi)$ iff   for each $\sigma \in PATH(\mathrm{L}, ds)$ such that $FirstS(\sigma) = v^{\#}(t)$, L, $v, \sigma \models \pi$
- L, $v, \sigma \models [\lambda\, x.\, \phi]$ iff   L, $v[FirstS(\sigma)/x] \models \phi$
- L, $v, \sigma \models \langle \lambda x.\, \phi \rangle$ iff $FirstL(\sigma)$ is defined and L, $v[FirstL(\sigma)/x] \models \phi$
- L, $v, \sigma \models \pi_1\, \mathcal{U}\, \pi_2$ iff   there exists $j \geq 0$ such that for all h, $0 < h < j$, L, $v, \sigma\lfloor_h \models \pi_1$ and L, $v, \sigma\lfloor_j \models \pi_2$
- $\neg\, \phi, \phi \Rightarrow \phi', \forall\, x.\, \phi, \neg\, \pi, \pi \Rightarrow \pi', \forall\, x.\, \pi$ as usual.

ϕ is *valid* in L (written L $\models \phi$) iff L, $v \models \phi$ for all evaluations v.

In the above definitions we have used a minimal set of combinators. However it is possible to define other, derived, combinators as for the classical logics of Section 13.2.3; plus $\bigtriangledown(t, \pi) =_{\mathrm{def}} \neg\, \triangle\, (t, \neg\, \pi)$ (which means at least in one case, i.e., the property represented by π holds at least on one path).

Example 13.9. (**Bit using** LTL (**requirements**)) As already done in Example 13.6, we give here some sample requirements on **Bit**; but in a different way to before, we specify the system modularly, by considering its components first and then how they are put together in order to cooperate. Furthermore, temporal LTL has also edge formulas, so we can also conveniently express properties concerning the interactions of the system with the environment. Since properties are anchored to processes (concurrent systems), we can relate properties of the system to properties of its components.

Below $\langle \lambda l. \, l = t \rangle$ is abbreviated to $\langle t \rangle$.

spec *USER* =
   enrich *INT* **by**
   dsorts *user* : _ $\overset{\cdot}{\longrightarrow}$ _
   preds *Terminated* : *user*
   opns *PUT, GET* : *int* → *lab_user*
   axioms *Terminated*(*u*) ⇒
         △ (*u*, [λ*u'*. *Terminated*(*u'*)])
      ** if the user is terminated, it remains so forever (**R1**)

spec *BUFFER* =
   enrich *INT* **by**
   dsorts *buffer* : _ $\overset{\cdot}{\longrightarrow}$ _
   opns *RECEIVE, RETURN* : *int* → *lab_buffer*
   axioms $b \xrightarrow{RECEIVE(i)} b' \Rightarrow \triangle(b', \Diamond [\lambda x. \exists x'. x \xrightarrow{RETURN(i)} x'])$
      ** after receiving *i* the buffer eventually will have
      ** the capability to return *i*
      $\triangledown(b, \langle RECEIVE(i) \rangle)$
      ** the buffer must be able to receive any integer (**R0**)

spec *SYSTEM* =
   enrich *BUFFER* + *USER* **by**
   dsorts *system* : _ $\overset{\cdot}{\longrightarrow}$ _
   opns *START, OK, ERROR*, τ : → *lab_system*
         _ | _ : *buffer* × *user* → *system*
   axioms ∃ *b, u. s* = *b* | *u*
         $\langle b, u \rangle \xrightarrow{l} \langle b', u' \rangle \Rightarrow$
            $(b = b' \wedge u \xrightarrow{l} u') \vee$
            $(u = u' \wedge b \xrightarrow{l} b') \vee$
            $(\exists i. l = \tau \wedge u \xrightarrow{GET(i)} u' \wedge b \xrightarrow{RETURN(i)} b') \vee$
            $(\exists i. l = \tau \wedge u \xrightarrow{PUT(i)} u' \wedge b \xrightarrow{RECEIVE(i)} b')$
      ** (**R5**)
      $\triangledown(s, \square \langle \lambda l. \neg l = ERROR \rangle)$
      ** there always exists a possible "correct" behavior (**R2**)
      Terminated(*u*) ⇒
      △ (*b* | *u*, [λ*s*. ∃ *b', u'*. (*s* = *b'* | *u'* ∧ *Terminated*(*u'*))] $\mathcal{U}$
         [λ *s*. ¬ ∃ *l, s'*. *s* $\xrightarrow{l}$ *s'*])
      ** if the user is terminated, it remains so until the system stops

$$s \xrightarrow{START} s' \Rightarrow \triangle (s', \Diamond (\langle OK \rangle \vee \langle ERROR \rangle))$$

** after the system has been started, in any case eventually
** it will send out either OK or $ERROR$ **(R3)**

$$(s \xrightarrow{OK} s' \vee s \xrightarrow{ERROR} s') \Rightarrow$$
$$\triangle (s', \square \langle \lambda l. \neg (l = OK \vee l = ERROR) \rangle)$$

** OK and $ERROR$ are sent at most once, and it cannot
** happen that both are sent **(R4)**

$$\nabla (b \mid u, \Diamond \square \langle \tau \rangle) \Rightarrow \nabla (b, \Diamond \square \langle \tau \rangle)$$

** if the system may eventually only do internal actions,
** then the buffer component has such a possibility, too

Some of the axioms of the above specifications are just to show the peculiarity of this logic. For example, the unique axiom of *USER* requires that it must remain terminated in isolation; while the axiom of *SYSTEM* about terminated requires something about the behavior of the user when put within the system. The last axiom of *SYSTEM* shows how properties of the components can be related to properties of the whole system.   ∎

13.3.2   Rewriting logic (RL)

RL is a formalism for the specification of concurrent systems developed by Meseguer in the recent years, sharing some of the ideas of LTL; moreover, its specifications are syntactically very similar to those of LTL. For both formalisms the behavior of processes is modeled by means of transition systems; the states of such systems are elements of some carriers of an algebra, given as the initial model of a conditional specification; the structure of a term representing one of such states models the concurrent structure of the system in that state; and the transitions are defined by conditional formulas in which the transition symbol (arrow) appears.

Clearly, there are also major differences between the two formalisms: the transitions are labeled in the case of LTL and nonlabeled for RL, and have special properties in the RL case, as to be closed by reflexivity, transitivity, and congruence w.r.t. the operations; and, most important, their intuitive interpretation is very different in the two cases:

LTL $t \xrightarrow{l} t'$ means that the system in the state represented by t has the "capability" of passing into the state represented by t' by performing some "atomic" activity, i.e., an activity that cannot be interrupted, where no information on the intermediate states is available, whose interaction with the environment is represented by l, and at each instant a system can perform only one of these activities.

RL $t \longrightarrow t'$ means that the system in the state represented by t can pass into the state represented by t' by performing some activity completely independently from the environment; such activity may be also the composition of several "smaller" activities of the same system, and so information on the intermediate states may be available (for a terminating system, for example, we may have transitions which correspond to whole evolutions of the system from the beginning till the termination).

A complete study of the relationship between RL and LTL can be found in [AR97b].

Below we give a short presentation of RL, the main references are [Mes92, MM93]; notice that in such papers Meseguer has used the language of category theory to present RL, while here, for clarity, we use a more logic-algebraic style.

A *rewrite theory*, i.e., an RL specification, is a 4-tuple $\mathcal{R} = \langle \Sigma, E, L, R \rangle$, where $\Sigma = \langle S, \Omega \rangle$ is a signature, E a set of equations on Σ, and R a set of *rewrite rules* of the form

$$r \colon [t] \longrightarrow [t'] \quad if \quad [u_1] \longrightarrow [v_1] \wedge \ldots \wedge [u_k] \longrightarrow [v_k]$$

with $r \in L$ and $[t], [t'], [u_1], [v_1], \ldots, [u_k], [v_k] \in T_\Sigma(X)/{\equiv_E}$.

The *entailment system* associated with $\mathcal{R}$ has the following rules:

1. Reflexivity For each $[t] \in T_\Sigma(X)/{\equiv_E}$ $\dfrac{}{[t] \longrightarrow [t]}$

2. Congruence For each $f \colon s_1 \times \cdots \times s_n \to s \in \Omega$

$$\dfrac{[t_1] \longrightarrow [t_1'] \ldots [t_n] \longrightarrow [t_n']}{[f(t_1, \ldots, t_n)] \longrightarrow [f(t_1', \ldots, t_n')]}$$

3. Replacement' For each rewrite rule

$$r \colon [t(\overline{x})] \longrightarrow [t'(\overline{x})] \quad if \quad [u_1(\overline{x})] \longrightarrow [v_1(\overline{x})] \wedge \ldots \wedge [u_k(\overline{x})] \longrightarrow [v_k(\overline{x})]$$

belonging to R, where $\overline{x}$ is the vector of all variables appearing in the rule and $\overline{w}$ a corresponding vector of elements in $T_\Sigma(X)/{\equiv_E}$

$$\dfrac{[u_1(\overline{w}/\overline{x})] \longrightarrow [v_1(\overline{w}/\overline{x})] \ldots [u_k(\overline{w}/\overline{x})] \longrightarrow [v_k(\overline{w}/\overline{x})]}{[t(\overline{w}/\overline{x})] \longrightarrow [t'(\overline{w}/\overline{x})]}$$

4. Transitivity $\dfrac{[t_1] \longrightarrow [t_2] \quad [t_2] \longrightarrow [t_3]}{[t_1] \longrightarrow [t_3]}$

Here, for simplicity, we use the entailment system above, which is a slightly modified version of the original [Mes92]: rule 3, shown below, has been changed to avoid the simultaneous rewriting of an element substituted for a variable. In [AR97b] we show that this entailment system is equivalent to the original, since the entailed sequents are the same and the structure of the proofs is preserved.

Now we define the models of the rewrite theories using the entailment system.

- An $\mathcal{R}$-*presystem* is a direct reflexive graph $G = (\theta_0, \theta_1 \colon Edges \to Nodes, id)$, where $id \colon Nodes \to Edges$ such that $\theta_0(id(n)) = \theta_1(id(n)) = n$ (id expresses the reflexivity of the graph) together with:

- a Σ-structure on *Nodes* such that it satisfies E and the edges respect the sorts (i.e., for each edge e, $\theta_0(e)$ and $\theta_1(e)$ have the same sort);
- for each $f\colon s_1 \times \cdots \times s_n \to s \in \Omega$,
  a partial operation $f\colon Edges^n \dashrightarrow Edges$ such that $f(e_1,\ldots,e_n)$ is defined iff for $i = 1,\ldots,n$, $e_1, \ldots, e_n$ are edges of sorts $s_1, \ldots, s_n$ respectively, and
  $\theta_0(f(e_1,\ldots,e_n)) = f(\theta_0(e_1),\ldots,\theta_0(e_n))$,
  $\theta_1(f(e_1,\ldots,e_n)) = f(\theta_1(e_1),\ldots,\theta_1(e_n))$;
- a partial operation $\_;\_\colon Edges^2 \dashrightarrow Edges$ such that $e;e'$ is defined iff e and e' have the same sort and $\theta_1(e) = \theta_0(e')$ and $\theta_0(e;e') = \theta_0(e)$, $\theta_1(e;e') = \theta_1(e')$;
- for each rewrite rule
  $r\colon [t(\overline{x})] \longrightarrow [t'(\overline{x})]$ *if* $[u_1(\overline{x})] \longrightarrow [v_1(\overline{x})] \wedge \ldots \wedge [u_k(\overline{x})] \longrightarrow [v_k(\overline{x})]$
  belonging to R, a partial operation[4] $r\colon VarEv \times Edges^k \dashrightarrow Edges$,
  where $VarEv$ is the set of variable evaluations from $\overline{x}$ into *Nodes*, such that $r(v,e_1,\ldots,e_k)$ is defined iff, for $i = 1,\ldots,k$, $\theta_0(e_i) = v^\#(u_i(\overline{x}))$ and $\theta_1(e_i) = v^\#(v_i(\overline{x}))$, and $\theta_0(r(v,e_1,\ldots,e_k)) = v^\#(t(\overline{x}))$, $\theta_1(r(v,e_1,\ldots,e_k)) = v^\#(t'(\overline{x}))$.

- A *morphism* ϕ between two $\mathcal{R}$-presystems P and P' is a graph morphism which preserves the Σ-structure on the nodes and the operations on the edges.
- An $\mathcal{R}$-*system* is an $\mathcal{R}$-presystem satisfying the following equations on edges (the adaptation of those of [Mes92] to our modified entailment system):
  1. **Category** $(e;e');e'' = e;(e';e'')$
  2. **Functoriality of the Σ-structure**
     for each $f\colon s_1 \times \cdots \times s_m \to s \in \Omega$
     $f(e_1;e'_1,\ldots,e_m;e'_m) = f(e_1,\ldots,e_m);f(e'_1,\ldots,e'_m)$
     $f(id(n_1),\ldots,id(n_m)) = id(f(n_1,\ldots,n_m))$
  3. **Axioms in E** For each $t(x_1,\ldots,x_n) = t'(x_1,\ldots,x_n) \in E$
     $t(e_1,\ldots,e_n) = t'(e_1,\ldots,e_n)$.
- $\mathcal{T}_\mathcal{R}$ is the initial element in the class of the $\mathcal{R}$-systems, i.e., the $\mathcal{R}$-system where *Nodes* is $T_\Sigma/{\equiv_E}$, the edges are generated by the operations with edge types, and the edges represented by two different terms are identified iff their identification follows from equations 1, 2, and 3 above.

The ground terms built by the edge operations bijectively correspond to the *proofs* of sequents in the entailment systems associated with $\mathcal{R}$; and the axioms on edges correspond to the required identifications on the proofs.

$\mathcal{R}$-systems, and therefore also $\mathcal{T}_\mathcal{R}$, can be seen as categories, where the objects are the nodes, the morphisms are the edges, and "$;$" is the composition operation on morphisms; axiom 1 and the fact that the graph is reflexive ensure that they really are categories.

---

[4] If several rules with the same label will result to operations of the same functionality, then we assume that the names of such operations are made different.

$\mathcal{T_R}$ is considered the standard semantics of a theory $\mathcal{R}$.

We can see a striking difference between RL and LTL: for LTL the precise form and number of the axioms of a specification is irrelevant, while the precise form and number of rules of a theory is extremely important for RL. If two RL theories $\mathcal{R}$ and $\mathcal{R}'$ have the same signature, equivalent sets of equations, and $\to_{\mathcal{R}} = \to_{\mathcal{R}'}$, then in general $\mathcal{T_R}$ and $\mathcal{T}_{\mathcal{R}'}$ are not isomorphic.

In LTL we have that for each specification there exist infinite isomorphic specifications with different sets of axioms (e.g., they can be obtained by adding derived axioms).

Furthermore, the rule labels are not relevant in RL, only the rules are; indeed the same label can be used for several rules but the edges and the operations over them are determined by the rules, not by the labels; and we have that applications of operations associated with different rules, labeled in the same way with premises and consequences of the same sorts, must be disambiguated.

Example 13.10. (**Bit using** RL)

> **spec** *BUFFER* =
>   **enrich** *INT_QUEUE*[*buffer*/*queue*] **by**
>   **rl** $\tau1$: $b \longrightarrow Dup(b)$
>   **rl** $\tau2$: $b \longrightarrow Put(10^{10}, Empty)$

> **spec** *USER* = *USER_STATUS*[*user*/*user_status*],

where *USER_STATUS* has been defined in Example 13.8; the latter is a static algebraic specification, and thus may be considered as a particular RL specification without proper transitions.

> **spec** *SYSTEM* =
>   **enrich** *BUFFER* + *USER* **by**
>   **sorts**   *system*
>   **opns**   _ | _: *buffer* × *user* → *system*
>   **rl**   *START* : *Empty* | *Initial* $\longrightarrow$ *Empty* | *Putting_0*
>   **rl** $\tau3$: b | *Putting_0* $\longrightarrow$ *Put*(0, *b*) | *Putting_1*
>   **rl** $\tau4$: b | *Putting_1* $\longrightarrow$ *Put*(1, *b*) | *Reading*
>   **rl** $\tau5$:   *Not_Empty*(b) $\Rightarrow$
> b | *Reading* $\longrightarrow$ *Remove*(b) | *Read*(*First*(b))
>   **rl**   *OK* : b | *Read*(0) $\longrightarrow$ b | *Terminated*
>   **rl**   *ERROR* :*NotEq*(i, 0) $\Rightarrow$ b | *Read*(i) $\longrightarrow$ b | *Terminated*

Notice the major differences compared with the LTL specification of Example 13.8, even though the two seem very similar. In this case the specification is structured by giving first the specifications of the two components; but now the activity of the user is not given in *USER*, and only part of that of the buffer is given in *BUFFER*. Indeed, if, e.g., we have the rule *Putting_0* $\longrightarrow$ *Putting_1* in *USER*, then the user in any case may perform such a transition also without synchronizing with the buffer.

Furthermore there is no provision for knowing that $START$, OK, and $ERROR$ are different interactions with the environment, while $\tau 1, \ldots, \tau 5$ correspond to internal activities; see [AR97b] for a detailed comparison.   ∎

13.4   Dynamic data-types (A4)

In recent literature various approaches have been proposed to extend the classical algebraic framework for the specification of data types to handle processes; the first one was Goguen and Meseguer's reflexive semantics for object–orientation in [GM87b]. All these approaches share some common features, which have been nicely summarized in [EO94] by Ehrig and Orejas, where they report informally a general schema for building an algebraic framework following the state-as-algebra style:

- states-as-algebras; thus, and implicitly, dynamics is modeled by a (labeled or not) transition system;
- all the statealgebras extend a fixed algebra of basic nondynamic values (static or value algebra);
- the elements of the carriers of the nonstatic sorts of a statealgebra are the components of the system at that moment, and the nonstatic operations represent how they are organized at that moment;
- state-transformations = transitions from a statealgebra to another statealgebra, correspond in most cases to operations of a special kind (dynamic operations) and in general are not homomorphisms (the organization and the number of components may change).
  Notice that dynamic operation calls are the common mechanism to express the interactions with the environment; but in this way the reaction to an external stimulus (a dynamic operation call) must be deterministic (except if we leave the classical algebraic frameworks for some nondeterministic framework).
  While the usual algebraic techniques may be used to define/specify the value and the statealgebras, there is no standard way to define/specify the dynamic operations (the transitions).
- there is no a general way to handle concurrency/distribution/cooperation among process components or to give in a structured way the specification of a concurrent system by composing the specifications of its components; usually each approach offers ad hoc techniques;
- most of these approaches have been developed with object-orientation in mind rather than concurrency generally.

Here, we briefly report on only some approaches; other can be found, for instance, in [Bau95, PP95]. Among them, evolving algebras are peculiar; since the emphasis is not on the data structure aspects, but more on the operational idea of state transitions; indeed, recently they have been called "Abstract State Machines" (ASM).

13.4.1 Evolving algebras (abstract state machines)

The basic idea of the "evolving algebras" (see, for instance, [Gur93, Gur95]) is perfectly summarized by their name. Essentially an evolving algebra (specification) consists of a description of a (nonlabeled) transition system, whose states are algebras on the same homogeneous signature built over the same universe (including Boolean values). Some of the operation symbols are qualified as "static" and their interpretation is the same in any (algebra which is a) state. The transitions are defined by rules of the following form:

$$econd \Rightarrow up_1, \ldots up_k$$

where, for each $j = 1, \ldots, k$, the *function update* up_j has form

$$f_j(e_1^j, \ldots, e_{n_j}^j) := e^j,$$

$econd$, $e_1^1, \ldots, e_{n_1}^1, e^1, \ldots, e_1^k, \ldots, e_{n_k}^k, e^k$ are "descriptions" (any possible mathematically intelligible expressions) of elements of the universe, the first describing a Boolean value, and for $j = 1, \ldots, k$, f_j is an operation of the signature of arity n_j.

The interpretation of one of such rules is that the system in a state A such that $econd$ holds on it, may pass to another algebra B where for each operation of the signature f, $f_B(x_1, \ldots, x_n) =$

- the interpretation of e^j in A, if $f = f_j$ and $x_1, \ldots, x_n$ coincide with the interpretations of $e_1^j, \ldots, e_{n_j}^j$ in A;
- $f_A(x_1, \ldots, x_n)$, otherwise.

Obviously, the function updates $up_1, \ldots up_k$ in a rule must be consistent in the sense that they do not simultaneously update the same function on the same arguments with different values.

Here, for simplicity, we consider only the basic evolving algebras, where there is no provision for concurrent structuring; concerning reactivity, some operations of the signature are classified "external" with the idea that they can change under the influence of (the nonfurther qualified) environment and cannot be modified by the rules.

Thus, essentially, to give an evolving algebra specification means to give a (nonlabeled) transition system where the states are finite tuples, which can also be of functional type, describing the transitions by saying only which components of the state tuples change and how, and for the components of functional type, only for which arguments change and how. In this way state transformations are described in a very economic way. Consider, for example, the conditional rules describing the modification of tuple-like states, following, e.g., an LTL-style (see Section 13.3.1), as

$$econd \Rightarrow \langle c_1, \ldots, c_i, \ldots, c_n \rangle \rightarrow \langle c_1, \ldots, c_i', \ldots, c_n \rangle$$

and

$$econd \Rightarrow$$
$$\langle c_1, \ldots, c_i, \ldots, c_n \rangle \rightarrow \langle c_1, \ldots, \lambda\ x.\ \text{if}\ x = a\ \text{then}\ b\ \text{else}\ c_i(x), \ldots, c_n \rangle$$

versus the corresponding simpler versions given in an evolving algebra style:

$$econd \Rightarrow C_i := c_i' \qquad \text{and} \qquad econd \Rightarrow C_i(a) := b.$$

After all, evolving algebras are nothing but lts where the states are algebras. As for other purely lts–based approaches, the treatment of liveness conditions and modular composition are less elegant.

Example 13.11. **(Bit using evolving algebras)**

** SIGNATURE
** Static operations
    Empty (0–ary)
    Remove, Dup, First, Not_Empty (1–ary)
    Put (2–ary)
    Initial, Terminated, Putting_0, Putting_1, Reading (0–ary)
    Read (1–ary)
    START, OK, ERROR (0–ary)
** Dynamic operations
    Buf_Cont, User_State, Output (0–ary)
** External operations
    Input (0–ary)
** definition of static operations

    .....
** RULES

$Input = START \land Buf\_Cont = Empty \Rightarrow$
    $User\_State := Putting\_0$

$User\_State = Putting\_0 \Rightarrow$
    $User\_State := Putting\_1$
    $Buf\_Cont := Put(0, Buf\_Cont)$

$User\_State = Putting\_1 \Rightarrow$
    $User\_State := Reading$
    $Buf\_Cont := Put(1, Buf\_Cont)$

$User\_State = Reading \land Not\_Empty(Buf\_Cont) \Rightarrow$
    $User\_State := Read(First(Buf\_Cont))$
    $Buf\_Cont := Remove(Buf\_Cont)$

true $\Rightarrow Buf\_Cont := Put(10^{10}, Empty)$

true $\Rightarrow Buf\_Cont := Dup(Buf\_Cont)$

$User\_State = Read(0) \Rightarrow$

   User_State := *Terminated*
   Output := *OK*

User_State ≠ *Read*(0) ⇒
   User_State := *Terminated*
   Output := *ERROR*   ∎

13.4.2   D–oids

D-oids [AZ95] are mathematical structures aimed at abstractly modeling concurrent systems by extending the algebraic approach for modeling data structures. In [AZ95] a very general approach is taken, since the definition of a d-oid is parameterized by the underlying static framework for (values and) statealgebras, but here for simplicity, we fix such framework to the usual algebras.

A d-oid has a signature, called a *dynamic signature*. In general a dynamic signature is a pair consisting of a signature Σ with a set of sorts S and a family of dynamic operation symbols DOP over S. A dynamic operation *dop* may have a functionality $dop\colon w \to [s]$, with $w \in S^*$ and $[s] \in S \cup \{\Lambda\}$. This corresponds to the idea that a dynamic operation may also return a null value. There are also constant dynamic operations *dop* having functionality $dop\colon \to [s]$.

A d–oid over $\langle \Sigma, DOP \rangle$ consists of a class $\mathcal{A}$ of Σ–algebras and an interpretation of the dynamic operations. If $dop\colon w \to [s]$, then an interpretation of *dop* is a partial function mapping $\langle A, \bar{a} \rangle$, with $A \in \mathcal{A}$ and $\bar{a} \in |A|_w$, to a transformation of A into an algebra $B \in \mathcal{A}$ and a returned value $v \in |B|_{[s]}$, when $[s]$ is not null. A *transformation* of A in B is a triple $\langle A, f, B \rangle$, where f is a partial map from the carrier of A into the carrier of B, called a *tracking map*. The tracking map is essential for keeping track of the identity of the elements of the system: if e is an element in A, then we can recover it in the new state B by applying f to e; tracking maps allow us to deal in a very abstract way with object creation (nonsurjective maps) and deletion (non total tracking maps). Tracking maps may be noninjective, to model the cases where some elements with different identities are glued together.

Static basic values may be provided by a value part, just an algebra, see [AZ95].

The interpretation of a constant dynamic operation $dop\colon \to [s]$ is an algebra $A \in \mathcal{A}$ and a returned value $v \in |A|_{[s]}$, whenever $[s]$ is not null.

Finally, it is possible to extend in a natural way the notion of term to the dynamic case. Terms define a syntactic structure, the *term d–oid*, which is, under some assumptions, a free structure for the appropriate categorical setting [AZ96].

Concerning specifications based on d-oids, [Zuc96] presents a general way to build institutions for dynamic data type specifications and shows an application to the d-oid case. The formulas of the proposed logic allow us to express static properties, precisely on the value part and the statealgebras (a kind of system invariants). Concerning the properties on dynamics, it allows formulas for expressing pre- and post-conditions on sequential compositions of dynamic operations, represented by the elements of the term d-oid, and for requiring that two compositions are the same, i.e., they represent the same transformation.

Example 13.12. (**Bit using d-oids**) The value part is given by the initial model of following algebraic specification, where *USER_STATUS* and *MESSAGE* have been given respectively in Examples 13.8 and 13.3, and *INT_QUEUE* is in Appendix A.

> **spec** *VALUE = INT_QUEUE + USER_STATUS + MESSAGE*

> **spec** *SYSTEM =*
> > **enrich** *VALUE* **by**   ** value part
> > > ** statealgebra part
> > > *Buf_Cont:* → *queue*
> > > *User_State:* → *user_status*
> > > *Output:* → *message*
> > > ** dynamic operations
> > > *START, INT1, INT2, INT3:* →
> > > *OUTPUT:* → *message*
> > > ** dynamic properties
> > > $\{Buf\_Cont = Empty \land User\_State = Initial\}START$
> > > > $\{User\_State = Putting\_0\}$
> > > $\{User\_State = Putting\_0\}$ *INT1*
> > > > $\{User\_State = Putting\_1 \land Buf\_Cont = Put(0, Buf\_Cont)\}$
> > > $\{User\_State = Putting\_1\}$ *INT1*
> > > > $\{User\_State = Reading \land Buf\_Cont = Put(1, Buf\_Cont)\}$
> > > $\{User\_State = Reading \land Not\_Empty(Buf\_Cont)\}$ *INT1*
> > > > $\{User\_State = Read(First(Buf\_Cont)) \land$
> > > > $Buf\_Cont = Remove(Buf\_Cont)\}$
> > > *INT2* $\{Buf\_Cont = Put(10^{10}, Empty)\}$
> > > *INT3* $\{Buf\_Cont = Dup(Buf\_Cont)\}$
> > > $\{User\_State = Read(0)\}$ *INT1*
> > > > $\{User\_State = Terminated \land Output = OK\}$
> > > $\{User\_State = Read(i) \land \neg (i = 0)\}$ *INT1*
> > > > $\{User\_State = Terminated \land Output = ERROR\}$
> > > $\{User\_State = Terminated\}$ $m \leftarrow OUTPUT$ ∎

13.4.3   Algebraic specifications with implicit state

The specification formalism of Dauchy and Gaudel [DG93, GDK96] is based on the notion of *elementary access function* and *elementary modifier*. Elementary access functions characterize the structure of the states of the system, as kinds of observation functions, while elementary modifiers allow us to perform updates of the elementary accesses without returning any value.

Elementary modifiers are built-in features of the specification language, associated with the elementary access functions.

A specification in this formalism is a 4–tuple

$$langle\langle \Sigma, AX \rangle, \langle \Sigma_{ac}, AX_{ac} \rangle, \langle \Sigma_{mod}, Def_{mod} \rangle, AX_{init} \rangle,$$

where:

- $\langle \Sigma, AX \rangle$ is the specification of the static values used.
- $\langle \Sigma_{ac}, AX_{ac} \rangle$ is the specification of the access functions and is a conservative extension (see, Chapter 6) of $\langle \Sigma, AX \rangle$ with no new sort (thus $\Sigma \subseteq \Sigma_{ac}$). Some of the access functions are elementary, while the others are defined in terms of the elementary ones by the access axioms of AX_{ac}.
- The admissible initial states are characterized by the set of axioms AX_{init}.
- The definition of the elementary access functions makes implicitly available the corresponding elementary modifiers in the following way: given an elementary access function f with functionality $s_1 \times \cdots \times s_n \to s$, the corresponding elementary modifier is $\mu\text{-}f$ with domain $s_1 \times \cdots \times s_n \, s$. Elementary modifiers are the tools for describing the statealgebras transformation.

  Given some terms with variables $t_1, \ldots, t_n$ of sorts $s_1, \ldots, s_n$ and a term t of sort s, the meaning of the statement $\mu\text{-}f(t_1, \ldots, t_n, t)$ is a modification of f. More precisely, it transforms a statealgebra A into a statealgebra B such that:

  * $f_B(v_1, \ldots, v_n) = (\sigma(t))^B$ if there exists a ground substitution σ such that for $i = 1, \ldots, n$ $v_i = (\sigma(t_i))^A$;
  * $f_B(v_1, \ldots, v_n) = f_A(v_1, \ldots, v_n)$ otherwise;
  * derived access functions which depend on f are changed accordingly;
  * any other operation and all carriers are unchanged.

$\langle \Sigma_{mod}, Def_{mod} \rangle$ define some composite modifiers, the functionalities of which have no range. The axioms in Def_{mod} are positive conditional and their premises are built on Σ_{ac}. They define the modifiers using statements built from the elementary modifiers and the following constructs: **nil** identity; **;** sequential composition; **and** composition in any order (it is responsibility of the specifier to check that the result of the composition does not depend on the order); • denotes modifications made on the same state, i.e., all preconditions and arguments of the involved modifiers must be evaluated in the initial state prior to doing all corresponding modifications.

Example 13.13. (**Bit using access functions and modifiers**) Let *VALUE* be the algebraic specification defined in Example 13.12.

    spec *SYSTEM* =
        enrich *VALUE* **by**   ******   basic static values
               **   elementary accesses
               Buf_Cont: $\to$ *queue*

$User\_State: \rightarrow user\_status$
$Output: \rightarrow message$
** modifiers
$START: \rightarrow$
$\tau: \rightarrow$
** modifier definitions
$User\_State = Initial \wedge Buf\_Cont = Empty \Rightarrow$
$\quad START = \mu\text{-}User\_State(Putting\_0)$

$\tau =$
cases
$User\_State = Putting\_0 \Rightarrow$
$\quad (\mu\text{-}User\_State(Putting\_1) \wedge \mu\text{-}Buf\_Cont(Put(0, Buf\_Cont)))$
$\quad \vee (\mu\text{-}Buf\_Cont(Put(10^{10}, Empty)))$
$\quad \vee (\mu\text{-}Buf\_Cont(Dup(Buf\_Cont)))$
$User\_State = Putting\_1 \Rightarrow$
$\quad (\mu\text{-}User\_State(Reading) \wedge \mu\text{-}Buf\_Cont(Put(1, Buf\_Cont)))$
$\quad \vee (\mu\text{-}Buf\_Cont(Put(10^{10}, Empty)))$
$\quad \vee (\mu\text{-}Buf\_Cont(Dup(Buf\_Cont)))$
$User\_State = Reading \wedge Not\_Empty(Buf\_Cont) \Rightarrow$
$\quad (\mu\text{-}User\_State(Read(First(Buf\_Cont)))) \wedge$
$\quad \mu\text{-}Buf\_Cont(Remove(Buf\_Cont)))$
$\quad \vee (\mu\text{-}Buf\_Cont(Put(10^{10}, Empty)))$
$\quad \vee (\mu\text{-}Buf\_Cont(Dup(Buf\_Cont)))$
$User\_State = Read(0) \Rightarrow$
$\quad (\mu\text{-}User\_State(Terminated) \wedge \mu\text{-}Output(OK))$
$\quad \vee (\mu\text{-}Buf\_Cont(Put(10^{10}, Empty)))$
$\quad \vee (\mu\text{-}Buf\_Cont(Dup(Buf\_Cont)))$
$User\_State = Read(i) \wedge NotEq(i, 0) = True \Rightarrow$
$\quad (\mu\text{-}User\_State(Terminated) \wedge \mu\text{-}Output(ERROR))$
$\quad \vee (\mu\text{-}Buf\_Cont(Put(10^{10}, Empty)))$
$\quad \vee (\mu\text{-}Buf\_Cont(Dup(Buf\_Cont)))$
end cases   ∎

13.4.4 Statealgebras with references

In [GR95, GR97] Große-Rhode presents a state-as-algebra approach based on
a general idea of "reference".

A statealgebra is a partial algebra which is an extension of a given algebra A, called the *base model* (static value algebra). More precisely, a state
is considered as a static data type where references are added. For some
sorts, say s, a special reference sort $ref(s)$ and a contents operation symbol
$!_s: ref(s) \rightarrow s$ are introduced in the signature. The base model is a model of a
normal partial equational specification, called the *base specification*, with existence equations and some minor restrictions concerning equations between
references.

For a list of (pairwise different) references $d_\omega = d_1 \ldots d_n$ with $d_i \in$
$|A|_{ref(s_i)}$ and a corresponding list of values $a_\omega = a_1 \ldots a_n$ with $a_i \in |A|_{s_i}$, the

state $A^{[d_\omega := a_\omega]}$ on the base model A, where the references d_i have contents a_i, is formally defined as a free extension of A by an existence equation of the form $!_{s_1}(d_1) = a_1 \wedge \ldots \wedge !_{s_n}(d_n) = a_n$.

The definition of states as free extensions of the base model allows one to formalize the notion of *persistent state*: a persistent state on A is a model of the base specification whose restriction to the signature of the value datatypes (disregarding the reference sorts) is isomorphic to A. Intuitively a persistent state can be regarded as an extension with the content functions of A, i.e., as a pair $\langle A, env \rangle$, where *env* is an environment which is a family of partial functions mapping references to values. It is proved that persistent states are in a one-to-one correspondence with the pairs $\langle A, env \rangle$.

On top of a base specification a *transition specification* can be defined. Dynamic operations are specified by a set of method definitions, which are conditional parallel assignments. It is possible to have several assignments for the same method with overlapping conditions, hence dynamic operations are nondeterministic. Finally, state transitions are specified by method expressions built by the application of methods to arguments and by sequential composition of them.

Some interesting results concern structured specifications; following the well-established theory of composition of specifications and of parametric specifications in an arbitrary institution, Große-Rhode proves that his specifications enjoy the properties needed for defining the usual structuring mechanisms for composing specifications.

Example 13.14. (**Bit using statealgebras with references**) Let *VALUE* be the specification defined in Example 13.12.

> **spec** $VALUE-REF =$
> > **enrich** $VALUE[buffer/queue]$ **by**
> > **sorts** $ref(user\_status), ref(buffer), ref(message)$
> > **refs** $U: \rightarrow ref(user\_status)$
> > > $B: \rightarrow ref(buffer)$
> > > $Output: \rightarrow ref(message)$

> **spec** $SYSTEM =$
> > **enrich** $VALUE-REF$ **by**
> > > ****** methods
> > > $START: \rightarrow ref(user\_status)$
> > > $!U = Initial \wedge !B = Empty \Rightarrow START(U) := Putting\_0 . O$
> > >
> > > $INT1: \rightarrow ref(buffer) \times ref(user\_status)$
> > > $!U = Putting\_0 \Rightarrow INT1(B, U) := < Put(0, !B), Putting\_1 > . O$
> > > $!U = Putting\_1 \Rightarrow INT1(B, U) := < Put(1, !B), Reading > . O$
> > > $!U = Reading \wedge Not\_Empty(!B) \Rightarrow$
> > > > $INT1(B, U) := < Remove(!B), Read(First(!B)) > . O$
> > >
> > > $INT2: \rightarrow ref(buffer)$

$$INT2(B) := Put(10^{10}, Empty)$$
$$INT2(B) := Dup(!B)$$

$$INT3: \ \to ref(user\_status) \times ref(message)$$
$$!U = Read(0) \ \Rightarrow \ INT3(U, Output) :=< Terminated, OK > . O$$
$$!U = Read(i) \ \wedge \ NotEq(i, 0) \ \Rightarrow$$
$$INT3(U, Output) :=< Terminated, ERROR > . O \qquad \blacksquare$$

13.5   Conclusion

We have distinguished four main approaches in the use of algebraic techniques for the specification of reactive, and concurrent systems and presented some methods illustrative of the different viewpoints. There has not been room to include all methods, in particular those more recent (like the coalgebraic) or requiring a deeper treatment (like hidden specifications), but we have provided pointers to the relevant literature. Nor we have made any attempt at comparing the different methods, since a thorough comparison should follow some rigorous criteria, which are still under discussion (see, e.g., some hints in [AR97a]). Instead we have provided a guided tour, hopefully stimulating further reading and research.

Looking back, we can now observe that, in order to handle properly the features typical of concurrent and reactive systems, the algebraic techniques need some kind of extension of a very different nature. First they all need an underlying model able to deal with the concurrency issues (like Petri nets or Labeled Transition Systems). Then there are specific adjustments either at the level of the specification language (**A2**), or of some basic technical point (generalized bisimulations, coalgebras instead of algebras, hidden specifications), or at the method level [BCPR96].

In general any really usable formalism for the specification of systems must be complemented by a specification formalism for data and in this respect algebraic techniques have the advantage of being very abstract and linked to languages supporting modularity . This is the rationale behind the success of methods following viewpoint **A2**, like LOTOS.

We can also observe that only the algebraic methods following approach **A3** keep the fully axiomatic flavor of the original algebraic specifications; this would apply to the hidden specification and coalgebraic methods too.

An issue which has only been mentioned as an aside, but of paramount importance, is the support of automatic tools both for development and verification. This is a fast developing field, which could provide one basic key to the successful use of algebraic techniques in the future. Another key could come from a standardization of the algebraic notation and of the associated methods, contrary to the direction of the current proliferation of notations. In this respect we are greatly looking forward to the outcome of CoFI, the ongoing Common Framework Initiative, sponsored by the IFIP WG 2.2 [Mos97].

Acknowledgements. We thank Davide Ancona and Elena Zucca for Section 13.4.

Work partially funded by HCM project Express and MURST 40% Modelli della computazione e dei linguaggi di programmazione.

A   Specifications of data types used by Bit

Below we report the algebraic specifications of the data types, integers and
finite queues of integers, which are used in various specifications of **Bit**.

> **spec** *INT* =
> > **sorts**   *int*
> > **opns** $0: \rightarrow int$
> > > $S, P: int \rightarrow int$
> > > $\_ + \_: int \times int \rightarrow int$
> >
> > **preds** $\_ < \_, NotEq: int \times int$
> > **axioms** $0 + i = i$
> > > $S(i) + i' = S(i + i')$
> > > $P(i) + i' = P(i + i')$
> > > $0 < S(0)$
> > > $P(0) < 0$
> > > $i < i' \Rightarrow (S(i) < S(i') \wedge P(i) < P(i') \wedge i < S(i') \wedge P(i) < i')$
> > > $(i < i' \vee i' < i) \Rightarrow NotEq(i, i')$
>
> **spec** *INT_QUEUE* =
> > **enrich** *INT* **by**
> > **sorts**   *queue*
> > **opns** $Empty: \rightarrow queue$
> > > $Put: int \times queue \rightarrow queue$
> > > $First: queue \rightarrow int$
> > > $Remove, Dup: queue \rightarrow queue$
> > > ∗∗   *Dup duplicates the first element of a queue*
> >
> > **preds** $Not\_Empty: queue$
> > **axioms** $Not\_Empty(Put(i, q))$
> > > $First(Empty) = 0$
> > > $First(Put(i, Empty)) = i$
> > > $First(Put(i, Put(i', q))) = First(Put(i', q))$
> > > $Remove(Empty) = Empty$
> > > $Remove(Put(i, Empty)) = Empty$
> > > $Remove(Put(i, Put(i', q))) = Put(i, Remove(Put(i', q)))$
> > > $Dup(Empty) = Empty$
> > > $Dup(Put(i, Empty)) = Put(i, Put(i, Empty))$
> > > $Dup(Put(i, Put(i', q))) = Put(i, Dup(Put(i', q)))$

14 Formalization of the Development Process

David Basin[1] and Bernd Krieg-Brückner[2]

[1] Department of Computer Science, University of Freiburg, Freiburg, Germany
`basin@informatik.uni-freiburg.de`
[2] Department of Computer Science, University of Bremen, Bremen, Germany
`bkb@informatik.uni-bremen.de`

14.1 Introduction

14.1.1 What is development?

Software development encompasses many phases including requirements engineering, specification, design, implementation, verification or testing, and maintenance. In this chapter we concentrate on the intermediate tasks: the transition from requirements specification to verified design and design optimization, in particular, techniques for developing correct designs as opposed to ad hoc or *a posteriori* methods in which a postulated design is later verified or tested.

We view software development as the composition of correctness preserving refinement steps. Hence, a development is meaningful relative to a notion of refinement and correctness. There are many such notions (e.g., Chapter 6), each defining a type of problem transformation or reduction. For example, specification refinement from a specification SP to SP' through a refinement map ρ might be defined as correct when the images of ρ over realizations of SP' are realizations of SP. Here a refinement step represents a reduction to a hopefully simpler design objective and results in an incomplete program that expresses design decisions made during refinement. These steps can be combined into a development that constructs programs satisfying the initial specification.

Our view of development is very general and there is a strong analogy between program development in our sense and proof, as a "development" of some logical formula; both are forms of deduction. When we refer to *development* below, we often mean both. While we suggest that there is no conceptual difference, there may be good reasons in practice to differentiate between them; proof systems are often based on a fixed set of context-free and rather simple deduction rules and fixed built-in tactics, while development rules may be quite complex and context-sensitive.

Previous chapters have considered concrete formalisms for specification and deduction. Instead of focusing on a particular development methodology, we consider how methodologies could be formalized, and the advantages of formalized development. While our use of *development* is quite general, we use *formal* in a strong and restrictive sense: A development (or assertion) is

$$[] \mathbin{+\!\!+} x = x \tag{14.1}$$
$$h{:}t \mathbin{+\!\!+} x = h{:}(t \mathbin{+\!\!+} x) \tag{14.2}$$
$$x \mathbin{+\!\!+} [] = x \tag{14.3}$$
$$x \mathbin{+\!\!+} (y \mathbin{+\!\!+} z) = (x \mathbin{+\!\!+} y) \mathbin{+\!\!+} z \tag{14.4}$$

Fig. 14.1. Defining axioms and properties of $\mathbin{+\!\!+}$

formal when it is carried out (or stated) in an explicitly formalized calculus and is *informal* otherwise. Formal mathematics, as pioneered by the project of Automath [dB80], has been taken up with considerable enthusiasm by computer scientists. However, most arguments about program correctness and, in general, mathematics are informal in our strong sense. Before we consider what extra effort is required to formalize developments, let us first motivate why it might be worth it.

14.1.2 A motivating example

For motivation, we employ a simple example from functional program transformation. A well-studied formalism is unfold–fold, developed by Burstall and Darlington [BD77]. It is a good example because the formalism is simple, but yet rich enough to illustrate many of the advantages of formalized development that we present later. Moreover, it is well-known and serves as a basis for many kinds of transformational development.

Unfold–fold consists of a set of rules that are applied to sets of recursion equations producing new equations. Among the rules are: *Definition* which introduces a new, distinctly named recursion equation; *Instantiation* which introduces a substitution instance of a previous equation; *Unfold* which replaces an instance of a function name by its appropriately instantiated body; *Fold* which replaces an instance of a function body by a corresponding function call; and *Laws* which rewrites an expression using known equations about primitive functions. A refinement step is the application of a rule. A development is then a sequence of these rule applications. At the end of a development, a subset of the resulting equations is taken as the new function definitions.

Development in this setting is best illustrated through an example. Consider a theory of lists where lists are constructed with *nil* and *cons*, written as $[]$ and infix ":" respectively. Here $[h]$ denotes the singleton list containing h, i.e., $h{:}[]$. The append function, $\mathbin{+\!\!+}$, is the infix concatenation function defined in the first two equations in Figure 14.1 by primitive recursion on its first argument.

1.	$qrev(l, m)$	$= rev(l) \mathbin{+\!\!\!+} m$	*definition (14.7)*
2.	$qrev([\,], m)$	$= rev([\,]) \mathbin{+\!\!\!+} m$	*instantiation with* $[\,]$
2a.		$= [\,] \mathbin{+\!\!\!+} m$	*unfold with rev, (14.5)*
2b.		$= m$	*unfold with* $\mathbin{+\!\!\!+}$*, (14.1)*
3.	$qrev(h{:}t, m)$	$= rev(h{:}t) \mathbin{+\!\!\!+} m$	*instantiation with h:t*
3a.		$= (rev(t) \mathbin{+\!\!\!+} [h]) \mathbin{+\!\!\!+} m$	*unfold with rev, (14.6)*
3b.		$= rev(t) \mathbin{+\!\!\!+} ([h] \mathbin{+\!\!\!+} m)$	*law (14.4), associativity of* $\mathbin{+\!\!\!+}$
3c.		$= rev(t) \mathbin{+\!\!\!+} h{:}([\,] \mathbin{+\!\!\!+} m)$	*unfold with* $\mathbin{+\!\!\!+}$*, (14.2)*
3d.		$= rev(t) \mathbin{+\!\!\!+} h{:}m$	*unfold with* $\mathbin{+\!\!\!+}$*, (14.1)*
3e.		$= qrev(t, h{:}m)$	*fold with qrev (14.7)*

Fig. 14.2. An unfold–fold transformation of *qrev* from *rev*

Furthermore, suppose we are given the following standard, but inefficient, program for list reversal.

$$rev([\,]) = [\,] \tag{14.5}$$

$$rev(h{:}t) = rev(t) \mathbin{+\!\!\!+} [h] \tag{14.6}$$

Figure 14.2 gives a development of *qrev*, an accumulator-based version of reverse. The justification for each rule is given in the right-hand column. Since *rev* is defined by embedding into *qrev* (with $m = [\,]$),

$$qrev(l, m) = rev(l) \mathbin{+\!\!\!+} m \tag{14.7}$$

$$qrev(l, [\,]) = rev(l) \tag{14.8}$$

the development yields a function for list reversal that runs in linear (instead of quadratic) time.

$$qrev([\,], m) = m \tag{14.9}$$

$$qrev(h{:}t, m) = qrev(t, h{:}m) \tag{14.10}$$

The rules of unfold–fold comprise a simple calculus. We use it to motivate which aspects of development can be formalized and the advantages there might be in formalizing them.

Correctness. First, there is the question of correctness: what is the relationship between the initial equations and the extracted program definition? In general, unfold–fold derivations guarantee only *partially correct* programs: on those inputs where the defined function terminates, it is equal to the original definition. On the other hand, many derivations result in totally correct (always terminating) functions. For example, *qrev* defines a total function.[1]

---

[1] Evaluated in the left-to-right direction, the defining rewrite rules terminate on any given pair of inputs and compute a unique output. On the other hand, a simple (and well-known) example of a partially, but not totally, correct derivation

However, within unfold–fold itself we cannot express the relationship (neither partial nor total correctness) between the original and the synthesized program.

An alternative to this informal notion of correctness is to formalize unfold–fold within a metalogic capable of expressing program equivalence and establishing sufficient conditions for proofs to yield correct programs. Such analysis has been done for unfold–fold: [Kot78, San95] present restrictions on unfold–fold derivations that guarantee total correctness. However the analysis and restrictions are not formalized. This is in contrast to, for example, the use of a programming logic like LCF [GMW79, Pau87] where a model of computation may be specified, and properties such as equality and termination of programs may be expressed and reasoned about, relative to a desired computational model. LCF is a particularly good example of such a framework. It was developed to experiment with Scott's Logic of Computable Functions and is based on a first-order logic for domain theory. This logic has been used, for example, to verify algebraic laws of functional programs [Les81], correctness of program schemes for recursion removal [Coh80], and relationships between denotational and Hoare-style axiomatic semantics [Sok87]. Recently, various alternatives have been suggested to verify the correctness of development rules or nontrivial transformation rules, using formal frameworks such as IsABELLE or LEGO [AB95b, AB96, Bas94, WK95].

The major advantage of formalizing development rules and formally proving that they constitute correct refinement steps is a separation of concerns and the resulting economy of verification effort. Rules can be verified once, as correctness-preserving, perhaps under some schematic applicability conditions. When applied, any applicability conditions are instantiated as verification conditions for the particular application. While considerable effort may have to be invested in the initial correctness proof, the proof of the applicability of the rule, if any, usually requires much less effort than a formal *a posteriori* proof for a development step.

By embedding and developing a formalism like unfold–fold within a metalogic like LCF, we would be able to state and verify formally that developed designs are partially or totally correct. Even more is possible. Rather than just considering which designs are total, we could formalize classes of development that guarantee total (as opposed to partial) correctness. The work of Kott, and more recently Sands [Kot78, San95], is relevant for our example. Both have examined sufficient conditions for restricted unfold–fold derivations to be totally correct. If we imagine a formalization of unfold–fold where derivations are formalized by, say, a recursive data type of proofs, then if the application conditions of rules in the restricted calculus are recursively

---

is to start with the program definition $f(x) = 42$ and observe that for any instance, say 7, $f(7) = 42$. Using this, and the folding rule, allows us to derive the diverging program $f(x) = f(7)$.

Defining equations for f (left-hand side of the transformation rule)

$$A(x) \Rightarrow f(x) = B(x) \tag{14.11}$$

$$\neg A(x) \Rightarrow f(x) = K(f(H(x)), J(x)) \tag{14.12}$$

Defining equations for f' (right-hand side of the transformation rule)

$$f'(x) = g'(x, N) \tag{14.13}$$

$$A(x) \Rightarrow g'(x, y) = K(B(x), y) \tag{14.14}$$

$$\neg A(x) \Rightarrow g'(x, y) = g'(H(x), K(J(x), y)) \tag{14.15}$$

Applicability conditions:

$$K(x, N) = x \tag{14.16}$$

$$K(x, K(y, z)) = K(K(x, y), z) \tag{14.17}$$

Correctness:

$$f(x) = f'(x) \tag{14.18}$$

Fig. 14.3. Transformation rule for tail recursion introduction, presented as a program schema

checkable, then it should be possible to define a subtype of derivations based on development rules that only produce totally correct designs.

Rule extensibility. We have indicated how restricting developments may enable correctness proofs. Alternatively, a formalization can provide a basis for *extending* refinement in a correctness-preserving manner. Extensibility is important for scalability. The rules of unfold–fold, for example, are appropriate for constructing small derivations but inappropriate for transforming large programs because development steps are too small. This problem is common since designers of logics usually aim for minimality at the cost of usability. Hence large developments often require some kind of user extensibility.

One of the simplest kinds of extensibility is to admit new rule schemata that preserve correct refinement. For example, one popular approach to program transformation is based on rule schemata (also called transformation templates) that specify the form of programs (or program fragments) before and after transformation [HL78b]. In particular, a *transformation rule* is specified by providing a pair of program schemata along with additional applicability conditions that must be satisfied. The schemata are applied using higher-order matching or unification and the conditions are checked on the generated instance. This may lead to more succinct derivations than are possible using only deductive transformation methods with fixed sets of rules, such as the unfold–fold rules.

Consider Figure 14.3, which presents a transformation rule taken from [HK93]. Under the appropriate semantic interpretation, we could formulate and prove the equality of the definitions for f and f' for all instances (for the schematic variables) that satisfy the stated applicability conditions. This approach was taken in the 1970s, for example, by Huet and Lang [HL78b] who investigated the correctness of similar schemata relative to a denotational semantics within LCF. Once formalized, this schema may be applied using second-order matching to transform instances of the schema for f to instances of f'.

Returning to the *rev* example, let us give an alternative formalization of *rev* using conditional equations.

$$x = [] \Rightarrow rev(x) = [] \tag{14.19}$$
$$\neg(x = []) \Rightarrow rev(x) = rev(tl(x)) +\!\!+ [hd(x)] \tag{14.20}$$

Now, (14.19) second-order matches (14.11) and (14.20) matches (14.12) under the substitutions $f(x) \equiv rev(x)$, $A(x) \equiv x = []$, $B(x) \equiv []$, $K(x, y) \equiv x +\!\!+ y$, $H(x) \equiv tl(x)$, and $J(x) \equiv [hd(x)]$. Note that the variable N takes no part in matching (in Section 14.2.3 we show how it can be computed by applying proof rules using unification; in Section 14.3.4, N is picked up from the context), but augmenting the substitution with $N \equiv []$ and applying it to (14.13)–(14.15) yields the following program.

$$f'(x) = g'(x, []) \tag{14.21}$$
$$x = [] \Rightarrow g'(x, y) = [] +\!\!+ y \tag{14.22}$$
$$\neg(x = []) \Rightarrow g'(x, y) = g'(tl(x), [hd(x)] +\!\!+ y) \tag{14.23}$$

The programs f and f' are equivalent as the applicability conditions are valid: $[]$ is a right-identity of $+\!\!+$ and $+\!\!+$ is associative (see (14.3) and (14.4)); in other words, $+\!\!+$ and $[]$ form a monoid. The resulting program for g' can be further simplified by observing (as in Figure 14.2) that $[hd(x)] +\!\!+ y = hd(x){:}y$ (see (14.2) and (14.1)) and when $x = []$ that $x +\!\!+ y = y$ (see (14.1)). Performing this simplification yields

$$x = [] \Rightarrow g'(x, y) = y \tag{14.24}$$
$$\neg(x = []) \Rightarrow g'(x, y) = g'(tl(x), hd(x){:}y) \tag{14.25}$$

which is a trivial reformulation of *qrev* as given in (14.9) and (14.10).

This example indicates that if we formalize enough metatheory, then new development rules may be proved correct and used in subsequent developments. In this example, "enough metatheory" means the appropriate semantics of recursion equations. Perhaps surprisingly, one can often get by with much less. In Section 14.2.3 we show that it is possible to derive a specialized form of this schema, without any formalized semantics, directly within a first-order theory of lists. There we formally develop both the program schema and the transformation of *qrev* from *rev*.

Tactic extensibility. Deriving new schematic rules is a powerful way of extending refinement and hence development. Often though, refinement strategies for constructing developments cannot be captured by rules. A strategy may involve iteration, case analysis, and indeed, arbitrary computation. Returning to our unfold–fold example, Feather, for instance, has looked at strategies defining common patterns of reasoning in unfold–fold and defined a class of programs, called design tactics, that implement them [Fea82]. In his work each tactic implements a new kind of transformation step that performs a specific and useful task; e.g., the *combine* tactic replaces nested function calls with a call to a new function definition that can subsequently be optimized.

This idea of using tactics to raise the level of refinement in development goes back to the LCF project. Tactics are programs that, given a specification (goal) SP, reduce it to subspecifications $SP_1, \ldots, SP_n$; that is, a tactic is simply a user-programmable rule. The correctness of tactics is guaranteed in an after-the-fact fashion: each tactic has an associated *validation*, which is a program that constructs a solution to SP from solutions to the SP_i, using only primitive rules of the underlying logic.

Correctness based on validation functions makes tactic construction sometimes unnatural and their execution slow. Consider, for example, proving two terms equal that contain some associative commutative operator such as $+$, e.g., $a + ((b + c) + d)$ and $(c + a) + (d + b)$. A formal proof might consist of applying axioms for associativity, commutativity, and congruence. Alternatively we can view terms as trees and test equality by checking if their fringes are permutations of each other. This could be done by sorting both fringes and checking that the results are identical, i.e., the lists $[a, b, c, d]$. A tactic must do the former; but often we want development rules like the latter. Their construction may be simpler and their execution faster.

If our development formalism is strong enough to express and reason about programs and their properties, then an alternative to the tactic paradigm is possible. Rather than having tactics construct validations with each application, a tactic could be verified correct once and for all. After all, the problem of developing trusted tactics (that do not require validations) is a special case of the general problem of developing trusted software. Returning to the above example we could verify that the "fringe sorting" program is a decision procedure for the appropriate theory. Once verified, it could then be executed without producing a primitive justification because the program has been formally proven correct.

The unfold–fold formalism is not strong enough to show that functional programs are equivalent to arbitrary, not necessarily executable, specifications. We will consider more powerful approaches in Section 14.3. However, it could be used for tactic optimization. Namely, if a tactic is a functional program represented as recursion equations, then unfold–fold could be used to develop a more efficient version of it. This kind of optimization will be discussed in Section 14.4.

14.1.3   Development rules, refinement

The discussion above may have given the reader the impression that formalizing development is difficult and time consuming. It need not be! The size of the task depends on what is to be formalized. It may be easy (e.g., hours or days as opposed to months) to derive new development rules within a suitable framework. Alternatively, more work may be required to formalize nonschematic rules computed by arbitrary functions, or to provide semantic justifications for new rules. Overall, the amount of effort depends on the kinds of rules employed and how their correctness is to be established.

Logicians have looked at the general question "What constitutes a rule?", and the answer is surprisingly complex (e.g., [Avr92, FHV92]). For our purpose, development means problem reduction and the nature of a development rule depends on the syntactic entities we reason about, and the semantic notion of refinement involved. This is an open ended concept; we can imagine heterogeneous kinds of developments where refinement rules are parameterized by data types, models, executable programs, and assertions or *judgements* [Mar82, Mar85] between objects of these syntactic categories. The following rule, taken from [SST92], is not untypical.

$$\frac{SP : \mathbf{Spec}(\varSigma) \quad \varPhi' \subseteq Sen(\varSigma') \quad \sigma : \varSigma \to \varSigma'}{\mathbf{abstract}\ SP\ \mathbf{wrt}\ \varPhi'\ \mathbf{via}\ \sigma : \mathbf{Spec}(\varSigma)} \qquad (14.26)$$

It is impressive for sheer syntax alone as it combines algebraic signatures (e.g., $\varSigma$), signature morphisms (σ), sets of sentences, judgements about set inclusion, and judgements about the satisfaction relation between objects and specifications.

Different kinds of rules require different kinds of metatheories to formalize and reason about them. In this chapter, we will consider two paradigms:

1. a rule is a *partial function*; the type of its domain and range is governed by the judgements of the theory, and
2. a rule is a *relation*; it is a (usually binary) relation between the terms (formulas, proofs, etc.) of the theory.

The first view is the traditional one in logic; it leads to a tree-like structure of developments, i.e., proofs seen as applications of proof rules. The second view is the predominant one in term rewriting, the λ-calculus, rewrite logics, program transformation, and, last but not least, the internal view in many theorem provers.

For example, consider the simple setting of first-order logic. Let $\mathcal{F}$ denote the set of first-order formulas and let pr be a judgement-valued function, where $pr(A)$ represents the judgement that $A \in \mathcal{F}$ is provable. Then, following [HS86, Tro82], a rule is a partial function ϕ from n-tuples of judgements

($n \geq 0$) to judgements. The n-tuple $\langle pr(A_1), \dots, pr(A_n) \rangle$ in the domain of ϕ contains the *premises*, and if $\phi(pr(A_1), \dots, pr(A_n)) = pr(B)$, then $pr(B)$ is the corresponding *conclusion* and the expression

$$\frac{pr(A_1), \dots, pr(A_n)}{pr(B)}$$

is an instance of the rule ϕ.

In the second view, similar to that of rewriting, a rule is specified by a pair written as $l \to r$. For a given context C, two terms t and s are *in the rewriting relation at C*, written $t \to_C s$, iff there is a substitution σ such that $t = C[\sigma(l)]$ and $s = C[\sigma(r)]$. Two terms are *in the rewriting relation* iff there is a context C such that they are in the rewriting relation at C. For example, we can convert the set of equations formed by (14.5) and (14.6) into a typical first-order rewrite system composed of the two rules:

$$rev([]) \to [] \tag{14.27}$$

$$rev(h{:}t) \to rev(t) +\!\!\!+ [h] \tag{14.28}$$

In Section 14.4.1 we will return to this equational/rewriting view of developments.

Sometimes rules, either expressed as functions or relations, involve applicability conditions that must be checked before the rule can be applied; Figure 14.3 contains one such example, namely showing that K is associative with right-identity N. Such preconditions can often be checked automatically (see Section 14.3.4). Rules may also be applied where matching or unification builds in equational or higher-order theories (see Section 14.3.3).

Taking the view of rules as functions or relations, a metalogic that supports the formalization of rules and the derivation of new rules must be a logic capable of formalizing and reasoning about these entities. In practice, however, many rules are of a particularly simple form where they are expressible as *schematic rules* given by (first or higher-order) rule schemata. As we will see in the following section, the standard proof rules of first-order logic can be so formulated. For example, as a natural deduction rule we might formulate $\wedge$-I (conjunction introduction) by the rule schema

$$\frac{pr(\phi) \qquad pr(\psi)}{pr(\phi \wedge \psi)} \tag{14.29}$$

where ϕ and ψ are not concrete formulas of the logic, but rather variables in a metalanguage, which range over the formulas in $\mathcal{F}$. These variables are often called *metavariables* or *schematic variables*. A more complicated example might be a rule expressing program equivalence. We can formalize (14.11)–(14.18) as the single schematic rule

$$\frac{pr(A_1) \quad pr(A_2) \quad \dots \quad pr(A_7)}{pr(\forall x. f(x) = f'(x))} \tag{14.30}$$

where the A_i are the universal closure of the equations (14.11)–(14.17).

14.1.4 Composition

In general, we identify an elementary development step with the application of a development rule. A compound development step corresponds to the application of a series of rules, possibly by a tactic. To illustrate this, let us continue working in the first-order setting above. According to the partial function view of rules, (14.29) corresponds to the function

$$andIntro : \{x : \mathcal{F}|pr(x)\} \rightarrow \{y : \mathcal{F}|pr(y)\} \rightarrow \{z : \mathcal{F}|pr(z)\}$$

with

$$andIntro = \lambda x.\lambda y.x \wedge y .$$

As a consequence, function composition gives meaning to the usual composition of proof trees:

$$\frac{pr(A) \quad \dfrac{pr(B) \quad pr(C)}{pr(B \wedge C)}}{pr(A \wedge (B \wedge C))}$$

Moreover, when extending to the higher-order case, this functional view (which is extensively used in constructive type theory, e.g., [CH88, Mar82]), gives a precise meaning to rules involving assumptions and their discharge, for example, *modus ponens*:

$$\frac{pr(A) \quad \begin{array}{c} [pr(A)] \\ \vdots \\ pr(B) \end{array}}{pr(B)}$$

which can be represented as the function $modusPonens$ of type

$$\{x : \mathcal{F}|pr(x)\} \rightarrow (\{y : \mathcal{F}|pr(y)\} \rightarrow \{z : \mathcal{F}|pr(z)\}) \rightarrow \{r : \mathcal{F}|pr(r)\}) ,$$

which is defined by

$$modusPonens = \lambda x.\lambda F.F(x) .$$

According to the relational view, relational composition is the analog of functional composition and a development is a linear sequence of the form

$$t_1 \rightarrow_{C_1} \cdots \rightarrow_{C_{n-1}} t_n .$$

An alternative representation, which makes explicit use of the relational composition operator $\circ$, is

$$t_1 (\rightarrow_{C_1} \circ \cdots \circ \rightarrow_{C_{n-1}}) t_n .$$

This representation makes explicit that under the relational view, a development is a simple kind of compound object.

There is often a good deal of flexibility as to whether developments are formulated functionally or relationally, with the different formalizations emphasizing different views or aspects of development. In Section 14.2.3 we will give a functional formulation of the *qrev* development based on a functional formalization of the transformation schema given in Figure 14.3. Under this view, unfold–fold-style development can be seen as constructing equality proofs using derived rules. Alternatively, Figure 14.2 suggests a relational formalization where each development step relates sets of equations and corresponds to an instantiation step, or the application of a single equation, interpreted as a rewrite rule from left to right as an unfold step, from right to left as a fold step, or either way as a law. In each transformation step, the focus of applicability of a rewrite rule is adjusted appropriately, e.g., by a matching process.

14.1.5 Organization of this chapter

The remainder of this chapter is organized as follows. In Section 14.2 we present how simple schematic rules can be formalized using first-order and higher-order syntax and be formally derived. To make this concrete, we work through a particular example in the ISABELLE system, first deriving a development rule and then using it for program transformation. This motivates a general discussion about the relationship between developments for verification, synthesis, and transformation. In Section 14.3 we consider development rules that cannot be directly expressed as simple schemata. The problem then becomes one of formalizing general functions or relations. We also consider how schematic rules can be generalized, for example, by building in equational theories. In Section 14.4 we consider how developments can be formalized and how formalization supports their verification and optimization. Finally, in Section 14.5, we draw conclusions.

14.2 Simple schematic development rules

14.2.1 Motivation

As indicated above, many rules can be expressed schematically; here we consider what this entails. We address both metatheoretic extensibility, where users may derive new schematic rules, and how such encodings support the use of a metalogic as a "logical framework". We also consider the possibility of *schematic developments* where not only are rules schematic, but developments themselves may contain metavariables. This allows certain kinds of program synthesis to be integrated into the development process.

Let us first consider what kind of metatheory is needed to support schematic rules and developments. Many different logics have been proposed as metalogics for formalizing rules and derivations. These include equational logic [GSHH92], rewrite logic [MM93], first-order logic [GT93], type theory [HHP93], and higher-order logic [Pau94b], among others. Where possible, we will avoid a commitment to a particular metalogic although we will indicate minimal requirements when necessary.

14.2.2   Formalization and supporting metatheory

Syntax. The idea of encoding the syntax of one language in another goes back at least to Gödel [Göd31] who proposed representing formulas of first-order logic as constants in a theory of arithmetic and Tarski [Tar36] who suggested encodings based on "structural descriptive names" where each formula is associated with a term that mimics its structure. For example, under Tarski's approach one might encode the formula $A \vee (B \wedge C)$ as the term $or(A, and(B, C))$. The importance of this approach has been highlighted by the development of logical frameworks on computers where users encode logics within a given metalogic. Most approaches in current use are refinements of Tarski's idea, usually couched within a typed logic.

The basic idea is simple. Given a specific object logic, for each of its syntactic categories we declare a type in the metalogic. Further, for each operator (i.e., connective or quantifier) in the object logic we declare a constant in the metalogic whose type reflects the syntactic categories of its arguments and the result. The following, for example, is a fragment of the syntax of a first-order theory of lists with equality over a given data type such as the naturals Nat.[2]

$$
\begin{array}{ll}
\textbf{datatype } \mathcal{T} = nil : \mathcal{T} & \textbf{datatype } \mathcal{F} = eq : \mathcal{T} \times \mathcal{T} \to \mathcal{F} \\
\quad | \ cons : Nat \times \mathcal{T} \to \mathcal{T} & \quad | \ not : \mathcal{F} \to \mathcal{F} \\
\quad | \ append : \mathcal{T} \times \mathcal{T} \to \mathcal{T} & \quad | \ and : \mathcal{F} \times \mathcal{F} \to \mathcal{F} \\
\quad \vdots & \quad \vdots
\end{array}
$$

Here we are using a simple data type syntax to declare that elements in $\mathcal{T}$ are lists of natural numbers built from constructors nil and $cons$ and functions like $append$ (there may be others, of course), and formulas in $\mathcal{F}$ are built from the standard constructors. Of course, different logics will have different syntax for declaring types and constants over them. As an example, $not(eq(nil, cons(0, nil)))$ is an element of $\mathcal{F}$. To simplify presentation we will use the standard syntax (used in the introduction) to represent these constructors, e.g., $\neg([\,] = 0\!:\![\,])$ or even $[\,] \neq [0]$. It should always be clear from the context whether we are referring to object or metalevel syntax.

---

[2] We will ignore the issue of parameterization here. A logical framework like IsABELLE supports parameterized types through polymorphic typing of formulas in the metatheory.

$\wedge\text{-}I \quad \forall \phi \, \psi : \mathcal{F}. \, [\![pr(\phi); pr(\psi)]\!] \Longrightarrow pr(\phi \wedge \psi)$
$\wedge\text{-}El \quad \forall \phi \, \psi : \mathcal{F}. \, pr(\phi \wedge \psi) \Longrightarrow pr(\phi)$
$\wedge\text{-}Er \quad \forall \phi \, \psi : \mathcal{F}. \, pr(\phi \wedge \psi) \Longrightarrow pr(\psi)$
$\rightarrow\text{-}I \quad \forall \phi \, \psi : \mathcal{F}. \, (pr(\phi) \Longrightarrow pr(\psi)) \Longrightarrow pr(\phi \Rightarrow \psi)$
$\rightarrow\text{-}E \quad \forall \phi \, \psi : \mathcal{F}. \, [\![pr(\phi \Rightarrow \psi); pr(\phi)]\!] \Longrightarrow pr(\psi)$

Fig. 14.4. Typical proof rules for FOL

Rules. Proof rules, which are to be represented in the metatheory, may be organized around the concept of a judgement or assertion [Mar85], where a logic dictates a notion of proof (or, more generally, development) that justifies the judgements. Since judgements are assertions over elements of the logic's syntactic categories, we may formalize them as appropriately typed propositions in the metalogic. In first-order logic, for example, there is only one kind of judgement, provability, and we might write $pr(\phi)$ to indicate that the formula ϕ is provable. Martin-Löf has four judgements in his type theory (typehood, equality of types, membership in a type, and equalities of members) [Mar82]. If we were reasoning about Hoare logic then our judgement might represent validity of a triple of arguments, e.g., $\{S\} \, P \, \{Q\}$ where S and Q are first-order formulas and P a "while-loop" program.

In the terminology of [HHP93], proof rules represent judgements, which are *hypothetical* and *schematic*. If our metalogic contains implication[3], $\Longrightarrow$, then if J_1 and J_2 are judgements, the hypothetical judgement $J_1 \Longrightarrow J_2$ expresses consequence or logical entailment: J_2 is provable under the assumption of J_1. Alternatively, if C is a type representing a syntactic category, x a variable ranging over C, and $J(x)$ a judgement, then if the metalanguage contains universal quantification, $\forall$, $\forall x : C. \, J(x)$ is a schematic judgement expressing generality in rules and axioms, that is, $J(a)$ for all $a \in C$.

Given this view, it is straightforward to translate most rules into axioms in the metalogic by expressing them as hypothetical and schematic judgements. For example, the $\wedge\text{-}I$ rule given by

$$\frac{pr(\phi) \qquad pr(\psi)}{pr(\phi \wedge \psi)} \, , \tag{14.31}$$

states that $pr(\phi \wedge \psi)$ follows from $pr(\phi)$ and $pr(\psi)$. Moreover, it is general because ϕ and ψ range over formulas. Hence we may formalize the rule as the first rule given in Figure 14.4. Other rules are axiomatized analogously.

We use $\forall$ and $\Longrightarrow$ to represent universal quantification and implication in the metalogic. These operators (in **boldface font**) should not be confused with $\forall$ and $\Rightarrow$ in the declared object logic. Further, to simplify syntax, we write $[\![\phi_1; \dots ; \phi_n]\!] \Longrightarrow \phi$ for $(\phi_1 \Longrightarrow (\dots \Longrightarrow (\phi_n \Longrightarrow \phi) \dots))$ when $n > 1$.

---

[3] This is not the only possibility, e.g., [GSHH92] encodes rules as equations in OBJ3.

Given a collection of axiomatized rules, new rules may be derived by deduction in the metatheory. For example, using the axioms in Figure 14.4 we might derive the new formula

$$\forall \phi \, \psi \, \gamma : \mathcal{F}. \, [\![pr(\phi); pr(\psi); pr(\gamma)]\!] \Longrightarrow pr(\phi \wedge (\psi \wedge \gamma)), \qquad (14.32)$$

which corresponds to adding

$$\frac{pr(\phi) \quad\quad pr(\psi) \quad\quad pr(\gamma)}{pr(\phi \wedge (\psi \wedge \gamma))} \qquad\qquad (14.33)$$

to the set of rules. Its addition preserves correctness (in first-order logic, validity) since it is a derived rule.

Explicit axiomatizations of variables and substitution. So far we have not committed ourselves to a particular metalogic. The encoding above could be carried out within a sorted metatheory based on conditional equations or, at the other extreme, higher-order logic. However the alert reader will have noticed that we have ignored variables and quantifiers. In the past, two main approaches to their formalization have been pursued: explicit encoding and higher-order abstract syntax. Each has its advantages and disadvantages.

Explicitly axiomatizing operations on variables has the advantage that it is possible within weak metalogics (e.g., equational logic). Moreover, it makes syntactic properties explicit and this enables formal metareasoning about syntax itself. Let us reconsider our list theory example. To begin with, we augment the term and formula data types given on page 532. First, we add a new type $\mathcal{V}$, corresponding to the syntactic category of variables. Second, we extend the data type definition $\mathcal{T}$ by adding a new term constructor $var : \mathcal{V} \to \mathcal{T}$. And finally, we extend $\mathcal{F}$ with quantifiers $\forall$ and $\exists$, each of type $\mathcal{V} \times \mathcal{F} \to \mathcal{F}$.

With our extended syntax, we can formalize proof rules for quantifiers, for example, $\forall$-E,

$$\frac{pr(\forall x. \, \phi)}{pr(\phi[t/x])} \, . \qquad\qquad (14.34)$$

This rule relates a judgement about ϕ to a substitution instance $\psi = \phi[t/x]$. To express this we must formalize substitution. This may be axiomatized as a ternary predicate, *subst*, in the metatheory over formulas, terms, and variables, where $subst(\phi, t, x) = \psi$ is provable precisely when $\phi[t/x] = \psi$. With this addition, $\forall$-E is axiomatizable as

$$\forall \phi, \psi : \mathcal{F}. \forall t : \mathcal{T}. \forall x : \mathcal{V}. [\![equal(subst(\phi, t, x), \psi); pr(\forall(x, \phi))]\!] \Longrightarrow pr(\psi) \, . \qquad (14.35)$$

Here we have used two judgements. The new judgement *equal* over pairs of formulas is true when the two are provably equal. Note that we cannot simply

$$FV_{\mathcal{T}}(var(x)) = \{x\} \qquad\qquad FV_{\mathcal{F}}(s = t) \ = FV_{\mathcal{T}}(s) \cup FV_{\mathcal{T}}(t)$$
$$FV_{\mathcal{T}}(\square) \quad\ = \{\} \qquad\qquad\ \ FV_{\mathcal{F}}(\neg(p)) \ = FV_{\mathcal{F}}(p)$$
$$FV_{\mathcal{T}}(h{:}t) \quad = FV_{\mathcal{T}}(h) \cup FV_{\mathcal{T}}(t) \quad FV_{\mathcal{F}}(\wedge(p,q)) = FV_{\mathcal{F}}(p) \cup FV_{\mathcal{F}}(q)$$
$$FV_{\mathcal{T}}(s \mathbin{+\!\!+} t) = FV_{\mathcal{T}}(s) \cup FV_{\mathcal{T}}(t) \quad FV_{\mathcal{F}}(\forall(x,p)) = FV_{\mathcal{F}}(p) - \{x\}$$

$$\vdots \qquad\qquad\qquad\qquad\qquad \vdots$$

Fig. 14.5. Axiomatization of free term and formula variables

restate this as $pr(subst(\phi, t, x) = \psi)$ using the equality term constructor for the type $\mathcal{F}$ since its arguments must be members of $\mathcal{T}$.

There are a variety of approaches to explicitly formalizing theories of substitution and binding. The most direct approach is to take a standard textbook account of basic concepts, such as free and bound variables, capture, and equivalence under bound variable renaming, and translate them into an equational setting. As these may be defined by structural recursion on terms and formulas, the definitions are more or less straightforward. Figure 14.5 gives a flavor of this by equationally defining functions that, given terms ($FV_{\mathcal{T}}$) or formulas ($FV_{\mathcal{F}}$), return the set of their free variables. We have assumed a prior axiomatization of sets where $\{\}$ represents the empty set, $\{x\}$ represents the singleton set with element x, and, if A and B are sets, then $A \cup B$ is their union and $A - B$ their difference. In [MM93], Martí-Oliet and Meseguer have given equational encodings of several logics involving binding operators based on direct encodings of this form.

Less direct kinds of encoding have also been explored. One may change the representation of terms to avoid bound variables entirely. This was originally suggested by de Bruijn [dB72]: his idea was to avoid explicitly named binding variables and, instead, replace variables by a kind of pointer to the operator that binds them. This idea has been adopted, for example, in the work of Shankar [Sha88]. Another approach has been proposed by Talcott [Tal93], who has axiomatized a general theory of contexts (terms with "holes"), binding structure, and substitutions; any logic requiring these concepts should be able to import and build directly upon this theory. Another alternative, which is gaining some popularity, is based on extensions of the lambda calculus that formalize operators for explicit substitutions [ACCL91].

Explicit axiomatization has the advantage that one can state and prove theorems that depend on properties of variables, e.g., their names or whether their occurrences are free or bound. This is useful, for example, in formalizing transformation rules since we may wish to state explicit conditions on variables occurring in terms and their number of occurrences. For example, we may wish to make sure in the transformation rule of Figure 14.3, that f is linearly recursive in the source schema, i.e., does not occur in the schema variables A, B, K, H, or J. Another example would be the transformation rule for variable abstraction ("common subexpression elimination" [BW82a, HK93]); a simplified version might be:

$$K(e) \rightarrow let\ x = e\ in\ K(x)$$

Here, we might want to insist that e occurs more than once in K.

Alternatively, there are decision procedures for formulas in restricted syntactic classes and these procedures must analyze variables and their occurrences. A simple example from propositional logic, due to Lukasiewicz, is that a propositional formula, built only from the equivalence connective ($\Leftrightarrow$), is provable if and only if every propositional variable in it occurs an even number of times. To formalize such a theorem requires a representation of syntax where we can reason about the names of propositional variables and their occurrences. This would not be possible under the approach we consider next.

Higher-order abstract syntax. An alternative to an explicit axiomatic theory of binding is the use of "higher-order abstract syntax" (henceforth HOAS), which was inspired by Church [Chu40] and Martin-Löf's system of arities [NPS90]. This formalization paradigm is supported by higher-order metalogics whose type system is at least as powerful as the simply typed λ-calculus, e.g., higher-order logic [Chu40], λ-Prolog [Mil90a], and type theories like the Edinburgh LF [HHP93], Pfenning's related ELF system [Pfe91], the Calculus of Constructions [CH88], and NUPRL [C$^+$86]. In this approach, variables in the metalogic are used to represent variables in the object logic. Moreover, as the metalogic is higher-order, operators that bind variables (e.g., $\forall$, $\exists$, λ) can be represented using constants whose domain is of function type. In this way, all binding in the object logic is reduced to binding in the λ-calculus and substitution is denoted by application and performed by β-reduction. This approach is very popular among designers of logical frameworks [HHP93, Pau94b]; its advantage is that, once α-conversion, substitution, and the like are implemented within the logical framework for the λ-calculus, these operations can be "inherited" in declared theories whose binding operators behave similarly. Hence there is a significant saving of effort over explicit encoding because a theory of binding and substitution has, operationally, already been implemented.

HOAS is best illustrated by returning to our list theory example: the $\mathcal{T}$ and $\mathcal{F}$ data types defined on page 532. The operators $\forall$ and $\exists$ are added to $\mathcal{F}$ by adding the new constructors $\forall : (\mathcal{T} \rightarrow \mathcal{F}) \rightarrow \mathcal{F}$ and $\exists : (\mathcal{T} \rightarrow \mathcal{F}) \rightarrow \mathcal{F}$. Binding is handled by representing the operators as higher-order functions with argument types $\mathcal{T} \rightarrow \mathcal{F}$. For example, the formula $\forall x. \exists y. x = y$ is encoded as $\forall(\lambda x : \mathcal{T}. \exists(\lambda y : \mathcal{T}. x = y))$. Observe that if our metalogic identifies α-convertible terms (in the λ-calculus), then the above term is equal, for example, to $\forall(\lambda z : \mathcal{T}. \exists(\lambda x : \mathcal{T}. z = x))$. This was not the case for our previously given formula type with explicitly named variables. There we would have had to define predicates on terms and formulas that axiomatize equality

$\forall$-E $\forall P : \mathcal{T} \rightarrow \mathcal{F}. \forall t : \mathcal{T}. pr(\forall x : \mathcal{T}. P(x)) \implies pr(P(t))$
$\forall$-I $\forall P : \mathcal{T} \rightarrow \mathcal{F}. (\forall x : \mathcal{T}.pr(P(x))) \implies pr(\forall x : \mathcal{T}. P(x))$
sub $\forall P : \mathcal{T} \rightarrow \mathcal{F}. \forall x\, y : \mathcal{T}. [pr(x = y); pr(P(x))] \implies pr(P(y))$
ind $\forall \phi : \mathcal{T} \rightarrow \mathcal{F}. [pr(\phi([\,])); \forall h : Nat. \forall x : \mathcal{T}. pr(\phi(x)) \implies pr(\phi(h{:}x))]$
$\implies pr(\forall x : \mathcal{T}. \phi(x))$

Fig. 14.6. Typical HOAS proof rules

under bound variable renaming. The HOAS approach to representing syntax is very flexible. Examples in [AHMP92, HHP93, Pau94b] include higher-order logic, constructive type theories, modal logics, Hoare logic, LCF, and various lambda calculi.

HOAS eases the formalization of proof rules in that substitution, α-conversion, and side conditions on variables can also be "internalized". Let us return to the rule $\forall$-E, (14.34). We may represent this with the $\forall$-E axiom given in Figure 14.6. Note that, as syntactic sugaring, when Q is a variable binding operator, we shall write $Qx : T. \phi$ instead of $Q(\lambda x : T. \phi)$ (recall that we use a different font to distinguish $\forall$ in the metalogic from $\forall$ in the object logic). The $\forall$-E axiom has two parameters: the proposition-valued function P and the term t that instantiates it. So, for example, if we instantiate P with $\lambda x : \mathcal{T}.x \neq [\,]$ and t with the list $[0]$, then, after β-reduction, this yields $pr([0] \neq [\,])$ under the assumption $pr(\forall x : \mathcal{T}. x \neq [\,])$. This indicates how substitution has been modeled by β-conversion. In the rule $\forall$-I, higher-order syntax has allowed us to model the variable occurrence condition (that x may not occur free in any undischarged assumptions), using universal quantification in the metalanguage and its associated side condition. The substitution rule indicates again how substitution can be modeled by application. As a final example, consider the following rule for structural induction over lists.

$$\frac{pr(\phi[[\,]])\qquad pr(\phi[h{:}x])}{pr(\forall x : \mathcal{T}. \phi[x])}\;\begin{array}{c}[pr(\phi[x])]\\ \vdots\end{array}$$

This rule has the side condition that x does not occur free in any assumption, other than those discharged by the application of the rule. This is axiomatized by the final rule in Figure 14.6. Reading this "backwards", it says that to prove some $\forall x.\phi[x]$, we must establish two assertions. First, we must prove $\phi[[\,]]$ and second, for some arbitrary (eigenvariables) h and x, we must show that $\phi[h{:}x]$ is provable under the assumption of $\phi[x]$.

With some experience, the translation of natural deduction presentations of development rules into HOAS becomes natural. However, the question of whether an encoding of syntax and rules actually represents the intended logic is a difficult question, especially since HOAS mixes the syntax of the object and metalogics. In [Gar93, HHP93], considerable attention is devoted to this problem and methodologies are provided for formally establishing the *adequacy* of a representation, i.e., there is the intended correspondence between the syntax and proofs of the object logic and the metatheory. A proof of adequacy for an encoding of a first-order theory similar to that sketched here (first-order arithmetic formalized within the LF-metalogic) can be found in [HHP93].

14.2.3   Developments in a logical framework

One advantage of schematic rules, especially those formalized using HOAS, is that, within a logical framework that supports this style of encoding, it is easy to construct developments. Moreover, program verification, program synthesis, and the construction of new development rules are similar activities. In order to demonstrate this clearly, and also to support our contention that formalization is not difficult, we will present some simple examples carried out in the logical framework ISABELLE [Pau89b, Pau94b]. This is not the only possible system for such work, e.g., ELF [Pfe91] and λ-Prolog [FM88] also come to mind.

ISABELLE **background.** Before illustrating developments in ISABELLE, we provide some background necessary to understand our examples. We will omit many details and [Pau94b] should be consulted for a full account of ISABELLE. ISABELLE is a tactic-based theorem prover whose (meta)logic is the fragment of higher-order logic based on $\forall$ and $\Longrightarrow$. Within this logic, object logics are encoded. The proof rules given in Figures 14.4 and 14.6 are almost exactly as they would be input to ISABELLE; the only difference is that in ISABELLE one often omits the outermost quantifiers and types as they can be reconstructed by the system. Developments may be constructed in ISABELLE in a top-down refinement style. A rule has the form

$$\forall v_1, \ldots, v_m. [\![\phi_1; \ldots ; \phi_n]\!] \Longrightarrow \phi \tag{14.36}$$

where the notation $[\![\phi_1; \ldots ; \phi_n]\!] \Longrightarrow \phi$ is shorthand for the iterated implication $\phi_1 \Longrightarrow \ldots \Longrightarrow (\phi_n \Longrightarrow \phi)$. This also represents a proof state where ϕ is the goal to be established and the ϕ_i represent the subgoals to be proven. Given this view, an initial proof state has the form $\phi \Longrightarrow \phi$, i.e., it has one subgoal, namely ϕ. The final proof state *is itself* the desired theorem.

ISABELLE supports proof construction through higher-order resolution, which is roughly analogous to resolution in Prolog. That is, given a proof state with subgoal ψ and a rule like (14.36), then (treating the v_i as variables for unification) we higher-order unify ϕ with ψ. If this succeeds, then the unification yields a substitution σ and the proof state is updated by applying σ to it, replacing ψ with the subgoals $\sigma(\phi_1), \ldots, \sigma(\phi_n)$. This resolution step can be justified by a sequence of proof steps in the metalogic. Note that, as unification is used to apply rules, the proof state may contain metavariables itself. We will show that this supports program transformation and synthesis during development. Also note that, although rules are formalized in a natural deduction style, (14.36) can be read as an intuitionistic sequent where the ϕ_i are the hypotheses. ISABELLE has resolution tactics that apply rules in a way that maintains this illusion of working with sequents.

All proofs presented in this section are, with the exception of minor printing enhancements, taken directly from an ISABELLE session (given in typewriter font). Note that ISABELLE's displays $\forall$ and $\Longrightarrow$ as !! and ==> respectively. Also note that, as our proofs are restricted to one kind of judgement (namely pr), we omit this "coercion" and display $pr(\phi)$ as simply ϕ; context will always disambiguate metalogic from object-logic syntax. Finally note that we have *lengthened* proofs to make them easier to follow.

Verification. A formalized metatheory may be straightforwardly applied to prove judgements. For example, in our first-order metatheory we can easily prove a theorem like the following.

$$pr(\forall A\,B\,C.\,A \Rightarrow B \Rightarrow C \Rightarrow (A \wedge B \wedge C))$$

This theorem can be proven automatically in ISABELLE using a simplification tactic that comes with the system. However, we shall go through an interactive proof to give a feel for such proof construction using a formalized metatheory, and also as a "warm-up" for more complicated examples.

We begin by giving this goal (in our first-order theory *thy*) to the system.

```
- goal thy "ALL A B C. A --> B --> C --> (A & B & C)";
ALL A B C. A --> B --> C --> A & B & C
 1. ALL A B C. A --> B --> C --> A & B & C
```

We have typed the first line. The following two are system output. On the second line ISABELLE prints the ultimate goal and the third line contains the subgoal (identical to the ultimate goal) required to establish it. We proceed by resolving the subgoal (numbered 1) against $\forall$-I given in Figure 14.6.

```
- by (resolve_tac [allI] 1);
ALL A B C. A --> B --> C --> A & B & C
 1. !!A. ALL B C. A --> B --> C --> A & B & C
```

In the new subgoal, the resolution step has converted quantification over A in the object level to quantification in the metalogic. ISABELLE is a tactic-based theorem prover and we can use standard LCF-style tacticals (combinators that combine tactics) to shorten proofs. For example, we use *REPEAT* to iterate application of $\forall$-I.

```
- by (REPEAT (resolve_tac [allI] 1));
ALL A B C. A --> B --> C --> A & B & C
 1. !!A B C. A --> B --> C --> A & B & C
```

This applies $\forall$-I twice. At this point we proceed by resolving with $\rightarrow$-I (Figure 14.4).

```
- by (REPEAT (resolve_tac [impI] 1));
ALL A B C. A --> B --> C --> A & B & C
 1. !!A B C. [| A; B; C |] ==> A & B & C
```

The effect has been to convert implication at the object level to implication at the metalevel. Read as a sequent, the subgoal states that we must prove the conjunction from the three hypotheses. We apply $\wedge$-I to decompose the leading conjunction, which yields two subgoals.

```
- by (resolve_tac [conjI] 1);
ALL A B C. A --> B --> C --> A & B & C
 1. !!A B C. [| A; B; C |] ==> A
 2. !!A B C. [| A; B; C |] ==> B & C
```

Subgoals in ISABELLE are maintained as a numbered list. We may solve the first subgoal using the rule *ba* (by assumption), which invokes the assumption rule of the metalogic.

```
- ba 1;
ALL A B C. A --> B --> C --> A & B & C
 1. !!A B C. [| A; B; C |] ==> B & C
```

The remaining goals (in this case only one) are renumbered. Here, after one more application of $\wedge$-I and two of *ba*, we have completed the proof. ISABELLE responds with the following proven theorem.

```
ALL A B C. A --> B --> C --> A & B & C
No subgoals!
```

This example is artificially simple, but it demonstrates that, with the right metalevel support, constructing developments using a formal metatheory is not much different from directly using a specially built implementation. A significant difference though is that, with a metatheory, we can formalize and derive new development rules.

Development rules. In Figure 14.3 we presented a schema from [HK93] and suggested that it could be formalized and derived as a development rule. To verify the exact schema given requires a formalized semantics of functions. Here we illustrate a less general, but much simpler alternative, which is often just as useful. Namely, we formulate and prove a specialized version within our metatheory of lists. With slight modification we can restate the schema as a formula that is directly derivable by list induction.

```
- val assums = goal thy
  "[| f([]) = B;
      ALL h t. f(h:t) = K(f(t),J(h,t));
      ALL x.f'(x) = g'(x,N);
      ALL y.g'([],y) = K(B,y);
      ALL h t y.g'(h:t,y) = g'(t,K(J(h,t),y));
      ALL x. K(x,N) = x;
```

```
        ALL x y z. K(K(x,y),z) = K(x,K(y,z))  |]
             ==> ALL x. f(x) = f'(x)";
ALL x. f(x) = f'(x)
 1. ALL x. f(x) = f'(x)
```

The first nine lines (up to the closing double quote and semicolon) are our input, which specifies the theorem to be proven. Lines 2–8 contain the assumptions that can be used in establishing the theorem: Lines 2–3 provide the schema for f, lines 4–6 provide the schema for f' (in terms of the auxiliary function g'), and lines 7–8 specify additional properties of K and N. The list of these assumptions are bound to a variable *assums* and may be used in proving the goal $\forall x.\, f(x) = f'(x)$.

   Proving this theorem is actually quite similar to proving the equivalence of *rev* and *qrev* (*assums* can be seen as axiomatizing the algebraic laws needed for that theorem) and we begin by proving a generalization: we resolve with $\forall$-I and then use *subgoal_tac* to apply *cut* in the metalogic to prove a more general theorem. Cut allows us to insert a new hypothesis, provided we first prove it.

```
- by (resolve_tac [allI] 1 THEN subgoal_tac "ALL y. K(f(x),y) = g'(x,y)" 1);
ALL x. f(x) = f'(x)
 1. !!x. ALL y. K(f(x), y) = g'(x, y) ==> f(x) = f'(x)
 2. !!x. ALL y. K(f(x), y) = g'(x, y)
```

This step results in two subgoals. First, we must show that the cut formula implies the original goal and second, we must prove the cut formula. We proceed with the first goal and apply $\forall$-E to the assumption (the universally quantified formula on the left-hand side of ==>) and instantiate y with N.

```
- by (inst_var [allE] ("y","N") 1);
ALL x. f(x) = f'(x)
 1. !!x. K(f(x), N) = g'(x, N) ==> f(x) = f'(x)
 2. !!x. ALL y. K(f(x), y) = g'(x, y)
```

Now using the second and sixth assumption (in *assums*) we simplify both the hypothesis and conclusion of the first goal. ISABELLE's rewriter completely solves this goal. This command uses a version of *simp_tac*, which performs first-order rewriting using a given "simplification set". The simplification set *list_ss* is a standard collection of previously proven rewrite rules for lists augmented with our assumptions *assums*.

```
- by (asm_full_simp_tac (list_ss addsimps assums) 1);
ALL x. f(x) = f'(x)
 1. !!x. ALL y. K(f(x), y) = g'(x, y)
```

We are left with our strengthened goal. We resolve against our structural induction rule for lists (*ind* in Figure 14.6) and then apply the rewriter to the resulting base and step case.

```
- by ((resolve_tac [ind] 1) THEN
       (ALLGOALS (simp_tac (list_ss addsimps assums)))));
ALL x. f(x) = f'(x)
 1. !!x xa xs.
       ALL y. K(f(xs), y) = g'(xs, y) ==>
       ALL y. K(f(xs), K(J(xa, xs), y)) = g'(xs, K(J(xa, xs), y))
```

We only see the (simplified) step case because the rewriter has completely proved the base case, which is $\forall y.K(f([]),y) = g'([],y)$. This case was proved automatically by simplifying both sides of the equality to $K(B,y)$. In the step case we proceed by using the induction hypothesis as a rewrite rule to simplify the conclusion. This completes the proof.

```
- by (asm_simp_tac list_ss 1);
ALL x. f(x) = f'(x)
No subgoals!
```

ISABELLE allows us to bind this derived rule to a constant (which we call *tail_opt*) for use in subsequent developments.

This example indicates that deriving new development rules in a logical framework like ISABELLE that supports proof construction in a metalogic (e.g., through resolution and simplification) is not difficult. With some experience, this kind of derivation can be carried out interactively in minutes.

Transformation. Let us now use our derived rule *tail_opt* to synthesize the *qrev* function from the *rev* specification. Our development will not only illustrate how we can apply derived development rules, but also how HOAS and higher-order unification support program synthesis.

Our stated goal is to prove the equivalence of *rev* and an unknown function *qrev* under assumptions (bound again to *assums*) that axiomatize *rev*.

```
- val assums = goal thy
   "[| rev([])=[]; ALL h t. rev(h:t)=rev(t) ++ [h] |]
       ==> ALL l. rev(l) = ?qrev(l)";
ALL l. rev(l) = ?qrev(l)
 1. ALL l. rev(l) = ?qrev(l)
```

The question-mark in front of *qrev* tells ISABELLE that *qrev* should be treated as a metavariable for unification (otherwise it would be interpreted as an undeclared constant, like *rev*). That is, we are allowed to prove an *instance* of this theorem and in our case we intend this instantiation to be the new program.

Our first step is to resolve against the previous derived *tail_opt* rule. The tactic *res_inst_tac* performs resolution after first partially instantiating *tail_opt* by assigning *rev* to f and *qrev* to f'.[4]

---

[4] Strictly speaking it is unnecessary to provide this assignment because higher-order unification would generate this instance. However, in this case there are many (inappropriate) unifiers so it is simpler to provide the substitution up front.

```
- by (res_inst_tac [("f", "rev"),("f'","qrev")] tail_opt 1);
ALL 1. rev(1) = ?qrev(1)
 1. rev([]) = ?B
 2. ALL h t. rev(h:t) = ?K(rev(t), ?J(h, t))
 3. ALL x. ?qrev(x) = ?g'(x, ?N)
 4. ALL y. ?g'([], y) = ?K(?B, y)
 5. ALL h t y. ?g'(h:t, y) = ?g'(t, ?K(?J(h, t), y))
 6. ALL x. ?K(x, ?N) = x
 7. ALL x y z. ?K(?K(x, y), z) = ?K(x, ?K(y, z))
```

The seven goals correspond to the seven premises of the *tail_opt* rule. Note
that the uninterpreted constants in the *tail_opt* rule (i.e., B, K, G', J, and
N) were generalized to variables by ISABELLE after the rule was derived, and
these can be instantiated here by unification; hence, unification will determine
an instance of the rule. Our plan is to discharge goals 1 and 2 by resolving
them with the assumptions that define *rev*. This will generate an instantiation
for most of the program variables (e.g., K will become append); afterwards
we prove 6 and 7 using properties of append. This leaves goals 3 to 5, which
after simplification will constitute the definition for *qrev*.

Let us begin by resolving the first two goals against the equations for *rev*
in our assumptions *assums*. These resolution steps produce the substitutions
$K(x,y) \equiv x ++ y$ and $J(h,t) \equiv [h]$. These substitutions are propagated to
the remaining subgoals, which are then renumbered.

```
- by (OnHyps (resolve_tac assums) [1,2] );
ALL 1. rev(1) = ?qrev(1)
 1. ALL x. ?qrev(x) = ?g'(x, ?N)
 2. ALL y. ?g'([], y) = [] ++ y
 3. ALL h t y. ?g'(h:t, y) = ?g'(t, [h] ++ y)
 4. ALL x. x ++ ?N = x
 5. ALL x y z. (x ++ y) ++ z = x ++ y ++ z
```

We now turn our attention to goals 4 and 5, which have been instantiated
to be assertions about append. Goal 5 may be proved by rewriting (the
simplification set *list_ss* contains the lemma, already proved, that append
is associative). Note that ++ is printed infix as a right-associative operator.
Goal 4 may be proved by resolving with the lemma *append_Nil2*, which states
that $x ++ [] = x$. We combine these steps as follows.

```
- by (simp_tac list_ss 5 THEN
      ((resolve_tac [allI] 4 THEN resolve_tac [append_Nil2] 4)));
ALL 1. rev(1) = ?qrev(1)
 1. ALL x. ?qrev(x) = ?g'(x, [])
 2. ALL y. ?g'([], y) = [] ++ y
 3. ALL h t y. ?g'(h:t, y) = ?g'(t, [h] ++ y)
```

Notice that resolution with the lemma $x ++ [] = x$ generates the substitution $N \equiv []$ that is propagated to the first subgoal. Now we are almost
done: we may take the first goal as the definition of *qrev* under the auxiliary
function g' given by the second and third definitions. Before doing this, we
simplify these remaining goals.

```
- by (ALLGOALS (simp_tac (list_ss addsimps assums)));
ALL 1. rev(1) = ?qrev(1)
 1. ALL x. ?qrev(x) = ?g'(x, [])
 2. ALL y. ?g'([], y) = y
 3. ALL h t y. ?g'(h:t, y) = ?g'(t, h:y)
```

Finally, (using *uresult*) we can discharge these goals by making them assumptions of the theorem we are proving. ISABELLE returns the following
proven theorem, which states the equality of *rev* and *qrev* under the given
recurrence equations.

```
- val qsynth = uresult();
val qsynth =
  "[| ?rev([]) = []; ALL h t. ?rev(h:t) = ?rev(t) ++ [h];
     ALL x. ?qrev(x) = ?g'(x, []);
     ALL y. ?g'([], y) = y;
     ALL h t y. ?g'(h:t, y) = ?g'(t, h:y) |] ==> ALL 1. ?rev(1) = ?qrev(1)"
```

Let us review what we have achieved. First, we began by giving a simple axiomatization of list theory in ISABELLE's metalogic. Next, within this
theory we formally derived a development rule for recursion optimization. Finally, we used this rule to prove a theorem that states the equality between
a given function *rev* and an unknown function *qrev*. During the development
we generated a specialized definition of *qrev*, which reverses lists in linear,
instead of quadratic, time. This example illustrates how higher-order unification can be used to synthesize programs during their correctness proofs. We
will explore this development paradigm further in the next section.

14.2.4   Relating verification and synthesis

Our last example illustrated a schematic development: the development itself
contained metavariables. Such developments facilitate a simple but powerful kind of program construction where unification can solve for unknown
functions. In this section we will try to give more insight into this kind of development by briefly surveying some other approaches to synthesis and transformation, and indicating how they can be understood or recast as schematic
development.

Our starting point is to observe that schematic development blurs the
distinction between *verification*, which is proving that some object (e.g., program) has some desired property (e.g., specification), and *synthesis*, which is
constructing an object with a desired property. The *rev* to *qrev* example suggests that a calculus designed for one of these activities can be used for both.
We could use the *tail_opt* schema either to verify the equality of two definitions or, using metavariables, to delay commitment to *qrev* and synthesize it
during the development.

We shall use an even simpler example to bring out this relationship between verification and synthesis. We shall make an analogy with Prolog

proofs. Consider a simple Prolog theory about "happy people". We might state that rich and healthy people are happy. Written as a Prolog clause this is

$$happy(X) \leftarrow rich(X), healthy(X).$$

Further, we might have facts in our theory that a certain individual, say *bob*, is both healthy and rich, i.e., *healthy(bob)* and *rich(bob)*. We might choose to verify in this theory an assertion like *happy(bob)*. Alternatively we might try to synthesize some expression X satisfying the specification *happy(X)*.

$$\frac{\overline{rich(bob)} \quad \overline{healthy(bob)}}{happy(bob)} \qquad \frac{\dfrac{}{rich(X)}\{bob/X\} \quad \dfrac{}{healthy(X)}\{X = bob\}}{happy(X)}\{X = bob\}$$

On the left we have given the verification proof that a Prolog interpreter would (implicitly) construct in answering the query *happy(bob)*. On the right is a synthesis proof. If we assume that goals are solved from left-to-right (as is the case in Prolog) then, in the synthesis proof, after resolving with the clause defining *happy*, we have a resolution step with the ground clause *rich(bob)*. This generates a substitution *bob* for X that is propagated to the remaining subgoal (as well as the original goal), which is then solved with a ground resolution step. This extremely simple example illustrates that these two proofs are essentially the same: from the perspective of resolution they are identical modulo the base case of unification! The only difference is that on the left we always unify ground terms against ground terms, whereas on the right we also unify variables against ground terms, which produces an answer substitution.

The situation is analogous for proof construction in our metalogic. In Prolog, one begins with an axiomatized Horn clause theory. In our case, we begin with an axiomatized metatheory of rules that we may formally extend. In Prolog, proofs are constructed using a uniform search procedure based on resolution (i.e., SLD resolution). In our case, proofs are constructed semi-interactively using higher-order resolution. In both, unification can be used either to verify properties of given objects or to construct them during proof. In our setting though, we require higher-order unification because we wish to solve for functions, which are second-order objects in our declared metatheory.

Relationship to resolution-based development. This idea of unifying verification and synthesis by using unification to apply rules is old. It goes back at least to Green's use of resolution, not only for checking answers to queries, but also for synthesizing programs. In [Gre69], Green suggests using the "answer predicate trick" in resolution[5] to find unifiers using resolution

---

[5] The resolution rule provides no means to keep track of substitution instances. The trick is to extend the goal clause with the literal *answer(X)* where X is

to answer questions. In this setting, as in Prolog, object-level variables stand in for metavariables and are assigned values, possibly incrementally, using resolution.

Let $R(x, y)$ be a first-order proposition. Green showed how resolution can serve as a basis for answering the following types of questions:

Problem	Question	Desired Answer
checking	$R(a, b)$	yes or no
simulation	$\exists x.\, R(a, x)$	yes $x = b$ or no
verifying	$\forall x.\, R(x, g(x))$	yes or no $x = c$
programming	$\forall x.\, \exists y.\, R(x, y)$	yes $y = f(x)$ or no $x = c$

The first three are straightforward, but the last may be surprising; it is closely related to our *qrev* synthesis example. To see if $\forall x.\, \exists y.\, R(x, y)$ follows from a set of formulas, Green negates it, turns it into a clause, and searches for a refutation. Negated and clausified, it becomes $\{\neg R(x_0, y)\}$, and if we can prove inconsistency we get a unifier for y, e.g., $y \equiv hd(x_0)$. Hence we have y as a "function" of x_0, and only first-order variables and unification have been used in the proof process. First-order variables and unification suffice because skolemizing removes binding operators and, at the same time, (implicitly) extends the signature (with x_0) in which the theorem is proved. Green is not really synthesizing a function, but rather a first-order term. However, since x_0 is arbitrary, he can generalize afterwards.

We see here the essence of the schematic development idea in a simple first-order setting: the same resolution steps can be used either to verify that a function is correct, or to construct the function.

Relationship to deductive synthesis. There are a number of approaches loosely classified under the heading of deductive synthesis that also have close ties to schematic developments, even if they are formalized otherwise. For example, in Manna and Waldinger's deductive tableau, one proves a $\forall/\exists$ goal like $\forall x.\exists y.R(x, y)$ by manipulating $R(x, y)$ in a tableau, where x is turned into an eigenvariable and y is placed in a special "output column" [MW80]. Although it is not expressed this way, y plays the role of a metavariable and is incrementally instantiated during subsequent proof steps. This is quite similar to Green's approach, except that the goal is not negated, since proof construction is not refutation-based. At the end of the proof, if variables have been instantiated to ground terms meeting certain side conditions, the tableau yields a completed program.

It is not difficult to draw a precise, formal correspondence between deductive tableau proofs and natural deduction proofs with metavariables. These

---

the variable whose instance we want. Then, rather than halting when resolution produces the empty clause, we may halt with a clause *answer(t)* where t contains the generated substitution instance for X. See also [CL73].

details have been worked out in [Aya95, AB96] and it is shown how, in the spirit of the derivation of *tail_opt*, one can derive program construction rules that can be used to simulate deductive tableau proofs. For example, corresponding to program generation by case splitting, we may derive the rule

$$[pr(A) \Longrightarrow pr(P(s)); pr(\neg A) \Longrightarrow pr(P(t))] \Longrightarrow pr(P(if(A, s, t))) .$$

Now, if we have a goal $pr(R(x, ?y))$ we can resolve this goal against the rule above, which instantiates $?y$ to $if(?A, ?s, ?t)$ and yields the subgoals $pr(?A) \Longrightarrow pr(R(x, ?s))$ and $pr(\neg ?A) \Longrightarrow pr(R(x, ?t))$. The metavariables $?s$, $?t$, and $?A$ will be further instantiated as the cases are proven.

A number of other researchers, e.g., [Cle92, Coe92, HRS90, KBB93], have also observed that different approaches to deductive synthesis may be implemented by providing support for metavariables in developments.

Relationship to constructive synthesis. A distinction is sometimes made in the literature between deductive and constructive synthesis methods (see e.g., [DL93]). The latter are typically based on some kind of constructive logic or type theory equipped with a notion of realizability, e.g., Martin-Löf's type theory [Mar82], NUPRL [C$^+$86], or the Calculus of Constructions [CH88]. In such logics one proves that a term t meets a specification T by demonstrating that t has the type T, denoted $t \in T$. For example, if we show that $t \in (\forall x : T_1 . \exists y : T_2 . R(x, y))$ then, according to the semantics of these logics, t must be an executable function where for any x in T_1, $t(x)$ evaluates to a pair, whose first component, $f(x)$, belongs to T_2, and second component belongs to $R(x, f(x))$.

Should one view a type theory as a theory for program verification, synthesis, or both? In the NUPRL system, for example, the program t may or may not be given *a priori*, and the system contains two similar sets of rules depending on whether t is present (i.e., the goal is $t \in T$) or not (i.e., the goal is simply T). In the former case, proofs verify programs. In the latter, they say how to construct or synthesize programs, and the system may *extract* the inhabiting term t. The extraction, however, may be understood in a very simple way: the term t may be represented initially by a metavariable stored "behind the scenes" and each development rule application elaborates the structure of t. In [Pau94b], Paulson documents an implementation in ISABELLE of a similar type theory, where metavariables are "first-class" objects and may appear directly in proofs. Rules are applied with higher-order resolution (as we have described) and one proves the judgement $t \in T$, where t may be a specific ground term of the type theory, a metavariable, or even a combination of the two. In the first case we have verification, in the second synthesis, and in the third a hybrid where some parts of the program structure are known and others left unspecified (see [HBS92] for uses of such hybrid verification/synthesis).

It is worth emphasizing that the idea of schematic developments suffices to explain program construction in such theories. A common explanation of program extraction is that it is possible because the logic is constructive: a proof yields a program that exhibits the implicit construction. This is correct but the emphasis can be misleading. Unification during schematic developments is enough to account for extraction in type theories like Martin-Löf's or NUPRL's. Constructivity does not play a role in the *extraction* of a program from a proof but rather in its *execution* and *meaning*. That is, because the logic is constructive, the terms extracted can be effectively executed and their evaluation behavior agrees with the semantics of the type theory. For example, see [How91] for a semantics for a classical type theory where programs correspond to oracles that cannot be executed.

Admittedly this discussion takes us a long way from the topic of formalizing developments. However it does help to illustrate that, by choosing the proper development framework, one can formalize developments in ways that support not just program verification, but also transformation and synthesis.

14.3   General development rules

14.3.1   Motivation

The previous section has examined developments based on schematic rules. Unfortunately, not all rules are so easily formalized. Recall from Section 14.1.3 the more general view of a rule as a partial function. Below we consider two examples of rules that are not formalizable as simple schematic rules because computing an instance requires more than just substitution into a schema.

"Regular" schemata. In our formalized first-order theory, for all sequences of formulas ϕ_i and ϕ, it is provable that

$$(\phi_1 \Rightarrow (\ldots \Rightarrow (\phi_n \Rightarrow \phi) \ldots)) \Leftrightarrow ((\phi_1 \wedge \ldots \wedge \phi_n) \Rightarrow \phi).$$

Now suppose we wished to formalize a rule that allowed us to reason backwards from a formula in uncurried form (the right-hand side) to curried form (the left). We might formulate this, say, for $n = 3$, as

$$\forall \phi_1 \, \phi_2 \, \phi_3 \, \phi : \mathcal{F}. \, pr(\phi_1 \Rightarrow (\phi_2 \Rightarrow (\phi_3 \Rightarrow \phi))) \Longrightarrow pr((\phi_1 \wedge \phi_2 \wedge \phi_3) \Rightarrow \phi).$$

Unfortunately, while we can formalize this rule for any particular n, we cannot capture this pattern for every n, e.g.,

$$\forall \phi_1 \ldots \phi : \mathcal{F}. \, pr(\phi_1 \Rightarrow (\ldots \Rightarrow (\phi_n \Rightarrow \phi) \ldots)) \Longrightarrow pr((\phi_1 \wedge \ldots \wedge \phi_n) \Rightarrow \phi).$$

The problem is that the " ... " have only an informal meaning: " ... " expresses a set of schemata. This set can in turn be described, for example, by a context-free language and recognized by a simple parsing algorithm. Hence, one way to formalize this rule is to formalize such an algorithm; but then establishing the correctness of the rule requires reasoning about the given algorithm or grammar (see Section 14.3.3).

Decision procedures. Another useful class of rules are those that implement decision procedures for logics or sublogics. A simple example of this is the theorem of Łukasiewicz that we alluded to earlier: a propositional formula built only from the equivalence connective is provable when all its propositional variables occur an even number of times.

Let us consider the formalization of this. This is a theorem about propositional logic, so let us assume we have a type $\mathcal{F}$ of propositions constructed from variables (over a type $\mathcal{V}$) with a constructor $var : \mathcal{V} \rightarrow \mathcal{F}$ and at least the connective (there could be others) $\mathit{Iff} : \mathcal{F} \times \mathcal{F} \rightarrow \mathcal{F}$ that builds equivalences. Further, assume we have a type $Bool$ in the metatheory with operators like $\&$ defining conjunction. Now, we may define a function $parity : \mathcal{F} \rightarrow Bool$,

$$parity(\phi) = \mathit{iff\_form}(\phi) \ \& \ even\_occ(\phi),$$

that checks that (1) ϕ is only built from variables and the Iff operator and that (2) all variables occurring in ϕ have an even number of occurrences. The first condition, $\mathit{iff\_form}$, might be formalized as follows.

$$\forall x : \mathcal{V}.\ \mathit{iff\_form}(var(x)) = true$$
$$\forall \phi_1 \ \phi_2 : \mathcal{F}.\ \mathit{iff\_form}(\mathit{Iff}(\phi_1, \phi_2)) = \mathit{iff\_form}(\phi_1) \ \& \ \mathit{iff\_form}(\phi_2)$$
$$\forall \phi_1 \ \phi_2 : \mathcal{F}.\ \mathit{iff\_form}(and(\phi_1, \phi_2)) = false$$

$$\vdots$$

The function $even\_occ$, which checks the second condition, could be formalized, for example, as the composition of a function which converts the fringe of the formula, viewed as a tree, to a list and one which checks that each variable in the list occurs an even number of times. Finally, let Eq be a judgement over Booleans, where $Eq(b_1, b_2)$ is true when b_1 and b_2 have the same Boolean value. Now, we may formalize Łukasiewicz's theorem in the metatheory as

$$\forall \phi : \mathcal{F}.\ Eq(parity(\phi), true) \implies pr(\phi)\,. \qquad (14.37)$$

This theorem may be proved by induction over the size of formulas as we now sketch. Given a $\phi \in \mathcal{F}$ such that $parity(\phi) = true$, then ϕ is a term built from Iff and must contain at least two occurrences of the same variable, call it a. One case is $\phi = \mathit{Iff}(a, a)$, but $pr(\mathit{Iff}(a, a))$ may be proved using rules for Iff. In the second case, ϕ must contain more than one Iff connective. Since Iff is associative and commutative, we may reorder variables in ϕ and "shuffle" together two occurrences of a from the remainder, ϕ'. Hence we must have

$$pr(\mathit{Iff}(\mathit{Iff}(a, a), \phi')) \implies pr(\phi).$$

Furthermore, we can "cancel" the formula $\mathit{Iff}(a, a)$ yielding $pr(\phi') \implies pr(\phi)$. Since $parity(\phi') = true$ and ϕ' is simpler (e.g., fewer variables) than ϕ, we have $pr(\phi')$ by the induction hypothesis. Hence, we can conclude $pr(\phi)$.

Once proved, this theorem can be used as a new development rule; its application reduces proving $pr(\phi)$ to proving $Eq(parity(\phi), true)$. The resulting goal can be proved by symbolic evaluation using the equations defining *parity*; this evaluates to *true* when ϕ meets the precondition *parity*. This evaluation can proceed either by rewriting in the metatheory using the equations defining *parity* or by evaluation in the theorem prover, if the prover implements a metatheory based on a programming language (e.g., type theories like [C+86] or pure Lisp like [BM79]). Evaluation is often more efficient than proving the original goal with a tactic. The metatheorem need not construct a validation since it has been formally verified to be correct, just like the rules from the previous section.

14.3.2   Supporting metatheory

In Section 14.2.2 we showed how the syntactic categories of a logic are specified as data types in a metatheory. Formalizing and reasoning about functions over these types is a special case of the more general problem examined in this book: how should programs be formalized (e.g., using equations, Horn clauses, etc.) and what formalisms are adequate to specify and establish their properties? Rather than surveying such possibilities again, we will restrict ourselves to pointing out some of the options that have been successfully applied to formal metatheory in the past.

First-order metatheories. In the last section we considered using first-order metatheories for explicitly encoding syntax. The same kinds of metatheories suffice to reason about functions represented as equations or Horn clauses. For example, the definition for *iff_form* used in *parity* could be directly formalized in a metalogic based on sorted equations. In the last section, rule definitions did not involve recursion. Here the difference is that to reason about rules defined by recursive functions, or containing recursively specified applicability conditions, we require induction principles for our data types. In the *parity* example, we performed induction over formulas, which may be semantically justified by restricting ourselves to considering the initial model of this data type [GTW78].

This approach to developing rules as verified metatheorems in an equational setting has been carried out using the OBJ3 theorem prover, which is based on a term-rewriting implementation of equational logic. [GSHH92] presents the use of OBJ3 as a metalogic where object logics are encoded in equational logic in the explicit style we have presented. There, a rule like the Łukasiewicz example can be directly formalized as we have done. Once proved, it may be applied, and OBJ3's rewriter can be used to evaluate the preconditions of *parity*. A similar investigation of the use of first-order metatheories for the derivation of rewrite rules has been carried out using Maude [MM94], which is a logical framework based on rewrite logic.

Higher-order logics and type theories. In higher-order settings, e.g., based on higher-order logic or some kind of type theory, we can carry out similar developments by defining data types and axiomatizing induction and computation principles for these types. For example, [BC93] presents a kind of higher-order abstract data type based on existential types in a type theory like Martin-Löf's. The definition of data type constructors is similar to the algebraic setting, however each data type is equipped with a structural induction rule and a rule for defining functions by primitive recursion. These principles suffice to define functions like *parity* and prove their properties using derived induction principles. Most type theories support some form of inductive definition suitable for these kinds of proofs (e.g., [BCMS89, Pau89a]). A notable exception is the type theory of the Edinburgh LF, which is very weak and neither supports the definition of functions by primitive recursion nor induction over defined signatures. However, Pfenning has been carrying out an ambitious program of formalized development, where programs are expressed as logic programs using LF signatures and induction is carried out indirectly through a kind of *schema checking* [PR92].

Within a sufficiently expressive metalogic (e.g., HOL, or Calculus of Constructions), an alternative to axiomatizing induction is formally to define data types as inductive definitions within the metatheory, and then derive rules like induction within this theory. One such approach, formalized within the ISABELLE system by Paulson [Pau94a] is based on an internalized account of the Knaster–Tarski fixpoint theorem within higher-order logic. This theorem is used to justify inductive definitions and recursion over definitions given by least and greatest fixpoint equations.

There are advantages in using a constructive type theory as a metalogic as opposed to classical logics. In particular, as proofs construct programs, the verification of rules often has useful computational content. Consider Łukasiewicz's theorem. A proof of this theorem yields a function f such that, for any formula ϕ that satisfies the *parity* predicate, $f(\phi)$ evaluates to a member of $pr(\phi)$, which constitutes a proof of ϕ. This function plays a role analogous to the validation of a tactic (except here proofs are terms within the metalogic, instead of derivations constructed using the metalogic). But f need not be executed to verify the correctness of the development step since the rule itself has been formally verified. On the other hand, f could be executed if we wanted to construct a proof term showing $pr(\phi)$. [BC93] contains a further discussion of these points. Related work also includes [CH90, How88a] in which term rewriters, matchers, and normalizers, are implemented and formally verified within the NUPRL type theory (see also Section 14.4.1) and the work of [Kre93] in formulating and verifying rules for developing "global search algorithms" based on schemata due to Smith [Smi91].

Specially designed frameworks. The final possibility we consider is the use of a metalogic explicitly designed as a formal metatheory, such as Fefer-

man's FS_0 [Fef88]. FS_0 was designed as a minimal logic for allowing formal
metareasoning (it is a conservative extension of primitive recursive arithmetic). The syntax of the object logic is explicitly encoded using S-expressions
(as in Lisp) in FS_0. For each syntactic category of the object logic, an inductive class is defined that contains all S-expressions representing members
of that category. FS_0 allows computation over S-expressions by primitive
recursion and reasoning by structural induction. In [MSB93] an example is
given showing how the predicate calculus may be formalized and how one
may reason about explicitly encoded bound variables in the context of a
prenex normal form theorem for predicate calculus. Note that proofs in the
object logic are also defined in FS_0 by inductively defined classes. This allows
the development of rules that are admissible, but not necessarily derivable
(for the distinction, see [HS86, Tro82]). Furthermore, this facilitates a formal
development of proof transformations such as cut elimination [Mat94].

14.3.3   Generalized unification and matching

The advantage of schematic rules over rules specified using arbitrary programs is that the former may be directly applied using matching or unification. With the latter, a development step requires the user to provide
appropriate instances, "execute" the program, and apply the result. There
is a useful middle ground between these two extremes: the limitations of
schemata in expressing rules may often be overcome by extending matching
and unification to incorporate equational theories. With this approach, the
meaning of functions may be declared equationally and incorporated into the
matching or unification algorithm used to apply rules.

The problem of equational matching and unification has been extensively
studied and various algorithms for $\mathcal{E}$-unification (unification modulo a set of
equations) have been developed. In particular, techniques based on *narrowing* have been developed and proved complete in solving unification problems
associated with convergent term-rewriting systems [BHS89, GS88, Hul80a].
Narrowing has been used as the major computational mechanism underlying
functional logic languages such as EQLOG [GM86b] and SLOG [Fri85b],
however it has rarely been used in program synthesis and program transformation, or for large scale theorem proving in systems like ISABELLE.

Let us illustrate how the equational approach can provide a direct solution
to the problem of formalizing ellipsis for the kind of schemata we considered
earlier, e.g.,

$$(\phi_1 \Rightarrow (\ldots \Rightarrow (\phi_n \Rightarrow \phi) \ldots)) \Leftrightarrow ((\phi_1 \wedge \ldots \wedge \phi_n) \Rightarrow \phi) . \tag{14.38}$$

We may specify these schemata by defining in Figure 14.7 equations that
define curried and uncurried formulas. We assume that our signature for
formulas contains functions *and* and *imp*, which build conjunctions and implications. These equations, oriented from left to right, define a canonical

$$curry(p, []) = p$$
$$curry(p, x{:}xs) = imp(x, curry(p, xs))$$
$$uncurry(p, []) = p$$
$$uncurry(p, x{:}xs) = imp(aux(x{:}xs), p)$$
$$aux(x{:}[]) = x$$
$$aux(x{:}x'{:}xs) = and(x, aux(x'{:}xs))$$

Fig. 14.7. Axioms defining *curry* and *uncurry*

rewrite system. Now (14.38) can be formalized as the following rule in our metalogic

$$\forall \phi : \mathcal{F}. \forall l : \mathcal{F}\ List.\ pr(curry(\phi, l)) \implies pr(uncurry(\phi, l)). \tag{14.39}$$

Now, given a particular goal like

$$pr((\psi_1 \wedge \psi_2 \wedge \psi_3) \Rightarrow \psi),$$

we can (equationally) match this against the right-hand side of our rule, which results in an instance where $\phi \equiv \psi$ and $l \equiv [\psi_1, \psi_2, \psi_3]$. If we had applied this proof rule using resolution incorporating the equations above, then with this substitution instance, our goal would be reduced to proving

$$pr(\psi_1 \Rightarrow (\psi_2 \Rightarrow (\psi_3 \Rightarrow \psi))).$$

It is actually possible to specify (14.39) in a more modular way because the desired schemata are simple and "regular" enough that we can express the rule using a higher-order recursive function *foldr* as a kind of control combinator:

$$foldr\ op\ z\ [] = z$$
$$foldr\ op\ z\ (x{:}xs) = op\ x\ (foldr\ op\ z\ xs)$$

Direct computation shows that the left-hand side of (14.38) is equal to

$$foldr\ imp\ \phi\ [\phi_1, \dots, \phi_n]$$

and the right-hand side is

$$imp\ (foldr\ and\ true\ [\phi_1, \dots, \phi_n])\ \phi$$

where *true* is the neutral element for conjunction. Hence, in a metatheory that contains the definition of *foldr*, our rule for uncurrying can be directly expressed as a formula in the metalogic, namely

$$\forall \phi : \mathcal{F}. \forall l : \mathcal{F}\ List.\ (foldr\ imp\ \phi\ l) \implies imp((foldr\ and\ true\ l), \phi).$$

The use of a combinator like *foldr* is possible since some of the parameters are known; *curry* is essentially the partial parameterization *foldr imp φ*. The use of such combinators represents a kind of *control abstraction* and facilitates the reuse of development rules and their verification.

It is well-known that $\mathcal{E}$-unification is complex and inefficient in general; there are often many solutions, even to simple unification problems, and a combination of equations usually requires a new unification algorithm. This is also true for the general narrowing approach. These problems have hindered the incorporation of general equational unification procedures into program synthesis and transformation systems. Recent work by Shi [Shi94] presents algorithms that can often efficiently solve matching problems formalized using a class of recursive functions, called *matching combinators*. The combinators *foldr* and *curry* are examples of higher-order and first-order matching combinators, respectively. Under this approach, second-order matching against formulas in the metalogic with *foldr* is decidable and also unitary if it contains a concrete function (e.g., $\Rightarrow$ in the example above). This work is currently being integrated into the ISABELLE system and promises considerable support for both formalizing and using schematic rules.

14.3.4   Context sensitivity

Many rules have applicability conditions that formalize assumptions about the context in which the rule must be applied. According to the partial function view of rules, which gives rise to tree-shaped deductions, such assumptions take the form of additional premises, e.g., the equations about the monoid property in our example in Figure 14.3 and its later variants. According to the relational view of rules, applicability conditions give rise to verification conditions, which are generated after matching the left-hand side of a transformation rule; these conditions must be discharged before the rule can be safely applied.

In both cases, we can roughly classify conditions based on how they may be checked.

1. Structural or static semantic conditions that can be discharged automatically by some inference mechanism built into the transformation system. Such mechanisms include type inference and incremental evaluation of attributes decorating the abstract syntax tree [HK93]. Examples of such conditions include information on types, occurrences, scope information, etc.
2. Conditions that can be discharged by user-provided tactics.
3. Conditions that have to be proved interactively by the user.

Depending on the representation of context information in the system, our rule for tail recursion may fall into any of the above categories:

1. The context may provide access to all locally applicable axioms and theorems (this "local theory" was provided as an extra attribute in [HK93]) such that associativity can be checked against an appropriate axiom and the value of the neutral element N can be picked up from an axiom as well.
2. The monoid axioms can be proved and the value of N deduced by some other verification tactic.
3. The user has to provide the value of N as an input parameter and prove associativity and that N is the identity.

Note that only if contextual conditions are computable (the first two cases) can rules be easily composed or used in complex application strategies, which, for example, apply a rule deep down in some recursion process (see Section 14.4).

14.4   Developments

We have seen how individual proof rules can be specified and derived. Here we shall consider how development steps can be composed, i.e., how we may reason about the composition of rules and the strategies that combine them. In particular, we consider how combinators can be used to build developments in a uniform way that facilitates formal specification of development strategies and their verification. In order to illustrate this concretely, we turn to the relational view of developments and examine developments based on equational reasoning and the formalization of rewrite proofs.

14.4.1   Formalizing equational developments

In the setting of algebraic development, equational reasoning forms a central component of formal correctness proofs (see Chapter 8) and is implemented in systems that support rewriting. Traditionally, rewriting is not formalized (although see [MM93]). System support may consist of some kind of completion procedure that converts a rule set into a canonical rewrite system that can be used as a decision procedure [Gan91, KSZ86]. Alternatively, in tactic-based theorem proving systems, the user typically provides the system with a set of rewrite rules, which are exhaustively applied (as was the case in our ISABELLE examples), perhaps under some user-specified strategy. The rewrite theory need not be confluent or even terminating.

Below we consider a standard approach for implementing rewriting in a tactic-based setting in which rewriting can be programmed in a structured and modular way. The idea goes back to the implementation of rewriting functions called *conversions* in the LCF system [Pau83, Pau87]. To ease our presentation, we will give a slightly simplified account of conversions; implementations in the spirit described here have been carried out in the LCF, HOL, NUPRL, and ISABELLE systems.

Primitive conversions. We shall assume that a tactic programming language, such as ML or some other functional programming language, is given. Conversions are programs that implement rewriting; they are partial functions of type $\mathcal{T} \to \mathcal{T} \times \textit{tactic}$. A conversion should, on input t, return a pair $\langle s, T \rangle$ where the tactic T proves the equation $pr(t = s)$ in the appropriate theory, or returns a failure token (an *exception* in the ML programming language). With conversions, to rewrite a term t, we execute $c(t)$ and afterwards execute the resulting tactic T to prove $pr(t = s)$.

We begin by describing primitive conversions, which correspond to simple kinds of rewriting steps. Initially, there is the identity conversion *IdConv*, defined by the function $\lambda t. \langle t, \textit{Refl} \rangle$, that simply returns the input term paired with a tactic *Refl* that proves $pr(t = t)$ for all terms t. Primitive conversions may also be defined from known equalities, i.e., $pr(l = r)$. We can write a conversion-valued function that, given an equation $pr(l = r)$, returns a conversion c such that $c(t)$ executes by matching l against t, which either returns the substitution σ or fails if the terms do not match, and afterwards returns the pair $\langle \sigma(r), T \rangle$ where the tactic T proves $pr(\sigma(l) = \sigma(r))$ using the given equation.

Composite conversions. Primitive conversions comprise basic development steps: rewriting at the root of a term with the identity or an equation. These basic steps can be combined by introducing choice (branching) between conversions, iteration, and extension to rewriting within contexts. Composition of steps is programmed by *conversionals*, which are analogous to tacticals in combining tactic-based developments. Indeed, the analogy is very close; for example, in the ISABELLE development in Section 14.2.3 we presented the tacticals *THEN* and *REPEAT* for composing proof steps and, here, the conversionals *ThenC* and *RepeatC* play an analogous role in composing rewrites. For both tactics and conversions, the use of combinators gives rise to a programming calculus that may in turn be formalized as an algebra. The modular construction of tactics and conversions raises their abstraction, eases reasoning about their correctness, and facilitates reuse.

There are two basic conversionals for sequencing conversions: *OrelseC* and *ThenC*. The former provides selective composition and the latter a method of sequentially composing conversions. *OrelseC* works by testing if conversions succeed or fail (return an exception). Suppose the expression $e_1 ? e_2$ computes the value of e_1 and, if that fails, it computes the value of e_2. We use this to define $(c_1 \ \textit{OrelseC} \ c_2)(t)$ as $(c_1 \ t) ? (c_2 \ t)$. This conversional can be used to pick a successful rewrite strategy out of a set of conversions, that is, it provides a way of trying different strategies. On the other hand, the (infix) conversional *ThenC* uses transitivity,

$$\forall t_1, t_2, t_3 : \mathcal{T}. [pr(t_1 = t_2); pr(t_2 = t_3)] \implies pr(t_1 = t_3),$$

to compose conversions sequentially. That is, the conversion c_1 *ThenC* c_2, when applied to a term t, first computes

$$c_1 t = \langle t_1, T_1 \rangle$$
$$c_2 t_1 = \langle t_2, T_2 \rangle$$

and returns $\langle t_2, T \rangle$, where the tactic T proves $pr(t = t_2)$ by backchaining through the transitivity rule above and proving the subgoals $pr(t = t_1)$ and $pr(t_1 = t_2)$ using T_1 and T_2, respectively.

Using sequencing we can directly program iteration. For example, we can define recursively the conversional *RepeatC*, where *RepeatC*(c) repeatedly applies the conversion c until failure, i.e.,

$$RepeatC(c) = (c \; ThenC \; (RepeatC(c))) \; OrelseC \; IdConv.$$

Subterm traversal. In the case of tactics, the objects operated upon are goal trees, and tacticals apply tactics to the fringes of these trees (with the exception of transformation tactics in the sense of NUPRL [C$^+$86]). In the case of rewriting, the objects are terms, and developments must specify a subterm or context where simplification occurs. The conversions we have previously described allow simplification at the root of the input term. We may extend them to subterms with *homomorphic extension* functionals, which extend the effect of a (local) transformation over larger contexts (see also [HK93, Kri89]).

Rewriting subterms is justified by the fact that equality is a congruence, i.e., for each n-ary function f, we have

$$\forall t_1 \ldots t_n \, s_1 \ldots s_n : \mathcal{T}. \, [pr(t_1 = s_1); \ldots ; pr(t_n = s_n)]$$
$$\Longrightarrow pr(f(t_1, \ldots, t_n) = f(s_1, \ldots, s_n))$$

The conversion $SubConv(c)$, when applied to a term $f(t_1, \ldots, t_n)$ produces a pair $\langle f(s_1, \ldots, s_n), T \rangle$ by first applying c to each subterm t_i, yielding the pair $\langle t_i', T_i \rangle$, or, if it fails, the pair $\langle t_i, Refl \rangle$. If all n conversions fail, then *SubConv* fails. The tactic T is constructed that applies the congruence rule above, which reduces proving the equation $pr(f(t_1, \ldots, t_n) = f(s_1, \ldots, s_n))$ to proving the subgoals $pr(t_i = t_i')$. T then proves each such subgoal by T_i.

SubConv provides the basis for traversing terms recursively. We may now write operators which recursively traverse and rewrite subterms. For example, the operator *Depth*(c) that recursively rewrites all subterms of a term in depth-first order may be recursively defined as follows:

$$Depth(c) = ((SubConv(Depth(c))) \; ThenC \; (RepeatC(c)))$$
$$OrelseC \; (RepeatC(c)).$$

Similarly, top-down rewriting is accomplished by *Top*(c) whose recursive definition is[6]

$$Top(c) = (ProgressC(RepeatC(c))) \; OrelseC \; (SubConv(Top(c))).$$

---

[6] *ProgressC*(c) is a conversion that fails when c behaves like *IdConv*.

With a small set of combinators, we can define powerful, indeed arbitrary, rewrite strategies for applying a given set of rewrite rules.

14.4.2 Verified development strategies

Tactics and conversions are programs that construct developments. They are used as flexible programming languages for expressing development strategies and how developments can be composed. In both cases, a validation is required to guarantee the correctness of these steps. A conversion, when given an input t, simultaneously computes a rewritten term s and a tactic T that justifies the equality. The execution of T is necessary since any function of type $\mathcal{T} \to \mathcal{T} \times tactic$ is a conversion and users can program (maliciously or by accident) a conversion where $pr(t = s)$ is not provable.

Call a conversion c correct if, when $c(t) = \langle s, T \rangle$, T will succeed in proving $pr(t = s)$. If we restrict the way in which conversions are built, we can establish the correctness of conversions and their composition, namely, only use conversions built from (the previously described) primitive conversions and the conversionals. If we denote the set of such conversions $\mathcal{C}$, then we can show that every $c \in \mathcal{C}$ is correct by induction on the structure of c: primitive conversions are easily seen to be correct and each conversional preserves correctness.

We have previously considered the formalization and verification of rules expressed by programs. If our metalanguage for writing conversions is part of our formalized metatheory, then we can formally establish the correctness of these development strategies. The advantage of formally establishing correctness is that for $c \in \mathcal{C}$, if $c(t) = \langle s, T \rangle$ we can assert $pr(t = s)$ without executing T. This is analogous to the way in which decision procedures can be verified, as discussed in Section 14.3. Actually, if conversions are verified, then, since T need not be executed, we need not construct it in the first place. Instead, conversions can be defined as partial functions of type $\mathcal{T} \to \mathcal{T}$ and primitive conversions and conversionals may be defined that do not build tactics. As before, a subset of these corresponding to $\mathcal{C}$ can be proved correct and serve as a basis for composing development steps (rewriting steps) in a flexible and verified way.

The kind of verification of rewriting developments described here has been carried out by Howe [How88b] in the NUPRL system. There Howe specified a type of correct conversions (analogous to the notion of correctness given above) and defined and showed correct a class of conversions similar to those considered here.[7] Term rewriting based on a similar idea has been formalized in the OBJ3 system [GSHH92]. In this work, OBJ3 is used as a metalanguage for itself, and a first-order algebra of rewriting combinators is defined.

---

[7] Note that Howe's work is significantly more involved because he was also concerned with *reflection*, that is, how verified rewriting in NUPRL could be applied to reason about NUPRL proofs themselves (see also [ACHA90]). If one maintains the meta/object logic distinction, the problem is considerably simpler.

14.4.3   Metaprogram transformation and optimization

As observed earlier, the development of correct metaprograms (i.e., tactics) is a special case of developing correct programs. Similarly, when metaprograms are defined in a modular way, for example, based on higher-order combinators, then we can reason about the combinator algebra to derive more efficient tactics using transformation rules at the metalevel.

As a concrete example, consider homomorphic extension. Many functions on lists can be described using the following homomorphic extension functional *hom*.

$$hom \; op \; e \; h \; [] = e$$
$$hom \; op \; e \; h \; (x{:}xs) = op \; (h \; x) \; (hom \; op \; e \; h \; xs)$$

Examples of this functional include the reverse function

$$rev = hom \; (\lambda x \, y.y \mathbin{+\!\!+} x) \; [] \; \lambda x.[x] \, ,$$

the right folding functional *foldr*, where *foldr op e = hom op e id*, or the mapping functional *map*, where *map f = hom* $\mathbin{+\!\!+}$ $[] \; f$.

Now, provided *op* and *e* form a monoid, we can optimize *hom* by introducing tail recursion.

$$hom \; op \; e \; h \; xs = h_2 \; op \; e \; h \; xs \; e$$
$$h_2 \; op \; e \; h \; [] \; r = op \; r \; e$$
$$h_2 \; op \; e \; h \; (x : xs) \; r = h_2 \; op \; e \; h \; xs \; (op \; r \; (h \; x))$$

In doing so, we have optimized all instantiations of *hom* (where *op* and *e* form a monoid); this includes those functions that are used at the metalevel.

This kind of optimization is very common; we can perform similar kinds of optimizations for homomorphisms over trees and this could be used to improve term-rewriting tactics requiring subterm traversal (e.g., *SubConv* in Section 14.4.1). In the case of rewriting, we can also imagine reasoning about the equivalence of rewrite strategies and showing that some particular inefficient rewrite strategy is equivalent to a more efficient one. For example, we might prove that repeated outermost rewriting with each rule is equivalent to one bottom-up sweep, where every rule is tried once at every applicable position [HK93, KKLT91, Kri89].

14.4.4   Development methods = rules + tactics

The construction and transformation of programs, either interactively or automatically, requires more than the simple blind application of some set of transformation rules. In practice, a designer of transformations must go further and build *development methods* that combine both verified development rules and tactics that embody strategies for applying them. Both aspects are necessary as a rule cannot formalize all aspects of program development and one rule often gives rise to various strategies. For example, in the relational view of transformation rules, a tactic may say how to automatically check context conditions, generate verification conditions, or request user input, such as how to instantiate metavariables. In the partial function view, when there are multiple subgoals, the tactic may chose to attack the subgoals in a particular order and this may correspond to different approaches to program development.

A concrete example of this is provided in the ISABELLE derivation we gave of *qrev* in Section 14.2.3. After applying the transformation rule we had seven goals. We chose to discharge them in a particular order: we proved goals 1 and 2 first, which provided instantiations for most of the metavariables in the resulting proof state, before moving on to goals 6 and 7, which expressed properties of some of these metavariables. As a result, 6 and 7 were almost fully instantiated when we proved them. If the order were reversed, we would have had to solve goals 6 and 7 with more degrees of freedom. In particular, there might be other possibilities for K and N in our background theory (or theories), which together form a monoid and we would have to try these various possibilities. This is, of course, wasteful since goal 2 forces a unique assignment for K, which further determines N. It is easy to imagine similar degrees of freedom in algebraic data type implementations, where it may be advantageous to state the target first, and then develop a corresponding abstraction homomorphism, or to develop the homomorphism first, or to start by describing the induced congruence relation, etc.

Our view of combining rules with tactics leads to a natural hierarchical structure of development theories (see, too, the discussion of *development graphs* that follows). At the foundation is a logic or set of related logics. The next layer consists of derived rules, which are proved correct in these logics relative to the intended semantics of the programs and data types. At the highest level, the rules are combined with tactics into methods, which specify strategies for applying the rules and resolving subsequent deductive tasks. This view leads to a methodology for structuring development theories and clarifies conceptually the roles of the underlying theory, derived rules, and tactics.

14.4.5 System support

Formal development requires adequate system support. The kinds of development and transformation methodologies described here can best be carried out with machine assistance for tasks such as theory organization, application of proof rules, checking applicability conditions, and the like. Moreover, support should not simply be for rule application, but should center around all aspects of development including specification, verification of rules, reuse, and modification.

It is possible that all of these development activities may be carried out within the same system; in this chapter we gave such an example using Is-ABELLE. However this is not the only possibility. Different activities (e.g., verifying, as opposed to simply applying, rules) require different skills and thus different kinds of user support. Moreover, considerable structuring facilities are required since numerous kinds of documents are related to each other in a formal development: specifications, program modules, proof obligations and proofs, proof and transformation rules, tactics, development scripts, etc. All these form part of a *project development graph* [KMR+94, PW95]; their relations are formal since they refer to the underlying semantics and methodology. Moreover, these relationships carry usable information; for example, inheritance of properties between specifications may aid reusability.

Orthogonal to a configuration composed of specification or program components, the most important relation in a development graph is the refinement relation between, say, a requirement specification SP and a design specification SP'. The edge in the graph carries, as specialized information, the incremental development script from SP to SP'; in turn, the script, which is a single application of a transformation rule, or tactic, or a composition of these, relates to verification conditions, their proofs, lemmata that arose during such proofs, and so on. Such information is important for the purposes of certification of a development and is vital for any kind of reuse. In fact, each development step corresponds to a development graph transformation that introduces such items and the relation between them [PW95].

For development "in the large", it is important to have a system architecture that manages global correctness and integration between all of the heterogeneous components. An interesting observation is that the notions of refinement, development, and their formalization arise for program development, transformation development ("metadevelopment"), and system development (i.e., integration of new tools, updates) in much the same way. Indeed their implementation can be reused at different levels for different users [KPO+95, HK93, KKLT91].

14.5 Conclusions

We have taken a general view of development and considered it as the composition of correctness-preserving refinement steps. Given this view, we have examined how, and to what degree, development can be formalized; this has included rules, compositions of rules, their combination with tactics and the like. We have also presented the kinds of advantages that formalized development offers. In particular, formalization offers not only some assurance of correctness of development calculi themselves, but also a concrete foundation for formally extending them and optimizing their rules.

Although much of this chapter was a survey of relevant ideas and literature, we have tried to use a concrete example of a classical transformation rule (linear to tail recursion) to illustrate many of the principles that we covered. One particular contribution, we hope, is to show that, with the appropriate foundations and support, formal development of rules and their application is not just a desirable goal, but a realistic proposition.

Bibliography

[AB92]   J. Avenhaus and K. Becker. Conditional rewriting modulo a built-in algebra. SEKI Report SR-92-11, University of Kaiserslautern, 1992.

[AB95a]   M. Aiguier and G. Bernot. Algebraic specification of object type specifications. In R.J. Wieringa and R.B. Feenstra, editors, *Information Systems—Correctness and Reusability, Selected Papers from the IS-CORE Workshop'94*, pages 16–32. World Scientific Publishers, 1995.

[AB95b]   Penny Anderson and David Basin. Deriving and applying logic program transformers. In *Algorithms, Concurrency and Knowledge (1995 Asian Computing Science Conference)*, volume 1023 of *Lecture Notes in Computer Science*, pages 301–318. Springer, 1995.

[AB96]   Abdelwaheb Ayari and David Basin. Generic system support for deductive program development. In *Tools and Algorithms for the Construction and Analysis of Systems (TACAS'96)*, volume 1055 of *Lecture Notes in Computer Science*, pages 313–328. Springer, 1996.

[Aba89]   M. Abadi. The power of temporal proofs. *Theoretical Computer Science*, 65:35–83, 1989.

[ABB+86]   E. Astesiano, C. Bendix Nielsen, N. Botta, A. Fantechi, A. Giovini, P. Inverardi, E. Karlsen, F. Mazzanti, J. Storbank Pedersen, G. Reggio, and E. Zucca. The draft formal definition of Ada. Deliverable 7 of the CEC-MAP project, 1986.

[Abr96]   J.-R. Abrial. *The B-Book*. Cambridge University Press, 1996.

[ABW88]   K.R. Apt, H.A. Blair, and A. Walker. Towards a theory of declarative knowledge. In J. Minker, editor, *Deductive Databases and Logic Programming*, pages 89–148. Morgan Kaufmann, 1988.

[AC92]   E. Astesiano and M. Cerioli. Partial higher-order specifications. *Fundamenta Informaticae*, 16(2):101–126, 1992.

[AC93]   E. Astesiano and M. Cerioli. Relationships between logical frameworks. In M. Bidoit and C. Choppy, editors, *Recent Trends in Data Type Specification, Proc. Workshop on Specification of Abstract Data Types ADT'91*, volume 655 of *Lecture Notes in Computer Science*, pages 126–143. Springer, 1993.

[AC95]   E. Astesiano and M. Cerioli. Free objects and equational deduction for partial conditional specifications. *Theoretical Computer Science*, 152(1):91–138, 1995.

[AC96]   E. Astesiano and M. Cerioli. Non-strict don't care algebras and specifications. *Mathematical Structures in Computer Science*, 6(1):85–125, 1996.

[ACCL91]   Martín Abadi, Luca Cardelli, Pierre-Louis Curien, and Jean-Jacques Lévy. Explicit substitutions. *Journal of Functional Programming*, 1(4):375–416, 1991.

[ACHA90]   Stuart F. Allen, Robert L. Constable, Douglas J. Howe, and William E. Aitken. The semantics of reflected proof. In *Symposium on Logic in Computer Science*. Computer Society Press of the IEEE, 1990.

[AEH94]   S. Antoy, R. Echahed, and M. Hanus. A needed narrowing strategy. In *Proc. 21st ACM POPL*, pages 268–279, 1994.

[AG91]   A.J. Alencar and J.A. Goguen. OOZE: An object-oriented z environment. In P. America, editor, *Proc. ECOOP'91*, pages 180–199. Springer, 1991.

[AG96]   T. Arts and J. Giesl. Termination of constructor systems. In H. Ganzinger, editor, *Proc. 7th Intl. Conference on Rewriting Techniques and Applications*, volume 1103 of *Lecture Notes in Computer Science*, pages 63–77. Springer, 1996.

[AG97]   T. Arts and J. Giesl. Automatically proving termination where simplification orderings fail. In M. Bidoit and M. Dauchet, editors, *Theory and Practice of Software Development (TAPSOFT'97), Proc. 7th Intl. Joint Conference CAAP/FASE*, volume 1214 of *Lecture Notes in Computer Science*, pages 261–272. Springer, 1997.

[AGM92]   S. Abramsky, D.M. Gabbay, and T.S.E. Maibaum. *Handbook of Logic in Computer Science, Vols. 1–6*. Clarendon Press, Oxford, 1992.

[AGR92]   E. Astesiano, A. Giovini, and G. Reggio. Observational structures and their logic. *Theoretical Computer Science*, 96(1):249–283, 1992.

[AHMP92]   Arnon Avron, Furio Honsell, Ian Mason, and Randy Pollack. Using typed lambda calculus to implement formal systems on a machine. *Journal of Automated Reasoning*, 9:309–352, 1992.

[AHS90]   J. Adámek, H. Herrlich, and G. Strecker. *Abstract and Concrete Categories*. Wiley, New York, 1990.

[AL94]   J. Avenhaus and C. Loría-Sáenz. On conditional rewrite systems with extra variables and deterministic logic programs. In *Proc. LPAR'94*, volume 822 of *Lecture Notes on Artificial Intelligence*, pages 215–229. Springer, 1994.

[ALN^{+}91] J.-R. Abrial, M. Lee, D. Neilson, P. Scharbach, and I. Sørenson. The B-method. In S. Prehn and W.J. Toetenel, editors, *VDM, Formal Software Development Methods, Proc. 4th Intl. Symposium of VDM Europe; Vol. 2: Tutorials*, volume 552 of *Lecture Notes in Computer Science*, pages 398–405. Springer, 1991.

[AM82]   M.A. Arbib and E.G. Manes. Parametrized data types do not need highly constrained parameters. *Information and Control*, 52:139–158, 1982.

[AM90]   J. Avenhaus and K. Madlener. Term rewriting and equational reasoning. In R.B. Banerji, editor, *Formal Techniques in Artifial Intelligence*, pages 1–43. Elsevier Science (North-Holland), 1990.

[AMRW85]   E. Astesiano, G.F. Mascari, G. Reggio, and M. Wirsing. On the parameterized algebraic specification of concurrent systems. In H. Ehrig, C. Floyd, M. Nivat, and J. Thatcher, editors, *Proc. TAPSOFT'85, Vol. 1*, volume 185 of *Lecture Notes in Computer Science*. Springer, 1985.

[Apt90]   K.R. Apt. Logic programming. In J. van Leeuwen, editor, *Handbook of Theoretical Computer Science*, pages 493–574. Elsevier Science (North-Holland), 1990.

[AR87]   E. Astesiano and G. Reggio. SMoLCS-driven concurrent calculi. In H. Ehrig, R. Kowalski, G. Levi, and U. Montanari, editors, *Proc. TAPSOFT'87, Vol. 1*, volume 249 of *Lecture Notes in Computer Science*. Springer, 1987.

[AR93a]   E. Astesiano and G. Reggio. Algebraic specification and concurrency. In M. Bidoit and C. Choppy, editors, *Recent Trends in Data Type Specification, Proc. Workshop on Specification of Abstract Data Types ADT'91*, volume 655 of *Lecture Notes in Computer Science*, pages 1–39. Springer, 1993.

[AR93b]   E. Astesiano and G. Reggio. Specifying reactive systems by abstract events. In *Proc. 7th Intl. Workshop on Software Specification and Design (IWSSD-7)*, Los Alamitos, CA, 1993. IEEE Computer Society.

[AR94]   J. Adámek and J. Rosický. *Locally Presentable and Accessible Categories*. Cambridge University Press, 1994.

[AR96a]   E. Astesiano and G. Reggio. A dynamic specification of the RPC-memory problem. In M. Broy, S. Merz, and K. Spies, editors, *Formal System Specification: The RPC-Memory Specification Case Study*, volume 1169 of *Lecture Notes in Computer Science*. Springer, 1996.

[AR96b]   E. Astesiano and G. Reggio. Labelled transition logic: An outline. Technical Report DISI-TR-96-20, DISI – Università di Genova, Italy, 1996.

[AR97a]   E. Astesiano and G. Reggio. Formalism and method. In M. Bidoit and M. Dauchet, editors, *Proc. TAPSOFT'97*, volume 1214 of *Lecture Notes in Computer Science*. Springer, 1997.

[AR97b]   E. Astesiano and G. Reggio. On the relationship between labelled transition logic and rewriting logic. Technical Report DISI-TR-97-23, DISI – Università di Genova, Italy, 1997.

[ARS96]   http://www-formal.stanford.edu/clt/ARS/systems.html, 1996. Links to most currently available automated reasoning systems.

[Asp97]   D. Aspinall. *Type Systems for Modular Programs and Specifications*. PhD thesis, University of Edinburgh, 1997.

[AT89]   J. Adamek and V. Trnkova. *Automata and Algebras in Categories*. Kluwer Academic Publishers, 1989.

[Avr92]   Arnon Avron. Simple consequence relations. *Information and Computation*, 92(1):105–140, 1992.

[AW89]   E. Astesiano and M. Wirsing. Bisimulation in algebraic specifications. In H. Aït-Kaci and M. Nivat, editors, *Resolution of Equations in Algebraic Structures 1*, pages 1–31. Academic Press, 1989.

[Aya95]   Abdelwaheb Ayari. A reinterpretation of the deductive tableaux system in higher-order logic. Master's thesis, Universität Saarbrücken, 1995.

[AZ95]   E. Astesiano and E. Zucca. D-oids: a model for dynamic data types. *Mathematical Structures in Computer Science*, 5(2):257–282, 1995.

[AZ96]   E. Astesiano and E. Zucca. A free construction of dynamic terms. *Journal of Computer and System Sciences*, 52(1), 1996.

[Bac88]   L. Bachmair. Proof by consistency in equational theories. In *Proc. 3rd IEEE Symposium on Logic in Computer Science, Edinburgh (UK)*, pages 228–233, 1988.

[Bac91]   L. Bachmair. *Canonical equational proofs*. Computer Science Logic, Progress in Theoretical Computer Science. Birkhäuser Verlag AG, 1991.

[Bar74]   J. Barwise. Axioms for abstract model theory. *Ann. Math. Logic*, 7:221–265, 1974.

[Bar77]   Jon Barwise, editor. *Handbook of Mathematical Logic*, volume 90 of *Studies in Logic and the Foundations of Mathematics*. North-Holland, 1977.

[Bar81]   H.P. Barendregt. *The Lambda Calculus: Its Syntax and Semantics*, volume 103 of *Studies in Logic and the Foundations of Mathematics*. North-Holland, revised edition, 1981.

[Bar84]   H.P. Barendregt. *The Lambda-Calculus, its syntax and semantics*. Studies in Logic and the Foundation of Mathematics. Elsevier Science (North-Holland), second edition, 1984.

[Bas94]   David A. Basin. Logic frameworks for logic programs. In *4th Intl. Workshop on Logic Program Synthesis and Transformation (LOPSTR'94)*, volume 883 of *Lecture Notes in Computer Science*, pages 1–16. Springer, 1994.

[Bau91]   H. Baumeister. Unifying initial and loose semantics of parameterized specifications in an arbitrary institution. In *Proc. TAPSOFT'91*, volume 493 of *Lecture Notes in Computer Science*, pages 103–120. Springer, 1991.

[Bau92]   H. Baumeister. Parameter passing in the typed λ-calculus approach to parameterized specifications. In *Proc. WADT'92, Caldes de Malavella (Spain)*, 1992.

[Bau93]   Hubert Baumeister. Parameter passing in the typed λ-calculus approach to parameterized specifications. Available at URL http://www.mpi-sb.mpg.de/~hubert/index/index6.html, 1993.

[Bau95]   H. Baumeister. Relations as abstract datatypes: An institution to specify relations between algebras. In P.D. Mosses, M. Nielsen, and M.I. Schwartzbach, editors, *Proc. TAPSOFT'95*, volume 915 of *Lecture Notes in Computer Science*. Springer, 1995.

[BB87]   T. Bolognesi and E. Brinksma. Introduction to the ISO specification language LOTOS. *Computer Networks and ISDN Systems*, 14, 1987.

[BBB+85]   F. Bauer, R. Berghammer, M. Broy, W. Dosch, F. Geiselbrechtinger, R. Gnatz, E. Hangel, W. Hesse, B. Krieg-Brückner, A. Laut, T. Matzner, B. Möller, F. Nickl, H. Partsch, P. Pepper, K. Samelson, H. Wössner, and M. Wirsing. *The Munich Project CIP. Volume I: The Wide Spectrum Language CIP-L*, volume 183 of *Lecture Notes in Computer Science*. Springer, 1985.

[BBC86a]   G. Bernot, M. Bidoit, and C. Choppy. Abstract data types with exception handling: an initial approach based on a distinction between exceptions and errors. *Theoretical Computer Science*, 46(1):13–45, 1986.

[BBC86b]   G. Bernot, M. Bidoit, and C. Choppy. Abstract implementations and correctness proofs. In *Proc. 3rd Symp. on Theoretical Aspects of Computer Science*, volume 210 of *Lecture Notes in Computer Science*, pages 236–251. Springer, 1986.

[BBD+81]   F. Bauer, M. Broy, W. Dosch, R. Gnatz, B. Krieg-Brückner, A. Laut, M. Luckmann, T. Matzner, B. Möller, H. Partsch, P. Pepper, K. Samelson, R. Steinbrüggen, H. Wössner, and M. Wirsing. Programming in a wide spectrum language: a collection of examples. *Science of Computer Programming*, 1:73–114, 1981.

[BBH94]   R. Barnett, D. Basin, and J. Hesketh. A recursion planning analysis of inductive completion. Report MPI-I-94-230, Max-Planck-Institut für Informatik Saarbrücken, 1994.

[BBTW81]   J. Bergstra, M. Broy, J.V. Tucker, and M. Wirsing. On the power of algebraic specifications. In *Proc. MFCS'81*, volume 118 of *Lecture Notes in Computer Science*, pages 193–204. Springer, 1981.

[BC93]   David Basin and Robert Constable. Metalogical frameworks. In Gérard
            Huet and Gordon Plotkin, editors, *Logical Environments, Proc. Work-
            shop on Logical Frameworks '91*, pages 1–29. Cambridge University
            Press, 1993.

[BCM88]   E. Battiston, F. De Cindio, and G. Mauri. OBJSA nets: a class of
            high-level nets having objects as domains. In G. Rozenberg, editor,
            Advances in Petri Nets, volume 340 of *Lecture Notes in Computer Sci-
            ence*. Springer, 1988.

[BCMS89]   Roland Backhouse, Paul Chisholm, Grant Malcolm, and Erik Saaman.
            Do-it-yourself type theory. *Formal Aspects of Computing*, 1(1):19–84,
            1989.

[BCPR96]   M. Bidoit, C. Chevenier, C. Pellen, and J. Ryckbosh. An algebraic spec-
            ification of the steam-boiler control system. In J.-R. Abrial, E. Borger,
            and H. Langmaack, editors, *Formal Methods for Industrial Applica-
            tions*, volume 1165 of *Lecture Notes in Computer Science*. Springer,
            1996.

[BD77]   Rod M. Burstall and John Darlington. A transformation system for de-
            veloping recursive programs. *Journal of the Association for Computing
            Machinery*, 24(1):44–67, 1977.

[BD86]   L. Bachmair and N. Dershowitz. Commutation, transformation and
            termination. In J. Siekmann, editor, *Proc. 8th Intl. Conference on Au-
            tomated Deduction*, volume 230 of *Lecture Notes in Computer Science*,
            pages 5–20. Springer, 1986.

[BD87]   L. Bachmair and N. Dershowitz. Completion for rewriting modulo a
            congruence. In *Proc. 2nd Conference on Rewriting Techniques and
            Applications*, volume 256 of *Lecture Notes in Computer Science*, pages
            192–203. Springer, 1987.

[BD89]   L. Bachmair and N. Dershowitz. Completion for rewriting modulo a
            congruence. *Theoretical Computer Science*, 67(2-3):173–202, 1989.

[BDD+92]   M. Broy, F. Dederichs, C. Dendorfer, M. Fuchs, T. Gritzner, and R. We-
            ber. The design of distributed systems: an introduction to FOCUS.
            Report TUM–I9203, Institut für Informatik, Technische Universität
            München, 1992.

[BDH86]   L. Bachmair, N. Dershowitz, and J. Hsiang. Orderings for equational
            proofs. In *Proc. 1st IEEE Symposium on Logic in Computer Science,
            Cambridge, Mass. (USA)*, pages 346–357. IEEE, 1986.

[BDP89]   L. Bachmair, N. Dershowitz, and D. Plaisted. Completion without fail-
            ure. In H. Aït-Kaci and M. Nivat, editors, *Resolution of Equations in
            Algebraic Structures, Vol. 2: Rewriting Techniques*, pages 1–30. Aca-
            demic Press Inc., 1989.

[Bee85]   Michael J. Beeson. *Foundations of Constructive Mathematics*, vol-
            ume 6 of *Ergebnisse der Mathematik und ihrer Grenzgebiete (3. Folge)*.
            Springer, revised edition, 1985.

[Bee86]   M.J. Beeson. Proving programs and programming proofs. In R.B.
            Marcus et al., editors, *Logic, Methodology and Philosophy of Science
            VII*, pages 51–82. North-Holland, 1986.

[Bee95]   C. Beeri. Recent trends in data type specification. In E. Astesiano,
            G. Reggio, and A. Tarlecki, editors, *Bulk Types and Query Language
            Design*, volume 906 of *Lecture Notes in Computer Science*. Springer,
            1995.

[BEP87]   E.K. Blum, H. Ehrig, and F. Parisi-Presicce. Algebraic specification of modules and their basic interconnections. *Journal of Computer and System Sciences*, 34(2-3):293–339, 1987.

[Ber89]   G. Bernot. Correctness proofs for abstract implementations. *Information and Computation*, 80:121–151, 1989.

[BF85]   J. Barwise and S. Feferman, editors. *Model-Theoretic Logics*. Springer, 1985.

[BFG⁺93]   M. Broy, C. Facchi, R. Grosu, R. Hettler, H. Hußmann, D. Nazareth, F. Regensburger, and K. Stølen (The Munich SPECTRUM Group). The requirement and design specification language SPECTRUM: An informal introduction, Version 1.0, Part I. Technical Report TUM–19311, TUM–19312, Institut für Informatik, Technische Universität München, 1993.

[BFLM94]   J.C. Bicarregui, J.S. Fitzgerald, P.A. Lindsay, and R. Moore. *Proofs in VDM, a practitioner's guide*. Springer, 1994.

[BG77]   R. Burstall and J. Goguen. Putting theories together to make specifications. In *Proc. 5th Intl. Joint Conference on Artificial Intelligence, Cambridge, Mass. (USA)*, pages 1045–1058, 1977.

[BG80]   R.M. Burstall and J.A. Goguen. The semantics of CLEAR, a specification language. In D. Björner, editor, *Proc. Copenhagen Winter School on Abstract Software Specification*, volume 86 of *Lecture Notes in Computer Science*, pages 292–332. Springer, 1980.

[BG89]   H. Bertling and H. Ganzinger. Completion-time optimization of rewrite-time goal solving. In N. Dershowitz, editor, *Proc. 3rd Conference on Rewriting Techniques and Applications (RTA'89)*, volume 355 of *Lecture Notes in Computer Science*, pages 45–58. Springer, 1989.

[BG94a]   L. Bachmair and H. Ganzinger. Rewrite-based equational theorem proving with selection and simplification. *Journal of Logic and Computation*, 4(3):1–31, 1994.

[BG94b]   L. Bachmair and H. Ganzinger. Rewrite techniques for transitive relations. In *Proc. 9th IEEE Symposium on Logic in Computer Science*, pages 384–393. IEEE Computer Society Press, 1994. Short version of TR MPI-I-93-249.

[BGLS92]   L. Bachmair, H. Ganzinger, C. Lynch, and W. Snyder. Basic paramodulation and superposition. In *Proc. 11th Intl. Conference on Automated Deduction, Saratoga Springs, N.Y. (USA)*, pages 462–476, 1992.

[BGLS95]   L. Bachmair, H. Ganzinger, C. Lynch, and W. Snyder. Basic paramodulation. *Information and Computation*, 121(2):172–192, 1995.

[BGM89]   M. Bidoit, M.-C. Gaudel, and A. Mauboussin. How to make algebraic specifications more understandable? An experiment with the PLUSS specification language. *Science of Computer Programming*, 12(1):1–38, 1989.

[BGM91]   G. Bernot, M.-C. Gaudel, and B. Marre. Software testing based on formal specifications: a theory and a tool. *Software Engineering Journal*, 6(6):387–405, 1991.

[BGT91]   R.M. Burstall, J.A. Goguen, and A. Tarlecki. Some fundamental algebraic tools for the semantics of computation, Part 3. Indexed categories. *Theoretical Computer Science*, 91:239–264, 1991.

[BH94]   J. Bowen and M. Hinchey. Formal methods and safety critical standards. *Computers*, 27(8):68–71, 1994.

[BH96]   Michel Bidoit and Rolf Hennicker. Behavioural theories and the proof
            of behavioural properties. *Theoretical Computer Science*, 165(1):3–55,
            1996.

[BHK89]   J.A. Bergstra, J. Heering, and P. Klint. The algebraic specification
            formalism ASF. In J.A. Bergstra, J. Heering, and P. Klint, editors,
            Algebraic Specification, ACM Press Frontier Series. Addison-Wesley,
            1989.

[BHK90]   J.A. Bergstra, J. Heering, and P. Klint. Module algebra. *Journal of
            the Association for Computing Machinery*, 37(2):335–372, 1990.

[BHS89]   Hans-Jürgen Bürckert, Alexander Herold, and Manfred Schmidt-
            Schauß. On equational theories, unification, and decidability. *Journal
            of Symbolic Computation*, 8:3–50, 1989.

[BHW95]   M. Bidoit, R. Hennicker, and M. Wirsing. Behavioural and abstractor
            specifications. *Science of Computer Programming*, 25:149–186, 1995.

[Bid81]   M. Bidoit. *Une Méthode de Présentation des Types Abstraits, Applica-
            tions*. Thèse de Doctorat de Troisième Cycle, Université de Paris-Sud,
            1981.

[Bid88]   M. Bidoit. The stratified loose approach. A generalization of initial and
            loose semantics. In *Recent Trends in Data Type Specification*, volume
            332 of *Lecture Notes in Computer Science*, pages 1–22. Springer, 1988.

[Bid89]   M. Bidoit. *Pluss, un langage pour le développement de spécifications
            algébriques modulaires*. Thèse d'Etat, Université de Paris-Sud, 1989.

[BJ97]   A. Bouhoula and J.-P. Jouannaud. Automata-driven automated induc-
            tion. In *Proc. 12th IEEE Symposium on Logic in Computer Science,
            Warsaw (Poland)*, pages 14–25. IEEE Comp. Soc. Press, 1997.

[BJM97]   A. Bouhoula, J.-P. Jouannaud, and J. Meseguer. Specification and
            proof in membership equational logic. In M. Bidoit and M. Dauchet,
            editors, *Proc. Theory and Practice of Software Development*, volume
            1214 of *Lecture Notes in Computer Science*, pages 67–92. Springer,
            1997.

[BK86a]   J.A. Bergstra and J.W. Klop. Conditional rewrite rules: Confluence and
            termination. *Journal of Computer and System Sciences*, 32(3):323–362,
            1986.

[BK86b]   J.A. Bergstra and J.W. Klop. Process algebra: Specification and verifi-
            cation in bisimulation semantics. In M. Hazewinkel, J.K. Lenstra, and
            L.G.L.T. Meertesen, editors, *Math. & Comp. Sci. II*, volume 4 of *CWI
            Monograph*. North-Holland, 1986.

[BK98]   M. Białasik and B. Konikowska. A logic for nondeterministic specifi-
            cations. In E. Orłowska, editor, *Logic at work. Essays dedicated to the
            memory of H. Rasiowa*. Kluwer, 1998.

[BKK96a]   P. Borovanský, C. Kirchner, and H. Kirchner. Controlling rewriting by
            rewriting. In J. Meseguer, editor, *Proc. 1st Intl. Workshop on Rewriting
            Logic*, volume 4 of *Electronic Notes in Theoretical Computer Science*,
            1996.

[BKK⁺96b]   P. Borovanský, C. Kirchner, H. Kirchner, P.-E. Moreau, and M. Vit-
            tek. ELAN: A logical framework based on computational systems. In
            J. Meseguer, editor, *Proc. 1st Intl. Workshop on Rewriting Logic*, vol-
            ume 4 of *Electronic Notes in Theoretical Computer Science*, 1996.

[BKK98]   P. Borovanský, C. Kirchner, and H. Kirchner. A functional view of rewriting and strategies for a semantics of ELAN. In *The Third Fuji International Symposium on Functional and Logic Programming, Kyoto (Japan)*, pages 143–167. World Scientific, 1998.

[BKR95]   A. Bouhoula, E. Kounalis, and M. Rusinowitch. Automated mathematical induction. *Journal of Logic and Computation*, 5(5):631–668, 1995.

[BL87]   F. Bellegarde and P. Lescanne. Transformation orderings. In *Proc. 12th Coll. on Trees in Algebra and Programming (TAPSOFT)*, volume 249 of *Lecture Notes in Computer Science*, pages 69–80. Springer, 1987.

[BLH92]   D. Bjørner, H. Langmaack, and C.A.R. Hoare. Provably correct systems. PROCOS I final deliverable, 1992.

[BM79]   Robert S. Boyer and J. Strother Moore, editors. *A Computational Logic*. Academic Press, 1979.

[BM84]   J.A. Bergstra and J.J. Meyer. On specifying sets of integers. *Elektronische Informationsverarbeitung und Kybernetik*, 20:531–541, 1984.

[BM86]   John L. Bell and Moshé Machover. *A course in mathematical logic*. North-Holland, 1986.

[BMPW86]   M. Broy, B. Möller, P. Pepper, and M. Wirsing. Algebraic implementations preserve program correctness. *Science of Computer Programming*, 7:35–53, 1986.

[BN98]   F. Baader and T. Nipkow. *Term Rewriting and all That*. Cambridge University Press, 1998.

[Boe81]   B.W. Boehm. *Software engineering economics*. Advances in Computing Science and Technology Series. Prentice Hall, 1981.

[Boo94]   G. Booch. *Object-Oriented Analysis and Design with Applications*. Benjamin Cummings, 2nd edition, 1994.

[Bor94]   F. Borceux. *Handbook of Categorical Algebra I – III*. Cambridge University Press, 1994.

[Bor98]   Tomasz Borzyszkowski. Completeness of a logical system for structured specifications. In F. Parisi Presicce, editor, *Recent Trends in Algebraic Development Techniques. Selected Papers, 12th Intl. Workshop, WADT'97*, volume 1376 of *Lecture Notes in Computer Science*, pages 107–121. Springer, 1998.

[Bou94]   A. Bouhoula. *Preuves automatiques par récurrence dans les théories conditionnelles*. Thèse de Doctorat d'Université, Université Henri Poincaré – Nancy 1, 1994.

[Bou96]   A. Bouhoula. Using induction and rewriting to verify and complete parameterized specifications. *Theoretical Computer Science*, 170(1-2), 1996.

[Bou97]   A. Bouhoula. Automated theorem proving by test set induction. *Journal of Symbolic Computation*, 23(1):47–77, 1997.

[BPW89]   T.B. Baird, G.E. Peterson, and R.W. Wilkerson. Complete sets of reductions modulo associativity, commutativity and identity. In N. Dershowitz, editor, *Proc. 3rd Conference on Rewriting Techniques and Applications*, volume 355 of *Lecture Notes in Computer Science*, pages 29–44. Springer, 1989.

[BR83]   K. Benecke and H. Reichel. Equational partiality. *Algebra Universalis*, 16:219–232, 1983.

[BR90]   W. Bousdira and J.-L. Rémy. On sufficient completeness of conditional specifications. In S. Kaplan and M. Okada, editors, *Proc. 2nd Intl. Workshop on Conditional and Typed Rewriting Systems*, Lecture Notes in Computer Science. Springer, 1990.

[BR93]   A. Bouhoula and M. Rusinowitch. Automatic case analysis in proof by induction. In R. Bajcsy, editor, *Proc. 13th Intl. Joint Conference on Artificial Intelligence, Chambéry (France)*, volume 1, pages 88–94. Morgan Kaufmann, 1993.

[BR95a]   A. Bouhoula and M. Rusinowitch. Implicit induction in conditional theories. *Journal of Automated Reasoning*, 14(2):189–235, 1995.

[BR95b]   A. Bouhoula and M. Rusinowitch. Spike: a system for automatic inductive proofs. In V. Alagar and M. Nivat, editors, *Proc. 4th Intl. Conference on Algebraic Methodology and Software Technology*, volume 936 of *Lecture Notes in Computer Science*. Springer, 1995.

[Bre91]   Ruth Breu. *Algebraic Specification Techniques in Object Oriented Programming Environments*, volume 562 of *Lecture Notes in Computer Science*. Springer, 1991.

[Bri89]   E. Brinksma. A theory for the derivation of tests. In *The Formal Description Technique LOTOS*, pages 235–247. Elsevier Science (North-Holland), 1989.

[Bro75]   T. Brown. *A structured design-method for specialized proof procedures*. PhD thesis, California Institute of Technology, Pasadena, California, 1975.

[Bro87]   M. Broy. Semantics of finite or infinite networks of communicating agents. *Distributed Computing*, 2, 1987.

[Bro93]   M. Broy. Interaction refinement: The easy way. In M. Broy, editor, *Program Design Calculi*, volume 118 of *NATO ASI Series F: Computer and System Sciences*. Springer, 1993.

[Bro96]   M. Broy. Specification and refinement of a buffer of length one. In M. Broy, editor, *Deductive Program Design*, volume 152 of *NATO ASI Series F: Computer and System Sciences*. Springer, 1996.

[Bro97]   M. Broy. Compositional refinement of interactive system. *Journal of the Association for Computing Machinery*, 44(6), 1997.

[BS93]   R. Berghammer and G. Schmidt. Relational specifications. In *Algebraic Methods in Logic and Computer Science*, volume 28 of *Banach Center Publications*, pages 167–190. Institute of Mathematics, Polish Academy of Sciences, Warsaw, 1993.

[BST99]   M. Bidoit, D. Sannella, and A. Tarlecki. Architectural specifications in CASL. In A.M. Haeberer, editor, *Proc. 7th Intl. Conference on Algebraic Methodology and Software Technology (AMAST'98)*, volume 1548 of *Lecture Notes in Computer Science*. Springer, 1999.

[BSvH⁺93]   A. Bundy, A. Stevens, F. van Harmelen, A. Ireland, and A. Smaill. Rippling: A heuristic for guiding inductive proofs. *Journal of Artificial Intelligence*, 62:185–253, 1993.

[BT87]   J.A. Bergstra and J.V. Tucker. Algebraic specifications of computable and semicomputable data types. *Theoretical Computer Science*, 50:137–181, 1987.

[Bur69]   R.M. Burstall. Proving properties of programs by structural induction. *Comput. J.*, 12:41–48, 1969.

[Bur82]   P. Burmeister. Partial algebras – survey of a unifying approach towards a two-valued model theory for partial algebras. *Algebra Universalis*, 15:306–358, 1982.

[Bur86]   P. Burmeister. *A Model Theoretic Oriented Approach to Partial Algebras*. Akademie-Verlag, Berlin, 1986.

[BW82a]   Friedrich L. Bauer and Hans Wössner. *Algorithmic Language and Program Development*. Texts and Monographs in Computer Science. Springer, 1982.

[BW82b]   M. Broy and M. Wirsing. Partial abstract types. *Acta Informatica*, 18(1):47–64, 1982.

[BW90]   J.C.M. Baeten and W.P. Weijland. *Process Algebra*. Cambridge University Press, 1990.

[BW95]   M. Barr and Ch. Wells. *Category Theory for Computing Science*. Prentice Hall, New York and London, second edition, 1995.

[C⁺86]   Robert L. Constable et al., editors. *Implementing Mathematics with the Nuprl Proof Development System*. Prentice Hall, 1986.

[CAB⁺94]   D. Coleman, P. Arnold, S. Bodoff, C. Dollin, H. Gilchrist, F. Hayes, and P. Jeremaes. *Object-oriented Development – The Fusion Method*. Prentice Hall, 1994.

[Cal96]   C. Caleiro. Distributed object communities. Master's thesis, Instituto Superior Técnico, Universidade Técnica de Lisboa, Av. Rovisco Pais, 1096 Lisboa Codex, Portugal, 1996. Supervised by A. Sernadas. In Portuguese.

[CELM96]   M. Clavel, S. Eker, P. Lincoln, and J. Meseguer. Principles of Maude. In J. Meseguer, editor, *Proc. 1st Intl. Workshop on Rewriting Logic*, volume 4 of *Electronic Notes in Theoretical Computer Science*. North-Holland, 1996.

[Cen94]   María Victoria Cengarle. *Formal Specifications with Higher-Order Parameterization*. PhD thesis, Institut für Informatik, Ludwig-Maximilians-Universität München, 1994.

[Cen95]   María Victoria Cengarle. Semantic typing for parametric algebraic specifications. In V.S. Alagar and M. Nivat, editors, *Proc. Algebraic Methodology and Software Technology (AMAST'95)*, volume 936 of *Lecture Notes in Computer Science*, pages 261–276. Springer, 1995.

[CEOB]   F. Cornelius, H. Ehrig, F. Orejas, and M. Baldamus. Abstract and behaviour module specifications. *Mathematical Structures in Computer Science*. To appear.

[Cer93]   M. Cerioli. *Relationships between Logical Formalisms*. PhD thesis, Universities of Genova, Pisa and Udine, 1993. Available as internal report of Pisa University, TD-4/93 or by anonymous ftp at ftp.disi.unige.it in /pub/cerioli/thesis92.ps.z.

[Cer95]   M. Cerioli. A lazy approach to partial algebras. In E. Astesiano, G. Reggio, and A. Tarlecki, editors, *Recent Trends in Data Type Specification: 10th Workshop on Specification of Abstract Data Types – Selected Papers*, volume 906 of *Lecture Notes in Computer Science*, pages 188–202. Springer, 1995.

[CEW93]   I. Claßen, H. Ehrig, and D. Wolz. *Algebraic Specification Techniques and Tools for Software Development*. AMAST Series in Computing, Vol. 1. World Scientific, 1993.

[CF92]   J.R.B. Cockett and T. Fukushima. About charity. Technical Report
           92/480/18, Department of Computer Science, University of Calgary,
           1992.

[CG91]   P.-L. Curien and G. Ghelli. On confluence for weakly normalizing sys-
           tems. In R.V. Book, editor, *Proc. 4th Conference on Rewriting Tech-
           niques and Applications*, volume 488 of *Lecture Notes in Computer Sci-
           ence*, pages 215–225. Springer, 1991.

[CGR93]   D. Craigen, S. Gerhart, and T. Ralston. On the use of formal methods
           in industry – an authoritative assessment of the efficacy, utility, and
           applicability of formal methods to systems design and engineering by
           the analysis of real industrial cases. Report to the us national institute
           of standards and technology, 1993.

[CGW95]   I. Claßen, M. Große-Rhode, and U. Wolter. Categorical concepts for
           parameterized partial specifications. *Mathematical Structures in Com-
           puter Science*, 5(2):153–188, 1995.

[CH88]   Thierry Coquand and Gérard Huet. The Calculus of Constructions.
           Information and Computation, pages 95–120, 1988.

[CH90]   Robert L. Constable and Douglas J. Howe. Implementing metamathe-
           matics as an approach to automatic theorem proving. In *A source book
           of formal approaches to A.I.* North-Holland, 1990.

[CHKM97]   M. Cerioli, A. Haxthausen, B. Krieg-Brückner, and T. Mossakowski.
           Permissive subsorted partial logic in CASL. In Michael Johnson, editor,
           Algebraic Methodology and Software Technology (AMAST'97), volume
           1349 of *Lecture Notes in Computer Science*, pages 91–107. Springer,
           1997.

[CHL96]   P.-L. Curien, T. Hardin, and J.-J. Lévy. Confluence properties of weak
           and strong calculi of explicit substitutions. *Journal of the Association
           for Computing Machinery*, 43(2):362–397, 1996.

[Chr92]   J. Christian. Some termination criteria for narrowing and E-narrowing.
           In D. Kapur, editor, *Proc. 11th Intl. Conference on Automated Deduc-
           tion*, volume 607 of *Lecture Notes in Computer Science*, pages 582–588.
           Springer, 1992.

[Chu40]   Alonzo Church. A formulation of the simple theory of types. *Symbolic
           Logic*, 5(1):56–68, 1940.

[CJ91]   J.H. Cheng and C.B. Jones. On the usability of logics which handle
           partial functions. In C. Morgan and J.C.P. Woodcock, editors, *Proc.
           3rd Refinement Workshop*, Workshops in Computing series, pages 51–
           69. Springer, 1991.

[CL73]   Chin-Liang Chang and Richard Char-Tung Lee, editors. *Symbolic logic
           and mechanical theorem proving*. Academic Press, 1973.

[CL87]   A. Ben Cherifa and P. Lescanne. Termination of rewriting systems by
           polynomial interpretations and its implementation. *Science of Com-
           puter Programming*, 9(2):137–160, 1987.

[CL92]   E.A. Cichon and P. Lescanne. Polynomial interpretations and the com-
           plexity of algorithms. In D. Kapur, editor, *Proc. 11th Intl. Conference
           on Automated Deduction*, volume 607 of *Lecture Notes in Computer
           Science*, pages 139–147. Springer, 1992.

[Cla89]   I. Claßen. Revised ACT ONE: categorical constructions for an algebraic
           specification language. In *Proc. Workshop on Categorical Methods in*

Computer Science with Aspects from Topology, volume 393 of *Lecture Notes in Computer Science*, pages 124–141. Springer, 1989.

[Cle91]   J.P. Cleave. *A Study of Logics*. Oxford University Press, 1991.

[Cle92]   Tim Clement. Using metavariables in natural deduction proofs. In Cliff B. Jones, Roger C. Shaw, and Tim Denvir, editors, *5th Refinement Workshop*, Workshops in computing, London, pages 255–271. Springer, 1992.

[CM93]   M. Cerioli and J. Meseguer. May I borrow your logic? In *Proc. 18th Intl. Symp. on Mathematical Foundations of Computer Science MFCS'93*, volume 711 of *Lecture Notes in Computer Science*, pages 342–351. Springer, 1993.

[CM97]   M. Cerioli and J. Meseguer. May I borrow your logic? (transporting logical structures along maps). *Theoretical Computer Science*, 173(2):311–347, 1997.

[CO88]   S. Clérici and F. Orejas. GSBL: an algebraic specification language based on inheritance. In *Proc. 1988 European Conference on Object Oriented Programming*, volume 322 of *Lecture Notes in Computer Science*, pages 78–92. Springer, 1988.

[Coe92]   Martin David Coen. Interactive program derivation. Technical Report 272, Cambridge University Computer Laboratory, 1992.

[CoF98]   The CoFI Task Group on Language Design. CASL – the common algebraic specification language – summary (version 1.0). Available at http://www.brics.dk/Projects/CoFI/Documents/CASL/Summary/, 1998.

[Coh80]   Avra Cohn. *Machine Assisted Proofs of Recursion Implementation*. PhD thesis, University of Edinburgh, 1980.

[Com86]   H. Comon. Sufficient completeness, term rewriting system and anti-unification. In J. Siekmann, editor, *Proc. 8th Intl. Conference on Automated Deduction*, volume 230 of *Lecture Notes in Computer Science*, pages 128–140. Springer, 1986.

[Com92]   H. Comon. Completion of rewrite systems with membership constraints. In W. Kuich, editor, *Proc. ICALP'92*, volume 623 of *Lecture Notes in Computer Science*. Springer, 1992.

[Com98a]   H. Comon. Completion of rewrite systems with membership constraints. Part I: Deduction rules. *Journal of Symbolic Computation*, 25(4):397–419, 1998.

[Com98b]   H. Comon. Completion of rewrite systems with membership constraints. Part II: Constraint solving. *Journal of Symbolic Computation*, 25(4):421–453, 1998.

[CR94]   M. Cerioli and G. Reggio. Institutions for very abstract specifications. In H. Ehrig and F. Orejas, editors, *Recent Trends in Data Type Specification*, volume 785 of *Lecture Notes in Computer Science*, pages 113–127. Springer, 1994.

[CR96]   E. Coscia and G. Reggio. Deontic concepts in the algebraic specification of dynamic systems: The permission case. In O.-J. Dahl, M. Haveraan, and O. Owe, editors, *Recent Trends in Data Types Specification, Proc. 11th Workshop on Specification of Abstract Data Types joint with the 8th General COMPASS Meeting*, volume 1130 of *Lecture Notes in Computer Science*, pages 182–199. Springer, 1996.

[CR97]   G. Costa and G. Reggio. Specification of abstract dynamic data types: A temporal logic approach. *Theoretical Computer Science*, 173(2), 1997.

[CS92]   J.R.B. Cockett and D. Spencer. Strong categorical data types i. In R.A.G. Seely, editor, *Intl. Meeting on Category Theory 1991*, Canadian Mathematical Society Proceedings, Montreal, 1992. AMS.

[CS95]   J.R.B. Cockett and D. Spencer. Strong categorical data types ii: A term logic for categorical programming. *Theoretical Computer Science*, 139:69–113, 1995.

[CSS94]   J.F. Costa, A. Sernadas, and C. Sernadas. Object inheritance beyond subtyping. *Acta Informatica*, 31:5–26, 1994.

[CW95]   María Victoria Cengarle and Martin Wirsing. A calculus of higher-order parameterization for algebraic specifications. *Bulletin of the Interest Group in Pure and Applied Logics (IGPL)*, 3(4):615–641, 1995. Special Issue 'Workshop on Logic, Language, Information and Computation 1994'.

[Dav65]   M. Davis, editor. *The Undecidable*. Raven Press, New York, 1965.

[Dav93]   A.M. Davis, editor. *Software requirements, objects, functions and states*. Prentice-Hall, 1993.

[dB72]   N.G. de Bruijn. Lambda calculus notation with nameless dummies, a tool for automatic formula manipulation, with application to the Church-Rosser theorem. *Indagationes Mathematicae*, 34:381–392, 1972.

[dB80]   N.G. de Bruijn. A survey of the project Automath. In *Essays in Combinatory Logic, Lambda Calculus, and Formalism*, pages 589–606. Academic Press, 1980.

[DDPW94]   E. Dubois, Ph. Du Bois, M. Petit, and S. Wu. ALBERT: A formal agent-oriented requirements language for distributed composite systems. In E. Dubois, P. Hartel, and G. Saake, editors, *Proc. Workshop on Formal Methods for Information System Dynamics (CAiSE'94-Workshop)*, pages 25–39. University of Twente, 1994. Technical Report.

[DE95]   G. Denker and H.-D. Ehrich. Action reification in object-oriented specification. In R.J. Wieringa and R.B. Feenstra, editors, *Information Systems—Correctness and Reusability, Selected Papers from the IS-CORE Workshop'94*, pages 103–118. World Scientific, 1995.

[Den80]   J.B. Dennis. Data flow supercomputers. *IEEE Computer*, 13(11), 1980.

[Den96a]   G. Denker. Reification—changing viewpoint but preserving truth. In O.-J. Dahl, M. Haveraan, and O. Owe, editors, *Recent Trends in Data Types Specification, Proc. 11th Workshop on Specification of Abstract Data Types joint with the 8th General COMPASS Meeting*, volume 1130 of *Lecture Notes in Computer Science*, pages 182–199. Springer, 1996.

[Den96b]   G. Denker, editor. *Verfeinerung in objektorientierten Spezifikationen: Von Aktionen zu Transaktionen*. Reihe DISDBIS, Band 6. infix-Verlag, Sankt Augustin, 1996.

[Der82]   N. Dershowitz. Orderings for term-rewriting systems. *Theoretical Computer Science*, 17:279–301, 1982.

[Der83]   N. Dershowitz. Applications of the Knuth-Bendix completion procedure. Technical Report ATR-83(8478)-2, The Aerospace Corporation, El Segundo, Calif. 90245, 1983.

[Der87]   N. Dershowitz. Termination of rewriting. *Journal of Symbolic Computation*, 3(1-2):69–116, 1987.

[Der89]   N. Dershowitz. Completion and its applications. In H. Aït-Kaci and M. Nivat, editors, *Resolution of Equations in Algebraic Structures, Vol. 2: Rewriting Techniques*, pages 31–86. Academic Press Inc., 1989.

[DF93]   J. Dick and A. Faivre. Automating the generation and sequencing of test cases from model-based specification. In *Proc. FME'93*, volume 670 of *Lecture Notes in Computer Science*, pages 268–284. Springer, 1993.

[DF98]   R. Diaconescu and K. Futatsugi. *CafeOBJ Report: The Language, Proof Techniques, and Methodologies for Object-Oriented Algebraic Specification*. World Scientific, 1998.

[DG93]   P. Dauchy and M.-C. Gaudel. Implicit state in algebraic specifications. In U.W. Lipeck and G. Koschorreck, editors, *Proc. Intl. Workshop Is-Core'93, Hannover (Germany)*. Universität Hannover, 1993.

[DGM93]   P. Dauchy, M.-C. Gaudel, and B. Marre. Using algebraic specifications in software testing: a case study on the software of an automatic subway. *Journal of Systems and Software*, 21(3):229–244, 1993.

[DGS91]   R. Diaconescu, J.A. Goguen, and P. Stefaneas. Logical support for modularisation. Report prog. res. group, Oxford University, 1991.

[DGS93]   Răzvan Diaconescu, Joseph Goguen, and Petros Stefaneas. Logical support for modularisation. In Gérard Huet and Gordon Plotkin, editors, *Logical Environments, Proc. Workshop on Logical Frameworks '91*, pages 83–130. Cambridge University Press, 1993.

[DH91]   C. Dimitrovici and U. Hummert. Composition of algebraic high-level nets. In H. Ehrig, K.P. Jantke, F. Orejas, and H. Reichel, editors, *Recent Trends in Data Type Specification, Proc. 7th Workshop on Specification of Abstract Data Types*, volume 534 of *Lecture Notes in Computer Science*. Springer, 1991.

[DH95]   N. Dershowitz and C. Hoot. Natural termination. *Theoretical Computer Science*, 142(2):179–207, 1995.

[DH97]   G. Denker and P. Hartel. TROLL—an object oriented formal method for distributed information system design: Syntax and pragmatics. Informatik-Berichte 97-03, Technische Universität Braunschweig, 1997.

[DHLT87]   M. Dauchet, T. Heuillard, P. Lescanne, and S. Tison. Decidability of the confluence of ground term rewriting systems. In *Proc. 2nd IEEE Symposium on Logic in Computer Science, Ithaca, N.Y. (USA)*, pages 353–359, 1987.

[DJ90]   Nachem Dershowitz and Jean-Pierre Jouannaud. Rewrite systems. In J. van Leeuwen, editor, *Handbook of Theoretical Computer Science, Chapter 6*, pages 244–320. Elsevier Science (North-Holland), 1990.

[DJ91]   N. Dershowitz and J.-P. Jouannaud. Notations for rewriting. *Bulletin of European Association for Theoretical Computer Science*, 43:162–172, 1991.

[DL93]   Yves Deville and Kung-Kiu Lau. Logic program synthesis. *Journal of Logic Programming*, 12, 1993.

[DMN67]   O.-J. Dahl, B. Myrhaug, and K. Nygaard, editors. SIMULA 67, *Common Base Language*. Norwegian Computer Center, Oslo, 1967.

[DO90]   N. Dershowitz and M. Okada. A rationale for conditional equational programming. *Theoretical Computer Science*, 75:111–138, 1990.

[DOS88a] N. Dershowitz, M. Okada, and G. Sivakumar. Canonical conditional rewrite systems. In E. Lusk and R. Overbeek, editors, *Proc. 9th Intl. Conference on Automated Deduction (CADE'88)*, volume 310 of *Lecture Notes in Computer Science*, pages 538–549. Springer, 1988.

[DOS88b] N. Dershowitz, M. Okada, and G. Sivakumar. Confluence of conditional rewrite systems. In *Proc. Intl. Conf. on Conditional Term Rewriting Systems (CTRS'87)*, volume 308 of *Lecture Notes in Computer Science*, pages 31–44. Springer, 1988.

[DS88] Nachem Dershowitz and G. Sivakumar. Goal-directed equation solving. In *Proc. 7th National Conference on Artificial Intelligence, St. Paul, Minn. (USA)*, pages 166–170, 1988.

[EBC91] H. Ehrig, M. Baldamus, and F. Cornelius. Theory of algebraic module specification including behavioural semantics, constraints and aspects of generalized morphisms. In *Proc. 2nd Intl. Conf. Algebraic Methodology and Software Technology AMAST'91, Iowa City, Io. (USA)*, pages 101–125, 1991.

[EBCO91] H. Ehrig, M. Baldamus, F. Cornelius, and F. Orejas. Theory of algebraic module specifications including behavioural semantics and constraints. In *Proc. AMAST'91*, 1991.

[Ech88] R. Echahed. On completeness of narrowing strategies. In M. Dauchet and M. Nivat, editors, *Proc. CAAP'88*, volume 299 of *Lecture Notes in Computer Science*, pages 89–101. Springer, 1988.

[ECSD98] H.-D. Ehrich, C. Caleiro, A. Sernadas, and G. Denker. Logics for specifying concurrent information systems. In G. Saake and J. Chomicki, editors, *Logics for Databases and Information Systems*. Kluwer Publ. Comp., 1998. To appear.

[EDS93] H.-D. Ehrich, G. Denker, and A. Sernadas. Constructing systems as object communities. In M.-C. Gaudel and J.-P. Jouannaud, editors, *Proc. Theory and Practice of Software Development (TAPSOFT'93)*, volume 668 of *Lecture Notes in Computer Science*, pages 453–467. Springer, 1993.

[EFH83] H. Ehrig, W. Fey, and H. Hansen. ACT ONE: An algebraic specification language with two levels of semantics. Technical Report 83-01, Technische Universität Berlin, 1983.

[EG94] H. Ehrig and M. Große-Rhode. Functorial theory of parameterized specifications in a general specification framework. *Theoretical Computer Science*, 135:221–226, 1994.

[EGS90] H.-D. Ehrich, J.A. Goguen, and A. Sernadas. A categorial theory of objects as observed processes. In W.-P. de Roever J.W.de Bakker and G. Rozenberg, editors, *Proc. REX/FOOL Workshop*, volume 489 of *Lecture Notes in Computer Science*, pages 203–228. Springer, 1990.

[EH96] H.-D. Ehrich and P. Hartel. Temporal specification of information systems. In A. Pnueli and H. Lin, editors, *Proc. Intl. Workshop in Honor of Chih-Sung Tang*, pages 43–70, Singapore, 1996. World Scientific.

[Ehr78] H.-D. Ehrich. Extension and implementation of abstract data types. In J. Winkowski, editor, *Proc. 7th Symp. on Math. Foundations of Computer Science*, volume 64 of *Lecture Notes in Computer Science*. Springer, 1978.

[Ehr82]   H.-D. Ehrich. On the theory of specification, implementation and parameterization of abstract data types. *Journal of the Association for Computing Machinery*, 29:209–277, 1982.

[EJD93]   H.-D. Ehrich, R. Jungclaus, and G. Denker. Object roles and phases. In U.W. Lipeck and G. Koschorreck, editors, *Proc. Intern. Workshop on Information Systems – Correctness and Reusability IS-CORE'93*, pages 114–121. University of Hannover, 1993. Technical Report No. 01/93.

[EJDS94]   H.-D. Ehrich, R. Jungclaus, G. Denker, and A. Sernadas. Object-oriented design of information systems: Theoretical foundations. In J. Paredaens and L. Tenenbaum, editors, *Advances in Database Systems, Implementations and Applications*, CISM Courses and Lectures No. 347, pages 201–218. Springer, 1994.

[EK83]   H. Ehrig and H.-J. Kreowski. Compatibility of parameter passing and implementation of parameterized data types. *Theoretical Computer Science*, 27(3):255–286, 1983.

[EKMP82]   H. Ehrig, H.-J. Kreowski, B. Mahr, and P. Padawitz. Algebraic implementation of abstract data types. *Theoretical Computer Science*, 20:209–263, 1982.

[EKO95]   H. Ehrig, H.-J. Kreowski, and F. Orejas. Correctness of actualization for parameterized implementation concepts based on constructors and abstractors. *EATCS Bulletin*, 56:79–85, 1995.

[EKO97]   H. Ehrig, H.-J. Kreowski, and F. Orejas. Correctness of horizontal and vertical composition for implementation concepts based on constructors and abstractors. *REVISTA MATEMATICA de la Universidad Complutense de Madrid*, 10(2):365–387, 1997.

[EKP78]   H. Ehrig, H.-J. Kreowski, and P. Padawitz. Stepwise specification and implementation of abstract data types. In *Proc. 5th Intl. Colloq. on Automata, Languages and Programming*, volume 62 of *Lecture Notes in Computer Science*, pages 205–226. Springer, 1978.

[EKT+84]   H. Ehrig, H.-J. Kreowski, J.W. Thatcher, E. Wagner, and J. Wright. Parameter passing in algebraic specification languages. *Theoretical Computer Science*, 28:45–81, 1984.

[EL80]   H.-D. Ehrich and U. Lipeck. Proving implementation correct – two alternative approaches. *Information Processing Letters*, 80:83–88, 1980.

[EM85]   Hartmut Ehrig and Bernd Mahr. *Fundamentals of Algebraic Specification 1: Equations and Initial Semantics*, volume 6 of *EATCS Monographs on Theoretical Computer Science*. Springer, 1985.

[EM90]   H. Ehrig and B. Mahr. *Fundamentals of Algebraic Specification 2: Module Specifications and Constraints*, volume 21 of *EATCS Monographs on Theoretical Computer Science*. Springer, 1990.

[Eme90]   A.E. Emerson. Temporal and modal logic. In J. van Leeuwen, editor, *Handbook of Theoretical Computer Science, Vol. B*. Elsevier Science (North-Holland), 1990.

[EO94]   H. Ehrig and F. Orejas. Dynamic abstract data types: An informal proposal. *EATCS Bulletin*, 53, 1994.

[EP91]   H. Ehrig and F. Parisi-Presicce. Algebraic specification graph-grammars: A junction between module specifications and graph grammars. In H. Ehrig, H.-J. Kreowski, and G. Rozenberg, editors, *Graph-*

Grammars and Their Application to Computer Science, volume 532 of *Lecture Notes in Computer Science*, pages 292–310. Springer, 1991.

[EPO89]   H. Ehrig, P. Pepper, and F. Orejas. On recent trends in algebraic specification. In *Proc. ICALP'89*, volume 372 of *Lecture Notes in Computer Science*, pages 263–288. Springer, 1989.

[ES90]   H.-D. Ehrich and A. Sernadas. Algebraic implementation of objects over objects. In J.W. de Bakker, W.-P. de Roever, and G. Rozenberg, editors, *Proc. REX Workshop "Stepwise Refinement of Distributed Systems: Models, Formalisms, Correctness"*, volume 430 of *Lecture Notes in Computer Science*, pages 239–266. Springer, 1990.

[ES95]   H.-D. Ehrich and A. Sernadas. Local specification of distributed families of sequential objects. In E. Astesiano, G. Reggio, and A. Tarlecki, editors, *Recent Trends in Data Type Specification: 10th Workshop on Specification of Abstract Data Types joint with the 5th COMPASS Workshop*, volume 906 of *Lecture Notes in Computer Science*, pages 218–235. Springer, 1995.

[ESD93]   Espírito Santo Data Informática (ESDI), Lisbon. OBLOG CASE V1.0 – *The User's Guide*, 1993.

[ETLZ82] H. Ehrig, J.W. Thatcher, P. Lucas, and S.N. Zilles. Denotational and initial algebra semantics of the algebraic specification language LOOK. Draft report, IBM Research, 1982.

[Eva51]   T. Evans. On multiplicative systems defined by generators and relations. In *Proc. Cambridge Philosophical Society*, pages 637–649, 1951.

[EWT83]   H. Ehrig, E.G. Wagner, and J.W. Thatcher. Algebraic specifications with generating constraints. In *Proc. ICALP'83*, volume 154 of *Lecture Notes in Computer Science*. Springer, 1983.

[Far91]   W.M. Farmer. A partial functions version of Church's simple type theory. *Journal of Symbolic Logic*, 55:1269–1291, 1991.

[Far92]   Jordi Farrés-Casals. *Verification in ASL and Related Specification Languages*. PhD thesis, University of Edinburgh, 1992. Report CST-92-92.

[Far93]   W.M. Farmer. A simple type theory with partial functions and subtypes. *Annals of Pure and Applied Logic*, 64:211–240, 1993.

[Far95]   W.M. Farmer. Reasoning about partial functions with the aid of a computer. *Erkenntnis*, 43:279–294, 1995.

[Fay79]   M. Fay. First order unification in equational theories. In *Proc. 4th Workshop on Automated Deduction, Austin, Tex. (USA)*, pages 161–167, 1979.

[Fea82]   Martin S. Feather. A system for assisting program transformation. *ACM Transactions on Programming Languages and Systems*, 4:1–20, 1982.

[Fef88]   Solomon Feferman. Finitary inductively presented logics. In *Logic Colloquium '88*. North-Holland, 1988.

[Fef92]   S. Feferman. A new approach to abstract data types, I informal development. *Mathematical Structures in Computer Science*, 2:193–229, 1992.

[Fef95]   S. Feferman. Definedness. *Erkenntnis*, 43:295–320, 1995.

[Fei89]   L.M.G. Feijs. The calculus $\lambda\pi$. In M. Wirsing and J.A. Bergstra, editors, *Algebraic Methods: Theory, Tools and Applications*, volume 394 of *Lecture Notes in Computer Science*, pages 307–328. Springer, 1989.

[Fey88]   W. Fey. *Pragmatics, Concepts, Syntax, Semantics, and Correctness Notions of ACT TWO: An Algebraic Module Specification and Interconnection Language.* PhD thesis, Technische Universität Berlin, 1988. Report 88/26.

[FHV92]   Ronald Fagin, Joeseph Y. Halpern, and Mosche Y. Vardi. What is an inference rule? *Journal of Symbolic Logic*, 57(3):1018–1045, 1992.

[FJ90]   J.S. Fitzgerald and C.B. Jones. *Modularizing the formal description of a database system*, volume 428 of *Lecture Notes in Computer Science*. Springer, 1990.

[FJ92]   L.M.G. Feijs and H.B.M. Jonkers. *Formal specification and design.* Cambridge University Press, 1992.

[FJL97]   J. Fitzgerald, C.B. Jones, and P. Lucas, editors. *Proc. FME'97, Industrial applications and strenghtened foundations of formal methods*, volume 1313 of *Lecture Notes in Computer Science*. Springer, 1997.

[FM88]   Amy Felty and Dale Miller. Specifying theorem provers in a higher-order logic programming language. In *9th Intl. Conference on Automated Deduction, Argonne, Ill. (USA)*, 1988.

[FM92]   J. Fiadeiro and T. Maibaum. Temporal theories as modularisation units for concurrent system specification. *Formal Aspects of Computing*, 4:239–272, 1992.

[Fri85a]   L. Fribourg. Handling function definitions through innermost superposition and rewriting. In J.-P. Jouannaud, editor, *Proc. 1st Conference on Rewriting Techniques and Applications (RTA'85)*, volume 202 of *Lecture Notes in Computer Science*, pages 325–344. Springer, 1985.

[Fri85b]   L. Fribourg. SLOG: A logic programming language interpreter based on clausal superposition and rewriting. In *IEEE Symposium on Logic Programming, Boston, Mass. (USA)*, 1985.

[Fri86]   L. Fribourg. A strong restriction of the inductive completion procedure. In *Proc. 13th Intl. Colloquium on Automata, Languages and Programming*, volume 226 of *Lecture Notes in Computer Science*, pages 105–115. Springer, 1986.

[FS88]   J. Fiadeiro and A. Sernadas. Structuring theories on consequence. In *Recent Trends in Data Type Specification, Proc. 5th Workshop on Specification of Abstract Data Types ADT'88*, volume 332 of *Lecture Notes in Computer Science*, pages 44–72. Springer, 1988.

[FS97]   M. Fowler and K. Scott. *UML Distilled: Applying the Standard Object Modeling Language.* Addison-Wesley, New York, 1997.

[FW93]   R.B. Feenstra and R. Wieringa. LCM 3.0: A language for describing conceptual models – syntax definition. Rapport IR-344, Vrije Universiteit Amsterdam, 1993.

[G⁺93]   D.M. Gabbay et al. *Handbook of Logic in Artificial Intelligence and Logic Programming, Vol. 1: Logical Foundations.* Oxford Science Publications, 1993.

[Gan83]   H. Ganzinger. Parameterized specifications: parameter passing and implementation with respect to observability. *ACM Transactions on Programming Languages and Systems*, 5(3):318–354, 1983.

[Gan89]   H. Ganzinger. Order-sorted completion: the many-sorted way. In *Proc. Intl. Joint Conference on Theory and Practice of Software Development: Colloquium on Software Engineering*, Lecture Notes in Computer Science. Springer, 1989.

[Gan91]   Harald Ganzinger. A completion procedure for conditional equations. *Journal of Symbolic Computation*, 11:51–81, 1991.

[Gar93]   Philippa A. Gardner. A new type theory for representing logics. In A. Voronkov, editor, *Proc. 4th Intl. Conference on Logic Programming and Automated Reasoning*, volume 698 of *Lecture Notes on Artificial Intelligence*. Springer, 1993.

[Gau86]   M.-C. Gaudel. Towards structured algebraic specifications. In *ESPRIT 85 Status Report of Continuing Work*, pages 493–510. North-Holland, 1986.

[Gau92]   M.-C. Gaudel. Structuring and modularizing algebraic specifications: the PLUSS specification language, evolutions and perspectives. In *9th Annual Symposium on Theoretical Aspects of Computer Science (STACS'92)*, volume 577 of *Lecture Notes in Computer Science*, pages 3–18. Springer, 1992.

[Gau95]   M.-C. Gaudel. Advantages and limits of formal approaches for ultra-high dependability. In Randell et al., editors, *Predictably Dependable Computing Systems, Chapter IV-A*, Springer Basic Research Series, pages 241–251. Springer, 1995.

[GB80]   J.A. Goguen and R.M. Burstall. CAT, a system for the structured elaboration of correct programs from structured specifications. Technical Report CSL-118, Computer Science Laboratory, SRI International, 1980.

[GB84]   J.A. Goguen and R.M. Burstall. Introducing institutions. In E. Clarke and D. Kozen, editors, *Logics of Programs*, volume 164 of *Lecture Notes in Computer Science*, pages 221–256. Springer, 1984.

[GB86]   J. Goguen and R. Burstall. A study in the foundations of programming methodology: specifications, institutions, charters and parchments. In *Proc. Intl. Workshop on Category Theory and Computer Programming*, volume 240 of *Lecture Notes in Computer Science*, pages 313–333. Springer, 1986.

[GB92]   J.A. Goguen and R.M. Burstall. Institutions: abstract model theory for specification and programming. *Journal of the Association for Computing Machinery*, 39(1):95–146, 1992.

[GD90]   D. Garlan and N. Delisle. *Formal specifications as reusable framework*, volume 428 of *Lecture Notes in Computer Science*. Springer, 1990.

[GD92]   J.A. Goguen and R. Diaconescu. A short survey of order-sorted algebra. *EATCS Bulletin*, 49:121–133, 1992.

[GD94a]   J. Goguen and R. Diaconescu. An Oxford survey of order sorted algebra. *Mathematical Structures in Computer Science*, 4:363–392, 1994.

[GD94b]   J.A. Goguen and R. Diaconescu. Towards an algebraic semantics for the object paradigm. In H. Ehrig and F. Orejas, editors, *Recent Trends in Data Type Specification*, volume 785 of *Lecture Notes in Computer Science*, pages 1–29. Springer, 1994.

[GDK96]   M.-C. Gaudel, P. Dauchy, and C. Khoury. A formal specification of the steam-boiler control problem by algebraic specification with implicit state. In J.-R. Abrial, E. Borger, and H. Langmaack, editors, *Formal Methods for Industrial Applications*, volume 1165 of *Lecture Notes in Computer Science*. Springer, 1996.

[GDLE84] M. Gogolla, K. Drosten, U. Lipeck, and H.-D. Ehrich. Algebraic and operational semantics of specifications allowing exceptions and errors. *Theoretical Computer Science*, 34:289–313, 1984.

[GG90] S.J. Garland and J.V. Guttag. Using lp to debug specifications. In *IFIP TC2 Working Conference on Programming Concepts and Methods*. North-Holland, 1990.

[GG94] Dov Gabbay and Franz Guenthner, editors. *What is a Logical System?* Oxford University Press, 1994.

[GGM76] V. Giarratana, F. Gimona, and U. Montanari. Observability concepts in abstract data type specifications. In A.W. Mazurkiewicz, editor, *Proc. Mathematical Foundations of Computer Science (MFCS'76)*, volume 45 of *Lecture Notes in Computer Science*, pages 576–587. Springer, 1976.

[GH78] J.V. Guttag and J.J. Horning. The algebraic specification of abstract data types. *Acta Informatica*, 10:27–52, 1978.

[GH86] J. Guttag and J. Horning. Report on the Larch shared language. *Science of Computer Programming*, 6:103–134, 1986.

[GH90] G. Guiho and C. Hennebert. SACEM software validation. In *12th IEEE-ACM Intl. Conference on Software Engineering*, pages 186–191, 1990.

[GH93] J.V. Guttag and J.J. Horning. *Larch: Languages and Tools for Formal Specification*. Springer, 1993.

[GHM78] J.V. Guttag, E. Horowitz, and D.R. Musser. Abstract data types and software validation. *Communications of the Association for Computing Machinery*, 21(12):1048–1063, 1978. Also as: Research Report ISI/RR-76-48, University of Southern California, 1976.

[GHM88] A. Geser, H. Hußmann, and A. Mück. A compiler for a class of conditional term rewriting systems. In *Proc. Intl. Conf. on Conditional Term Rewriting Systems (CTRS'87)*, volume 308 of *Lecture Notes in Computer Science*, pages 84–90. Springer, 1988.

[GHW85] J.V. Guttag, J.J. Horning, and J.M. Wing. Larch in five ease pieces. Technical report, Digital Systems Research Center, Palo Alto, CA, 1985.

[Gie95] J. Giesl. Generating polynomial orderings for termination proofs. In J. Hsiang, editor, *Proc. 6th Intl. Conference on Rewriting Techniques and Applications (RTA'95)*, volume 914 of *Lecture Notes in Computer Science*, pages 426–431. Springer, 1995.

[GJM85] J.A. Goguen, J.-P. Jouannaud, and J. Meseguer. Operational semantics for order-sorted algebra. In W. Brauer, editor, *Proc. 12th Intl. Colloquium on Automata, Languages and Programming*, volume 194 of *Lecture Notes in Computer Science*, pages 221–231. Springer, 1985.

[GKK90] I. Gnaedig, C. Kirchner, and H. Kirchner. Equational completion in order-sorted algebras. *Theoretical Computer Science*, 72:169–202, 1990.

[GL97] R.D. Gumb and K. Lambert. Definitions in nonstrict positive free logic. *Modern Logic*, 7:25–55, 1997.

[GLT89] J.-Y. Girard, Y. Lafont, and P. Taylor. *Proofs and Types*. Cambridge University Press, 1989.

[GM] J. Goguen and G. Malcolm. More higher order programming in OBJ. In J. Goguen and G. Malcolm, editors, *Software Engineering with OBJ: Algebraic Specification in Action*. To appear.

[GM82]   J.A. Goguen and J. Meseguer. Universal realization, persistent inter-
           connection and implementation of abstract modules. In *Proc. 9th Intl.
           Colloq. on Automata, Languages and Programming*, volume 140 of *Lec-
           ture Notes in Computer Science*, pages 265–281. Springer, 1982.

[GM85]   J. Goguen and J. Meseguer. Completeness of many-sorted equational
           logic. *Houston Journal of Mathematics*, 11(3):307–334, 1985.

[GM86a]   J.A. Goguen and J. Meseguer. Remarks on remarks on many-sorted
           equational logic. *EATCS Bulletin*, 30:66–73, 1986.

[GM86b]   Joseph A. Goguen and José Meseguer. EQLOG: Equality, types, and
           generic modules for logic programming. In D. DeGroot and G. Lind-
           strom, editors, *Logic Programming: Functions, Relations and Equa-
           tions*, pages 259–363. Prentice Hall, 1986. An earlier version appeared
           in *Journal of Logic Programming*, 1(2):179–210, 1984.

[GM87a]   J.A. Goguen and J. Meseguer. Order-sorted algebra solves the con-
           structor selector, multiple representation and coercion problems. In
           Proc. Second Symposium on Logic in Computer Science, pages 18–29.
           IEEE Computer Society, 1987. Also Report CSLI-87-92, Center for the
           Study of Language and Information, Stanford University, March 1987;
           revised version in *Information and Computation, 103*, 1993.

[GM87b]   J.A. Goguen and J. Meseguer. Unifying functional, object-oriented
           and relational programming with logical semantics. In B. Shriver anf
           P. Wegner, editor, *Research Direction in Object-Oriented Programming*,
           Computer System Series, pages 417–477. MIT Press, 1987.

[GM92]   J.A. Goguen and J. Meseguer. Order-sorted algebra I: equational de-
           duction for multiple inheritance, overloading, exceptions and partial
           operations. *Theoretical Computer Science*, 105(2):217–273, 1992.

[GM93]   M.J. Gordon and T.F. Melham. *Introduction to HOL: a theorem prov-
           ing environment for higher-order logic*. Cambridge University Press,
           1993.

[GM94]   G. Guiho and F. Meija. Operational safety critical software methods in
           railways. In *Information Processing '94, Vol. 3*, pages 262–269. North-
           Holland, 1994.

[GM97]   J.A. Goguen and G. Malcolm. A hidden agenda. Technical Report
           CS97-538, UCSD, 1997. `http://www-cse.ucsd.edu/users/goguen/`
           `ps/ha.ps.gz`.

[GMJ85]   J.A. Goguen, J. Meseguer, and J.-P. Jouannaud. Operational semantics
           of order-sorted algebra. In Wilfried Brauer, editor, *Proc. 1985 Intl.
           Conference on Automata, Languages and Programming*, volume 194 of
           Lecture Notes in Computer Science, pages 221–231. Springer, 1985.

[GMSB96]   M.-C. Gaudel, B. Marre, S. Schlienger, and G. Bernot. *Précis de génie
           logiciel*. Masson, Enseignement de l'Informatique, 1996.

[GMW79]   Michael J. Gordon, Robin Milner, and Christopher P. Wadsworth. *Ed-
           inburgh LCF: A Mechanized Logic of Computation*, volume 78 of *Lec-
           ture Notes in Computer Science*. Springer, 1979.

[GMW⁺93] J. Goguen, J. Meseguer, T. Winkler, K. Futatsugi, P. Lincoln, and
           J.-P. Jouannaud. Introducing OBJ3. Technical Report SRI-CSL-88-8,
           Computer Science Lab, SRI International, August 1988; to appear in
           Applications of Algebraic Specification using OBJ, 1993.

[Göd31]   Kurt Gödel. Über formal unentscheidbare Sätze der *Principia Mathematica* und verwandter Systeme I. *Monatshefte für Mathematik und Physik*, 38:173–98, 1931. Translation by Elliot Mendelsohn printed in [Dav65].

[Gog78]   J.A. Goguen. Abstract errors for abstract data types. In *Proc. IFIP Working Conference on the Formal Description of Programming Concepts*. North-Holland, 1978.

[Gog80]   J.A. Goguen. How to prove algebraic inductive hypotheses without induction, with applications to the correctness of data type implementation. In W. Bibel and R. Kowalski, editors, *Proc. 5th Intl. Conference on Automated Deduction*, volume 87 of *Lecture Notes in Computer Science*, pages 356–373. Springer, 1980.

[Gog84]   M. Gogolla. Partially ordered sorts in algebraic specifications. In *Proc. 9th Colloquium on Trees in Algebra and Programming*, pages 139–153. Cambridge University Press, 1984.

[Gog87]   M. Gogolla. On parametric algebraic specifications with clean error handling. In H. Ehrig, R. Kowalski, G. Levi, and U. Montanari, editors, *Proc. TAPSOFT'87*, volume 249 of *Lecture Notes in Computer Science*, pages 81–95. Springer, 1987.

[Gog90]   J. Goguen. Higher-order functions considered unnecessary for higher-order programming. In D. Turner, editor, *Research Topics in Functional Programming*, pages 309–352. Addison-Wesley, 1990.

[Gor95]   A.D. Gordon. A tutorial on co-induction and functional programming. In Springer, editor, *Proc. Functional Programming, Glasgow 1994*, pages 78–95, 1995.

[GPSS80]   D. Gabbay, A. Pnueli, S. Shelah, and J. Stavi. The temporal analysis of fairness. In *Proc. 7th ACM Symp. on Principles of Programming Languages*, pages 163–173, 1980.

[GR83]   A. Goldberg and D. Robson. *Smalltalk-80: The Language and Its Implementation*. Addison-Wesley, Reading, MA, 1983.

[GR95]   M. Große-Rhode. *Specification of Transition Categories – An Approach to Dynamic Abstract Data Types*. PhD thesis, Technische Universität Berlin, 1995.

[GR97]   M. Große-Rhode. Transition specifications for dynamic abstract data types. *Applied Categorical Structures*, 5, 1997.

[Grä79]   G. Grätzer. *Universal Algebra*. Springer, 2nd edition, 1979.

[Gre69]   Cordell Green. Application of theorem proving to problem solving. In *Intl. Joint Conference on Artificial Intelligence '69*, pages 219–239, 1969.

[GS88]   Jean Gallier and Wayne Snyder. Complete sets of transformations for general e-unification. *Theoretical Computer Science*, 67:203–260, 1988.

[GS95a]   J.A. Goguen and A. Socorro. Module composition and system design for the object paradigm. *Journal of Object Oriented Programming*, 7(14):47–55, 1995.

[GS95b]   David Gries and Fred B. Schneider. Avoiding the undefined by underspecification. In J. van Leeuwen, editor, *Computer Science Today: Recent Trends and Developments*, volume 1000 of *Lecture Notes in Computer Science*, pages 366–373. Springer, 1995.

[GSHH92]   Joseph A. Goguen, Andrew Stevens, Keith Hobley, and Hendrik Hilberdink. 2OBJ, a metalogical framework based on equational logic. *Philosophical Transactions of the Royal Society, Series A*, 339:69–86, 1992. Also in *Mechanized Reasoning and Hardware Design*, edited by C.A.R. Hoare and M.J.C. Gordon, Prentice-Hall, 1992, pages 69–86.

[GT77]   J.A. Goguen and J. Tardo. OBJ-0 preliminary users manual. Semantics and Theory of Computation Report No. 10, UCLA Computer Science Dept., 1977.

[GT93]   Fausto Giunchiglia and Paolo Traverso. Reflective reasoning with and between a declarative metatheory and the implementation code. In *Intl. Joint Conference on Artificial Intelligence '91, Sydney (Australia)*, pages 111–117, 1993.

[GTW76]   Joseph A. Goguen, James W. Thatcher, and Eric G. Wagner. An initial algebra approach to the specification, correctness, and implementation of abstract data types. Report RC 6487, IBM T.J. Watson Research Center, Yorktown Heights, 1976.

[GTW78]   Joseph A. Goguen, James W. Thatcher, and Eric G. Wagner. An initial algebra approach to the specification, correctness, and implementation of abstract data types. In Raymond Yeh, editor, *Current Trends in Programming Methodology, IV*, pages 80–149. Prentice-Hall, 1978. Also as Report RC 6487, IBM T.J. Watson Research Center, Yorktown Heights, 1976.

[GTWW77]   J. Goguen, J. Thatcher, E. Wagner, and J. Wright. Initial algebra semantics and continuous algebras. *Journal of the Association for Computing Machinery*, 24(1):68–95, 1977.

[Gur93]   Y. Gurevich. Evolving algebras, an attempt to discovery semantics. In G. Rozenberg and A. Salomaa, editors, *Current Trends in Theoretical Computer Science*. World Scientific, 1993.

[Gur95]   Y. Gurevich. Evolving algebras 1993: Lipari guide. In E. Börger, editor, *Specification and Validation Methods*. Oxford University Press, 1995.

[GV92]   J.F. Groote and F. Vaandrager. Structured operational semantics and bisimulation as a congruence. *Information and Computation*, 100:202–260, 1992.

[GW88]   J.A. Goguen and T. Winkler. Introducing OBJ3. Research Report SRI-CSL-88-9, SRI International, 1988.

[GW93]   H. Ganzinger and U. Waldmann. Termination proofs of well-moded logic programs via conditional rewrite systems. In *Proc. CTRS'92*, volume 656 of *Lecture Notes in Computer Science*, pages 430–437. Springer, 1993.

[GW96]   M.-C. Gaudel and J. Woodcock, editors. *Proc. FME'96, Industrial benefits and advances in formal methods*, volume 1051 of *Lecture Notes in Computer Science*. Springer, 1996.

[GWM+92]   J.A. Goguen, T. Winkler, J. Meseguer, K. Futatsugi, and J.-P. Jouannaud. Introducing OBJ3. Technical Report SRI-CSL-92-03, SRI International, 1992.

[Hag87]   T. Hagino. *A categorical programming language*. PhD thesis, Edinburgh University, 1987. Report CST-47-87.

[Hai53]   T. Hailperin. Quantification theory and empty individual-domains. *Journal of Symbolic Logic*, 18:197–200, 1953.

[Han93]   M. Hanus. The integration of functions into logic programming: from theory to practice. Technical report, Max-Planck-Institut für Informatik, September 1993.

[Han94]   M. Hanus. The integration of functions into logic programming: From theory to practice. *Journal of Logic Programming*, 19/20:583–628, 1994.

[Han95]   M. Hanus. Analysis of residuating logic programs. *Journal of Logic Programming*, 24:219–245, 1995.

[Har89]   T. Hardin. Confluence results for the pure strong categorical combinatory logic CCL: λ-calculi as subsystems of CCL. *Theoretical Computer Science*, 65:291–342, 1989.

[Har97]   P. Hartel. *Konzeptionelle Modellierung von Informationssystemen als verteilte Objektsysteme*. Reihe DISDBIS, Band 6. infix-Verlag, Sankt Augustin, 1997.

[HBS92]   Jane Hesketh, Alan Bundy, and Alan Smaill. Using middle-out reasoning to transform naive programs into tail recursive ones. In *Proceedings of CADE-11*, 1992.

[HD83]   J. Hsiang and N. Dershowitz. Rewrite methods for clausal and non-clausal theorem proving. In *Proc. 10th Intl. Colloquium on Automata, Languages and Programming*, volume 154 of *Lecture Notes in Computer Science*, pages 331–346. Springer, 1983.

[Hen50]   L. Henkin. Completeness in the theory of types. *Journal of Symbolic Logic*, 15:81–91, 1950.

[Hen88]   M. Hennessy. *An Algebraic Theory of Processes*. MIT Press, 1988.

[Hen89]   R. Hennicker. Observational implementations. In *Proc. 6th Symp. on Theoretical Aspects of Computer Science*, volume 349 of *Lecture Notes in Computer Science*. Springer, 1989.

[Hen91a]   R. Hennicker. Context induction: A proof principle for behavioural abstractions. *Formal Aspects of Computing*, pages 326–345, 1991.

[Hen91b]   R. Hennicker. Observational implementation of algebraic specifications. *Acta Informatica*, 28:187–230, 1991.

[Hen97]   Rolf Hennicker. Structured specifications with behavioural operators: Semantics, proof methods and applications. Habilitation Thesis, Ludwig-Maximilians-Universität München, 1997.

[Her73]   H. Hermes. *Introduction to mathematical logic*. Springer, 1973.

[HG86]   H. Hußmann and A. Geser. Experiences with the RAP system – a specification interpreter combining term rewriting and resolution. In *European Symposium on Programming*, volume 213 of *Lecture Notes in Computer Science*, pages 339–335. Springer, 1986.

[HH82]   G. Huet and J.-M. Hullot. Proofs by induction in equational theories with constructors. *Journal of Computer and System Sciences*, 25(2):239–266, 1982. Preliminary version in Proc. 21st Symposium on Foundations of Computer Science, IEEE, 1980.

[HHP93]   Robert Harper, Furio Honsell, and Gordon Plotkin. A framework for defining logics. *Journal of the Association for Computing Machinery*, 40(1):143–184, 1993.

[HJ95]   P. Hartel and R. Jungclaus. Modeling business processes over objects. *Intl. Journal of Intelligent and Cooperative Information Systems*, 4:165–188, 1995.

[HK93]   Berthold Hoffmann and Bernd Krieg-Brückner, editors. *Program Development by Specification and Transformation*, volume 680 of *Lecture Notes in Computer Science*. Springer, 1993.

[HKK91]   M. Hermann, C. Kirchner, and H. Kirchner. Implementations of term rewriting systems. *Computer Journal*, 34(1):20–33, 1991.

[HKK94a]   C. Hintermeier, C. Kirchner, and H. Kirchner. Dynamically-typed computations for order-sorted equational presentations –extended abstract–. In S. Abiteboul and E. Shamir, editors, *Proc. 21st Intl. Colloquium on Automata, Languages, and Programming*, volume 820 of *Lecture Notes in Computer Science*, pages 450–461. Springer, 1994.

[HKK94b]   C. Hintermeier, C. Kirchner, and H. Kirchner. Order-sorted completion with dynamic types. In *Proc. 10th WADT - 6th Compass Workshop*, May 1994.

[HKK98]   C. Hintermeier, C. Kirchner, and H. Kirchner. Dynamically-typed computations for order-sorted equational presentations. *Journal of Symbolic Computation*, 25(4):455–526, 1998.

[HKLR92]   J. Hsiang, H. Kirchner, P. Lescanne, and M. Rusinowitch. The term rewriting approach to automated theorem proving. *Journal of Logic Programming*, 14(1-2):71–99, 1992.

[HKM98]   A.E. Haxthausen, B. Krieg-Brückner, and T. Mossakowski. Extending CASL with higher-order functions – design proposal. CoFI note: L-8, available at **http://www.brics.dk/Projects/CoFI/Notes/L-8/ index.html**, 1998.

[HL78a]   G. Huet and D.S. Lankford. On the uniform halting problem for term rewriting systems. Technical Report 283, Laboria, France, 1978.

[HL78b]   Gérard Huet and Bernard Lang. Proving and applying program transformations expressed with second-order patterns. *Acta Informatica*, pages 31–55, 1978.

[HL91a]   G. Huet and J.-J. Lévy. Computations in orthogonal rewriting systems, I. In J.-L. Lassez and G. Plotkin, editors, *Computational Logic, Chapter 11*, pages 395–414. The MIT Press, 1991.

[HL91b]   G. Huet and J.-J. Lévy. Computations in orthogonal rewriting systems, II. In J.-L. Lassez and G. Plotkin, editors, *Computational Logic, Chapter 12*, pages 415–443. The MIT Press, 1991.

[HO80]   G. Huet and D. Oppen. Equations and rewrite rules: A survey. In R.V. Book, editor, *Formal Language Theory: Perspectives and Open Problems*, pages 349–405. Academic Press, New York, 1980.

[Hoa72]   C.A.R. Hoare. Proofs of correctness of data respesentations. *Acta Informatica*, 1:271–281, 1972.

[Hoa85]   C.A.R. Hoare, editor. *Communicating Sequential Processes*. Prentice Hall, 1985.

[How88a]   Douglas J. Howe. *Automating Reasoning in an Implementation of Constructive Type Theory*. PhD thesis, Cornell University, 1988.

[How88b]   Douglas J. Howe. Computational metatheory in Nuprl. In *9th Intl. Conference on Automated Deduction, Argonne, Ill. (USA)*, pages 238–257, 1988.

[How91]   Douglas J. Howe. On computational open-endedness in Martin-Löf's type theory. In *Sixth Annual Symposium on Logic in Computer Science, Amsterdam (The Netherlands)*, 1991.

[HPW92]   P. Hudak, S. Peyton Jones, and P. Wadler, editors. *Report on the programming language Haskell, a non-strict, purely functional language (version 1.2)*. SIGPLAN Notices 27(5). 1992.

[HR86]   J. Hsiang and M. Rusinowitch. A new method for establishing refutational completeness in theorem proving. In J. Siekmann, editor, *Proc. 8th Intl. Conference on Automated Deduction*, volume 230 of *Lecture Notes in Computer Science*, pages 141–152. Springer, 1986.

[HR87]   J. Hsiang and M. Rusinowitch. On word problem in equational theories. In T. Ottmann, editor, *Proc. 14th Intl. Colloquium on Automata, Languages and Programming*, volume 267 of *Lecture Notes in Computer Science*, pages 54–71. Springer, 1987.

[HRS90]   Maritta Heisel, Wolfgang Reif, and Werner Stephan. Tactical theorem proving in program verification. In *10th Intl. Conference on Automated Deduction, Kaiserslautern (Germany)*, pages 117–131, 1990.

[HS86]   J.Roger Hindley and Jonathan P. Seldin. *Introduction to Combinators and λ-Calculus*. Cambridge University Press, 1986.

[HS96]   M. Hofmann and D. Sannella. On behavioural abstraction and behavioural satisfaction in higher-order logic. *Theoretical Computer Science*, 167, 1996.

[HST94]   Robert Harper, Donald Sannella, and Andrzej Tarlecki. Structured theory presentations and logic representations. *Annals of Pure and Applied Logic*, 67(1):113–160, 1994. An earlier version: Structure and representation in LF. *Proc. 4th IEEE Symposium on Logics in Computer Science LiCS'89*, Asilomar, Cal. (USA) 1989, 226–237.

[Hue72]   G. Huet. *Constrained Resolution: A Complete Method for Higher Order Logic*. PhD thesis, Case Western Reserve University, 1972.

[Hue80]   G. Huet. Confluent reductions: Abstract properties and applications to term rewriting systems. *Journal of the Association for Computing Machinery*, 27(4):797–821, 1980. Preliminary version in 18th Symposium on Foundations of Computer Science, IEEE, 1977.

[Hul80a] J.-M. Hullot. Canonical forms and unification. In W. Bibel and R. Kowalski, editors, *Proc. 5th Intl. Conference on Automated Deduction (CADE'80)*, volume 87 of *Lecture Notes in Computer Science*, pages 318–334. Springer, 1980.

[Hul80b] J.-M. Hullot. *Compilation de Formes Canoniques dans les Théories équationelles*. Thèse de Doctorat de Troisième Cycle, Université de Paris Sud, Orsay (France), 1980.

[Hum90]   W.S. Humphrey. *Managing the software process*. SEI series in Software Engineering. Addison-Wesley, 1990.

[Huß89]   H. Hußmann. *Nichtdeterministische algebraische Spezifikation*. PhD thesis, Universität Passau, 1989.

[Huß94a] H. Hußmann. Formal foundations for pragmatic software engineering methods. In [Wol94], pages 27–34. 1994.

[Huß94b] H. Hußmann. *Formal Foundations for SSADM*. Habilitation thesis, Technische Universität München, 1994.

[Hut90]   D. Hutter. Guiding induction proofs. In M.E. Stickel, editor, *Proc. 10th Intl. Conference on Automated Deduction (CADE'90)*, volume 449 of *Lecture Notes in Computer Science*, pages 147–161. Springer, 1990.

[HWB97]   Rolf Hennicker, Martin Wirsing, and Michel Bidoit. Proof systems
           for structured specifications with observability operators. *Theoretical
           Computer Science*, 173:393–443, 1997.

[ISO89]   ISO 8807. information processing systems – open systems interconnec-
           tion – LOTOS – a formal description technique based on the temporal
           ordering of observational behaviour. IS, International Organization for
           Standardization, 1989.

[Jac92]   I. Jacobson. *Object-Oriented Software Engineering*. Addison-Wesley,
           Reading, MA, 1992.

[JK86a]   J.-P. Jouannaud and H. Kirchner. Completion of a set of rules modulo a
           set of equations. *SIAM Journal of Computing*, 15(4):1155–1194, 1986.
           Preliminary version in Proc. 11th ACM Symposium on Principles of
           Programming Languages, Salt Lake City (USA), 1984.

[JK86b]   J.-P. Jouannaud and E. Kounalis. Proof by induction in equational
           theories without constructors. In *Proc. 1st IEEE Symposium on Logic
           in Computer Science, Cambridge, Mass. (USA)*, pages 358–366, 1986.

[JK89]   J.-P. Jouannaud and E. Kounalis. Automatic proofs by induction in
           theories without constructors. *Information and Computation*, 82:1–33,
           1989.

[JKK83]   J.-P. Jouannaud, C. Kirchner, and H. Kirchner. Incremental construc-
           tion of unification algorithms in equational theories. In *Proc. Intl.
           Colloquium on Automata, Languages and Programming*, volume 154 of
           Lecture Notes in Computer Science, pages 361–373. Springer, 1983.

[JKKM92]   J.-P. Jouannaud, C. Kirchner, H. Kirchner, and A. Mégrelis. Pro-
           gramming with equalities, subsorts, overloading and parameterization
           in OBJ. *Journal of Logic Programming*, 12(3):257–280, 1992.

[JLR82]   J.-P. Jouannaud, P. Lescanne, and F. Reinig. Recursive decomposition
           ordering. In D. Bjørner, editor, *Formal Description of Programming
           Concepts 2*, pages 331–348. Elsevier Science (North-Holland), 1982.

[JM90]   J.-P. Jouannaud and C. Marché. Completion modulo associativ-
           ity, commutativity and identity (AC1). In A. Miola, editor, *Proc.
           DISCO'90*, volume 429 of *Lecture Notes in Computer Science*, pages
           111–120. Springer, 1990.

[JOE95]   R. Jiménez, F. Orejas, and H. Ehrig. Compositionality and compati-
           bility of parameterization and parameter passing in specification lan-
           guages. *Mathematical Structures in Computer Science*, 5(2):283–313,
           1995.

[Jon89]   H.B.M. Jonkers. Description algebra. In M. Wirsing and J.A. Bergstra,
           editors, *Algebraic Methods: Theory, Tools and Applications*, volume 394
           of *Lecture Notes in Computer Science*, pages 283–328. Springer, 1989.

[Jon90]   C.B. Jones. *Systematic Software Development Using VDM*. Prentice
           Hall, 1990.

[Jou83]   J.-P. Jouannaud. Confluent and coherent equational term rewriting
           systems. Applications to proofs in abstract data types. In G. Ausiello
           and M. Protasi, editors, *Proc. 8th Colloquium on Trees in Algebra and
           Programming*, volume 159 of *Lecture Notes in Computer Science*, pages
           269–283. Springer, 1983.

[JR91]   K. Jensen and G. Rozenberg, editors. *High-Level Petri Nets – Theory
           and Application*. Springer, 1991.

[JR97]   B. Jacobs and J. Rutten. A tutorial on (co)algebras and (co)induction. *EATCS Bulletin*, 62:222–259, 1997.

[JSHS96]   Ralf Jungclaus, Gunter Saake, Thorsten Hartmann, and Cristina Sernadas. TROLL—A language for object-oriented specification of information systems. *ACM Transactions on Information Systems*, 14:175–211, 1996.

[JW86]   J.-P. Jouannaud and B. Waldmann. Reductive conditional term rewriting systems. In M. Wirsing, editor, *3rd IFIP Conference on Formal Description of Programming Concepts, Ebberup (Denmark)*, pages 223–244. Elsevier Science (North-Holland), 1986.

[KA84]   S. Kamin and M. Archer. Partial implementations of abstract data types: A dissenting view on errors. In *Proc. Intl. Symposium on Semantics of Data Types*, volume 173 of *Lecture Notes in Computer Science*, pages 317–336. Springer, 1984.

[Kah74]   G. Kahn. The semantics of a simple language for parallel programming. In J.L. Rosenfeld, editor, *Information Processing 77*. North-Holland, 1974.

[Kah87]   G. Kahn. Natural semantics. Technical Report 601, INRIA Sophia-Antipolis, 1987.

[Kam83]   S. Kamin. Final data type specifications: A new data type specification method. *ACM TOPLAS*, 5:97–123, 1983.

[Kap83]   S. Kaplan. *Un langage de spécification de types abstraits algébriques*. Thèse de Doctorat de Troisième Cycle, Université d'Orsay, France, 1983.

[Kap84]   S. Kaplan. Conditional rewrite rules. *Theoretical Computer Science*, 33:175–193, 1984.

[Kap87]   S. Kaplan. Simplifying conditional term rewriting systems: Unification, termination and confluence. *Journal of Symbolic Computation*, 4(3):295–334, 1987.

[KB70]   D.E. Knuth and P.B. Bendix. Simple word problems in universal algebras. In J. Leech, editor, *Computational Problems in Abstract Algebra*, pages 263–297. Pergamon Press, Oxford, 1970.

[KB91]   S. Krischer and A. Bockmayr. Detecting redundant narrowing derivations by the lse-sl reducibility test. In R.V. Book, editor, *Proc. 4th Conference on Rewriting Techniques and Applications*, volume 488 of *Lecture Notes in Computer Science*, pages 74–85. Springer, 1991.

[KBB93]   Ina Kraan, David Basin, and Alan Bundy. Middle-out reasoning for logic program synthesis. In *10th Intl. Conference on Logic Programming (ICLP'93), Budapest (Hungary)*, pages 441–455, 1993.

[Kei71]   H. Jerome Keisler. *Model theory for infinitary logic*, volume 62 of *Studies in Logic and the Foundations of Mathematics*. North-Holland, 1971.

[KG76]   A. Kay and A. Goldberg. *Smalltalk-76 Instruction Manual*. Xerox PARC, 1976.

[KGS88]   L. Kreiser, S. Gottwald, and W. Stelzner. *Nichtklassische Logik – Eine Einführung*. Akademie Verlag, Berlin, 1988.

[Kir95a]   H. Kirchner. On the use of constraints in automated deduction. In A. Podelski, editor, *Constraint Programming: Basics and Trends*, volume 910 of *Lecture Notes in Computer Science*, pages 128–146. Springer, 1995.

[Kir95b]   H. Kirchner. Some extensions of rewriting. In H. Comon and J.-P. Jouannaud, editors, *Term Rewriting*, volume 909 of *Lecture Notes in Computer Science*, pages 54–73. Springer, 1995.

[KK67]   G. Kreisel and J.L. Krivine. *Eléments de Logique Mathématique*. Dunod (Paris), 1967.

[KKLT91]   B. Krieg-Brückner, E.W. Karlsen, J. Liu, and O. Traynor. The PROSPECTRA methodology and system: Uniform transformational (meta-) development. In S. Prehn and W.J. Toetenel, editors, *VDM, Formal Software Development Methods, Proc. 4th Intl. Symposium of VDM Europe; Vol. 2: Tutorials*, volume 552 of *Lecture Notes in Computer Science*, pages 363–397. Springer, 1991.

[KKM88]   C. Kirchner, H. Kirchner, and J. Meseguer. Operational semantics of OBJ-3. In *Proc. 15th Intl. Colloquium on Automata, Languages and Programming*, volume 317 of *Lecture Notes in Computer Science*, pages 287–301. Springer, 1988.

[KKM98]   Kolyang, B. Krieg-Brückner, and T. Mossakowski. Static semantic analysis of CASL. In F. Parisi Presicce, editor, *Recent Trends in Algebraic Development Techniques. Selected Papers, 12th Intl. Workshop, WADT'97*, volume 1376 of *Lecture Notes in Computer Science*, pages 333–348. Springer, 1998.

[KKR90]   C. Kirchner, H. Kirchner, and M. Rusinowitch. Deduction with symbolic constraints. *Revue d'Intelligence Artificielle*, 4(3):9–52, 1990. Special issue on Automatic Deduction.

[KKV95]   C. Kirchner, H. Kirchner, and M. Vittek. Designing constraint logic programming languages using computational systems. In P. Van Hentenryck and V. Saraswat, editors, *Principles and Practice of Constraint Programming. The Newport Papers, Chapter 8*, pages 131–158. The MIT Press, 1995.

[KL82]   S. Kamin and J.-J. Lévy. Attempts for generalizing the recursive path ordering. Technical report, Inria, Rocquencourt, 1982.

[Kle52]   S.C. Kleene. *Introduction to Metamathematics*. North-Holland, 1952.

[Klo92]   Jan Willem Klop. Term rewriting systems. In S. Abramsky, Dov M. Gabbay, and T.S.E. Maibaum, editors, *Handbook of Logic in Computer Science, Vol. 2, Chapter 1*, pages 1–116. Oxford University Press, 1992.

[KM87]   D. Kapur and D.R. Musser. Proof by consistency. *Artificial Intelligence*, 31(2):125–157, 1987.

[KM95]   H.-J. Kreowski and T. Mossakowski. Equivalence and difference of institutions: Simulating Horn clause logic with based algebras. *Mathematical Structures in Computer Science*, 5:189–215, 1995.

[KMR+94]   B. Krieg-Brückner, W. Menzel, W. Reif, H. Ruess, Th. Santen, D. Schwier, G. Schellhorn, K. Stenzel, and W. Stephan. System architecture framework for KORSO. In Bernd Krieg-Brückner, editor, *Programmentwicklung durch Spezifikation und Transformation - Bremer Beiträge zum Verbundprojekt KORSO (Korrekte Software)*, Informatik Bericht Nr. 1/94. Universität Bremen, 1994.

[KN85]   D. Kapur and P. Narendran. A finite Thue system with decidable word problem and without equivalent finite canonical system. *Theoretical Computer Science*, 35:337–344, 1985.

[KN94]   P. Knijnenburg and F. Nordemann. Partial hyperdoctrines: categorical models for partial function logic and Hoare logic. *Mathematical Structures in Computer Science*, 4:117–146, 1994.

[KNO90]   Deepak Kapur, Paliath Narendran, and Friedrich Otto. On ground-confluence of term rewriting systems. *Information and Computation*, 86(1):14–31, 1990.

[Kot78]   Laurent Kott. About a transformation system: A theoretical study. In *3rd Intl. Symposium on Programming, Paris (France)*, pages 232–247, 1978.

[Kou85]   E. Kounalis. Completeness in data type specifications. In B. Buchberger, editor, *Proc. EUROCAL Conference*, volume 204 of *Lecture Notes in Computer Science*, pages 348–362. Springer, 1985.

[Kou90]   E. Kounalis. Testing for inductive (co)-reducibility. In A. Arnold, editor, *Proc. 15th CAAP*, volume 431 of *Lecture Notes in Computer Science*, pages 221–238. Springer, 1990.

[KPO⁺95]   Bernd Krieg-Brückner, Jan Peleska, Ernst-Rüdiger Olderog, D. Balzer, and Alexander Baer. Universelle Entwicklungsumgebung für formale Methoden (uniform workbench). Informatik Bericht Nr. 8/95, Universität Bremen, 1995.

[KR90a]   E. Kounalis and M. Rusinowitch. Mechanizing inductive reasoning. In *Proc. American Association for Artificial Intelligence Conference, Boston*, pages 240–245. AAAI Press and MIT Press, 1990.

[KR90b]   E. Kounalis and M. Rusinowitch. Mechanizing inductive reasoning. *EATCS Bulletin*, 41:216–226, 1990.

[Kre87]   H.-J. Kreowski. Partial algebras flow from algebraic specifications. In T. Ottmann, editor, *Proc. ICALP'87*, volume 267 of *Lecture Notes in Computer Science*, pages 521–530. Springer, 1987.

[Kre93]   Christoph Kreitz. Meta-synthesis: Deriving programs that develop programs. Technical Report AIDA-93-03, Technische Hochschule Darmstadt, 1993. Habilitation.

[Kri89]   B. Krieg-Brückner. Algebraic specification and functionals for transformational program and meta-program development. In J. Diaz and F. Orejas, editors, *Proc. TAPSOFT'89, Vol. 2*, volume 352 of *Lecture Notes in Computer Science*, pages 36–59. Springer, 1989.

[Kri94]   S. Krischer. *Méthodes de vérification de circuits digitaux*. PhD thesis, Institut National Polytechnique de Lorraine, 1994.

[KS91]   B. Krieg-Brückner and D. Sannella. Structuring specifications in-the-large and in-the-small: higher-order functions, dependent types and inheritance in SPECTRAL. In *Proc. Colloq. on Combining Paradigms for Software Development, Joint Conf. on Theory and Practice of Software Development (TAPSOFT)*, volume 494 of *Lecture Notes in Computer Science*, pages 313–336. Springer, 1991.

[KS98]   S. Kahrs and D. Sannella. Reflections on the design of a specification language. In *Proc. Intl. Colloq. on Fundamental Approaches to Software Engineering. European Joint Conferences on Theory and Practice of Software (ETAPS'98)*, volume 1382 of *Lecture Notes in Computer Science*, pages 154–170. Springer, 1998.

[KST94]   S. Kahrs, D. Sannella, and A. Tarlecki. The definition of Extended ML. Report ECS-LFCS-94-300, University of Edinburgh, 1994.

[KST97]   S. Kahrs, D. Sannella, and A. Tarlecki. The definition of Extended ML: a gentle introduction. *Theoretical Computer Science*, 173:445–484, 1997.

[KSZ86]   Deepak Kapur, G. Sivakumar, and H. Zhang. RRL: A rewrite rule laboratory. In *8th Intl. Conference on Automated Deduction, Oxford (UK)*, pages 691–692, 1986.

[Küc89]   W. Küchlin. Inductive completion by ground proof transformation. In H. Aït-Kaci and M. Nivat, editors, *Resolution of Equations in Algebraic Structures, Vol. 2: Rewriting Techniques*, pages 211–244. Academic Press, 1989.

[Lam83]   L. Lamport. Specifying concurrent program modules. *ACM TOPLAS*, 3, 1983.

[Lam91]   K. Lampert, editor. *Philosophical Applications of Free Logic*. Oxford University Press, New York, 1991.

[Lan75]   D.S. Lankford. Canonical inference. Technical report, Louisiana Tech. University, 1975.

[Lan79]   D.S. Lankford. On proving term rewriting systems are Noetherian. Technical report, Louisiana Tech. University, Mathematics Dept., Ruston LA, 1979.

[Lar88]   K.G. Larsen. Proof systems for Hennessy-Milner logic with recursion. In M. Dauchet and M. Nivat, editors, *Proc. CAAP'88*, volume 299 of *Lecture Notes in Computer Science*, pages 215–230. Springer, 1988.

[LcP90]   Huimin Lin and Man chi Pong. Modelling multiple inheritance with colimits. *Formal Aspects of Computing*, 2:301–311, 1990.

[Les81]   J. Leszczyłowski. Theory of FP systems in Edinburgh LCF. In *Formalization of Programming Concepts*, volume 107 of *Lecture Notes in Computer Science*, pages 374–386. Springer, 1981.

[Les90]   P. Lescanne. On the recursive decomposition ordering with lexicographical status and other related orderings. *Journal of Automated Reasoning*, 6:39–49, 1990.

[LEW96]   J. Loeckx, H.-D. Ehrich, and B. Wolf. *Specification of Abstract Data Types*. Wiley, 1996.

[LFA93]   R. Loogen, F.L. Fraguas, and M.R. Artalejo. A demand driven computation strategy for lazy narrowing. In *Proc. PLILP'93*, volume 714 of *Lecture Notes in Computer Science*, pages 184–200. Springer, 1993.

[LG86]   B. Liskov and J.V. Guttag. *Abstraction and Specification in Program Development*. MIT Press, McGraw Hill, 1986.

[LG91]   R. Loogen and U. Goltz. Modelling nondeterministic concurrent processes with event structures. *Fundamenta Informaticae*, 14:39–74, 1991.

[Lip83]   U. Lipeck. *Ein algebraischer Kalkül für einen strukturierten Entwurf von Datenabstraktionen*. PhD thesis, Abteilung Informatik, Universität Dortmund, 1983.

[Llo87]   J.W. Lloyd. *Foundations of Logic Programming*. Springer, 1987.

[LLT90]   A. Lazrek, P. Lescanne, and J.-J. Thiel. Tools for proving inductive equalities, relative completeness and ω-completeness. *Information and Computation*, 84(1):47–70, 1990.

[LMRT91]   K. Lodaya, M. Mukund, R. Ramanujam, and P.S. Thiagarajan. Models and logics for true concurrency. In P.S. Thiagarajan, editor, *Some Models and Logics for Concurrency. Advanced School on the Algebraic, Log-*

> *ical and Categorical Foundations of Concurrency, Gargnano del Garda,* 1991.

[LNS82]   J.-L. Lassez, V.L. Nguyen, and E.A. Sonenberg. Fixed point theorems and semantics: A folk tale. *Information Processing Letters,* 14:112–116, 1982.

[LOT97]   *Enhanced LOTOS Documentation,* 1997. Available at `ftp://ftp.dit.upm.es/pub/lotos/elotos/`.

[LR96]   S. Limet and P. Réty. Conditional directed narrowing. In *Proc. 5th Intl. Conference on Algebraic Methodology and Software Technology,* volume 1101 of *Lecture Notes in Computer Science,* pages 637–640. Springer, 1996.

[LRT92]   K. Lodaya, R. Ramanujam, and P.S. Thiagarajan. Temporal logics for communicating sequential agents: I. *International Journal of Foundations of Computer Science,* 3:117–159, 1992.

[LS93]   C. Lynch and W. Snyder. Redundancy criteria for constrained completion. In C. Kirchner, editor, *Proc. 5th Conference on Rewriting Techniques and Applications,* volume 690 of *Lecture Notes in Computer Science,* pages 2–16. Springer, 1993.

[LS95]   F. Laroussinie and Ph. Schnoebelen. A hierarchy of temporal logics with past. *Theoretical Computer Science,* 148:303–324, 1995.

[LT87]   K. Lodaya and P.S. Thiagarajan. A modal logic for a subclass of event structures. In Th. Ottmann, editor, *Proc. 14th Intl. Colloq. on Automata, Languages and Programming,* volume 267 of *Lecture Notes in Computer Science,* pages 290–303. Springer, 1987.

[Luo90]   Z. Luo. *An Extended Calculus of Constructions.* PhD thesis, University of Edinburgh, 1990. Report CST-65-90.

[LvF67]   K. Lampert and B. v. Fraassen. On free description theory. *Zeitschrift für Logik und Grundlagen der Mathematik,* 13:225–240, 1967.

[LW98]   Gary T. Leavens and Jeannette M. Wing. Protective interface specifications. *Formal Aspects of Computing,* 10:59–75, 1998.

[Lyn94]   C. Lynch. Local simplification. In J.-P. Jouannaud, editor, *Constraints in Computational Logics,* volume 845 of *Lecture Notes in Computer Science,* pages 3–18. Springer, 1994.

[Lyn95]   C. Lynch. Paramodulation without duplication. In D. Kozen, editor, *Proc. 10th IEEE Symposium on Logic in Computer Science, San Diego, Cal. (USA),* pages 167–177. IEEE, 1995.

[LZ74]   B. Liskov and S. Zilles. Programming with abstract data types. *SIGPLAN Notices,* 9(4):50–60, 1974.

[Mac71]   S. MacLane. *Categories for the Working Mathematician.* Springer, 1971.

[Maj77]   M.E. Majster. Limits of the "algebraic" specification of abstract data types. *ACM SIGPLAN Notices,* 12:37–42, 1977.

[Maj79]   M.E. Majster. Data types, abstract data types and their specification problem. *Theoretical Computer Science,* 8:89–127, 1979.

[Mar82]   Per Martin-Löf. Constructive mathematics and computer programming. In *6th Intl. Congress for Logic, Methodology, and Philosophy of Science,* pages 153–175. North-Holland, 1982.

[Mar85]   Per Martin-Löf. On the meaning of the logical constants and the justifications of the logical laws. Technical report, Scuola di Specializzazione in Logica Matematica, Dipartimento di Matematica, Universita di Siena, 1985.

[Mar94]   C. Marché.   Normalised rewriting and normalised completion.   In S. Abramsky, editor, *Proc. 9th IEEE Symposium on Logic in Computer Science, Paris (France)*, pages 394–403. IEEE, 1994.

[Mat94]   Sean Matthews. A theory and its metatheory in FS_0. In Gabbay and Guenthner [GG94].

[May85]   B. Mayoh. Galleries and institutions. Technical Report DAIMI PB-191, Aarhus University, 1985.

[McD91]   J. McDermid. Formal methods: Use and relevance for the development of safety critical systems. In *Safety Aspects of Computer Control*. Butterworth/Heineman, 1991.

[MdR94]   S. Muggleton and L. de Raedt. Inductive logic programming: Theory and methods. *Journal of Logic Programming*, 19/20:629–679, 1994.

[Mei92]   K. Meinke. Universal algebra in higher types. *Theoretical Computer Science*, 100(2):385–417, 1992.

[Men87]   Elliot Mendelson. *Introduction to Mathematical Logic*. Mathematics Series. Wadsworth & Brooks/Cole, third edition, 1987.

[Mes89]   José Meseguer. General Logics. In H.-D. Ebbinghaus, J. Fernández-Prida, M. Garrido, D. Lascar, and M. Rodríguez Artalejo, editors, *Logic Colloquium '87*, pages 275–329. North-Holland, 1989.

[Mes92]   J. Meseguer. Conditional rewriting logic as a unified model of concurrency. *Theoretical Computer Science*, 96(1):73–155, 1992.

[Mes93]   J. Meseguer. A logical theory of concurrent objects and its realization in the Maude language. In G. Agha, P. Wegner, and A. Yonezawa, editors, *Research Directions in Concurrent Object-Oriented Programming*, pages 314–390. MIT Press, 1993.

[MG85]   J. Meseguer and J.A. Goguen. Initiality, induction and computability. In M. Nivat and J. Reynolds, editors, *Algebraic Methods in Semantics*, pages 459–541. Cambridge University Press, 1985.

[MG93]   J. Meseguer and J.A. Goguen. Order-sorted algebra solves the constructor selector, multiple representation and coercion problems. *Information and Computation*, 103(1):114–158, 1993.

[Mil80]   R. Milner. *A Calculus of Communicating Systems*, volume 92 of *Lecture Notes in Computer Science*. Springer, 1980.

[Mil89]   R. Milner. *Communication and concurrency*. Prentice Hall, London, 1989.

[Mil90a]   Dale Miller. Abstractions in logic programs. In P. Odifreddi, editor, *Logic and Computer Science*, pages 329–359. Academic Press, 1990.

[Mil90b]   R. Milne. The semantic foundations of the RAISE specification language. Technical Report REM/11, RAISE project, STC Technology Ltd, 1990.

[MK98]   P.-E. Moreau and H. Kirchner. A compiler for rewrite programs in associative-commutative theories. In C. Palamidessi, H. Glaser, and K. Meinke, editors, *Proc. ALP/PLILP: Principles of Declarative Programming*, volume 1490 of *Lecture Notes in Computer Science*, pages 230–249. Springer, 1998.

[MM93]   Narciso Martí-Oliet and José Meseguer. Rewriting logic as a logical and semantic framework. Technical Report SRI-CSL-93-05, Computer Science Laboratory, SRI International, 1993.

[MM94]   Narciso Martí-Oliet and José Meseguer. General logics and logical frameworks. In Gabbay and Guenthner [GG94].

[MM95]   J. Meseguer and N. Martí-Oliet. From abstract data types to logical frameworks. In E. Astesiano, G. Reggio, and A. Tarlecki, editors, *Recent Trends in Data Type Specification: 10th Workshop on Specification of Abstract Data Types joint with the 5th COMPASS Workshop*, volume 906 of *Lecture Notes in Computer Science*, pages 48–80. Springer, 1995.

[MN70]   Z. Manna and S. Ness. On the termination of Markov algorithms. In *Proc. 3rd Hawaii Intl. Conference on System Science, Honolulu, Haw. (USA)*, pages 789–792, 1970.

[MN90]   U. Martin and T. Nipkow. Ordered rewriting and confluence. In M.E. Stickel, editor, *Proc. 10th Intl. Conference on Automated Deduction*, volume 449 of *Lecture Notes in Computer Science*, pages 366–380. Springer, 1990.

[Mog88]   E. Moggi. *The Partial Lambda-Calculus*. PhD thesis, University of Edinburgh, 1988. Report CST-53-88.

[Möl87]   B. Möller. Algebraic specification with higher-order operations. In *Proc. IFIP TC 2 Working Conference on Program Specification and Transformation*. North-Holland, 1987.

[Mor71]   J.H. Morris. Another recursion induction principle. *Communications of the Association for Computing Machinery*, 14:351–354, 1971.

[Mos89a]   P. Mosses. Unified algebras and institutions. In *Proc. 4th Annual IEEE Symposium on Logic in Computer Science*, pages 304–312, 1989.

[Mos89b]   P. Mosses. Unified algebras and modules. In *Proc. 16th ACM Symp. on Principles of Programming Languages*, pages 329–343, 1989.

[Mos93]   Peter D. Mosses. The use of sorts in algebraic specifications. In M. Bidoit and C. Choppy, editors, *Recent Trends in Data Type Specification, Selected Papers from the 8th Workshop on Specification of Abstract Data Types*, volume 655 of *Lecture Notes in Computer Science*, pages 66–91. Springer, 1993.

[Mos96a]   T. Mossakowski. Equivalences among various logical frameworks of partial algebras. In H. Kleine Büning, editor, *Proc. 9th Workshop on Computer Science Logic (CSL'95). Selected Papers*, volume 1092 of *Lecture Notes in Computer Science*, pages 403–433. Springer, 1996.

[Mos96b]   T. Mossakowski. Using limits of parchments to systematically construct institutions of partial algebras. In M. Haveraaen, O.-J. Dahl, and O. Owe, editors, *Recent Trends in Data Type Specifications. Selected Papers, 11th Workshop on Specification of Abstract Data Types ADT'95*, volume 1130 of *Lecture Notes in Computer Science*, pages 379–393. Springer, 1996.

[Mos97]   Peter D. Mosses. CoFI: The common framework initiative for algebraic specification and development. In M. Bidoit and M. Dauchet, editors, *Proc. TAPSOFT'97*, volume 1214 of *Lecture Notes in Computer Science*, pages 115–137. Springer, 1997.

[Mos98a]   T. Mossakowski. Colimits of order-sorted specifications revisited. In F. Parisi Presicce, editor, *Recent Trends in Algebraic Development Techniques. Selected Papers, 12th Intl. Workshop, WADT'97*, volume 1376 of *Lecture Notes in Computer Science*, pages 316–332. Springer, 1998.

[Mos98b]   T. Mossakowski. Institution-independent semantics for CASL-in-the-large. Technical Report S-8, The Common Framework Initiative, 1998. Available from http://www.brics.dk/Projects/CoFI/Notes/S-8.

[MP89]   Z. Manna and A. Pnueli. The anchored version of the temporal framework. In J.W. de Bakker, W.-P. de Roever, and G. Rozenberg, editors, *Linear Time, Branching Time and Partial Order in Logics and Models for Concurrency*, volume 354 of *Lecture Notes in Computer Science*. Springer, 1989.

[MP92]   Z. Manna and A. Pnueli, editors. *The Temporal Logic of Reactive and Concurrent Systems*. Springer, 1992.

[MP95]   Z. Manna and A. Pnueli, editors. *Temporal Verification of Reactive Systems—Safety*. Springer, 1995.

[MPS93]   B. Möller, H. Partsch, and S. Schuman, editors. *Formal Program Development, IFIP WG 2.1 State-of-the-Art Report*, volume 755 of *Lecture Notes in Computer Science*. Springer, 1993.

[MPT97]   T. Mossakowski, W. Pawłowski, and A. Tarlecki. Combining and representing logical systems. In *Proc. 7th Intl. Conference on Category Theory and Computer Science CTCS'97*, volume 1290 of *Lecture Notes in Computer Science*, pages 177–198. Springer, 1997.

[MR92]   J. Moreno-Navarro and M. Rodriguez-Artalejo. Logic programming with functions and predicates: the language BABEL. *Journal of Logic Programming*, 12(3):191–223, 1992.

[MS97]   O. Mueller and K. Slind. Treating partiality in a logic of total functions. *The Computer Journal*, 40(10):640–651, 1997.

[MSB93]   Sean Matthews, Alan Smaill, and David Basin. Experience with FS_0 as a framework theory. In Gérard Huet and Gordon Plotkin, editors, *Logical Environments, Proc. Workshop on Logical Frameworks '91*, pages 61–82. Cambridge University Press, 1993.

[MSS90]   V. Manca, A. Salibra, and G. Scollo. Equational type logic. *Theoretical Computer Science*, 77:131–159, 1990. Special Issue dedicated to AMAST'89.

[MSS92]   V. Manca, A. Salibra, and G. Scollo. On the expressiveness of equational type logic. In C.M.I. Rattray and R.G. Clark, editors, *The Unified Computation Laboratory*. Oxford University Press, 1992.

[MT92a]   K. Meinke and J. Tucker. Universal algebra. In S. Abramsky, D. Gabbay, and T. Maibaum, editors, *Handbook of Logic in Computer Science, Vol. 1*, pages 189–411. Oxford University Press, 1992.

[MT92b]   M. Mukund and P.S. Thiagarajan. A logical characterization of well-branching event structures. *Theoretical Computer Science*, 96:35–72, 1992.

[MT93]   J. Meinke and J.V. Tucker, editors. *Many-sorted Logic and its Applications*. Wiley, 1993.

[MTH90]   R. Milner, M. Tofte, and R. Harper. *The Definition of Standard ML*. MIT Press, 1990.

[MTW88a]   B. Möller, A. Tarlecki, and M. Wirsing. Algebraic specification with built-in domain constructions. In M. Dauchet and M. Nivat, editors, *Proc. CAAP'88*, volume 299 of *Lecture Notes in Computer Science*, pages 132–148. Springer, 1988.

[MTW88b] B. Möller, A. Tarlecki, and M. Wirsing. Algebraic specifications of reachable higher-order algebras. In *Recent Trends in Data Type Specification, Selected Papers from the 5th Workshop on Specification of Abstract Data Types*, volume 332 of *Lecture Notes in Computer Science*, pages 154–169. Springer, 1988.

[Mus80]   D.R. Musser. On proving inductive properties of abstract data types. In *Proc. 7th ACM Symp. on Principles of Programming Languages*, pages 154–162. Association for Computing Machinery, 1980.

[MV89]   S. Mauw and G.J. Veltink. An introduction to psf$_d$. In J. Diaz and F. Orejas, editors, *Proc. TAPSOFT'89, Vol. 2*, volume 352 of *Lecture Notes in Computer Science*, pages 375–389. Springer, 1989.

[MV90]   S. Mauw and G.J. Veltink. A process specification formalism. *Fundamenta Informaticae*, XIII, 1990.

[MV91]   S. Mauw and G.J. Veltink. A proof assistant for PSF. In K. Larsen and A. Skou, editors, *Proc. 3rd Workshop on Computer Aided Verification, Vol. 1*, Aalborg, Denmark, 1991. The University of Aalborg.

[MV93]   S. Mauw and G.J. Veltink. *Algebraic Specification of Communication Protocols*, volume 36 of *Cambridge Tracts in Theoretical Computer Science*. Cambridge University Press, Aalborg, Denmark, 1993.

[MVS85]   T.S.E. Maibaum, P.A.S. Veloso, and M.R. Sadler. A theory of abstract data types for program development: bridging the gap? In H. Ehrig, C. Floyd, M. Nivat, and J. Thatcher, editors, *Proc. Joint Conference on Theory and Practice of Software Development (TAPSOFT'85)*, volume 186 of *Lecture Notes in Computer Science*, pages 214–230. Springer, 1985.

[MW77]   J.H. Morris and B. Wegbreit. Subgoal induction. *Communications of the Association for Computing Machinery*, 20:209–222, 1977.

[MW80]   Zohar Manna and Richard Waldinger. A deductive approach to program synthesis. *ACM Transactions on Programming Languages and Systems*, 2(1):90–121, 1980.

[Neu89]   P.G. Neumann. Flaws in specifications and what to do about them. In *IEEE Intl. Workshop on Software Specification and Design*, 1989.

[New42]   M.H.A. Newman. On theories with a combinatorial definition of equivalence. *Annals of Math*, 43:223–243, 1942.

[NH84]   R. De Nicola and M. Hennessy. Testing equivalence for processes. *Theoretical Computer Science*, 34:83–133, 1984.

[NHWG89] M. Nielsen, K. Havelund, K.R. Wagner, and C. George. The RAISE language, method and tools. *Formal Aspects of Computing*, 1:85–114, 1989.

[Nip86]   T. Nipkow. Non-deterministic data types: models and implementations. *Acta Informatica*, 22:629–661, 1986.

[Niv87]   P. Nivela. *Semantica de comportamiento para especificaciones algebraicas*. PhD thesis, Universitat Politecnica de Catalunya, Barcelona, 1987.

[NM88]   G. Nadathur and D. Miller. An overview of λprolog. In *Proc. 5th Intl. Logic Programming Conference, Seattle*, pages 810–827. MIT Press, 1988.

[NO88]   P. Nivela and F. Orejas. Initial behaviour semantics for algebraic specifications. In D.T. Sannella and A. Tarlecki, editors, *Recent Trends* in *Data Type Specification, Selected Papers from the 5th Workshop on Specification of Abstract Data Types*, volume 332 of *Lecture Notes in Computer Science*, pages 184–207. Springer, 1988.

[NOS95]   M. Navarro, F. Orejas, and A. Sanchez. On the correctness of modular systems. *Theoretical Computer Science*, 140(1):139–178, 1995.

[NPS90]   Bengt Nordström, Kent Petersson, and Jan M. Smith. *Programming in Martin-Löf's Type Theory: An Introduction*, volume 7 of *The International Series of Monographs on Computer Science*. Oxford Science Publications, 1990.

[NPW81]   M. Nielsen, G. Plotkin, and G. Winskel. Petri nets, event structures and domains, Part I. *Theoretical Computer Science*, 13:85–108, 1981.

[NR92a]   R. Nieuwenhuis and A. Rubio. Basic superposition is complete. In B. Krieg-Brückner, editor, *Proc. ESOP'92*, volume 582 of *Lecture Notes in Computer Science*, pages 371–389. Springer, 1992.

[NR92b]   R. Nieuwenhuis and A. Rubio. Theorem proving with ordering constrained clauses. In D. Kapur, editor, *Proc. 11th Intl. Conference on Automated Deduction*, volume 607 of *Lecture Notes in Computer Science*, pages 477–491. Springer, 1992.

[NR94]   R. Nieuwenhuis and A. Rubio. AC-superposition with constraints: no AC-unifiers needed. In A. Bundy, editor, *Proc. 12th Intl. Conference on Automated Deduction*, volume 814 of *Lecture Notes in Computer Science*, pages 545–559. Springer, 1994.

[NRS89]   W. Nutt, P. Réty, and G. Smolka. Basic narrowing revisited. *Journal of Symbolic Computation*, 7(3-4):295–318, 1989. Special issue on unification. Part one.

[NRT92]   M. Nielsen, G. Rozenberg, and P.S. Thiagarajan. Elementary transition systems. *Theoretical Computer Science*, 96:3–33, 1992.

[Obe62]   A. Oberschelp. Untersuchungen zur mehrsortigen Quantorenlogik. *Mathematische Annalen*, 145(1):297–333, 1962.

[ONS93]   F. Orejas, M. Navarro, and A. Sanchez. Implementation and behavioural equivalence: A survey. In M. Bidoit and C. Choppy, editors, *Recent Trends in Data Type Specification, Proc. Workshop on Specification of Abstract Data Types ADT'91*, volume 655 of *Lecture Notes in Computer Science*, pages 93–125. Springer, 1993.

[ONS96]   F. Orejas, M. Navarro, and A. Sanchez. Algebraic implementation of abstract data types: A survey of concepts and new compositionality results. *Mathematical Structures in Computer Science*, 6:33–67, 1996.

[Ore81]   F. Orejas. On the representation of data types. In *Formalization of Programming Concepts*, volume 107 of *Lecture Notes in Computer Science*, pages 419–431. Springer, 1981.

[Ore85]   F. Orejas. On implementability and computability in abstract data types. In *Algebra, Logics and Combinatorics in Computer Science*. North-Holland, 1985.

[ORS92]   S. Owre, J. Rushby, and N. Shankar. PVS, a prototype verification system. In D. Kapur, editor, *Proc. 11th Intl. Conference on Automated Deduction*, volume 607 of *Lecture Notes in Computer Science*, pages 748–752. Springer, 1992.

[OSC89]   F. Orejas, V. Sacristan, and S. Clérici. Development of algebraic specifications with constraints. In *Proc. Workshop on Categorical Methods in Computer Science with Aspects from Topology*, volume 393 of *Lecture Notes in Computer Science*, pages 102–123. Springer, 1989.

[Pad84]   Peter Padawitz. Towards a proof theory of parameterized specifications. In G. Kahn, D.B. MacQueen, and G. Plotkin, editors, *Intl. Symposium on Semantics of Data Types*, volume 173 of *Lecture Notes in Computer Science*, pages 375–391. Springer, 1984.

[Pad87]   Peter Padawitz. Parameter preserving data type specifications. *Journal of Computer and System Sciences*, 34:179–209, 1987.

[Pad88a]   P. Padawitz. *Computing in Horn Clause Theories*. Springer, 1988.

[Pad88b]   Peter Padawitz. The equational theory of parameterized specifications. *Information and Computation*, 76:121–137, 1988.

[Pad91]   P. Padawitz. Inductive expansion: A calculus for verifying and synthesizing functional and logic programs. *Journal of Automated Reasoning*, 7:27–103, 1991.

[Pad92]   P. Padawitz. *Deduction and Declarative Programming*. Cambridge University Press, 1992.

[Pad94]   P. Padawitz. *Expander: A system for testing and verifying functional-logic programs*. FB Informatik, University of Dortmund, 1994.

[Pad96a]   P. Padawitz. Inductive theorem proving for design specifications. *Journal of Symbolic Computation*, 21:41–99, 1996.

[Pad96b]   P. Padawitz. Swinging data types: Syntax, semantics, and theory. In M. Haveraaen, O.-J. Dahl, and O. Owe, editors, *Recent Trends in Data Type Specification. Selected Papers, 11th Workshop on Specification of Abstract Data Types, WADT'95*, volume 1130 of *Lecture Notes in Computer Science*, pages 409–435. Springer, 1996. Extended version: `http://ls5.cs.uni-dortmund.de/~peter/Swing.ps.gz`.

[Pad97]   P. Padawitz. Theory of programming. German course notes, FB Informatik, University of Dortmund, 1997. `http://ls5.cs.uni-dortmund.de/~peter/TdP96.ps.gz`.

[Pad98]   P. Padawitz. Towards the one-tiered design of data types and transition systems. In F. Parisi Presicce, editor, *Recent Trends in Algebraic Development Techniques. Selected Papers, 12th Intl. Workshop, WADT'97*, volume 1376 of *Lecture Notes in Computer Science*, pages 365–380. Springer, 1998.

[Par69]   David Park. Fixpoint induction and proofs of program properties. In B. Meltzer and D. Michie, editors, *Machine Intelligence 5*, pages 59–78. Elsevier, 1969.

[Par81]   D. Park. Concurrency and automata on infinite sequences. In *Proc. 5th GI Conference*, volume 104 of *Lecture Notes in Computer Science*. Springer, 1981.

[Par89]   H. Partsch. From informal requirements to a running program: a case study in algebraic specification and transformational programming. *Science of Computer Programming*, 11(3):263–297, 1989.

[Par95]   D.L. Parnas. A logic for describing, not verifying, software. *Erkenntnis*, 43:321–338, 1995.

[Pau83]   Lawrence C. Paulson. A higher-order implementation of rewriting. *Science of Computer Programming*, 3:119–149, 1983.

[Pau87]   Lawrence C. Paulson. *Logic and Computation: Interactive Proof with Cambridge LCF*. Cambridge University Press, 1987.

[Pau89a] Christine Paulin-Mohring. Inductive definitions in the calculus of con-
          structions. In *The Calculus of Constructions, Documentation and
          users's guide*. Project Formel, 1989.

[Pau89b] Lawrence C. Paulson. The foundation of a generic theorem prover.
          Journal of Automated Reasoning, 5:363–397, 1989.

[Pau94a] Lawrence C. Paulson. A fixedpoint approach to implementing
          (co)inductive definitions. In *Proc. 12th Intl. Conference on Automated
          Deduction (CADE-12)*, volume 814 of *Lecture Notes on Artificial In-
          telligence*. Springer, 1994.

[Pau94b] Lawrence C. Paulson. *Isabelle: a generic theorem prover*, volume 828
          of *Lecture Notes in Computer Science.* Springer, 1994.

[Pau96]   L.C. Paulson. *ML for the Working Programmer*. Cambridge University
          Press, 2nd edition, 1996.

[Pep92]   P. Pepper. Transforming algebraic specifications. In *Proc. AMAST'91,
          Workshop in Computing*. Springer, 1992.

[Pet83]   G. Peterson. A technique for establishing completeness results in theo-
          rem proving with equality. *SIAM Journal of Computing*, 12(1):82–100,
          1983.

[Pet90]   G.E. Peterson. Complete sets of reductions with constraints. In M.E.
          Stickel, editor, *Proc. 10th Intl. Conference on Automated Deduction*,
          volume 449 of *Lecture Notes in Computer Science*, pages 381–395.
          Springer, 1990.

[Pet94]   H. Peterreins. A natural deduction calculus for structured specifica-
          tions. Report 9410, LMU München, 1994.

[Pfe91]   Frank Pfenning. Logic programming in the LF logical framework. In
          Logical Frameworks, pages 149–181. Cambridge University Press, 1991.

[Pla93]   D. Plaisted. Equational reasoning and term rewriting systems. In
          D. Gabbay, C. Hogger, J.A. Robinson, and J. Siekmann, editors, *Hand-
          book of Logic in Artificial Intelligence and Logic Programming, Vol. 1*,
          pages 273–364. Oxford University Press, 1993.

[Plo70]   G.D. Plotkin. A note on inductive generalization. In B. Meltzer and
          D. Michie, editors, *Machine Intelligence 5*, pages 153–163. Elsevier,
          1970.

[Plo72]   G.D. Plotkin. Building-in equational theories. In B. Meltzer and
          D. Michie, editors, *Machine Intelligence 7*, pages 73–90. Elsevier, 1972.

[Plo83]   Gordon D. Plotkin. An operational semantics for CSP. In D. Bjørner,
          editor, *Formal Description of Programming Concepts – II, Proc. IFIP
          TC-2 Working Conference*, pages 199–225. North-Holland, 1983.

[Pnu77]   A. Pnueli. The temporal logic of programs. In *Proc. 18th Annual
          Symposium on Foundations of Computer Science*, pages 46–57, New
          York, 1977. IEEE Computer Society Press.

[Pnu86]   A. Pnueli. Applications of temporal logic to the specification and ver-
          ification of reactive systems: a survey of current trends. In *Current
          Trends in Concurrency*, volume 224 of *Lecture Notes in Computer Sci-
          ence*. Springer, 1986.

[Poi84]   A. Poigné. Another look at parameterization using algebraic specifica-
          tions with subsorts. In *Proc. 11th Symp. on Mathematical Foundations
          of Computer Science*, volume 176 of *Lecture Notes in Computer Sci-
          ence*, pages 471–479. Springer, 1984.

[Poi86]   A. Poigné. On specifications, theories, and models with higher types. *Information and Control*, 68:1–46, 1986.

[Poi87]   A. Poigné. Partial algebras, subsorting, and dependent types: Prerequisites of error handling in algebraic specifications. In *Recent Trends in Data Type Specification: 5th Workshop on Specification of Abstract Data Types – Selected Papers*, volume 332 of *Lecture Notes in Computer Science*, pages 208–234. Springer, 1987.

[Poi90]   A. Poigné. Parameterization for order-sorted algebraic specification. *Journal of Computer and System Sciences*, 40:229–268, 1990.

[PP95]   F. Parisi-Presicce and A. Pierantonio. Dynamical behaviour of object systems. In E. Astesiano, G. Reggio, and A. Tarlecki, editors, *Recent Trends in Data Type Specification: 10th Workshop on Specification of Abstract Data Types – Selected Papers*, volume 906 of *Lecture Notes in Computer Science*. Springer, 1995.

[PR92]   Frank Pfenning and Ekkehard Rohwedder. Implementing the metatheory of deductive systems. In *Proc. 11th Intl. Conference on Automated Deduction (CADE-11)*, pages 537–551. Springer, 1992.

[PS81]   G. Peterson and M.E. Stickel. Complete sets of reductions for some equational theories. *Journal of the Association for Computing Machinery*, 28:233–264, 1981.

[PS87]   A. Pettorossi and A. Skowron. Higher-order generalization in program derivation. In *Proc. TAPSOFT'87*, volume 250 of *Lecture Notes in Computer Science*, pages 182–196. Springer, 1987.

[PSF97]   PSF toolkit manual pages. Technical report, WINS, University of Amsterdam, 1997. Available at http://www.wins.uva.nl/~bobd/work/.

[PW95]   P. Pepper and M. Wirsing. A method for the development of correct software. In M. Broy and S. Jähnichen, editors, *KORSO: Correct Software by Formal Methods*, Lecture Notes in Computer Science. Springer, 1995.

[Qui59]   W.V. Quine. On cores and prime implicants of truth functions. *American Math. Monthly*, 66:755–760, 1959.

[Ram96]   R. Ramanujam. Locally linear time temporal logic. In *Proc. 11th Annual IEEE Symposium on Logics in Computer Science*, pages 118–127, New York, 1996. IEEE Computer Society.

[RBP+91]   J. Rumbaugh, M. Blaha, W. Premerlani, F. Eddy, and W. Lorensen. *Object-Oriented Modeling and Design*. Prentice-Hall, Englewood Cliffs, NJ, 1991.

[Red90]   U.S. Reddy. Term rewriting induction. In M.E. Stickel, editor, *Proc. 10th Intl. Conference on Automated Deduction (CADE 10)*, volume 449 of *Lecture Notes in Computer Science*, pages 162–177. Springer, 1990.

[Reg91]   G. Reggio. Entities: an institution for dynamic systems. In H. Ehrig, K.P. Jantke, F. Orejas, and H. Reichel, editors, *Recent Trends in Data Type Specification, Proc. 7th Workshop on Specification of Abstract Data Types*, volume 534 of *Lecture Notes in Computer Science*. Springer, 1991.

[Reg93]   G. Reggio. Event logic for specifying abstract dynamic data types. In M. Bidoit and C. Choppy, editors, *Recent Trends in Data Type Specification, Proc. Workshop on Specification of Abstract Data Types ADT'91*, volume 655 of *Lecture Notes in Computer Science*. Springer, 1993.

[Reg94]   F. Regensburger. *HOLCF: Eine konservative Erweiterung von HOL um LCF*. PhD thesis, Technische Universität München, 1994.

[Rei80]   H. Reichel. Initially restricting algebraic theories. In *Proc. MFCS'80*, volume 88 of *Lecture Notes in Computer Science*, pages 504–514. Springer, 1980.

[Rei81]   H. Reichel. Behavioural equivalence – a unifying concept for initial and final specification methods. In *Proc. 3rd Hungarian Comp. Sci. Conference*, pages 27–39, 1981.

[Rei85]   W. Reisig. *Petri Nets: an Introduction*, volume 4 of *EATCS Monographs on Theoretical Computer Science*. Springer, 1985.

[Rei87]   H. Reichel. *Initial Computability, Algebraic Specifications, and Partial Algebras*. Oxford University Press, 1987.

[Rei91]   W. Reisig. Petri nets and algebraic specifications. *Theoretical Computer Science*, 80(1):1–34, 1991.

[Rei95]   H. Reichel. An approach to object semantics based on terminal coalgebras. *Mathematical Structures in Computer Science*, 5:129–152, 1995.

[Rei98]   W. Reisig. *Elements of Distributed Algoritms: Modelling and Analysis with Petri Nets*. Springer, 1998.

[Rém82]   J.-L. Rémy. *Etude des systèmes de Réécriture Conditionnels et Applications aux Types Abstraits Algébriques*. Thèse de Doctorat d'Etat, Institut National Polytechnique de Lorraine, Nancy (France), 1982.

[Rét87]   P. Réty. Improving basic narrowing. In P. Lescanne, editor, *Proc. 2nd Conference on Rewriting Techniques and Applications*, volume 256 of *Lecture Notes in Computer Science*, pages 228–241. Springer, 1987.

[Río93]   A. Ríos. *Contributions à l'étude des λ-calculs avec des substitutions explicites*. Thèse de Doctorat d'Université, Université Paris VII, 1993.

[RKKL85]   P. Réty, C. Kirchner, H. Kirchner, and P. Lescanne. Narrower: A new algorithm for unification and its application to logic programming. In J.-P. Jouannaud, editor, *Proc. 1st Conference on Rewriting Techniques and Applications*, volume 202 of *Lecture Notes in Computer Science*, pages 141–157. Springer, 1985.

[RS92]   L. Rapanotti and A. Socorro. Introducing FOOPS. Report PRG-TR-28-92, Programming Research Group, Oxford University Computing Laboratory, 1992.

[RT86]   G. Rozenberg and P.S. Thiagarajan. Petri nets: Basic notions, structure, behaviour. In *Current Trends in Concurrency*, volume 224 of *Lecture Notes in Computer Science*. Springer, 1986.

[Rus87]   M. Rusinowitch. *Démonstration automatique par des techniques de réécriture*. Thèse de Doctorat d'Etat, Université Henri Poincaré – Nancy 1, 1987. Also published by InterEditions, Collection Science Informatique, directed by G. Huet, 1989.

[Rus89]   M. Rusinowitch. *Démonstration automatique-Techniques de réécriture*. InterEditions, 1989.

[Rus93]   J. Rushby. Formal methods and the certification of critical system. Report SRI-CSL-93-07, SRI International, 1993. Available at http://www.csl.sri.com.

[Rut96]   J.J.M.M. Rutten. Universal coalgebra: A theory of systems. Report CS-R9652, CWI, SMC Amsterdam, 1996.

[RvH93]   J. Rushby and F. von Henke. Formal verification of algorithms for criti-
          cal systems. *IEEE Transactions on Software Engineering*, SE 19(1):13–
          23, 1993.

[RW69]   G.A. Robinson and L.T. Wos. Paramodulation and first-order theorem
          proving. In B. Meltzer and D. Mitchie, editors, *Machine Intelligence 4*,
          pages 135–150. Edinburgh University Press, 1969.

[San91]   D. Sannella. Formal program development in Extended ML for the
          working programmer. In *Proc. 3rd BCS/FACS Workshop on Refine-
          ment, Hursley Park*, Workshops in Computing, pages 99–130. Springer,
          1991.

[San92]   A. Sanchez. *Implementacion de Especificaciones Algebraicas*. PhD the-
          sis, Depto. de Leng. y Sist. Inf. Basque Country University, 1992.

[San95]   David Sands. Correctness of recursion-based automatic program trans-
          formations. In *TAPSOFT'95*, volume 915 of *Lecture Notes in Computer
          Science*, pages 681–965. Springer, 1995.

[SB83]   Donald Sannella and R.M. Burstall. Structured theories in LCF. In
          G. Ausiello and M. Protasi, editors, *Proc. 8th Colloquium on Trees in
          Algebra and Programming (CAAP)*, volume 159 of *Lecture Notes in
          Computer Science*, pages 377–391. Springer, 1983.

[Sch70]   J. Schmidt. A homomorphism theorem for partial algebras. *Coll. Math.*,
          21:5–21, 1970.

[Sch87a] M. Schmidt-Schauß. *Computational Aspects of an Order-Sorted Logic
          with Term Declarations*. PhD thesis, Universität Kaiserslautern, 1987.

[Sch87b] O. Schoett. *Data abstraction and correctness of modular programming*.
          PhD thesis, University of Edinburgh, 1987. Report CST-42-87.

[Sch92]   O. Schoett. Two impossibility theorems on behavioural specifications
          of abstract data types. *Acta Informatica*, 29:595–621, 1992.

[SCL70]   J.R. Slagle, C.L. Chang, and R.C.T. Lee. A new algorithm for gener-
          ating prime implicants. *IEEE Transactions on Computing*, 19(4):304–
          310, 1970.

[Sco76]   D. Scott. Data types as lattices. *SIAM Journal of Computing*, 5:522–
          587, 1976.

[Sco79]   D.S. Scott. Identity and existence in intuitionistic logic. In M.P. Four-
          man, C.J. Mulvey, and D.S. Scott, editors, *Application of Sheaves*, vol-
          ume 753 of *Lecture Notes in Mathematics*, pages 660–696. Springer
          Verlag, 1979.

[Sel72]   A. Selman. Completeness of calculi for axiomatically defined classes of
          algebras. *Algebra Universalis*, 2:20–32, 1972.

[SFSE88] A. Sernadas, J. Fiadeiro, C. Sernadas, and H.-D. Ehrich. Abstract
          object types: A temporal perspective. In B. Banieqbal, H. Barringer,
          and A. Pnueli, editors, *Proc. Colloq. on Temporal Logic in Specifica-
          tion*, volume 398 of *Lecture Notes in Computer Science*, pages 324–350.
          Springer, 1988.

[SG96]   M. Shaw and D. Garlan. *Software Architecture. Perspectives on an
          emerging discipline*. Prentice Hall, 1996.

[Sha88]   Natarayan Shankar. A mechanical proof of the Church-Rosser theorem.
          Journal of the Association for Computing Machinery, 35(3):475–522,
          1988.

[Shi94]   Hui Shi. *Extended Matching with Applications to Program Transfor-
          mation*. PhD thesis, Universität Bremen, 1994.

[SHJE94]   G. Saake, T. Hartmann, R. Jungclaus, and H.-D. Ehrich. Object-oriented design of information systems: TROLL language features. In J. Paredaens and L. Tenenbaum, editors, *Advances in Database Systems, Implementations and Applications*, CISM Courses and Lectures No. 347, pages 219–245. Springer, 1994.

[Siv96]   T. Sivertsen. A case study on the formal development of a reactor safety system. In M.-C. Gaudel and J. Woodcock, editors, *Proc. FME'96, Industrial benefits and advances in formal methods*, volume 1051 of *Lecture Notes in Computer Science*, pages 18–38. Springer, 1996.

[SJE92]   G. Saake, R. Jungclaus, and H.-D. Ehrich. Object-oriented specification and stepwise refinement. In J. de Meer, V. Heymer, and R. Roth, editors, *IFIP Transactions C: Communication Systems, Vol. 1: Proc. Open Distributed Processing*, pages 99–121. North-Holland, 1992.

[SJH93]   G. Saake, R. Jungclaus, and T. Hartmann. Application modelling in heterogeneous environments using an object specification language. *Intl. Journal of Intelligent and Cooperative Information Systems*, 2:425–449, 1993.

[Sla74]   J.R. Slagle. Automated theorem-proving for theories with simplifiers, commutativity and associativity. *Journal of the Association for Computing Machinery*, 21(4):622–642, 1974.

[Smi91]   Doug R. Smith. KIDS – a knowledge-based software development system. In Michael R. Lowry and Robert D. Mcartney, editors, *Automating Software Design*, pages 483–514. AAAI Press/The MIT Press, 1991.

[SMI95]   T. Suzuki, A. Middeldorp, and T. Ida. Level-confluence of conditional rewrite systems with extra variables in right-hand sides. In J. Hsiang, editor, *Proc. 6th Intl. Conference on Rewriting Techniques and Applications (RTA'95)*, volume 914 of *Lecture Notes in Computer Science*, pages 179–193. Springer, 1995.

[Smo86]   G. Smolka. Order-sorted Horn logic: semantics and deduction. SEKI report SR-86-17, FB Informatik, Universität Kaiserslautern, 1986.

[SNGM89]   G. Smolka, W. Nutt, J.A. Goguen, and J. Meseguer. Order-sorted equational computation. In H. Aït-Kaci and M. Nivat, editors, *Resolution of Equations in Algebraic Structures, Vol. 2: Rewriting Techniques*, pages 297–367. Academic Press Inc., 1989.

[SNW93]   V. Sassone, M. Nielsen, and G. Winskel. A classification of models for concurrency. In E. Best, editor, *Proc. CONCUR'93*, pages 82–96, 1993.

[Soc91]   R. Socher-Ambrosius. Boolean algebra admits no convergent term rewriting system. In R.V. Book, editor, *Proc. 4th Conference on Rewriting Techniques and Applications*, volume 488 of *Lecture Notes in Computer Science*, pages 264–274. Springer, 1991.

[Sok87]   Stefan Sokołowski. Soundness of Hoare's logic: an automatic proof using LCF. *ACM Transactions on Programming Languages and Systems*, 9:100–120, 1987.

[Som92]   I. Sommerville. *Software Engineering*. Addision Wesley, 4th edition, 1992.

[Spi92]   J.M. Spivey. *The Z notation, a reference manual*. Prentice Hall, 1992.

[SR94]   A. Sernadas and J. Ramos. The GNOME language: Syntax, semantics and calculus. Technical report, Instituto Superior Técnico, Lisboa, 1994.

[SS93]   A. Salibra and G. Scollo. A soft stairway to institutions. In M. Bidoit
           and C. Choppy, editors, *Recent Trends in Data Type Specification, Proc.
           Workshop on Specification of Abstract Data Types ADT'91*, volume 655
           of *Lecture Notes in Computer Science*, pages 310–329. Springer, 1993.

[SSC95]   A. Sernadas, C. Sernadas, and J.F. Costa. Object specification logic.
           Journal of Logic and Computation, 5:603–630, 1995.

[SSE87]   A. Sernadas, C. Sernadas, and H.-D. Ehrich. Object-oriented specifi-
           cation of databases: An algebraic approach. In P. Hammerslay, editor,
           Proc. 13th Intl. Conference on Very Large Databases (VLDB'87), pages
           107–116, Palo Alto, 1987. Morgan–Kaufmann.

[SST92]   Donald Sannella, Stefan Sokołowski, and Andrzej Tarlecki. Toward
           formal development of programs from algebraic specifications: Param-
           eterisation revisited. *Acta Informatica*, 29(8):689–736, 1992.

[ST]   D. Sannella and A. Tarlecki. *Foundations of Algebraic Specifications
           and Formal Program Development*. Cambridge University Press. To
           appear.

[ST86]   Donald Sannella and Andrzej Tarlecki. Extended ML: An institution-
           independent framework for formal program development. In *Proc.
           Workshop on Category Theory and Computer Programming*, volume
           240 of *Lecture Notes in Computer Science*, pages 364–389. Springer,
           1986.

[ST87]   D.T. Sannella and A. Tarlecki. On observational equivalence and alge-
           braic specification. *Journal of Computer and System Sciences*, 34:150–
           178, 1987.

[ST88a]   D. Sannella and A. Tarlecki. Specifications in an arbitrary institution.
           Information and Computation, 76:165–210, 1988.

[ST88b]   Donald Sannella and Andrzej Tarlecki. Toward formal development
           of programs from algebraic specifications: Implementations revisited.
           Acta Informatica, 25(3):233–281, 1988.

[ST89]   D.T. Sannella and A. Tarlecki. Toward formal development of ML
           programs: foundations and methodology. In J. Diaz and F. Orejas,
           editors, *Proc. TAPSOFT'89, Vol. 2*, volume 352 of *Lecture Notes in
           Computer Science*, pages 375–389. Springer, 1989.

[ST91]   D. Sannella and A. Tarlecki. Extended ML: past, present and future.
           In H. Ehrig, K.P. Jantke, F. Orejas, and H. Reichel, editors, *Recent
           Trends in Data Type Specification, Proc. 7th Workshop on Specifica-
           tion of Abstract Data Types*, volume 534 of *Lecture Notes in Computer
           Science*, pages 297–322. Springer, 1991.

[ST92]   D.T. Sannella and A. Tarlecki. Towards formal development of pro-
           grams from algebraic specifications: parameterisation revisited. *Acta
           Informatica*, 29:689–736, 1992.

[ST96]   Donald Sannella and Andrzej Tarlecki. Mind the gap! Abstract ver-
           sus concrete models of specifications. In *Proc. 21st Intl. Symp. on
           Mathematical Foundations of Computer Science*, volume 1113 of *Lec-
           ture Notes in Computer Science*, pages 114–134. Springer, 1996.

[ST97]   D. Sannella and A. Tarlecki. Essential concepts of algebraic specifica-
           tion and program development. *Formal Aspects of Computing*, 9:229–
           269, 1997. An earlier short version: Toward formal development of pro-
           grams from algebraic specifications: model-theoretic foundations. *Proc.*

19th Intl. Coll. on Automata, Languages and Programming ICALP'92, LNCS 623, 656–671, Springer 1992.

[Ste90]   J. Steinbach. AC-termination of rewrite systems — A modified Knuth-Bendix ordering. In H. Kirchner and W. Wechler, editors, *Proc. 2nd Intl. Conference on Algebraic and Logic Programming*, volume 463 of *Lecture Notes in Computer Science*, pages 372–386. Springer, 1990.

[Sti85]   M. Stickel. Automated deduction by theory resolution. *Journal of Automated Reasoning*, 1(4):333–355, 1985.

[Sti92]   C. Stirling. Modal and temporal logics. In S. Abramsky, Dov M. Gabbay, and T.S.E. Maibaum, editors, *Handbook of Logic in Computer Science, Vol. 2*, pages 477–563. Clarendon Press, Oxford, 1992.

[SW82]   Donald Sannella and Martin Wirsing. Implementation of parameterized specifications. In *Proc. 9th Intl. Colloq. on Automata, Languages and Programming (ICALP)*, volume 140 of *Lecture Notes in Computer Science*, pages 473–488. Springer, 1982.

[SW83]   Donald Sannella and Martin Wirsing. A kernel language for algebraic specification and implementation. In M. Karpinski, editor, *Proc. 11th Colloquium on Foundations of Computation Theory*, volume 158 of *Lecture Notes in Computer Science*, pages 413–427. Springer, 1983.

[Tal93]   Carolyn Talcott. A theory of binding structures and applications to rewriting. *Theoretical Computer Science*, 112:99–143, 1993.

[Tan94]   C.S. Tang. A temporal logic language oriented toward software engineering – an introduction to the XYZ sytem. *Chinese Journal of Advanced Software Research*, 1:1–29, 1994.

[Tar36]   Alfred Tarski. Der Wahrheitsbegriff in den formalisierten Sprachen. *Studia Philosophica*, 1:261–405, 1936.

[Tar85]   A. Tarlecki. On the existence of free models in abstract algebraic institutions. *Theoretical Computer Science*, 37(3):269–304, 1985.

[Tar86a]   A. Tarlecki. Bits and pieces of the theory of institutions. In *Proc. Intl. Workshop on Category Theory and Computer Programming*, volume 240 of *Lecture Notes in Computer Science*, pages 334–363. Springer, 1986.

[Tar86b]   A. Tarlecki. Quasi-varieties in abstract algebraic institutions. *Journal of Computer and System Sciences*, 33:333–360, 1986.

[Tar87]   A. Tarlecki. Institution representation. Unpublished note, Dept. of Computer Science, University of Edinburgh, 1987.

[Tar96]   A. Tarlecki. Moving between logical systems. In M. Haveraaen, O.-J. Dahl, and O. Owe, editors, *Recent Trends in Data Type Specifications. Selected Papers, 11th Workshop on Specification of Abstract Data Types ADT'95*, volume 1130 of *Lecture Notes in Computer Science*, pages 478–502. Springer, 1996.

[Tar98]   A. Tarlecki. Towards heterogeneous specifications. In D. Gabbay and M. van Rijke, editors, *Proc. 2nd Intl. Workshop on Frontiers of Combining Systems, FroCoS'98*. Kluwer, 1998.

[TBG91]   A. Tarlecki, R. Burstall, and J. Goguen. Some fundamental algebraic tools for the semantics of computation. Part III: Indexed categories. *Theoretical Computer Science*, 91:239–264, 1991.

[Thi84]   J.-J. Thiel. Stop losing sleep over incomplete data type specifications. In *Proc. 11th ACM Symp. on Principles of Programming Languages*, pages 76–82. Association for Computing Machinery, 1984.

[Thi94]   P.S. Thiagarajan. A trace based extension of linear time temporal logic. In *Proc. 9th Annual IEEE Symposium on Logic in Computer Science*, pages 438–447. IEEE Computer Science Press, 1994.

[Tro82]   Anne S. Troelstra. *Metamathematical Investigation of Intuitionistic Arithmetic and Analysis*, volume 344 of *Lecture Notes in Mathematics*. Springer, 1982.

[Tse91]   T.H. Tse. *A Unifying Framework for Structured Systems Analysis and Design Models*. Cambridge University Press, 1991.

[TW86]   A. Tarlecki and M. Wirsing. Continuous abstract data types. *Fundamenta Informaticae*, 9:95–126, 1986.

[TWW81]   J. Thatcher, E.G. Wagner, and J.B. Wright. Specification of abstract data types using conditional axioms. Technical Report RC 6214, IBM Yorktown Heigths, 1981.

[TWW82]   J.W. Thatcher, E.G. Wagner, and J.B. Wright. Data type specification: parameterization and the power of specification techniques. *ACM TOPLAS*, 4:711–732, 1982.

[UML97]   UML notation guide, version 1.1. Technical Report ad/97-08-05, Object Management Group, 1997. Available from http://www.omg.org.

[Vaa89]   F. Vaandrager. A simple definition for parallel composition of prime event structures. Report CS-R8903, CWI Amsterdam, 1989.

[Var92]   P. Varnish. Proof obligations in Extended ML. Master's thesis, University of Edinburgh, 1992.

[Vau87]   J. Vautherin. Parallel system specifications with coloured Petri nets and algebraic data types. In G. Rozenberg, editor, *Advances in Petri Nets*, volume 266 of *Lecture Notes in Computer Science*. Springer, 1987.

[vE91]   P. van Eijk. Tools for LOTOS, a lotosfere overview. Memoranda Informatica 91-25, Universiteit Twente - Faculteit der Informatica, Enschede, 1991.

[vG90]   R.J. van Glabbeek. The linear time – branching time spectrum. In J.C.M. Baeten and J.W. Klop, editors, *Proc. CONCUR'90, Theories of Concurrency: Unification and Extension*, volume 458 of *Lecture Notes in Computer Science*. Springer, 1990. Extended abstract.

[Vig94]   L. Vigneron. Associative-commutative deduction with constraints. In A. Bundy, editor, *Proc. 12th Intl. Conference on Automated Deduction*, volume 814 of *Lecture Notes in Computer Science*, pages 530–544. Springer, 1994.

[Vit96]   M. Vittek. A compiler for nondeterministic term rewriting systems. In H. Ganzinger, editor, *Proc. 7th Intl. Conference on Rewriting Techniques and Applications*, volume 1103 of *Lecture Notes in Computer Science*, pages 154–168. Springer, 1996.

[Wal92]   U. Waldmann. Semantics of order-sorted specifications. *Theoretical Computer Science*, 94(1):1–33, 1992.

[Wan79]   M. Wand. Final algebra semantics and data type extensions. *Journal of Computer and System Sciences*, 19:27–44, 1979.

[WBK94]   A. Werner, A. Bockmayr, and S. Krischer. How to realize LSE narrowing. In *Proc. 4th Intl. Conference on Algebraic and Logic Programming*, Lecture Notes in Computer Science. Springer, 1994.

[WDC⁺94]   U. Wolter, K. Didrich, F. Cornelius, M. Klar, R. Wessaly, and H. Ehrig. How to cope with the spectrum of SPECTRUM. Report 94-22, Technische Universität Berlin, 1994.

[WE87]   E.G. Wagner and H. Ehrig. Canonical constraints for parameterized data types. *Theoretical Computer Science*, 30:323–351, 1987.

[Weg89]   P. Wegner. Learning the language. *Byte*, 14:245–253, 1989.

[Wer93]   A. Werner. A semantic approach to order-sorted rewriting. In C. Kirchner, editor, *Proc. 5th Conference on Rewriting Techniques and Applications*, volume 690 of *Lecture Notes in Computer Science*, pages 47–61. Springer, 1993.

[Wer98]   A. Werner. A semantic approach to order-sorted rewriting. *Journal of Symbolic Computation*, 25(4):527–569, 1998.

[WG94]   C.-P. Wirth and B. Gramlich. On notions of inductive validity for first-order equational clauses. In A. Bundy, editor, *Proc. 12th Intl. Conference on Automated Deduction*, volume 814 of *Lecture Notes in Computer Science*, pages 162–176. Springer, 1994.

[Win84]   G. Winskel. Synchronization trees. *Theoretical Computer Science*, 34:33–82, 1984.

[Wir86]   M. Wirsing. Structured algebraic specifications: a kernel language. *Theoretical Computer Science*, 42:123–249, 1986.

[Wir90]   Martin Wirsing. Algebraic specification. In J. van Leeuwen, editor, *Handbook of Theoretical Computer Science, Vol. B*, chapter 13, pages 675–788. Elsevier Science (North-Holland), 1990.

[Wir93]   Martin Wirsing. Structured specifications: Syntax, semantics and proof calculus. In F.L. Bauer, W. Brauer, and H. Schwichtenberg, editors, *Logic and Algebra of Specification*, volume 94 of *NATO ASI Series F: Computer and Systems Sciences*, pages 411–442. Springer, 1993.

[Wir95]   M. Wirsing. Algebraic specification languages: an overview. In E. Astesiano, G. Reggio, and A. Tarlecki, editors, *Recent Trends in Data Type Specification: 10th Workshop on Specification of Abstract Data Types – Selected Papers*, volume 906 of *Lecture Notes in Computer Science*, pages 81–115. Springer, 1995.

[WK95]   Burkhart Wolff and Kolyang. Development by refinement revisited: Lessons learnt from a case study. In *Proc. Softwaretechnik '95*, Software-Technik Trends. Springer, 1995.

[WN95]   G. Winskel and M. Nielsen. Models for concurrency. In [AGM92], volume 4, pages 1–148. 1995.

[Wol87]   U. Wolter. The power of behavioural validity. Preprint 1-87, TU Magdeburg, Section Mathematik, 1987.

[Wol90]   U. Wolter. An algebraic approach to deduction in equational partial Horn theories. *J. Inf. Process. Cybern. EIK*, 27(2):85–128, 1990.

[Wol94]   B. Wolfinger, editor. *Innovationen bei Rechnern und Kommunikationssystemen*. Springer, 1994.

[WPP+83]   M. Wirsing, P. Pepper, H. Partsch, W. Dosch, and M. Broy. On hierarchies of abstract data types. *Acta Informatica*, 20:1–33, 1983.

[WRCS67]   L. Wos, G.A. Robinson, D.F. Carso, and L. Shalla. The concept of demodulation in theorem proving. *Journal of the Association for Computing Machinery*, 14(4):698–709, 1967.

[Yan93]   Han Yan. *Theory and Implementation of Sort Constraints for Order Sorted Algebra*. PhD thesis, Oxford University, 1993.

[You89]   J.-H. You. Enumerating outer narrowing derivations for constructor-based term rewriting systems. *Journal of Symbolic Computation*, 7(3-4):319–342, 1989. Special issue on unification. Part one.

[ZHR92]   C.C. Zhou, C.A.R. Hoare, and A.P. Ravn. A calculus of durations. *Information Processing Letters*, 40:269–276, 1992.

[ZKK88]   H. Zhang, D. Kapur, and M.S. Krishnamoorthy. A mechanizable induction principle for equational specifications. In E. Lusk and R. Overbeek, editors, *Proc. 9th Intl. Conference on Automated Deduction*, volume 310 of *Lecture Notes in Computer Science*, pages 162–181. Springer, 1988.

[Zuc96]   E. Zucca. From static to dynamic abstract data types. In W. Penczek and A. Szlas, editors, *Proc. MFCS'96*, volume 1113 of *Lecture Notes in Computer Science*. Springer, 1996.

Subject Index

Author Index